Lecture Notes in Networks and Systems

Volume 102

The series "Lecture Notes in Networks and Systems" publishes the latest developments in Networks and Systems—quickly, informally and with high quality. Original research reported in proceedings and post-proceedings represents the core of LNNS.

Volumes published in LNNS embrace all aspects and subfields of, as well as new challenges in, Networks and Systems.

The series contains proceedings and edited volumes in systems and networks, spanning the areas of Cyber-Physical Systems, Autonomous Systems, Sensor Networks, Control Systems, Energy Systems, Automotive Systems, Biological Systems, Vehicular Networking and Connected Vehicles, Aerospace Systems, Automation, Manufacturing, Smart Grids, Nonlinear Systems, Power Systems, Robotics, Social Systems, Economic Systems and other. Of particular value to both the contributors and the readership are the short publication timeframe and the world-wide distribution and exposure which enable both a wide and rapid dissemination of research output.

The series covers the theory, applications, and perspectives on the state of the art and future developments relevant to systems and networks, decision making, control, complex processes and related areas, as embedded in the fields of interdisciplinary and applied sciences, engineering, computer science, physics, economics, social, and life sciences, as well as the paradigms and methodologies behind them.

**** Indexing: The books of this series are submitted to ISI Proceedings, SCOPUS, Google Scholar and Springerlink ****

More information about this series at http://www.springer.com/series/15179

Mustapha Hatti
Editor

Smart Energy Empowerment in Smart and Resilient Cities

Renewable Energy for Smart and Sustainable Cities

Springer

Editor
Mustapha Hatti
EPST-CDER
Unité de Développement
des Equipements Solaires
Bou-Ismail, Algeria

ISSN 2367-3370 ISSN 2367-3389 (electronic)
Lecture Notes in Networks and Systems
ISBN 978-3-030-37206-4 ISBN 978-3-030-37207-1 (eBook)
https://doi.org/10.1007/978-3-030-37207-1

This Springer imprint is published by the registered company Springer Nature Switzerland AG
The registered company address is: Gewerbestrasse 11, 6330 Cham, Switzerland

Contents

Smart Grid, Energy Control and Management

Particle Swarm Optimization and Artificial Neural Network

Frequency Control in Microgrid Power System with Renewable Power Generation Using PID Controller Based on Particle Swarm Optimization

M. Regad[1(✉)], M. Helaimi[1], R. Taleb[1], Ahmed M. Othman[2], and Hossam A. Gabbar[3]

[1] Electrical Engineering Department, Laboratoire Génie Electrique et Energie Renouvelable (LGEER), Hassiba Benbouali University of Chlef, BP. 78C, Ouled Fares, 02180 Chlef, Algeria
m.regad@univ-chlef.dz
[2] Electrical Power and Machine Department, Faculty of Engineering, Zagazig University, Zagazig, Egypt
[3] Faculty of Energy Systems and Nuclear Science, University of Ontario Institute of Technology (UOIT), 2000 Simcoe St. N, Oshawa, ON L1H 7K4, Canada

Abstract. This paper addresses an application of proportional-integral-derivation (PID) controller based Particle Swarm Optimisation for frequency control of a microgrid power system with the integration of wind power generation and PV generator. Fuel cell and diesel engine generator are used as secondary sources. The Energy Storage System is often applied in the microgrid system for supply energy to the connected load. Particle Swarm Optimization is used to tune the gains of the PID controller through revolving of an objective function. The PSO is robust and more efficient optimization method. The main objective of this work is to reduce the fluctuations of the system frequency and power system. The reults shows that the PID controller based PSO given better performances of system for frequency and power regulation in comparison with PID controller based Genetic Algorithm.

Keywords: Renewable energy · Hybrid power system · PID controller · Frequency control · Particle Swarm Optimization

1 Introduction

In the last decades, the increase of the depletion and environment impacts of the fossil fuel cell to using new renewable green energy in order to limit the power demand and avoid the greenhouse caused by the emission of gas from fossil fuel sources [1]. However, the energy from renewable energy sources such as wind and photovoltaic is not constant and varies according to the weather conditions. The stochastic and intermittent nature results in some fluctuations in power and frequency system that must be controlled using an adequate control strategy [2, 3]. One of the most existing

M. Hatti (Ed.): ICAIRES 2019, LNNS 102, pp. 3–13, 2020.
https://doi.org/10.1007/978-3-030-37207-1_1

solutions is to combine more than one of this sources with a storage system and conventional sources called hybrid energy system which is largely used to overcome the need to energy and decrease the effect of the greenhouse. The reason for using the energy storage system is to absorb the fluctuations from wind and photovoltaic systems and supply this stored power to load later. This hybrid energy system can operate in connected mode or in island mode in rural areas [1]. The main challenges that face the use of hybrid energy system are to promise the electricity supply to customers and providing for the long term energy security [4]. Control system based PID controller is comely used in industry due to its simplicity and clear functionality amongst all others control scheme [8]. Several works are done using the PID controller for control of a hybrid energy system such reported in [5–7].

Frequency and power fluctuation present a big challenge in the operation of such a hybrid energy system. Many researchers investigate the optimal design of PID controller parameters using various optimizations techniques such as Genetic Algorithm, Particle Swarm Optimization, and Mine Blast Algorithm for frequency and power control in a hybrid energy system. In [9] the authors discuss the frequency control of stand-alone microgrid with a battery energy storage system using particle swarm optimization (PSO). In [10, 11] the authors studied the frequency control of the hybrid energy system using PID/PI based GA. Another optimization method called MBA is reported in [12] in order to optimize the PID controller parameters for hybrid microgrid system.

For tuning the optimal values for the PID controller parameters, we propose to use the PSO technique proposed to be applied. This technique developed by Kennedy and Eberhart in 1995 [13], is a stochastic search, robust and flexible in solving of optimization problem due to its high-quality of solution within shorter calculation time and stable convergence characteristic than other stochastic methods [14].

The rest of the paper is summarized as fellow: the configuration of the proposed system is presented in Sect. 2. In Sect. 3 the controller strategy and optimization method are addressed. The simulation results and discussions are reported in Sect. 4. This paper is ended by a conclusion in the last section followed by a reference.

2 Configurations of Microgrid Power System

The various microgrid energy system components are presented by first-order transfer function as shown in Fig. 1. In this study, power is generated by a WTG, a PV, a FC, and a DEG generator. The integration of different energy storage system devices can facilitate the reliable power supply to the connected load. The storage systems along with, PV and FC are to be connected through suitable converters but have not been considered here in order to avoid the complexities in the modeling. The energy storage systems BESS and FESS store energy during the surplus generation and release efficiently during the peak-load demand. The DEG is to be taken as the standby generator which starts automatically to make up the deficit power demand. The parameters of different components of the proposed system are given in the Table 1 [11, 15–17].

Microgrid system conists of various distributed generation connected together for forming independent units which can operate in two modes; isolated mode in remote

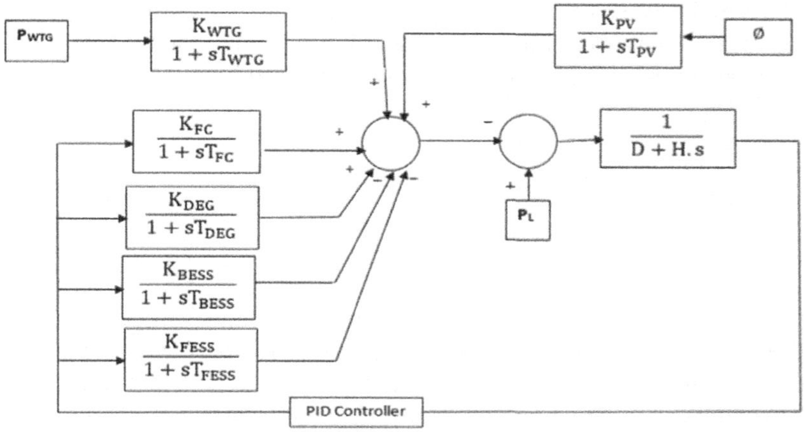

Fig. 1. Hybrid energy generations with the storage energy system

Table 1. Microgrid parameters

Component	Gain (K)	Time constant (T)
Wind turbine generator	$K_{WTG} = 1$	$T_{WTG} = 1.5$
Photovoltaic generator	$K_{PV} = 1$	$T_{IN} = 0.04$ $T_{I/C} = 0.004$
Fuel Cell (FC)	$K_{FC} = 0.01$	$T_{FC} = 4$
Diesel engine generator	$K_{DEG} = 0.003$	$T_{DEG} = 2$
Battery energy storage system	$K_{BESS} = 1$	$T_{BESS} = 0.1$
Flywheel energy storage system	$K_{FESS} = 1$	$T_{FESS} = 0.1$

areas or in conneted mode to support and enhance the microgrid security and reability. The microgrid can increase the reliability and efficiency of the power system.

3 Controller Design and Optimization Technique

In this, the proposed controller and optimization method are discussed.

3.1 PID Controller

The conventional Proportional Integral Derivative Controller (PID) is considered the most popular controller used in almost all the industries processes. It can provide excellent control performances due to its quite structure and robustness. PID is the most simple and easy understood controller despite the varied dynamic characteristics of process plant [6].

A proportional controller has the effect of reducing the rise time, but cannot eliminate the steady-state error. An integral mode has the ability to eliminate the steady-state error [7].

PID controller is a control scheme generally used in an industrial process system. A standard PID controller structure is also known as the "three terms" controller, whose can be presented by a transfer function in the ideal form in (1) or in the parallel form in (2) [8].

$$G(s) = K_P\left(1 + \frac{1}{T_I s} + T_D s\right) \tag{1}$$

$$G(s) = K_P + \frac{K_I}{s} + K_D s \tag{2}$$

Where K_P, K_I and K_D are the proportional, integral and derivative gain respectively, T_I and T_D are integral, and the derivative time constant respectively (Fig. 2).

PID controller composed of three types of control actions, Proportional, Integral, and Derivative control action.

The "three-term" functionalities are highlighted below [5, 6].

- The proportional term – gives a control action proportional to the error signal through the all-pass gain factor.
- The integral term – mismatches' steady state errors through low-frequency compensation by an integrator.
- The derivative term – enhancing the transient response through high-frequency compensation by a differentiator.

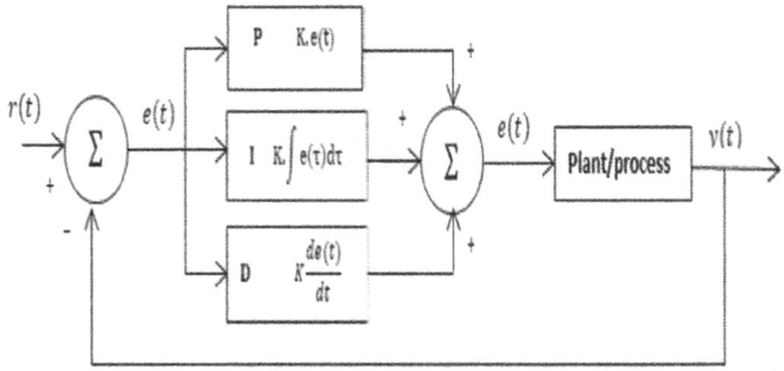

Fig. 2. Block of PID controller model

The objective function (J) used for optimizing the controller parameter's has been considered an integral performance index over the simulation period using the weighted sum of squared frequency deviation (Δf) and the deviation of the control signal (Δu) as given by (3).

$$= J_{opt} = \int_{T_{min}}^{T_{max}} [w(\Delta f)^2 + \left(\frac{1-w}{Kn}\right)(\Delta u)^2]dt \tag{3}$$

Where w dictates the relative importance of the two objectives (i.e., Integral of Squared Error—ISE and Integral of squared Deviation of Control Output—ISDCO), and its value is taken as 0.7. $Kn = 10^4$ is the normalizing constant to scale ISE and ISDCO in uniform scale. The model of the objective function is presented using Matlab/Simulink.

3.2 Interview on PSO algorithm

Many problems have not an exact solution that gives the results in a reasonable time. For overcoming these problems some metaheuristics methods which offer an approached solution after many iterations are recently proposed. Among these methods Particle Swarm Optimization algorithm that has generic principle to be applied in many fields of optimization problems. PSO is a stochastic optimization algorithm developed by Eberhart and Kennedy, inspired by the social behavior and fish schooling of bird flocking. Each particle in the swarm is a different possible set of the unknown parameters of the objective function to be optimized. The swarm consists of N particles moving around in a D-dimensional search space. Each particle is initialized with a random position and a random velocity [9]. The new velocity can be calculated by the fellow formula.

$$V_{i+1} = w.V_i + C_1.r_1(P_{best} - Xi_k) + C_2.r_{2.}(G_{best} - X_i) \tag{4}$$

$$X_{i+1} = X_i + V_{i+1} \tag{5}$$

Where V_i is the component in the dimension of the particle velocity in iteration, X_i is the component in the dimension of the particle position in iteration, $C_1\, and\, C_2$ are constant weight factors, *Pbest* is the best position achieved so far by particle, *Gbest* is the best position found by the neighbors of particle, and are random factors in between 0 and 1 interval, and w is inertia weight which is started from a positive initial value (w0) and decreases during the iterations by $W_{k+1} = \beta.W_k$.

The algorithms of PSO can be described as follows:

Step1: Initialize a population of particles with random positions and velocities on D-dimensions in the problem space.

Step2. Evaluation of desired optimization fitness function in D variables for each particle,

Step3. Comparison of particle's fitness evaluated with its best previous position. If the current value is better, then set the best previous position equal to the current value, and pi equals to the current location xi in D dimensional space.

Step4. Identifying the particle in the neighborhood with the best fitness so far, and assign its index to the variable g,

Step5. Change velocity and position of the particle according to Eqs. (4) and (5).

Step6. Return to step 2 until a criterion is met or end of iterations.

The flowchart of this algorithm is presented by Fig. 3 as follow:

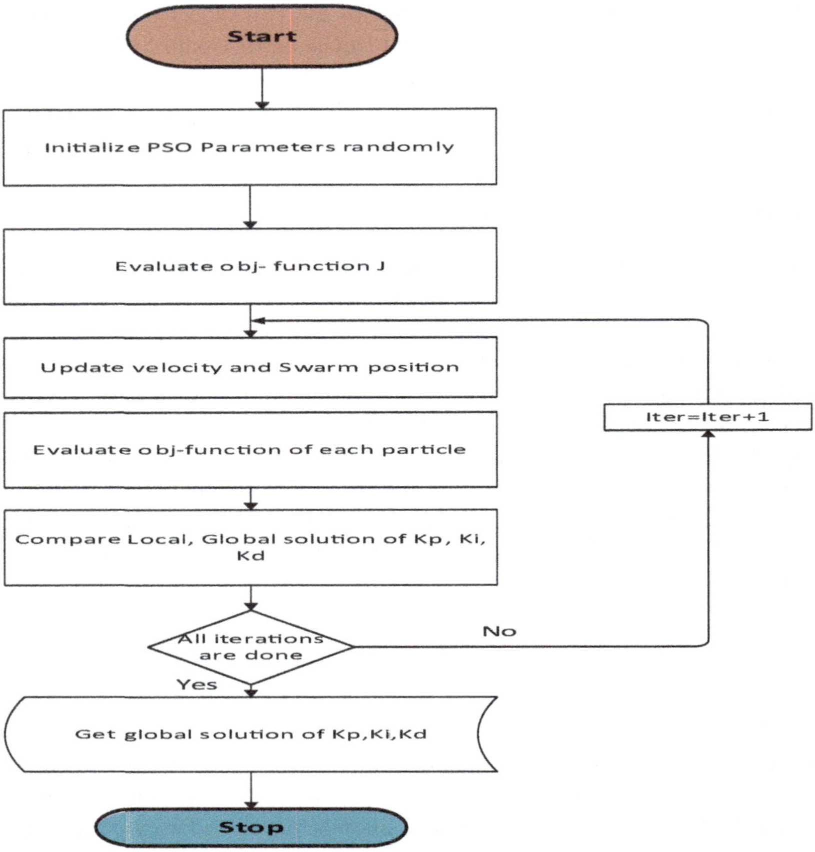

Fig. 3. Flowchart of PSO algorithm

4 Results and Discussion

The proposed configuration of the microgrid is simulated under nominal condition during 120 s using Matlab/Simulink interface. A PID controller is introduced in order to eliminate the frequency and power fluctuations provoked by the integration of renewable power generation such as PV and wind which have intermittent nature and stochastic changing during the simulation time. The simulation results are showed in followed figures (Figs. 4, 5, 6 and 7).

The change in power demand and generation causes fluctuations in frequency and power deviations which are settled down after few second due to the coordination with generation sources and storage devices through the controller. With a thorough analysis of results, it can be observed that small deviation of frequency and power is achieved by using of PID controller based PSO compared to the results given using GA based PID controller. The proposed controller based PSO appeared better than the controller based GA in performances and stability of the system.

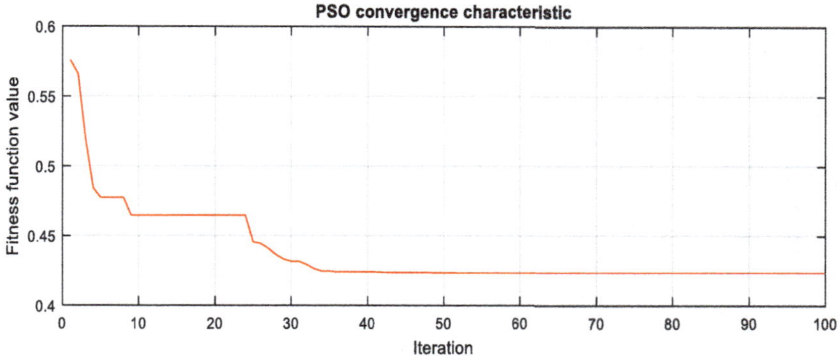

Fig. 4. PSO convergence characteristic

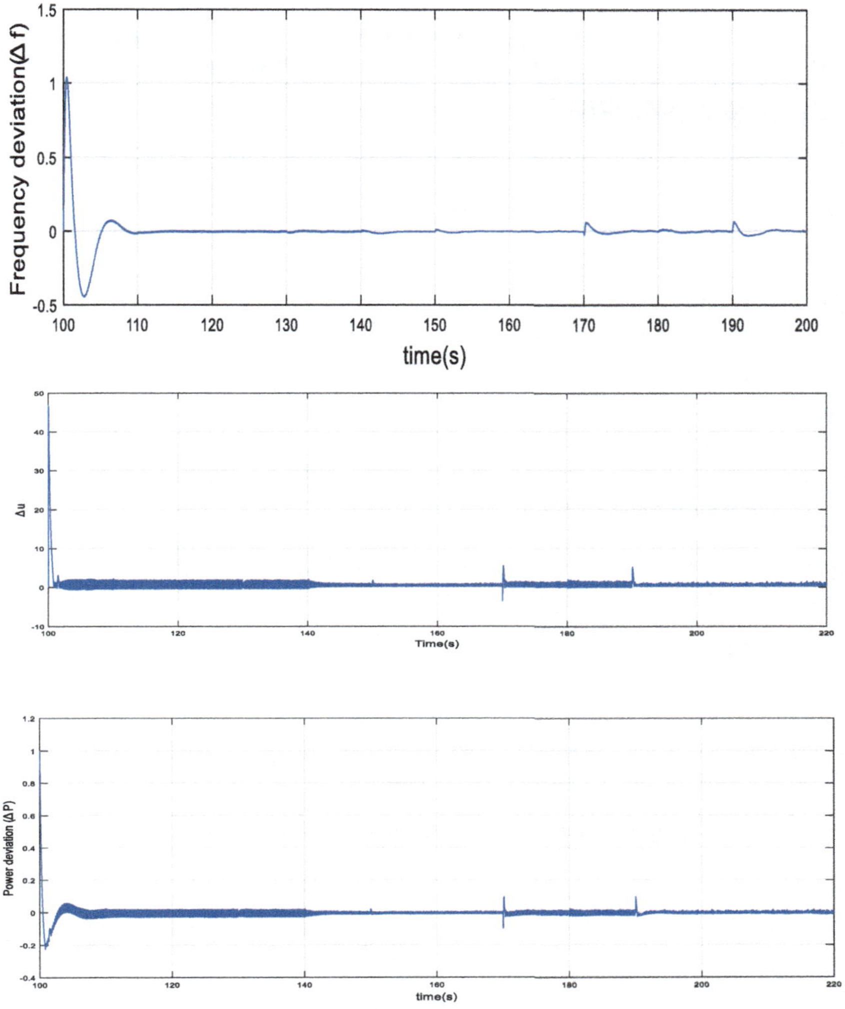

Fig. 5. Frequency and power deviation with signal control

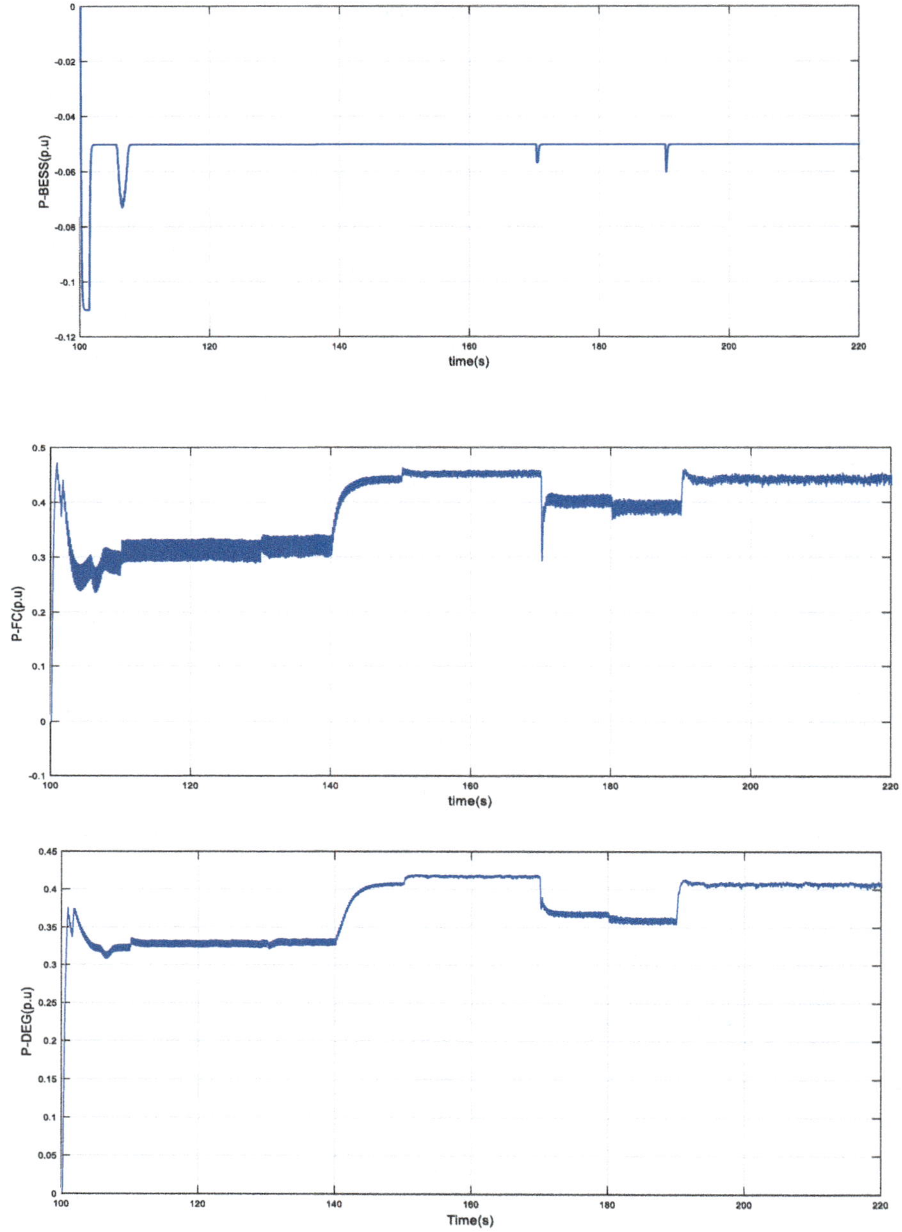

Fig. 6. Generated power by each component of a microgrid

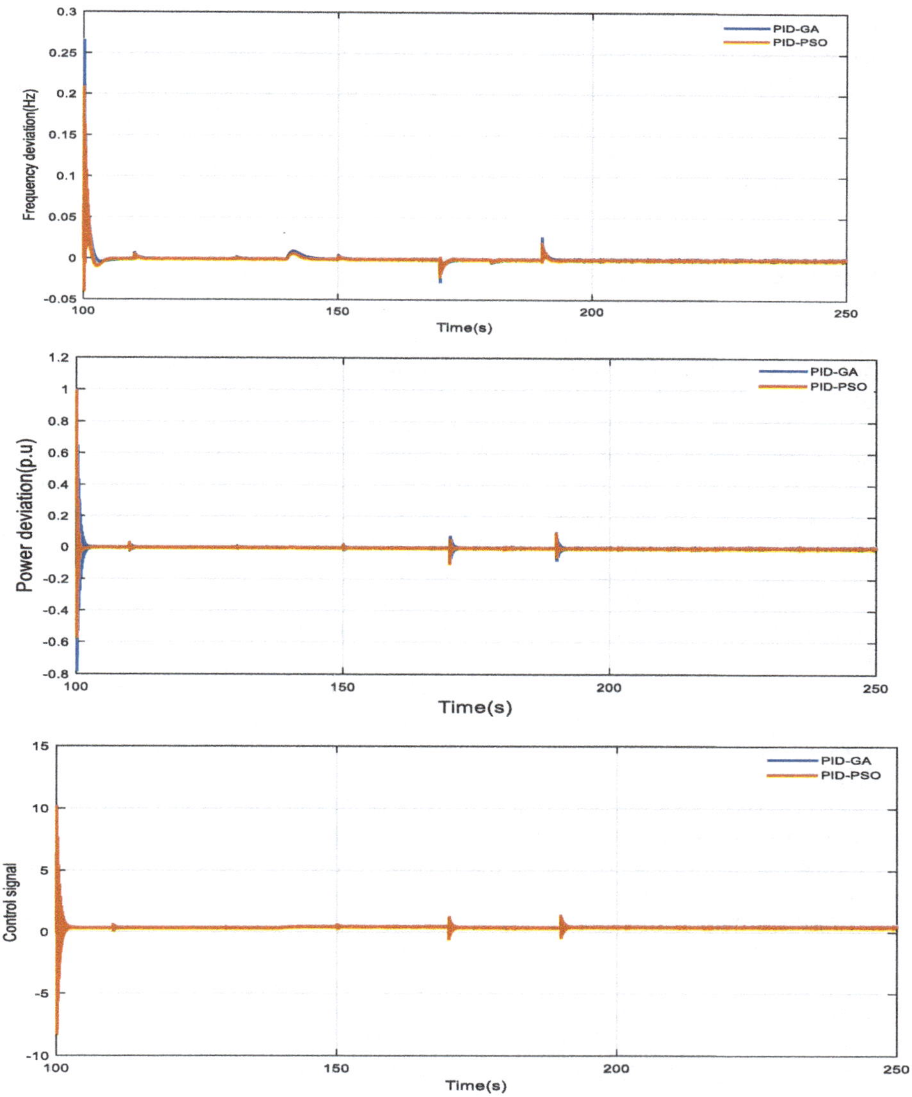

Fig. 7. Power and frequency deviations with control signal using the best PID based GA and PSO.

5 Conclusion

The proposed hybrid system consisting of PV, WTG, FC, DEG, BESS, and FESS is presented, controlled using PID and simulated autonomously. The system components are modeled using the first order transfer function considering reasonable approximation. Different types of energy storage such as BESS and FESS have been used to absorb the fluctuation in output power of photovoltaic and wind systems. However, a control scheme based on the PID controller is used to provide power and, frequency deviations. The PID parameters are optimized using the Particle Swarm Optimization algorithm. The obtained results are compared with the Genetic Algorithm based PID controller. This proposed method is a good choice for application in control of the hybrid energy system based on the renewable energy system in comparison with Genetic Algorithm.

References

1. Bouzid, A.M., Guerrero, J.M., Cheriti, A., Bouhamida, M., Sicard, P., Benghanem, M.: A survey on control of electric power distributed generation systems for microgrid applications. Renew. Sustain. Energy Rev. **44**, 751–766 (2015)
2. Anwar, M.N., Pan, S.: A frequency response model matching method for PID controller design for processes with dead-time. ISA Trans. **55**, 175–187 (2015)
3. Senjyu, T., Nakaji, T., Uezato, K., Funabashi, T.: A hybrid power system using alternative energy facilities in isolated Island. IEEE Trans. Energy Convers. **20**(2), 406–414 (2005)
4. Mahmoud, M.S., Hussain, S.A., Abido, M.A.: Modeling and control of microgrid: an overview. J. Franklin Inst. **351**(5), 2822–2859 (2014)
5. Das, D.C., Roy, A.K., Sinha, N.: GA based frequency controller for solar thermal–diesel–wind hybrid energy generation/energy storage system. Int. J. Electr. Power Energy Syst. **43** (1), 262–279 (2012)
6. Das, D.C., Roy, A.K., Sinha, N.: Genetic algorithm based PI controller for frequency control of an autonomous hybrid generation system. In: World Congress on Engineering 2012, London, UK., 4–6 July 2012, vol. 2189, pp. 953–958. International Association of Engineers, March (2010)
7. Shayanfar, H.A., Shayeghi, H., Younesi, A.: Optimal PID controller design using Krill Herd algorithm for frequency stabilizing in an isolated wind-diesel system. In: Proceedings on the International Conference on Artificial Intelligence (ICAI), p. 516. The Steering Committee of The World Congress in Computer Science, Computer Engineering and Applied Computing (WorldComp) (2015)
8. Shayanfar, H.A., Shayeghi, H., Younesi, S.A.A.: Design of optimal PID controller using Jaya algorithm for frequency stabilizing. In: An Isolated Wind-Diesel System (2016)
9. Iruthayarajan, M.W., Baskar, S.: Evolutionary algorithms based design of multivariable PID controller. Expert Syst. Appl. **36**(5), 9159–9167 (2009)
10. Kerdphol, T., Fuji, K., Mitani, Y., Watanabe, M., Qudaih, Y.: Optimization of a battery energy storage system using particle swarm optimization for stand-alone microgrids. Int. J. Electr. Power Energy Syst. **81**, 32–39 (2016)
11. Rajesh, K.S., Dash, S.S.: Load frequency control of autonomous power system using adaptive fuzzy based PID controller optimized on improved sine cosine algorithm. J. Ambient Intell. Humaniz. Comput. **10**(5), 2361–2373 (2019)
12. Ranjan, S., Das, D.C., Latif, A., Sinha, N.: LFC for autonomous hybrid micro grid system of 3 unequal renewable areas using mine blast algorithm. Int. J. Renew. Energy Res. (IJRER) **8** (3), 1297–1308 (2018)
13. Debbarma, S., Bhattacharya, M., Meena, B. K., Datta, A.: Frequency control of autonomous hybrid power system using smart controllable load. In: 2015 International Conference on Robotics, Automation, Control and Embedded Systems (RACE), pp. 1–7. IEEE, February 2015

14. Maiti, D., Biswas, S., Konar, A.: Design of a fractional order PID controller using particle swarm optimization technique. arXiv preprint arXiv:0810.3776 (2008)
15. Lee, D.J., Wang, L.: Small-signal stability analysis of an autonomous hybrid renewable energy power generation/energy storage system part I: time-domain simulations. IEEE Trans. Energy Convers. **23**(1), 311–320 (2008)
16. Pan, I., Das, S.: Kriging based surrogate modeling for fractional order control of microgrids. IEEE Trans. Smart Grid **6**(1), 36–44 (2014)
17. Ray, P., Mohanty, S., Kishor, N.: Small-signal analysis of autonomous hybrid distributed generation systems in presence of ultracapacitor and tie-line operation. J. Electr. Eng. **61**(4), 205–214 (2010)

Five PV Model Parameters Determination Through PSO and Genetic Algorithm, a Comparative Study

M. Rezki(✉), S. Bensaid, I. Griche, and H. Houassine

Electrical Engineering Department, Faculty of Sciences and Applied Sciences, Bouira University, Bouira, Algeria
mohamedrezki197@yahoo.fr

Abstract. The main goal of this paper is the application of PSO (Particle Swarm Optimization) and Genetic Algorithm (GA) in Renewable energy in general and particularly photovoltaics (PV) in order to extract the five parameters that governs the PV module (the photocurrent, the serial resistance, the saturation current, the parallel resistance and the ideality factor). Indeed, PSO and GA are intelligent post-analytic global optimization algorithms that give a minimal error. The application of these algorithms aimed at comparing the experimental results of a fairly well known photovoltaic module with is the MSX 60 has given good results. This is confirmed by the calculation of statistical performance measurement factors such as RMSE (root-mean-square error) and MAPE (mean absolute percentage error).

Keywords: Optimization · Five PV parameters · PSO · GA

1 Introduction

The Photovoltaic solar energy which is a clean energy comes from the conversion of sunlight into electricity through semiconductor materials such as silicon or composite materials. These photosensitive materials have the property of releasing their electrons under the influence of an external energy (light and temperature). This junction constituting the solar cell is based on solar modules constructed by manufacturers. Modeling and simulation of PV module helps in better understanding in terms of the behavior and characteristics [1]. Many models have been developed to reflect the true behavior of the solar module such as Ideal Photovoltaic model (model of three parameters), 1D-2R model (five parameters model) and the two-diode model (the seven parameters model) [2]. The most used common model is the five parameters model for its offering a closer representation of the solar cell [3]. These five parameters are the: photocurrent (Iph)t, serial resistance (R_S), saturation current, (I_0), parallel resistance (R_{sh}) and the ideality factor (n).

There were different optimizing algorithms For evaluating and optimizing the PV model [4–8]. In general, we cal classify these algorithms in three groups: (a) analytical methods such as Newton-raphson method [9], (b) iterative methods like search fitting curves [10, 11], (c) intelligent algorithms (heuristic and metaheuristic). Among the heuristic methods it can be found the PSO algorithm and the genetic algorihms [12–14].

M. Hatti (Ed.): ICAIRES 2019, LNNS 102, pp. 14–21, 2020.
https://doi.org/10.1007/978-3-030-37207-1_2

The remainder of the paper is organized as follows. In Sect. 2, the problem of solar cell modeling is presented.

Section 3 describes the GA algorithm as well as PSO algorithm. The different results and discussion with validation are exposed in Sect. 4. Finally, Sect. 5 gives a summary and conclusions.

2 The Five Parameters PV Cell/Module Modeling

Due to its simplicity and acceptable accuracy, the single-diode (five parameters) model has been selected (see Fig. 1).

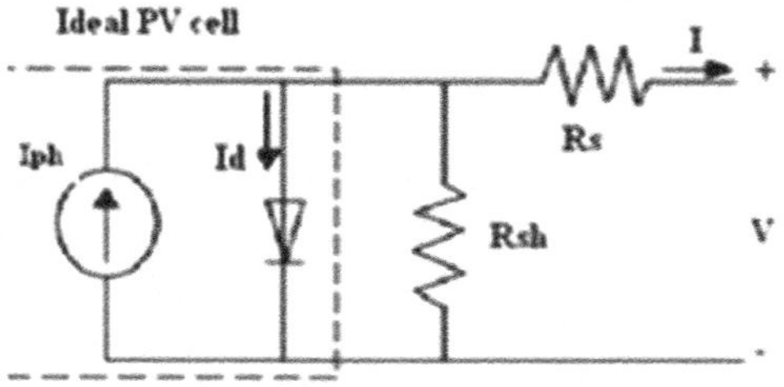

Fig. 1. Equivalent model of five parameters solar cell

From Fig. 1, it can be shown that the output current of the solar cell can be given as follows:

$$I = I_{ph} - I_{rs}\left[e^{\left(\frac{q(v+R_S.I)}{A.k.T}\right)} - 1\right] - \left(\frac{q(v+R_S I)}{R_{sh}}\right) \quad (1)$$

PV cells are connected together in series and parallel solar cell combinations to form a module the terminal equation of the PV module can be written as follows [15]:

$$I = N_p I_{ph} - N_p I_{rs}\left[e^{\left(\frac{q(v+R_S.I)}{A.k.T.N_S}\right)} - 1\right] - N_p\left(\frac{q(v+R_S I)}{N_S.R_{sh}}\right) \quad (2)$$

Where:

V is the cell output voltage; q is the electron charge ($1.60217646 \times 10^{-19}$C); k is the Boltzmann's constant ($1.3806503 \times 10^{-23}$ J/K); T is the temperature in Kelvin; I_{rs} is the cell reverse saturation current; A is the diode ideality constant; N_p is the number of PV cells connected parallel; N_s is the number of PV cells connected in series.

The generated photocurrent I_{ph} depends on solar irradiation and it's by the following equation:

$$I_{ph} = [I_{sc} + k_i(T - T_r)]\frac{G}{1000} \quad (3)$$

Where:

k_i is the short-circuit current temperature coefficient; G is the solar irradiation in W/m^2; I_{sc} is the cells short-circuit current at reference temperature; T_r is the cell reference temperature.

The cell's saturation current i_s varies with temperature according to the following equation

$$I_{rs} = I_{rr}\left[\frac{T}{T_r}\right]^3 \exp\left(\frac{q.E_G}{k.A}\left[\frac{1}{T_r}-\frac{1}{T}\right]\right) \tag{4}$$

Where:

E_G is the band-gap energy of the semiconductor used in the cell. I_{rr} is the reverse saturation at T_r.

3 PV Module Parameters Extraction Based on GA & PSO Algorithms

Using the objective function well defined by the Eq. (1), it can be easily implemented on the heuristic algorithms such as genetic algorithm (GA) and particle swarm optimization (PSO) for determining the five parameters of our chosen model 1D–2R.

3.1 Genetic Algorithm (GA)

The genetic algorithm is a search heuristic inspired by the process of natural selection (well reputed as Darwin's theory). Its principal application is to generate useful solutions to optimization and search problems. The Genetic Algorithm was originally proposed by Holland (Holland 1975). After that, many authors (Goldberg 2000; Michalewicz 1994) have modified the existing one and improved genetic algorithms are proposed) [16]. Procedure of the GA starts from a seed and generates a set of individuals. Each of these individuals can extract a group of parameters, the best-fitted parameters are selected to form the new population when the process has been repeated 50 000 times until this iterations procedure been accomplished [17].

The basic components common to almost all genetic algorithms are [18]:

- a fitness function for optimization
- a population of chromosomes
- selection of which chromosomes will reproduce
- crossover to produce next generation of chromosomes
- random mutation of chromosomes in new generation.

For the implementation of GA to perform PV module parameters extraction the fitness function has the role of optimizing the objective function defined above (I–V curve) and the population of chromosomes express the five electrical PV cell parameters (Iph, I0, A, Rs and Rsh). The task of selection and crossover is to promote chromosome with high fitness. On the other hand, the random mutation ensures diversification of solutions by creating another generation.

3.2 Particle Swarm Optimization (PSO)

Particle Swarm Optimization (PSO) is an heuristic optimization (sometimes called metaheuristic), invented by Russel Eberhart and James Kennedy in 1995, inspired by the behaviour of social organisms in groups, such as bird and fish schooling or ant colonies [19]. This optimization method is based on the collaboration of individuals with each other. Elsewhere this method has many similarities with the genetic algorithm except for the use of a group (swarm of particles) instead of an individual (chromosome) in the search.

Just the algorithm starts each particle is positioned randomly in the search space of the problem. Each iteration moves the particles according to 3 components: (1) Its current velocity $v_i = (v_{i1}, v_{i2}, \ldots, v_{id})$ with i is the rank of the particle. (2) Its best positions (or solution) namely p_i. (3) The best solution obtained before (previous best position) which is p_g. This gives the following equations [4]:

$$v_i^{n+1} = w * v_i^n + C_1 * r_1 * \left(Pbest_i - x_i^n\right) + C_2 * r_2 * \left(Gbest - x_i^n\right) \tag{4}$$

$$x_i(j+1) = v_i(j+1) + x_i(\mathrm{j}) \tag{5}$$

Where: x_i is the position of the particle, C_1 and C_2 are acceleration factors r_1 and r_2 are two uniform random numbers between 0 and 1. w is an inertia weight (had a high value for searching global solution). v_i is the initial velocity. j is the iteration index. *Pbest* and *Gbest* are respectively: personal and global best fitness of each particle. The global solution is indeed represents the desired solution of the solar cell equation.

4 Results and Discussions

The GA and PSO algorithms were implemented on the MSX 60 PV module (polycrystalline) by using the Matlab software (see Table 1).

Table 1. Manufacturing datasheet of MSX 60

Characteristics	Value
Open-Circuit Voltage (V_{oc})	21.1 V
Optimum Operating Voltage (V_{mp})	17.1 V
Short-Circuit Current (I_{sc})	3.8 A
Optimum Operating Current (I_{mp})	3.5 A
Maximum Power at STC* (P_{max})	60 W
Number of cells	36
Temp. coefficient of V_{oc}	−80 m V/°C
Temp. coefficient of I_{sc}	0.0024 A/°C

For measuring the efficiency of the proposed algorithms, we opted for statistical tools by calculating the errors: *RMSE* (root-mean-square error) and *MAPE* (mean absolute percentage error).

RMSE is defined by [20]:

$$RMSE = \sqrt{\frac{1}{N}\sum\nolimits_{i=1}^{N} f_i(V_m, I_m, x)^2} \quad (6)$$

Where, N is the number of the experimental data.

The formula of the mean absolute percentage error (MAPE) is as follows:

$$MAPE = \frac{1}{N}\sum\nolimits_{i=1}^{N} \left|\frac{A_i - F_i}{A_i}\right| .100 \quad (7)$$

Where A_i is the actual value and F_i is the forecast value, N is the number of the experimental data.

In this section, the proposal methods (PSO & GA) are applied to extract the five parameters governing the MSX60 solar module under standard test conditions (STC: 1000 W/m^2, 25 C°). Referring to the experimental I–V curve and the manufacturer's given datasheet, a statistical study performed by calculating the RMSE, the MAPE and the execution time was done. The different results can be shown in Table 2.

Table 2. Extracted parameters for the MSX 60 PV module under STC

Parameters	GA algorithm	PSO algorithm
Rs [Ω]	0.4843	0.4821
R_{sh} [Ω]	2.0001×10^4	2×10^4
A (ideality factor)	1.0001	1
I_{ph} [A]	3.7999	3.7999
I_0 [A]	4.6314×10^{-10}	4.6121×10^{-10}
RMSE	0.0723	0.0707
MAPE	1.8035	1.7734
Time execution [s/run]	4529.5486	1468.2126

The results in Table 2 show that statistically the results of the PSO are much better than those of the GA (see values of RMSE and MAPE). Another advantage of the PSO is its relatively low execution time compared to the algorithm GA.

The Figs. 2 and 3 shows experimental model (curve I–V) and the estimated characteristics (computed model with PSO and GA) applied on the MSX 60 PV module.

It can be seen from Figs. 2 and 3 the acceptable matching curves between experimental and proposal models. It's due also for the integration of Nelder-Mead algorithm in the main PSO-GA programms in order to resolve the non linear objective function. That's the reason of the enhancement of results.

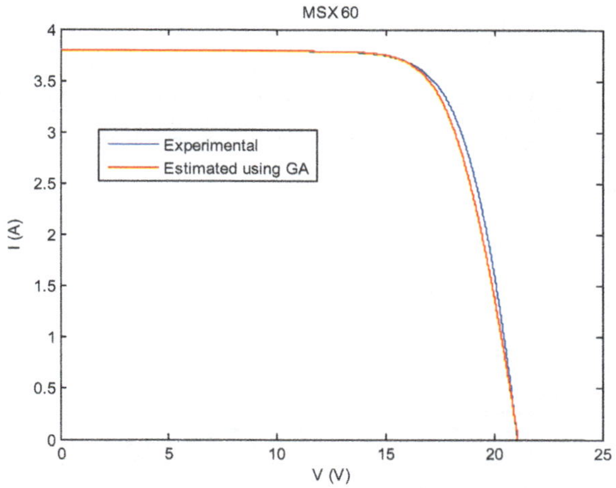

Fig. 2. GA model of the five parameters PV MSX60 module

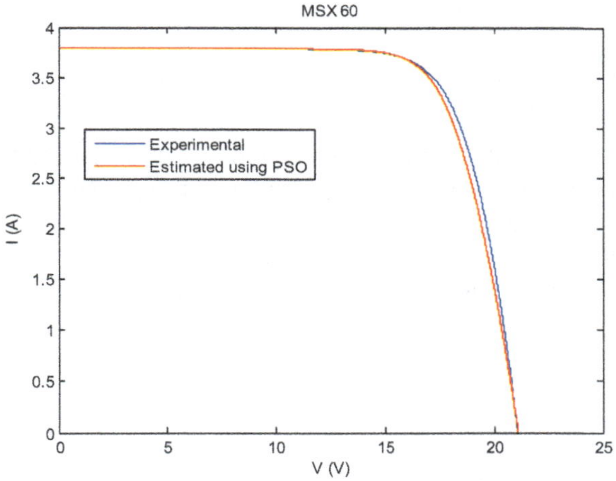

Fig. 3. PSO model of the five parameters PV MSX60 module

Figure 4 depicts the comparison between the convergence characteristics of GA and PSO algorithms depending on number of iterations. As an initial solution GA algorithm is the best but in terms of accuracy and speed of convergence it's very clear that the PSO algorithm is better than GA. In general both PSO and GA methods converge towards the global solution, which is an advantage.

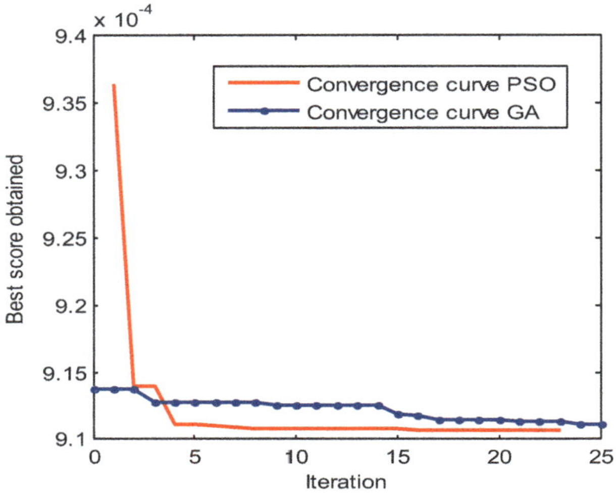

Fig. 4. Comparison between the convergence characteristics of GA and PSO algorithms

5 Conclusion

In this paper, a comparison study between the GA algorithm and PSO for extraction of PV parameters was done. this comparison confirms the power of PSO compared to GA as a modeling tool. The speed of convergence towards a global optimum is clearly in favor of the PSO. The integration of the Nelder-Mead method in order to solve the nonlinearity of the objective function solve the non-linearity of the objective function helped us to avoid falling into the local solutions and it has contributed to have precise results.

References

1. Zainal, N.A., et al.: Modelling of photovoltaic module using matlab simulink. IOP Conf. Series Mater. Sci. En. **114**, 1–9 (2016)
2. Bonkoungou, D., et al.: Modelling and simulation of photovoltaic module considering single-diode equivalent circuit model in MATLAB. Int. J. Emerg. Technol. Adv. Eng., 493–502 (2008)
3. de Blas, M.A., Torres, J.L., Prieto, E., García, A.: Selecting a suitable model for characterizing photovoltaic devices. Renew. Energy **25**, 371–380 (2002)
4. Ye, M., et al.: Parameter extraction of solar cells using particle warm optimization. J. Appl. Phys. **105**, 0945021–0945028 (2009)
5. Ismail, M.S., Moghavvemi, M., Mahlia, T.M.I.: Characterization of PV panel and global optimization of its model parameters using genetic algorithm. Energy Convers. Manag. **73**, 10–25 (2013)
6. Chan, D.S.H., Phang, J.C.H.: Analytical methods for the extraction of solar-cell single- and double-diode model parameters from I–V characteristics. IEEE Trans. Electron Devices **34**, 286–293 (1987)

7. Wolf, P., Benda, V.: Identification of PV solar cells and modules parameters by combining statistical and analytical methods. Sol. Energy **93**, 151–157 (2013)
8. AlHajri, M.F., et al.: Optimal extraction of solar cell parameters using pattern search. Renewable Energy **44**, 238–245 (2012)
9. Reis, L.R.D., Camacho, J.R., Novacki, D.F.: The Newton Raphson method in the extraction of parameters of PV modules. Renew. Energy Power Qual. J. (RE&PQJ) **1**, 634–639 (2017)
10. Khezzar, R., Zereg, M., Khezzar, A.: Comparative study of mathematical methods for parameters calculation of current-voltage characteristic of photovoltaic module. In: IEEE International Conference on Electrical and Electronics Engineering 'ELECO', pp. 24–28, November 2009
11. Villalva, M., Gazoli, J.: Comprehensive approach to modeling and simulation of photovoltaic arrays. IEEE Trans. Power Electron. **24**, 1198–1208 (2009)
12. Lodhi, E., et al.: Application of particle swarm optimization for extracting global maximum power point in PV system under partial shadow conditions. Int. J. Electron. Electrical Eng. **5**, 223–229 (2017)
13. Amokrane, Z., Haddadi, M.: An improved technique based on PSO to estimate the parameters of the photovoltaics cell/module. In: The 5th International Conference on Electrical Engineering – Boumerdes (ICEE-B), Boumerdes, Algeria, pp. 1–9, 29–31 October 2017
14. Zagrouba, M., Sellami, A., Bouaicha, M., Ksouri, M.: Identification of PV solar cells and modules parameters using the genetic algorithms: application to maximum power extraction. Sol. Energy **84**, 860–866 (2010)
15. AlRashidi, M.R., et al.: A new estimation approach for determining the I–V characteristics of solar cells. Sol. Energy **85**, 1543–1550 (2011)
16. Rajasekar, N., et al.: Bacterial foraging algorithm based solar PV parameter estimation. Sol. Energy **97**, 255–265 (2013)
17. Peng, W., et al.: Evolutionary algorithm and parameters extraction for dye-sensitized solar cells one-diode equivalent circuit model. Micro Nano Lett. **8**, 86–89 (2013)
18. Carr, J.: An introduction to genetic algorithms. Senior Project **16**, 1–40 (2014)
19. Gopalakrishnan, K.: Particle swarm optimization in civil infrastructure systems: state-of-the-art review. In: Metaheuristic Applications in Structures and Infrastructures, pp. 49–76 (2013)
20. Askarzadeh, A., Rezazadeh, A.: Parameter identification for solar cell models using harmony search-based algorithms. Sol. Energy **86**, 3241–3249 (2012)

Maximum Power Point Tracking Based on the Bio Inspired BAT Algorithm

Sabrina Titri[1(✉)], Karim Kaced[1], and Cherif Larbes[2]

[1] Division Microélectronique et Nanotechnologie, Centre de Développement des Technologies Avancées (CDTA), Baba Hassen, Algiers, Algeria
{stitri,kkaced}@cdta.dz

[2] Laboratoire des Dispositifs de Communication et de Conversion Photovoltaïque (LDCCP), Ecole Nationale Polytechnique (ENP), El-Harrach, Algeria
Cherif.larbes@g.enp.edu.dz

Abstract. Nowadays, there is an increasing trend in the use of solar energy by using photovoltaic system (PVS). The power generated by a PVS highly relies on solar intensity. Therefore, a Maximum Power Point Tracker (MPPT) is one of the key components of solar electricity generation. It is used to extract the maximum power point (MPP) produced by a PVS. In this paper, we present a bio inspired Bat Swarm Optimization (BSO) algorithm to track the MPP thereby increasing the performance of the PVS. The proposed BSO algorithm is developed in Matlab/Simulink environment. Furthermore, the results obtained from the BSO algorithm are compared with the well-known conventional Perturb and Observe (P&O) algorithm.

Keywords: Maximum power point tracking · Photovoltaic system · Bio inspired algorithm · Bat algorithm · Perturb and Observe

1 Introduction

Since the output characteristics of a photovoltaic panel (PV) depends on the solar radiation, the temperature and the load, the maximum power point (MPPT) is not constant. Therefore, tracking the MPPT of a photovoltaic panel is usually an essential part of photovoltaic system (PVS) since it maximizes the power output of the PVS, and therefore maximizes the PV panel efficiency. Thus, to improve the conversion efficiency of the electric power generation, an MPPT controller is integrated with the PVS so that the PV panel will be able to deliver the maximum power under all variable atmospheric conditions. Therefore, a significant number of MPPT controller have been proposed in the literature and industry [1, 2], starting with conventional method (CM) to soft computing (SC) [3, 4]. The main goal of these MPPT controllers is to extract maximum output power from the PV panel under different temperature and sunlight radiation. Despite the fact that these methods are designed for the same objectives, they differ in terms of complexity, flexibility, convergence speed, cost, hardware implementation and effectiveness [5]. Thus, MPPT based conventional methods are simple, easy to

M. Hatti (Ed.): ICAIRES 2019, LNNS 102, pp. 22–29, 2020.
https://doi.org/10.1007/978-3-030-37207-1_3

implement and capable of tracking the MPP efficiently in normal conditions. However, they have a drawback related to the continuous oscillations that occurs around the MPP resulting in considerable loss of power during steady state. In addition, none of these methods are capable of handling the problem of partial shading conditions.

In recent years, MPPT based SC are attracting huge interest from research communities. These MPPT are developed and used both to improve the energy conversion efficiency under uniform and non-uniform irradiation and to track the MPP accurately. A large number of MPPT based on these methods has been proposed. These methods can be categorized as Artificial Intelligent Methods (AIM) and Bio inspired Methods (BIM) [6]. The MPPT based AIM [7–9], such as Artificial Neural Networks (ANN) requires huge computation time and months of training to ensure the tracking of the MPP. Regarding the BIM, they ensure optimal searching ability without involving excessive mathematical computations [10, 11]. In addition, their implementation simplicity makes them very attractive for solving the MPPT problem.

In this study, an MPPT based BIM is developed, this MPPT is based on a swarm intelligent algorithm namely Bat Swarm Optimization [12, 13] used for maximum power point tracking which is analyzed and compared with the well-known conventional Perturb and Observe (P&O) [14] MPPT controller. The paper is organized as follows, in Sect. 2, we present the system overview of a PVS. Section 3 deals with the techniques of MPPT, where the proposed BSO algorithm is presented, followed in Sect. 4 by the simulation results and Finally, a conclusion is given.

2 System Overview

Figure 1 illustrates the whole architecture of a PV system which contains a solar panel for energy extraction from the sun, a DC/DC converter, a resistive load and an MPPT controller.

The equivalent circuit of a solar panel, which is composed of several photovoltaic cells employing parallel, series or series/parallel is depicted in Fig. 2. The model consists of a current source ipv, a diode d and a couple of resistances rs and rp.

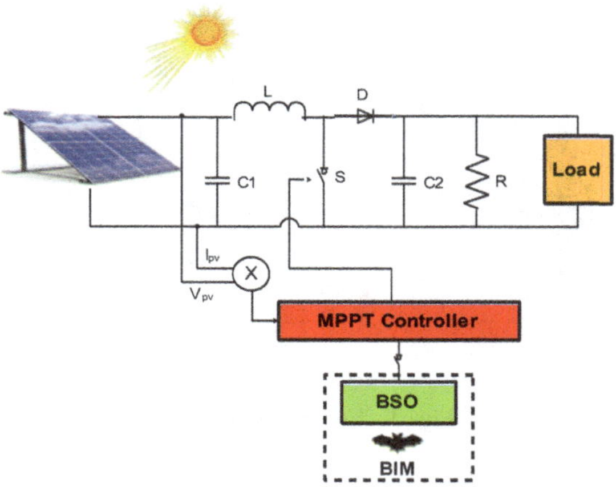

Fig. 1. Photovoltaic system architecture

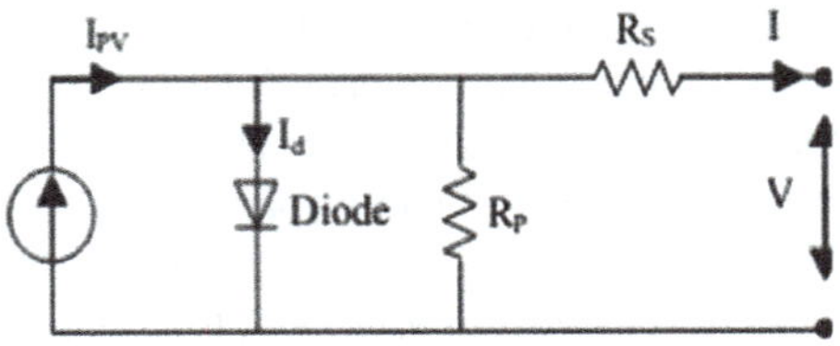

Fig. 2. Equivalent circuit of a PV cell

Equation (1) show the current produced by the solar cell.

$$I = Ipv - I_0\left[e^{\frac{(V+I.R_S)}{aVt}} - 1\right] - \frac{V + IR_s}{R_P} \tag{1}$$

The PV module operation depends strongly on the load characteristics to which it is connected, under constant uniform irradiance the current–voltage (I–V) characteristic has a unique point on the curve, called the maximum power point (MPP), at which the array operates with maximum efficiency and produces maximum output power. Furthermore the characteristics of a PV system vary with temperature (Tc) and insolation (S). Thus, MPPT controller is required to track the new modified MPP in its corresponding curve whenever temperature and/or insolation variation occurs.

3 Bat Swarm Optimization Algorithm

3.1 Bat Swarm Optimization Basic Concepts

The standard BAT search algorithm is a Bio Inspired algorithm developed by Yang [13] in 2010, and used for solving various optimization problems. As depicted in Fig. 3, the algorithm is inspired by the echolocation behavior of natural bats in locating their foods and avoids obstacles.

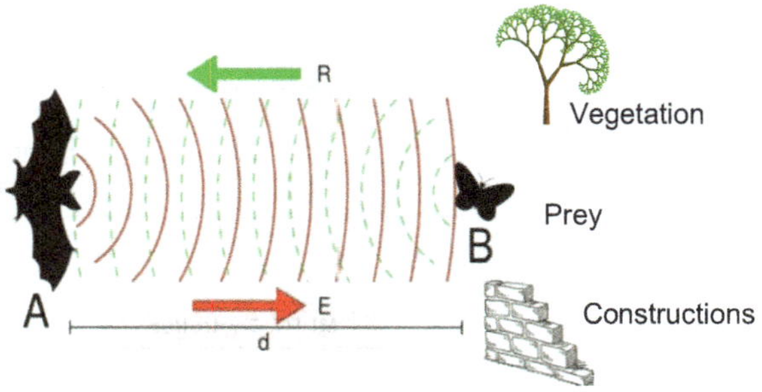

Fig. 3. Echolocation of bat.

When searching for prey, these bats emit a loud and short pulse of sound (wave), wait a fraction of time for the echo to return back to their ears. With this echo, a bat can decide an object's dimensions, shape, direction, and movement. The loudness increases and decreases when approaching towards prey. Accordingly, bats can determine how far they are from the surrounding objects. In order to transform these behaviors of bats to algorithm, Yang introduced the following rules [13]:

1 All bats use echolocation to sense distance, and also can differentiate between food, prey and barriers;
2 All bats fly randomly with velocity $\boldsymbol{v_i}$ at position $\boldsymbol{x_i}$ with a fixed frequency $\boldsymbol{f_{min}}$, varying wavelength $\boldsymbol{\lambda}$ and loudness $\boldsymbol{A_0}$ to search for prey. They can automatically adjust the wavelength (frequency) of their emitted pulses and adjust the rate of pulse emission $\boldsymbol{r}$ in range of [0,1], depending on the proximity of their target;
3 The loudness $\boldsymbol{A}$ can varies from a large (positive) $\boldsymbol{A_0}$ to a minimum constant value $\boldsymbol{A_{min}}$.

For each bat (i), the position $\boldsymbol{x_i}$ and velocity $\boldsymbol{v_i}$ should be defined and update during the iterations. The mathematical equations for updating the locations $\boldsymbol{x_i}$ and velocities $\boldsymbol{v_i}$ can be written as:

$$x_i^t = x_i^{t-1} + v_i^t \tag{2}$$

$$v_i^t = v_i^{t-1} + \left(x_i^t - x^*\right) f_i \tag{3}$$

$$f_i = f_{min} + \left(f_{max} - f_{min}\right) \alpha \tag{4}$$

where α in the range of [0,1] is a random vector drawn from a uniform distribution, x^* is the current global best location, which is achieved after comparing all the locations among all the n bats.

3.2 The Proposed Bat Swarm Optimization MPPT Implementation

This section describes the implementation of the BSO algorithm for solving the problem involved to MPPT controller in PV system. The flow chart of the proposed BSO based MPPT method is depicted in Fig. 4.

The PV power serves as the target function. The function to optimize is $\mathbf{P_{PV}} = \mathbf{f}\ (\mathbf{V_{PV}})$. The algorithm starts by having no previous information about the position of the best value (bat).

In the beginning, bats are generated randomly in the interval $\mathbf{[0,\ V_{co}]}$. For each solution there is a corresponding fitness function. Based on these solutions, the archive is constructed.

Depending on the region strength, the transition probability of bats is calculated. By successive iteration bats move towards the optimized point i.e. MPP. Figure 5 a–b and c, depicts the different steps according to the fitness function, the restrict search space and bats movement towards the optimized point, i.e. the MPP.

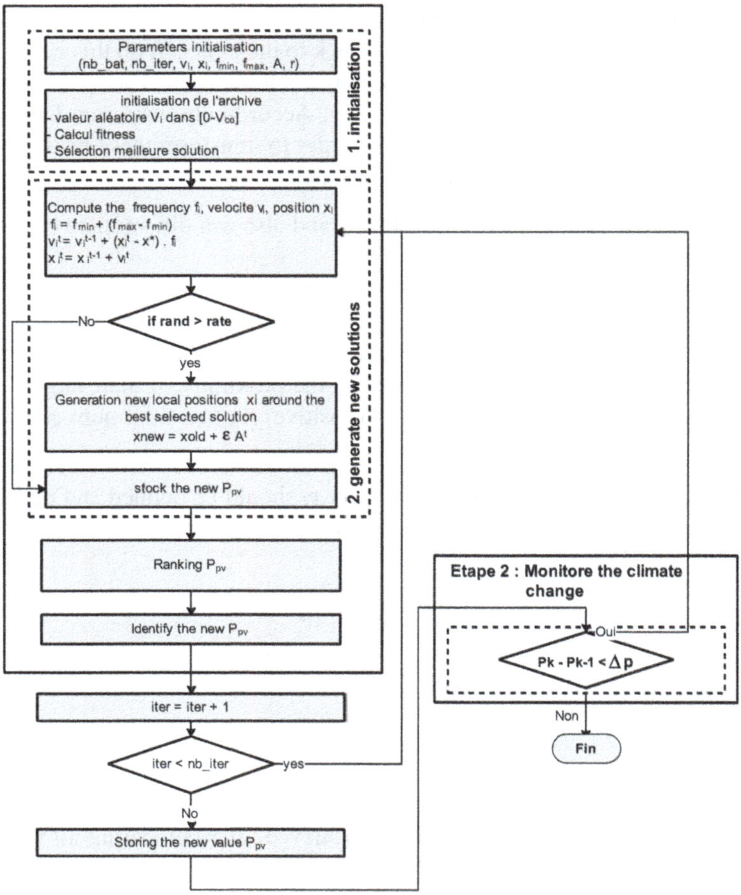

Fig. 4. Flow chart of the BSO MPPT controller.

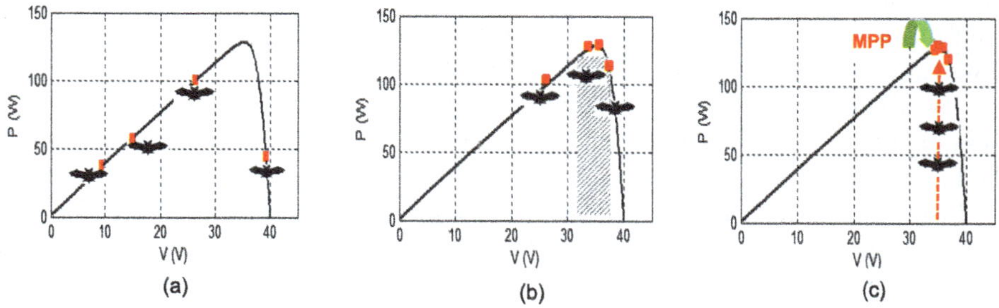

Fig. 5. (a) Distribution of the bat population; (b) Echolocation of bat; (c) Bats move towards the optimized point, i.e. the MPP.

4 Simulation Results

For the simulation, we used the Matlab/Simulink environment tools [15]. These facilitate the change of the atmospheric conditions (T°, S) to assess the MPP tracking trajectory, thus several scenarios are developed in order to evaluate the MPPT respond under dynamic operating conditions (Fig. 6).

The MPPT based BSO algorithm was tested and compared to the well-known Perturb and Observe algorithm P&O.

Firstly, the BSO MPPT controller is simulated at standard conditions STCD (1000 W/m^2, 25 °C), then at rapidly changing atmospheric conditions.

The temperature is constant at 25 °C and the irradiance is varying from 500 to 1000 W/m^2. The performance of each MPPT controller was evaluated at steady state condition.

Figures 7 and 8 depicts respectively the progression of the output power delivered to the load for the P&O and BSO algorithms under standard conditions (1000 W/m^2, 25 °C).

As depicted in Fig. 9, the BSO MPPT controller converges rapidly to the MPP. This latter is reached after only 0.7 s with a zero oscillations around the MPP.

Regarding to the MPPT based P&O, results show that the P&O MPPT controller converge slowly, the MPP is reached after 9.8 s with a continuous oscillations around the MPP which cause energy losses.

In the case of rapidly changing atmospheric conditions regarding to irradiance, the condition of changing solar insolation was modeled as depicted in Fig. 9. The temperature is constant at 25 °C and the irradiance is varying from 500 to 1000 W/m^2. From these results, it is shown that the BSO controller gives better results compared to P&O controller in terms of power value, time response and oscillations around the MPP. In Fig. 10, it is clearly shown that at 20 s, in the transition of the irradiance from 500 to 1000 W/m^2, the BSO MPPT controller follow the variation of the irradiance and track the MPP rapidly, it take negligible time to stabilize with no oscillation around the MPP, whereas for the P&O controller, as illustrate in Fig. 11, its take more time (5 s) to track the MPP and adjust the value of the power, where the oscillations around the MPP are still not eliminated.

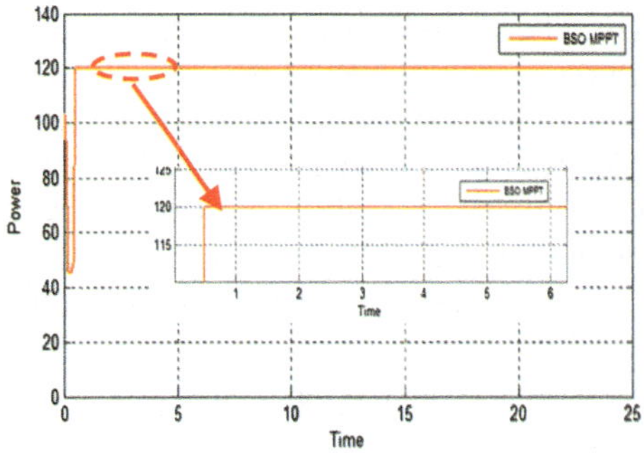

Fig. 6. BSO Power curve under STDC.

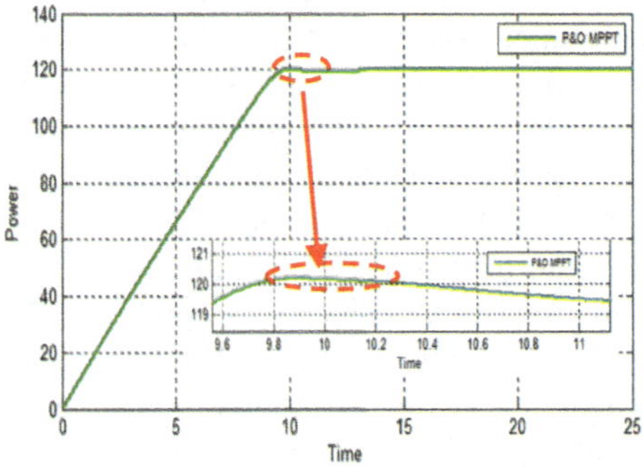

Fig. 7. P&O Power curve under STDC.

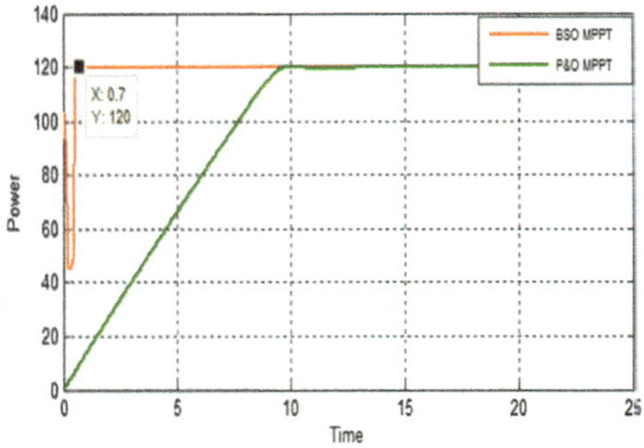

Fig. 8. Power curves of the P&O and BSO under STDC.

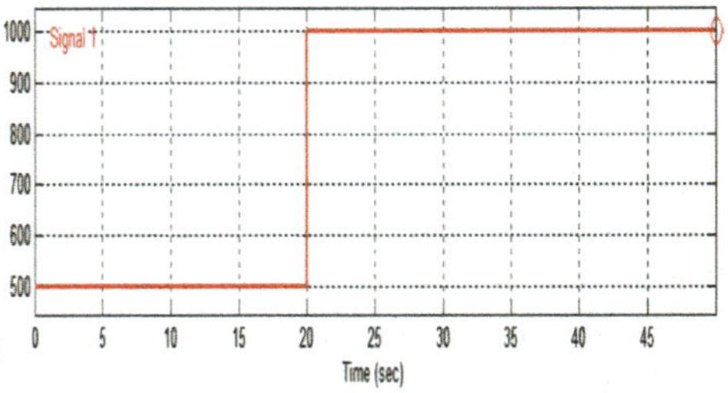

Fig. 9. Variation of the irradiance level from 500 to 1000 W/m^2

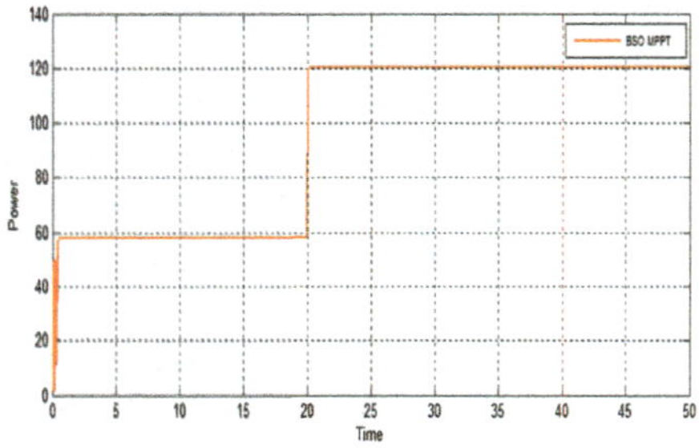

Fig. 10. BSO Power curve under rapid variation of irradiance.

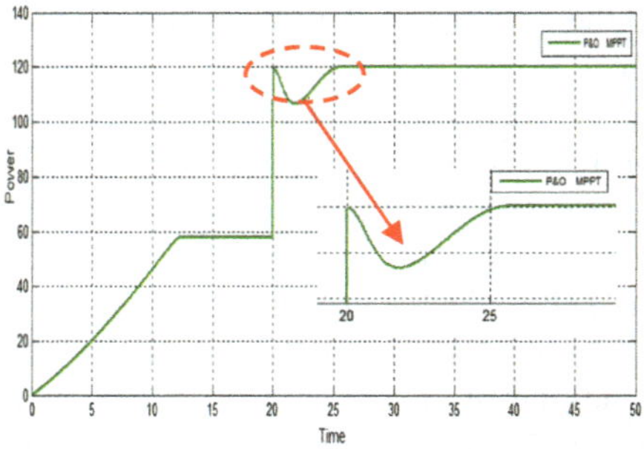

Fig. 11. P&O Power curve under rapid variation of irradiance

5 Conclusion

In this study, a new bio inspired algorithm namely bat swarm optimization algorithm (BSO) used for maximum power point tracking is developed, analyzed and compared to the well-known conventional method with the Perturb and Observe (P&O) MPPT controller. The results show that the BSO gives better performances in term of convergence speed, accuracy, stability and robustness; where the BSO controller reached the MPP speedily with a good accuracy and no oscillation around the point at steady state even for standard condition or when a rapid variation of irradiance occurs compared to the P&O MPPT controller. Regarding to the P&O controllers, the biggest drawback of the controller is due to the multiple oscillations around the MPP, especially under rapid variation of the atmospheric condition. Thus, the BSO can be considered as one of the main competitive algorithms for tracking the maximum power point of a Photovoltaic system under rapid variation of weather (Solar irradiance and Temperature) where it is able to readjust rapidly to the new conditions compared to P&O where it take more time. As future work we aim to test the proposed algorithm to solve the problem of partial shading then propose a hardware implementation for the proposed method.

References

1. Eason, G., Noble, B., Sneddon, I.N.: A review of improved maximum peak power tracking algorithms for photovoltaic. Int. J. Electr. Eng. Technol. (IJEET) **1**, 85–107 (2010). ISSN 0976-6553
2. Salas, V., Olias, E., Barrado, A., Lazaro, A.: A review of the maximum power point tracking algorithms for stand alone photovoltaic systems. Sol. Energy Mater. Sol. Cells **90**, 1555–1578 (2006)
3. Mellit, A., Kalougirou, S.A.: Artificial intelligence techniques for photovoltaic application: a review. Prog. Energy Combust. Sci. **34**, 574–632 (2008)
4. Mellit, A.: A soft computing technique modeling and simulation of stand alone photovoltaic power supply system. Int. Rev. Electr. Eng. IREE **2**(1), 480–488 (2007)
5. Titri, S., Larbes, C., Youcef Toumi, K.: Rapid prototyping of PVS into FPGA: from model based design to FPGA/ASICs implementation. In: 2014 9th International Design and Test Symposium (2014)
6. Titri, S., Larbes, C., Youcef Toumi, K., Benatchba, K.: A new MPPT controller based on the ant colony optimization algorithm for photovoltaic systems under partial shading conditions. Appl. Soft Comput. J. http://dx.doi.org/10.1016/j.asoc.2017.05.017
7. Larbes, C., Ait Cheikh, S.M., Obeidi, T., Zerguerras, A.: Genetic algorithms optimized fuzzy logic control for the MPPT in photovoltaic system. Renew. Energy **34**(10), 2093–2100 (2009)
8. Ait Cheikh, M.S., Larbes, C., Tchoketch Kebir, G.F., Zergueras, A.: Maximum power point tracking using fuzzy logic scheme. Revue des Energies Renouvelables **10**(3), 387–395 (2007)
9. Chekired, F., Mahrane, A., Chikh, M., Smara, Z.: Optimization of energy management of a photovoltaic system by the fuzzy logic technique. Energy Procedia **6**, 513–521 (2011)
10. Hanif Halim, A., Ismail, I.: Bio inspired optimization method: a review. Int. J. Artif. Intell. **1** (2014)
11. Binitha, S., Siva Sathya, S.: A survey of bio inspired optimization algorithms. Int. J. Soft Comput. Eng. IJSCE **2**(2), 137–150 (2012)
12. Yang, X.S.: BAT algorithm for multiobjective optimization. Int. J. Bio Inspired Comput. **3**(5) (2011)
13. Yang, X.S.: A new metaheuristic bat inspired algorithm. In: Nature Inspired Cooperative Strategies for Optimization, vol. 284. Springer Press (2010)
14. Liu, X., Lopez, L.A.C.: An improved perturb and observe maximum power point tracking algorithm for PV arrays. In: IEEE 35th Power Electronics Specialists Conference, PESC 2004, pp. 2005–2010 (2004)
15. www.mathworks.com

New Design of an Optimized Synergetic Control by Hybrid BFO-PSO for PMSG Integrated in Wind Energy Conversion System Using Variable Step HCS Fuzzy MPPT

M. Beghdadi(✉), K. Kouzi, and A. Ameur

Laboratory of Semi-Conductors and Functional Materials, University Amar Telidji Laghouat, Laghouat, Algeria
beghdadi.elbayadh@gmail.com, k.kouzi@lagh-univ.dz

Abstract. The aim of this study is to improve the performance of wind energy conversion system (WECS) based on permanent magnet synchronous generator (PMSG) using an optimized synergetic control by hybrid BFO-PSO. For this, it is synthesized on one hand a new synergetic controller model of PMSG which can guarantee fast response, asymptotic stability of the closed-loop system in the all range operating condition, and system robustness in presence of parameter variation. On the other one, to overcome the problem of synergetic controller parameter tuning, it is proposed in this work, hybrid BFO-PSO the algorithm which allow obtaining the optimal parameter of proposed regulator and consequently improve the performance of control system. Moreover, in order to make sure that the speed generator continues effectively the variation of the wind speed and guarantee the maximum extracted power from the wind energy the hill-climb search (HCS) fuzzy-based MPPT was proposed.

The obtained simulation results illustrate clearly the efficiency of proposed method witch work effectively under very variable wind speed and with high-performances compare to the conventional HCS.

Keywords: PMSG · BFO-PSO · Fuzzy-based wind MPPT · Synergetic control · HCS (P&O) · Wind energy conversion system WECS

1 Introduction

With the fast-growing demand on energy in the world because the awesome advance technology in industry field and for the many drawbacks of conventional energy like fossil energy especially the pollution problem, it is be more than necessary to go towards renewable energy source. The wind turbine installed around the world has grown fast in the last two decades to be over than 440 GW and will reach 760 GW in the horizon of 2020 [1].

In the wind turbine there are three mains types of generator which are the squirrel cage inductor generator (SCIG), the double fed inductor generator (DFIG) and the permanent magnet synchronous generator. The using of PMSM in the WECS offer many advantages as the high efficiency, the possibility of working without the gearbox

M. Hatti (Ed.): ICAIRES 2019, LNNS 102, pp. 30–40, 2020.
https://doi.org/10.1007/978-3-030-37207-1_4

who rise up the reliability of the system and the effective control methods available for this machine.

The synergetic control is very powerful method where it takes distinguished place and acceptance in the control community because its advantages as the robustness against the parameters variating and the high-performances in the stability, precision and the settling time. But there are control parameters in this technique need to be set up.

The choice of parameters control of SC is the master key of performance of this method where inappropriate values can give control with large settling time or with overshoot and that inacceptable especially when we use in the WECS where we need very fast techniques to keep up the quick variation of wind speed.

To select the right parameters of any control method the best way is using optimization methods. In optimization of speed control technique, the error between the reference and the measured is formulated as fitness function where the objective of these methods is to minimize this function to assure the right selecting of control parameters and that mean a control with the best possible performances [2]. There are many optimization algorithms as PSO, GA, ACO, NA and BFO where each algorithm has advantages and drawbacks comparing to others.

To guarantee the maximum generated power in WECS under variable wind speed we must tracking the optimal power according to wind speed as we call maximum power point tracking MPPT. Many MPPT was invented which are regrouped in two big families the first methods need the pre-knowing of wind turbine characteristics of power in function of wind speed and it's defined by experiences or simulations, then we have MPPT with no need to wind turbine characteristics. In the first category we find the tip speed ratio (TSR) MPPT, power signal feedback (PSR) and the torque optimal control (TOC). In the second category we have many methods but the famous one is hill-climb search (HCS) MPPT which knows also by perturb and observe (P&O).

This paper is constructed as follows: in Sect. 2, the modeling of the wind generator is presented. Section 3 deals with the design of synergetic control of PMSG. The Hybrid BFO-PSO Algorithm based on synergetic control is presented in Sect. 4. In Sect. 5 the HCS fuzzy-based MPPT is proposed. In Sect. 6 the performances of the proposed control are illustrated by some simulation.

2 Mathematical Model of PMSG

For reason of simplicity of mathematical model, we will assume that the air gap is perfectly constant so $L_d = L_q = L_s$ and the rotor permanent magnetic field is distributed in sine wave form in this air gap.

By using of dq (Park base on Concordia transformer) synchronous system, the two components of stator voltage become:

$$\begin{cases} V_d = R_s I_d + L_s \dot{I}_d - p\Omega L_s I_q \\ V_q = R_s I_q + L_s \dot{I}_q + p\Omega L_s I_d + p\Omega \psi_{d0} \end{cases} \tag{1}$$

By using the second low of Newton we find:

$$\dot{\Omega} = \frac{T_{em} - T_r - f\Omega}{Jp}, \quad T_{em} = p\psi_{d0} I_q \tag{2}$$

In the Eq. (2) the torque is in function of only the quadrature component of stator current I_q ($p\psi_{d0} = ct$) that guide us to the principle of vector control which is the separating of the control of machine torque and the rotor magnetic flux in other word we can control the torque (speed) by controlling I_q and the rotor flux by I_d. In the case of PMSM the rotor magnetic field is permanent and constant so there is no need to I_d and generally it is controlled to be none and that to reduce the Joule losses [3].

3 Synergetic Control

In the past, the regulation in industry was done strictly by using conventional regulator as PI but this regulation is weak against the parameters changing of the machines like the changing of the stator resistance because of short-circuit. So many researches done for solving this problem by using adaptation mechanism in first time then the appearance of the sliding mode control who was an interesting solution but with this control method another problem appear which is the chattering phenomenon causing serious problems in the control of motors and generators.

For better performances and reducing the chattering problem Kolesnikov propose new control approach [4] called the synergetic control becoming an alternative method of the SMC (sliding mode control).

Recently the synergetic control earns its place in the robust control community as well the industrial partners and that after the successful implementation in power electronics and its industrial application in battery charging [5, 6].

Any n-order nonlinear dynamic system can be described by:

$$\dot{x} = f(x, u, t) \tag{3}$$

where x is the state vector, u is the control vector, and t is the time.

The synergetic control can be defined in these steps:

Step 1: selection of macro-variables ψ_s who can be set as a linear relation or combinations of the variables of the system [7] and that reduce significantly the order of our system.

The main objective of the control is forcing system to operate on the manifold $\Psi s = 0$.

Step 2: put the dynamic evolution for macro-variable as next:

$$T\dot{\Psi}_s + \varphi(\Psi_s) = 0 \tag{4}$$

where **T** is a control parameter and ϕ is function of Ψ_s but it must be choosing according to two conditions: $\phi(0) = 0$ and $\phi(\Psi_s).\Psi s > 0$.

The simplest way to define ϕ it to set ϕ = Ψs and that assure respect of previous conditions and the dynamic evolution become:

$$T\dot{\Psi}_s + \Psi_s = 0 \tag{5}$$

Step 3: it's can be writing $\dot{\Psi}\mathrm{s} = \frac{d\Psi S}{dt} = \frac{d\Psi S}{dx}\frac{dx}{dt}$ (8) and we know that $\frac{dx}{dt} = \dot{x}$ so by substitution of (6) and (3) in (5) we find:

$$T\frac{d\Psi_s}{dx}f(x,u,t) + \Psi_s = 0 \tag{6}$$

Step 4: solving (6) give us the general solution u:

$$u = g(x, \Psi_s, t, T) \tag{7}$$

by finding the solution 'u' we can easily control the system to follow the reference value by the control vector.

3.1 Synergetic Control on PMSG

First thing for build the SC (synergetic control) it's defining the macro-variables, the direct component of stator current can be control just with simple PI regulator because it's will set to be zero and it has weak depending on the machine parameters.

As we have now one channel of control, we must choose one macro-variable Ψs. For better performances of our control and because we want to control the speed of the machine the macro-variable will be [7]:

$$\Psi_s = K_1(\Omega^* - \Omega) + K_2\int(\Omega^* - \Omega)dt \tag{8}$$

with Ω* is the reference speed, K_1 and K_2 are two control parameters to change the performances of the SC to make it fastest and more stable.

By the substitution of (8) in (5) give us:

$$T\left(-K_1\dot{\Omega} + K_2(\Omega^* - \Omega) + K_1(\Omega^* - \Omega)\right) + K_2\int(\Omega^* - \Omega)dt \tag{9}$$

Finally, by substitution and after simplifying the V_q become:

$$\begin{aligned} Vq = R_s\left(\frac{J}{p\psi_{d0}}\left(\frac{T_r + f\Omega}{J} + \left(\frac{1}{T} + \frac{K_2}{K_1}\right)(\Omega^* - \Omega)\right) + \frac{K_2}{TK_1}\int(\Omega^* - \Omega)dt\right) \\ + L_s\dot{I}_q + p\Omega L_s I_d + p\Omega\psi_{d0} \end{aligned}$$

After the analysis of Eq. (10) we know that the control of V_q allows to control speed but the system answer depend on control parameters T, K_1 and K_2. Those parameters effect the performances of the system like the stabilisation, precision and the speed of system answer. Wrong value of these parameters can make the system slow, with big overshoot or even instable so it's must be choosing carefully and the best way to do that is by using optimization algorithms who provide for us appropriate values with satisfy performances and that what will see in the next section.

4 Proposed Hybrid BFO-PSO Algorithm

The BFO is a powerful optimization method but there is one problem which can affects our results which is the weak information exchange capacity between the bacteria of the population and this can lead us to slow or even non-convergence [8]. So, the best solution for this problem is to combine the BFO with the PSO because the last one has an excellent information exchange between the particles of swarm by determination the global best position gbest (Fig. 1).

The proposed algorithm for searching the best values of T, K_1 and K_2 is detailed as next:

a. set all the initial characteristics as the size of the swarm, random x_i^k, v_i^k, bacteria number and its position and finally determinate the three parts of the BFO: Chemotaxis, Reproduction, Elimination and Dispersal of initial population of bacteria for the 3-dimension.
b. like the PSO for each vector of bacteria (T, K_1 and $K_{2)}$ define and evaluate the fitness function.
c. turn over the bacteria for searching of the best position.
d. if the fitness of bacteria improves continue with the rest of the algorithm else comeback to step c.
e. if the bacteria do the maximum chemotactic step number complete with reproduction, elimination and dispersing operation else comeback to c.
f. do the division of the best fitness bacteria for make a new generation and comeback to c.
g. if we arrive at the maximum number of iterations the final vector of T, K_1 and K_2 is the best value to make sure that the synergetic control can be done with high-performances.

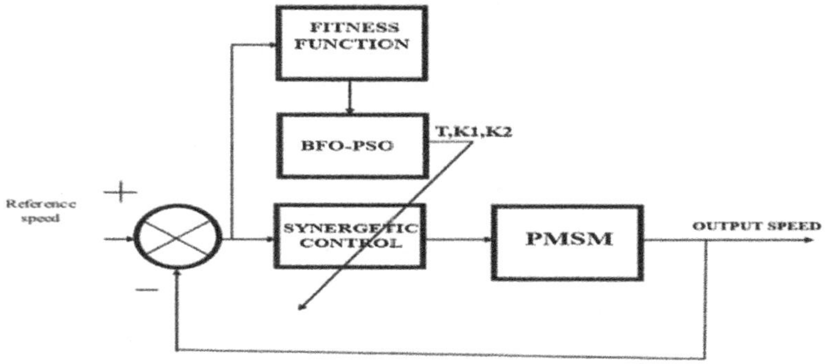

Fig. 1. Tuning SC parameters by hybrid BFO-PSO.

5 Maximum Power Point Tracking MPPT

The principle of HLC is very simple, first we have to measure the generator speed and the generated power. After that step will added to previous speed to increase it and measure the generated power again, if $\Delta P > 0$ that's mean we are in the right way and we must continue to increase the speed until ΔP become negative. In the opposite case we must decrease the speed to reach the pick point of power P_{opt} (Fig. 2). So, the control starts with x_i then x_{i+1} and care on to reach the peak point, if we pass this point the step speed change the sign to comeback to it and this is the reason of called it hill-climb trying keep $dP/d\Omega = 0$.

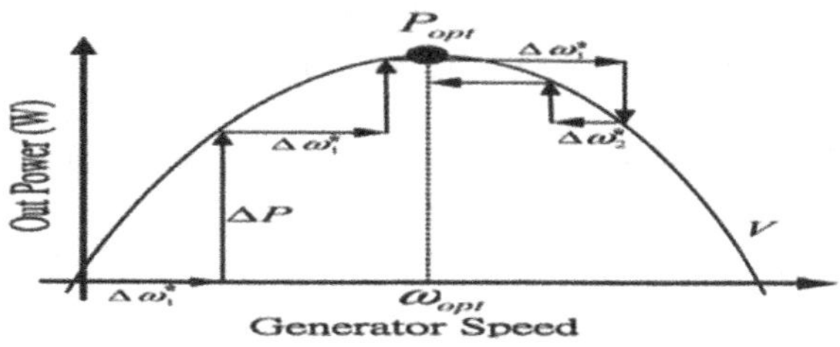

Fig. 2. HLC (P&O) principle.

So, the HTC use the principle of search-remember-reuse technique where in the beginning the process start with empty memory and modest performances and after the search is improved and we came to the peak point quickly and that allow to us to use the possible maximum power from the wind [9].

But there are two main disadvantages in this method which are: HCS become very slow if we choose a small step and this it's mean weak efficiency under high variable wind speed and the second is the serious oscillations around the MPPT point in case we select a big step [10].

5.1 Proposed HTC Fuzzy-Based MPPT

Selecting adaptive step is the optimal solution to overcome the drawbacks of the HCS which mean clearly set a large step when we are far from the peak point and a small one when we are arriving near to it. There are many ways to make the step variable but, in this paper, we are interesting of the fuzzy-based HCS.

In the case of PMSG there are three way to control the machine, controlling machine via the grid side converter, by the chopper in the DC bus and finally by the generator side converter.

If the two first ones are selected, the control of machine is done by the controlling of the DC voltage but with the third one, we control the speed directly by the gate signal of the generator side converter (Fig. 3).

In our work we choose the control by the generator side converter because two reasons:

- the possibility of adapting this mode easy with the synergetic control mentioned in Sect. 3.1;
- in this case we can use a matrix converter which give us many advantages and suppression of the DC bus stage.

The principle of HCS fuzzy-based is the same as the standard HCS the different is in the step size, where in fuzzy is variable according to the rules define by the user.

There are three main steps in any fuzzy:

- fuzzification where the inputs convert to linguistic variable, in our case N for negative, P: positive, S: small, M: middle, B: big and EZ: equal to zero;
- Table of rule which is the relation between the linguistic variable of inputs and outputs;
- The defuzzification which is the opposite operation of fuzzification to have numerical outputs.

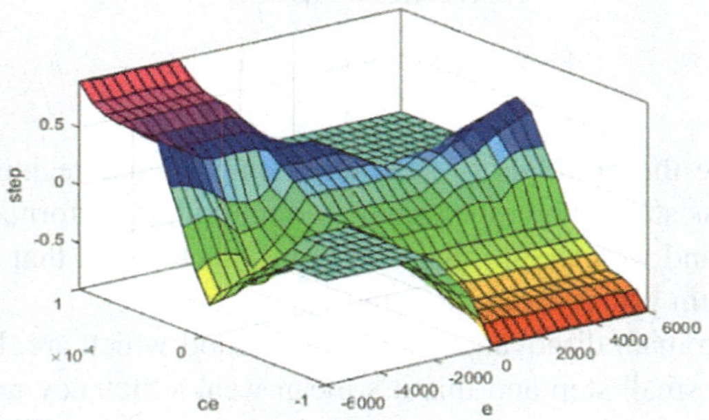

Fig. 3. Surface of rules.

6 Simulation Results and Discussion

The simulation of the system under study will be done with MATLAB/SIMULINK 2015a, where we have two major axes of simulation which are: optimization of the synergetic control by using hybrid BFO-PSO method and integration of this system in wind energy conversion system using variable step HCS fuzzy-based MPPT.

All the parameters of the simulation like PMSM parameters, synergetic control, BFO-PSO and the wind turbine is mentioned in Table 1.

6.1 Simulation of Synergetic Control of PMSM Optimized by BFO-PSO

The first thing to do for better evaluating of the performances of the optimized control is simulate the system under negative fixe load torque (generator mode) Tr = −30 N.m and step speed 1800 tr/mn at t = 0 s, 1200 tr/mn at t = 1.5 s and 200 tr/mn at t = 3 s. the period of the simulation is 4 s. We will use ITSE as fitness function.

After the appliqueing of BFO-PSO optimization on SC it provides the parameters T, K_1 and K_2 mention in Table 1. The simulation provides the results in Fig. 4.

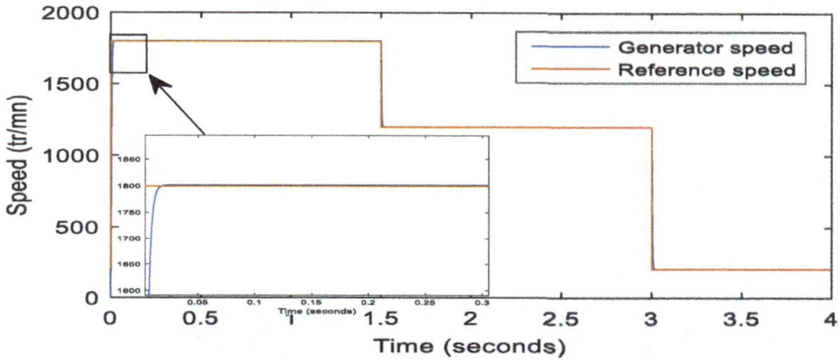

Fig. 4. Generator and reference speed.

It's clear in the figure that the performances of the control are excellent with high quality, very fast where is the settling time less than 0.02 s, no overshoot speed, high precision (static error ε_s = 0.08%) and very weak steady state oscillations.

With these results we confirm that the BFO-PSO is very powerful optimization method to tuning the synergetic control parameters by giving parameters T, K_1 and K_2 to become very fast, robust, stable, have a big precision and without overshoot.

6.2 Integration the System in Wind Energy System with HLC Fuzzy-Based MPPT

Now after evaluating the performances of the optimal control, the system integrated in wind energy conversion system with fuzzy-based HCS (P&O). The speed of wind is making by Sampled Gaussian Noise with mean speed = 11 m/s (Zone 2) and standard deviation = 1.

The period of the simulation is 10 s but these time present 10 h in real time (1 s → 1 h) to complete the simulation with reasonable time.

To confirm the performances and the speed answer of our MPPT algorithm, the wind turbine will expose to quick variable wind. The simulation gives the next results (Fig. 5):

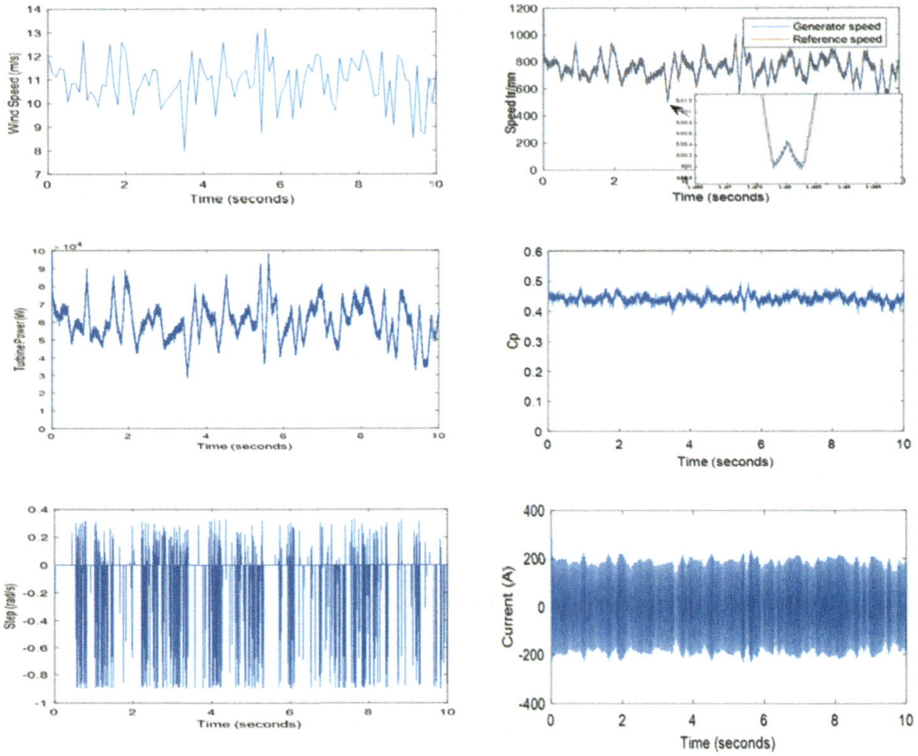

Fig. 5. Simulation results of WECS with synergetic control and HCS fuzzy-based MPPT.

The system although is it training by very variable wind speed, but it works very well and give us satisfy performances. We see that the reference speed provides by HCS fuzzy-based MPPT follow the variation of the wind speed effectively, fast and with very low oscillations proving that by using fuzzy in HCS assure almost the disappear of the major drawbacks of the standard HCS which are the big oscillation and the slowness in its answer. The generator speed also follows the reference speed very effectively, without overshoot and with high precision and this confirm the robustness and the power of synergetic control without forget the role of hybrid BFO-PSO optimization algorithm because it gives us a very appropriate values of control parameters to make our SC as good enough.

All the measured values of the turbine as the turbine power P_T, Cp, λ and stator currents follow exactly the variation of wind speed to reach the biggest value when the wind speed become over 13 m/s where the power is near to rated power (100 KW), Cp $\approx$ 0.48, $\lambda \approx 2.7$, $I_s \approx 155.56$ A. Cp and λ varying in tight band (around 0.43 for Cp and 2.3 for λ) to exploit the maximum possible power from the wind and rise up the efficiency of the WECS.

Table 1. Simulation's parameters.

Type of parameters	Designation	Symbol	Value
Turbine	Blade radius, gain de gearbox, air density, inertia, cut-out wind speed	R, G, ρ, J_T, v_{max}	20 m, 60, 1.225 kg/m^3, 64 N.m.s/rad, 16 m/s
PMSM	Rated power, stator resistor, stator inductor, inertia, rotor permanent magnetic flux	P_n, R_s, L_s, J, ψ_{d0}	100 KW, 0.6 Ω, 1.4 mH, 0.00417 N.m.s/rad, 0.9 Wb
BFO-PSO	Dimension of search space, number of bacteria, number of chemotactic steps, limits the length of a swim	p, s, Nc, Ns	3, 300, 100, 3
	PSO constants	c_1, c_2	1.5, 0.5
Synergetic control $_{(BFO\text{-}PSO)}$		T, K_1, K_2	8.42e-8, 13.2349, 2.5356

7 Conclusion

In this work an optimized synergetic control optimized by hybrid BFO-PSO for PMSG integrated in WECS which is driving by hill-climb search HCS fuzzy based MPPT is presented.

By using a BFO-PSO for the optimization of SC we have a right control parameter (T, K_1 and K_2) to make the control robust, fast, stable, without overshoot and with very high-precision.

After the optimal SC was testing, we integrated it in WECS all using an HCS fuzzy based MPPT to guarantee the maximum generator power without serious oscillations and fast enough to keep up the variation of wind speed and that by the adaptation of the speed step according to the distance from the peak point.

The simulation results prove the power of the optimal SC by BFO-PSO and also the many benefits of working with HCS fuzzy-based MPPT to make the wind turbine operate at its optimum power point for a wide range of wind speed and can be used in the environments where the wind speed varying quickly and continuously.

References

1. Blaabjerg, F., Ma, K.: Application of power electronics. In: Proceedings of the IEEE Wind Energy System, vol. 105, pp. 2116–2131 (2017)
2. Awadallah, A.: Parameter estimation of induction machines from nameplate data using particle swarm optimization and genetic algorithm techniques. In: Electric Power Components and Systems, pp. 801–814 (2008). https://doi.org/10.1080/15325000801911393
3. Retif, J.M.: Vector control of permanent magnetic synchronous machine. In: Vector Control of Asynchronous and Synchronous Machines, INSA, Lyon, p. 19 (2008)
4. Kolesnikov, A.: Introduction of synergetic control. In: American Control Conference, Portland, Oregon, USA, June, pp. 3013–3016 (2014)

5. Kolesnikov, A., Veselov, G.: A synergetic approach to the modeling of power electronic systems. In: Proceedings of COMPEL, Blacksburg, VA (2000)
6. Laribi, M., Ait Cheikh, M.S., Larbes, C., Essounbouli, N., Hamzaoui, A.: A sliding mode and synergetic control approach applied to induction motor. In: Proceedings of the 3rd International Conference on Systems and Control (2013)
7. Mingliang, Z., Tao, W.: A sliding mode and synergetic control approaches applied to permanent magnet synchronous motor. J. Phys. Conf. Ser. **1087** (2018). https://doi.org/10.1088/1742-6596/1087/4/042012
8. Wang, Y.K., Wang, J.S.: Optimization of PID controller based on PSO-BFO algorithm. In: IEEE 2016 28th Chinese Control and Decision Conference (CCDC), vol. 978, pp. 4633–4638 (2016)
9. Jogendra, S., Thongam, M.O.: MPPT control methods in wind energy conversion systems. In: Fundamental and Advanced Topics in Wind Power (2011). ISBN: 978-953-307-508-2. http://www.intechopen.com/books/fundamental-and-advancedtopics-in-wind-power/mppt-control-methods-in-wind-energy-conversion-systems
10. Mousa, H., Youssef, A., Mohamed, E.: Adaptive P&O MPPT algorithm based wind generation system using realistic wind fluctuations. Int. J. Electr. Power Energy Syst. (2019). https://doi.org/10.1016/j.ijepes.2019.04.038

ANFIS Technique to Estimate Daily Global Solar Radiation by Day in Southern Algeria

Abdeldjbbar Babahadj[1,2](✉), Lakhdar Rahmani[2], Kada Bouchouicha[1], Berbaoui Brahim[1], Ammar Necaibia[1], and Bellaoui Mebrouk[1]

[1] Unité de Recherche en Energies Renouvelables en Milieu Saharien (URERMS), Centre de Développement des Energies Renouvelables (CDER), 01000 Adrar, Algeria
Babahadj@urerms.dz, k.bouchouicha@gmail.com, berbaoui.brahim@gmail.com, necaibia.ammar@gmail.com, bellaoui_meb@hotmail.com

[2] Arids Zones Energetic Laboratory (ENERGARID), TAHRI Mohamed University of Béchar, BP 417, 08000 Béchar, Algeria
rahmani08@gmail.com

Abstract. In this paper, the performance and accuracy of the adaptive neuro-fuzzy inference system (ANFIS) technique is assessed for estimating the daily horizontal global solar radiation H in South Algeria. 21 years of experimental data for three cities in South Algeria Adrar, Timimoun and Bordj Badji Mokhtar (BBM) were used to train and test the ANFIS model. The results verified that the ANFIS model provides accurate predictions based on statistical formulas such as: the root mean square error (RMSE), mean absolute percentage error (MAPE and the coefficient of determination (R). In a nutshell, the results highly encouraged the implementation of ANFIS to estimate the daily horizontal global solar radiation using only the number of the day of years n_d.

Keywords: Radiation · Daily global · ANFIS · Solar · Adrar · Timimoun · Bordj Badji Mokhtar

1 Introduction

Currently, Algeria has an intention for utilizing further alternative energy resources due to different economic reasons and more importantly other environmental protection goals. Solar radiation is considered as the most important renewable energy sources in the world [1]. Algeria South has a greatly advantageous from geographic position viewpoint; which is characterized by a large global sun-belt and important solar radiation. Estimation of the solar radiation using ANFIS is one of such the adopted techniques for Assessment solar energy at the interest site. The barriers in the measurement of solar radiation have resulted in the evolution of so many models and algorithms for its estimation from some routinely measured meteorological parameters such as; sunshine hour, minimum, maximum and average air temperatures, relative humidity, and cloud factor [2]. According to that, many empirical models for prediction

M. Hatti (Ed.): ICAIRES 2019, LNNS 102, pp. 41–50, 2020.
https://doi.org/10.1007/978-3-030-37207-1_5

and estimating daily global solar radiation has been developed [3, 4]. In the last years, Artificial intelligence is an alternative to the old statistical methods, which have the ability to track complicated dependencies between different variables, where conventional methods have their limits [6]. Among the best techniques is the Adaptive Neuro-Fuzzy Inference System (ANFIS). It is a hybrid smart system that merges methodologies of the learning power of the artificial neural networks with the knowledge representation of fuzzy logic [5]. Computationally adaptability and efficiency are the most two main advantages of the ANFIS [8], which can be used as a tool for estimating solar radiation data [9]. Mellit et al. [6] used ANFIS for predicting the global solar radiation based on sunshine duration and air temperature in Algeria. Tymvios et al. [7] proposed a comparative study between the Angstrom-Prescott models and artificial neural network (ANN) in Athalassa for estimating daily global solar radiation. In Cyprus. They utilized the parameters of maximum possible sunshine hours, sunshine hours, the number of months and maximum air temperature and they developed five different ANN models; two models with 2 inputs and three models with 3 inputs. Benghanem et al. [8] developed six ANN-based models to estimate daily global solar radiation in Saudi Arabia. They used different combinations of the input variable of day number of the year, sunshine hours, relative humidity and ambient temperature. Their results demonstrated that the most significant element is the sunshine duration. Mohammadi et al. [9] developed a hybrid approach, which included generalized fuzzy models and hidden Markov models, to estimate solar irradiation in India. They considered 15 different sets of meteorological variables to assess their influence on solar radiation estimation. Their results displayed that the most appropriate model was dependent on sunshine duration. Other parameters such as air temperature, relative humidity, atmospheric pressure and wind speed were ranked in the next places. Mohammadi et al. [10] employed the adaptive neuro-fuzzy inference system (ANFIS) to chosen the most significant input parameters for estimation of daily H_d in Iran. In this study, an used of adaptive neuro-fuzzy inferences system (ANFIS) is suggested to develop a program computing-based model for estimation of daily global solar radiation by day of the year. The prime aim is evaluating the sufficiency of the ANFIS scheme to provide a convenient way for accurately predicting the daily global solar radiation using only one simple input. The potential of developed ANFIS model is verified by providing statistical comparisons between its estimation with those of three DYB models established by Aouna and Bouchouicha [11].

2 Data and Methodology

2.1 Study Area

Algeria is the big largest country in Africa with a total area of 2,381.741 km^2 of which 87″ are desert. Algeria extends between the longitudes 9° W and 12° E and the latitudes 19° and 37° N, as shown in Fig. 1. Figure 1 shows the location of the three selected cities, the Algeria frontier.

The National Aeronautical and Space Administration (NASA) is considered as one of the best important data sources through its earth science research program [13].

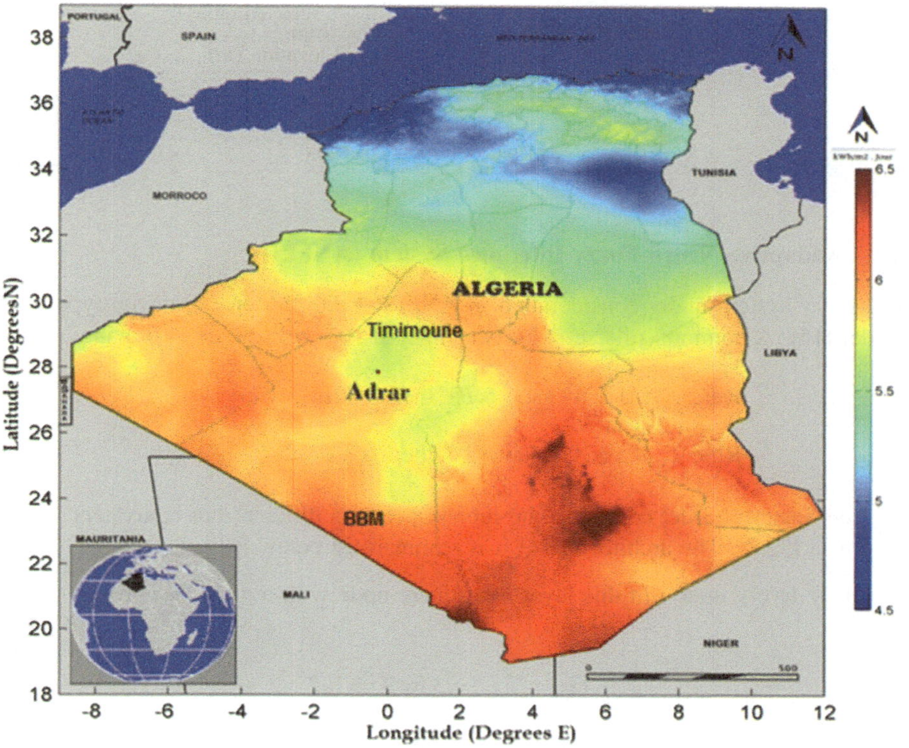

Fig. 1. The Annual average of the daily GHI (1983–2005) [12]

NASA has long supported both satellite systems and research providing data that is important for the study of weather and climate processes [14]. This data includes a large archive of over 200 satellite-derived meteorology and solar energy parameters for more than 21 years, from 1 July, 1983 to 30 June, 2005. The data is available on a 1° longitude by 1° latitude equal-angle grid covering the entire glob. Bouchouicha et al. [12] are validated the NASA-SSE solar data against historical ground measurements made in four Algerian National Office of Meteorology stations (Algiers, Tamanrasset, Oran, and Bechar) for over 10 years and another two radiometric stations (Adrar and Ghardaia) for more than 18 years. Regression analysis of this data with the monthly mean values of global solar irradiation shows a relative root mean square error of 12.4% and a relative mean bias error of (−4.6%). These results indicate that NASA's solar radiation data is acceptable for the development of the empirical models. 50% of the collected data were used for training and the subsequent 50% served for testing.

The geographical locations (latitude, longitude, elevation) for each of the three selected cities are showed in Table 1.

Table 1. Geographical locations of the selected cities

N	Location	Latitude	Longitude	Elevation (m)
1	Adrar	27.88	−0.28	263
2	Timimoun	29.25	0.28	312
3	BBM	21.33	0.95	398

2.2 Aadaptive Neuro-Fuzzy Inference System (ANFIS)

The fuzzy system under consideration in ANFIS is the first order Sugeno type fuzzy model [15]. A common rule set with two fuzzy if-then rules is the following:

$$\text{Rule 1 : If } x \text{ is } A1 \text{ and } y \text{ is } B1, \text{then} f_1 = c_{11}x + c_{12}y + c_{10} \tag{1}$$

$$\text{Rule 2 : If } x \text{ is } A2 \text{ and } y \text{ is } B2, \text{then} f_2 = c_{21}x + c_{22}y + c_{20} \tag{2}$$

The Sugeno Model could also be represented in a different but equivalent way as shown in Fig. 3. The architecture of it is summarized below [5]:

Layer 1. Every node i in this layer is a square node with a node function

$$O_{l,i} = \mu_{A,i}(x), \text{ for i} = 1,2 \tag{3}$$

Where $\mu_{A,i}(x)$, is the membership function of the fuzzy concept A_i. The usual choice of the membership function is of bell-shaped with maximum equal to 1 and minimum equal to 0, such as the generalized bell function or the Gaussian function

$$\mu_{A,i}(x) = \frac{1}{1 + \left[\left(\frac{x - c_i}{a}\right)^2\right]^{b_i}} \tag{4}$$

$$\mu_{A,i}(x) = exp\left[-\left(\frac{x - c_i}{a}\right)^2\right] \tag{5}$$

Where $\{a_i, b_i, ci\}$ (or $\{a_i, c_i\}$ in the latter case) is called the premise parameter set. The generalized bell function type is chosen for the present application.

Layer 2. Every node in this layer is a circle node labeled π which multiplies the incoming signals and sends the product out [16, 17]. For example,

$$\bar{w}_i = \mu_{A,i}(x) \times \mu_{B,i}(x) \; i = 1,2 \tag{6}$$

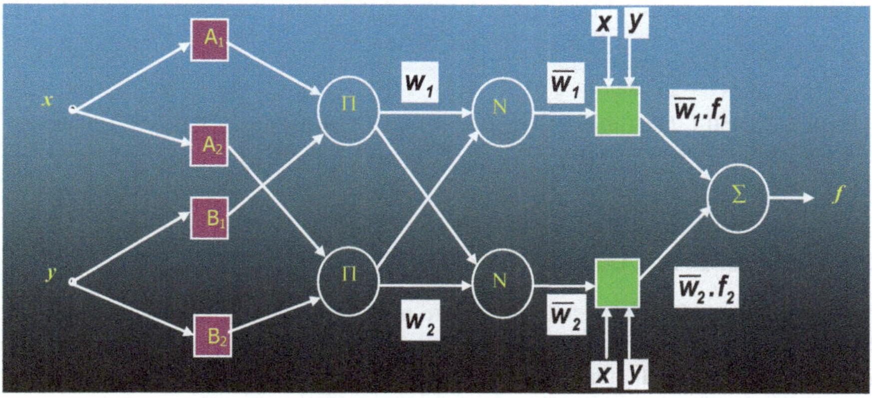

Fig. 2. ANFIS structure with two inputs

Layer 3. Every node in this layer is a circle node labeled N. The ith node calculates the ratio of the ith rule's firing strength to the sum of all rules' firing strengths (Fig. 2):

$$\bar{w}_i = \frac{w_i}{w_1 + w_2} \quad i = 1, 2. \tag{7}$$

Layer 4. Every node i in this layer is a square node with a node function

$$o_1^4 = \bar{w}_i f_i = \bar{w}_i(p_i x + q_i y + r_i). \tag{8}$$

Where $\bar{w}_i$ is the output of layer 3 and $\{p_i, q_i, r_i\}$ is called the consequent parameter set.

Layer 5. The single node in this layer is a circle node labeled Σ. It computes the overall output as the summation of all incoming signals [18], i.e.

$$o_1^5 = \sum\nolimits_i \bar{w}_i f_i \tag{9}$$

$$f_i = (\bar{w}_1 x)p_1 + (\bar{w}_1 y)q_1 + (\bar{w}_1)r_1 + (\bar{w}_2 x)p_2 + (\bar{w}_2 y)q_2 + (\bar{w}_2)r_2 \tag{10}$$

The learning method of ANFIS is a hybrid learning algorithm which consists of two passes, namely the forward pass and the backward pass. In the forward pass (the first pass), the premise parameters will be fixed and the consequent parameters are identified by the least square estimate. In the backward pass (the second pass), the consequent parameters will be fixed and the premise parameters will be updated by the gradient descent which is a supervised learning [19].

Table 2. The established DYB empirical models for Adrar, Timimoun and BBM [11]

N	Equation	Location
1	$H = 3.2847 + 267.86n_d + 2.67E - 4n_d^2 - 2.08E - 6n_d^3 + 3.14E - 9n_d^4$	Adrar
2	$H = 3.2950 + 2.66\sin\left(\frac{1.668\pi}{365}n_d - 0.8598\right) - 0.25cos\left(\frac{5.22\pi}{365}n_d - 1.127\right)$	Timimoun
3	$H = 6.17 + 1.907\sin\left(\frac{2.15\pi}{365}n_d - 1.558\right) - 0.7041cos\left(\frac{3.535\pi}{365}n_d + 0.8983\right)$	BBM

2.3 Statistical Error Analysis of Model ANFIS

The performance of the selected models are analyzed and evaluated with three statistical parameters. These are root mean square error (RMSE), mean absolute percentage error (MAPE) and coefficient of determination (R) test indicators. The mathematical formulae of the two statistical parameters are described below [20]:

$$RMSE = \left[\frac{1}{N}\sum\nolimits_{i=1}^{N}\left(\bar{H}_{i,c} - \bar{H}_{i,m}\right)^2\right]^{\frac{1}{2}} \tag{11}$$

$$MAPE = \frac{1}{N}\sum\nolimits_{i=1}^{N}\left|\frac{\bar{H}_{i,m} - \bar{H}_{i,c}}{\bar{H}_{i,m}}\right| \times 100 \tag{12}$$

$$R = \frac{\sum_{i=1}^{N}\left(\bar{H}_{i,c} - \bar{H}_{c,av}\right).\left(\bar{H}_{i,m} - \bar{H}_{m,av}\right)}{\sqrt{\sum_{i=1}^{N}\left(\bar{H}_{i,m} - \bar{H}_{m,av}\right)^2 \cdot \sum_{i=1}^{N}\left(\bar{H}_{i,c} - \bar{H}_{c,av}\right)^2}} \tag{13}$$

Where N is the total number of available data points, $H_{m,av}$, $H_{c,av}$ are the average of measured and estimated values of solar radiation (KWh/m^2) and $H_{i,c}$, $H_{i,m}$ are i^{th} estimated and measured daily global solar radiation (KWh/m^2) respectively.

Table 3. Statistical indicators obtained for the established DYB models for Adrar, Timimoun and BBM [11]

Model number	MAPE	RMSE (KWh/m^2)	R	Location
1	3.2034	0.2391	0.9892	Adrar
2	2.8499	0.2108	0.9924	Timimoun
3	2.6647	0.1998	0.9877	BBM

3 Results and Discussion

In this study, the potential of ANFIS technique to estimate the daily horizontal global solar radiation by day of the year as a single input was appraised. To achieve further reliability in the evaluations, the developed ANFIS model was tested by a data set that was not used during the training process. The suitability of the proposed ANFIS system was assessed statistically using different well-known indicators. Then to ensure the accuracy level of the ANFIS model, its performance was compared against DYB empirical models. Following offers the most significant results obtained in this research work. Figures 3, 4 and 5 show the scatter plots between the measured and computed global solar radiation values via the developed ANFIS model for Adrar, timmimoun and BBM respectively. At the beginning, the ANFIS networks were trained with the long-term averaged measured data. Three bell-shaped membership Functions were used to fuzzily the ANFIS input. Various types of membership function and also number of membership function were tested to recognize the most favorable type and number of membership functions. After training process the ANFIS networks were tested to calculate the daily horizontal global solar radiation (H) based on days of the year (n_d). Long-term measured daily global solar radiation (H) as the output parameter and the number of days (n_d) as the input parameter were collected and defined for the learning techniques. For this aim, one year averaged global radiation data achieved from the period of 10 years from January 1985 to December 1994 was used to train the samples and the one year averaged daily data set obtained from the remaining period of January 1995 to December 2004 was served to test the samples.

Table 4. Performance of the proposed ANFIS model based upon different statistical indicators

Location	Phase	MAPE	RMSE (KWh/m^2)	R
Adrar	Training	2.7593	0.2011	0.99558
	Testing	2.9584	0.2051	0.99396
Timimoun	Training	2.0879	0.1904	0.99341
	Testing	2.3478	0.1932	0.99301
BBM	Training	2.1493	0.1795	0.98993
	Testing	2.3986	0.1803	0.98983

The superiority of the developed ANFIS model can be justified by providing some comparisons with DYB empirical models. To achieve this, the performance of the ANFIS-based model is verified against the DYB empirical models previously established by Aouna and Bouchouicha [11] for Adrar, Timimoun and BBM. Table 2 shows these DYB models established based upon the long-term measured daily data set for three sites. The values RMSE, MAPE and R obtained by the study of Aouna and Bouchouicha [11] to evaluate the performance of three established DYB models are presented in Table 3. Nevertheless, by making a comparison between the statistical results listed in Table 3 with those of Table 4 it is apparently found that the ANFIS model enjoys greater performance compared to calibrated DYB empirical models. Thus, the developed ANFIS model can be introduced as the superior model to estimate the daily horizontal global solar radiation by the day of the year, from the results presented in Table 3 [11] and the obtained in this manuscript, it's clearly obvious that the better in terms of statistical indicators especially for the coefficient of determination (R) which was better compared to the empirical model for the 3 different locations (Adrar, Timmimoun, BBM) and the results for the comparison taken for the testing phase.

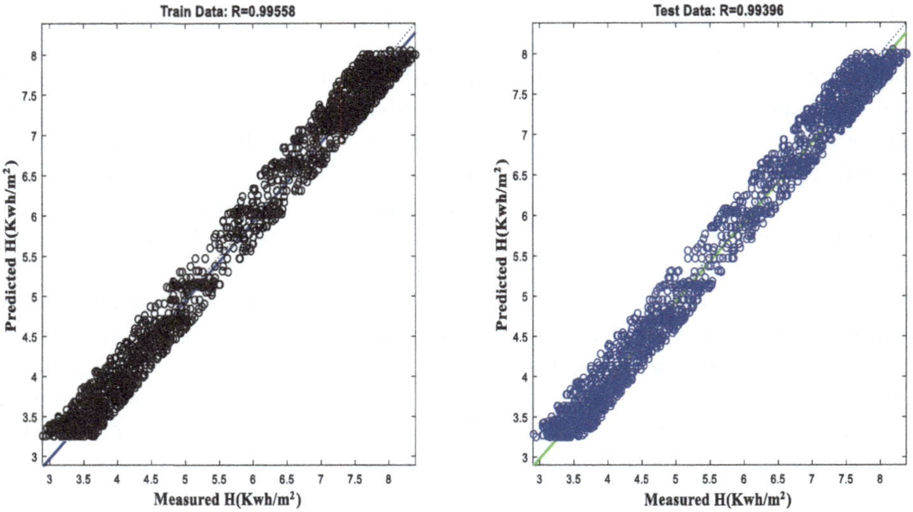

Fig. 3. Scatter plots of the measured data against the predicted daily global solar radiation by ANFIS model for Adrar

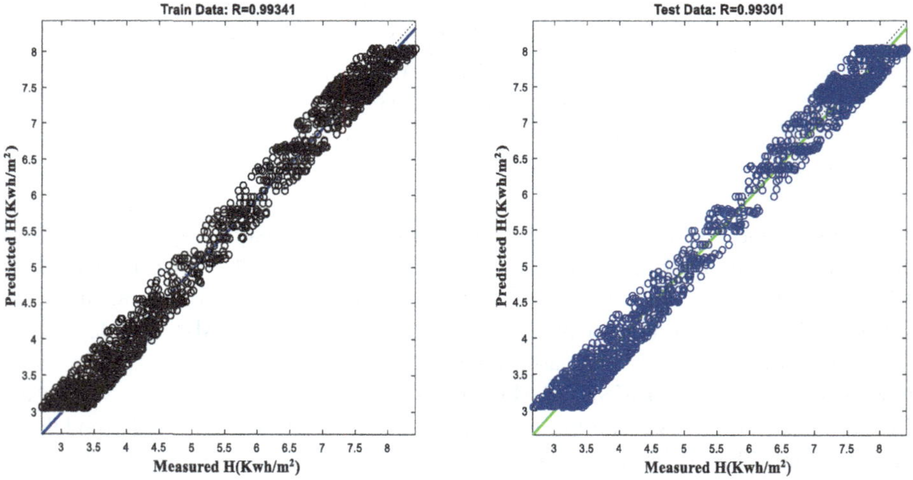

Fig. 4. Scatter plots of the measured data against the predicted daily global solar radiation by ANFIS model for Timimoun

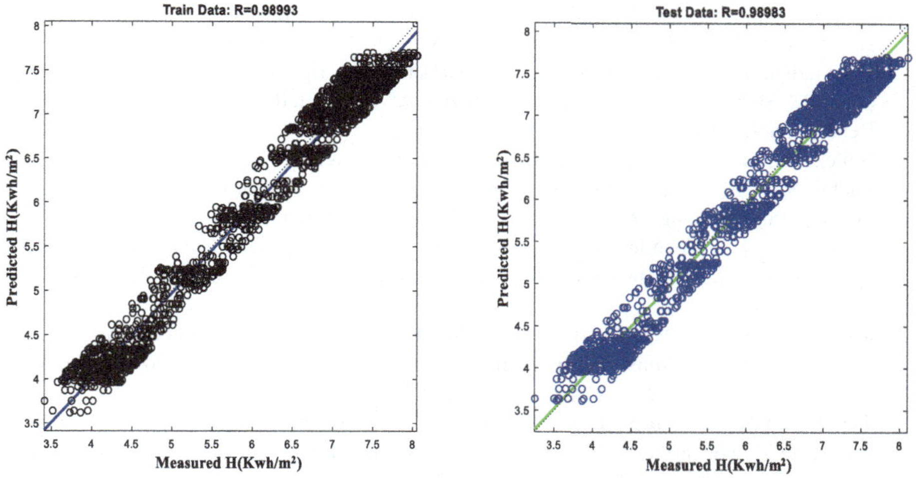

Fig. 5. Scatter plots of the measured data against the predicted daily global solar radiation by ANFIS model for BBM

4 Conclusions

In this work, an adaptive neuro-fuzzy inference system (ANFIS) was utilized to estimate the daily horizontal global solar radiation by day of the year. Basically, the prediction of global solar radiation based upon day of the year offers 2 advantages. First, there is no dependency to any specific input element such as meteorological data. Second, there is no need to any pre-calculation analysis. As a matter of fact, this study aimed at identifying the potential of ANFIS technique to predict the global solar radiation by day of the year. The predictions accuracy of the developed ANFIS model was evaluated using three statistical indicators such RMSE, MAPE and R. Thereafter to validate the adequacy of the developed ANFIS model its performance was compared against the DYB empirical models previously established by Aouna and Bouchouicha [11]. As a conclusion, the utilized ANFIS model can be used to estimate the horizontal global solar radiation with favorable level of reliability and precision. Generally, the developed ANFIS model in this study enjoys a series of merits including the simplicity, easy usage as well as high accuracy. As a result, the suggested ANFIS model would play a notable role in various applications such as designing, simulating and monitoring the solar energy technologies, particularly in isolated areas.

References

1. Bouraiou, A., Hamouda, M., Chaker, A., Lachtar, S., Neçaibia, A., Boutasseta, N., Mostefaoui, M.: Experimental evaluation of the performance and degradation of single crystalline silicon photovoltaic modules in the saharan environment. Energy **132**, 22–30 (2017)

2. Kassem, A.S., Aboukarima, A.M., El Ashmawy, N.M., Zayed, M.F.: Comparison of empirical models and an adaptive neuro fuzzy inference system for estimating hourly total solar radiation on horizontal surface at Alexandria City, Egypt, pp. 1–17, January 2016
3. Said, R., Mansor, M., Abuain, T.: Estimation of global and diffuse radiation at Tripoli. Renew. Energy **14**, 221–227 (1998)
4. Bailek, N., Bouchouicha, K., Al-Mostafa, Z., El-Shimy, M., Aoun, N., Slimani, A., Al-Shehri, S.: A new empirical model for forecasting the diffuse solar radiation over Sahara in the Algerian big south. Renew. Energy **117**, 530–537 (2018)
5. Olatomiwa, L., Mekhilef, S., Shamshirband, S., Petković, D.: Adaptive neuro-fuzzy approach for solar radiation prediction in Nigeria. Renew. Sustain. Energy Rev. **51**, 1784–1791 (2015)
6. Mellit, A., Arab, A.H., Khorissi, N., Salhi, H.: An ANFIS-based forecasting for solar radiation data from sunshine duration and ambient temperature. In: 2007 IEEE Power Engineering Society General Meeting PES (2007)
7. Tymvios, F.S., Jacovides, C.P., Michaelides, S.C., Scouteli, C.: Comparative study of Ångström's and artificial neural networks' methodologies in estimating global solar radiation. Solar Energy **78**(6), 752–762 (2005)
8. Benghanem, M., Mellit, A., Alamri, S.N.: ANN-based modelling and estimation of daily global solar radiation data: a case study. Energy Convers. Manage. **50**(7), 1644–1655 (2009)
9. Mohammadi, K., Shamshirband, S., Tong, C.W., Arif, M., Petković, D., Sudheer, C.: A new hybrid support vector machine-wavelet transform approach for estimation of horizontal global solar radiation. Energy Convers. Manage. **92**, 162–171 (2015)
10. Mohammadi, K., Shamshirband, S., Petković, D., Khorasanizadeh, H.: Determining the most important variables for diffuse solar radiation prediction using adaptive neuro-fuzzy methodology; case study: city of Kerman, Iran. Renew. Sustain. Energy Rev. **53**, 1570–1579 (2016)
11. Aoun, N., Bouchouicha, K.: Estimating daily global solar radiation by day of the year in Algeria. Eur. Phys. J. Plus **132**, 216 (2017)
12. Bouchouicha, K., Razagui, A., Bachari, N.E.I., Aoun, N.: Mapping and geospatial analysis of solar resource in Algeria. Int. J. Energy Environ. Econ. **23**, 735–751 (2015)
13. Bakirci, K.: Prediction of global solar radiation and comparison with satellite data. J. Atmos. Solar Terr. Phys. **152**, 41–49 (2017)
14. Olomiyesan, B.M., Oyedum, O.D.: Comparative study of ground measured, satellite-derived, and estimated global solar radiation data in Nigeria. J. Solar Energy, 1–7 (2016)
15. Moghaddamnia, A., Remesan, R., Kashani, M.H., Mohammadi, M., Han, D., Piri, J.: Comparison of LLR, MLP, Elman, NNARX and ANFIS models-with a case study in solar radiation estimation. J. Atmos. Solar Terr. Phys. **71**, 975–982 (2009)
16. Mohammadi, K., Shamshirband, S., Kamsin, A., Lai, P.C., Mansor, Z.: Identifying the most significant input parameters for predicting global solar radiation using an ANFIS selection procedure. Renew. Sustain. Energy Rev. **63**, 423–434 (2016)
17. Benmouiza, K., Cheknane, A.: Clustered ANFIS network using Fuzzy C-means, subtractive clustering, and grid partitioning for hourly solar radiation forecasting. Theor. Appl. Climatol. **137**, 1–13 (2018)
18. Quej, V.H., Almorox, J., Arnaldo, J.A., Saito, L.: ANFIS, SVM and ANN soft-computing techniques to estimate daily global solar radiation in a warm sub-humid environment. J. Atmos. Solar Terr. Phys. **155**, 62–70 (2017)
19. Jang, J.S.R.: ANFIS: adaptive-network-based fuzzy inference system. IEEE Trans. Syst. Man Cybern. **23**(3), 665–685 (1993)
20. Rensheng, C., Shihua, L., Ersi, K., Jianping, Y., Xibin, J.: Estimating daily global radiation using two types of revised models in china. Energy Convers. Manage. **47**, 865–878 (2006)

Impact of Artificial Intelligence Using Multilevel Inverters for the Evolution the Performance of Induction Machine

Lahcen Lakhdari[1](✉) and Bousmaha Bouchiba[2]

[1] Department of Electrical Engineering, Laboratory (COASEE), Tahri Mohamed Bechar University Algeria, Bechar, Algeria
l.lakhdari@hotmail.com

[2] Department of Electrical Engineering, Tahri Mohamed Bechar University Algeria, Bechar, Algeria
bouchiba_bousmaha@yahoo.fr

Abstract. This paper presents a study of Impact of artificial intelligence using multilevel inverters for the evolution the performance of induction machine.

Artificial intelligence it is a scientific discipline related to the processing of knowledge and reasoning in order to allow a machine to perform functions normally associated with the human being. Artificial intelligence tries to reproduce cognitive processes for the purpose of carrying out 'smart' actions; it is like 'building computer programs that perform tasks that are currently performed more satisfactorily by human beings because they require high-level mental processes.

This work presents a detailed analysis of the comparative advantage of two-level Inverter and the Three Level Diode Clamped Inverter fed Induction Motor. The comparison is based on the evaluation of harmonic distortion THD, on the other hand and to show the performance of the proposed Fuzzy Mode Controller, has been compared with the classical PI control method.

Keywords: Artificial intelligence · PI control · Fuzzy Mode Controller · Induction machine · NPC inverter

1 Introduction

All Artificial intelligence (AI) is the intelligence of machines and the branch of computer science that aims to create it [1]. Artificial intelligence has been the subject of optimism, but has also suffered setbacks and, today, has become an essential part of the technology industry, providing the heavy lifting for many of the most difficult problems in computer science. AI research is highly technical and specialized, deeply divided into subfields that often fail to communicate with each other. Subfields have grown up around particular institutions, the work of individual researchers, the solution of specific problems, longstanding differences of opinion about how AI should be done and the application of widely differing tools [2]. Artificial intelligence concerns itself with intelligent Behavior – the things that make us seem intelligent. In an ultimate view, engineers are about re-creating a perception of man and building a machine using

M. Hatti (Ed.): ICAIRES 2019, LNNS 102, pp. 51–59, 2020.
https://doi.org/10.1007/978-3-030-37207-1_6

the human framework [3]. The use of induction motors has increased tremendously since the day of its invention. They are being used as actuators in various industrial processes, robotics, house appliances (generally single phase) and other similar applications. The reason for its day by day increasing popularity can be primarily attributed to its robust construction, simplicity in design and cost effectiveness. These have also proved to be more reliable than DC motors [4, 5]. Induction Motors have many applications in the industries, because of the low maintenance and robustness. The speed control of induction motor is more important to achieve maximum torque and efficiency [6].

There were also advances in control methods and Artificial Intelligence (AI) techniques, including expert system, fuzzy logic, neural networks and genetic algorithm. Researchers soon realized that the performance of induction motor drives can be enhanced by adopting Artificial Intelligent based methods with the use of multilevel inverters [7].

The use of multilevel converters in industry has become an extremely large field because industrial equipment is using more and more variable speed drive, Multilevel inverters are widely used for induction machine control.

Multilevel inverters are widely used and marketed in various industrial areas of high voltage and high power. It's possible to increase the power delivered to the load from its topology. Thus, it enables not only the most sinusoidal voltage to be generated but also to improve the harmonic ratio to the high number of voltage levels offered by the structure of this type of converter [8]. The mathematical model of the induction machine is developed and presented in Sect. 2, Sect. 3 shows the NPC multilevel inverter, control techniques of multi-level inverter (spwm) is shows in Sect. 4. Section 5 shows the development of PI controller and the application to induction machine; indirect field-oriented control induction machine is given in Sect. 6, Sect. 7 shows the fuzzy logic controller, Sect. 8 shows the simulation results using matlab Simulink, finally the conclusion drawn in Sect. 10.

2 Mathematical Model of the Induction Machine

The modeling step of the induction machine is essential for the development of control laws.

In order to lighten the mathematical notations, we will use the indices s, r to designate the stator and rotor spectively, and d, q for the direct and quadrature axes.

A. ***Electrical Equations***

$$\begin{cases} v_{ds} = R_s.i_{ds} - \dot{\theta}_s.\varphi_{qs} + \frac{d\varphi_{ds}}{dt} \\ v_{qs} = R_s.i_{qs} + \dot{\theta}_s.\varphi_{ds} + \frac{d\varphi_{qs}}{dt} \\ v_{dr} = 0 = R_r.i_{dr} - \dot{\theta}_r.\varphi_{qr} + \frac{d\varphi_{dr}}{dt} \\ v_{qr} = 0 = R_r.i_{qr} + \dot{\theta}_r.\varphi_{dr} + \frac{d\varphi_{qr}}{dt} \end{cases} \quad (1)$$

$$\begin{cases} \varphi_{ds} = L_s \cdot i_{ds} + M \cdot i_{dr} \\ \varphi_{qs} = L_s \cdot i_{qs} + M \cdot i_{qr} \\ \varphi_{dr} = M \cdot i_{ds} + L_r \cdot i_{dr} \\ \varphi_{qr} = M \cdot i_{qs} + L_r \cdot i_{qr} \end{cases} \tag{2}$$

B. ***Mechanical Equation***

$$c_e = \frac{3}{2} . p . \frac{M}{L_r} . \left(\varphi_{dr} . i_{qs} - \varphi_{qr} . i_{ds} \right) \tag{3}$$

3 Clamped Multilevel Inverter

The three-level NPC inverter is shown in Fig. 1. The input DC bus is composed of two capacitors in series (C1 and C2), forming a midpoint marked (0) which allows the inverter to access a supplementary voltage level. Each of the three arms (a, b and c) of the inverter is composed of four controlled switches (K1, K2, K3 and K4 for arm a) and two holding diodes connected to the DC bus midpoint. The controlled switches are unidirectional in voltage and bidirectional in current there are classic combinations of a transistor and an antiparallel diode [9].

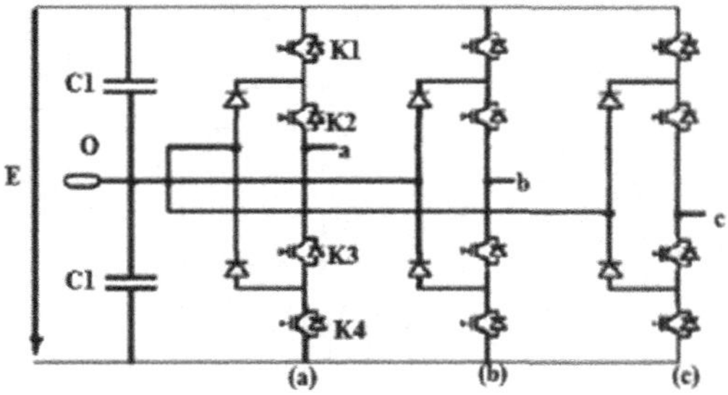

Fig. 1. Three level diode clamped multilevel inverter

4 Control Techniques of Multi-level Inverter (SPWM)

SPWM is realized by comparing a low frequency modulated wave (reference voltage) with a high carrier wave Frequency of triangular shape [10]. The switching times are determined by the points of intersection between the carrier wave and the modulated wave, the switching frequency of the switches is set by the carrier wave [10]. In three-phase system, the three sinusoidal references are out of phase with $2\pi/3$ to the same frequency f [10] (Fig. 2).

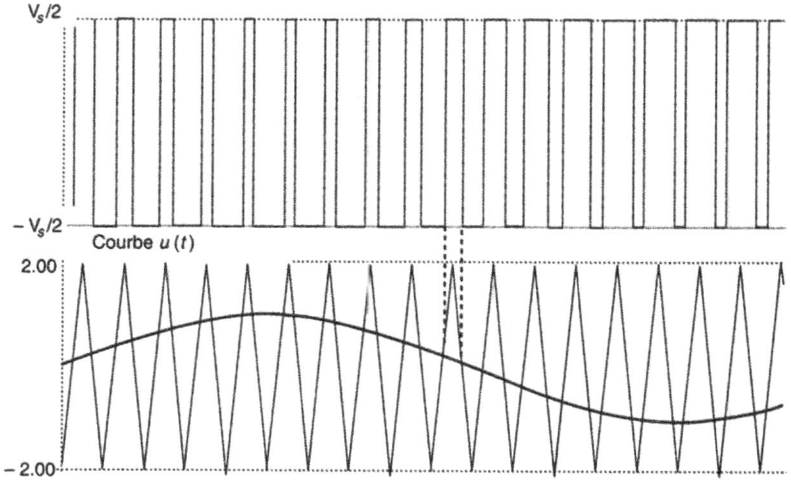

Fig. 2. SPWM waveform generation

5 Indirect Field-Oriented Control Induction Machine

The principle of speed control by the indirect field oriented control is presented in Fig. 3.

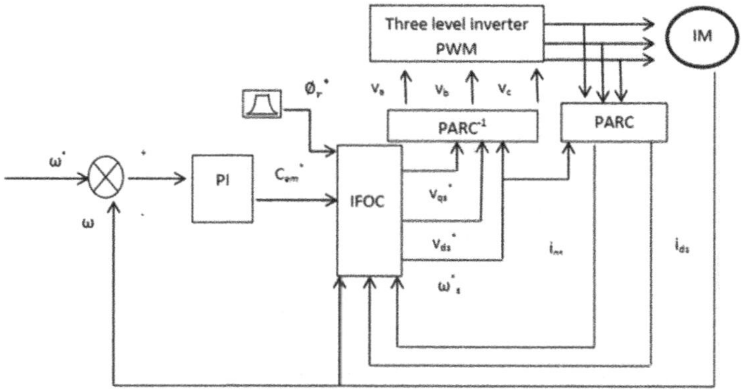

Fig. 3. The indirect field oriented control of an induction machine

In this method, the rotor flux is not regulated, so there is no need for a sensor, no estimation or a flow observer. If the amplitude of the actual rotor flux is not used, its position must be known to effect the coordinate changes. This requires the presence of a rotor position sensor.

6 PI Controller Design for Speed of an Induction Machine

The speed controller determines the reference torque in order to maintain the corresponding speed. For the cascade to be justified, it is necessary that the loop intern is very fast compared to that speed. The mechanical equation gives:

$$\frac{\omega}{C_{em}} = \frac{P}{f_c + J.s} \tag{4}$$

The block diagram of the speed control is therefore carried out as indicated in Fig. 4.

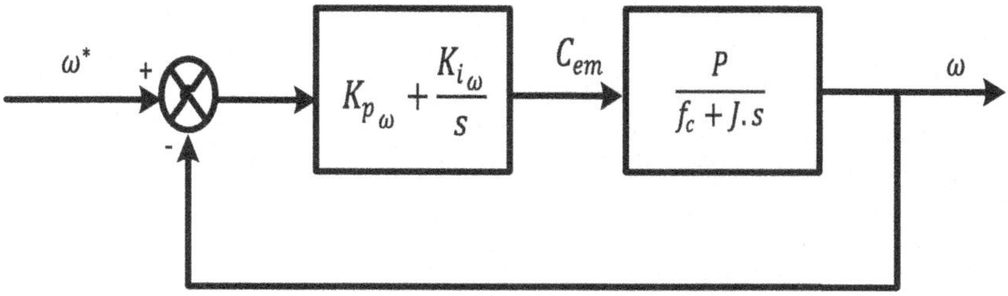

Fig. 4. Block diagram of the speed control.

7 Fuzzy Logic Controller

The fuzzy logic control algorithm is a series of IF……THEN rules. Every fuzzy control rule in the rulebase is a relationship between the input variables, membership functions and an output action or command [11]. Table 1 show the Fuzzy Controller Rule Base for the input and output variable. The proposed controller uses following linguistic labels BN (big negative), MN (medium negative), SN (small negative), EZ (zero), SP (small positive), MP (medium positive), BP (big positive). Each of the inputs and output contain membership function with all these seven linguistics.

Table 1. Units for magnetic properties.

$\Delta\dot{\omega}_s$	$\Delta\omega_s$						
	BN	MN	SN	EZ	SP	MP	BP
BN	*BN*	*BN*	*BN*	*BN*	*MN*	*SN*	*EZ*
MN	*BN*	*BN*	*BN*	*MN*	*SN*	*EZ*	*SP*
SN	*BN*	*BN*	*MN*	*SN*	*EZ*	*SP*	*SP*
EZ	*BN*	*MN*	*SN*	*EZ*	*SP*	*MP*	*MP*
SP	*MN*	*SN*	*EZ*	*SP*	*MP*	*BP*	*MP*
MP	*SN*	*EZ*	*SP*	*MP*	*BP*	*BP*	*MP*
BP	*EZ*	*SP*	*MP*	*BP*	*BP*	*BP*	*MP*

The internal architecture of a fuzzy controller is given by Fig. 5.

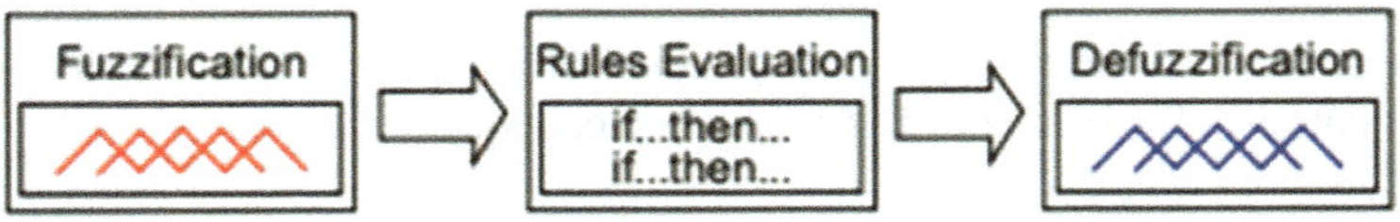

Fig. 5. Internal architecture of a fuzzy controller

In this system, the inputs of the fuzzy controller are $\Delta\omega_s$ (Speed error) and $\Delta\dot{\omega}_s$ (The integral of the speed error) they are defined as follows:

$$\Delta\omega_s = \omega_s^* - \omega_s \tag{5}$$

$$\Delta\dot{\omega}_s = \dot{\omega}_s^* - \dot{\omega}_s \tag{6}$$

Figure 6 shows the input and output membership functions.

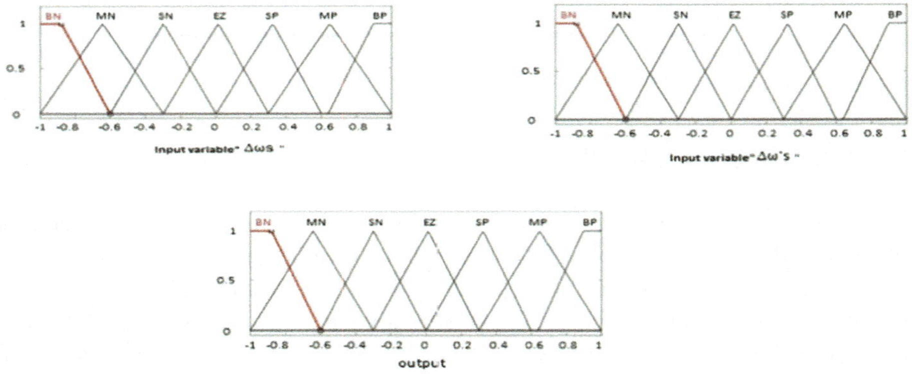

Fig. 6. Membership functions of fuzzy controller

8 Simulation Results

In this section, simulation results are presented to show the performance of the Multilevel inverter compared with the conventional inverter.

The Fig. 7 present the Vab compound tension and its harmonic spectrum for two-level inverter.

The Fig. 8 present the output currents and its harmonic spectrum for two-level inverter.

The Fig. 9 present the Vab compound tension and its harmonic spectrum for three-level inverter.

The Fig. 10 present the output currents and its harmonic spectrum for three-level inverter.

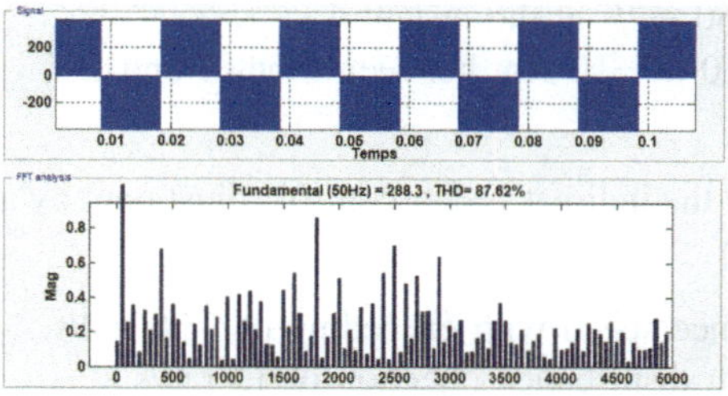

Fig. 7. The Vab compound tension and its harmonic spectrum for two-level inverter.

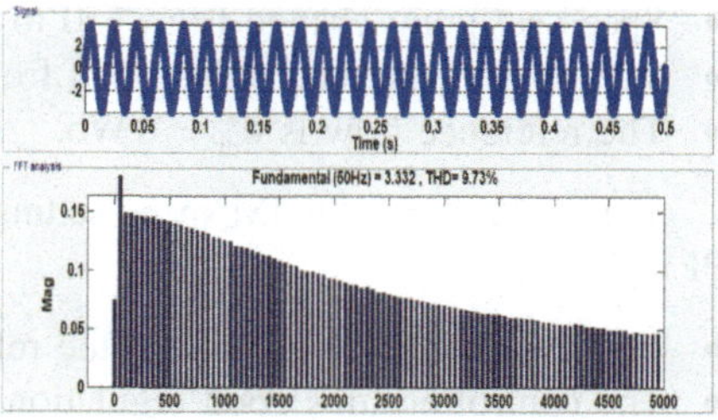

Fig. 8. The output currents and its harmonic spectrum for two-level inverter.

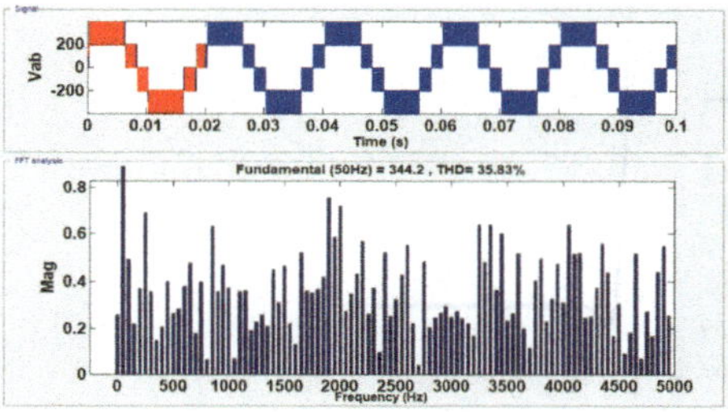

Fig. 9. The Vab compound tension and its harmonic spectrum for Three-level inverter.

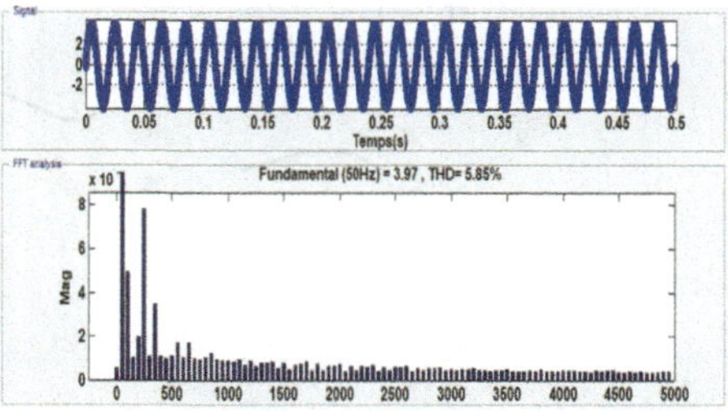

Fig. 10. The output currents and its harmonic spectrum for Three-level inverter.

9 Comparison Between Different Inverters

Table 2 shows well and gives a general idea about the actual waveform quality with its fundamental component. The distortion factor [THD (%)] clearly shows the advantage of the level. Note also that increasing the level of the inverter improves the output signal of the inverter.

Table 2. Distortion factors of each type of inverter

Inverter	Structure	Output voltage THD (%)	
		Compound voltage Vab	Output current
2 Levels	Classical	87.62	9.73
3 Levels	NPC	35.83	5.85

In order to evaluate the performance of the indirect speed vector control with adjustment by a PI regulator, and Fuzzy mode regulator we performed numerical simulations under the following conditions:

- Speed set point change from 150 to −150 rad/s at the instant 4 s
- Variation of the mechanical load from 10 to −10 Nm between times 1 and 2 s
- The reference flow is $\emptyset_{dr}^{*} = 1\text{Wb}$.

Figure 11 shows the IM speed setting by the indirect vector control adjustment by a PI regulator.

- The speed of rotation follows the reference speed with exceeding of 7.20 rad/s.
- The control ensures good regulation with disturbance rejection of 13 rad/s.

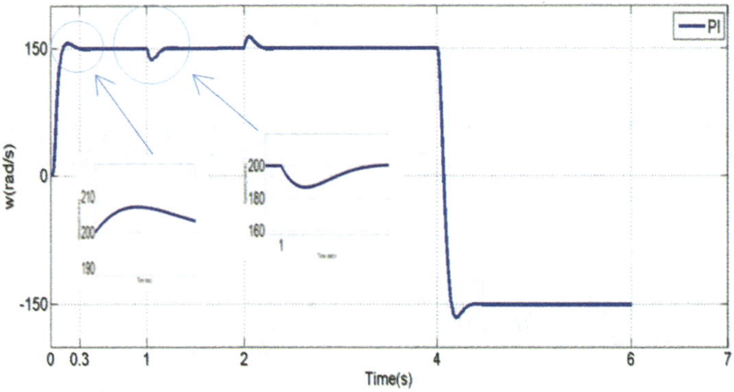

Fig. 11. Speed of IM with PI controller

Figure 12 shows the IM speed setting by the indirect vector control adjustment by Fuzzy logic regulator.

The results show that the regulation by a fuzzy logic regulator gives satisfactory results:

- The speed of rotation follows the reference speed without exceeding.
- The control ensures good regulation with disturbance rejection of 2.4 rad/s.
- A response time of 0.20 ms to reach the balanced state.

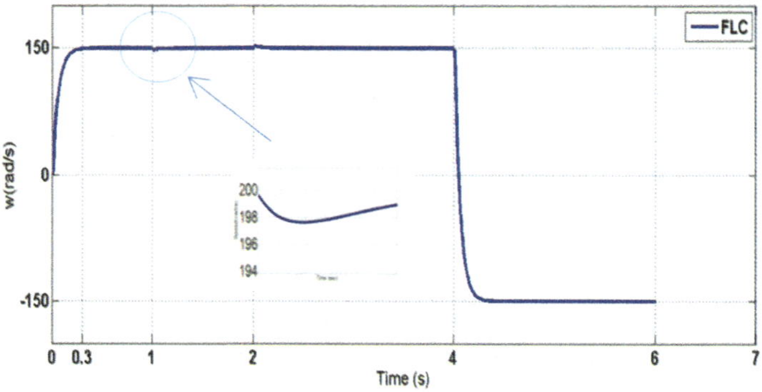

Fig. 12. Speed of IM with Fuzzy logic regulator

10 Conclusion

In this work, the PI and FLC have been tested in simulation; Fuzzy mode is a controller for non-linear systems with non-constant parameters; this leads to accuracy and the robustness, and solves the problems caused by the PI controller. The results of the simulation indicate that the intelegent control gives better results compared to the conventional control and secondly the use of the multilevel inverter produce less total harmonic distortion (THD) who Contribute at the performance evolution of the induction machine.

References

1. Shekar, K.C., Chandra, P., Rao, K.V.: Fault diagnostics in industrial application domains using data mining and artificial intelligence technologies and frameworks. In: IEEE International Advance Computing Conference (IACC), pp. 538–543 (2014)
2. Sivadasan, B.: Application of artificial intelligence in electrical engineering. In: National Conference on Emerging Research Trend in Electrical and Electronics Engineering (ERTEE 2018), March 2018. e-ISSN 2455-5703
3. Ferreira, J., Lobo, J., Bessiere, P., Castelo-Branco, M., Dias, J.: A Bayesian framework for active artificial perception. IEEE Trans. Cybern. **43**, 699–711 (2013)
4. Rao, M.N., Rajani, A.: Speed control of induction motor using fuzzy logic approach. Int. J. Eng. Res. Technol. (IJERT) **2**(8) (2013). ISSN 2278-0181
5. Lakhdari, L., Bouchiba, B.: Fuzzy sliding mode controller for induction machine feed by three level inverter. Int. J. Power Electron. Drive Syst. (IJPEDS) **9**(1), 55–63 (2018). https://doi.org/10.11591/ijpeds.v9n1.pp55-63. ISSN 2088-8694
6. Shi, K.L., Chan, T.F., Wong, Y.K., Ho, S.L.: Modeling and simulation of the three phase induction motor using SIMULINK. Int. J. Electr. Eng. Educ. **36**, 163–172 (1999)
7. Menghal, P., Laxmi, A.J.: Artificial intelligent control of induction motor drives. i-Manager's J. Instrum. Control Eng. **2**(1), 9 (2013). ISSN-2321-113X
8. Lakhdari, L., Bouchiba, B., Bechar, M.: Comparative analysis of SVPWM and the standard PWM technique for three level diode clamped inverter fed induction motor. Int. J. Electr. Comput. Eng. **12**(1), 1–10 (2018)
9. Daniel depernet, these présentée pour l'obtention du grade de docteur de l'université de reims champagne-ardenne, "optimisation de la commande d'un onduleur mli a trois niveaux de tension pour machine asynchrone"
10. ELOUED university Elearning. Cours Electronique de puissance avancée. http://elearning.univeloued.dz/courses/EPA01/document/Cours_Master2_ChapitreI.pdf?cidReq=EPA01. Accessed 26 Apr 2017

Machine Learning Techniques for Road Traffic Automatic Incident Detection Systems: A Review

S. Hireche(✉) and A. Dennai

Department of Exact Sciences, CS&SD Team, SGRE Laboratory, University of TAHRI Mohamed, Bechar, Algeria
hireche.sam@gmail.com, de_selam@yahoo.fr

Abstract. Recently, non-recurrent congestion caused by road traffic incidents has become a critical concern of road Traffic Management System (TMS). However, incidents can't be predicted. Hence, modern cities deployed Automatic Incidents Detection Systems (AIDSs) to early detect incidents and to improving road traffic flow efficiency and safety. For this, many AIDS approaches based on Machine Learning (ML) techniques are proposed. Although several reviews about AIDS have been written, a review of ML techniques based incident detection systems is required.

The purpose of this paper is to discuss the recent research contributions in automatic incidents detection systems based on ML techniques. To achieve this goal, a review and a comparison of data sources, datasets, techniques and detection performances in both freeway and urban roads are provided. Finally, the paper concludes by addressing the critical open issues for conducting research in the future as a proposal framework.

Keywords: Machine Learning · Automatic incident detection systems · Detection Rate · False alarm rates · Mean Time to Detect

1 Introduction

Machine Learning (ML) is a branch of artificial intelligence that has been in use since the 1959s, when Arthur Samuel defined machine learning, as "the field of study that gives computers the ability to learn without being explicitly programmed" [1]. ML include a set of techniques which allows computer systems to learn from the data [2]. These techniques can be classified into four categories: (i) Supervised learning; (ii) Semi-supervised learning; (iii) Unsupervised learning; and (iii) Reinforcement learning based on the type of learning. From these categories, ML algorithms are designed to "identify and exploit hidden patterns in data for (i) describing the outcome as a grouping of data for clustering problems, (ii) predicting the outcome of future events for classification and regression problems, and (iii) evaluating the outcome of a sequence of data points for rule extraction problems" [1].

Over the last decades, there has been increased interest among transportation researchers in exploring the feasibility of applying ML algorithms to address and solve different transportation problems by applying Intelligent Transportation System

M. Hatti (Ed.): ICAIRES 2019, LNNS 102, pp. 60–69, 2020.
https://doi.org/10.1007/978-3-030-37207-1_7

(ITS) applications. In this context, automatic incidents detection system (AIDS) is one of the main ITSs that was established since the 1970s [3] to identify the time and the location of traffic incidents. Traffic incident is defined as "any non- recurring event that causes a reduction of roadway capacity or an abnormal increase in demand" [4]. These events can be classified as planned or unplanned events, which includes accidents, disabled vehicles, temporary maintenance, truck load spills, adverse weather conditions or others events that occurs at a random time and location [5]. However, "traffic incidents usually cannot be predicted which poses great challenge to Traffic Management Centers" (TMCs) [6]. Therefore, early detection of such incidents reduces the effects of non recurent congestion, the probability of other collisions, and even enhances the traffic safety and efficiency in road network. Specifically, each AIDS is composed of two main components. The first one is the traffic detection system that is used to provide the traffic data necessary for detecting an incident by identifying traffic flow measures or derived features gathered from traffic data collection technologies. These technologies includes Inductive Loop Detectors (ILDs), Video Image Processing detectors (VIPs), mobile sensors, probe vehicle systems, Wifi and Bluetooth vehicle re-identification readers or even fusion sensors. The second one is the Automatic Incident Detection Algorithm (AIDA) that is employed to prove the presence or the absence of incidents [7]. Here, researchers applied a rich of techniques to identify abnormal changes in these data. These techniques covered in literature can be categorized into several groups, namely statistical, times series, image processing, filter and comparative based techniques. More recently, artificial intelligence, data mining, machine learning, and others techniques are also used.

To enhance the AIDA performances, researchers adopt various techniques to distinguish traffic incident and non-incident traffic patterns. Up to now, most research on AIDS mainly deals with the use of classification techniques, especially advanced ML-based techniques that have demonstrated its usage to handle the most traditional techniques limitation, i.e. the improvement of incident detection performances. For such reason, several studies investigate the combination of ML methods for processing real time traffic data. So, from this point it is clear that the improvement of incident detection performances still remain a great research. Hence, very little attention has been paid to examined the incident detection performances obtained when applying these ML techniques in this emerging field.

In fact, there are several analysis studies and reviews that are summarized different techniques used to detect incidents in freeway roads. Parkany and Xie [8] gave a detailed review about AIDAs. They categorized them according to traffic data collection technologies. They also reported the common measures that have been used to evaluate these AIDAs. Moreover, they included consideration of some AIDSs developed before 2001 on arterial roads. Deniz et al. [9] compared three AIDAs with different road condition and incident location scenarios. Moreover, they gave an overall evaluation on the results obtained from this comparison. Nikolaev et al. [5] evaluated seven discussed AIDAs based on different techniques. Further, they presented the results obtained from this analysis. Therefore, these existing reviews take two objectives different. In the first objective, the reviews are presented the freeway AIDS in general; they are concentrated on citing categories, techniques, and evaluation metrics.

In the second objective, they are evaluated different AIDAs based on the evaluation performance metrics.

However, the rapid developments in the road traffic field domain have presented a new requirement to present comprehensive reviews of such studies. This paper is a significant extension of these existing reviews in literature. It attempts to review and to compare the recent ML based AIDS studies in the last five years (from 2014 to 2018). Also, this study aims at bridging this gap by encapsulating all the literature related to ML-based techniques for incidents detection, and identifying their data sources, datasets, and evaluation performances. In the end, the open research shortcomings found through these ML based AIDSs selected papers will be provided as a proposed framework.

The remainder of this paper is organized as follows. Section 2 provides a review of ML based AIDS studies. Section 3 conducts to analyze and to compare the reviewed AIDSs and to discuss the results obtained from this comparison. The open research issues found through these selected papers are proposed as framework in Sect. 4. Finally, Sect. 5 provides the conclusions and future works.

2 Machine Learning Techniques for Road Traffic Incident Detection Systems

This section covers a review of the key machine learning techniques used in literature for AIDS in the five past years (from 2014 to 2018). These ML techniques include Support Vector Machines (SVMs), Artificial Neural Networks (ANNs), Decision Trees (DT), Naïve Bayes (NB) and ensemble learning. These techniques covered in this paper are based on supervised learning to classify traffic patterns as incidents or non-incidents patterns. The Fig. 1 shows the taxonomy of all ML techniques used in these AIDSs. We discuss all these propositions in detail in this section.

Qingchao et al. in [10] treated the incident detection as a pattern classification problem. For such goal, they applied the NB classifiers to classify traffic patterns as an incident or non-incident traffic pattern. Also, they combined the Naive Bayes and decision tree (NBTree) to detect incidents based on ILD traffic data. As results, authors provided that NBE enhance the stability of an AIDA performances. Gakis et al. in [11] presented a SVM-based approach for detecting traffic incidents. In this approach, researchers used traffic speed data retrieved from ILDs as feature to train the SVM. A very good performances are reported after using this SVM classification approach.

To improve the adaptability of existing AIDAs in signalized urban networks, Ghosh and Smith in [12] developed the CAIDA (Customization of Automatic Incident Detection Algorithms) scheme. Here, authors used three types of ANNs: Probabilistic Neural Network (PNN), Multilayer Feed-Forward Neural Network (MLFFNN) and Radial Basis Function Neural Network (RBFNN) with a SVM method to detect incidents. After evaluated all of these chosen algorithms, their results indicated that the PNN and RBFNN in conjunction with wavelet based denoising and fuzzy-logic-based algorithms (FWRBFNN) provided less consistent and accurate results compared to MLFNN algorithm. Based on real-time traffic data collected by wireless sensors, Zhou et al. in [13] proposed another freeway SVM based traffic incident detection method.

The I-880 data set is used to validate and evaluate the performance of their AIDS. As result, they confirmed that their features selected give a better detection performance.

In 2016, another AIDS classification approach based on logistic regression (LR) technique is developed by Agarwal et al. in [14]. Authors through this contribution proved that using the wavelet based denoising technique, before feeding the data in logistic regression model, can improve the performances compared to using LR as alone technique.

In the same year, Chlyah et al. in [15] used both Multi Agents System and Support Vector Machine (MAS-SVM) algorithm to detect incidents on signalized urban arterial roads. In this work, an incident detection agent is doing globally the incident detection, while the SVM model with RBF kernel is selected for the classification of traffic data collected from ILDs. Good performances are presented by applying this SVM technique. Another contribution based on a binary-logit regression model was applied in a pre-timed intersection incident detection system in [16]. Hawas and Ahmed in this model generated simulated traffic incidents based on NETSIM simulator. Coupled with some predefined threshold values, this logit model are used to estimate the incident status probability at each cycle time. Compared analysis to others models was demonstrated the promising fusibility of applying this model for urban roads.

In 2017, El Hatri and Boumhidi in [6] used a hybrid techniques that are Fuzzy Logic and Deep Neural Network Learning (FLDNNL) for detecting traffic incident in urban signalized intersection area. Parameters of the DNN is initialized using a Stacked Auto-Encoder (SAE) model, while the back-propagation algorithm is used to precisely adjust the parameters in the deep network. The FL system is employed to control the learning parameters. The simulation results confirmed the accuracy of the proposed approach compared to others advanced ANN methods.

Detect incidents in freeway roads based on Extreme Learning Machine (ELM) is another ML proposition by Li et al. in [17]. Through the experiment analysis with I-880 dataset, researchers indicted the high performances compared to others AID models. Implement this ELM in urban areas is the main future research required.

In more recently studies, Li et al. [18] combined two classification techniques i.e. the decision Tree and Augmented Naive Bayesian (TAN) to detect incidents. Reduce the dependency on experts' knowledge is the goal of this model. The results of this AIDA indicated that TAN have a better performances to NB and a similar performance to MLFNN.

An important study presented by Li & all in [19] compared four classification methods to detect traffic incidents. These methods included SVM, NB, Cart, and AdaBoost-Cart (ACT). After evaluated these classification methods, the results indicated that AdaBoost-Cart and NB models performed quite well compared to SVM and Cart models.

In 2018, Dardor et al. [20] tryed to resolve the problem of incident detection on signalized intersection urban areas based on SVM coupled with Genetic Algorithm (GA-SVM) model. In this proposition, the Radial Basis Function (RBF) was selected as the kernel function of SVM to classify the signal and determine the event type, while GA is selected as the optimization algorithm to maximize classification accuracy of SVM. Based on SUMO simulation traffic data, researchers confirmed that GA-SVM incident detection algorithm outperform those AIDAs based only on SVM.

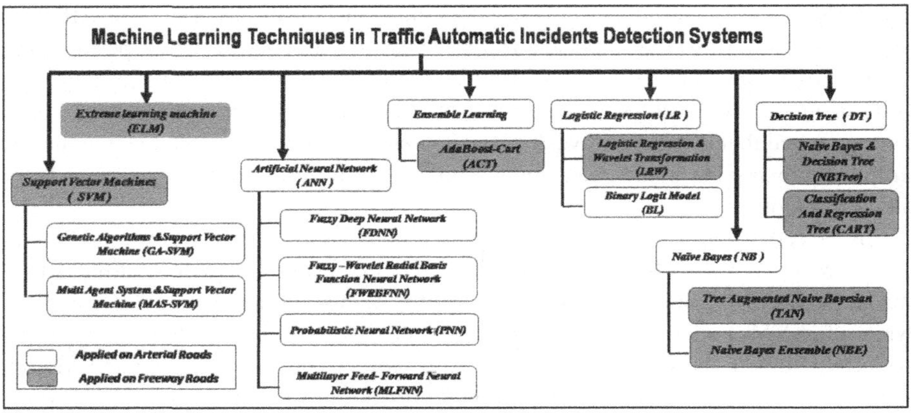

Fig. 1. Taxonomy of machine learning techniques used in traffic automatic incident detection systems.

3 Synthesis and Discussion

Having described our review of advanced ML techniques used in AIDS approaches as shown throughout Sect. 2. In this section, we synthesize all these studies reviewed. Table 1 presents a comparative analysis of all ML based AIDSs in terms: data sources, traffic inputs, techniques and performances with the scope area where these AIDS are applicable.

A. *Data Sources and Data Sets*

Only one data source is used in these studies i.e. the Inductive Loop Detector (ILD). Using this type of fixed intrusive traffic data gathering technology is due to its advantages. First, it is characterized by its accuracy and error level that is fairly good and known. Secondly, it is the most widely deployed low cost sensor and has been in use for decades. Thirdly, ILDs can gather the most of traffic parameters which include speed, volume, occupancy, density, queue, location, etc. An important issue in the field of AIDS research is the availability of datasets. Through this review, four data sets applied only on freeway roads are used to evaluate the AIDAs performances. I-880 Freeway in San Francisco Bay area, The Ayer Rajah Expressway (AYE) in Singapore, Freeways US-95 and I-15 in the Las Vegas area are the main traffic incidents datasets used. In the other hand, all the urban AIDSs are used simulated datasets to validate their algorithms based on VISSIM, SUMO or NETSSIM microscopic simulators.

B. *Machine Learning Methods*

Our analyses indicate that the three important ML methods the widely used by researchers are SVM based technique used alone or as hybrid techniques, ANN technique, followed by the Naïve Bays. Researchers typically used them to categorize or classify the gathered traffic data given inputs based on the training data set as incident or non incident outputs in order to solve the incident non-linear classification problem, and to improve the performance metrics.

Table 1. Comparative analysis of ML based traffic automatic incident detection systems.

Authors – (Year) – [Reference]	Data sources	Traffic inputs				Scope areas	Machine learning techniques	Data sets or simulators	AIDS Performances		
		V	S	O	R				DR (%)	FAR (%)	MTTD (S)
Qingchao et al. [10]	ILDs	√	√	√		Freeway	SNB NBE NBTree	I-880/AYE Dataset	82.28 89.63 81.43	3.98 2.97 0.89	1.2991 1.6193 1.3622
Gakis et al. [11]	ILDs		√			Freeway	SVM	I-880 Dataset	98.6	1.29	27.6
Ghosh and Smith [12]	ILDs	√		√		Urban	MLFNN PNN SVM FWRBFNN	VISSIM Simulator	87.57 88.82 87.31 61.88	0.30 2.96 0.90 3.72	98.09 134.67 98.61 168.39
Zhou et al. [13]	ILDs	√	√			Freeway	SVM	I-880 Dataset	≈90	≈2.5	≈8
Agarwal et al. [14]	ILDs	√	√	√		Freeway	LRW	US-95 & I-15 Dataset	98.78	6.5	N
Chlyah et al. [15]	ILDs	√	√	√	√	Urban	SVM-MAS	SUMO Simulator	98.21	2.3	103.5
Hawas and Ahmed [16]	ILDs	√	√	√		Urban	BL	NETSIM Simulator	80.7 47.8	0.75 0.47	130 190
El Hatri and Boumhidi [6]	ILDs	√	√	√		Urban	FDNN	SUMO Simulator	98.23	0.24	192.44
Li et al. [17]	ILDs	√	√	√		Freeway	ELM	I-880 dataset	86	1	0.81
Li et al. [18	ILDs	√	√	√		Freeway	TAN	AYE dataset	95.71	4.54	75
Li et al. [19]	ILDs	√	√	√		Freeway	SVM NB Cart ACT	I-880 dataset	72.5 83.8 71.0 88.2	0.2 6 0.5 0.2	190.73 0.150 1.030 12.650
Dardor et al. [20]	ILDs	√	√			Urban	GA-SVM	SUMO Simulator	95.95	0.38	40.86

V: Volume – S: Speed – O: Occupancy – R: otheRs – (S): Seconds – N: Not mentioned
DR: Detection Rate - FAR: False Alarms Rate - MTTD: Mean Time to Detection

C. *Evaluation of Performances*

There are numerous measures used to evaluate the performance of different AIDAs. Detection Rate (DR), False Alarms Rate (FAR), and Mean Time to Detection (MTTD) are the most frequently applied measurements to investigate the accuracy of AIDAs. The DR and FAR measures the effectiveness of an AID algorithm, while the MTTD reflects its efficiency. Their definitions are presented as in formulas (1), (2), (3) as the same as mentioned in the following references [15] and [16]. All the compared results in term DR, FAR and MTTD for these ML are described in Fig. 2.

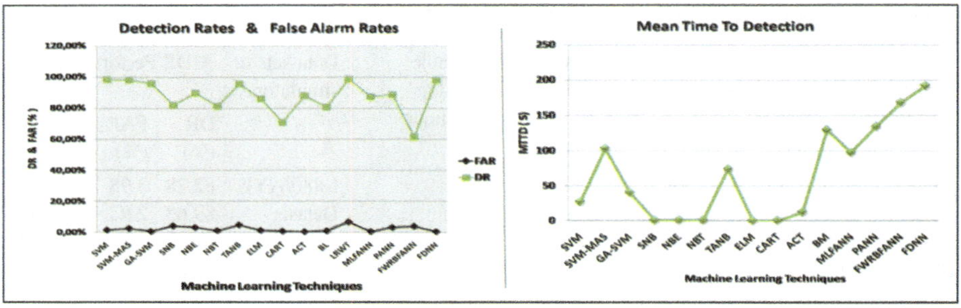

Fig. 2. Compared DR, FAR and MTTD of ML techniques used in AIDS.

$$\mathrm{DR} = \frac{\text{Number of detected incidents}}{\text{Total number of genrated incidents}} \times 100\% \tag{1}$$

$$\mathrm{FAR} = \frac{\text{Number of false alarm incidents}}{\text{Total number of declared incidents cases}} \times 100\% \tag{2}$$

$$\mathrm{MTTD} = \frac{1}{n}\sum\nolimits_{i=1}^{n} t_{ialg} - t_{iocc} \tag{3}$$

(t_{iocc}) is the time to incident actually occurred and (t_{ialg}) is the time that it was detected by the algorithm of a set of (n) incident. It's measured by minutes (min) or by seconds (s).

For DR, the LRW, SVM, and FDNN techniques reported the best detection accuracy by a ratio of 98.78%, 98.6% and 98.23% respectively. For FAR, the ACT, FDNN and MLFANN reported the best false alarm rates by a ratio of 0.2%, 0.24% and 0.3% respectively. In term MTTD, NB, ELM, CART techniques reported the lowest time to detect incidents by a time of 0.15 s, 0.81 s and 1.03 s respectively. In summary, a high DR and a minimal FAR and MTTD are the desired AIDA performances. So, although these ML models provide acceptable performance results, few of them are proved reasonable performances. As result, continuous research is recommended to improve these obtained results.

4 Proposal Framework

After analyzing these ML based AIDS propositions, we observe that the most approaches do not explain the weak and strong points for their obtained results. Authors just cited some of future research possibilities, which can be applied to enhance their results. In this section, we regroup most of them as three steps. The first step provides the recent traffic sensor technologies might be used to collect real time traffic data in the AIDS traffic detection component. The second step regroups all the mechanisms that must be applied in the core system of an AIDA from storage of incidents data to its processing by ML techniques. In the third and last step, we suggest some of solutions needed to apply in order to give incidents information to end users in road. Figure 3 presents the framework of the future research directions in ML based AIDS that are summarized based on research shortcomings presented in the reviewed proposed approaches.

Fig. 3. Proposal framework for conducting future research in AIDS based on ML techniques.

5 Conclusion and Future Works

In this paper, we have reviewed the recent research contributions of ML techniques used in literature for detecting road traffic incidents in both arterial and freeway roads. The literature covered is in the five past years (i.e. 2014–2018). The review reveals that the Support Vector Machine, the Naïve Bayes and the Artificial Neural Networks have been the most frequently used techniques in the AIDS field research domain. The reviewed approaches are compared in term data sources, input traffic variables, traffic data sets and the various ML techniques that are applied in urban or freeways scope road areas with the different measures used to evaluate the incident detection performance. Finally, a qualitative synthesis was investigated after analyzing these reviewed AIDSs.

The best reported detection accuracy is of the wavelet based Logistic Regression, the Support Vector Machine, and Fuzzy Deep Neural Network with a detection accuracy values of 98.78%, 98.6% and 98.23% respectively. These techniques are characterized by their higher detection rates, acceptable false alarm rates and time to detection. It is remarked also that all studies are based on fixed traffic data collection technology i.e. inductive loop detectors. A proposal framework to discuss the open research issues in ML based AIDSs are also presented to further look into.

Finally, although this review presents significant results to apply ML techniques in this research field, future work is recommended to focus towards on field testing validation as part of smart traffic incident management systems.

References

1. Boutaba, R., et al.: A comprehensive survey on machine learning for networking: evolution, applications and research opportunities. J. Internet Serv. Appl. **9**(1), 16 (2018)
2. Dey, A.: Machine learning algorithms: a review. Int. J. Comput. Sci. Inf. Technol. **7**, 1174–1179 (2016)
3. Khorashadi, B., Liu, F., Ghosal, D., Zhang, M., Chuah, C.: Distributed automated incident detection with vgrid. IEEE Wirel. Commun. **18**, 64–73 (2011)
4. Farradyne, P.B.: Traffic Incident Management Handbook, FHAOTM, November 2000
5. Nikolaev, A.B., Sapego, Y.S., Ivakhnenko, A.M., Mel'nikova, T.E., Stroganov, V.Y.: Analysis of the Incident detection technologies and algorithms in intelligent transport systems. Int. J. Appl. Eng. Res. **15**(12), 4765–4774 (2017)
6. El Hatri, C., Boumhidi, J.: Fuzzy deep learning based urban traffic incident detection. Cogn. Syst. Res. **50**, 206–213 (2017)
7. Zou, Y., Shi, G., Shi, G., Wang, Y.: Image sequences based traffic incident detection for signaled intersections using HMM. In: Proceedings of the Ninth International Conference on Hybrid Intelligent Systems, pp. 257–261 (2009)
8. Parkany, E., Xie, C.: A complete review of incident detection algorithms & their deployment: what works and what doesn't. Technical report, University of Massachusetts, Transportation Center. 214 Marston Hall, February 2005
9. Deniz, O., Celikoglu, H.B., Gurcanli, G.E.: Overview to some incident detection algorithms: a comparative evaluation with Istanbul freeway data. In: Proceedings of the 12th International Conference "Reliability and Statistics in Transportation and Communication" (RelStat 2012), Riga, Latvia, 17–20 October 2012, Transport and Telecommunication Institute, Lomonosova 1, LV-1019, Riga, Latvia, pp. 274–284 (2012)
10. Qingchao, L., Lu, J., Chen, S., Zhao, K.: Multiple naive bayes classifiers ensemble for traffic incident detection. In: Mathematical Problems in Engineering. Hindawi Publishing Corporation (2014)
11. Gakis, E., Kehagias, D., Tzovaras, D.: Mining traffic data for road incidents detection. In: Proceedings of 17th International Conference on Intelligent Transportation Systems, Qingdao, China, 8–11 October 2014, pp. 930–935 (2014)
12. Ghosh, B., Smith, D.P.: Customization of automatic incident detection algorithms for signalized urban arterials. J. Intell. Transp. Syst. **4**(18), 426–441 (2014)
13. Zhou, B., Lv, H., Ren, T., Chen, Y., Qiu, N.: Intelligent traffic incidents detection method in freeway corridors. In: Proceedings of International Conference on Mechatronics, Electronic, Industrial and Control Engineering (MEIC 2015), pp. 1623–1626. Atlantis Press (2015)

14. Agarwal, S., Kachroo, P., Regentova, E.: A hybrid model using logistic regression and wavelet transformation to detect traffic incidents. IATSS Res. **40**, 56–63 (2016)
15. Chlyah, M., Dardor, M., Boumhidi, J.: Multi-agent system based on support vector machine for incident detection in urban roads. In: Proceedings of 2016 11th International Conference on Intelligent Systems: Theories and Applications, Mohammedia, Morocco, 19–20 October 2016. IEEE (2016)
16. Hawas, Y.E., Ahmed, F.: A binary logit-based incident detection model for urban traffic networks. Transp. Lett. **9**(1), 49–62 (2016)
17. Li, L., Qu, X., Zhang, J., Ran, B.: Traffic incident detection based on extreme machine learning. J. Appl. Sci. Eng. **4**(20), 409–416 (2017)
18. Li, D., Hu, X., Jin, C., Zhou, J.: Learning to detect traffic incidents from data based on tree augmented naive bayesian classifiers. Discrete Dyn. Nat. Soc. **2017**, 1–9 (2017). https://doi.org/10.1155/2017/8523495. Article ID 8523495
19. Li, L., Zhang, J., Zheng, Y., Ran, B.: Real-time traffic incident detection with classification methods. In: Green Intelligent Transportation Systems, pp. 777–788. Springer, Singapore (2018)
20. Dardor, M., Chlyah, M., Boumhidi, J.: Incident detection in signalized urban roads based on genetic algorithm and support vector machine. In: Proceedings of the 2018 International Conference on Intelligent Systems and Computer Vision, Fez, Morocco, 2–4 April 2018. IEEE (2018)

Investigate of Different MPPT Algorithm Based on P&O, INC and Second Order Sliding Mode Control Applied to Photovoltaic System Conversion Under Strong Conditions

L. Baadj[1(✉)], K. Kouzi[1], M. Birane[1], and M. Hatti[2]

[1] Laboratory of Semiconductors and Functional, Materials, Electrical Engineering Department, University of Laghouat, Laghouat, Algeria
{l.baadj,m.birane}@lagh-univ.dz, kouzi.univ@gmail.com
[2] Unité de Développement des Equipments Solaires, UDES/Centre de Développement des Energies Renouvelables, CDER, Bou Ismaïl, Tipasa, Algeria
mustapha.hatti@ieee.org

Abstract. This study investigate different robust maximum power point tracker (MPPT) schemes applied to improve energy conversion efficiency of photovoltaic system based on super twisting sliding mode control (second order). To do this, in first step the two classical control techniques (Perturb and Observe 'P&O', Incremental Conductance 'INC') were presented. In second step the robust sliding mode controller 'SMC', and Second order sliding mode controller 'SOSMC' were proposed. In final step a Comparative study of various schemes (MPPT) was done. The main goal of the suggested MPPT algorithm is to achieve an optimum operation without the need of atmospheric conditions measurements and to enhance the efficiency of the PV power system. The validation of the comparative study is shown by MATLAB/SIMULINK simulation for the same Photovoltaic power system.

Keywords: MPPT · Sliding mode controller · SMC · Second order sliding mode controller · SOSMC · Perturb and Observe · P&O · Incremental Conductance · INC · Photovoltaic · PV

1 Introduction

A photovoltaic (PV) system converts sun light into electricity. The PV module is the result of connecting a group of PV cells in series and/or parallel, they are the conversion unit in this generation system [1].

Limitations of the PV system such as low efficiency and non-linearity of the output characteristics make it necessary to obtain the MPP process. Differences in solar radiation levels, ambient temperatures and dust accumulation on the surface of the PV panel affect the output of the PV system [2]. A complete analysis of 30 different MPPT algorithms can be found. Among these techniques: Perturb and Observe [3] and Incremental Conductance [4] are the most common algorithms used in the literature because of their simplicity in the implementation and autonomy of PV array parameters.

M. Hatti (Ed.): ICAIRES 2019, LNNS 102, pp. 70–81, 2020.
https://doi.org/10.1007/978-3-030-37207-1_8

Other techniques that differ in duration have the same goal are neural network, fuzzy-logic [5] and sliding mode control [1].

The papers contain the following order: The characteristic of PV array is presented in Sect. 2. In Sect. 3, we introduce the DC-DC boost converter inserted between the PV array and the load. The MPPT searching algorithms in Sect. 4. Numerical simulations under varying climate parameters are given in Sect. 5. Finally, some conclusions in Sect. 6.

2 Photovoltaic Generator

2.1 How a PV Cell Working

Photovoltaic cell is basically a p-n junction which converts directly sunlight to electricity. When the cell exposed to the sunlight, the cell Photons are absorbed by semiconductor atoms, releasing electrons from the negative layer. These electrons feed their way through an outer circle towards the positive layer giving an electric current. Mono crystalline, polycrystalline and thin film technologies are the major families of PV cells. The absorption depends on the cell surface, semiconductor or band gap, the temperature and the solar radiation [6].

2.2 PV Panel Modelling

To model the PV Panel, the scientific community offer several models. The single diode model is the classical one described in literature. The equivalent circuit (Fig. 1) consists of current source to model the incident luminous flux, a diode for cell polarisation phenomena, a parallel resistance due to leakage current and a series resistance representing various contacts [6].

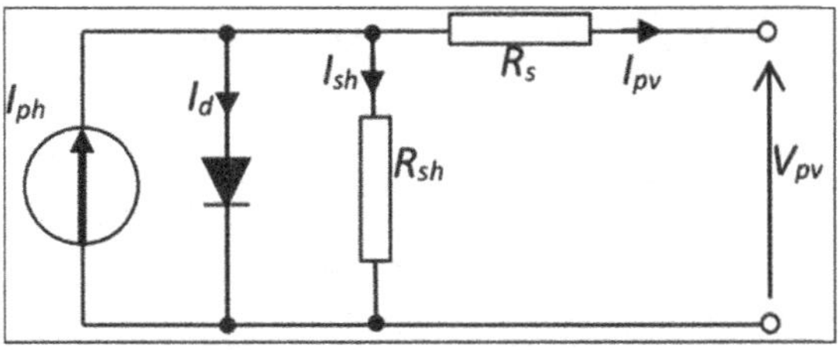

Fig. 1. Equivalent circuit of PV array

Consequently the nonlinear characteristics of PV cell can be represented by following equation:

$$I_{pv} = I_{ph} - I_0 \left[\exp\left(\frac{q(V_{pv} + R_s I_{pv})}{N_s n k_b T} \right) - 1 \right] - \frac{V_{pv} + R_s I_{pv}}{R_{sh}} \tag{1}$$

$$I_{ph} = G/1000(I_{sc} + K_i(T - Tr)) \tag{2}$$

$$I_0 = I_{rr}\left(\frac{T}{T_r}\right)^3 \exp\left(\left[\frac{qE_g}{k_b n}\right] \times \left[\left(\frac{1}{T_r}\right) - \left(\frac{1}{T}\right)\right]\right) \tag{3}$$

An ideal PV cell Rs ≈ 0 and Rsh ≈ + ∞
The Eq. (1) becomes:

$$I_{pv} = I_{ph} - I_d = I_{ph} - I_0\left[\exp\left(\frac{qV_{pv}}{N_s n k_b T}\right) - 1\right] \tag{4}$$

where Ipv is the output current (A), Vpv the voltage (V), I0 is reverse saturation current, q the electronic charge, kb is Boltzmann's constant, T is ambient temperature in Kelvin, Tr is reference temperature, Irr is the saturation current at the reference temperature, Isc is the short-circuit current of PV cell under standard conditions, Eg is the energy of the band gap for silicon, n is the P-N junction's idealist factor, Ki is the short-circuit current temperature coefficient, G is solar irradiance (W/m^2).

2.3 PV Characteristic

The PV characteristic is plotted in Figs. 2 and 3 under different irradiance levels, and PV characteristic under different temperatures is plotted in Figs. 4 and 5.

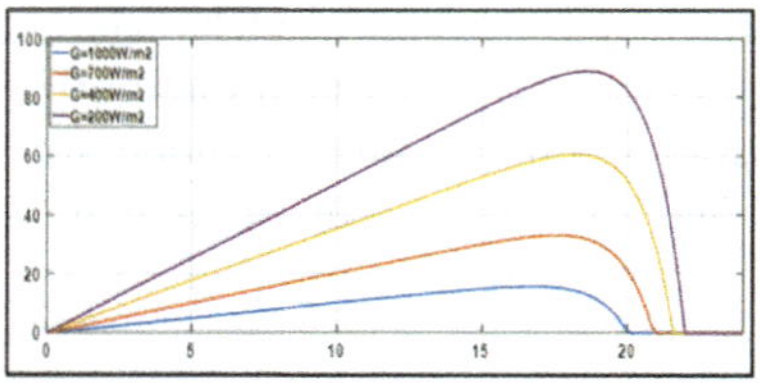

Fig. 2. P-V Curve in differents irradiantes (T = 25 °C)

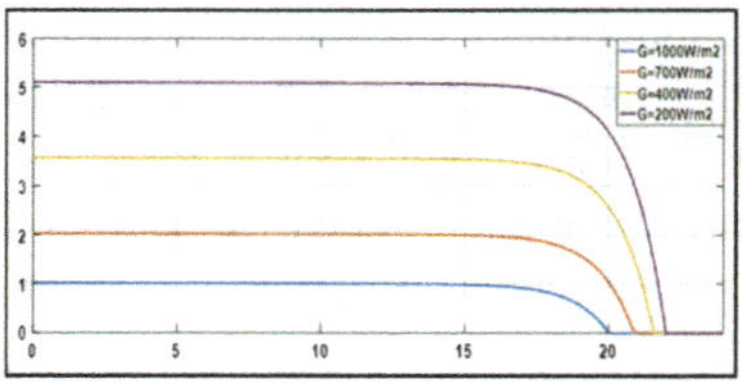

Fig. 3. I-V Curve in differents irradiantes (T = 25 °C)

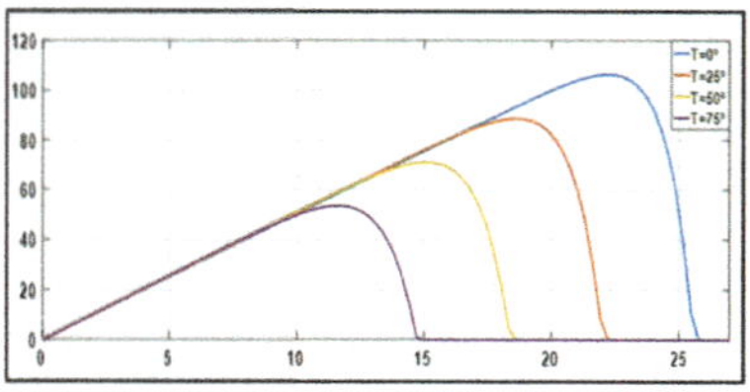

Fig. 4. P-V Curve in different temperature (G = 1000 W/m^2)

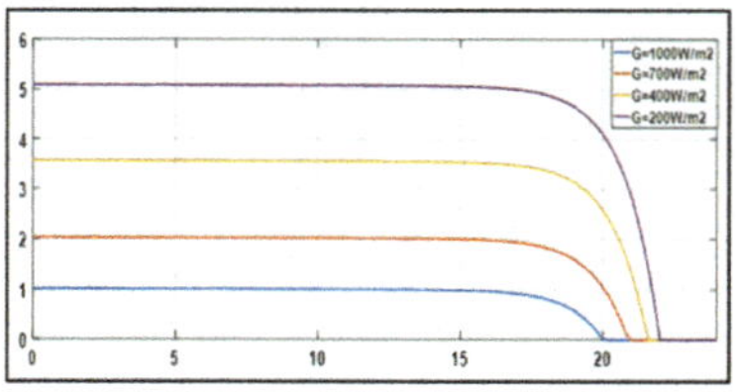

Fig. 5. I-V Curve in different temperature (G = 1000 W/m^2)

3 DC-DC Boost Converter Modeling

The diagrams in Fig. 6 show the structure of this converter with the switching period of T and duty cycle d. For this converter, state equation of voltage results in the Eq. (5).

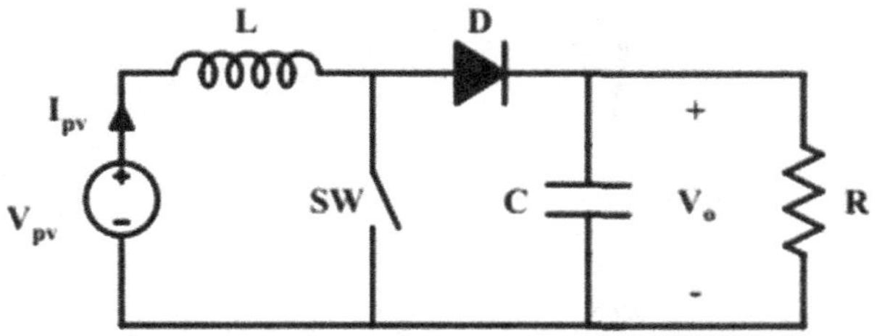

Fig. 6. DC-DC Boost converter

$$\begin{cases} \dfrac{di_L}{dt} = \dfrac{V_{pv} - V_0}{L} + \dfrac{V_0}{L}u \\ \dfrac{dV_0}{dt} = \left(-\dfrac{V_0}{RC_2} + \dfrac{i_L}{C_2}\right) - \dfrac{i_L}{C_2}u \end{cases} \tag{5}$$

Where VO is the load voltage and iL is the current across the inductor. The output voltage of the boost converter VO can be expressed in function of the input voltage Vpv and its duty cycle d:

$$\frac{V_0}{V_{pv}} = \frac{1}{1-d} \tag{6}$$

4 Maximum Power Point Tracking

4.1 Classic MPPT Commands

- Perturb and Observe (P&O)
  The most commonly used MPPT algorithm is P&O method. This algorithm uses simple ordering of notes and slightly measured parameters. In this approach, the array voltage is given periodically for the disturbance and the corresponding output energy is compared with those in the previous turbulence cycle. The P&O method measures the instantaneous operating area and then according to the area, the reference voltage is increased or increased so that the systems operate near the mpp. Since the method only increases or decreases the reference voltage, the implementation is simple. However, the method can track immediate and rapid changes in environmental conditions. The algorithm described in Fig. 8 [7] (Fig. 7).

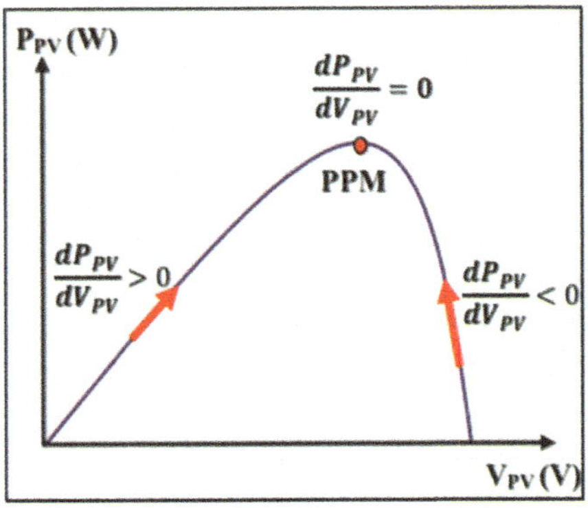

Fig. 7. Operations of P&O MPPT [8]

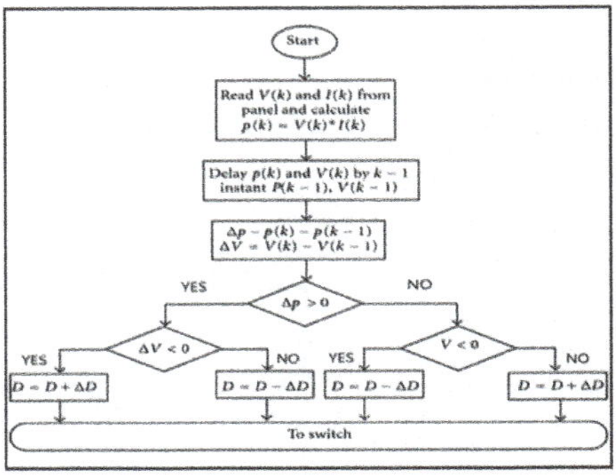

Fig. 8. Flowchart of Perturb and Observe (P&O) method algorithm

- Incremental Conductance (INC)
  Under rapidly changing climatic conditions, INC overcomes problems that have emerged after the P&O method. It can access MPP and stop disturbing the operating point This algorithm determines when MPPT arrives at MPP and does not fluctuate significantly unlike P&O. This is clearly an advantage over P&O. INC method is better than P&O in terms of accurate tracking of rapidly increasing and decreasing radiation conditions. The operations of INC MPPT is shown in Fig. 9, and the Fig. 10 shown the INC method algorithm [7].

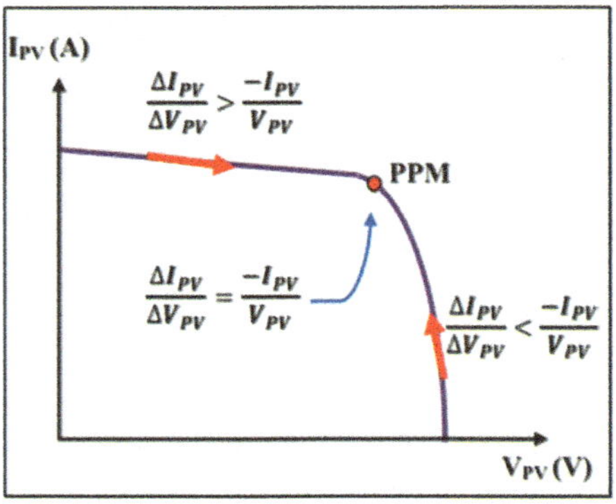

Fig. 9. Operations of INC MPPT [8]

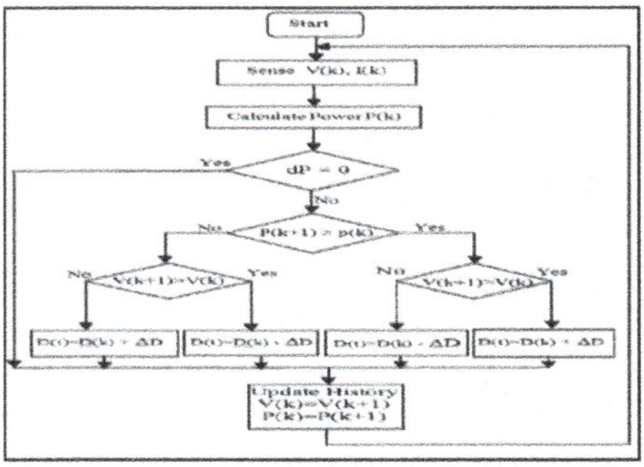

Fig. 10. Flowchart of incremental conductance (INC) method algorithm

4.2 Modern MPPT Commands

- Sliding mode control method
  Its main advantages are simplicity, stability, durability, high precision and good stability, etc.

It has three steps [9]:

1. Surface selection.

For the system determined by the following equation, the surface vector has the same dimension as the vector control (u).

$$x = A(x,t).x + B(x,t).u \tag{7}$$

The nonlinear model is the error function on the controlled variable (x), it is given by:

$$S(x) = \left(\frac{\partial}{\partial t} + \lambda_x\right)^{r-1} \times e(x) \tag{8}$$

Where

$$e(x) = \dot{x} - x$$

2. Establish conditions of stability

Stability and convergence conditions include different dynamics that allow the system to meet with the sliding surface and stay there regardless of turbulence: there are two considerations to ensure convergence.

The discrete function switching

This is the first condition for convergence. We have to give the dynamic surface converge to zero. Provided by:

$$\begin{cases} \dot{S}(x) > 0 \rightarrow si \triangleright S(x) > 0 \\ \dot{S}(x) < 0 \rightarrow si \triangleright S(x) < 0 \end{cases} \tag{9}$$

It can be written as:

$$\dot{S}(x).S(x) < 0 \tag{10}$$

– Lyapunov function:

The Lyapunov function is a positive scalar function for the state variables of the system. The idea is to choose a numerical function to ensure that the variable to be controlled is attracted to its reference value. We define the function Lyapunov as follows:

$$V(x) = \frac{1}{2} S^2(x) \tag{11}$$

$$\dot{V}(x) = S(x).\dot{S}(x) \tag{12}$$

3. Determination of the control law

The structure of a sliding mode controller consists of two parts. The first one concerns the exact linearization (u_{eq}) and the second one concerns the stabilization (u_n).

$$u = u_{eq} + u_n \tag{13}$$

Where ueq corresponds to the control. It serves to maintain the variable control on the sliding surface. u_n is the discrete control determined to check the convergence condition (Eq. 10). We consider a system defined in state space by (Eq. 7) and we have to find analogical expression of the control (u).

$$\dot{S} = \frac{\partial S}{\partial t} = \frac{\partial S}{\partial x} \cdot \frac{\partial x}{\partial t} \tag{14}$$

Substituting Eqs. 7 and 13 in Eq. 14, we get:

$$S(x) = \frac{\partial S}{\partial t} \cdot \left(A(x,t) + B(x,t) \cdot u_{eq}\right) + \frac{\partial S}{\partial x} \cdot B(x,t) \cdot u_n$$

We conclude the equivalent control expression:

$$u_{eq} = -\left(\frac{\partial S}{\partial t} \cdot B(x,t)\right)^{-1} \cdot \frac{\partial S}{\partial t} \cdot A(x,t) \tag{15}$$

For the equivalent control that can take a limited value, it must:

$$\frac{\partial S}{\partial x} \cdot B(x,t) \neq 0 \tag{16}$$

In the convergence mode and replacing the equivalent command by its expression in Eq. 15, we find the new expression of the surface derivative:

$$\dot{S}(x,t) = \frac{\partial S}{\partial t} \cdot B(x,t) \cdot u_n \tag{17}$$

And the condition expressed by Eq. 10 becomes:

$$S(x,t) \cdot \frac{\partial S}{\partial t} \cdot B(x,t) \cdot u_n < 0 \tag{18}$$

The simplest form can take separate control as follows:

$$u_n = ks \cdot sign(S(x,t)) \tag{19}$$

Where the ks tag must be different from that of $\frac{\partial S}{\partial t} \cdot B(x,t)$

- **Application of sliding mode control to Track Photovoltaic System**

To develop an MPPT controller based on the sliding mode control, it is necessary to use the system model and choose a sliding surface, so that if this is zero the maximum power point PPM is reached, then the control law is calculated to control the converter. We select the sliding surface as:

$$\frac{dP_{pv}}{dV_{pv}} = 0 \tag{20}$$

$$I_{pv} + V_{pv}\frac{dI_{pv}}{dV_{pv}} = 0 \tag{21}$$

The sliding surface is given as:

$$S = I_{pv} + V_{pv}\frac{dI_{pv}}{dV_{pv}} \tag{22}$$

We have to observe the work cycle d. Its control choose as follows:

$$u = u + \Delta u \triangleright \; if \; \triangleright S > 0$$
$$u = u - \Delta u \triangleright \; if \; \triangleright S < 0$$

By the non-linear time invariant system, Eqs. 17 and 18 can be written:

$$\dot{X} = f(X) + g(X) \cdot u \tag{23}$$

The equivalent control is determined from the following condition [9]:

$$\dot{S} = \left[\frac{dS}{dX}\right]^T \cdot \dot{X} = 0 \tag{24}$$

$$\dot{S} = \left[\frac{dS}{dX}\right]^T \cdot (f(X) + g(X) \cdot u_{eq}) = 0 \tag{25}$$

We obtain the equivalent control ueq:

$$u_{eq} = -\frac{\left[\frac{dS}{dX}\right]^T \cdot f(X)}{\left[\frac{dS}{dX}\right]^T \cdot g(X)} = 1 - \frac{V_{pv}}{V_0} \tag{26}$$

The equivalent duty cycle must lies in $0 < u_{eq} < 1$

The real control signal u is proposed as [9]: $u = 1$ if $u_{eq} + k_s.S \geq 1$, $u = u_{eq} + k_s.S$ if $0 \leq u_{eq} + ks.S \leq 1$ and $u = 0$ if $u_{eq} + k_s.S \leq 0$. Where the control saturates if $u_{eq} + k_s.S$ are out of range, ks is a positive scaling constant (Fig. 11).

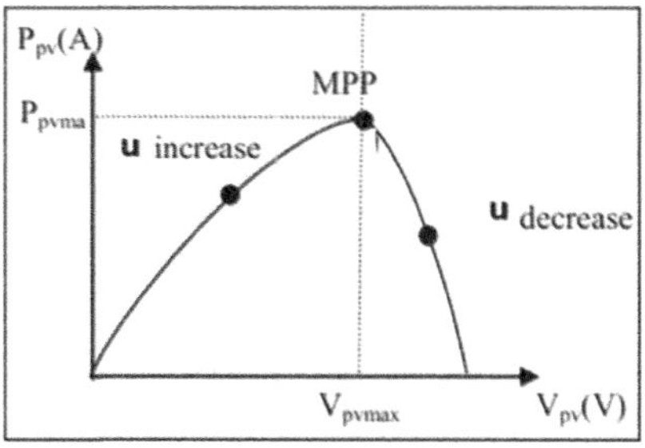

Fig. 11. Operating point according to the sign u

Note

The main disadvantage of the increasing mode control lies in the phenomenon of chatter, due to the absolute stop of the control law. Among the approaches that are proposed to reduce this brewing phenomenon, on the second order sliding mode control.

- **Application of SOSMC to Track Photovoltaic System**

The disadvantage of SMC is the chattering phenomenon for reduce this phenomenon a SOSMC is developed. According to the theory of the SMC, We must define the sliding surface S and design the control code in order to attract the path of the state S = 0 and maintain it there [10]. With model of DC converter we have:

$$\begin{cases} \dot{x} = f_1(x,t) + f_2(x,t)u \\ S = S(x,t) = e(x,t) + \lambda\dot{e}(x,t) \\ e(x,t) = (x_2 - x_{2ref}) \end{cases} \tag{27}$$

Where x϶R is state of the system, u϶R is the control, f1, f2 are functions easily identified from Eq. (5). S϶R is the sliding surface. If we differentiate the sliding surface S, we can write:

$$\ddot{S} = \varphi_1(t, S, \dot{S}) + \varphi_2(t, S, \dot{S}) \cdot \dot{u} \tag{28}$$

The control u is bonded ($0 \le u \le 1$). We assume that the Eq. (27) satisfy the following conditions [10]:

$$\begin{cases} 0 < \gamma_i \le |\varphi_2(t, S, \dot{S})| \le \gamma_e \\ |\varphi_2(t, S, \dot{S})| \le \varpi_0 \end{cases} \tag{29}$$

Where γ_i, γ_e and ϖ_0 are positive gains. The used control law is the super twisting algorithm given as follow [10]:

$$u = u_1 + u_2 \tag{30}$$

Where

$$\begin{cases} \dot{u}_1 = -\delta_1 sign(S) \\ u_2 = -\delta_2 |S|^{\rho} sign(S) \end{cases}$$

And σ_1, σ_2 are constant positive [11]

$$\begin{cases} \delta_1 > 0 \\ \delta_2 > 0 \\ 0 < \rho \le \frac{1}{2} \end{cases}$$

The use of the second order sliding mode guarantees the finite time convergence [10].

5 Simulation Results

The Simulink model of PV with dc-dc boost converter using INC, P&O, SMC and SOSMC MPPT methods. And its corresponding result is shown in Figs. 13, 14, 15 and 16, consist on the waveforms of PV current Ipv (A), voltage Vpv (V), power Ppv (W) and output voltage V0 (V) (Fig. 12).

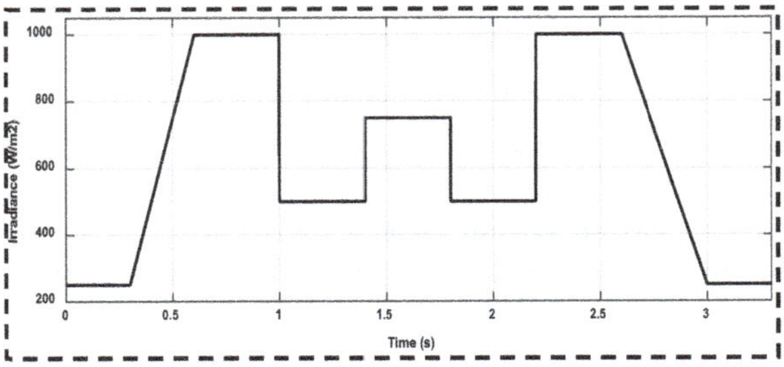

Fig. 12. Irradiance profile

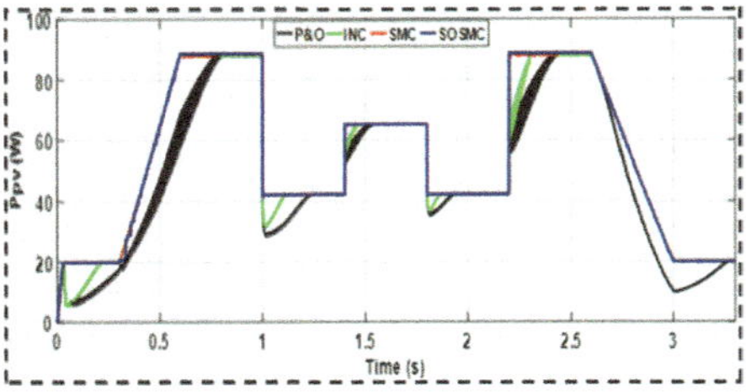

Fig. 13. PV power P_{pv} (W)

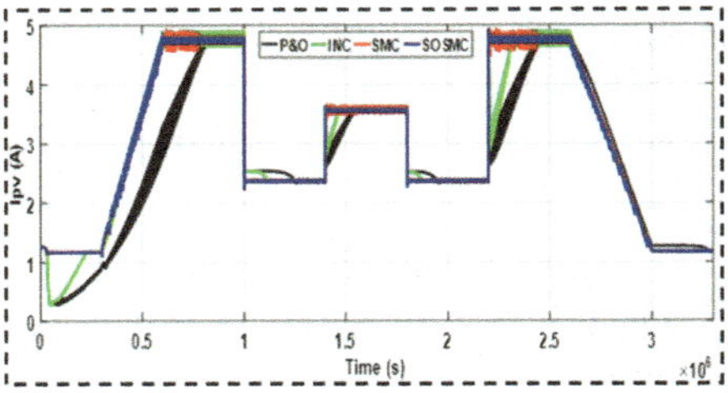

Fig. 14. PV current Ipv (A)

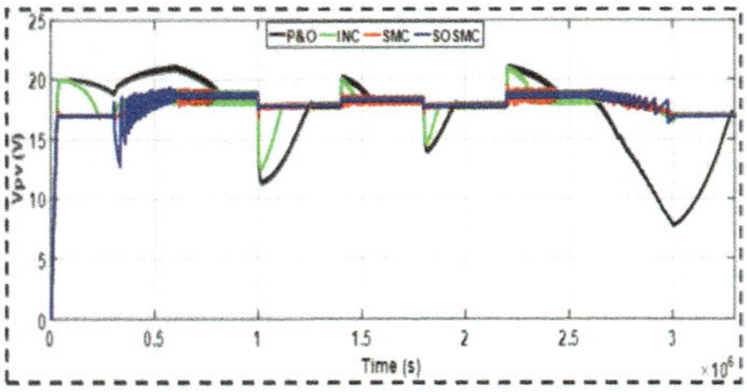

Fig. 15. PV voltage V_{pv} (V)

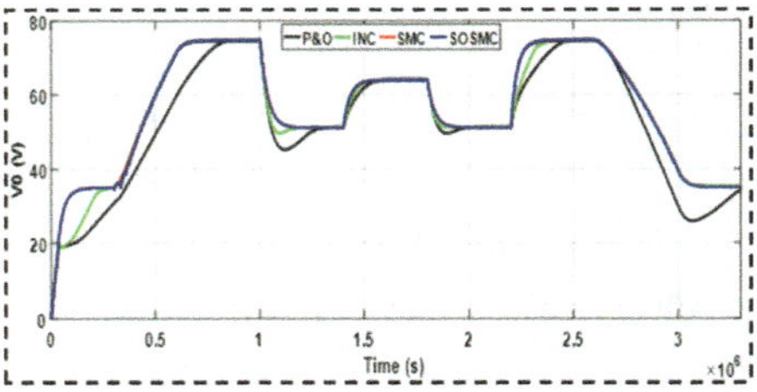

Fig. 16. Output voltage V_0 (V)

From the results we observed the significant influence of solar radiation change on photoelectric properties, Depicts the comparison of the PPT between the classical techniques (P&O, INC) and the modern techniques (SMC, SOSMC). It can be seen that the SMC exhibits better performance than the P&O and INC, in dynamic response and in the stable state. But when we use the latter, we notice the emergence of chattering effects and because of that we used the SOSMC which reduces chattering effects and robust to abrupt change of solar radiation.

6 Conclusion

In this paper, a study of several valid methods in the search for the maximum power point of a photovoltaic system was presented. Various points were discussed on the optimization of energy conversion which increases the overall electrical efficiency of a photovoltaic system. From the obtained results, it can be say: that Second order sliding mode controller 'SOSMC' can improve greatly the performance of the photovoltaic power system under strong operating condition. The special merit of proposed scheme control based on SOSMC is can be competitive with regard to the existing control strategies.

Prameters

PV parameters

ki = 0.0017 mA/°C, q = 1.6E–19, k = 1.38E–23, N = 1.3, Eg0 = 1.1, Rs = 0.0017Ω, Rsh = 1000 Ω, T_N = 298 K, Voc = 22 V, Isc = 5.1A, Ns = 36.

Boost converter parameters:

f = 10 kHz, L = 0.4 mH, C1 = C2 = 1000 µF.

References

1. Alsayid, B., Jallad, J.: Modeling and simulation of photovoltaic cells/module/arrays. J. Res. Rev. **2**(6), 1327–1331 (2012)
2. Alsumiri, M.A., Jiang, L., Tang, W.H.: MPPT Controller for Photovoltaic System Using SMC (2014)
3. Kollimalla, S.K., Mishra, M.K.: A new adaptive P&O MPPT algorithm based on FSCC method for photovoltaic system. In: Proceedings of IEEE International Conference on Circuit, Power Computing Technology ICCPCT 2013 (2013)
4. Belkaid, A., Colak, I., Isik, O.: Photovoltaic maximum power point tracking under fast varying of solar radiation. Appl. Energy **179**, 523–530 (2016)
5. Ben Salah, C., Ouali, M.: Comparison of fuzzy logic and neural network in maximum power point tracker for PV systems. Electr. Power Syst. Res. **81**(1), 43–50 (2011)
6. Mars, N., Grouz, F., Essounbouli, N., Sbita, L.: Synergetic MPPT controller for photovoltaic system. J. Electr. Electron. Syst. **6**(232), 2332–0796 (2017)
7. Rana, M.M., Ali, M.R., Ajad, A.K., Moznuzzaman, M.: Analysis of P&O and INC MPPT techniques for PV array using MATLAB. IOSR J. Electr. Electron. Eng. **11**(04), 80–86 (2016)
8. Attoui, H.: Contribution à la synthèse de nouvelles stratégies de commande des systèmes d'énergie renouvelable. UNIVERSITE FERHAT ABBAS — SETIF 1 (2017)
9. Rekioua, D., Achour, A.Y., Rekiouaa, T.: Tracking power photovoltaic system with sliding mode control strategy. Energy Procedia **36**, 219–230 (2013)
10. Sahraoui, H., Drid, S., Chrifi-Alaoui, L., Bussy, P.: Second order sliding mode control of DC-DC converter used in the photovoltaic system according an adaptive MPPT. Int. J. Renew. Energy Res. **6**(2), 375–383 (2016)
11. Massaoudi, Y.: A new backstepping sliding mode controller applied to a DC-DC boost converter. Int. J. Power Electron. Drive Syst. **7**(3), 759 (2016)

Optimum Design of a Hybrid Photovoltaic/Diesel/Battery/System Using Water Cycle Algorithm

F. Fodhil[1(✉)], A. Hamidat[2], O. Nadjemi[1], Z. Alliche[1], and L. Berkani[3]

[1] Faculté de Technologie, Département d'Electronique, Labset, Université Saad Dahlab Blida 1, Route de Soumaa, 09000 Blida, Algeria
fadfodl@yahoo.fr
[2] Centre de Développement des Energies Renouvelables, CDER, Route de l'Observatoire Bouzareah, 16340 Algiers, Algeria
[3] Faculté de Technologie, Département d'Electronique, LATSI, Université Saad Dahlab Blida 1, Route de Soumaa, 09000 Blida, Algeria

Abstract. This paper proposes a novel approach for the optimal design of an autonomous hybrid energy system. The hybrid system consists of photovoltaic panels, diesel generator and battery bank. The constrained water cycle algorithm is applied to minimize the cost of energy (COE) while the CO_2 emission and unmet load are considered as constraints. The proposed method is tested and evaluated to find the optimal configuration of hybrid system, which is designed to electrify 25 households located in a rural Saharan village. The optimization results stated the efficiency and robustness of the proposed method. Moreover, based on the values of the unmet load and the minimum quantity of CO_2 emissions constraints, an optimal solution (15 kWp of PV, 74.8 kWh of battery bank, 5 kW of DG) is chosen and analyzed. The COE of the optimal configuration is 0.45 \$/kWh with a PV penetration of 97%.

Keywords: Water cycle algorithm · Optimization · Hybrid energy system

1 Introduction

Integration of renewable energies in the rural electrification process has recently witnessed a promising trend worldwide. However, to keep this positive trend, the process requires reliable, accurate and environmentally friendly approaches. Given that standalone hybrid energy systems are the most common systems for rural and remote areas, the optimal design of these systems becomes more and more an attractive subject to researchers. Many studies have been conducted to enhance the efficiency and the energy performance of hybrid energy systems especially the hybrid PV-diesel-battery system [1–3] which is the most common and mature electrification system in rural regions [4]. In the last decade, there has been much research into the utilization of heuristic and artificial intelligence algorithms such as neural network (NN), genetic algorithm (GA), evolutionary algorithms (EA), scatter search (SS), ant colony optimization (ACO), cuckoo search algorithm (CSA), artificial bee colony optimization

M. Hatti (Ed.): ICAIRES 2019, LNNS 102, pp. 82–93, 2020.
https://doi.org/10.1007/978-3-030-37207-1_9

(ABCO) and particle swarm optimization (PSO) in the design of a reliable hybrid PV-Diesel system for rural sites worldwide [5–11]. In this context, this paper involves the optimal design of a standalone hybrid energy system using the water cycle algorithm (WCA). A case study was carried out in a rural Saharan village. The village is located at latitude 24°42′ N 4°39′ E far from the utility grid in the province of Tamanrasset. Moreover, it has 25 un-electrified households, with an average household power consumption about 2.4 kWh/day.

2 Problem Description

The chosen system for this study is the standalone hybrid PV-diesel-battery energy system. It consists basically of PV modules, a diesel generator, an inverter, and a battery bank, Fig. 1 illustrates the diagram of the hybrid system. The PV subsystem is the primary source in the system. The cycle charging dispatch strategy is considered as the system control strategy which means when the PV energy is more than the hourly load, the energy excess is stored in the battery bank based on the state of charge of batteries. When the energy produced by the PV fails to supply the load, the battery bank releases energy to assist the PV in supplying the load demand based also on the state of charge of batteries. When the batteries are not able to meet the load, the diesel generator is operated to satisfy the load and charge the batteries simultaneously.

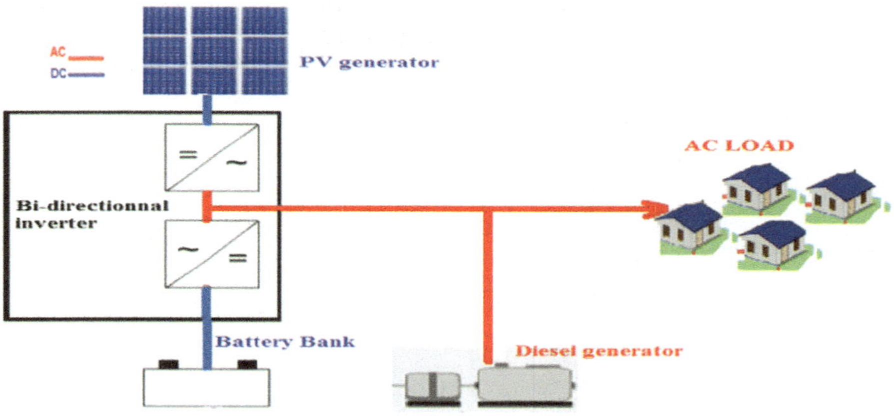

Fig. 1. The basic schematic of the hybrid PV-diesel-battery energy system.

3 Materials and Methods

3.1 The Proposed Approach

The aim of this study is the optimal sizing of a standalone hybrid PV-Diesel-battery system by using the water cycle algorithm (WCA). The cost of the energy (COE) is the main objective of the optimization problem, the total CO2 emissions produced by the diesel generator, and the loss of load probability (LPP) are considered as constraints. The design variables included in the optimization process are the size of the PV panels N_{pv}, the size of the battery bank N_{bat}, and the rated capacity of DG unit P_{dg}, these variables are defined in a vector named stream. Each stream represents a different configuration of the system. The optimization flowchart is illustrated in Fig. 2. After entering the system parameters such as the hourly load demand, hourly meteorological data, lifetime of the project, cost of the components…etc. the next step is the initialization phase, where the first population of streams is generated randomly. Afterward, in each iteration and for each stream, the objective function and the constraints are calculated and checked. If the particle does not meet the constraints, it is modified and sent back to simulation. Each iteration feasible particles are evaluated by the WCA algorithm based on their fitness until stopping criterion is checked. After terminating the cycle, the optimal solution is returned.

3.2 Water Cycle Algorithm

The water cycle algorithm (WCA) is first proposed by Eskander et al. [12] to solve various constrained optimization problems. The WCA is metaheuristic algorithm inspired from nature and based on water cycle observation. It mimics the flow of rivers and streams up into the sea. Similar to other metaheuristics, the primary objective of this algorithm is to find the optimum solution or near-optimum solution through effective exploration and exploitation [13]. In the initialization phase, a matrix of initial streams of size $N_{pop} \times D$ are generated randomly, where N_{pop} represents the population size, and D is the number of design variables [12].

$$Total\ population = \begin{bmatrix} Sea \\ River_1 \\ River_2 \\ River_3 \\ \vdots \\ Stream_{N_{sr}+1} \\ Stream_{N_{sr}+2} \\ Stream_{N_{sr}+3} \\ \vdots \\ Stream_{N_{pop}} \end{bmatrix} = \begin{bmatrix} x_1^1 & x_2^1 & x_3^1 & \cdots & x_D^1 \\ x_1^2 & x_2^2 & x_3^2 & \cdots & x_D^2 \\ \vdots & \vdots & \vdots & \vdots & \vdots \\ x_1^{N_{pop}} & x_2^{N_{pop}} & x_3^{N_{pop}} & \cdots & x_D^{N_{pop}} \end{bmatrix} \quad (01)$$

where, N_{sr} is the addition of the number of rivers and a single sea as follows [12]:

$$N_{sr} = Number\ of\ rivers + \underbrace{1}_{sea} \quad (02)$$

The rest of the population ($N_{streams}$) is given in Eq. (03):

$$N_{streams} = N_{pop} - N_{sr} \tag{03}$$

The amount of water entering a river or the sea depends on the magnitude of the flow. The assigned streams for each river and the sea are calculated using the following equation [12]:

$$NS_n = round\left\{ \left| \frac{Cost_n - Cost_{N_{sr}+1}}{\sum_{n=1}^{N_{sr}} C_n} \right| \times N_{streams} \right\}, n = 1, 2, 3, \ldots, N_{sr} \tag{04}$$

where NS_n is the number of streams which flow into specific rivers and the sea.

In the exploitation phase of WCA, the new positions of streams and rivers are calculated as follows [12]:

$$\vec{X}_{stream}(t+1) = \vec{X}_{stream}(t) + rand \times C \times \left(\vec{X}_{sea}(t) - \vec{X}_{stream}(t)\right) \tag{05}$$

$$\vec{X}_{stream}(t+1) = \vec{X}_{stream}(t) + rand \times C \times \left(\vec{X}_{river}(t) - \vec{X}_{stream}(t)\right) \tag{06}$$

$$\vec{X}_{river}(t+1) = \vec{X}_{river}(t) + rand \times C \times \left(\vec{X}_{sea}(t) - \vec{X}_{river}(t)\right) \tag{07}$$

Equations (05) and (06) are for streams which flow into the sea and their corresponding rivers, respectively. where, t is the iteration index, $1 < C < 2$, and *rand* is a uniformly distributed random number between 0 and 1. If a stream has a cost function better than its connecting river, the position of the river and stream are exchanged. The same happens for a river and the sea [12].

The raining process happens after satisfying the evaporation process. New population forms streams in different locations. The locations of the new streams are formulated as follows [12]:

$$X_{stream}^{new} = LB + rand \times (UB - LB) \tag{08}$$

where LB and UB are the lower and upper bounds, respectively. For constrained problems, the following equation is used to encourage the generation of streams which directly flow to the sea leading to improve the exploration near the sea (the optimum solution) [12]:

$$X_{stream}^{new} = X_{sea} + \sqrt{\mu} \times randn(1, D) \tag{09}$$

3.3 Mathematical Model of the Hybrid System Components

- PV Model

The following equation is used to calculate the hourly output PPV(t) (W) the PV array [14]:

$$P_{PV}(t) = f_{pv}.P_{STC}.\frac{G_h(t)}{G_{STC}}.\left[1 + \frac{\alpha}{100}(T_C(t) - T_{STC})\right]. \quad (10)$$

where f_{pv} is the PV derating factor (around 0.6). $G_h(t)$ (kWh/m^2) is the solar radiation incident over the tilted surface of the PV panels during time step of hour t. P_{STC}(Wp) is the output power of the panel in STC (standard test conditions). G_{STC} is the incident radiation at standard test conditions (G_{STC} = 1 kW/m2). T_{STC} is the PV cell temperature under standard test conditions (T_{STC} = 25 °C) and $T_c(t)$ (°C) is the PV cell temperature in the present time step, which can be calculated as:

$$T_C(t) = T_a(t) + \left(\frac{NOCT - 20}{0.8}\right).\frac{G_{h_{year}}(t)}{1kWh/m^2} \quad (11)$$

where $T_a(t)$ is the ambient temperature (°C) and NOCT is the nominal operation cell temperature (°C).

- Battery Bank

The power of battery bank $P_B(t)$ at hour t is expressed using the following equation [15]:

$$P_B(t) = P_B(t-1).(1-\sigma) + \left(P_{PV}(t) - \frac{P_{load}(t)}{\eta_{inv}}\right).\eta_{bat} \quad (12)$$

when:

$P_{PV}(t) - \frac{P_{load}(t)}{\eta_{inv}} > 0$ *and* $P_B(t-1) < P_{Bmax}$ the charging process.

$P_{PV}(t) - \frac{P_{load}(t)}{\eta_{inv}} < 0$ *and* $P_B(t-1) > P_{Bmin}$ the discharging process.

$$P_{Bmin} \le P_B(t) \le P_{Bmax} \quad (13)$$

where $P_B(t-1)$ is the availability power of the battery bank at the hour (t−1). The term σ is the self-discharge rate of battery bank, in the study it is assumed 0.002. $P_{load}(t)$ t is the load demand at the hour t. η_{bat} is the efficiency of the battery (η_{bat} = 0.8) and η_{inv} is the efficiency of the inverter. P_{Bmin} and P_{Bmax} denote the minimum allowable energy level remained in the battery bank and the maximum allowable energy level respectively.

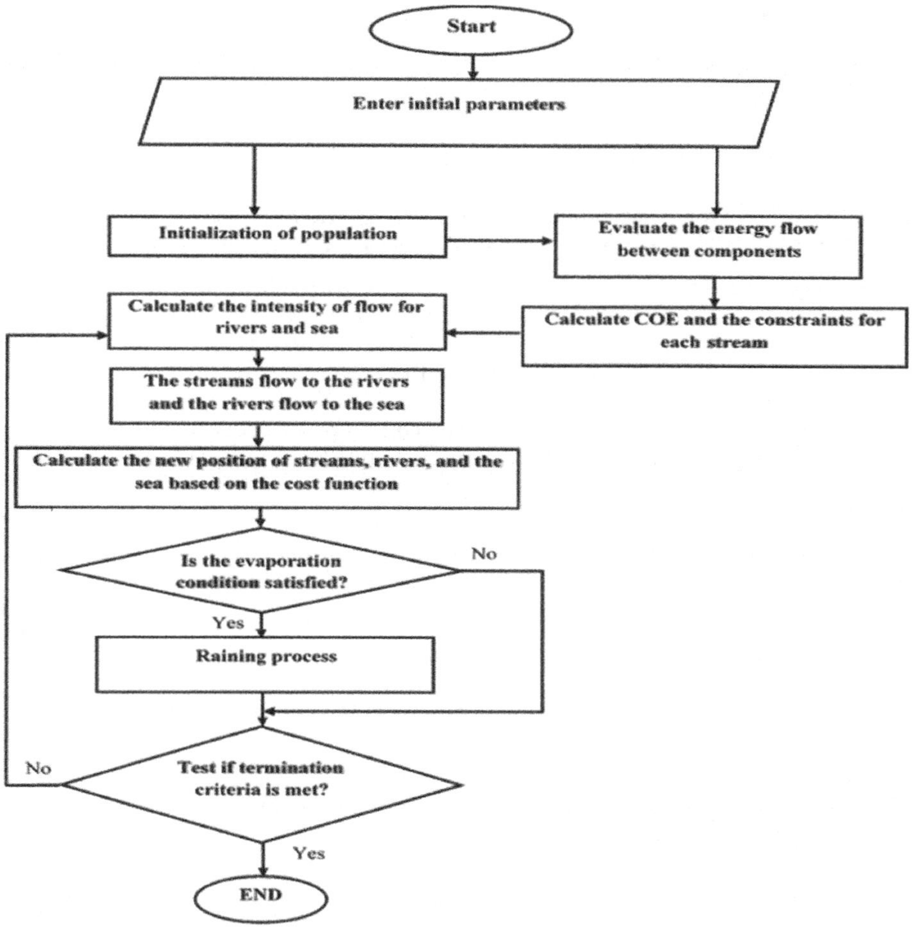

Fig. 2. The WCA optimization flowchart of the hybrid PV-diesel-battery system.

The maximum state of charge (SOC_{max}) and the minimum state of charge (SOC_{min}) of the battery bank during discharging are formulated as follows [16]:

$$SOC_{max} = P_B \mathrm{N}_{\mathrm{bat}} \tag{14}$$

$$SOC_{min} = SOC_{max}(1 - DOD_{max}), 30\% \leq DOD_{max} \leq 50\% \tag{15}$$

where N_{bat} is the total number of batteries, P_B is the nominal capacity of each battery and $DOD_{max}(\%)$ denotes the maximum depth of discharge. DOD = 30% is chosen for this study.

- Diesel Generator

The fuel consumption of the diesel generator, $Cons_G$ (L/h) is considered and expressed as follows [17]:

$$Cons_G = \alpha_{DG} P_{n-DG} + \beta_{DG} P_{a-DG} \tag{16}$$

where P_{n-DG} (kW) is the nominal power, P_{a-DG} (kW) is the electrical produced power of the diesel generator, α_{DG} and β_{DG} are the coefficients of the consumption curve, defined by the user (L/kWh). The diesel efficiency η_G is given as:

$$\eta_G = \frac{P_{a-DG}}{Cons_G LHV_{fuel}} \tag{17}$$

where LHV_{fuel} is the heating value of fuel consumption between $10 \leq LHV_{fuel} \leq 11.6$ kWh/L, $\alpha_{DG} = 0.246$ L/kWh and $\beta_{DG} = 0.08145$ L/kWh [17].

- Inverter

The power going from the inverter to cover the load demand is calculated as follows:

$$P_{in} = \frac{P_{load}(t)}{\eta_{inv}} \tag{18}$$

where η_{inv} is the inverter efficiency, and $P_{load}(t)$ is the hourly demand (W).

- Objective Function and Constraints

The cost of energy (COE) is the objective function of the optimization problem. The equation of COE is given as follows:

$$COE = \frac{C_{ann,tot}}{P_{TLoad}} \tag{19}$$

where P_{TLoad} is the total annual load consumption, and $C_{ann,tot}$ is the total annualized cost. The annualized cost includes annual capital cost, the annual operating & maintenance cost, the annual replacement cost plus the annual fuel cost.

As mentioned before, loss of load probability LLP and CO_2 emissions are considered as the problem constraints.

The following equation is used to formulate the LLP of the system [18]:

$$LLP = \frac{\sum_{t=1}^{8760} shortage(t)}{\sum_{t=1}^{8760} D(t)} \tag{20}$$

where $D(t)$ is electricity demand, and $shortage(t)$ is the annual unmet load.

The quantity of CO_2 emissions emitted by diesel generator is calculated as follows [18, 19]:

$$CO_{2_{emission}} = \sum_{t=1}^{8760} fuel_{cons}(t)EF \tag{21}$$

where *EF* denotes the emission factor of the diesel generator. 2.4 kg/L $\leq EF \leq$ 2.8 kg/L [17].

4 Case Study

A rural Saharan and un-electrified village namely Moulay Lahcen is chosen as a case study, the village is located at latitude 24°42′ N 4°39′ E and an altitude of 981 meters far from the electrical grid in the province of Tamanrasset. The local meteorological data (the global horizontal radiation, clearness index and temperature) are presented in Fig. 3. The solar radiation was sourced from the NASA Surface Meteorological dataset and the temperature was sourced from Meteonorm 6.0. A load pattern is assumed based on an investigation nearby the inhabitants of the rural villages in the southern of Algeria, two hourly load patterns are used for summer and winter months. The household has a basic load profile. Figure 4 shows the daily load profiles of household. Furthermore, the technical and economic details of the system components are summurized in Table 1.

Table 1. Hybrid system technical and economic parameters.

Parameters	Value
Project lifetime (years)	20
Reliability of inverter	0.95
Reliability of battery banks	0.85
Emission factor (kg/L)	2.4
Fuel cost ($/L)	0.3
Cost of PV panel ($/W)	2.3
Cost of diesel generator ($/kW)	375
Cost of a battery ($/kWh)	400
Cost of an inverter ($/kW)	700

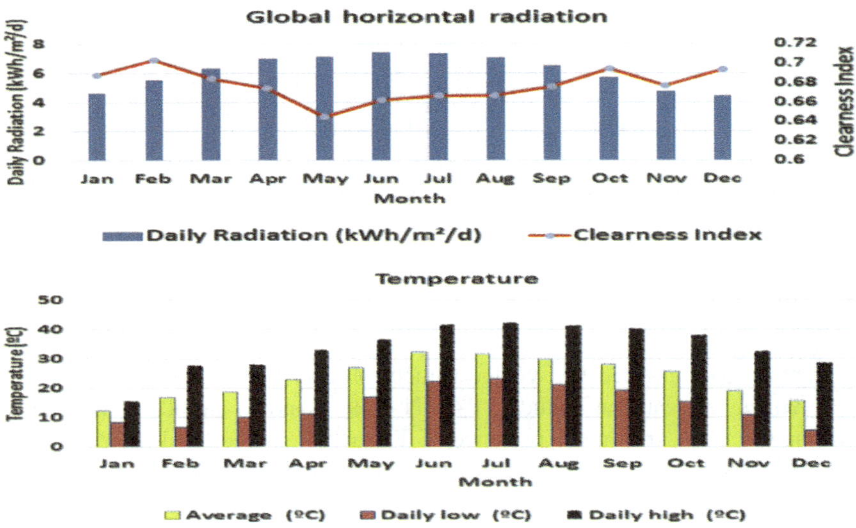

Fig. 3. Meteorological characteristics of the site.

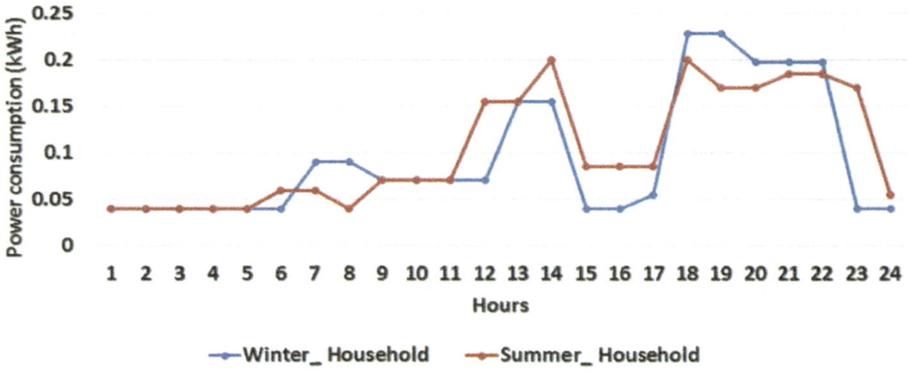

Fig. 4. Daily load profiles of a household.

Table 2. Technical, economic and CO_2 emissions of the optimal system.

PV (kWp)	Battery (kWh)	DG (kW)	LLP (%)	CO_2 (kg/year)	PV penetration (%)	ACS ($/year)	ACC ($/year)	AFC ($/year)	AOM ($/year)	ARC ($/year)	Energy produced by the DG (kW/year)	Energy produced by the PV (kW/year)
15	74.8	5	0.2	898.75	97	12872.06	7274.23	280.86	1832.05	3484.93	862.5	27792.22

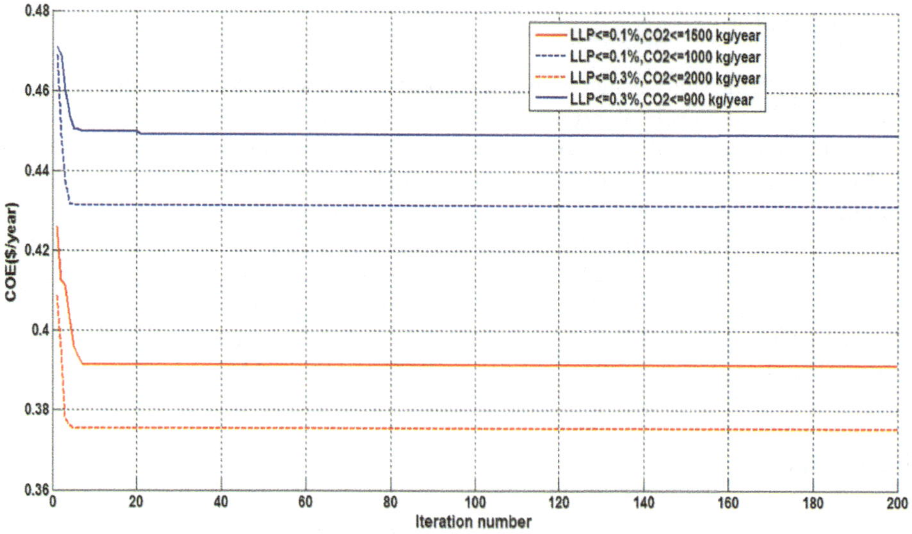

Fig. 5. Convergence curves of WCA for different cases.

5 Results and Discussion

The water cycle algorithm (WCA) is applied to find the optimal design of hybrid PV-diesel-battery system. The cost of the system (COE) is the objective function of this design optimization problem, LLP and CO_2 emission are considered as constraints bounds. The system modeling and optimization program are coded in MATLAB software on a PC with i3 CPU 2.5 GHz, 3 GB RAM. The number of population is 50 and the number of iterations is 200. The convergence curves of WCA for different values of LLP and CO_2 emissions constraints is depicted in Fig. 5. It can be seen that WCA reached the optimum solution in all cases after maximum of 20 iterations. Which means that the WCA has a highly efficient performance in solving this optimization problem. From the different cases, the designer can choose the most appropriate solution. In this study the optimal solution with the minimum CO_2 emissions is picked to be analyzed ($constr_{\mathrm{LLP}}$ <= 0.3%, $constr_{CO_2}$ <= 900 kg/year). The results of this solution are provided in Table 2. The optimal configuration of the hybrid system has COE of 0.45 \$/kWh. It involves 15 kWp of PV, 74.8 kWh of battery bank, 5 kW of DG, and produced 97% of the total energy from PV (27792.22 kWh/year) and 3% (862.5 kWh/year) from the diesel generator. The annual capital cost ACC is the dominant cost of the system with 56.5% of the ACS, followed by the annual replacement cost (ARC) with 27.07%, the annual operating and maintenance cost with 14.23% and the annual fuel cost (AFC) with 2.18%.

6 Conclusion

The purpose of the study is the optimal design of a standalone hybrid PV-diesel-battery system using water cycle algorithm (WCA) to supply power to rural Saharan village. The cost of energy (COE) is chosen as the objective function along with considering the loss of load probability (LLP) and CO_2 emissions as the problem constraints. The optimization results confirmed the efficiency and the adequacy of the water cycle algorithm (WCA) in solving the design problem of hybrid energy systems. Furthermore, the optimal solution (15 kWp of PV, 74.8 kWh of battery bank, 5 kW) with ($constr_{\mathrm{LLP}} <= 0.3\%$, $constr_{CO_2} <= 900$ kg/year) was intended for analysis. The obtained COE of the hybrid system is 0.45 $/kWh, the PV penetration is 97% and the annual capital cost of the system represents 56.5% of the total cost of the system.

References

1. Kadri, A., Hamidat, A.: Technico–economic assessment of hybrid diesel-photovoltaic power plant in the south of Algeria. In: AIP Conference Proceedings. AIP Publishing (2016)
2. Fodhil, F., Hamidat, A., Nadjemi, O.: Energy control strategy analysis of hybrid power generation system for rural Saharan community in Algeria. In: International Conference in Artificial Intelligence in Renewable Energetic Systems. Springer (2017)
3. Rodríguez-Gallegos, C.D., Gandhi, O., Bieri, M., Reindl, T., Panda, S.K.: A diesel replacement strategy for off-grid systems based on progressive introduction of PV and batteries: an Indonesian case study. Appl. Energy **229**, 1218–1232 (2018)
4. Chaurey, A., Kandpal, T.C.: Assessment and evaluation of PV based decentralized rural electrification: an overview. Renew. Sustain. Energy Rev. **14**(8), 2266–2278 (2010)
5. Gan, L.K., Shek, J.K.H., Mueller, M.A.: Optimised operation of an off-grid hybrid wind-diesel-battery system using genetic algorithm. Energy Convers. Manag. **126**, 446–462 (2016)
6. Al-Alawi, A., Al-Alawi, S.M., Islam, S.M.: Predictive control of an integrated PV-diesel water and power supply system using an artificial neural network. Renew. Energy **32**(8), 1426–1439 (2007)
7. Movahediyan, Z., Askarzadeh, A.: Multi-objective optimization framework of a photovoltaic-diesel generator hybrid energy system considering operating reserve. Sustain. Cities Soc. **41**, 1–12 (2018)
8. Eteiba, M.B., Barakat, S., Samy, M.M., Wahba, W.I.: Optimization of an Off-Grid PV/Biomass hybrid system with different battery technologies. Sustainable Cities and Society (2018)
9. Fodhil, F., Hamidat, A., Nadjemi, O.: Potential, optimization and sensitivity analysis of photovoltaic-diesel-battery hybrid energy system for rural electrification in Algeria. Energy **169**, 613–624 (2019)
10. Nadjemi, O., Nacer, T., Hamidat, A., Salhi, H.: Optimal hybrid PV/wind energy system sizing: Application of cuckoo search algorithm for Algerian dairy farms. Renew. Sustain. Energy Rev. **70**, 1352–1365 (2017)
11. Berbaoui, B., Dehini, R., Hatti, M.: An applied methodology for optimal sizing and placement of hybrid power source in remote area of South Algeria. Renewable Energy (2019)

12. Eskandar, H., Sadollah, A., Bahreininejad, A., Hamdi, M.: Water cycle algorithm – A novel metaheuristic optimization method for solving constrained engineering optimization problems. Comput. Struct. **110–111**, 151–166 (2012)
13. Sadollah, A., Eskandar, H., Lee, H.M., Yoo, D.G., Kim, J.H.: Water cycle algorithm: a detailed standard code. Software **5**, 37–43 (2016)
14. Dufo-López, R., Cristóbal-Monreal, I.R., Yusta, J.M.: Stochastic-heuristic methodology for the optimisation of components and control variables of PV-wind-diesel-battery stand-alone systems. Renew. Energy **99**, 919–935 (2016)
15. Wang, H., Zhang, D.: The stand-alone PV generation system with parallel battery charger. In: 2010 International Conference on Electrical and Control Engineering, pp. 4450–4453 (2010)
16. Sobu, A., Wu, G.: Optimal operation planning method for isolated micro grid considering uncertainties of renewable power generations and load demand. In: IEEE PES Innovative Smart Grid Technologies, Tianjin, China (2012)
17. Skarstein, Ø., Uhlen, K.: Design considerations with respect to long-term diesel saving in wind/diesel plants. Wind Engineering, pp. 72–87 (1989)
18. Abedi, S., Alimardani, A., Gharehpetian, G.B., Riahy, G.H., Hosseinian, S.H.: A comprehensive method for optimal power management and design of hybrid RES-based autonomous energy systems. Renew. Sustain. Energy Rev. **16**(3), 1577–1587 (2012)
19. Dufo-López, R., Bernal-Agustín, J.L.: Influence of mathematical models in design of PV-Diesel systems. Energy Convers. Manag. **49**(4), 820–831 (2008)

Enhancement of Extracted Power from Photovoltaic Systems Through Accelerated Particle Swarm Optimisation Based MPPT

Karim Kaced[1(✉)], Sabrina Titri[1], and Cherif Larbes[2]

[1] Division Microélectronique et Nanotechnologie, Centre de Développement des Technologies Avancées (CDTA), Baba Hassen, Algeria
{kkaced, stitri}@cdta.dz

[2] Laboratoire des Dispositifs de Communication et de Conversion Photovoltaïque (LDCCP), Ecole Nationale Polytechnique (ENP), El-Harrach, Algeria
Cherif.larbes@g.enp.edu.dz

Abstract. This paper presents a maximum power point tracking (MPPT) technique for photovoltaic (PV) system based on accelerated particle swarm optimisation (APSO) algorithm. The main purpose is to handle the multi-modal P-V characteristic curve under partial shading conditions. The APSO based MPPT has the advantages to be simple and accurate to track the global maximum power point (GMPP) under severe partial shading patterns. The simulation results have shown that the proposed technique can find the GMPP with high speed and efficiency and presents good dynamic behaviour. In addition, this technique is superior than the conventional Perturb and Observe (P&O) based MPPT in global peak tracking.

Keywords: Maximum power point tracking (MPPT) · Accelerated particle swarm optimisation (APSO) · Partial shading conditions · Photovoltaic (PV) system

1 Introduction

Solar photovoltaic energy is an important renewable energy that can be an alternative to other conventional sources of energy in order to meet the large energy demands in the future [1]. An important feature of photovoltaic panels is that the maximum available power is provided only at a single operating point, located by a known voltage and current, called the maximum power point. The problem is that the position of this point is not fixed but it changes according to the solar irradiance and temperature of the solar cells as well as the load. Because of the relatively expensive cost of this kind of energy, we must extract the maximum available power from the photovoltaic panels.

The photovoltaic module is an association of solar cells. When one or more cells of a PV module are illuminated differently from others cells (partial shading), a degraded behaviour appears for the latter. The cells receiving less energy must dissipate the

M. Hatti (Ed.): ICAIRES 2019, LNNS 102, pp. 94–102, 2020.
https://doi.org/10.1007/978-3-030-37207-1_10

surplus of the current delivered by those who are most enlightened, thus creating local heating up to the destruction of a part of the PV module if the defect persists. To protect the photovoltaic cells, a bypass diode is connected in parallel to the terminals of an elementary group of cells in series. When the photovoltaic panel is subjected to uniform irradiance conditions, the P-V characteristic is uni-modal and characterized by a single maximum power point. This situation is still not guaranteed and the photovoltaic panel is often subject to partially shaded conditions. Under these conditions, multiple peak power points appear on the P-V characteristic, several local maximum power points (LMPPs), and one global maximum power point (GMPP). The conventional maximum power point tracking (MPPT) techniques do not perform well under these conditions. For this, several improved techniques are proposed to take into account the multimodal character of the P-V characteristic curve in the presence of partial shading [2].

Due to their ability to handle multi-modal functions, swarm-intelligence-based algorithms, including particle swarm optimisation (PSO) [3], ant colony optimisation (ACO) [4], differential evolution (DE) [5], bat algorithm (BA) [6], cuckoo search (CS) [7] are proposed to tracks the GMPP. The PSO technique is proposed for MPPT by many researchers thanks to its simple structure and high accuracy to find the global MPP under partial shading condition. [3, 8] employed the conventional PSO whereas [9–11] proposed improved versions of this technique. For instance, [9] proposed a deterministic PSO where the stochastic factors are removed, which make the MPPT simpler. [10, 11] adopted adaptive PSO variants in order to enhance the tracking speed and the accuracy of GMPP tracking. Indeed, the control parameter are updated linearly according to iteration in [10] and with exponential form in [11]. In the basic variants of particle swarm optimization algorithm, the movements of the particles are guided by both their own best location and the global swarm's best known position. One of the reasons for using the individual best position is probably to increase the diversity in the quality solutions; however, this diversity can be simulated using some randomness. Subsequently, there is no compelling reason for using the individual best location unless the optimization problem of interest is highly nonlinear and multimodal. Yang [12] developed the so called accelerated particle swarm optimization (APSO), a simplified version that could accelerate the convergence toward the optimum using only the global best. Furthermore, the update strategy of particles doesn't use the velocity, and therefore, the APSO is much simpler to implement than the standard PSO algorithm.

The remainder of this paper is structured as follows. Section 2 presents the modelling of the photovoltaic panel and the P-V characteristic under uniform and partial shading condition. Section 3 gives a brief introduction of P&O and PSO algorithm followed by the description of the APSO technique. Section 4 shows the performance of the APSO algorithm under static and dynamic partial shading conditions. Finally, a conclusion is made in Sect. 5.

2 Modelling of the Photovoltaic System

The equivalent circuit of a photovoltaic cell is shown in Fig. 1. It includes a current source, two diodes D1 and D2, a series resistance R_s and a shunt resistance R_p. The photovoltaic module is composed of several photovoltaic cells connected in series. Based on the circuit, the current generated by the photovoltaic module can be presented by the following equation.

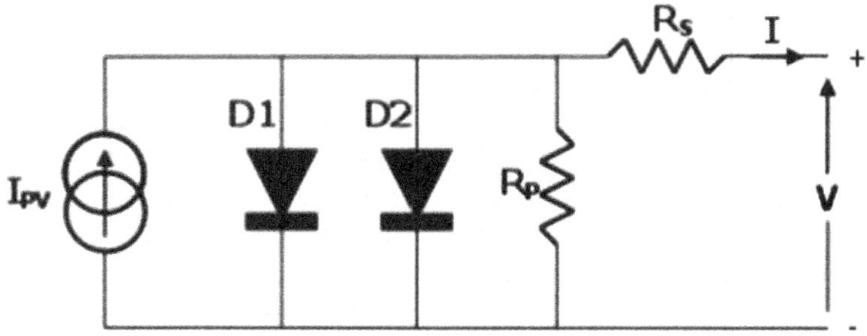

Fig. 1. Equivalent circuit of a solar cell.

$$I = I_{PV} - I_{01}\left[\exp\left(\frac{q(V + IR_sN_s)}{a_1kTN_s}\right) - 1\right] - I_{02}\left[\exp\left(\frac{q(V + IR_sN_s)}{a_2kTN_s}\right) - 1\right] - \left(\frac{V + IR_sN_s}{R_PN_s}\right) \quad (1)$$

In this equation, I_{PV} is the photo-current of the module, I_{01} and I_{02} are the inverse saturation currents of the diode D1 and D2, respectively. The parameters q, a_1, a_2, k and T respectively denote the charge of the electron, the ideality factor of the diode D1, factor of ideality of the diode D2, Boltzmann constant and the temperature of the cell. V is the voltage at the terminals of the PV module whereas I is the current of the cell.

Figure 2 shows the influence of solar irradiance on the P-V characteristic curve of three PV modules connected in series. Under uniform conditions, the P-V characteristic curve is uni-modal and has a single maximum power point. In practice, shadows are often projected on portions of the PV panel at a time of day, which could be caused by dust, clouds, trees, buildings or other objects around. When the photovoltaic panel is subjected to partially shaded conditions, irregularities appear on the P-V characteristic curve. Figure 2 shows the magnitude of the power deficit that can be caused by two partial shading patterns as well as the possibility of occurrence of several local power maxima.

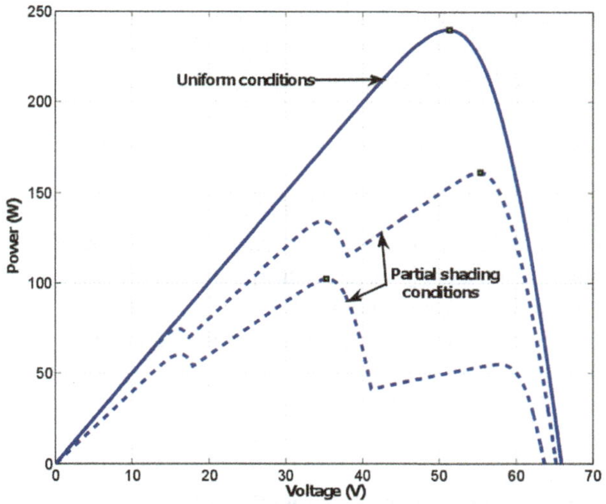

Fig. 2. P-V characteristics of PV panel under uniform and partial shading conditions.

3 APSO Based MPPT Controller

3.1 Conventional P&O Method

Fixed perturbation size based perturb and observation (P&O) method is the most popular MPPT algorithm due to its simplicity and easy implementation opportunity. The algorithm starts by sensing voltage or current from the PV array. Then it provides a perturbation in one direction by using either voltage (V) or current (I) or duty cycle (d). If the perturbation results in an increase in PV power increases in that direction, then the P&O algorithm will provide another perturbation in the same direction and continues to do so as the power keeps increasing. On the other hand if the change of power is negative, then the algorithm alters direction and provides perturbation in the opposite direction.

3.2 Particle Swarm Optimisation (PSO)

Particle swarm optimization (PSO) is a population based meta-heuristic which was developed by Kennedy and Eberhart in 1995. This technique is based on swarm behaviour in nature, such as fish and bird schooling. The PSO algorithm works on a population of agents called particles. To apply the PSO algorithm, it is necessary to define a search space consisting of particles and an objective function to optimize. The principle of the algorithm is to move these particles so that they find the optimum. Each of these particles has a position, a velocity and a neighbourhood. The position changes according to the particle best neighbour, the particle best position, and particle previous position.

At each iteration, the new particles in the swarm are determined taking into account their best positions and the best particle in the swarm. The new velocity of the particle i is calculated from the following equation:

$$v_i^{k+1} = \omega v_i^k + c_1 r_1 (Pbest_i - x_i^k) + c_2 r_2 (Gbest - x_i^k) \quad (2)$$

where: v_i^{k+1} and v_i^k are velocities of particle i at iteration $k+1$ and k, $Pbest_i$ is the best position of particle i, $Gbest$ is the best particle of the swarm, d_i^k is the position of particle i at iteration k. ω, c_1, c_2 are constant parameters. r_1, r_2 are uniform random numbers generated in $[0, 1]$ at each iteration.

The new positions are calculated according to the equation:

$$x_i^{k+1} = x_i^k + v_i^{k+1} \quad (3)$$

3.3 Accelerated Particle Swarm Optimisation (APSO)

The standard particle swarm optimization uses both the current global best Gbest and the individual best particles $Pbest_i$. However, this diversity can be simulated using some randomness. A simplified version that could accelerate the convergence of the algorithm toward the GMPP is to use only the global best. Therefore, the accelerated particle swarm optimization (APSO) algorithm was developed by Xin-She Yang. In the standard APSO, the velocity vector is generated by a simpler formula

$$v_i^{k+1} = v_i^k + \alpha(\varepsilon - 1/2) + \beta(Gbest - x_i^k) \quad (4)$$

where ε is a random variable with values from 0 to 1. We can also use a standard normal distribution, where ε_t is drawn from $N(0, 1)$ to replace the second term. Now we have

$$v_i^{k+1} = v_i^k + \beta(Gbest - x_i^k) + \alpha\,\varepsilon_t \quad (5)$$

The update of the position is the same as Eq. 3. However, we can see later that the velocity is redundant in accelerated PSO. In order to increase the convergence even further, we can also write the update of the location in a single step

$$x_i^{k+1} = (1 - \beta)x_i^k + \beta\, Gbest + \alpha\,\varepsilon_t \quad (6)$$

This simpler version will give the same order of convergence. It is worth pointing out that the velocity does not appear in Eq. 6, and thus there is no need to deal with the initialization of velocity vectors. Therefore, the APSO is much simpler to implement for MPPT.

4 Simulations

Figure 3 presents the Block diagram of the simulated PV system. The architecture of the PV system is built in MATLAB/Simulink in order to assess the effectiveness of the proposed algorithm. The duty cycle of the PWM signal is chosen to be the optimization variable. Thus, it is adjusted directly by the MPPT controller. Three particles are used and the algorithm continues the optimisation process until constraint on convergence is satisfied. The condition shown in the Eq. 7 is used as a convergence criterion.

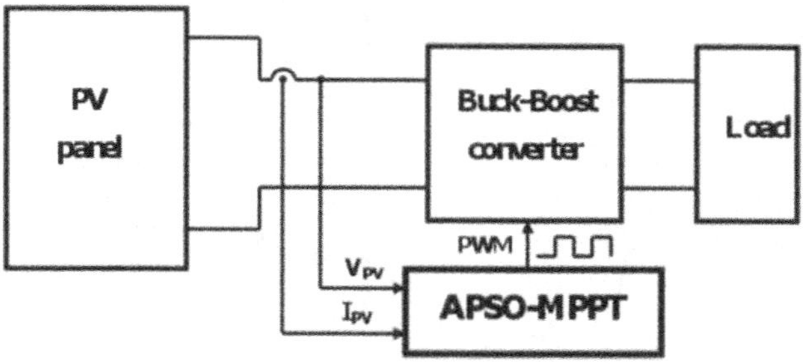

Fig. 3. Schematic diagram of the PV system.

$$\left|x_i^{k+1} - x_j^{k+1}\right| \leq \Delta x;\ i,j = 1,2,3;\ i \neq j \tag{7}$$

Due to varying weather and loading conditions, the global MPP is usually changing. The MPPT algorithm should have the ability to detect the variation of shading pattern and to search for the new global MPP. The search process is initialised when the following condition is satisfied

$$\frac{\left|P_{PVnew} - P_{PVlast}\right|}{P_{PVlast}} > \Delta P \tag{8}$$

In order to test and describe our proposed technique, firstly three static controversial conditions are selected and simulated. The PV array is composed of five modules connected in series arrangement.

The performance of the proposed technique is tested under the first condition. Figure 4 shows the output P-V characteristic curve of the PV panel along with the detailed simulation results of the proposed APSO method. The P-V curve is characterised by four MPPs. It is clearly observed that the APSO technique can ignore the local peaks and has converged accurately the GMPP.

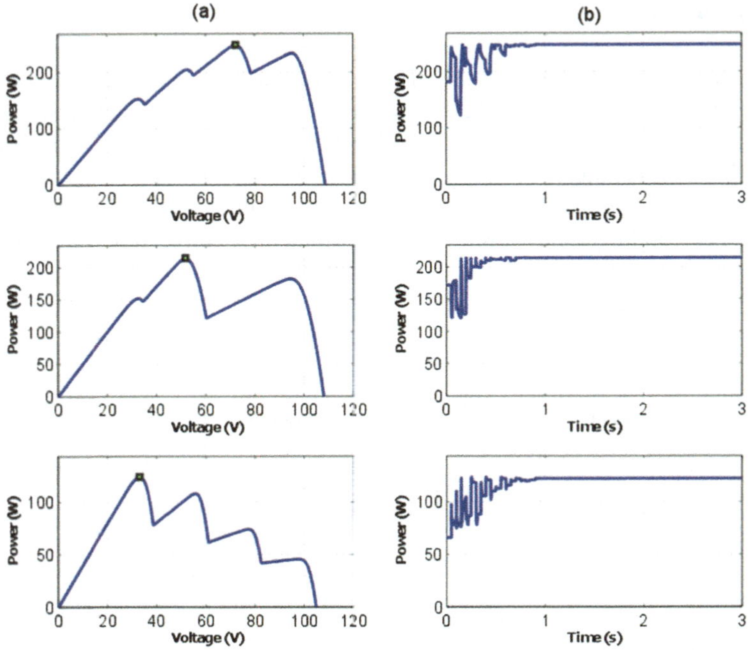

Fig. 4. APSO tracking results under different partial shading patterns.

For further verification, the system is tested under partial shading conditions. In this case, the first and second modules receive an equally distributed irradiance level of 1000 W/m^2, the third module receives irradiance level of 850 W/m^2 whereas the firth and fifth modules receives irradiance level of 400 W/m^2. In this case, the P-V characteristic curve contains three power peaks and the GMPP is located in the middle between the two local peaks. Most conventional MPPT methods cannot find the actual GMPP in such conditions and are normally trapped in the first MPP, which is not obviously the real GMPP. Thus, the efficiency of the PV systems drops dramatically. Figure 4 shows the simulation result of the proposed APSO technique applied to the second condition. The actual maximum power of the system is found and the operating point is maintained around the optimum voltage.

In the third shading pattern, the GMPP is located at left side of the P-V curve. In this condition, the PV panel is exposed to severe partial shading. Figure 4 presents the corresponding P-V characteristic and the power tracking results. The simulated power locus proves that the controller accurately tracked the actual GMPP.

Figures 5 and 6 asses the dynamic tracking capability of the proposed technique. Figure 5 shows the changing sequence of the test shading patterns. Figure 6 shows the measured waveforms for the dynamic tracking test. From Fig. 6, the proposed algorithm successfully detects the change in shading pattern and reinitializes the MPPT process accordingly. We can notice that the proper GMPP is captured accurately. On the other side, Fig. 6 reports also the results of tracking for the conventional P&O algorithm. It is clear that the P&O could not differentiate between the LMPPs and GMPP and get trapped in a LMPP.

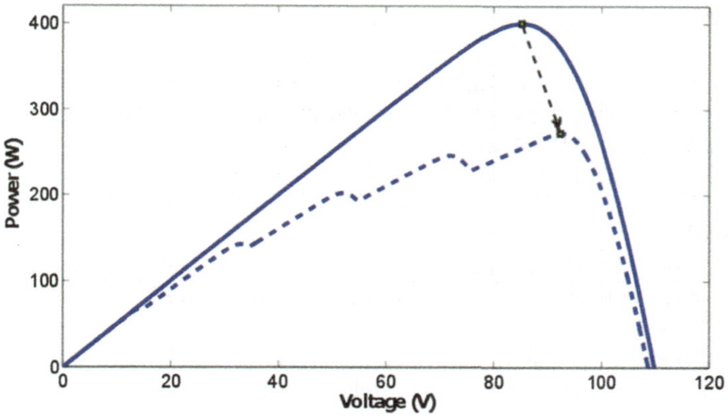

Fig. 5. P-V curves used for the dynamic tracking test.

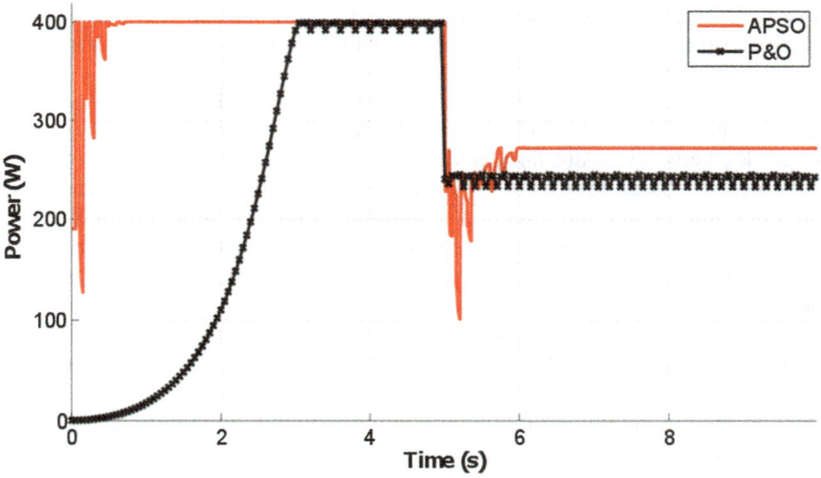

Fig. 6. Dynamic power tracking results for the proposed APSO and P&O techniques.

5 Conclusion

In this paper, the accelerated particle swarm optimisation (APSO) algorithm is proposed to track the GMPP of PV systems under partially shaded conditions. Under such situation, the P-V characteristic curve exhibits multiples peaks and the conventional MPPT techniques like the P&O algorithm don't perform well. The performances of the APSO algorithm are verified by simulation under extreme partial shading configurations. Despite its simple structure, the APSO algorithm based MPPT can track the GMPP with high accuracy and presents a fast convergence.

References

1. Mueller, A., Orosz, M., Narasimhan, A.K., Goswami, Y., Hemond, H.F.: Evolution and feasibility of decentralized concentrating solar thermal power systems for modern energy access in rural areas. MRS Energy Sustain. Rev. **3**(4), 32 (2016)
2. Jiang, L.L., Srivatsan, R., Maskell, D.L.: Computational intelligence techniques for maximum power point tracking in PV systems: a review. Renew. Sustain. Energy Rev. **85**, 14–45 (2018)
3. Miyatake, M., Veerachary, M., Toriumi, F., Fujii, N., Ko, H.: Maximum power point tracking of multiple photovoltaic arrays: a PSO approach. IEEE Trans. Aerosp. Electron. Syst. **47**(1), 367–380 (2011)
4. Titri, S., Larbes, C., Toumi, K.Y., Benatchba, K.: A new MPPT controller based on the ant colony optimization algorithm for photovoltaic systems under partial shading conditions. Appl. Soft Comput. **58**, 465–479 (2017)
5. Tajuddin, M.F.N., Ayob, S.M., Salam, Z., Saad, M.S.: Evolutionary based maximum power point tracking technique using differential evolution algorithm. Energy Build. **67**, 245–252 (2013)
6. Kaced, K., Larbes, C., Ramzan, N., Bounabi, M., elabadine Dahmane, Z.: Bat algorithm based maximum power point tracking for photovoltaic system under partial shading conditions. Sol. Energy **158**, 490–503 (2017)
7. Ahmed, J., Salam, Z.: A maximum power point tracking (MPPT) for PV system using cuckoo search with partial shading capability. Appl. Energy **119**, 118–130 (2014)
8. Kaced, K., Larbes, C., Ait-Chikh, S.M., Bounabi, M., Dahmane, Z.E.: FPGA implementation of PSO based MPPT for PV systems under partial shading conditions. In: 6th International Conference on Systems and Control (ICSC), pp. 150–155. IEEE (2017)
9. Ishaque, K., Salam, Z.: A deterministic particle swarm optimization maximum power point tracker for photovoltaic system under partial shading condition. IEEE Trans. Ind. Electron. **60**(8), 3195–3206 (2013)
10. Liu, Y.-H., Huang, S.-C., Huang, J.-W., Liang, W.-C.: A particle swarm optimization-based maximum power point tracking algorithm for PV systems operating under partially shaded conditions. IEEE Trans. Energy Convers. **27**(4), 1027–1035 (2012)
11. Chao, K.-H., Lin, Y.-S., Lai, U.-D.: Improved particle swarm optimization for maximum power point tracking in photovoltaic module arrays. Appl. Energy **158**, 609–618 (2015)
12. Yang, X.S., Deb, S., Fong, S.: Accelerated particle swarm optimization and support vector machine for business optimization and applications. In: Networked Digital Technologies. Communications in Computer and Information Science, vol. 136, pp. 53–66. Springer, Heidelberg (2011)

The Pursuit of the Maximum Power Point of a Photovoltaic System Using Artificial Neural Network

F. Saadaoui[1(✉)], K. Mammar[2], and A. Hazzab[3]

[1] Research Laboratory Control Analysis and Optimization of Systems Electro-Energetic (CAOSEE), University of Tahri Mohamed, Bp 417, Bechar, Algeria
fsaadaoui@gmail.com

[2] Department of Electrical Engineering, University of Tahri Mohamed, Bp 417, Bechar, Algeria
mammar.univ.dz@gmail.com

[3] CAOSEE Research Laboratory Control, Analysis and Optimization of Systems Electro- Energetic Systems, University of Tahri Mohamed Bechar, Bp 417, Bechar, Algeria
a_hazzab@yahoo.com

Abstract. The use of maximum power point tracking techniques in photovoltaic systems attracts particular attention to research and ensures that the photovoltaic energy system delivers as much as possible of the output power available to the load, regardless of the climatic conditions (variation in temperature and solar radiation), the choice and development are made to implement a more appropriate and effective maximum power point tracking controller using neural networks.

In order to obtain maximum power point tracking, the importance has also been given in this paper to the photovoltaic panel, these inputs (temperature and solar radiation) and also to the control of the power converter.

The results obtained using the Matlab/Simulink environment; confirm the effectiveness of the proposed method in terms of efficiency, fast calculation time of its robustness and the precision of its outputs which allow giving decisions correct, reliable and immediate.

Keywords: Maximum power point tracking (MPPT) · Photovoltaic (PV) system · Artificial neural network (ANN) · Boost converter · Perturbation and observation (P & O)

1 Introduction

The objective of this study is based on the application of neural networks in three independent parts of the photovoltaic (PV) chain (incoming radiation and temperature control, PV panel and command MPPT), in order to study the influence of artificial intelligence on this behavior.

M. Hatti (Ed.): ICAIRES 2019, LNNS 102, pp. 103–114, 2020.
https://doi.org/10.1007/978-3-030-37207-1_11

The exploitation and necessity of solar energy has become very useful over the last decades in such a way that it has been able to solve the problems of the energy crisis caused by the significant global energy demand and the global warming caused by the use of fossil fuels.

Renewable energy sources have a necessary role in the production of electricity. Solar energy is one of these important alternative sources depending on climatic conditions, such as temperature and solar radiation, or its optimal use requires taking into account different types of loads. The photovoltaic system must operate at its maximum power (MPPT) despite these climate changes. That's why for the sake of exploiting it, we have to assure him a better installation which can give a large production of solar energy takes into account the problem of the degradation of the power supplied by the solar generator due to the radiation variation of up to 50% [1].

In order to minimize these disadvantages through the operation of conversion systems and to increase the efficiency of the photovoltaic field, several studies are under development to optimize the performance of these energy systems.

Solar photovoltaic energy is a major area of competition for other conventional energy sources because of its high installation cost and the low energy conversion efficiency of PV cells [2].

For this purpose a DC/DC converter is placed between the PV panel and the load, to ensure the control and implementation of the photovoltaic generator at its maximum power point (MPPT) as a function of the temperature of the PV cell and the variations in the solar radiation of the PV generator and the state of the charge [3].

Several techniques are applied for the follow-up of the MPPT such as the P & O method and the incremental conductance method.

But recently, intelligent methods such as fuzzy logic and artificial neural networks have been adopted for photovoltaic applications, mainly because of their flexibility, symbolic reasoning and explanatory capabilities, which are useful for processing strong nonlinearities and complex systems [4] (Fig. 1).

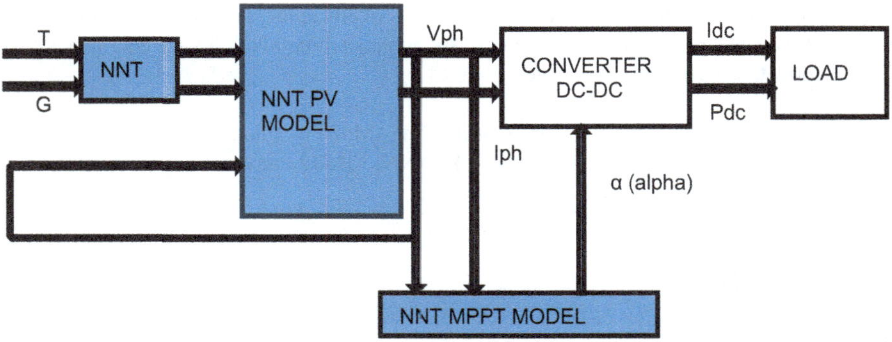

Fig. 1. Overview of a maximum power point tracking, MPPT-controlled photovoltaic, PV generator system with neural network ANN

2 Photovoltaic Generator Modeling

2.1 Photovoltaic Cell Modeling

The photovoltaic panel, receives solar energy and transform it into electric current, it consists of several cells connected in series and in parallel Fig. 2.

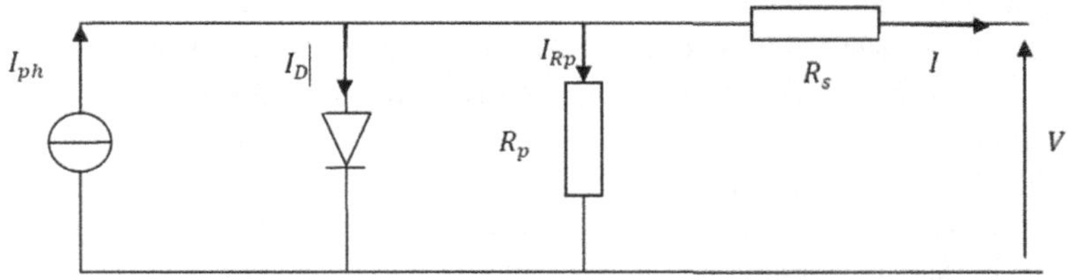

Fig. 2. Equivalent circuit of solar cell

$$I_{Ph} = I_D + I + I_{Rp} \tag{1}$$

from Eq. (1) we obtain the expression of characteristic I-V of the chosen model

$$I = I_{Ph} - I_s\left[\exp\left(\frac{V + IR_S}{nV_t}\right) - 1\right] - \frac{V + IR_s}{Rp} \tag{2}$$

$$I_{Ph} = (I_{sc} + K_i dT)\frac{G}{G_n} \tag{3}$$

G: Solar radiation, and G_n: Nominal solar radiation

$$I_s = \frac{I_{sc} + K_i dT}{\exp\left(\frac{V_{OC} + K_V dT}{V_t a}\right) - 1} \tag{4}$$

a: diode ideality factor, Is: is the saturation current of the diode, and V_t: is the thermic voltage at temperature T

$$I = N_{PP} I_{Ph} - N_{PP} I_s\left[\exp\left(\frac{V + (R_S N_{SS}/N_{PP})I}{V_t a}\right) - 1\right] - \frac{V + (R_S N_{SS}/N_{PP})I}{R_P N_{SS}/N_{PP}} \tag{5}$$

N_{PP}: is the number of photovoltaic modules in parallel = 1
N_{SS}: is the number of photovoltaic modules in series = 36.

The choice of a PV model generally depends on the use that we want to make and the information available to determine the parameters. Each model must be based on classical principles [5] Table 1.

Table 1. The electrical characteristics of photovoltaic module

Parameters	Values
Gr: Solar radiation reference	1000 (W/m^2)
Rs: Equivalent series resistance of the resistive contacts	0.5 Ohms
Rsh: Shunt resistance representing the leaking currents	500 Ohms
q: Electron charge	$1.602*10^{-19}$C
K: Boltzmann constant	$1.38*10^{-23}$ J/° K
T: is the effective temperature of the cell	°K
Tinit: Initial temperature of the PV Cell	298°K
Isc: Short-circuit current	3.33 A
Voc: Initial open-circuit voltage	21.3 V
Ki: Temperature versus current coefficient	0.0025 A/°C
Eg: Gap energy silicium	1,12 eV

2.2 Artificial Neural Network Model

The choice of ANN is made according to: the type of neurons, the number of neuron layers and the study methods, their advantages in broadband used in industry is the storage of an enormous mass of information and the direct storage in real time of this information, allowing making logical decisions, efficiency to solve problems of non-linearities. The ANN is noted for sensitivity to noise. Generally the neuron output will be one or more reference signal(s) [5].

3 Photovoltaic Generator Characteristics

3.1 Characteristic I-V & P-V

The optimal power of the cell is proportional to the radiation and the current Iph produced by the cell is proportional to the solar radiation G, with a small variation of voltage and the open circuit voltage will only decrease slightly with the radiation, and the power points maximum are at approximately the same voltage [6, 7] Figs. 3 and 4.

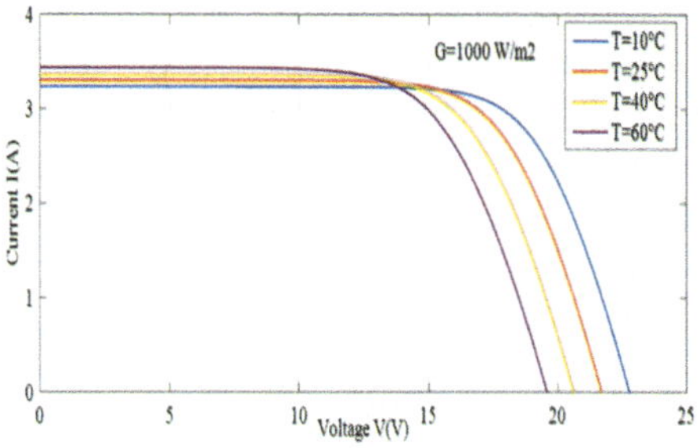

Fig. 3. Voltage current characteristic, I-V of a photovoltaic module, PV for different temperatures and constant radiation

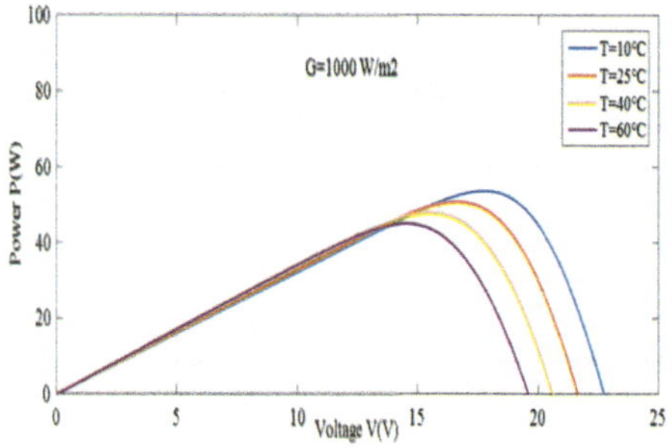

Fig. 4. Voltage power characteristic, P-V of a photovoltaic module, PV for different temperatures and constant radiation

As the temperature increases, the short circuit current increases, but the open-circuit voltage decreases. Thus, the output power of the cell is also decreased Figs. 3 and 4. But if the radiation increases the maximum power is higher and also the current also increases Figs. 5 and 6.

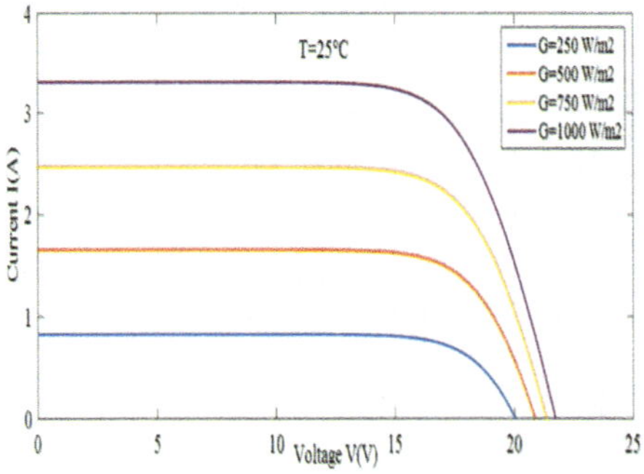

Fig. 5. Voltage current characteristic I-V of a photovoltaic module PV for different radiations and constant temperature

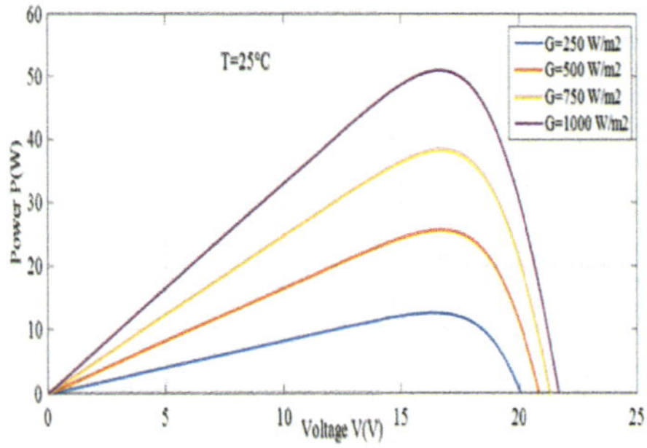

Fig. 6. Voltage power characteristic P-V of a photovoltaic module PV for different radiations and constant temperature

4 Photovoltaic System Control by ANN

4.1 PV System Based on Neuron-Network

The following figure shows the climatic variations measured on the 06/01/2011 by the meteorological services of some Saharan cities of Algeria over a period of 24 h of which Bechar was chosen as a sample.

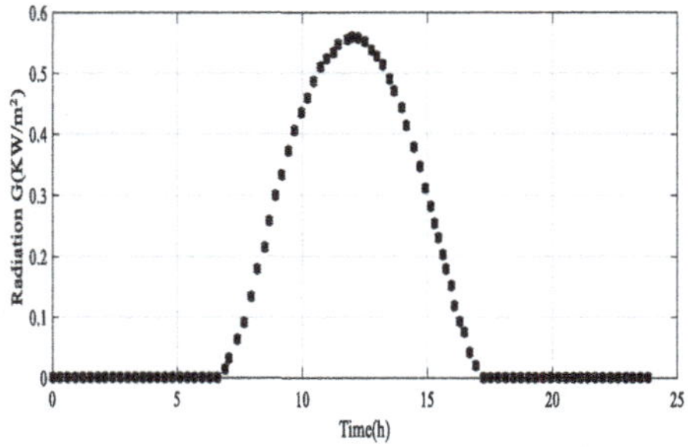

Fig. 7. Data base radiation variation of Bechar city measured by the weather services on January 06, 2011 over a period of 24 h without neural network.

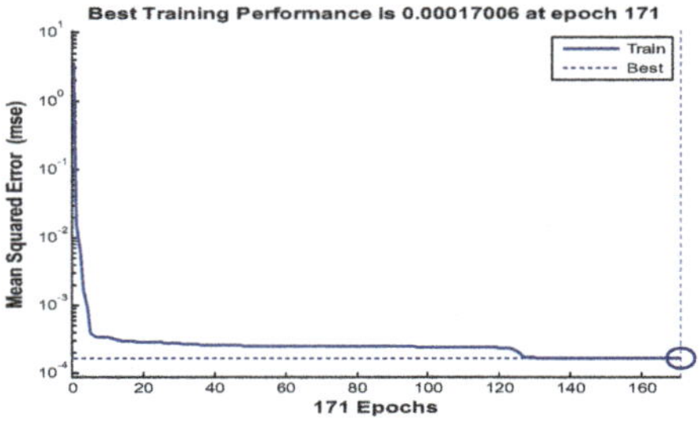

Fig. 8. The squared error of the ANN training of radiation with Matlab/Simulink

4.2 Application of ANN on Radiation

The quadratic solar radiation error is very negligible; it is on the order of 10^{-4} which explains the importance of ANN as input of the PV panel Fig. 8.

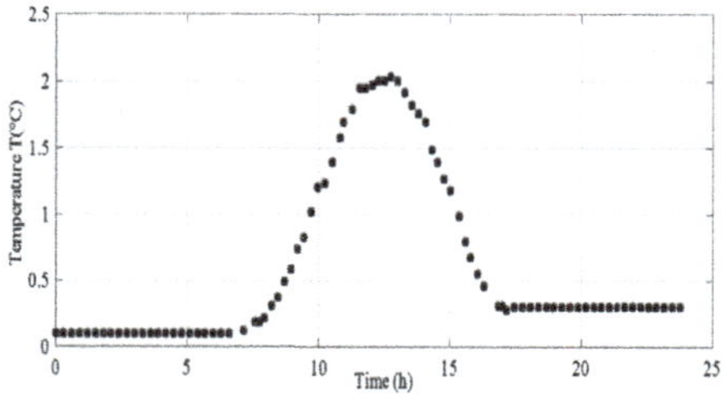

Fig. 9. Data base temperature variation of Bechar city measured by the weather services on January 06, 2011 over a period of 24 h without neural network

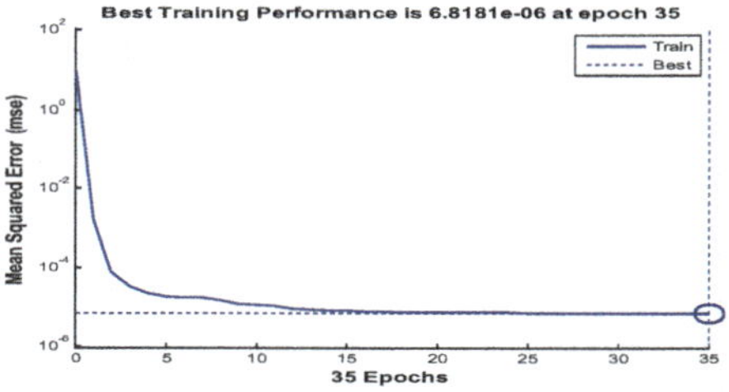

Fig. 10. The squared error of the ANN training of temperature with Matlab/Simulink

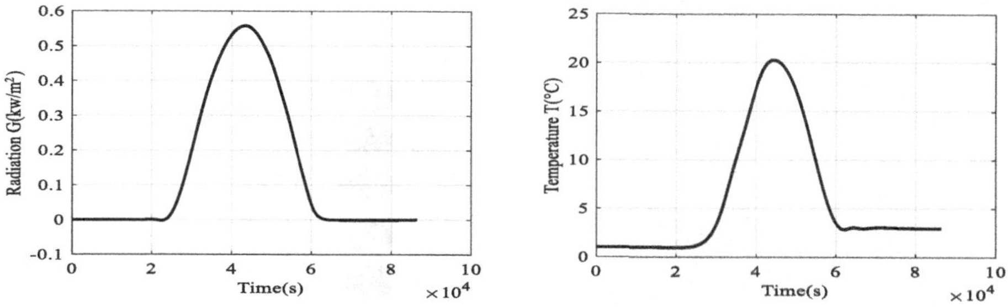

Fig. 11. Schema of radiation and temperature variation Bechar city as January 06, 2011 in duration of 24 h with neural network

4.3 Application of ANN on Radiation

The squared error of the temperature is very negligible, it is on the order of 10^{-4} which explains the importance of ANN and the following schema shows the equivalence with Figs. 6 and 9.

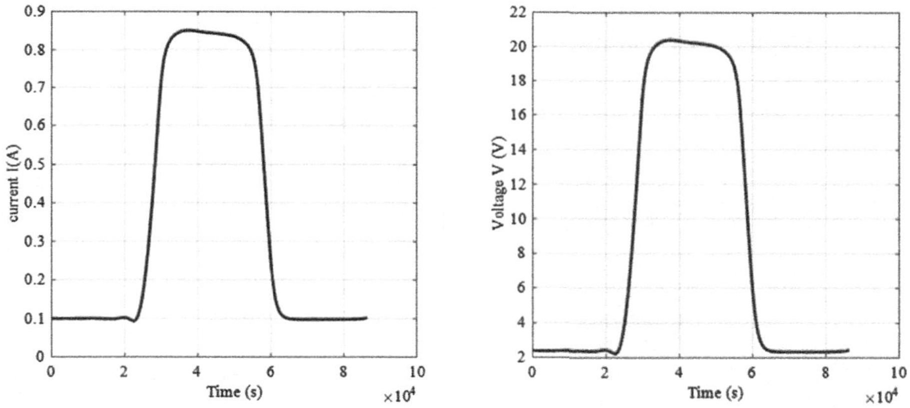

Fig. 12. Current and voltage variation supplied by the PV panel without neural network

4.4 Application of ANN on the PV Panel

To make the conversion of solar energy into electrical power effective at the output of the PV panel, it is preferable and indispensable to operate it close to a point of maximum power, as this power varies constantly according to the climatological conditions (radiation level solar, temperature of surface and the voltage of module.), or the operating power that is determined from these three inputs directly affects the design and performance of the system (Fig. 7).

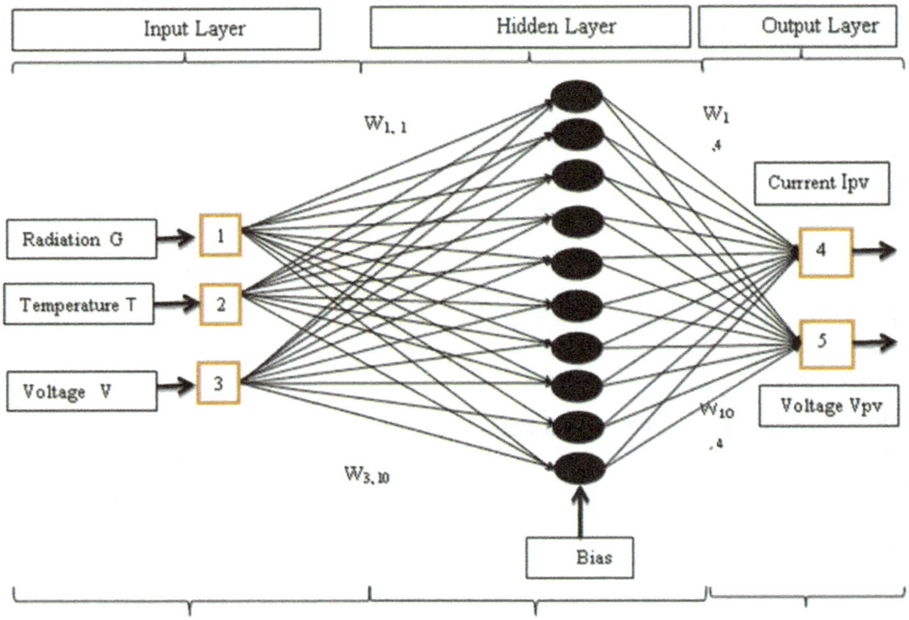

Fig. 13. Architectural synoptic of the proposed neural network (3-10-2)

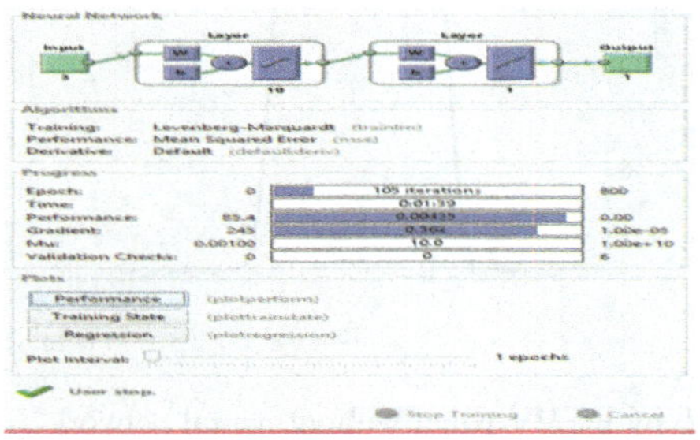

Fig. 14. ANN training of radiation and temperature with Matlab/Simulink

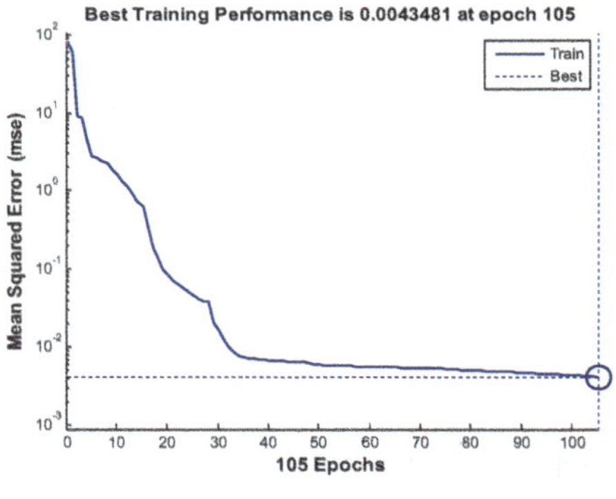

Fig. 15. The squared error of the ANN training of radiation and temperature with Matlab/Simulink

The quadratic error is always negligible; it is on the order of 10^{-4} which also explains the importance of ANN for the PV panel (Figs. 10 and 11).

The variation of the temperature considering the voltage of Fig. 16 with ANN

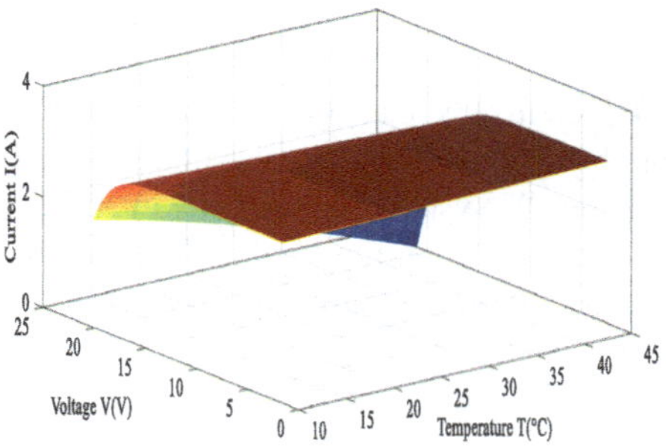

Fig. 16. Voltage current characteristic, I-V of a photovoltaic module for the different temperatures and constant radiation with neural network

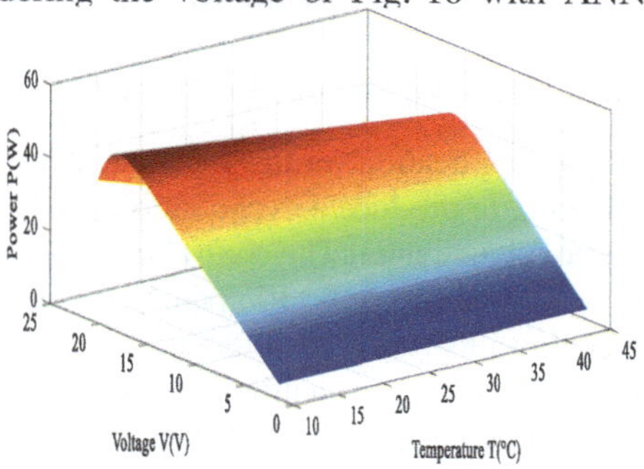

Fig. 17. Voltage power characteristic, P-V of a photovoltaic module for the different temperatures and constant radiation with neural network

directly influences the PV system and gives the same PV current values as shown in Fig. 3, which results in the performance use of the ANN technique for the system (Figs. 12, 13 and 14).

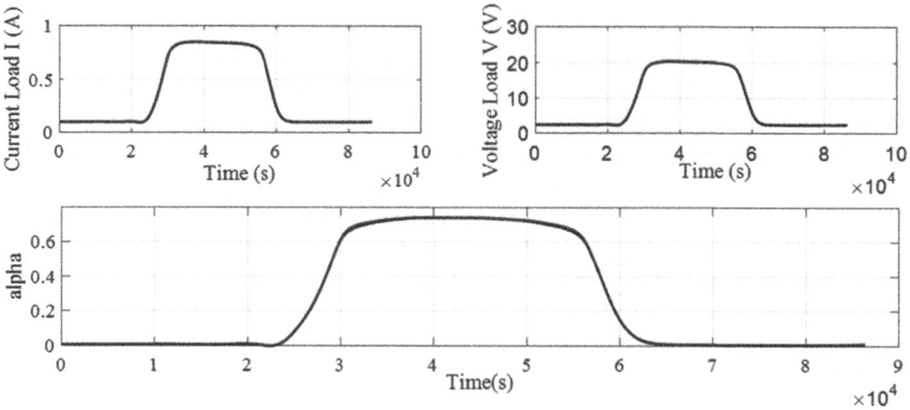

Fig. 18. DC load current and voltage variation with MPPT control based on ANN

The variation of the temperature taking into account the voltage of Fig. 17, it has a direct influence on the PV system in Fig. 4, which results in the use performance of the ANN technique for the system (Fig. 15).

These figures clearly show the dependence of the current and the output power of the PV as a function of temperature and radiation.

5 ANN Application on MPPT Controller

5.1 P&O Algorithm Based on ANN

For an application on the PV system, an MPPT controller based on neural networks allowing to produce a cyclic ratio from an input layer included (current, voltage) of the photovoltaic panel. The controller MPPT varies the PV system operating its maximum power by shifting the PV panel voltage to its identified optimum value. The control signal results from the difference between the measured operating power of the PV and the optimal power identified by the ANN [8, 9].

In a period of 24 h, MPPT control by ANN oscillates the operating point around the MPPT point.

In the different cases of change of radiation and temperature of the day of 06/01/2011, it appears that. During this climate change that lasts from 07 h to 18 h, the system converges towards the desired MPPT. So the MPPT by ANN command used controls the operation well of the PV generator around the optimal conditions. Figure 16 shows that the controlled current and charge voltage have the same values as those delivered by the PV cell Fig. 5, which shows the reliability and importance of the MPPT command by ANN [10–12].

6 Results and Discussions

This section is intended to give the reader a comparative view of the results compared to P & O's conventional MPPT method. The ANN command generates a reference voltage which is compared to the instantaneous output voltage of the panel by producing a signal of error that will be passed to a PI controller for the duty cycle adjustment for MPPT tracking. Or the performance of the behavior is evaluated in terms of system response time and effectiveness monitoring Fig. 18.

The Boost DC-DC converter has the function of increasing the DC voltage. The maximum power is reached when the MPPT algorithm changes and adjusts the duty cycle of the Boost DC-DC converter [13–15].

7 Conclusion

Artificial neural networks are tools for specific calculations given their ability to adapt to complex problems through learning. They have come together to solve several cases where conventional methods have shown their limits. The implementation of the energy system with its maximum power by the ANN approach has been declined or more precise calculation solutions have been proposed, with a better response time and good performance. Moreover, a very good agreement is obtained between the model of photovoltaic cells and the neuronal approach. The results demonstrate that the proposed MPPT method track very fast the MPP with negligible oscillations.

The main advantages of the developed MPPT control method are the rapid convergence to the MPPT, its robustness and high efficiency, and its ease of implementation.

References

1. Sourov, M.R., Tania Ahmed, U., Golam Rabbani, M.A.: High performance maximum power point tracker for photovoltaic power system using DC-DC boost converter. IOSR J. Eng. **2** (12), 12–20 (2012)
2. Mahjoub-Essefi, R., Souissi, M., Hadj-Abdallah, H.: Maximum power point tracking control technique for photovoltaic systems using neural networks. In: Proceedings of the 5th Annual International Renewable Energy Congress, IREC'2014, Hammamet, 27 March, pp. 22–427 (2014)
3. Mahjoub Essefi, R., Souissi, M., Hadj Abdallah, H.: Maximum power point tracking control using neural networks for stand-alone photovoltaic systems. Int. J. Modern Nonlinear Theory Appl. **3**, 53–65 (2014)
4. Almonacid, F., Fernandez, E.F., Rodrigo, P., Pérez-Higueras, P.J., Rus-Cascas, C.: Estimating the maximum power of a high concentrator photovoltaic (HCPV) module using an artificial neural network. Energy **53**, 165–172 (2013)
5. Mahanta, J., Sharma, B., Sarmah, N.: A review of maximum power point tracking algorithm for solar photovoltaic applications. J. Electr. Electron. Eng. (IOSR-JEEE) **13**(4), 01–13 (2018)
6. Amoozadeh, M., Gholamian, S.A.: Active and reactive power control of photovoltaic systems connected to the network for maximum power point tracking. Int. J. Mechatron. Electr. Comput. Technol. (IJMEC) **4**(12), 857–885 (2014)
7. Kishor, N., Mohanty, S.R., Villalva, M.G., Ruppert, E.: Simulation of PV array output power for modified PV cell model. In: IEEE, International Conference on Power and Energy (PEC), Kuala Lumpur, Malaysia (2010)
8. Guingane, T.T., Koalaga, Z., Simonguy, E., Zougmore, F., Bonkoungou, D.: Modélisation et simulation d'un champ photovoltaïque utilisant un convertisseur élévateur de tension (boost) avec le logiciel MATLAB/SIMULINK. Journal International de Technologie, de l'Innovation, de la Physique, de l'Energie et de l'Environnement JITIPEE. **2**(1), 1–14 (2016)
9. Aouchiche, N., Aït Cheikh, M.S., Malek, A.: Poursuite du point de puissance maximale d'un système photovoltaïque par les méthodes de l'incrémentation de conductance et la perturbation & observation. Revue des Energies Renouvelables **16**(3), 485–498 (2013)
10. Kumari, J.S., Sai Babu, Ch., Babu, K.: Design and Analysis of P&O and IP&O MPPT techniques for photovoltaic system. Int. J. Modern Eng. Res. (IJMER) **2**(4), 2174–2180 (2012)
11. Remy, Gh., Bethoux, O., Marchand, C., Dogan, H.: Review of MPPT Techniques for Photovoltaic Systems. Laboratoire de Génie Electrique de Paris (LGEP)/SPEE-Labs, CNRS UMR 8507, SUPELEC; Université Pierre et Marie Curie, France, pp. 1–14
12. Mammar, K., Chacker, A.: Adaptive neuro-fuzzy controller of PEM fuel cell system power generation. Int. J. Comput. Sci. (IJCSI) **9**(6), 1694–081 (2012)
13. Bendib, B., Krim, F., Belmili, H., Almi, M., Bolouma, F.S.: An intelligent MPPT approach based on neural-network voltage estimator and fuzzy controller, applied to a stand-alone PV system. In: IEEE 23rd International Symposium on Industrial Electronics (ISIE), Istumbul, Turkey, pp. 404–409 (2014)

14. Manas, M., Kumari, A., Das, S., et al.: An artificial neural network based maximum power point tracking method for photovoltaic system. In: IEEE International Conference on Recent Advances and Innovations in Engineering (ICRAIE), Jaipur, India (2016)
15. Chen, M., Ma, S., Wu, J., Huang, L.: Analysis of MPPT failure and development of an augmented nonlinear controller for MPPT of photovoltaic systems under partial shading conditions. J. Appl. Sci. (MDPI) **7**(95), 1–22 (2017)

Modified Particle Swarm Optimization Based MPPT with Adaptive Inertia Weight

Hadjer Azli[1(✉)], Sabrina Titri[2], and Cherif Larbes[1]

[1] Ecole Nationale Polytechnique, ENP, LDCCP, El Harrach, Algerie
az.hadjer.li@gmail.com, cherif.larbes@g.enp.edu.dz
[2] Centre de Développement des Technologies Avancées, CDTA Division Microélectronique DMN, Baba Hassen, Algerie
stitri@cdta.dz

Abstract. In this paper we study Maximum Power Point Tracking of PV system under partial shading condition using PSO with a new approach of adaptive weight. The proposed approach has a rapid convergence and high efficiency over the simple PSO and the conventional P&O algorithm. The exploitation and exploration phase are maintained by choosing the right values for c1 and c2 and by integrating a random inertia weight with exponential decrease over iterations, this leads to high speed convergence and maintains a zero steady state oscillation compared to P&O. Also, we used a statistical characteristics (Variance and Mean) to detect the step change in the irradiance. The simulation was carried out on MATLAB Simulink using two diode model of MSX60 solar power system under uniform and partial shading condition with step change in irradiance. To see the effectiveness of the method we compared it with the conventional P&O. The results shows that our approach is more effective in finding the Global peak and in avoiding the local peak stagnation.

Keywords: Maximum Power Point Tracking (MPPT) · Bio-Inspired Algorithms · Partial shading · Photovoltaic system · Particle swarm optimization · Perturb and Observe

1 Introduction

Energy is a fundamental element for a for building a civilization, and renewable energies are the best sources to produce a clean, free, endless and unlimited energy for us to use in our daily life. Among the renewable energies there is the solar power that nowadays is becoming more famous and eco friendliness. The photovoltaic system (PV) is the responsible one for transforming the solar energy into electricity but their efficiency is low. According to [1] the efficiency may vary from 6% for silicon–based to 42.8% for multiple junction research lab cells. In addition, the PV has a non-linear characteristics, his P-V curve shows a maximum peak at uniform irradiation and many local peaks for partial shading condition. Thus, recent literature focused their studies in finding methods to extract the maximum power MPP of the PV under different environment conditions. In order to extract the maximum power, a DC-DC converter is connected in between the PV and the load to control the seeking of the MPP by the use

M. Hatti (Ed.): ICAIRES 2019, LNNS 102, pp. 115–123, 2020.
https://doi.org/10.1007/978-3-030-37207-1_12

of MPPT controlled that measures by itself the outputs of the PV and send the corresponding duty cycle to the converter.

Many conventional methods was developed for MPPT tracking, such as Perturb and Observe (P&O) [2], Incremental conductance (IC) [3], and Hill Climbing (HC). CM methods have advantages like simplicity, low cost, few measured parameters. In the other side they have a major problem of locating the global optimum, also suffer from the steady state oscillation. Besides, various soft computing methods was used to overcome CM drawbacks such us Fuzzy Logic Controller (FLC) [4], and Artificial Neural Network (ANN) [5]. Despite their advantages they are harder, slower and need expensive hardwares.

Another field of algorithms was born called Meta-heuristic algorithms or Nature-Based Methods. They are a lot, and has been classified by many authors by the field of inspiration as: Physical/Chemical Based Algorithms, Bio-Inspired Algorithms (BIM). The BIM also is divided into: Evolutionary Algorithms (EA), and Swarm Intelligence Algorithms (SI). The BI are becoming better known in different areas of application since they are: easy to implement, can find the optimal solution and can be applied to solve different and complex problems. Additionally they can find the global maximum without stagnation in local peaks also they are capable of reducing the steady state oscillation to practically zero. Some of the bio-inspired algorithms the most popular in MPPT are: Particle swarm algorithm (PSO) [6] Ant Colony Algorithm (ACO) [7], Artificial Bee Colony (ABC), Firefly Algorithm (FA), Gray Wolf Optimization (GWO), Salp Swarm Algorithm (SSA), and many of them. Among BIM algorithm, PSO was the most used one in the field of MPPT because it is simple to implement, have few parameters to adjust and have short computation time. many authors employed this technique for MPPT, and they added many modification for better results like [8–11]. In our paper we used a random inertia weight with decreased value over iteration using exponential function. A brief comparison with P&O was made to see the effectiveness of the modified PSO over the conventional method.

This paper is organized as follows, Sect. 2 Generality on PV system, Sect. 2.3 PSO overview and application on MPPT, in Sect. 3 simulation and results, and the last Sect. 5 the conclusion.

2 Generality on PV System

2.1 PV Cell Modelisation

The PV cell is a combined layers of semiconductors material (p-layer and n-layer) therefore the realistic PV model is introduced as a current generator associated with a parallel diode and two resistor R_p and R_s. The R_s resistance is the sum of several structural resistances of the device and The R_p resistance exists mainly due to the leakage current of the p-n junction [12]. A two diode model presented by [13] is used in this paper due to the high accuracy of the modeled PV. The schematic of the two diode model is given in Fig. 1:

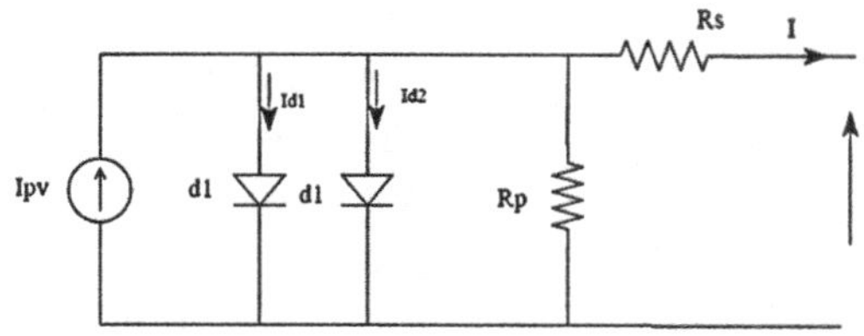

Fig. 1. Two diode model

The output current of the Cell after simplification [13] is given as:

$$I = Ipv - I_0\left[exp\left(\frac{V + IR_s}{V_T}\right) + exp\left(\frac{V + IR_s}{(p-1)V_T}\right) + 2\right] - \left(\frac{V + IR_s}{R_p}\right), \; and \; p \geq 2.2 \tag{1}$$

The PV current I_{pv} as function of the temperature and irradiance, also the saturation current I_o are given as:

$$I_{pv} = (I_{sc} + K_I\Delta T)\frac{G}{G_{STC}}, \tag{2}$$

$$I_o = \frac{(I_{sc} + K_I\Delta T)}{exp\left(\frac{(V_{oc} + K_V\Delta T)}{V_T}\right) - 1} \tag{3}$$

And G is the irradiance $G_{STC} = 1000\frac{W}{m^2}$.
Where:

V: The terminal voltage.
k: Boltzmann's constant, 1.3865×10^{-23}J/K.
q: Electron charge, 1.6021×10^{-19}C.
T: Temperature in Kelvin.
$p = a_1 + a_2$: a_1, a_2 are Diffusion and recombination current component of the diodes , $a_1 = 1$.
$V_T = \frac{N_sKT}{q}$: The thermal voltage of the PV having N_s cells connected in series;
R_p, R_s: are shunt resistor and series resistor.
K_V: Open circuit voltage coefficient.
K_I: short-circuit current coefficient.

2.2 PV System Architecture

The architecture of the PV system illustrated in Fig. 2 contains the (MSX-60) module configured in (4×1) PV array, a Buck boost converter, a load and an MPPT controller. Table 1 shows the parameters of the two diode model of MSX-60 module.

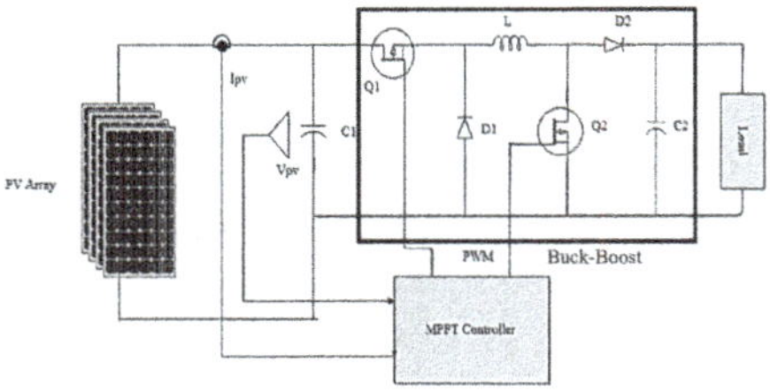

Fig. 2. Photovoltaic system architecture

Table 1. Parameters of Solar MSX-60 PV Module at STC

Parameters	Value
Peak power (W), P_{mpp}	60
Peak power voltage $V_{mpp}(V)$	17.1
Peak power current $I_{mpp}(A)$	3.5
Short circuit current $I_{sc}(A)$	3.8
Open circuit voltage $V_{oc}(V)$	21.1
Temperature coefficient of current, $K_I(mA/°C)$	3
Temperature coefficient of voltage, $K_V(mV/°C)$	−80
Number of series cells, N_s	36
Parallel resistor, $R_p(\Omega)$	176.4
Series resistor, $R_s(\Omega)$	0.35

The Solar cells of Photovoltaic system is influenced by the insolation, temperature and partial shading. In the PSC we used 4 × 1 PV arrays connected in series and accompanied by a bypass diode connected in parallel with each PV module to avoid the hotspot problem. The impact of those three factors is illustrated in Fig. 3.

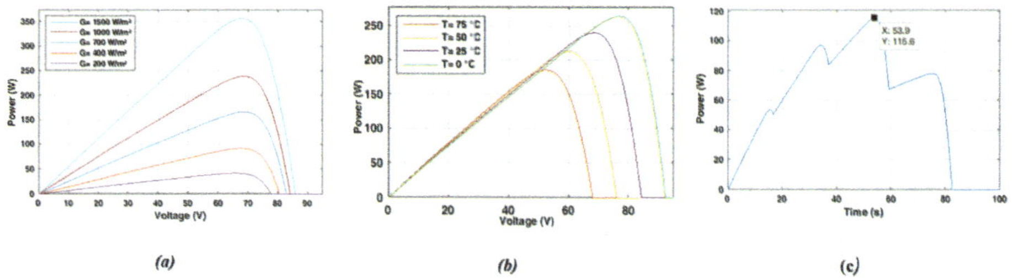

Fig. 3. (a) P-V curve under different insulation, (b) P-V curve under different temperature, (c) P-V curve under partial shading condition.

3 Particle Swarm Optimisation: Overview, and Application on MPPT

3.1 PSO Concept

Particle swarm algorithm PSO was firstly developed by [14] that imitates the behavior of flocks of birds searching for food. In this algorithm, a number of random particles, candidates or agent is sent for the search. Each particle's position is influenced by the previous velocity v_i^{k-1}, its best position found so far $P_{i,best}$ and the best position found by all the swarm G_{best}. Figure 4 shows the movement of the particle in the search space:

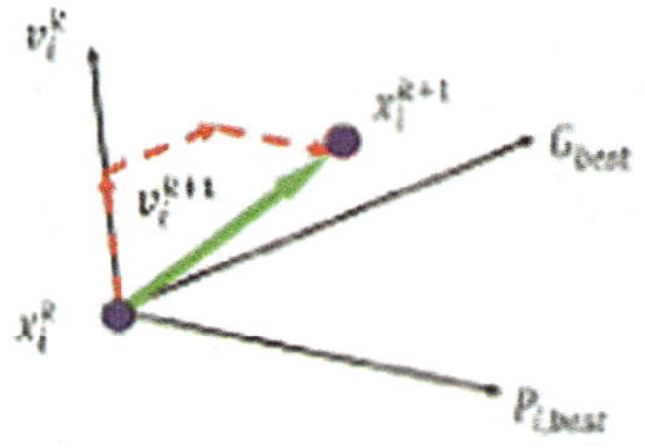

Fig. 4. Movement of the particles

The particle's position x_i^k is updated by:

$$x_i^{k+1} = x_i^k + v_i^{k+1} \quad (4)$$

And the velocity v_i^k is updated by:

$$v_i^{k+1} = w.v_i^k + c1.r1.\left(P_{i,best} - x_i^k\right) + c2.r2.\left(G_{best} - x_i^k\right) \quad (5)$$

v_i^k Represent the velocity of the ith particle in the kth iteration, w is the inertia weight, $c1$ and $c2$ are the individual and the social coefficient, $r1, r2$ are random numbers in $[0, 1]$.

3.2 Particle Swarm Optimization Based MPPT Controller

In this section we applied the PSO algorithm on MPPT controller to find the maximum power point. A projection of the PSO is made on the PV system by replacing the position x_i^k by the duty cycle d_i^k, and the velocity by the perturbation step size Δd_i^{k+1}. The PV output power serves as the objective function.

$$d_i^{k+1} = d_i^k + \Delta d_i^{k+1} \quad (6)$$

$$P_{pv}^k = f(d_i^k) \quad (7)$$

At first the system sends four particles $d = \{d_1, d_2, d_3, d_4\}$ to the DC-DC converter, and then it calculate the corresponding power. The MPPT controller interpret the P_{pv}, it determines the $G_{best}\, and\, P_{best}$, it calculate the velocity then it generate an updated duty cycle to the converter. This process repeats until it found the duty cycle that gives the maximum power of the PV system.

3.3 The Modified PSO

In the original work of [14] the inertia weight wasn't included in the velocity equation, it was firstly introduced by [15] as a constant value. Further, an adaptive weight was used with different mathematical approaches seeking for better solutions like in [16, 17]. The constant weight has two major problems on MPPT, if w is large the MPP is well located but the accuracy is very low, contrary if w is small the accuracy is higher but the chance of stagnating in local peak is big. Thus we need a modification on w formula for better chance finding the global peak with high accuracy and in short time. A modified PSO is proposed by using a random values of inertia weight with a non-linear decrease using exponential function. Figure 5 shows the pseudo code of the algorithm:

```
-Initialize the parameters: MaxIter, c1, c2, tau, n_swarm, and the bounds of search space
-Initialize the PSO population d_i and velocity Δd_i (i=1, 2... n_swarm)
While (item<MaxIter)
    For i=1:n_swarm
        Calculate power Ppv(k)
        Find the current best solution P_i,best for each particle
    End for
-Find the global best G_best
-Calculate the new weight with Eq.8 and Eq.9
-Update the velocity Δd_i^k and the position d_i^k with eq. 6 and eq.5
-Enforce the positions into the search space bounds
%% detecting the step change of irradiance
-Calculate the mean difference Δm of the current and the previous population mean m^k, m^(k-1)
-Calculate the variance σ^k
    If σ^k ≤ 2 && Δm ≤ 2
                        Etat=1;
    Else  if Δm > 2 && Etat == 1
              Etat=0;
              Set the velocity, global bests and the best position to zero
              Initialize the population again
          End
    End
End while
```

Fig. 5. Pseudo code of modified PSO

The interval of w is taken from $[0.2, 0.9]$, then this interval is divided into sub-intervals with exponential function over iterations. At each iteration random value of w is selected from the corresponding sub-interval. The randomness is used here to help for the exploration phase and also to mutate the stochastic process of the bird's behavior.

$$w^{iter} = A * \exp\left(-\frac{iter - 1}{\tau - 1}\right) + B, \quad \begin{array}{c} A = (w_{ub} - w_{lb}) * \left(1 - \exp\left(-\frac{MaxIter-1}{\tau-1}\right)\right)^{-1} \\ B = w_{ub} - A \end{array} \tag{8}$$

Where w_{ub}, w_{lb} are the upper and lower bound of the inertia weight, w^{iter} is the upper bound of the sub-interval and τ is the time constant of the exponential function. The selected weight W^{*iter} is calculated by the Eq. (9), and a is a constant parameter to adjust the width of the sub-intervals, and it is settled to $a = 0.8$.

$$W^{*iter} = \left(w^{iter} - a * w^{iter}\right) * rand() + a * w^{iter} \tag{9}$$

4 Simulation and Results

4.1 Selection of PSO Parameters

We tested the performance of the PV system in finding the MPP under PSC with above insulations: [1000 800 600 300] and at $T = 25°C$. We carried the tests under PSC and not under STC because we need to find the values that drive the algorithm to find the Global peak when the local peaks power are closer to the global one. At first we settled the values of $c1, c2$ to 1.2 *and* 1.6 respectfully, we varied the time constant τ then we selected the best one that gives the best results in term of finding the global peak, convergence speed and reduced steady state oscillation. Secondly: we settled τ value to its best then changed $c1$ and $c2$, the swarm population is fixed to N = 4 and the sampling time = 0.01 s.

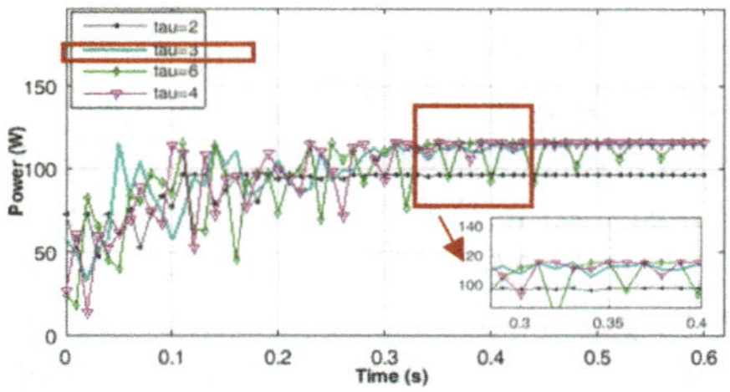

Fig. 6. PSO algorithm results with different τ

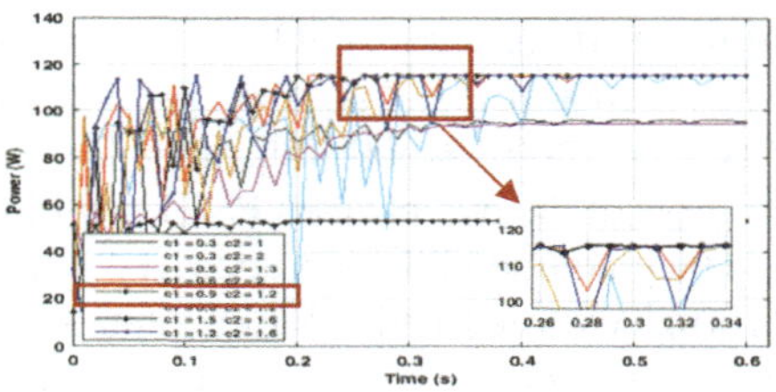

Fig. 7. PSO algorithm results when changing c1 and c2

From the test in Fig. 6 it's clear that $\tau = 3$ gives better results in term of convergence speed and reduced oscillation. And for the second test in Fig. 7, $c1 = 0.9$ *and* $c2 = 1.2$ drives good results compared to other values. Thus we have chosen those values to be used in our tests.

4.2 Performance of PSO And P&O Under Uniform Condition

We compared the performance of PSO with P&O under uniform irradiation with a step change of irradiance from $STC1 = 1000W/m^2$ to $STC2 = 700W/m^2$ at t = 1 s, see Fig. 8. As it is clear in the PSO and P&O graph, PSO has reached the MPP at t = 0.3 s with 99% of the actual MPP value and with almost zero oscillations, while P&O took 0.4 s to find the MPP with oscillation around it. At t = 1 s, P&O and PSO both took the same amount of time to find the MPP but PSO has more accuracy than P&O.

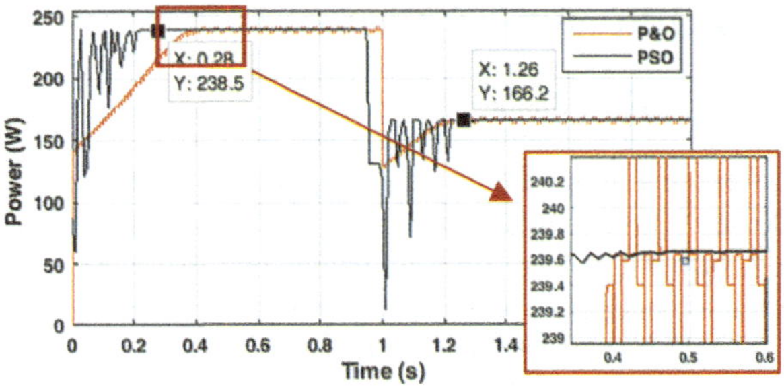

Fig. 8. Performance of PSO and P&O under step change of irradiance with uniform condition

4.3 Performance of Modified PSO Under Partial Shading Condition With Step Change of Irradiance

In this part we tested PSO with 3 patterns, at t = 0 the PV under standard condition $STC = 1000\,W/m^2$, at t = 1 s the insolation changes to the partial shading $PSC1 = [1000\,800\,600\,300]$, at t = 2 s it becomes $PSC2 = [1000\,700\,600\,400]$. The P-V curve of the two patterns PSC1 and PSC2 is illustrated in Fig. 9. From the results we see that modified PSO has approved its ability to find the MPP under rapid changement of irradiance and under hard condition, where the global max at PSC1 is equal to the global max of the PSC2 but in different position see Fig. 10. The modified PSO took about 0.3 s to find the solution, which is good result.

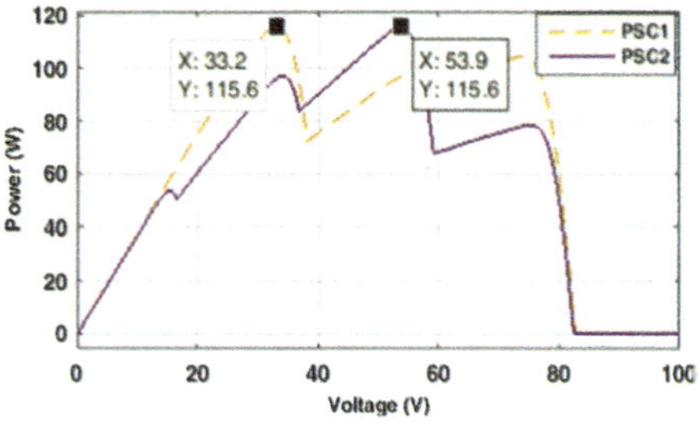

Fig. 9. P-V curve under patterns: PSC1 and PSC2

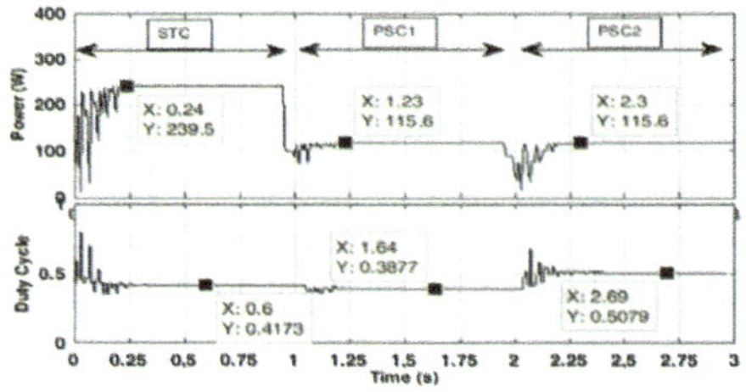

Fig. 10. Performance of PSO under step change of irradiance with PSC

5 Conclusion

In this paper, a PSO algorithm with an exponential decreased weight is used for tracking the MPP. Previous literature shows that the constant value of the weight can't balance between the exploration and exploitation of the algorithms thus it fails in finding the MPP. In this study, we've found that using an adaptive weight with decreased values leads to better results in term of finding the global maximum and in term of accuracy and convergence speed. This work was compared with P&O

algorithm and the results shows that our algorithm performed better than P&O. In further work we would like to implement the proposed algorithm on a Hardware device like FPGA, then we test it under real condition of temperature and irradiance.

References

1. Petreu, D.: Modelling and simulation of photovoltaic cells, vol. 49, no. 1, p. 7 (2008)
2. Ahmed, J., Salam, Z.: An improved perturb and observe (P&O) maximum power point tracking (MPPT) algorithm for higher efficiency. Appl. Energy **150**, 97–108 (2015)
3. Putri, R.I., Wibowo, S., Rifa'i, M.: Maximum power point tracking for photovoltaic using incremental conductance method. Energy Procedia **68**, 22–30 (2015)
4. Cheikh, M.S.A., Larbes, C., Kebir, G.F.T., Zerguerras, A.: Maximum power point tracking using a fuzzy logic control scheme. Revue des Energies Renouvelables **10**(3), 9 (2007)
5. Yaichi, M., Fellah, M.-K., Mammeri, A.: A neural network based MPPT technique controller for photovoltaic pumping system. Int. J. Power Electron. Drive Syst. (IJPEDS) **4**, 241–255 (2014)
6. Kennedy, J., Eberhart, R.: Particle swarm optimization. In: Proceedings of ICNN'95 - International Conference on Neural Networks, vol. 4, pp. 1942–1948 (1995)
7. Titri, S., Larbes, C., Toumi, K.Y., Benatchba, K.: A new MPPT controller based on the Ant colony optimization algorithm for Photovoltaic systems under partial shading conditions. Appl. Soft Comput. **58**, 465–479 (2017)
8. Gavhane, P.S., Krishnamurthy, S., Dixit, R., Ram, J.P., Rajasekar, N.: EL-PSO based MPPT for Solar PV under partial shaded condition. Energy Procedia **117**, 1047–1053 (2017)
9. Ishaque, K., Salam, Z., Amjad, M., Mekhilef, S.: An Improved Particle Swarm Optimization (PSO)–Based MPPT for PV with reduced steady-state oscillation. IEEE Trans. Power Electron. **27**(8), 3627–3638 (2012)
10. Zhang, L., Fu, Q., Chen, J., Bai, H., Zhou, X.: A modified particle swarm optimization algorithm — CPSODE. In: 29th Chinese Control and Decision Conference (CCDC), Chongqing, China, pp. 6659–6663 (2017)
11. Ishaque, K., Salam, Z.: A deterministic particle swarm optimization maximum power point tracker for photovoltaic system under partial shading condition. IEEE Trans. Ind. Electron. 1 (2012)
12. Dhar, S., Scholar, P., Sridhar, R.: Modeling and Simulation of Photovoltaic Arrays, p. 5
13. Ishaque, K., Salam, Z., Taheri, H.: Simple, fast and accurate two-diode model for photovoltaic modules. Sol. Energy Mater. Sol. Cells **95**(2), 586–594 (2011)
14. Kennedy, J., Eberhart, R.: Particle Swarm Optimization, p. 7 (1995)
15. Shi, Y., Eberhart, R.: A modified particle swarm optimizer. Presented at the Proceedings of the IEEE Conference on Evolutionary Computation, ICEC, vol. 6, pp. 69–73 (1998)
16. Nickabadi, A., Ebadzadeh, M., Safabakhsh, R.: A novel particle swarm optimization algorithm with adaptive inertia weight. Appl. Soft Comput. **11**, 3658–3670 (2011)
17. Xin, J., Chen, G., Hai, Y.: A particle swarm optimizer with multi-stage linearly-decreasing inertia weight. In: 2009 International Joint Conference on Computational Sciences and Optimization, vol. 1, pp. 505–508 (2009)

MPPT Based Fuzzy-Logic Controller for Grid Connected Residential Photovoltaic Power System

A. Abbadi[1(✉)], F. Hamidia[1], A. Morsli[1], O. Benbouabdellah[2], and Y. Chiba[3]

[1] Electrical Engineering and Automatic LREA, Electrical Engineering Department, Faculty of Technology, University of Medea, Medea, Algeria
amel.abbadi@yahoo.fr
[2] Chargée d'études HT, BE/DG/KAHRIF, Medea, Algeria
[3] Faculty of Technology, Department of Mechanical Engineering, University of Medea, Medea, Algeria

Abstract. This paper focuses on control strategy of the direct current voltage reference of a voltage source converter based on fuzzy logic controller. A PV array is connected to residential system via an inverter. The inverter control system uses three controllers. The first one is a Maximum Power Point Tracking (MPPT) controller. Based on fuzzy logic system scheme, this MPPT system automatically varies the VDC reference signal of the inverter VDC regulator in order to obtain the DC voltage which will extract maximum power from the PV string. The second controller regulates the DC link voltage to its reference value and the third controller regulates active and reactive grid current components. The active current reference is the output of the DC voltage controller. Result from the simulation shows that the fuzzy controller is able to varies the VDC reference signal of the inverter VDC regulator in various irradiation condition to extract maximum power from the PV string.

Keywords: Fuzzy controller · Residential photovoltaic system · PV array · Inverter control

1 Introduction

Traditional power generations that are on a basic of fossil fuel resource are considered to be unsustainable in long-term national strategies. This has been one of the main driving forces for an increasing installation of renewable energies like wind power, solar photovoltaic power, hydropower, biomass power, geothermal power, and ocean power, etc. into the public grids [1].

Renewable energy, headed by photovoltaic power generation, has attracted worldwide attention [2]. The decline in PV module prices accompanied by financial incentives, such as feed in tariffs, as well as increased price of electricity, have caused a rapid increase in grid connected residential photovoltaic systems around the world [3].

M. Hatti (Ed.): ICAIRES 2019, LNNS 102, pp. 124–131, 2020.
https://doi.org/10.1007/978-3-030-37207-1_13

Photovoltaic has good public acceptance, so it can be sited in a residential areas usually on roof top of domestic, commercial buildings and can also be mounted on ground. It can be used to provide much or all of electric power needed by any residential building occupants [4].

Photovoltaic system has no moving part so the operation is completely noiseless and it is easy to install. PV power station is environmental friendly, no pollutant on air, land and water, no cost of working fluid or fuel since sunlight is freely available from nature. It has durability of more than twenty five years with a very minimal maintenance and does not require highly skill person for its maintenance.

The residential photovoltaic applications as single-phase systems are different in configuration. The major functions of the power converters are the same, including PV power maximization [5–7], DC to AC power conversion and transfer, synchronization, grid code compliance, islanding detection and protection, etc.

It also requires advanced and intelligent controls to perform these PV features and also to meet customized demands, to enhance the PV integration in the residential photovoltaic applications.

This paper discusses the control of the VDC reference signal of the VDC controller of the inverter based on fuzzy control scheme. The proposed MPPT controller automatically varies the VDC reference signal of the inverter's VDC controller to obtain a DC voltage that will extract the maximum power of the PV array.

This paper is organized as follows. In Sect. 2, the system under study is described. The fuzzy logic controller design is presented in Sect. 3. In Sect. 4, the proposed control is validated by means of simulation. Finally, the conclusions are summarized in Sect. 5.

2 System Structure

Grid-connected residential photovoltaic power system is combined with PV array, inverter control module, residential load and distribution network. Figure 1 shows the system structure.

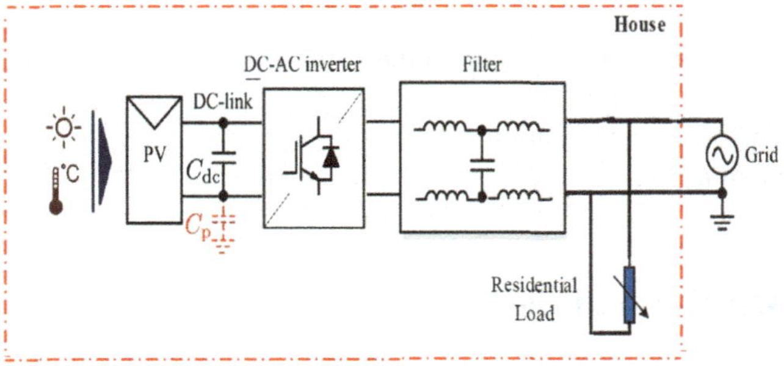

Fig. 1. General block diagram of a single-stage single-phase PV topology

The curve of Fig. 2 shows the I-V and P-V characteristics for the whole array.

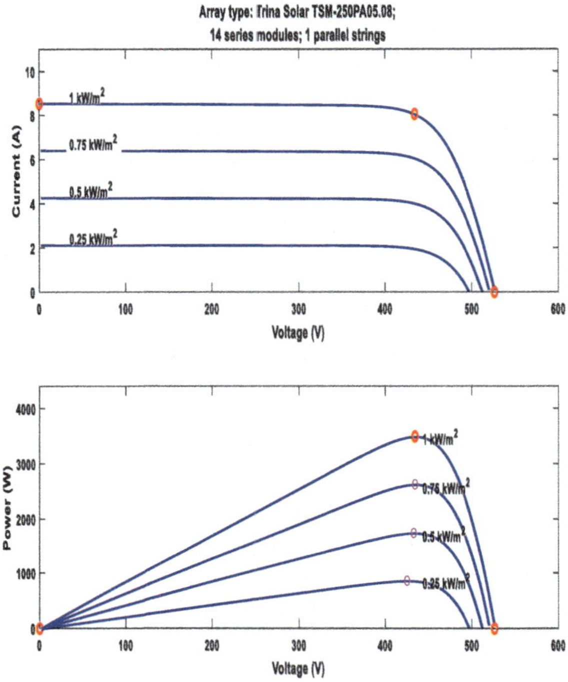

Fig. 2. The I-V and P-V characteristics for the Trina Solar TSM-250PA05.08

The control system contains five major subsystems:

- MPPT Controller: The Maximum Power Point Tracking (MPPT) controller is based on fuzzy logic control scheme. This MPPT system allows to changes the VDC reference signal of the inverter VDC regulator in order to draw maximum power from the PV array.
- VDC Regulator: Determine the required Id (active current) reference for the current regulator.
- Current Regulator: The regulation defines the intended reference voltages for the inverter.
- PLL is used for synchronization.
- PWM Generator.

3 Fuzzy MPPT Controller

In this paper a fuzzy control strategy is proposed to track the Maximum Power Point (MPP). The fuzzy logic system is composed of fuzzification, rule-base, inference and defuzzification as shown in Fig. 3. In the proposed system, the input variables of the FLC are variation of the photovoltaique voltage and variation of the power of the photovoltaique PV.

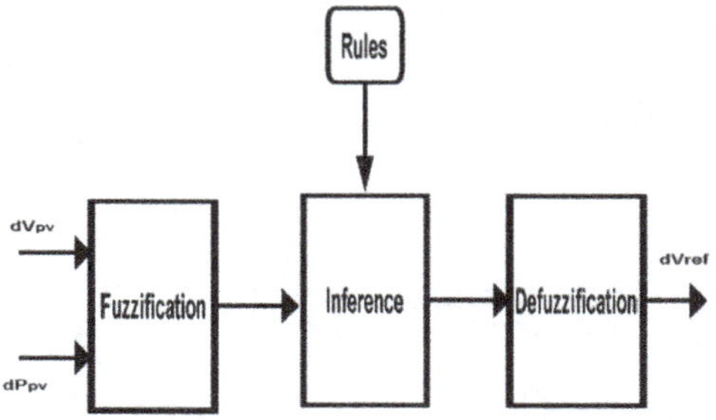

Fig. 3. Fuzzy controller diagram

The VDC reference signal is generated by the proposed MPPT technique in order to extract maximum power from the PV array (Table 1).

Table 1. FLC rules base for MPPT controller

Fuzzy rule		dVpv			
		NB	NS	PS	PB
dPpv	NB	PB	PS	NS	NB
	NS	PB	PS	NS	NS
	PS	NS	NS	PS	PS
	PB	NB	NS	PS	PB

4 Simulation Results

The simulation is achieved under various sun irradiation conditions (Fig. 4) to exhibit the performances of the fuzzy MPPT controller.

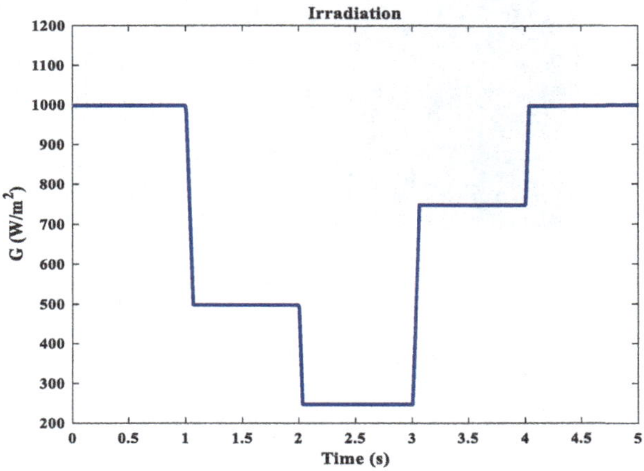

Fig. 4. Different sun irradiation conditions

Figure 5 shows the VDC link Voltage and the VDC reference signal generated by the proposed controller. Compared to the characteristics of the PV string (Fig. 2), these values correspond well to the expected values.

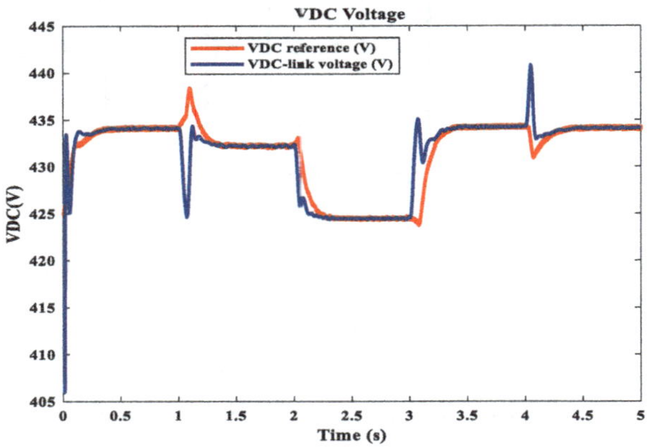

Fig. 5. VDC reference and VDC-link voltage

Figure 6 depicts the residential load voltage, and the Fig. 7 represents the residential load current.

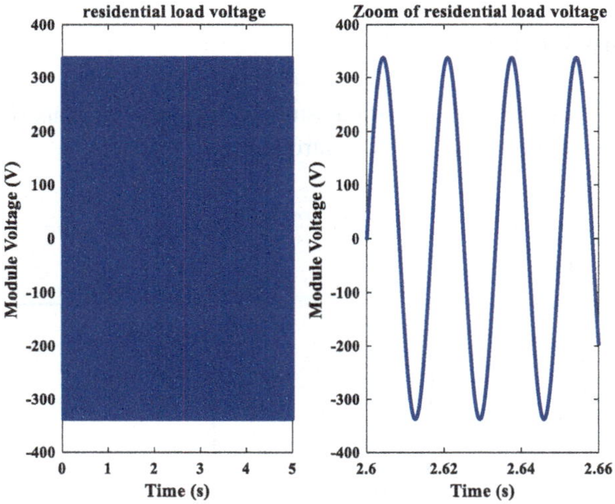

Fig. 6. Residential load voltage

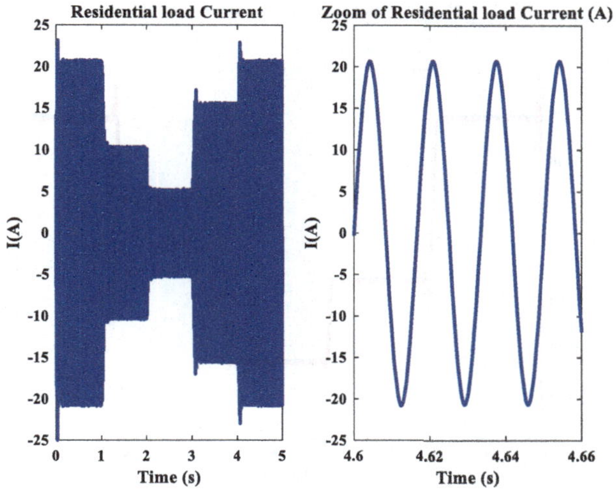

Fig. 7. Residential load current

The total harmonic distortions (THD) of the residential load current is presented by Fig. 8.

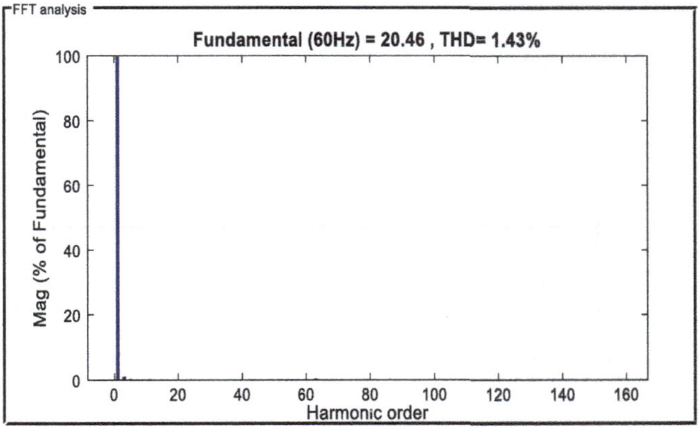

Fig. 8. Current harmonics spectra of residential load current

The Fig. 9 shows the power generated by the PV array and active power of the residential load. The difference in the curves represents the losses in the converter.

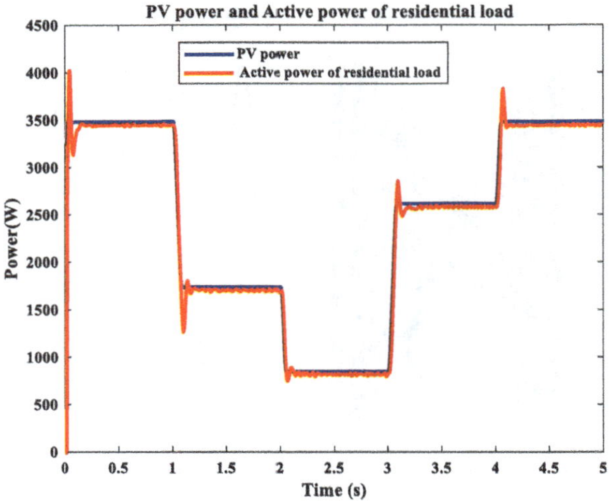

Fig. 9. Active power of the residential load

It is also shown in Fig. 10 that there is no leakage current flowing through the stray capacitance of the PV modules in the transformerless topology adopted.

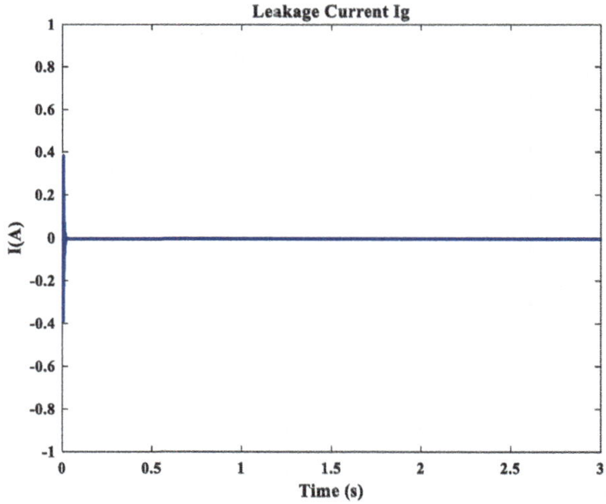

Fig. 10. Leakage current

5 Conclusion

In this paper, fuzzy logic system is proposed as control strategy to regulate the VDC reference signal of the inverter VDC structure. The developed control strategy is checked via simulation studies on residential photovoltaic system. The obtained results showed that the control system employing fuzzy MPPT controller is very efficient in the regulation of the DC reference voltage when irradiance levels change. The THD of the residential load current is very satisfying. The simulated results, obtained for different operating conditions, have shown good performances of the residential photovoltaic system with the fuzzy controller proposed.

References

1. Yang, Y., Blaabjerg, F.: Overview of single-phase grid-connected photovoltaic systems. Electric Power Compon. Syst. **43**(12), 1352–1363 (2015)
2. Li, D.H., Wang, H.X., Zhu, X.D., Li, Z.: Research on several critical problems of photovoltaic grid-connected generation system. Power System Prot. Control **38**(21), 208–214 (2010)
3. Ranaweera, I., Sanchez, S., Midtgård, O.M.: Residential photovoltaic and battery energy system with grid support functionalities. In: 6th International Symposium on Power Electronics for Distributed Generation Systems (PEDG), pp. 1–7 (2015)
4. Johnson, O., Ogunseye, A.A.: Grid connected photovoltaic system design for local government offices in Nigeria. Nigerian J. Technol. **36**(2), 571–581 (2017)
5. Shiau, J.-K., Wei, Y.-C., Chen, B.-C.: A study on the fuzzy-logic-based solar power MPPT algorithms using different fuzzy input variables. Algorithms **8**, 100–127 (2015)
6. Hohm, D.P., Ropp, M.E.: Comparative study of maximum power point tracking algorithms. Prog. Photovoltaics Res. Appl. **11**, 47–62 (2003)
7. Hussein, K.H., Muta, I., Hoshino, T., Osakada, M.: Maximum photovoltaic power tracking: an algorithm for rapidly changing atmospheric conditions. IEE Proc. Gener. Transm. Distrib. **142**, 59–64 (1995)

Control of the Energy Produced by Photovoltaic System Using the Fuzzy PI Controller

Mohammed Kendzi[1(✉)], Abdelghani Aissaoui[2], Ahmed Hasnia[2], and Ahmed Tahour[3]

[1] Department of Electrical Engineering, Smart Grids and Renewable Energies Laboratory, University of Tahri Mohammed Bechar, Béchar, Algeria
kendzimohammed@gmail.com

[2] Department of Electrical Engineering, University of Tahri Mohammed Bechar, Béchar, Algeria
irecom_aissaoui@yahoo.fr, hasnia_ahmed_ett@yahoo.fr

[3] Ecole superieur en science techniques, Tlemcen, Algeria
tahourahmed@yahoo.fr

Abstract. This paper presents a comparison between tow control strategy applied in a solar power system. The first is classical PI controller and the second is a fuzzy PI controller. The photovoltaic conversion system extracts the maximum power from the sun. This power is injected in the grid. The connection between the photovoltaic generator and the grid is made by means of static converters (chopper/inverter). In the photovoltaic conversion chain, a controller for tracking the maximum power point is designed using the direct search approach the Perturb and observe method. The fuzzy PI controller is included firstly to adapted the reference current used in the direct search approach the Perturb and observe method and secondly, to control the continues voltage generated with the chopper converter to fed the boost inverter. The simulation was done to prove the validity of the strategy and of the control method used for the control of the power generated by the photovoltaic system.

Keywords: Photovoltaic · Chopper · Boost · Fuzzy PI controller · PI controller

1 Introduction

The Solar PV technology is becoming popular as a renewable source as it is free, non-polluting and requires little maintenance. Regardless of the intermittency of sunlight, solar energy is widely available and completely free of cost [1, 2]. Recently, PV array systems have been used in several electric power applications [3–5]. When the solar cell is exposed to light, the electric energy is produced by the photoelectric effect, which is proportional to the intensity of sunlight [6]. For better utilization of available solar energy, the photovoltaic cells must be operated at their maximum power point (MPP). The boost converter (DC-DC converter) is employed in the PV array model to step up the voltage of the PV array to the required level of voltage. To control the stability and efficacy of the proposed MPPT algorithm with the boost converter and changes the duty

M. Hatti (Ed.): ICAIRES 2019, LNNS 102, pp. 132–142, 2020.
https://doi.org/10.1007/978-3-030-37207-1_14

cycle of the converter according to variations in irradiance and temperature we have used The conventional proportional integral controller (PI) and the fuzzy PI. The conventional proportional integral controller (PI) is widely used in the control of PV because of its simple structures and good performances. The problem that arises with the PI controller is that his parameters are initially calculated according to the photovoltaic generator parameters [7]. To overcome the disadvantages this paper proposes a comparison between fuzzy PI and PI controller, where the PI parameters can be adjusted online by an adaptive mechanism based on a fuzzy PI. A parameter adjustment method of PI controller with fuzzy PI was developed and used to find the optimal values of PI parameters. The simulation results show that the proposed controller provides better performance and good robustness compared with conventional PI controller.

The organization of this paper is as follows: In the 2nd section, the proposed photovoltaic System is presented. The 3rd section is dedicated to the Modeling a real cell. The 4th section was designed to the chopper booster. The 5th section, design of adaptive fuzzy gain scheduling of PI controller. The calculation of the reference current of the MPPT command in the method of disturbance and observation «p & o» will be presented in the 7th section. The results of the simulation and the conclusions are summarized in the last sections.

2 Presentation of the Photovoltaic System

The photovoltaic energy conversion system contains two converters, the first is a continuous-continuous converter (chopper) and the second is a continuous-alternative converter (inverter). The all system is connected to the Electrical network. Figure 1 shows the photovoltaic energy conversion system.

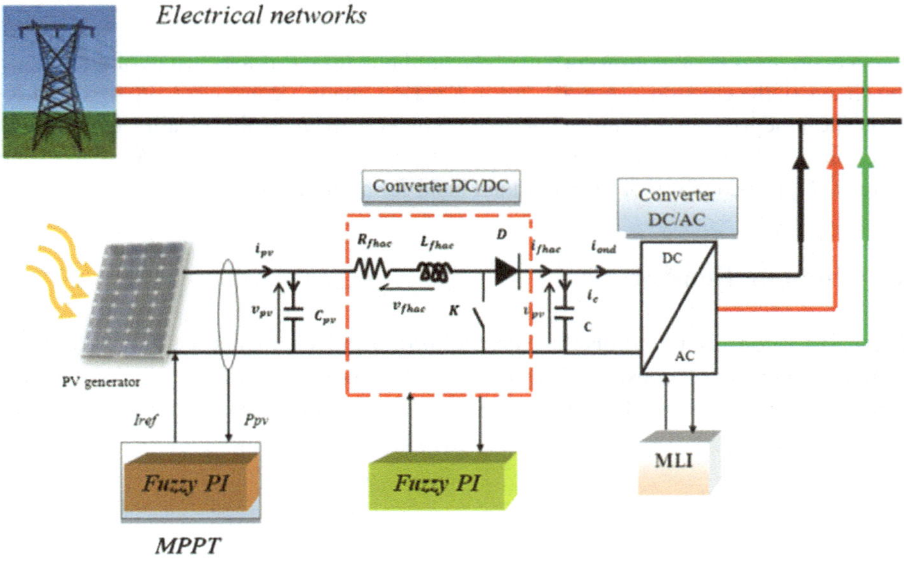

Fig. 1. Photovoltaic energy conversion system.

3 Modeling a Real Cell

The equivalent diagram of a photovoltaic cell is presented in Fig. 2. The current source generates a current I_{ph} proportional to the illumination G. Rs represents series resistance of PN junction cell and R_p represents shunt resistance which is inversely in relation with leakage current to the ground. The current Ipv supplied by the cell is the difference between the current I_{ph} and the current of the diode I_d.

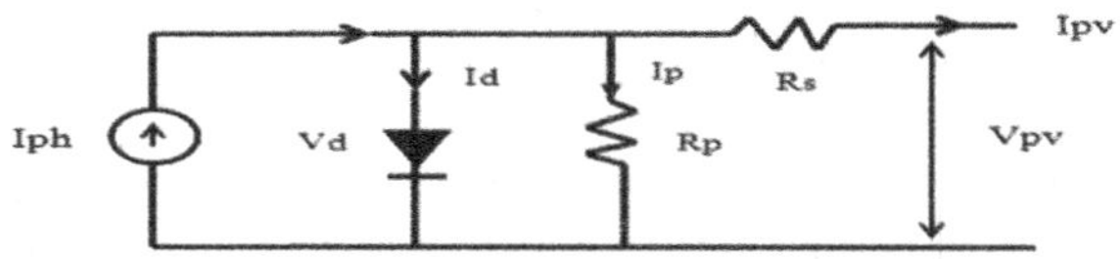

Fig. 2. Equivalent diagram of a real photovoltaic cell [8].

According to Kirchhoff's current law, the PV current equation in the single-diode model is [9]:

$$I_{pv} = I_{ph} - I_d - I_p \tag{1}$$

And

$$I_{pv} = I_{ph} - I_d = I_{ph} - I_0 \times \left(exp\left(\frac{e(V_{pv} + I_{pv}R_s)}{mKT_c}\right) - 1\right) - \frac{V_{pv} + I_{pv}R_s}{R_p} \tag{2}$$

$$I_0 = 1.5 \times 10^5 exp\left(\frac{-E_g}{KT_c}\right)$$

Where: I_0 is the saturation current of the unlit junction; I_c is the current of the cell; e is the charge of the electron (1.6×10^{-19} C); K is Boltzmann's constant (1.38×10^{-23} J/K); T_c is the junction temperature of the photovoltaic cell; m is the ideality factor of the junction; V_{pv} is the voltage across the cell; R_s is the series resistance; R_p is the shunt resistor. E_g is the energy of the forbidden band for a semiconductor material.

4 Chopper Booster

The booster converter controls the electrical power in DC circuits with great flexibility and high efficiency. It consists of capacitors, inductance, and switches; all these devices consume no active power [10]. Figure 3 presents an electrical boost converter scheme.

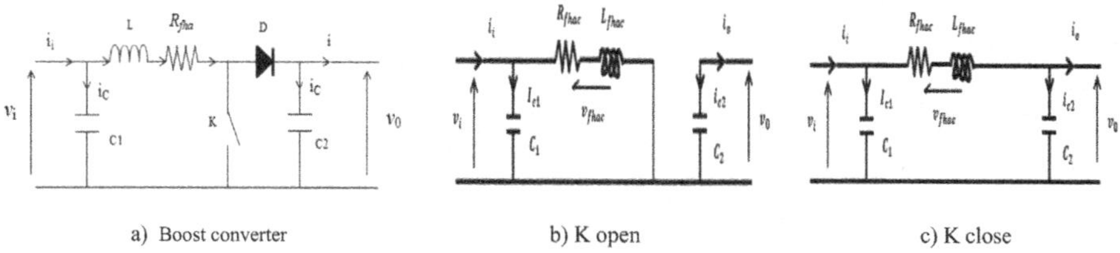

Fig. 3. Electrical boost converter scheme.

4.1 Mathematical Model of the Chopper Boost

Applying Kirchhoff's laws to the scheme of Fig. 3, the following systems of equations are obtained:

$0 < t < \alpha Ts$:

$$\begin{cases} i_{c_1}(t) = c_1 \frac{dv_i(t)}{dt} = i_i(t) - i_L(t) \\ i_{c_2}(t) = c_2 \frac{dv_0(t)}{dt} = -\, i_0(t) \\ v_L(t) = L\frac{di_L}{dt} = -v_i(t) - R_L i_L \end{cases} \tag{3}$$

$\alpha\, Ts < t < Ts$:

$$\begin{cases} i_{c_1}(t) = c_1 \frac{dv_i(t)}{dt} = i_i(t) - i_L(t) \\ i_{c_2}(t) = c_2 \frac{dv_0(t)}{dt} = i_L(t) - i_0(t) \\ v_L(t) = L\frac{di_L}{dt} = v_i(t) - v_0(t) - R_L i_L \end{cases} \tag{4}$$

4.2 Approximate Model of Chopper Boost

To find a valid dynamic representation for the whole period Ts, the following expression is generally used:

$$\frac{dx}{dt}T_s = \frac{dx}{dt_{(\alpha T_s)}}\alpha T_s + \frac{dx}{dt_{((1-\alpha)T_s)}}(1-\alpha)T_s \tag{5}$$

Applying the relation (5) to the systems of Eqs. (3) and (4), the equations that govern the system over an entire period are obtained:

$$\begin{cases} c_1 \frac{dv_i}{dt} T_s = \alpha T_s(i_i - i_L) + (1-\alpha)T_s(i_i - i_L) \\ c_2 \frac{dv_0}{dt} T_s = -\alpha T_s i_0 + (1-\alpha)T_s(i_L - i_0) \\ L\frac{di_L}{dt} T_s = \alpha T_s(v_i - R_L i_L) + (1-\alpha)T_s(v_i - v_0 - R_L i_L) \end{cases} \tag{6}$$

By arranging the terms of the preceding equations, (so that the Boost can be interconnected with the other simulation blocks), the dynamic modeling of the Boost converter is obtained:

$$\begin{cases} i_L = i_i - c_1 \frac{dv_i(t)}{dt} \\ i_0 = (1-\alpha)i_L - c_2 \frac{dv_0(t)}{dt} \\ v_i = L\frac{di_L}{dt} + (1-\alpha)v_0 + R_L i_L \end{cases} \tag{7}$$

4.3 Chopper Control

In this study, the switch of the converter is controlled by a Pulse Wave Large Modulation (PWM) signal, with a fixed frequency *fs* and a cyclic ratio α variable. To extract the control law of the GPV power output, it is necessary to study the two operating phases of the switch S:

- Shac is closed (0 < t < αTs):

$$\frac{di_{fhac}}{dt} = \frac{1}{L_{fhac}}\left(v_{pv} - v_{dc} - R_{fhac}\, i_{fhac}\right) \tag{8}$$

- Shac is open (αTs < t < Ts):

$$\frac{di_{fhac}}{dt} = -\frac{1}{L_{fhac}}\left(v_{dc} + R_{fhac}\, i_{fhac}\right) \tag{9}$$

For this and used of Eq. (5) the variation of the dynamic variable ifhac is considered to be of linear form, so the derivative of the dynamic variable ifhac can be defined according to Eq. (5) over the two time periods αTs and (1 − α)Ts:

$$\frac{di_{fhac}}{dt}T_s = \frac{di_{fhac}}{dt_{(\alpha T_s)}}\alpha T_s + \frac{di_{fhac}}{dt_{((1-\alpha)T_s)}}(1-\alpha)T_s \tag{10}$$

By replacing Eqs. (8) and (9) with their values in relation (10), the equation that governs the system for the entire period is obtained:

$$\frac{di_{fhac}}{dt} = \frac{1}{L_{fhac}}\left(\alpha v_{pv} - v_{dc} - R_{fhac}\, i_{fhac}\right) \tag{11}$$

The control law is given by the following expression:

$$\alpha = \frac{v_{fhac} + v_{dc}}{v_{pv}} \tag{12}$$

Where: Rf_{hac} is the chopper filter resistance; L_{fhac} is chopper filter inductance; v_{dc} is the continues voltage.

5 Design of Adaptive Fuzzy Gain Scheduling of PI Controller

In this section we have improved the performances of our system (precision, rapidity and robustness). To answer these requires, we are interested to develop a control strategy based on the online adaptation of the PI controller parameters by the fuzzy logic technique. The adaptive fuzzy gain scheduling of PI controller in a predefined way allows the use this controller for the nonlinear systems control [7].

Figure 4 show the block diagram of the adaptive fuzzy gain scheduler.

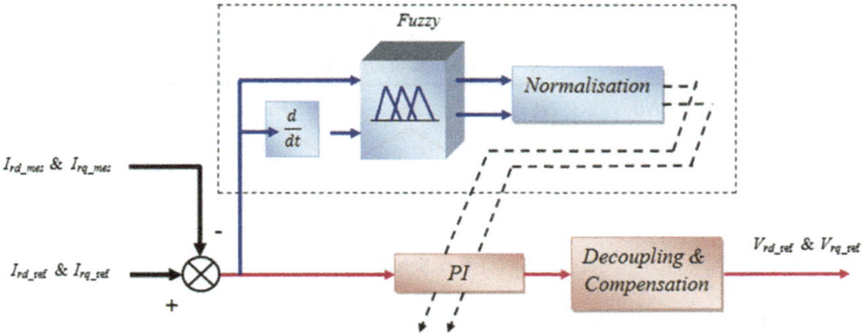

Fig. 4. Block diagram of the Fuzzy PI controller [7]

For good adjustment of the adaptive fuzzy gain scheduler we supposed the fuzzy controller with five fuzzy subsets and constructed from the following condition rule: Rule (i): if s is F is, then uf is F iuf, i = 1,…, 5

Table 1 shows one of possible control rules based on five membership functions, where: NB is negative big, NS is negative small, ZE is zero, PS is positive small and PB is positive big. These acronyms are labels of fuzzy sets and their corresponding membership functions are depicted in Fig. 5 [11].

Table 1. One of possible control rules based on five membership functions.

e	Δ*e*				
	NB	*NS*	*ZE*	*PS*	*PB*
NB	*ZE*	*ZE*	*PB*	*PB*	*PB*
NS	*ZE*	*ZE*	*PS*	*PS*	*PS*
ZE	*PS*	*ZE*	*ZE*	*ZE*	*NS*
PS	*NS*	*NS*	*NS*	*ZE*	*ZE*
PB	*NB*	*NB*	*NB*	*ZE*	*ZE*

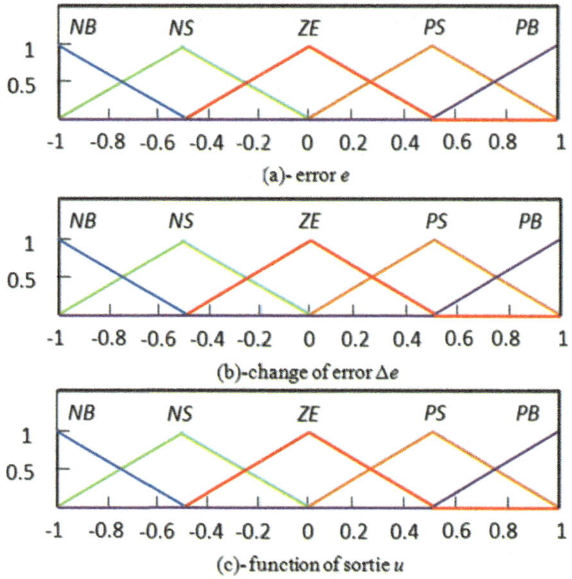

Fig. 5. Membership functions of: (a) error e; (b) change of error Δe; (c) function of sortie u

6 Calculation of the Reference Current of the MPPT Command in the Method of Disturbance and Observation «P & O»

In this paper, proposes a P & O control algorithm to track the maximum power delivered by the photovoltaic generator. The P & O method consists to disrupt the panel voltage vpv and to observe its impact on the change in PVG power. It only requires measurements on the panel voltage vpv and its current ipv. It can immediately detect the maximum power point of the panel. The general algorithm of this method is given in Fig. 6.

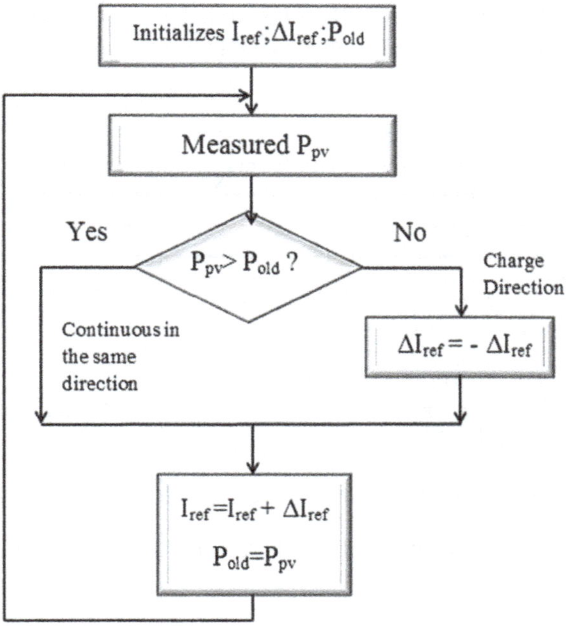

Fig. 6. Flowchart of the perturbation and observation (P & O) algorithm.

7 Simulations and Results

In this section, we simulated the system described in Fig. 1 in Matlab-Simulink. The first part of simulation consists of the regulation system of the DC voltage obtained by the chopper (regular at 330 V) and which feeds the inverter. To achieve the control of reference current and continues voltage and powers generated (P_{pv}), the Fuzzy PI controller used as described in Fig. 4.

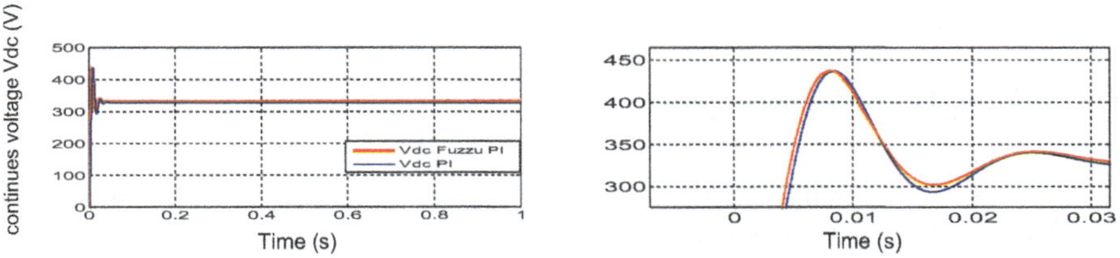

Fig. 7. The obtained chopper voltage vdc with PI and Fuzzy PI controller

The voltage of the chopper v_{dc} is maintained at 330 V as shown in Fig. 7. A transition period can be distinguished in the interval t = [0 0.04] s. After this period, the chopper (vdc =330 V). The curve in blue represents the controller PI, and the curve in red the Fuzzy PI controller. Both controllers have the same stabilization time, but the Fuzzy PI controller has a smaller maximum overshoot.

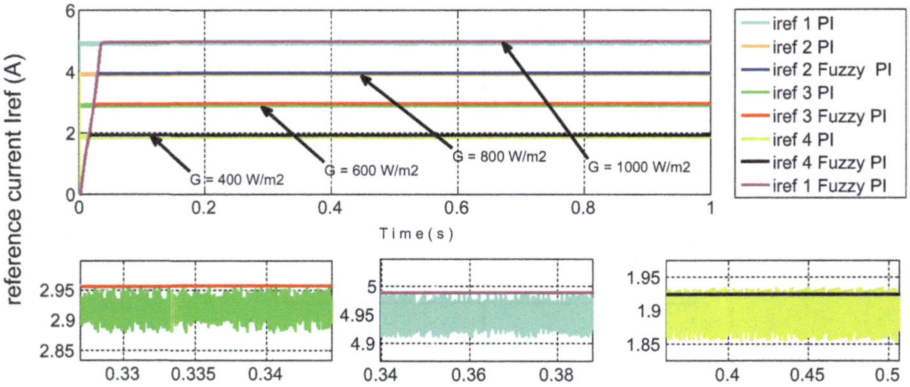

Fig. 8. Regulation of reference current of MPPT Iref with variation of radiation G = (1000-800-600-400) W/m^2

In Fig. 8, we have Regulation of reference current of MPPT I_{ref} with variation of radiation G = (1000-800-600-400) W/m^2. The Regulation with the Fuzzy PI controller is more precision then the PI controller.

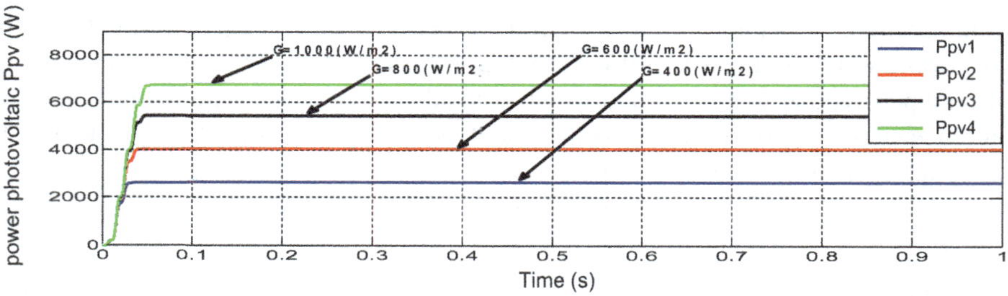

Fig. 9. Regulation of PV power generated with variation of G = (1000-800-600-400) W/m^2.

The Fig. 9 shows the Regulation of PV power generated in deferent value of radiation G = (1000-800-600-400) W/m^2

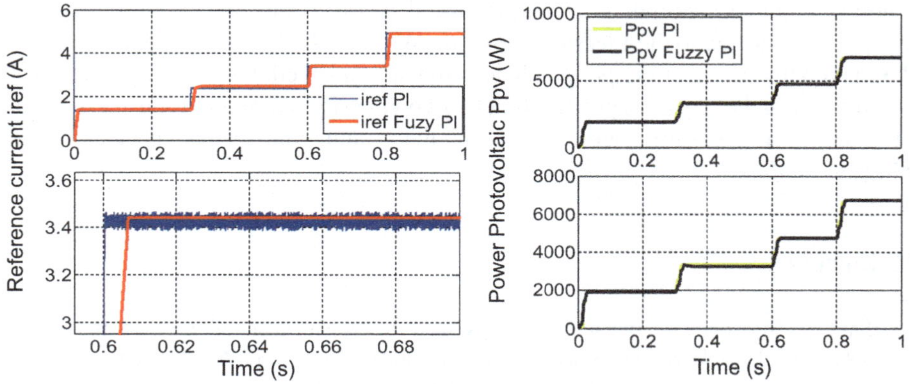

Fig. 10. Variation of reference current and PV power generated with variation of radiation G = (1000-800-600-400) W/m^2

Figure 10, shows the precision and the speed of the regulation of the reference current iref and the power P_{pv} as a function of time in any variation of G.

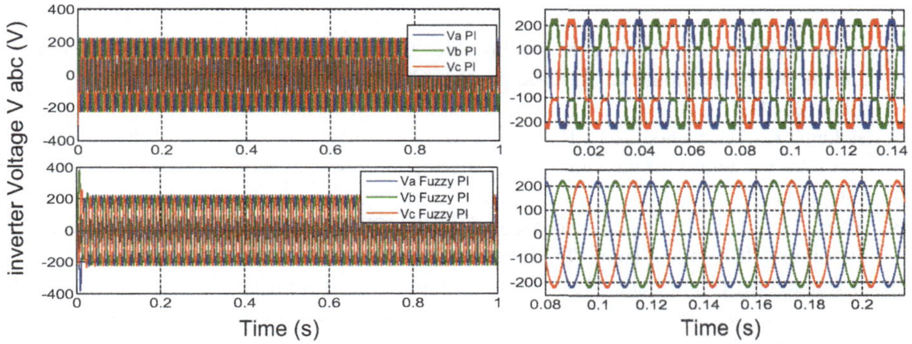

Fig. 11. Control of inverter voltage Vabc with PI and fuzzy PI controller

In Fig. 11, that Fuzzy PI controller gives a good form and good control of Inverter voltage V_{abc} to the classical PI controller. This voltage goes directly to the grid.

8 Conclusion

In this paper, we present the Control of the Energy Produced in Photovoltaic System by Using of tow control Strategy. The first is classical PI controller and the second is a fuzzy PI controller. the comparison between them is used to show the robustness of the control system of PV power with good precision. The photovoltaic energy conversion system equipped with two converters, the first is a continuous-continuous converter (chopper) and the second is a continuous- alternative converter (inverter). The all system is connected to the Electrical network. We used the Fuzzy PI controller to track the maximum power In the photovoltaic cell and to adapted the gain scheduling of PI controller in the chopper converter. Finely a good performance and robustness quality of Fuzzy PI controller are shown in the simulation results using software Matlab/Simulink.

References

1. Said, S., Massoud, A., Benammar, M., Ahmed, S.: A matlab/simulink based photovoltaic array model employing simpower systems. J. Energy Power Eng. **6**, 1965–1975 (2012)
2. Ahlawat, A., Gupta, I., Gupta, S.K.: Modeling of a PV array & implementation of an efficient mppt based control mechanism in stand-alone photovoltaic systems. i-Manager's J. Circuits Syst. **5**(2), 51–61 (2017)
3. Patel, H., Agarwal, V.: MATLAB-based modeling to study the effects of partial shading on PV array characteristics. IEEE Trans. Energy Convers. **23**, 302–310 (2008)
4. Ishaque, K., Salam, Z., Taheri, H.: Accurate MATLAB simulink PV system simulator based on a two-diode model. J. Power Electron. **11**, 179–187 (2011)
5. Hassan, A.A., Fahmy, F.H., Nafeh, A.A., El-Sayed, M.A.: Modeling and simulation of a single phase grid connected photovoltaic system. WSEAS Trans. Syst. Control. **5**, 16–25 (2010)
6. Sinha, D., Das, A.B., Dhak, D.K., Sadhu, P.K.: Equivalent circuit configuration for solar PV cell. In: 1st International Conference on Non-Conventional Energy (ICONCE), pp. 58–60. IEEE (2014)
7. Bedoud, K., Ali-rachedi, M., Bahi, T., Lakel, A.: Adaptive fuzzy gain scheduling of PI controller for control of the wind energy conversion systems. Energy Procedia **74**, 211–225 (2015)
8. Wang, M., Li, H., Wang, S.: Photovoltaic cell MPPT simulation system based on hybrid algorithm. Chem. Eng. Trans. **71**, 169–174 (2018)
9. Moltames, R., Boroushaki, M.: Sensitivity analysis and parameters calculation of PV solar panel based on empirical data and two-diode circuit model. Energy Equipment Syst. **6**(3), 235–246 (2018)
10. Umashankar, S.K., Aparna, P., Priya, R., Suryanarayanan, S.: Modeling and simulation of a PV system using DC-DC converter. Int. J. Latest Res. Eng. Technol. (IJLRET) **1**, 9–16 (2015)
11. Kendzi, M., Aissaoui, A., Abid, M., Tahour, A.L.: Control of the photoelectric generator for used in feeding of the independent wind turbine system. Int. J. Power Electron. Drive Syst. (IJPEDS) **3**, 1613–1627 (2018)

Application of Artificial Neural Network for Modeling Wastewater Treatment Process

A. Sebti[1,2(✉)], B. Boutra[2], M. Trari[1,2], L. Aoudjit[2], and S. Igoud[2]

[1] Chemical Engineering Department, National Polytechnic School, Algiers, Algeria
sebti.aicha@udes.dz, aicha_seb@yahoo.fr

[2] Unité de Développement des Equipements Solaires, UDES, Centre de Développement des Energies Renouvelables, CDER, 42415 Tipaza, Algeria

Abstract. Development of reliable model in chemical engineering facilitates all subsequent steps in process optimization and monitoring operation. Many chemical process intended to wastewater treatment can exhibit complex non-linear behaviour. In this paper, three layered feed forward neural network is used to predict the photo-catalytic degradation yield of solophenyl red, an azo dye widely used in textile industry. The approach adopted to find the optimal topology of the network is based on finding the architectural parameters (the hidden nodes number, the activation function and the training algorithm) that minimize the prediction error. Experimental data required for the development of the network are extracted from the study performed by [1]. In this study a new photo-catalyst have been used to eliminate under solar light the solophenyl red. The result show that the predicted data from the designed optimal neural network architecture was in good agreement with the experimental data. The excellent value of the correlation coefficient attested the accuracy of the model and proved its ability to fit this complex system.

Keywords: Photo-catalysis · Wastewater treatment · Artificial intelligence · Solar energy

1 Introduction

Development of reliable model in chemical engineering facilitates all subsequent steps in process optimization and monitoring operation. Many chemical process intended to wastewater treatment can exhibit complex nonlinear behavior. The use of analytical approach to model such system frequently have their limitations. Various papers have been proved that Artificial Neural Networks (ANNs) can overcome these limitations. ANNs are computational models that mimic the way in which the human brain deal with enormous amounts of sensory information. These empirical models are able to establish the unknown relationship between the input and output data without considering the complex physical and chemical laws governing the process under study [2–4]. They offer fast and accurate solutions to various problems in wide range of disciplines, particularly areas involving prediction, classification and data filtering [5]. An ANN always consists of at least three hierarchical layers of neurons fully

M. Hatti (Ed.): ICAIRES 2019, LNNS 102, pp. 143–154, 2020.
https://doi.org/10.1007/978-3-030-37207-1_15

interconnected and arranged in parallel structure: one input layer, one output layer and one or more hidden layers. Each neuron is linked to certain of its neighbors with weights that represent the strengths of these connections. According to the propagation direction of information, neural networks are categorized into: feed-forward (FFNN) networks and recurrent networks (RNN). In FFNN information flow only in one direction towards the output layer and there is no feedback between neurons. In RNN information pass both forward and backward by introducing loops between certain neurons in the network [3, 6, 7].

Neuron is the basic block in a network structure. It is a processing element which has usually one or more inputs and a single output (Fig. 1). Each neuron has an associated activation function (*f*) and weight (w_i). The purpose of the processing element is to perform simple computations (summation and multiplication) in order to determine the output signal (*y*) using the input values received from many other neurons (x_i) as a linear combination [8, 9].

$$y = f(\upsilon) \qquad with \qquad \upsilon = b_1 + \sum_{i=1}^{p-1} x_i * w_i$$

In fact, it applies the activation function to the sum of the weights and sends this signal in a forward direction layer by layer until last layer is reached [2]. The output delivered by the last layer is a prediction of the neural network that is compared to the target.

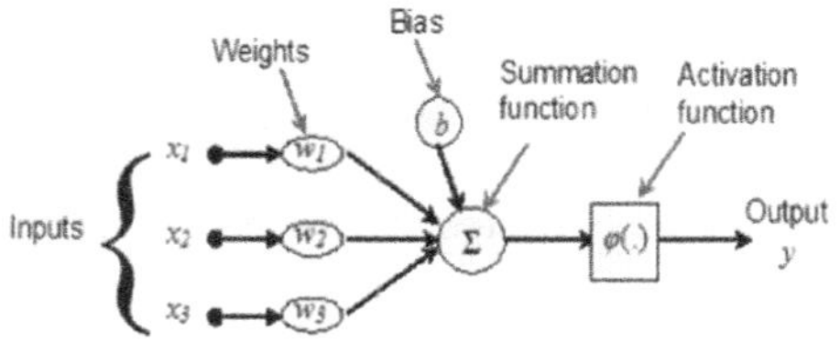

Fig. 1. Model of an artificial neuron [10].

When the predicted result is different of the target then the weight associated to each neuron is updated in an iterative procedure called training or learning. The most commonly learning method used to adjust weights is back propagation (BP). BP consists on calculating the error between the observed and the predicted responses and propagates this error signal in backward direction from the output layer to the input layer. This step is repeated until a convergence criterion is reached.

The purpose of this study is to demonstrate the ability of a three layered feed-forward neural network to model the complex behavior of a photo-catalytic wastewater treatment process.

2 Artificial Neural Network Modeling of Solophenyl Red Photo-Catalytic Degradation

2.1 Experimental Data

In this study, a three layered feed-forward neural network was used to model the photo-catalytic degradation of solophenyl red, an azo dye widely used in textile industry [1]. This organic pollutant is not biodegraded because of the complexity of the chemical structure and its presence in water is harmful to human health. The degradation experiments have been performed under solar light at different experimental conditions using a new catalyst (ZnO/Bentonite). According to the obtained results, the elimination of the SR, using the composite material, was successfully achieved within 160 min and the degradation yield of the azo dye was considered a function of: initial RS concentration, pH, ZnO/Bentonite dosage and irradiation time. The series of experiments conducted in batch reactor by [1] were collected and used for the development of the neural network. In fact, the 116 data were extracted from the experimental study. The four experimental variables (initial RS concentration, pH, ZnO/Bentonite dosage and irradiation time) were selected as inputs of the network whereas the degradation yield of the azo dye was chosen as output. The model variables and their ranges are summarized in Table 1.

Table 1. Neural network model variables and their ranges

Variables	Corresponding range
Initial RS concentration (mg/L)	5–75
pH	2,5–9
ZnO/Bentonite dosage (g/L)	0,25–1
Irradiation time (min)	0–210

In a neural network, the input layer accepts only independent variables. That's why, it's recommended to check whether the four variables are independent or not. In fact, this test reduces the size of the data set and feeds the network with only the most significant inputs by discarding those that are highly correlated. We use the Pearson's correlation coefficient that ranges from -1 (strong negative correlation) and 1 (strong positive correlation) with an insignificant correlation when the coefficient is close to 0 [11, 12].

The correlation coefficients for the SR photo-catalysis input variables are presented in the Fig. 2. The plot figure shows the Pearson coefficient for each pair of variables. The obtained values are close to 0 which indicate quite small correlation that can be neglected. Hence, all experimental variables are retained.

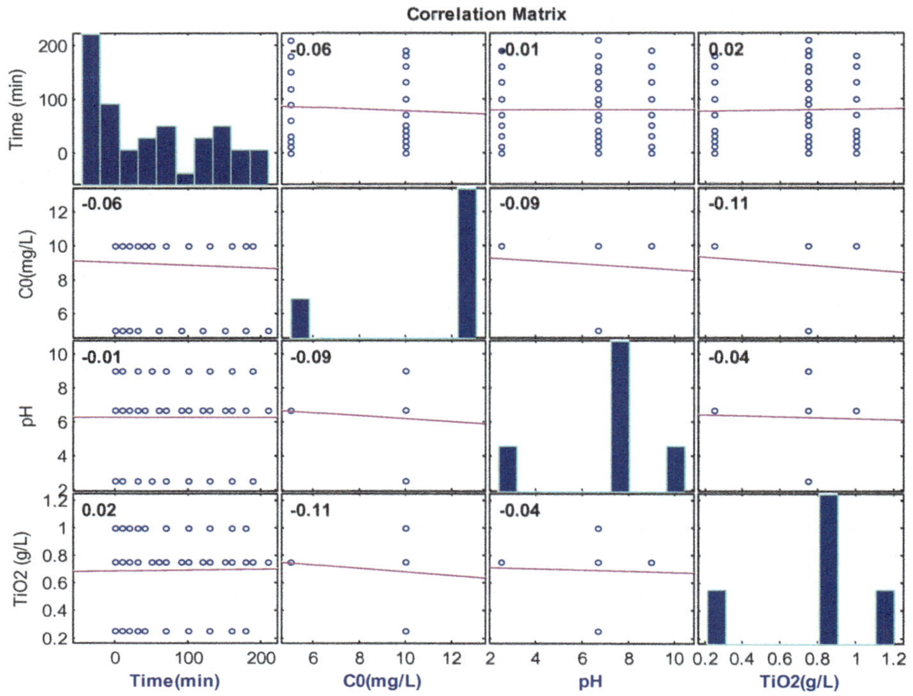

Fig. 2. Correlation coefficients for the solophenyl red photo-catalysis input variables.

In the training phase, the importance of each input variable is related to the magnitude of its range. Therefore, it is indispensable to normalize the raw data in the same range prior to feeding network. We normalize data in order to: equalize the importance of the input variables, minimize the bias and weight values within the network and reduce the training time.

Many techniques of normalization are proposed in the literature. The min-max method is adopted in this study and we normalize the data to fall in the range [−1, 1] [12].

2.2 ANN Model Development

The 116 experimental data were divided into three subsets training (60%), validation (20%) and test (20%). The training subset allowed the network to learn the linear and nonlinear relationships between the inputs and output vectors. The validation subset was employed to evaluate the generalization capability of the network and avoid over-fitting. The test data were fed to the network to assessing its prediction performance.

The number of hidden nodes, the activation function and learning algorithm are architectural parameters that affect the performance of the network model. The strategy adopted in this paper to optimize these parameters and determine the best feed forward network topology consists on:

- Varying the number of hidden nodes between 1 and 10.
- Testing four activation functions (logsig, tansig, hardlim and purelin) and ten training algorithms (trainlm, trainscg, trainbr, trainbfg, trainrp, traincgb, traincgf, traincgp, trainoss and traingdx).

This approach has allowed to select the optimal hidden nodes number and the suitable activation function and training algorithm that minimizes the prediction error. The script developed in Matlab has generated 400 topologies. Their performance to simulate the photo-catalysis process under study was evaluated by calculating the validation Root Mean Squared Error (RMSE) and the Perason's correlation coefficient between experimental and predicted data.

$$RMSE = \sqrt{\sum_{p=1}^{P}\sum_{j=1}^{n_L}\left(y_{jp}^{L} - t_{jp}\right)^2}$$

Where:

t_{jp}: Expected target at the jth output neuron for the input *p*.
y_{jp}^{L}: The output of the jth neuron at layer *L*

2.3 Results

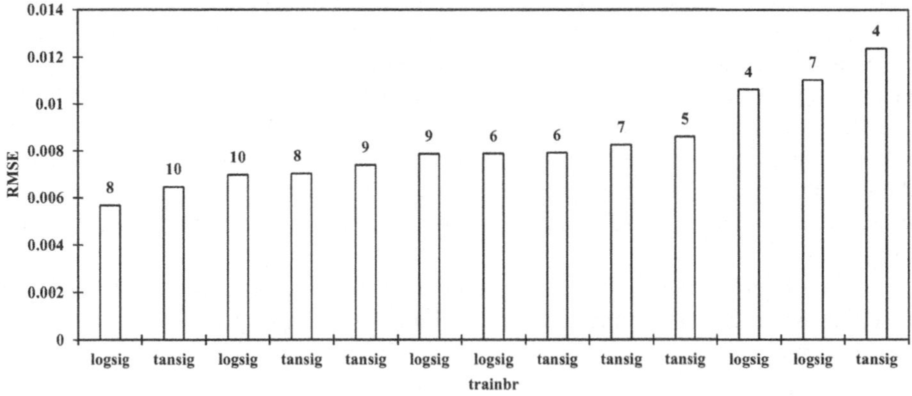

Fig. 3. Variation of RMSE as function of activation function for Bayesian regularization algorithm.

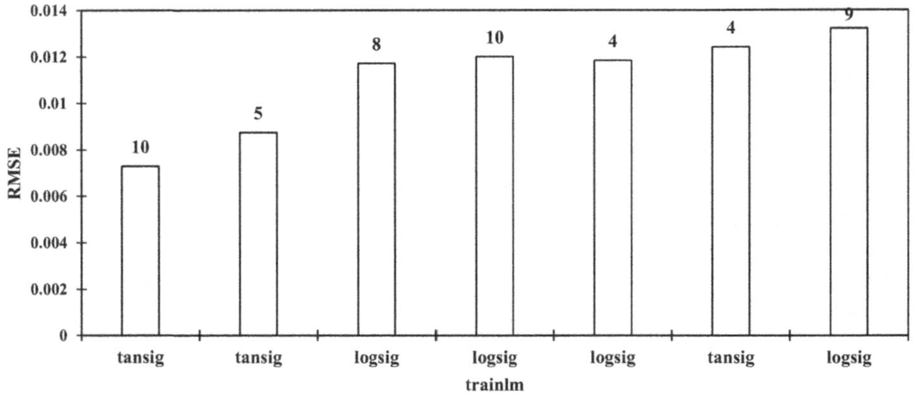

Fig. 4. Variation of RMSE as function of activation function for Levenberg-Marquardt algorithm.

It's important to report that among the 400 networks generated by the Matlab program, 21 topologies predict the photo-catalytic removal efficiency of the pollutant with high accuracy. In fact, the corresponding determination coefficients are over 0.97 and the validation RMSE are ranged between 0.005 and 0.013.

According to the values of the cross-validation mean squared errors and the determination coefficients between measured and predictable data, Bayesian regularization back-propagation (trainbr) and Levenberg-Marquardt (trainlm) show excellent ability to forecast the process efficiency compared to other algorithms (Figs. 3 and 4).

In addition, the two transfer functions namely hyperbolic log sigmoid (logsig) and tangent sigmoid (tansig) give the best fitting models.

The combined effect of the training algorithm and the activation function to improve the performance of the network is well confirmed and the best simulation of the complex system is performed by the network with Bayesian regularization back-propagation training algorithm and hyperbolic log sigmoid transfer function for the hidden layer.

After choosing the suitable combination of training algorithm and activation function, we discuss the neural network optimization with respect to the hidden neurons number.

The Fig. 5 shows the variation of the root mean squared error as function of the hidden neurons number and activation function for the best learning algorithm (trainbr). In fact, the Bayesian regularization algorithm (trainbr) has demonstrated an excellent ability to predict the degradation yield of the azo dye among the ten algorithms tested. In deed the logsig and tansig functions are the most suitable to model this complex and nonlinear system. For the logsig activation function, the best forecasting performance (least RMSE = 0.005 and high $R^2 = 0.99$) is achieved with eight neurons (4-8-1). Hence, the optimal topology of the feed forward network has four input neurons, eight hidden neurons and one output neuron.

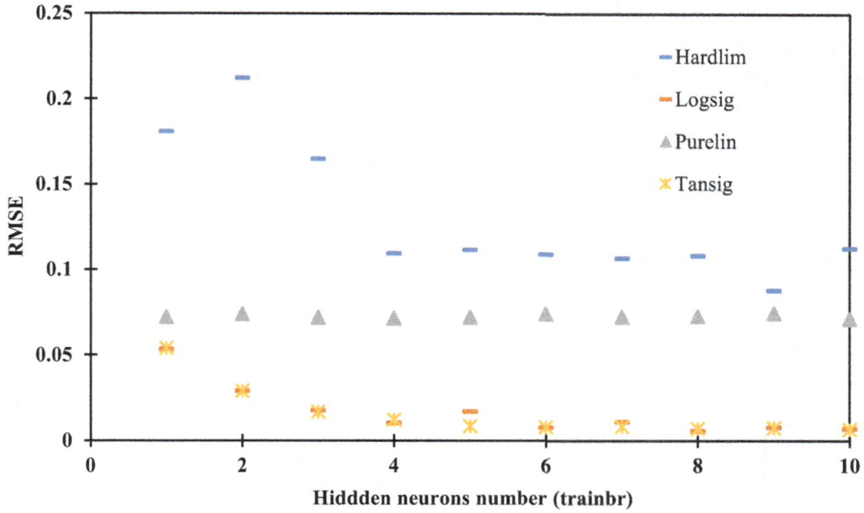

Fig. 5. RMSE as function of hidden nodes number for the trainbr learning algorithm.

In Fig. 6, the variation of the RMSE during the training step is reported as function of the epoch for the optimal topology (4-8-1). 100 training epoch was sufficient to reach the desired performance.

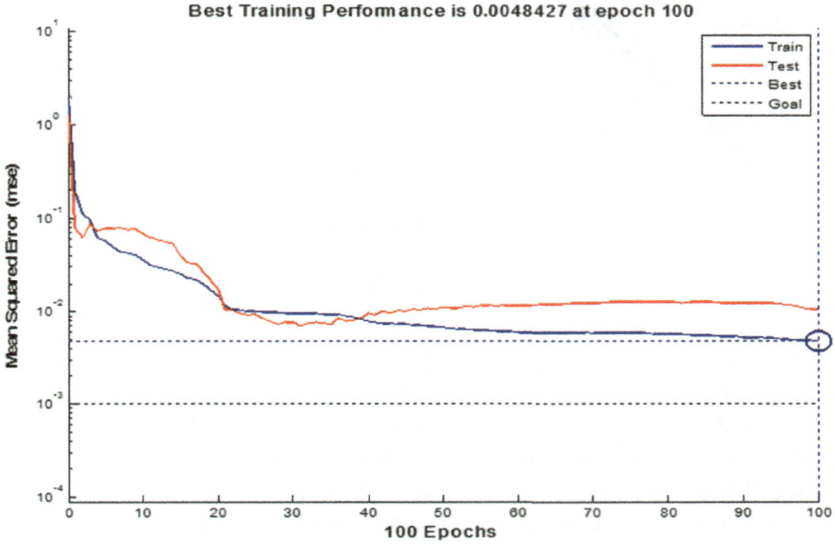

Fig. 6. RMSE as function of training epoch for the optimal network (4-8-1).

For further explanation, the prediction accuracy of the optimal neural network is investigated. Hence, Figs. 7, 8 and 9 illustrate the plots of the experimental results against the predicted ones for training, validation and test sets.

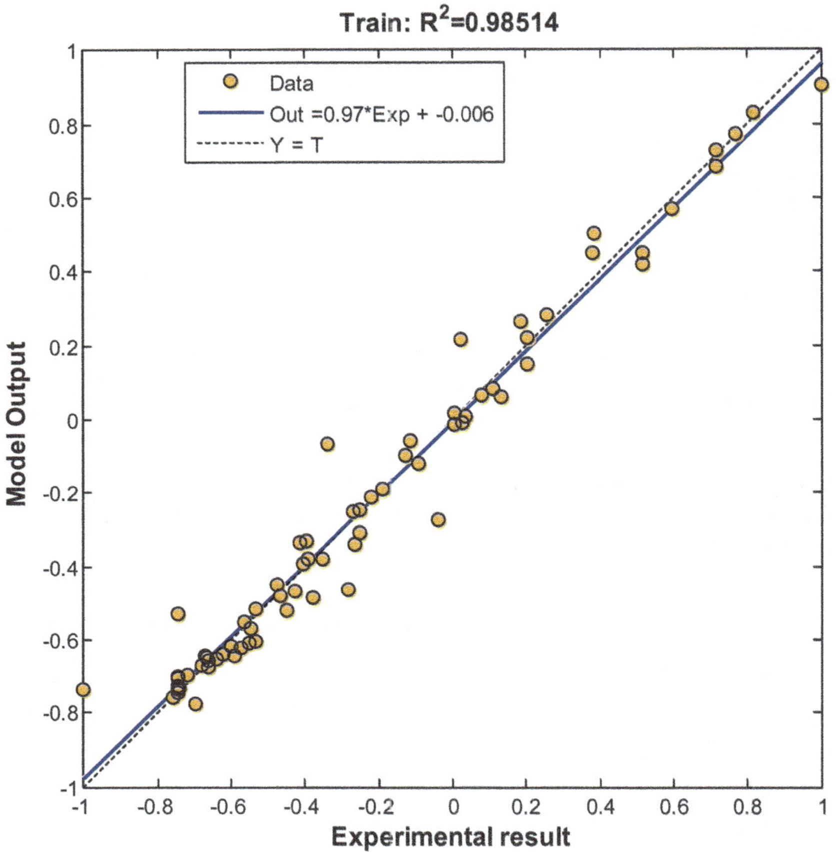

Fig. 7. Experiment results versus predicted ones for training set.

The coefficients R^2 for training, validation and test are 0.985, 0.995 and 0.983 respectively. These high values indicate a strong linear correlation between observed and predicted data. This result confirm the reliability of the developed model to predict the photo-catalytic degradation yield of solophenyl red (Fig. 10).

Artificial neural network are more and more frequently applied in various fields essentially to resolve forecasting problems. Nevertheless, these empirical models are generally regarded to behave as black box systems, unable to clarify the contribution of the independent variables to dependent one. Several authors have thus focused on the analysis of the relative influence of the input variables on the neural network response in order to make ANNs more interpretable [13]. In this paper, the algorithm proposed by Garson is performed. It's a sensitive analysis method which uses combinations of

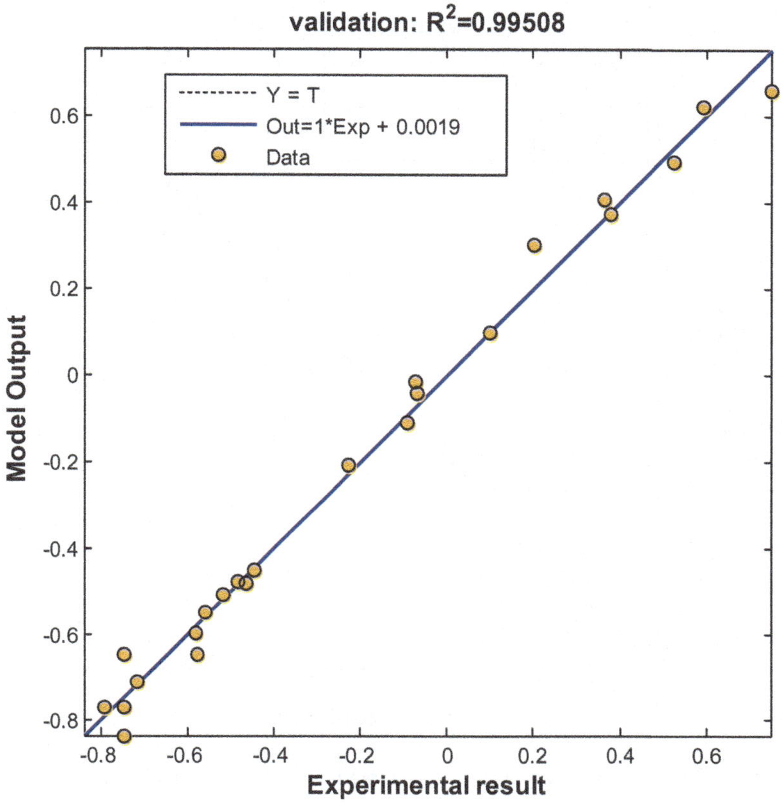

Fig. 8. Experiment results versus predicted ones for validation set.

the absolute values of the weights between layers of the neural network as shown in the formula (1) to obtain the relative importance of the input variable i with respect to the output variable k:

$$GRI_{ik} = \frac{\sum_{j=1}^{m}\left(\frac{|w_{ij}|}{\sum_{i=1}^{n}|w_{ij}|} * |v_{jk}|\right)}{\sum_{i=1}^{n}\left(\sum_{j=1}^{m}\left(\frac{|w_{ij}|}{\sum_{i=1}^{n}|w_{ij}|} * |v_{jk}|\right)\right)}$$

where i, j, k, respectively, refers to input layer, hidden layer and output layer neurons; w_{ij} is the connection weights between input layer and hidden layer neurons, v_{ij}, is the connection weights between hidden layer and output layer neurons, m is the total number of input neurons and n is the total number of hidden layer neurons.

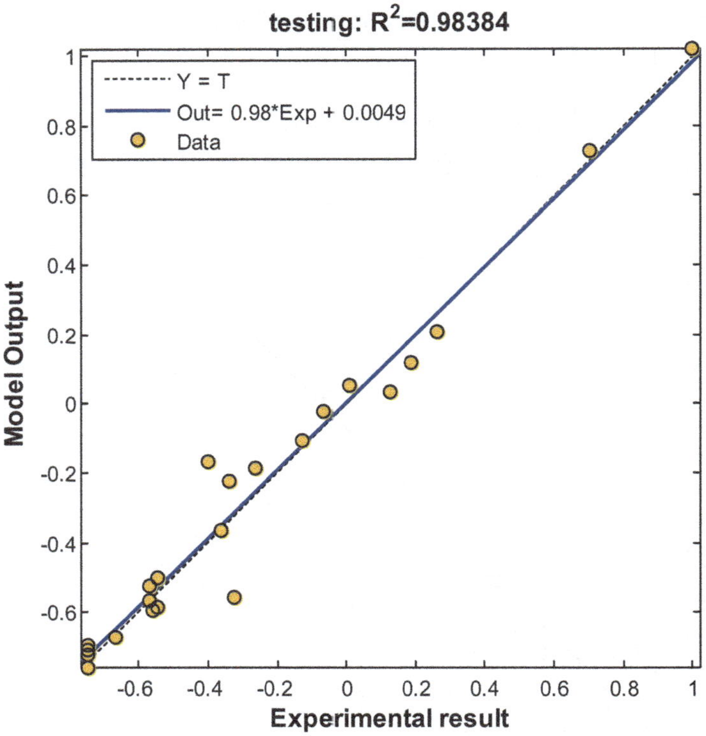

Fig. 9. Experiment results versus predicted ones for validation set.

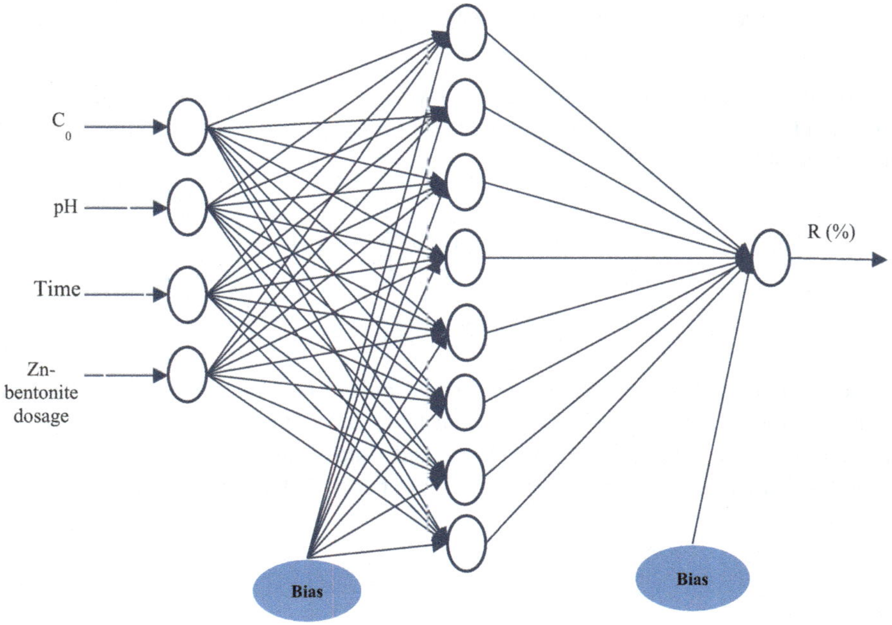

Fig. 10. Structure of the optimal neural network (4:8:1).

The weights matrix of the optimal neural network is given in Table 2.

Table 2. Weights matrix of the optimal neural network.

Weights and bias : Input layer – Hidden layer						Weights and bias: Hidden layer – Output layer		
Neuron	Inputs				Bias	Neuron	Weights	bias
	Irradiation time (h)	Initial solophenyl red concentration (mg/L)	Initial pH	Catalyst amount ZnO-Bentonite (g/L)				
1	−3,35	2,14	−1,31	2,07	−4,13	**1**	−3,46	−0,28
2	−0,44	−1,45	3,65	1,62	−0,76	**2**	3,99	
3	1,19	0,01	−1,78	−0,28	0,53	**3**	2,17	
4	3,86	−1,99	3,21	−1,22	4,02	**4**	−3,19	
5	1,54	1,34	0,40	0,05	3,20	**5**	2,54	
6	−0,28	−2,52	2,30	1,45	0,23	**6**	−2,53	
7	−1,64	1,57	1,64	3,94	1,26	**7**	−2,09	
8	−1,56	−0,31	−0,15	1,49	0,34	**8**	1,59	

The main objective of applying the Garson algorithm is to rank the four variables in order of the relative share of their contribution to the prediction of the solophenyl red removal efficiency by photo-catalytic under solar radiation. The obtained results are summarized in the Table 3.

Table 3. Relative importance of the process input variables

Input variable	Relative importance (%)	Rank
Irradiation time (h) (g/L)	7,10	4
Initial solophenyl red concentration (mg/L)	**58,78**	**1**
Initial pH	8,27	3
ZnO-Bentonite dosage (g/L)	25,84	2

Through the results of the sensitive analysis, it is possible to observe that the considered input variables possess strong effect on the pollutant removal. The initial concentration of solophenyl red is the most influential variable with relative importance of 58%. This parameter is followed by the catalyst loading (25%) and initial pH (8%).

3 Conclusion

The complexity of the photo-catalytic wastewater treatment process makes difficult the use of analytical approach for modelling purpose. Hence, in this work development of artificial neural network model was performed to predict the degradation yield of the solophenyl red azo dye. The strategy based on varying the three architectural parameters namely: hidden neurons number, activation function and learning algorithm has been adopted to optimize the network topology able to reproduce the experimental data. The best prediction performance was achieved with eight hidden neurons, logsig activation function and Bayesian regularization algorithm. This optimal architecture allowed to forecast the observed data with correlation coefficient of 0.995.

References

1. Boutra, B., Trari, M.: Solar photodegradation of a textile azo dye using synthesized ZnO/Bentonite. Water Sci. Technol. **75**(5), 1211–1220 (2017)
2. Dreyfus, G., Martinez, J.M., Samuelides, M., Gordon, M.B., Badran, F.: Réseaux de neurones, Méthodologie et applications, France (2004)
3. Bolanca, T., Ukic, S., Peternel, I., Kusic, H., Bozic, A.L., et al.: Artificial neural network models for advanced oxidation of organics in water matrix-comparison of applied methodologies. Indian J. Chem. Techn. **21**(1), 21–29 (2014)
4. Bagheri, M., Mirbagheri, S.A., Bagheri, Z., Kamarkhani, A.M.: Modeling and optimization of activated sludge bulking for a real wastewater treatment plant using hybrid artificial neural networks-genetic algorithm approach. Process Saf. Environ. Prot. **95**, 12–25 (2015)
5. Croall, I.F., Mason, J.P. (eds.): Industrial Applications of Neural Networks: Project ANNIE Handbook. Springer, Berlin; New York (1992)
6. Al Shamisi, M.H., Assi, A.H., Hejase, H.A.: Using MATLAB to Develop Artificial Neural Network Models for Predicting Global Solar Radiation in Al Ain City-UAE. INTECH Open Access Publisher (2011)
7. Burney, S.M.A., Jilani, T.A., Ardil, C.: A comparison of first and second order training algorithms for artificial neural networks. In: International Conference on Computational Intelligence, pp. 12–18 (2004)
8. Touzet, C.: les réseaux de neurones artificiels, introduction au connexionnisme. EC2 (1992)
9. Ammar, M.Y.: Mise en ø euvre de réseaux de neurones pour la modélisation de cinétiques réactionnelles en vue de la transposition batch/continu. Ph.D. thesis (2007)
10. Magoulas, G.D., Vrahatis, M.N.: Adaptive algorithms for neural network supervised learning: a deterministic optimization approach. Int. J. Bifurcat. Chaos **16**(7), 1929–1950 (2006)
11. Mjalli, F.S., Al-Asheh, S., Alfadala, H.E.: Use of artificial neural network black-box modeling for the prediction of wastewater treatment plants performance. J. Environ. Manag. **83**(3), 329–338 (2007)
12. Sebti, A., Souahi, F., Mohellebi, F., Igoud, S.: Experimental study and artificial neural network modeling of tartrazine removal by photocatalytic process under solar light. Water Sci. Technol. **76**(2), 311–322 (2017)
13. Olden, J.D., Joy, M.K., Death, R.G.: An accurate comparison of methods for quantifying variable importance in artificial neural networks using simulated data. Ecol. Model. **178**(3), 389–397 (2004)

Renewable Energy Conversion

Daily Global Solar Radiation Based on MODIS Products: The Case Study of ADRAR Region (Algeria)

M. Bellaoui(✉), K. Bouchouicha, and B. Oulimar

Unité de Recherche en Energies renouvelables en Milieu Saharien, UERMS, Centre de Développement des Energies Renouvelables, CDER, 01000 Setif, Adrar, Algeria
bellaoui_meb@hotmail.com

Abstract. The Measurements of the solar radiation is important in choosing sites to install projects with solar systems. Adrar region is classified from the high potential solar areas in the world. In this study, we using the atmospheric products of satellite MODIS (Moderate Resolution Imaging Spectroradiometer) and bird model to estimate global solar radiation. The results of estimation were validated with observed data of Adrar station (lat = 36.6, lon = 028). The correlation is very significant and the linear correlation coefficient between estimated and measured values of global solar radiation equal 0.78.

Keywords: Global solar radiation · Satellite MODIS · Bird model · Estimated value · Measured value · Correlation

1 Introduction

The Measurements of the solar radiation is important in choosing sites to install projects with solar systems. the availability of measurements of observed solar radiation has proven spatially and temporally inadequate for many applications [1]. Many approaches were proposed to estimate surface irradiance using satellite measurements, geostationary and polar satellite. Taking advantage of the high spatial-temporal resolutions of this type of data, which being the best option for constructing accurate estimation of the variability of radiation at high spatial resolution.

The most important methods were summarized and reviewed in several studies [2–4].

MODIS sensors aboard the Terra and Aqua satellite platforms, is combined to derive daily integrated PAR and mapped to a local coordinate system. The results was Compared to field observations [5].

Wang et al. [6] with products from the Moderate Resolution Imaging Spectroradiometer (MODIS) sensor implemented a model at one kilometer spatial resolution. The derived surface fluxes are evaluated against the globally distributed Baseline Surface Radiation Network (BSRN) measurements and compared with products from independent sources.

Porfirio et al. present a method for estimating Direct Normal Irradiance and daily direct normal irradiation that uses a minimal set of regional meteorological information

M. Hatti (Ed.): ICAIRES 2019, LNNS 102, pp. 157–163, 2020.
https://doi.org/10.1007/978-3-030-37207-1_16

and avoids empirical adjustment with ground-based radiometric data [7]. A model to estimate global solar radiation under cloudless conditions is presented. Atmospheric perceptible water vapour content is the only experimental inputs to the model [8]. In other study, instantaneous solar irradiances on a horizontal surface at 10:30 and 13:30 local time (LT) were calculated from Mcderate Resolution Imaging Spectroradiometer (MODIS) atmospheric data products with relatively high spatial resolution using a solar radiation model [9]. In order to estimates daily Surface shortwave net radiation, Wang et al. combined Terra and Aqua Moderate Resolution Imaging Spectroradiometer (MODIS) data [10]. The results were validated against measurements at seven stations of the Surface Radiation Budget Network.

Houborg et al. [11] presents a satellite-based scheme for the retrieval of all-sky solar irradiance components, which links a physically based clear-sky model with a neural network version of a rigorous radiative transfer model. The scheme exploits the improved cloud characterization and retrieval capabilities of the MODerate resolution Imaging Spectroradiometer (MODIS), and employs a cloud motion tracking scheme for the production of hourly solar irradiance data throughout the day. An algorithm was developed to calculate surface PAR by combining a clear-sky PAR model and the parameterizations for cloud transmittances. In the algorithm, the transmittances for water vapor, ozone, Rayleigh, aerosol, and cloud are each handled across the whole PAR band (400–700 nm) [12].

Artificial neural network (ANN) is utilized to build the mathematical relationship between measured monthly-mean daily GSR and several high-level remote sensing products available for the public [13].

In this study daily mean solar radiation is estimated using Moderate resolution Imaging Spectroradiometer (MODIS) satellite data. The Bird model was used to derive the solar radiation reaching the Earth's surface in clear skies condition. MODIS data sets are freely available via the MODIS webservice tool. The spatial resolution of this data varies by band from 250 m to 1 km. in this study the data used has a spatial resolution of 1 km. The model is classified into the best and the results confirmed the good performance of the proposed model.

2 Data and Method

2.1 Model

In order to calculate instantaneous solar irradiances at overpass times of satellite MODIS on horizontal surface we using Bird model. With this model we calculated the direct and diffuse radiation [14–16].

Direct irradiance (I_b, W/m^2)

$$I_b = I_{sc} \times \cos(\theta) \times T_r \times T_a \times T_w \times T_o \times T_u / E_r \quad (1)$$

$$T_r = exp\left\{-0.0903 \times (m')^{0.84} \times \left(1 + m' - (m')^{1.01}\right)\right\} \quad (2)$$

$$T_a = exp\{-\tau_A^{0.873} \times \left(1 + \tau_A - \tau_A^{0.7088}\right) \times m^{0.9108}\} \quad (3)$$

$$\tau_A = 0.2758 \times \tau_{A0.38} + 0.35 \times \tau_{A0.50} \quad (4)$$

$$T_w = 1 - 2.4959 \times w \times m \times \left[(1 + 79.034 \times w \times m)^{0.6828} + 6.385 \times w \times m\right]^{-1} \quad (5)$$

$$T_o = 1 - 0.1611 \times O_3 \times m \times (1 + 139.48 \times O_3 + m)^{-0.3035} - 0.002715$$
$$\times \left[1 + 0.044 \times O_3 \times m + 0.0003 + 6.385 \times (O_3 \times m)^2\right]^{-1} \quad (6)$$

$$T_u = exp\left\{-0.0127 \times (m')^{0.26}\right\} \quad (7)$$

Where

I_{sc} solar constant equal 1366.1 w/m^2,
θ solar zenith angle (in degrees)
T_r transmittance of Rayleigh scattering,
T_u transmittance of absorptance of mixed gases,
T_w transmittance of water vapor absorption,
T_o transmittance of ozone absorption
T_a transmittance of aerosol absorptance and scattering
E_r the correction factor for the earth-sun distance
m is the relative air mass and m' is the pressure-corrected air mass.
O_3 is the ozone amount (atm-cm). w is the amount of precipitable water in a certical column from the surface (cm)

Diffuse irradiance (Is, W/m^2)

$$I_s = I_{as} + I_G \quad (8)$$

$$I_{as} = I_{sc} \times \cos(\theta) \times T_o \times T_u \times T_w \times T_{AA} \times \frac{[0.5 \times (1 - T_r) + B_a \times (1 - T_a/T_{AA})]}{\left[1 - m + (m)^{1.02}\right]/E_R} \quad (9)$$

$$T_{AA} = 1 - K_1 \times \left(1 - m + m^{1.06}\right) \times (1 - T_a) \quad (10)$$

$$I_G = r_g \times r_s \times (I_b + I_{as})/\left(1 - r_g \times r\right) \quad (11)$$

$$r_s = 0.0685 + (1 - B_a) \times (1 - T_{as}) \quad (12)$$

$$T_{as} = 10^{-0.045 \times (m')^{0.7}} \quad (13)$$

$$Q_b = I_b \times (1 - TCF) \quad (14)$$

$$Q_s = I_s \times (1 - TCF) + C \times K \times TCF \times (I_b + I_s) \quad (15)$$

$$C = \begin{cases} 1, TCF < 0.95 \\ exp(-0.03 \times P_p \times t), TCF \geq 0.95 \end{cases} \quad (16)$$

$$Q_g = Q_b + Q_s \quad (17)$$

Where,

I_{as} is the atmospheric scattering of diffuse irradiance (w/m^2)
I_G the solar irradiance under clear sky.
T_{AA} is the transmittance of aerosol absorptance, $B_a = 0.84$
r_g is the ground albedo, r_s the sky albedo, T_{as} is the transmittance of dry air absorptance and scattering.
TCF is the total cloud fraction.

The Global solar radiation Q_g is the sum of direct and diffuse irradiances on horizontal surface under real conditions.

2.2 Data

The ground measurements of global solar radiation used in this study are obtained from pyranometers recorded in Adrar Site (27.88°N, 0.28°W), possessed and maintained by the Renewable Energy Research Unit in Saharian Medium (Table 1).

Table 1. Geographic and data records period of the studied station.

Station	Adrar
Latitude (°N)	27.88
Longitude (°E)	−0.27
Elevation (m)	269
Data series period	2010–2015
Mean GH (MJ/m^2.day)	6.89

Bird model used in this study requires many input data, including mod05_l2 (MODIS Atmosphere level 2 precipitation water vapor product), mod06_l2 (MODIS Atmosphere level 2 cloud product) and mod07_l2(MODIS Atmosphere profile product).

These products data contains the needed inputs for the model. The first product is using to obtain the values of: the amount of precipitation water in a vertical column from the surface, latitude and longitude. From the second product we acquire the total cloud fraction (TCF), solar zenith angle, solar azimuth and land surface elevation. And from the third we acquire the ozone amount.

In order to facilitate the calculation, because in Adrar region there is no change in ground situation over the year, so the ground albedo accept mean value equal 3.2 in all time of the year.

2.3 Data Measurement

The estimated global solar radiations are validated by comparing with ground measurements recorded in URERMS station, adrar region. In this study the data from January to December 2016 are used. In this station the data recorded hourly, but we need only the values in the time of overpass of the satellite in the pixel ***Structure of the Text***

3 Results and Discussion

Modis data products are downloaded and registered. The global solar radiation calculated in four steps summarized in (Fig. 1).

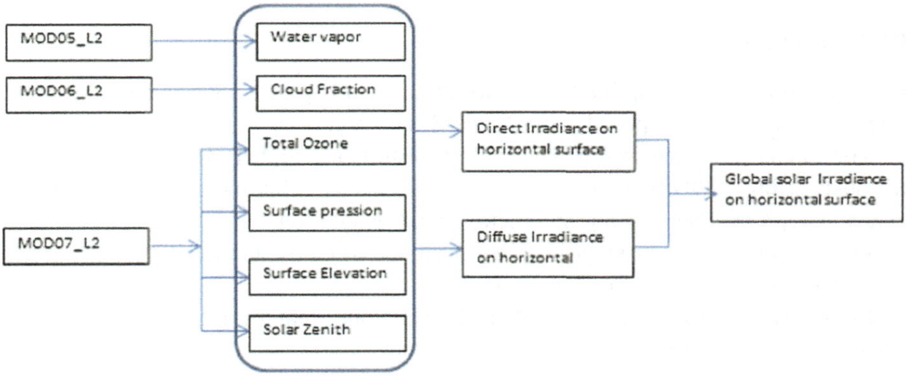

Fig. 1. Estimation of global solar radiation using MODIS products

A time series of instantaneous global solar radiation form 2016 year are estimated at one kilometer resolution. Some pixels presents well value of global soar radiation but others accept extremes values.

The validation results are presented in the scatter plot in (Fig. 2).

To deepen this validation the linear correlation coefficient between estimated and measured values equal 0.7863 so, the correlation between instantaneous solar radiation is generally significant.

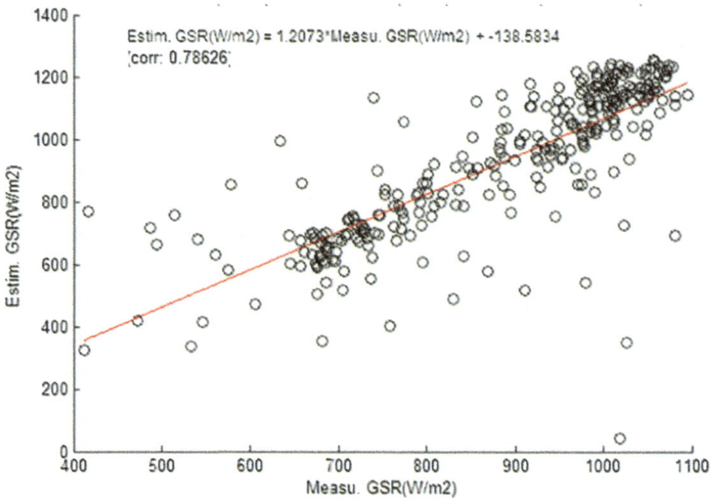

Fig. 2. Correlation between estimated and measured values of global solar radiation in Adrar station year 2016.

3.1 Conclusion

This study proposes a simplified model to estimate daily global solar radiation form the polar satellite images, especially the MODIS data products.

The model proposed in this study allows estimating global solar radiation instantaneously using MODIS Atmospheric Profiles product. This method provides global solar radiation estimates at one kilometre spatial resolution.

Using MODIS data products, offers advantage of freely and available data; which are widely used and continuously improved.

In our study, the model is evaluated using one year of global solar radiation measurements at Adrar site.

The correlation between estimate and measured values of global solar radiation is very significant, but some pixels in different period present extremes values. May be the inputs data require some corrections. The linear correlation coefficient equal 0.78.

The experimental results demonstrate that this model is applicable for estimating daily global solar radiation from polar satellite observations with acceptable performance.

References

1. Journée, M., Bertrand, C.: Remote sensing of environment improving the spatio-temporal distribution of surface solar radiation data by merging ground and satellite measurements. Remote Sens. Environ. **114**(11), 2692–2704 (2010)
2. Raphael, C., Hay, J.E.: An assessment of models which use satellite data to estimate solar irradiance at the earth's surface. J. Clim. Appl. Meteorol. **23**(5), 832–844 (1984)

3. Pinker, R.T., Frouin, R., Li, Z.: A review of satellite methods to derive surface shortwave irradiance. Remote Sens. Environ. **51**(1), 108–124 (1995)
4. Niemelä, S., Räisänen, P., Savijärvi, H.: Comparison of surface radiative flux parameterizations. Atmos. Res. **58**(2), 141–154 (2001)
5. Van Laake, P.E., Sanchez-Azofeifa, G.A.: Mapping PAR using MODIS atmosphere products. Remote Sens. Environ. **94**(4), 554–563 (2005)
6. Wang, H., Pinker, R.T.: Shortwave radiative fluxes from MODIS: model development and implementation. J. Geophys. Res. **114**(D20), D20201 (2009)
7. Porfirio, A.C.S., Ceballos, J.C.: A method for estimating direct normal irradiation from GOES geostationary satellite imagery: validation and application over Northeast Brazil. Sol. Energy **155**, 178–190 (2017)
8. López, G., Javier Batlles, F.: Estimating solar radiation from MODIS Data. Energy Procedia 49, 2362–2369 (2014)
9. Xu, X., et al.: Energy, vol. 111, no. C Pergamon (2016)
10. Wang, D., Liang, S., He, T., Shi, Q.: Estimation of daily surface shortwave net radiation from the combined MODIS data. IEEE Trans. Geosci. Remote Sens. **53**(10), 5519–5529 (2015)
11. Houborg, R., Soegaard, H., Emmerich, W., Moran, S.: Inferences of all-sky solar irradiance using Terra and Aqua MODIS satellite data, vol. 28, no. 20, pp. 4509–4535 (2007)
12. Tang, W., Qin, J., Yang, K., Niu, X., Min, M., Liang, S.: An efficient algorithm for calculating photosynthetically active radiation with MODIS products. Remote Sens. Environ. **194**, 146–154 (2017)
13. Qin, J., Chen, Z., Yang, K., Liang, S., Tang, W.: Estimation of monthly-mean daily global solar radiation based on MODIS and TRMM products. Appl. Energy **88**(7), 2480–2489 (2011)
14. Xu, X., et al.: A method for daily global solar radiation estimation from two instantaneous values using MODIS atmospheric products. Energy **111**(September), 117–125 (2016)
15. Chen, R., Kang, E., Ji, X., Yang, J., Wang, J.: An hourly solar radiation model under actual weather and terrain conditions: a case study in Heihe river basin. Energy **32**(7), 1148–1157 (2007)
16. Status Report: Application of Monte Carlo Techniques to Insolation Characterization and Prediction, July 1979

Integration of Direct Contact Membrane Distillation and Solar Thermal Systems for Production of Purified Water: Dynamic Simulation

A. Remlaoui(✉) and D. Nehari

Smart Structure Laboratory, University Center of Ain-Témouchent, Ain-Témouchent, Algeria
donremlaoui@gmail.com, nehari2746@gmail.com

Abstract. This paper investigates the integration of solar thermal (FPC) energy powering direct contact membrane desalination (DCMD) to produce clean water. The carried out system is modelized and simulated by using the commercial code TRNSYS. Doing this was possible by including a novel component able to simulate the physical behaviour of the DCMD. The simulation of the solar distillation system has been done during the 21st June under the meteorological conditions of Ain Témouchent city (Algeria). The results showed that the present model has a good agreement with the experimental data of the literature. The present desalination system allows getting a daily distillate production around 56 l/d. Furthermore, concerning the performance parameters, it was found that the energy collected, energy delivered by solar coil and to load reach to 953,25 kJ.h^{-1}, 929,43 kJ.h^{-1} and 7542 kJ.h^{-1} respectively. The solar fraction ranged from 0 to1 and the collector efficiencies was assessed 36%.

Keywords: Solar desalination · Direct contact membrane distillation · Flat plate collector · TRNSYS · Solar fraction

1 Introduction

Southern Mediterranean countries are facing a growing water scarcity. There are opportunities to address the problem of water scarcity in rural and remote areas through sustainable saltwater desalination technologies. According to the little scale seawater desalination, the membrane distillation (MD) can be a great option especially in view of the possibility to use the solar thermal and low-grade heat directly as the primary source of energy [1, 2]. Membrane Distillation is a half process that consolidates both thermal and membrane process. The Membrane has a direct contact with a seawater on the feed side and a fluid or vaporous stage on the permeate side. Therefore, can be characterized as a procedure for expelling water vapor from aqueous feed solution heated to a temperature under 100 °C. The transfer force of the process is the difference in partial pressures between two sides of the membrane, which causes evaporation on the feed side [1–4]. The present work focuses on using TRNSYS to analyze a combination of solar energy used to produce simultaneously thermal energy for heating

M. Hatti (Ed.): ICAIRES 2019, LNNS 102, pp. 164–172, 2020.
https://doi.org/10.1007/978-3-030-37207-1_17

seawater and electrical energy for desalinate seawater by FPC and DCMD (Direct Contact Membrane Distillation) for SW (seawater) desalination.

2 System Description

The proposed model for freshwater production from brackish water and thermal energy is a combination of direct Contact Membrane Distillation (DCMD) and flat plate solar collector (FPC) as shown in Fig. 1(a). The basic components like solar collector, storage tank, pumps, controller, heat exchanger, DCMD module.

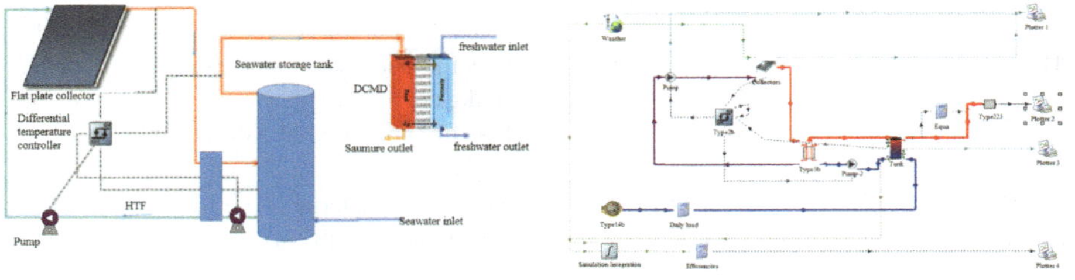

(a) (b)

Fig. 1. (a)Schematic diagram of the FPC-DCMD system and (b) Solar thermal and DCMD systems modeled in TRNSYS 17.

3 TRNSYS Simulation Model

The system described in the previous section was dynamically simulated by transient systems simulation (TRNSYS 17) [5]. Every component model is subroutines ("Type") that exists in the standard library of TRNSYS. TRNSYS software is utilized for analyzing the performance of the system, as the whole system is modeled as shown in Fig. 1(b). Table 1 demonstrates the devices and the corresponding simulation subroutine (Type). Once all the components of the system have been identified and a mathematical description of each component is available, the main components of this model are described and shown in Table 1.

3.1 Solar Collectors and Thermal Storage

Solar thermal collector is simulated using type 1b model. The following equations are the basic equations for the useful collector energy Q_u ($kJ.h^{-1}$) [6, 7] and for the collector efficiency η_{coll} [8, 9]:

$$Q_u = \dot{m}_{fs} C_{p,fs} (T_{outcol} - T_{incol}) \quad (1)$$

$$\eta_{coll} = \frac{Q_u}{A_c G_T} = \eta_0 - a_1 \frac{(T_{av} - T_{amb})}{G_T} - a_2 \frac{(T_{av} - T_{amb})^2}{G_T} \quad (2)$$

Table 1. The components and input design parameters in Transys17 simulation programs

Component	Type	Input parameter	Value
Flat plate solar thermal collector	TYPE 1b	Number in series Collector absorber area (m^2) Fluid specific heat (kj/kgk) Tested flow rate (kg/h) Intercept efficiency First order efficiency coefficient (kj/hm^{2}k) Second order efficiency coefficient (kj/hm^{2}k^2)	2 5 3.708 40 0.8 13 0.05
Pump	TYPE 3b	Rated flow rate (kg/h) Rated power (kj/h)	200 240
Heat exchanger	TYPE 5b	Specific heat of hot side fluid (kj/kgk) Specific heat of cold side fluid (kj/kgk)	3.708 4.19
DCMD	TYPE 223	Membrane material Membrane length (m) Pore size (μm) Contact angle (Degrees) Membrane area (m^2) Membrane Width (m) Specific heat of freshwater (J. kg^{-1} K^{-1}) Specific heat of feedwater (J. kg^{-1} K^{-1}) Feed water speed ($m.s^{-1}$) Salinity ($g_{NaCl}.L_{water}^{-1}$)	PTFE 0.145 1 126 0,0136 0.1 4190 4190 0.4 10
Thermal storage tank	TYPE 4c	Tank volume (m^3) Fluid specific heat (kJ kg^{-1} K^{-1}) Fluid density (kg m^{-3}) Tank loss coefficient (Wm^{-2} K^{-1}) Set point temperature for element (C) Dead band for heating element 1 (C) Maximum heating rate of element 1 (kJ h^{-1})	0.3 4.19 1000 0.6944 60 5 9000
Weather data reading and processing Mains seawater supply profile Online plotter Integrator Controller	TYPE1 5-6TM2 TYPE 14 h TYPE 65c TYPE 24 TYPE 2b	Ain Temouchent Flow rate (kg/h)	

Where T_{incol} and T_{outcol} are the inlet and outlet temperature of the solar fluid (K),$\dot{m}_{fs}$ is the solar fluid mass flow rate (kg.h^{-1}) and $C_{p,fs}$ is the specific heat capacity of solar fluid (kJ.kg^{-1} K^{-1}), G_T is the incident total solar radiation (kJ.h^{-1}.m^{-2}), A_c is the collector gross area (m^2), η_0 is the optical efficiency,a_1 is the global heat loss coefficient (kJ.h^{-1}.m^{-2}.K^{-1}) and a_2 is the temperature dependence of the global heat loss coefficient (kJ.h^{-1}.m^{-2}.K^{-2});T_{amb} Ambient temperature. The thermal storage tank is subjected to thermal stratification and modeled using type 4c.

3.2 Heat Exchanger and Membrane Distillation

Counter flow heat exchanger of Type 91 is used in the model. The simulation run using TRNSYS17 software has been executed via injected a new-programed component; in particular, TYPE 223 is added to the standard library. This new TYPE is committed to a desalination unit DCMD. This component is written by FORTRAN language. The membrane distillation process is governed by different heat and mass transfer mechanisms that occur at both the feed side, the membrane and the permeate side. Mass transfer occurs through the pores of the membrane while heat is transferred through both the membrane and its pores. The mass flux J_w (L/(m^2.h)) of water can be written as a linear function of the vapor pressure difference (P_{mf} and P_{mp} as a function of the temperature on the feed (T_{mf}) and permeate (T_{mp}) at the membrane surface (C)) across the membrane and the membrane mass transfer coefficient B_m (L/(m^{2}h.Pa)) [10], given by:

$$J_w = B_m\left(P_{mf} - P_{mp}\right) \tag{3}$$

The heat transfer involved in DCMD the heat flux can be written as follows [9, 11]:

$$Q = \left(\frac{1}{h_f} + \frac{1}{h_m + \frac{j_w \Delta H_v}{T_{mf} - T_{mp}}} + \frac{1}{h_p}\right)^{-1} \left(T_{bf} - T_{bp}\right) \tag{4}$$

4 The Energy Performance Indices

The energy performance indices evaluated in this study include: energy collected (Q_u), energy delivered by solar coil (Q_d) and delivered to load (Q_{load}), auxiliary energy (Q_{aux}), solar fraction (SF), collector efficiency (η_{coll}) and system efficiency (η_{sys}). The useful energy delivered by the solar coil to the hot water tank is given as:

$$Q_d = \dot{m}_{fs} C_{p,fs}\left(T_{outcoil} - T_{incoil}\right) \tag{5}$$

Where T_{incoil} solar fluid temperature at inlet to the solar coil, $T_{outcoil}$ solar fluid temperature at the outlet from the solar coil (°C).

4.1 Solar Fraction (SF)

Ratio of net utilized solar energy for SSH to total SSH demand including auxiliary energy [6, 7] is in this way:

$$SF = \frac{Q_d}{Q_d + Q_{aux}} \tag{6}$$

4.2 System Efficiency (η_{sys})

The efficiency of the FPC system is calculated as:

$$\eta_{sys} = \frac{Q_d}{A_c G_T} \tag{7}$$

5 Results and Discussion

5.1 Validation of the DCMD Model

In order to validate our numerical predictions based on the new TRNSYS component (DCMD unit); we have studied the system designed by Jianhua Zhang [11]. Figure 2 shows the comparison between the given experimental results [8] and the numerical results obtained from the model used in the present investigation. It can be noticed that the permeate flux predicted by the present model has a good agreement with the experimental data.

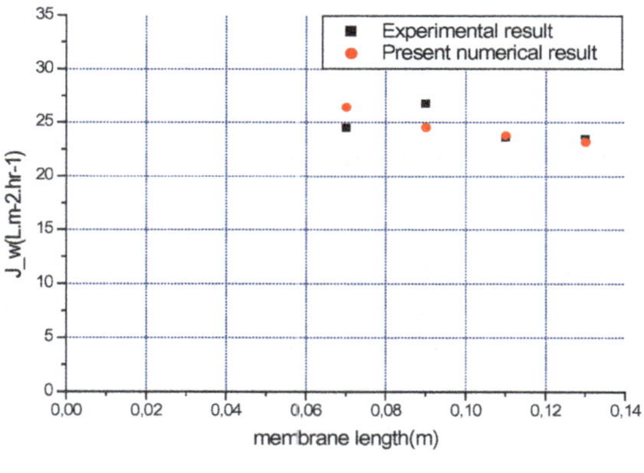

Fig. 2. Validation of the numerical model at (Tcin = 20 °C, Thin = 60 °C, V = 0.4 m.s^{-1}) with Jianhua Zhang [11] experimental result.

5.2 FPC Outlet Temperature

Figure 3(a) shows the plots of the solar fluid temperature in the outlet of FPC. It is seen that the modelled values follow has reached the uttermost value of 140 °C at the outlet of FPC in the tested days.

5.3 Hot Water Outlet Temperature

Figure 3(b) shows the hot water tank outlet fluid temperature for the FPC system measured along the hot water discharge pipe. The peaks show the outlet fluid temperatures during water draw-offs. It is seen that during the clear sky day the tank outlet fluid temperature stayed above 60 °C since energy was added to its content throughout the day thereby raising its temperature.

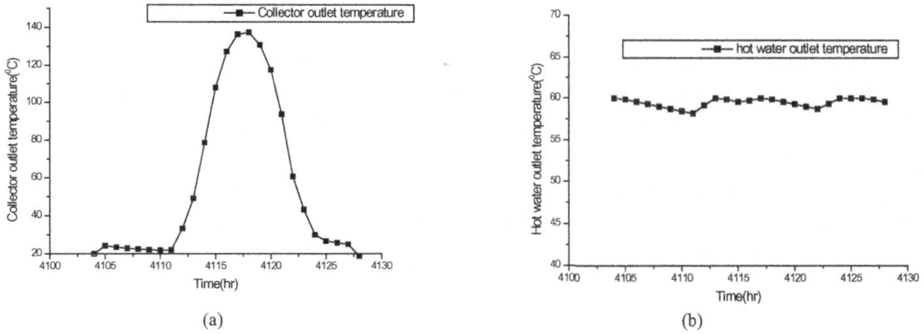

Fig. 3. Modelled (a) collector outlet temperature and (b) hot seawater outlet temperature for FPC system

5.4 The Permeate Flow of the DCMD System

The Fig. 4(a) presents the total mass transfer of the membrane during a summer day (June 21st). At feed side, the seawater inlet temperature is in the range 60–79.22 °C (not shown here), at permeate side, freshwater inlet temperature was about 25 °C and. The maximum production flux of the permeate is 12 L.h^{-1}.

5.5 Energy Collected (Q_u)

Figure 4(b) illustrates the energy supplied by the solar collector (Qu) that is provided to HTF for each hour during 21st June. The results of the simulation indicate during the time interval 8 am to 6 pm that the energy supplied by the solar collector is between 35,47 kJ.h^{-1} and 953,25 kJ.h^{-1}.

5.6 Energy Delivered by Solar Coil (Q_d) and Delivered to Load (Q_{load})

Figure 5(a) shows the total heat transfer rate across heat exchanger Q_d and energy delivered to load Q_{load} for each hour during 21 June. It can be seen in Fig. 5(a) hat the

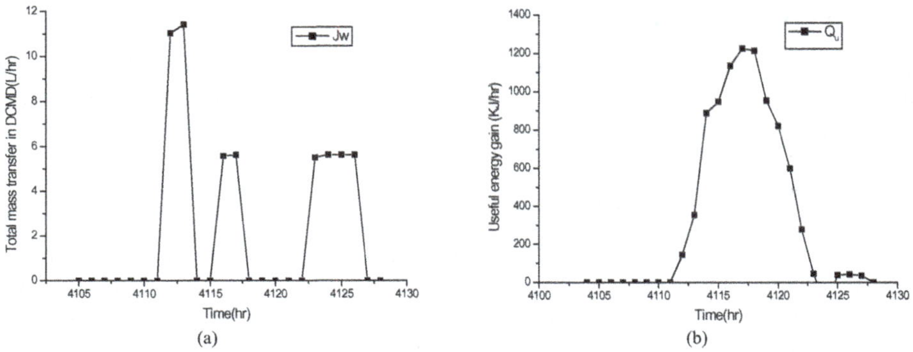

Fig. 4. Modelled (a) permeate flow of the DCMD system and (b) energy collected by FPC collector.

total heat transfer rate across heat exchanger Q_d is ranged between 72,59 kJ.h^{-1} and 929,43 kJ.h^{-1}, this parameter is proportionally to the solar radiation variation. In addition, in the first half daylight between 8 am to 14 pm (4110 h to 4116 h), the Q_d increasing until reach a hight value and this due to the rise of Qu of solar fluid. Whereas, for the second half daylight from the time 4117 h, the Q_d decreases due to diminution of solar radiation and Qu. It is seen in Fig. 5(a) that the FPC system delivered to load is ranged between 3771 kJ.h^{-1} and 7542 kJ.h^{-1}.

5.7 Auxiliary Energy (Q_{aux})

The results of the simulation indicate during the time interval 8 am to 6 pm that the auxiliary heating consumed a highest value of 3859.88 kJ.h^{-1} at the time 4113 h. This means that the high value of the solar radiation leads to minify the intervention of the auxiliary heater to augment the brackish water temperature until 60 °C.

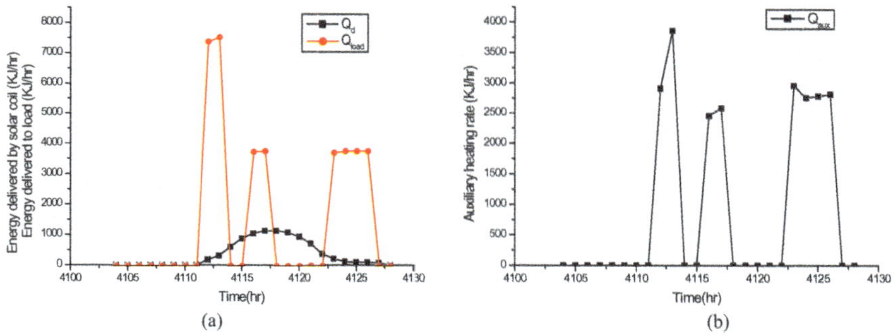

Fig. 5. Modelled (a) the total heat transfer rate across heat exchanger and energy delivered to load and (b) auxiliary energy in 21st June

5.8 Solar Fraction (SF)

According to the Fig. 6(a), the solar fraction SF for the FPC systems is presented. The SF is in the range $0 \leq SF \leq 1$, the solar savings fraction get zero '0' value for no solar energy utilization, and have the value 1 when the energy is provided only via solar way. For intermediate values different from 0 and 1, the pump and auxiliary heater work together.

5.9 Collector Efficiency (η_{coll}) and System Efficiency (η_{sys})

Concerning solar collector thermal efficiency η_{coll}, its maximum value in our case is 0.36 (i.e. 36%). This value is related to one collector with area of 1 m^2. The FPC system efficiency η_{sys} is also confined between 0.23 and 0.35

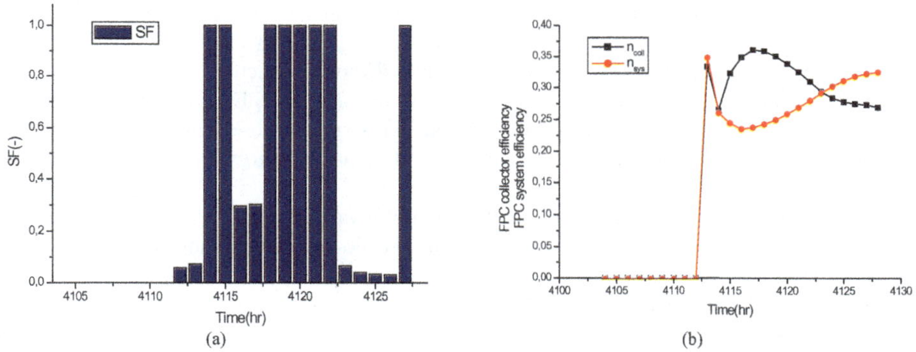

Fig. 6. Modelled (a) Solar fraction and (b) solar collector thermal efficiency and FPC system efficiency for n = 1FPC and S = 1 m^2

6 Conclusion

In this study, a new type of solar-energy-integrated DCMD system for seawater desalination was tested under actual environmental conditions in Ain Témouchent, Algeria. The system was simulated using TRNSYS program. A new TRNSYS component type of DCMD system has been built and added in TRNSYS library as type 223. The type was been included in the scheme with other additional components required for the solar desalination system. The simulation during the 21st June as a typical day in order to evaluate the thermal behaviors of the desalination system. The proposed solar DCMD system has shown favorable potential application in desalination of seawater. The HTF temperatures reached the highest value of 140 °C at the outlet of FPC. Solar thermal driven DCMD system can produce 12 kg/h in June 21st of drinking water. A TRNSYS simulation result was showed that the system was able to generated between 35,47 kJ.h^{-1} and 953,25 kJ.h^{-1} of heat whereas the auxiliary heating consumed a highest value of 3859,88 kJ.h^{-1} to augment the seawater temperature from the 60 °C. The total heat transfer rate across the heat exchanger ranges between 72,59 $kJ.h^{-1}$ and 929,43 $kJ.h^{-1}$. The performance parameters, which are the solar fractions, ranged from 0 to1 and the collector efficiencies was assessed 36%.

References

1. Schwantes, R., Cipollina, A., Gross, F., Koschikowski, J., Pfeifle, D., Rolletschek, M., Subiela, V.: Membrane distillation: solar and waste heat driven demonstration plants for desalination. Desalination **323**, 93–106 (2013)
2. Khayet, M.: Solar desalination by membrane distillation: dispersion in energy consumption analysis and water production costs (a review). Desalination **308**, 89–101 (2013)
3. Bahmanyar, A., Asghari, M., Khoobi, N.: Numerical simulation and theoretical study on simultaneously effects of operating parameters in direct contact membrane distillation. Chem. Eng. Process. **61**, 42–50 (2012)
4. Ashoor, B.B., Mansour, S., Giwa, A., Dufour, V., Hasan, S.W.: Principles and applications of direct contact membrane distillation (DCMD): a comprehensive review. Desalination **398**, 222–246 (2016)
5. TRNSYS: Transient System Simulation tool. http://www.trnsys.com/
6. Vargas-Bautista, J.P., García-Cuéllar, A.J., Pérez-García, S.L., Rivera-Solorio, C.I.: Transient simulation of a solar heating system for a small-scale ethanol-water distillation plant: thermal, environmental and economic performance. Energy Convers. Manag. **134**, 347–360 (2017)
7. Cao, F., Zhao, L., Zhang, F., Guo, L.: Redesign of a water heating system using evacuated tube solar collectors: TRNSYS simulation and techno-economic evaluation. Heat Transf. Eng. **35**, 556–566 (2014)
8. Mohan, G., Kumar, U., Pokhrel, M.K., Martin, A.: A novel solar thermal polygeneration system for sustainable production of cooling, clean water and domestic hot water in United Arab Emirates: Dynamic simulation and economic evaluation. Appl. Energy **167**, 173–188 (2016)
9. Bui, V.A., Vu, L.T.T., Nguyen, M.H.: Modelling the simultaneous heat and mass transfer of direct contact membrane distillation in hollow fibre modules. J. Membr. Sci. **353**, 85–89 (2010)
10. Eleiwi, F., Ghaffour, N., Alsaadi, A.S., Francis, L., Laleg-Kirati, T.M.: Dynamic modeling and experimental validation for direct contact membrane distillation (DCMD) process. Desalination **384**, 1–11 (2016)
11. Zhang, J., Li, J.-D., Gray, S.: Researching and modelling the dependence of MD flux on membrane dimension for scale-up purpose. Desalination Water Treat. **31**, 144–150 (2011)

Numerical Simulation of Shallow Solar Pond Operating Under Open and Closed Cycle Modes to Extract Heat, in the Medea Area, Algeria

Abdelkrim Terfai(✉), Younes Chiba, and Mohamed Nadjib Bouaziz

LBMPT, Mechanical Engineering Department, Faculty of Technology, University Yahia Fares, Medea, Algeria
terfaiabdelkrim@gmail.com, chibayounes@gmail.com, mn_bouaziz@email.com

Abstract. The aim of this work is to do a numerical simulation to study the thermal performance of a shallow solar pond operating under open and closed cycle mode of heat extraction. The pond is equipped with two glass covers and a surface equal to 1 (m^2) (1.3 m * 0.76 m). The bottom of the pond is painted black to increase the absorption of solar radiation. It is also equipped with a serpentine to extract heat through the heat-carrying fluid represented by water. taking into account the climatic and geographical conditions of the region of Medea by date 80/08/2019, and based on the differential equation solutions resulting from the pond heat balance equations, and by using the MATLAB software, a numerical code was developed to perform a numerical simulation To predict the temperature of the various components of the pond, The results obtained showed an increase in the temperature of the pond water and the temperature of the heat transfer fluid during closed mode 10.4173% and 11.5926% respectively, this compared to the open cycle mode.

Keywords: Solar thermal · Shallow solar pond · Numerical simulation · Performance

1 Introduction

The thermal energy provided by the sun can be used to heat water for use in various household and industrial needs, for this purpose we need solar collectors, for example Flat panel solar collectors, evacuated tube solar collectors, Plastic collector In addition to the shallow solar pond, which has been touched by many scientific researches during the last century, A clarification was provided of the use of the shallow solar pond as an effective means of producing electric power using water to collect and store heat, the collected heat is stored during the day inside an insulated water tank at night, this hot water heats the freon 11, which recycles the turbine and an electric generator. A shallow solar pond was built and tested as a portable device used to heat water for daily use such as camping or military sites (Kudish et al. 1978 and Ibrahim et al. 1995). The daily thermal performance of four shallow solar ponds varied among themselves in either the

M. Hatti (Ed.): ICAIRES 2019, LNNS 102, pp. 173–183, 2020.
https://doi.org/10.1007/978-3-030-37207-1_18

transparent upper cover or the glazing angle, and the economic feasibility of these ponds was analyzed as compared to oil, natural gas and electricity (Kudish 1981). A mathematical model was proposed and compared to previous experimental results of a shallow solar pond with a single cover, the inner and outer surfaces were painted and the exterior and lower surfaces were insulated to increase absorption and storage, the theoretical results correspond well to the experimental results. The thermal performance of the shallow solar pond was studied theoretically and practically, the change in the number of transparent covers of the shallow solar pond is from zero to three covers, and it was found that the best performance is the presence of two covers (Kamiuto et al., 1991). The thermal performance of a shallow solar pond operating under constant heat extraction was verified. The pond is equipped with two glass covers and a mirror installed above the pond to increase the intensity of solar radiation. A mathematical model was proposed and a software program was developed and the results obtained were compared with experimental results (Aboul-Enein et al. 2004). Performed a theoretical and experimental simulation of a shallow solar pond operating under the open and closed cycle of heat extraction. It was found that the daily efficiency is 59% and 33% when the pond is operated under closed and open cycle modes, respectively. The objective of this work is to do a numerical simulation to compare the thermal performance of a shallow solar pond operating under the open cycle on the one hand and the closed cycle mode to extract the heat on the other (El-Sebaii et al. 2013). An artificial neural network of a types multilayered feed-forward pack-propagation consists of three layer and ten neurons was created, fed by a set of inputs represented by the experimental data collection including solar radiation, ambient air temperature, pond water temperature, rate of thermal energy collected by heat exchanger, area of the glass cover was used in order to predict energy of the solar radiation, exergy output and exergy efficiency of a shallow solar pond. On the other hand, An artificial neural network of a types multilayered feed-forward pack-propagation consists of one layer and ten neurons was created, fed by a set of inputs represented by the experimental data collection including wind speed, solar radiation, ambient air temperature, inlet temperature of fluid and mass flow rate of the heat transfer fluid was used in order to predict pond water temperature, outlet temperature of the fluid, rate of heat the heat transfer fluid and instantaneous collection efficiency of a shallow solar pond (Terfai et al. 2018a, b)

2 Material and Method

Figure 1 represents the studied system, which is a Solar collector represented in a shallow solar pond, connected to the storage water tank and cold heat transfer tank by means of a heat exchanger in the form of serpentine welded to the bottom of the pond i.e. the absorbent plate, the pond length is 1.3 (m) and the width is 0.76 (m) With an area of 1 (m^2), and the height of the pond is 0.08 (m), the shallow solar pond was covered with two transparent glass that allows the passage of solar radiation and increase the warming of the pond, and also reduce the loss of heat reasonably, the distance between the two glass cover is estimated at 0.02 (m), The depth of the pond water is 0.06 (m), i.e. it can accommodate up to 60 L of water, In order to improve the efficiency of the pond, the bottom of the pond was painted black to increase the absorption of solar radiation. The pond was thermally insulated from the sides and bottom using polyester material. In order to extract the heat collected by the shallow solar pond, the heat-carrying fluid represented by water is pumped by a flow of 0.006 (Kg/S) in a heat exchanger with a length of 7 (m) and a diameter of 0.008 (m). In order to simulate the performance of the shallow solar pond, a computer program was developed using MATLAB based on the solution of energy balance equations for the pond. In this simula under open and closed cycle modes of heat extractiontion, the Bird and Hulstrom model (Mesri-Merad et al. 2012) was used to calculate the solar radiation using the climatic conditions of the city of Medea city in Algeria were used under latitude 36.28° and longitude 2.73°. The simulation was conducted on 08/08/2019 from 5 am to 6 am Next day.

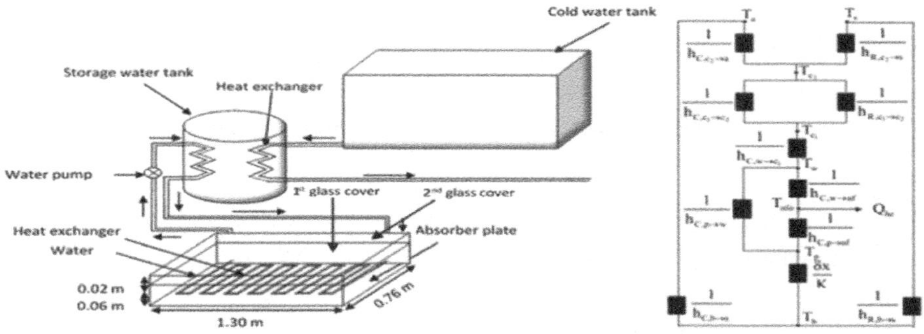

Fig. 1. A diagram showing the dimensions and equivalent electrical diagram of heat transfers in a shallow solar pond with two glass covers

3 Mathematical Model

Assuming that the solar pond is thermally insulated, that the lower glass cover touches the surface of the pond water and there is no thermal gradient of the pond water, the following energy balance equations can be written:

Energy balance equation for upper glass cover (C2):

$$I\alpha_g A_g + h_{cul} A_g (T_{gl} - T_{gu}) + h_{rlu} A_g (T_{gl} - T_{gu}) = h_{cua} A_g (T_{gu} - T_a) + h_{rus} A_g (T_{gu} - T_s) \quad (1)$$

Energy balance equation for lower glass cover (C1):

$$I\tau_g \alpha_g A_g + h_{cwl} A_w (T_w - T_{gl}) = h_{clu} A_g (T_{gl} - T_{gu}) + h_{rlu} A_g (T_{gl} - T_{gu}) \quad (2)$$

Energy balance equation for absorber plate:

$$I\tau_g^2 \tau'_w \alpha_p \left[A_p - \left(\frac{A_{he}}{2}\right)\right] = h_{cpw} \left[A_p - \left(\frac{A_{he}}{2}\right)\right] (T_p - T_w) + U_b A_p (T_p - T_a) \quad (3)$$

Temperature of the upper glass cover:

$$T_{c_2} = \frac{G\alpha_c + \left(h_{C,c_1\to c_2} + h_{R,c_1\to c_2}\right)T_{c_1} + h_{R,c_2\to s}T_s + h_{C,c_2\to a}T_a}{\left(h_{C,c_1\to c_2} + h_{R,c_1\to c_2} + h_{R,c_2\to s} + h_{C,c_2\to a}\right)} \quad (4)$$

Temperature of the lower glass cover:

$$T_{c_1} = \frac{h_{C,w\to c_1}A_wT_w + A_c\left[G\tau_c\alpha_c + T_{c_2}\left(h_{C,c_1\to c_2} + h_{R,c_1\to c_2}\right)\right]}{h_{C,w\to c_1}A_w + A_c\left(h_{C,c_1\to c_2} + h_{R,c_1\to c_2}\right)} \quad (5)$$

Temperature of the absorber plate:

$$T_p = \frac{\left(A_p - \frac{A_{he}}{2}\right)\left(G\tau_c^2\tau_w'\alpha_p + h_{C,p\to w}T_w\right) + U_bA_pT_a - \dot{Q}}{h_{C,p\to w}\left(A_p - \frac{A_{he}}{2}\right) + U_bA_p} \quad (6)$$

Rate of thermal energy transferred from the pond water to the pond HE's fluid.

$$\dot{Q}_{he} = \dot{m}_fC_w(T_{fo} - T_{fi}) \quad (7)$$

Energy balance equation for pond water:

$$\begin{aligned} &G\tau_c^2\alpha_wA_w + h_{C,p\to w}\left(A_p - \frac{A_{he}}{2}\right)(T_p - T_w) \\ &= m_wC_{p_w}\left(\frac{dT_w}{dt}\right) + h_{C,w\to c_1}A_w(T_w - T_{c1}) \\ &+ h_{C,w\to nf}\left(\frac{A_{he}}{2}\right)(T_w - T_{nf}) + U_sA_s(T_w - T_a) \end{aligned} \quad (8)$$

$$M\left(\frac{dT_w}{dt}\right) = f(t) - aT_w \quad (9)$$

$$T_w = \frac{\overline{f(t)}}{a}\left(1 - \exp\left(\frac{-at}{M}\right)\right) + T_{wi}\exp\left(\frac{-at}{M}\right) \quad (10)$$

Energy balance equation for heat exchanger tube welded on the absorber plate:

$$I\tau_g^2\tau_w'\alpha_{he}\left(\frac{P}{2}\right)\Delta x + h_{cwf}P(T_w - T_f)\Delta x = \dot{m}_fC_w\left(\frac{\partial T_f}{\partial_x}\right)\Delta x + A_t\rho_wC_w\left(\frac{\partial T_f}{\partial t}\right)\Delta x \quad (11)$$

$$\left(\frac{\partial T_f}{\partial x}\right) + a_1\left(\frac{\partial T_f}{\partial t}\right) + b_1T_f = f_1(t) \quad (12)$$

$$T_{fo} = T_f(x,t)_{x=L_{he}} = T_{fi} \exp\left[-\left(\frac{b_1}{2}\right)\left(L_{he} + \frac{t}{a_1}\right)\right] + \left(\frac{f_1(t)}{b_1}\right)\left\{1 - \exp\left[-\left(\frac{b_1}{2}\right)\left(L_{he} + \frac{t}{a_1}\right)\right]\right\} \quad (13)$$

Under the closed cycle mode.

The energy balance equation for the storage water tank

$$\dot{m}_{ft}C_w(T_{wt} - T_{ft}) + \dot{Q}_{loss} = \frac{m_{wt}C_w}{\Delta t}(T_{fo} - T_{wt}) \quad (14)$$

The total rate of energy losses from the storage tank

$$Q_{loss} = U_{tank}A_{tank}(T_{wt} - T_a) \quad (15)$$

The energy balance equation for the storage tank heat exchanger tube-heat exchanger's fluid assembly may be written as:

$$P_t h_{cwft}(T_{wt} - T_{ft})\Delta x = A_{tank}\rho_w C_w\left(\frac{\partial T_{ft}}{\partial t}\right)\Delta x + \dot{m}_{ft}C_w\left(\frac{\partial T_{ft}}{\partial x}\right)\Delta x \quad (16)$$

The outlet temperatures of the flowing water are obtained as:

$$T_{fot} = T_{ft}(x,t)_{x=L_{het}} = T_{fit} \exp\left[-\left(\frac{b_2}{2}\right)\left(L_{het} + \frac{t}{a_2}\right)\right] + \left(\frac{f_2(t)}{b_2}\right)\left\{1 - \exp\left[-\left(\frac{b_2}{2}\right)\left(L_{het} + \frac{t}{a_2}\right)\right]\right\} \quad (18)$$

The rate of energy collected

$$\dot{Q}_{coll} = \frac{m_w C_w(T_{fo} - T_{fi})}{\Delta t} \quad (19)$$

4 Results and Discussions

The thermal performance of the Shallow solar pond under open and close cycle mode for heat extraction was theoretically investigated under different configurations of dimensions of the pond. Different internal and external heat transfer coefficients: h_{cpw}, h_{cwl}, h_{cwf}, h_{cpf}, h_{cc1c2}, h_{rc1c2}, h_{cc2a}, h_{rc2s} and h_{cwft} were calculated using the correlations given in the literature (Duffie 2013) and (Kreith 2012) (Table 1).

Figure 2 shows the temporal variation of the coolant temperature for different heat exchanger diameter values. The value of the temperature of the heat transfer fluid reaches its maximum at 15:00 PM at the points; 45.3411 (°C), 45.5981 (°C), 46.0272 (°C), 46.3444 (°C), 46.5808 (°C), 46.6951 (°C), 46.8230 (°C) for different diameter values of heat exchanger 0.004 (m), 0.006 (m), 0.008 (m), 0.01 (m), 0.012 (m), 0.014 (m), 0.016 (m) respectively. The temperature of the heat transfer fluid increases as the diameter of the heat exchanger increases, due to an increase in the heat exchange area.

Table 1. Numerical parameters which were used for numerical calculations

Parameters	Value	Parameters	Value
α_C	0.05	$\rho_a\ (Kg/m^3)$	1.164
α_{he}	0.95	$K_a\ (W/mK)$	0.0251
α_p	0.95	$C_{P_a}\ (J/Kg\ K)$	1012
$\tau_w^{'}$	calculated by (A15)	$\mu_a\ (Kg/ms)$	18.64×10^{-6}
τ_C	0.9	$x_P = x_s\ (m)$	0.04
α_W	$1 - \tau_w^{'}$	$K_P = K_s\ (W/mK)$	0.157
ε_C	0.88	$x_{he}\ (m)$	0.001
$\beta_a\ (K^{-1})$	3.4×10^{-3}	$K_{he}\ (W/mK)$	59
$\sigma\ (W/m^2K^4)$	5.67×10^{-8}	V (m/s)	2

Figure 3 shows the temporal variation of the coolant temperature for different heat exchanger length values. The value of the temperature of the heat transfer fluid reaches its maximum at 15:00 PM at the points; 34.2330 (°C), 37.8365 (°C), 39.9819 (°C), 41.4338 (°C), 42.4557 (°C), 43.1711 (°C), 43.6553 (°C) for different length values of heat exchanger 1 (m), 2 (m), 3 (m), 4 (m),5 (m), 6 (m), 7 (m) respectively. The temperature of the heat transfer fluid increases as the length of the heat exchanger increases, due to an increase in the heat exchange area.

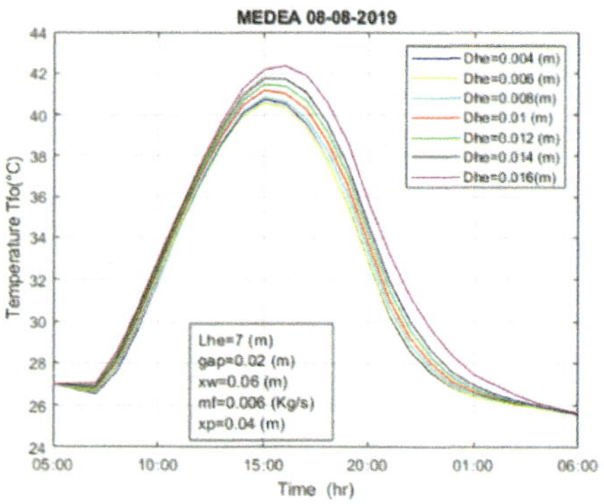

Fig. 2. Evolution of outlet temperature of the heat-carrying fluid as a function of time for different diameter heat exchanger (Dhe) under the open cycle mode of heat extraction.

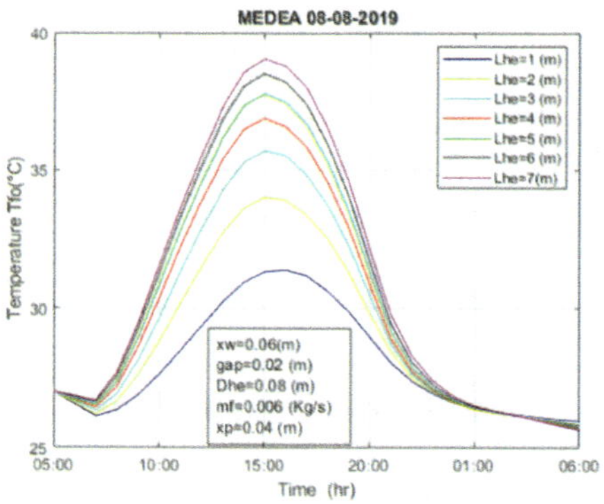

Fig. 3. Evolution of outlet temperature of the heat-carrying fluid as a function of time for different length of heat exchanger (Lhe) under the open cycle mode of heat extraction.

Figure 4 shows the temporal variation of the coolant temperature for different flow of heat transfer fluid values. The value of the temperature of the heat transfer fluid reaches its maximum at 15:00 PM at the points; 56.7071 (°C), 56.1790 (°C), 55.6558 (°C), 52.9014 (°C), 48.6540 (°C), 46.3644 (°C), 43.4622 (°C) for different diameter values of heat exchanger 0.006 (Kg/s), 0.008 (Kg/s), 0.01 (Kg/s), 0.02 (Kg/s),0.03 (Kg/s), 0.04 (Kg/s), 0.05 (Kg/s) respectively.

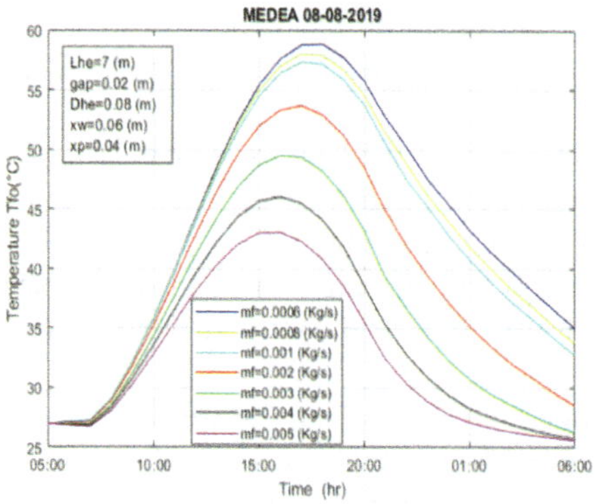

Fig. 4. Evolution of outlet temperature of the heat-carrying fluid as a function of time for different mass flow rate (mf) under the open cycle mode of heat extraction.

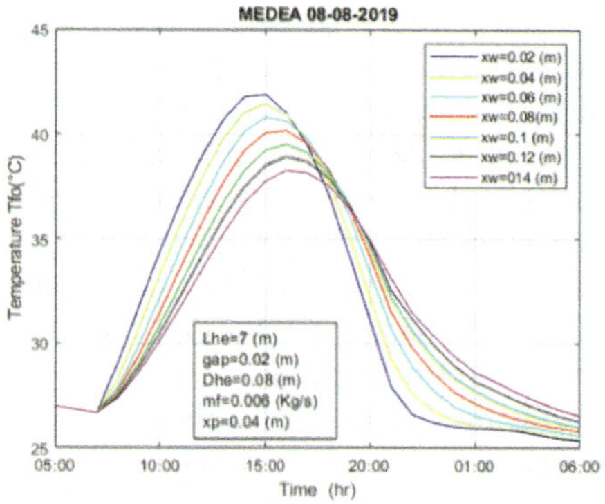

Fig. 5. Evolution of outlet temperature of the heat-carrying fluid as a function of time for different depth of the pond (xw) under the open cycle mode of heat extraction.

The temperature of the heat transfer fluid increases as the flow of heat transfer fluid decreases, due to the increased time needed to contact the heat transfer fluid with hot pond water.

Figure 5 shows the temporal variation of the coolant temperature for different depth of pond values. The value of the temperature of the heat transfer fluid reaches its maximum at 15:00 PM at the points; 48.1163 (°C), 47.3837 (°C), 46.3644 (°C), 45.2479 (°C), 44.1399 (°C), 43.6924 (°C), 42.1263 (°C) for different diameter values of heat exchanger 0.02 (m), 0.04 (m), 0.06 (m), 0.08 (m), 0.1 (m), 0.12 (m), 0.14 (m) respectively.

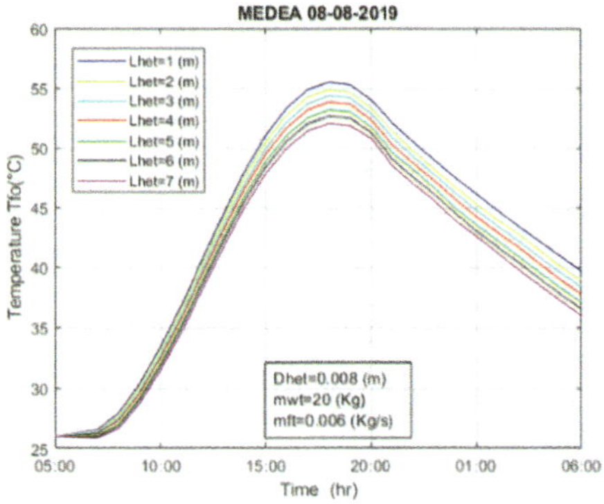

Fig. 6. Evolution of outlet temperature of the storage tank heat-carrying fluid as a function of time for different length of heat exchanger tank (Lhet) under the closed cycle mode of heat extraction.

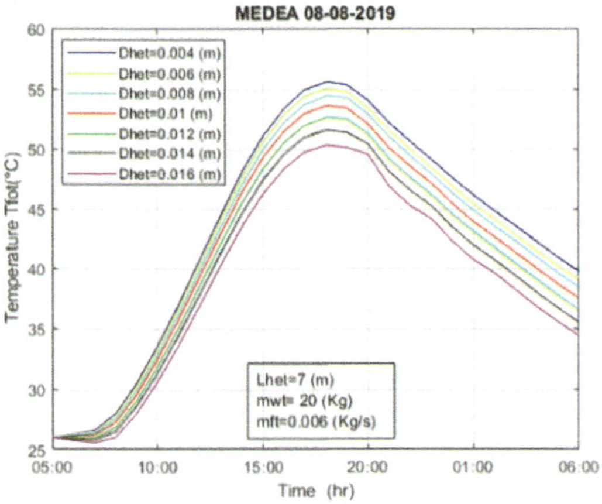

Fig. 7. Evolution of outlet temperature of the storage tank heat-carrying fluid as a function of time for different diameter of heat exchanger tank (Dhet) under the closed cycle mode of heat extraction.

The temperature of the heat transfer fluid decreases as the depth of the pond increases, due to the increase of water in the pond, which takes longer and the amount of heat is greater to heat.

Figure 6 shows the temporal change of temperature of the heat transfer fluid at the outlet of the storage tank values. The temperature of the heat transfer fluid reaches a maximum at 17:00 PM at the points; 54.8418 (°C), 54.1000 (°C), 53.1150 (°C), 51.8853 (°C), 50.4074 (°C), 48.6775 (°C), 46.6933 (°C) for different diameter values of heat exchanger 0.004 (m), 0.006 (m). 0.008 (m), 0.01 (m), 0.012 (m), 0.014 (m) and 0.016 (m) respectively.

Figure 7 shows the temporal change of temperature of the heat transfer fluid at the outlet of the storage tank values. The temperature of the heat transfer fluid reaches a maximum at 17:00 PM at the points; 54.7332 (°C), 53.9003 (°C), 53.0426 (°C), 52.1625 (°C), 51.2618 (°C), 50.3424 (°C), 49.4058 (°C) for different length values of heat exchanger tank 1 (m), 2 (m), 3 (m), 4 (m), 5 (m), 6 (m), 7 (m) respectively.

The temperature of the heat transfer fluid at the outlet of the heat exchanger of the tank decreases as the diameter and length of the heat exchanger increases, due to the increasing surface area of the heat exchanger that contributes to heat loss from the heat transfer fluid.

Figure 8 represents the temporal change of the solar radiation temperature of the heat transfer fluid in the open cycle and the closed cycle, the temperature of the pond water in the open and closed cycle, as well as the temperature of the tank water and heat transfer fluid at the outlet of the tank, the maximum value of solar radiation 940 (W) at midday. The maximum temperature of the pond water is 43, 6843 (°C) and 46, 7903 (°C) respectively. The reservoir water is 41, 3629 (°C). The temperature of the heat transfer fluid at the outlet of the tank is 39, 6060 (°C).

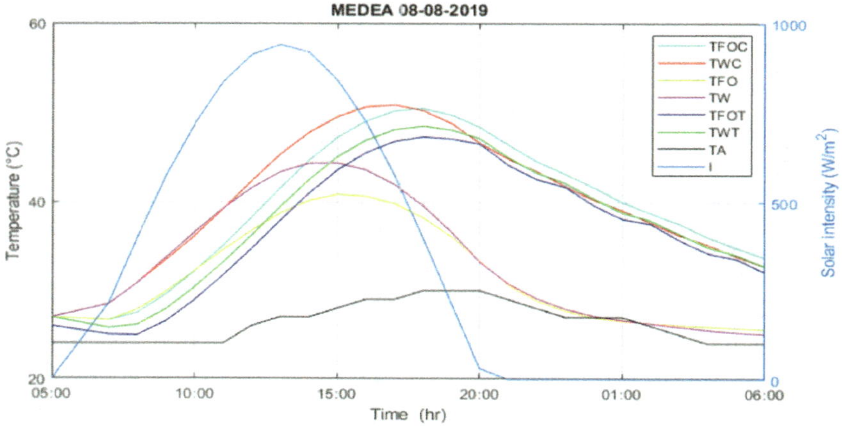

Fig. 8. Hourly variations of solar radiation, temperature of heat transfer fluid, pond temperature, storage tank temperature and outlet temperature of the storage tank in the open and closed cycle mode.

5 Conclusions

The results obtained from the numerical simulation of the shallow solar pond under open and closed flow heating conditions for heat extraction showed that the temperature of the pond elements during the closed cycle is greater than the temperature during the open cycle. The increase in the temperature of the pond water and the temperature of heat transfer fluid during the closed mode will be 10.4173% and 11.5926% respectively, larger than the open mode. Therefore, the pond under closed cycle mode will be more efficient as a source of hot water required for domestic industrial applications and low heat.

Appendix

$$a = h_{cpw}A_p + h_{cwl}A_g + U_sA_s - \frac{h_{cpw}^2 A_p}{h_{cpw} + U_b} - \frac{h_{cwl}^2 A_W^2}{h_{cwl}A_W + h_{clu}A_g + h_{rlu}A_g} \tag{A1}$$

$$\begin{aligned} f(t) = {} & G\tau_g^2 \alpha_w A_w + \frac{G\tau_g^2 \tau'_w \alpha_p h_{cpw}\left[A_p - \left(\frac{A_{he}}{2}\right)\right]}{h_{cpw} + U_b} \\ & + \frac{G\tau_g \alpha_w h_{cwl} A_g A_w}{h_{cwl}A_w + h_{clu}A_g + h_{rlu}A_g} \\ & + T_a\left(U_sA_s + \frac{h_{cpw}U_bA_p}{h_{cpw} + U_b}\right) T_{gu}\left(\frac{h_{cwl}A_gA_w(h_{clu} + h_{rlu})}{h_{cwl}A_w + h_{clu}A_g + h_{rlu}A_g}\right) - \dot{Q}_{he} \end{aligned} \tag{A2}$$

$$a_1 = \left(\frac{A_t \rho_w}{\dot{m}_f}\right) \tag{A3}$$

$$b_1 = \frac{(h_{cwf}P)}{\dot{m}_f C_w} \tag{A4}$$

$$M = m_w C_w \tag{A5}$$

$$f_1(t) = \frac{\left[\left(\frac{P}{2}\right) I \tau_g^2 \tau'_w \alpha_{he} + h_{cwf} T_w P\right]}{\dot{m}_f C_w} \tag{A6}$$

$$a_2 = \frac{A_{tank} \rho_w}{\dot{m}_{ft}} \tag{A7}$$

$$b_2 = \frac{(P_t h_{c \to wft} T_{fot})}{(\dot{m}_{ft} C_W)} \tag{A8}$$

$$f_2(t) \frac{P_t h_{cwf} T_{wt}}{\dot{m}_f C_w} \tag{A9}$$

Nomenclature

A	area (m^2)
a	air
At	Cross-sectional area of the heat exchanger (m^2)
C	Glass cover
Cp	Specific heat (J/kg K)
D	Diameter (m)
I	Solar radiation (W/m^2)
h_C	onvective heat transfer coefficient (W/m^2 K)
he	heat exchanger
h_R	radiation heat transfer coefficient (W/m^2 K)
K	thermal conductivity (W/m K)
L	Length (m)
m	mass flow rate (Kg/S)
N	Nusselt number
P	perimeter (m)
P	plaque absorbent
Q	Rate of thermal energy (W)
Ub	Bottom loss coefficient (W/m^2 K)
US	Sides loss coefficient (W/m^2 K)
T	Temperature (K)
T	Time (S)
V	Wind speed (m/s)
W	Width (m)
α	Absorptivity
β	Temperature coefficient of volume expansion (1/K)
x	Thickness (m)
τ	Transmissivity
σ	Stefan–Boltzmann constant (W/m^2 K^4)
μ	Absolute viscosity (N s/m^2)
ρ	Mass density (kg/m^3)
ε	Emissivity for radiation

References

Aboul-Enein, S., El-Sebaii, A.A., Ramadan, M.R.I., Khallaf, A.M.: Parametric study of a shallow solar-pond under the batch mode of heat extraction. Appl. Energy **78**(2), 159–177 (2004)

Duffie, J.A., Beckman, W.A.: Solar engineering of thermal processes. Wiley, Hoboken (2013)

El-Sebaii, A.A., Aboul-Enein, S., Ramadan, M.R.I., Khallaf, A.M.: Thermal performance of shallow solar pond under open and closed cycle modes of heat extraction. Sol. Energy **95**, 30–41 (2013)

Ibrahim, S.M., El-Reidy, M.K.: Performance of a mobile covered shallow solar pond. Renew. Energy **6**(2), 89–100 (1995)

Kamiuto, K., Oda, T.: Thermal performance of a shallow solar-pond water heater with semitransparent, multilayer surface insulation. Energy **16**(10), 1239–1245 (1991)

Kreith, F., Manglik, R.M., Bohn, M.S.: Principles of Heat Transfer: Cengage Learning (2012)

Kudish, A.I., Wolf, D.: A compact shallow solar pond hot water heater. Sol. Energy **21**(4), 317–322 (1978)

Kudish, A.: Sede boqer shallow pond project. Energy **6**(3), 277–292 (1981)

Mesri-Merad, M., Rougab, I., Cheknane, A., Bachari, N.I.: Estimation du rayonnement solaire au sol par des modèles semi-empiriques. Revue des Energ. Renouvelables **15**(3), 451–463 (2012)

Terfai, A., Chiba, Y., Bouaziz, M.N.: Numerical study of shallow solar pond by using neural networks method. In: 2018 International Conference on Applied Smart Systems (ICASS), pp. 1–4. IEEE (2018a)

Terfai, A., Chiba, Y., Bouaziz, M.N.: Artificial neural networks modeling of a shallow solar pond. In: International Conference in Artificial Intelligence in Renewable Energetic Systems, pp. 491–496. Springer, Cham (2018b)

Super Twisting High Order Sliding Mode Control of Vertical Axis Wind Turbine with Direct Attack Based on Doubly Fed Induction Generators

Lakhdar Saihi[1,2(✉)], Brahim Berbaoui[1], Fateh Ferroudji[1], Youcef Bakou[1], Khaled Koussa[1], Khayra Roummani[1], Farouk Meguelati[1], Abdeldjalil Slimani[1,2], Abd Elhaq Boutera[1,2], and Khaled Toumi[1,2]

[1] Centre de Développement des Energies Renouvelables CDER, Unité de Recherche en Energies Renouvelables en Milieu Saharien URERMS, 01000 Adrar, Algeria
saihi_lakhdar@urerms.dz

[2] Department of Electrical and Computer Engineering, University of Tahri Mohammed Bechar, P.O.B. 417, 08000 Bechar, Algeria

Abstract. In this paper we present a nonlinear control using super twisting high order sliding mode for a vertical axis wind turbine (VAWT) with direct attack based on doubly-fed induction generator (DFIG) supplied by an AC-AC converter. In the first place, we carried out briefly a study of modeling on the whole system (H-darrieus wind turbine). In order to control the power flowing between the stator of the DFIG and the grid, a proposed control design uses high order sliding mode technique is applied for implementing to remove completely the chattering phenomenon on a conventional sliding mode control. The use of this method provides very satisfactory performance for the DFIG control, and the chattering effect is also reduced by this technique. The machine is tested in association with a wind turbine. Simulations results are presented and discussed for the whole system.

Keywords: DFIG · Vertical axis wind turbine (VAWT) · Super twisting high order sliding mode control · Chattering phenomenon

1 Introduction

Wind energy conversion systems are becoming increasingly popular because of the demand on renewable energy resources, consequently wind power generation technique is being developed rapidly, and the wind energy systems using a doubly fed induction generator (DFIG) have some advantages due to variable speed operation [1]. Doubly fed induction generator (DFIG) is one of the most popular wind turbines, which includes an induction generator with slip ring, a partial scale power electronic converter and a common DC-link capacitor. Power electronic converter, which encompasses a back-to-back AC-DC-AC voltage source converter, has two main parts

M. Hatti (Ed.): ICAIRES 2019, LNNS 102, pp. 184–194, 2020.
https://doi.org/10.1007/978-3-030-37207-1_19

grid side converter (GSC) that rectifies grid voltage and rotor side converter (RSC) which feeds rotor circuit. Power converter only processes slip power therefore it's designed in partial scale and just about 30% of generator rated power which makes it attractive from economical point of view [2, 3].

In the context of control, several robust control methods of DFIG appeared; among the Sliding Mode control is one of them. This technique is a relatively new control method for nonlinear systems [2, 3]. It allows sequentially and systematically, by the choice of a Lyapunov function, to determine the system's control law. Its principle is to establish in a constructive manner the control law of the nonlinear system by considering some state variables as virtual drives and develop intermediate control laws [4, 5], however, the SMC has a major inconvenience which is the chattering effect created by the discontinuous part of control. In order to resolve this problem, one way to improve sliding mode controller performance is to high order sliding mode, which can be applied to reduce the chattering phenomenon of the SMC controller. The principal objective of this paper is show the advantages of high order sliding mode for control of DFIG based on vertical axis wind turbine (H-darrieus) with direct attack. The paper is organized as follow: in the second section, VAWT and the model of DFIG are presented. The third section is present to the high order sliding mode control strategy of the DFIG. In Sect. 4 devoted to simulation results and finally conclusions are summarized in the last section.

2 Wind Turbine Modeling

The Darrieus wind turbine is a type of vertical axis wind turbine (VAWT) used to generate electricity from the energy carried in the wind [3, 4] (Fig. 1).

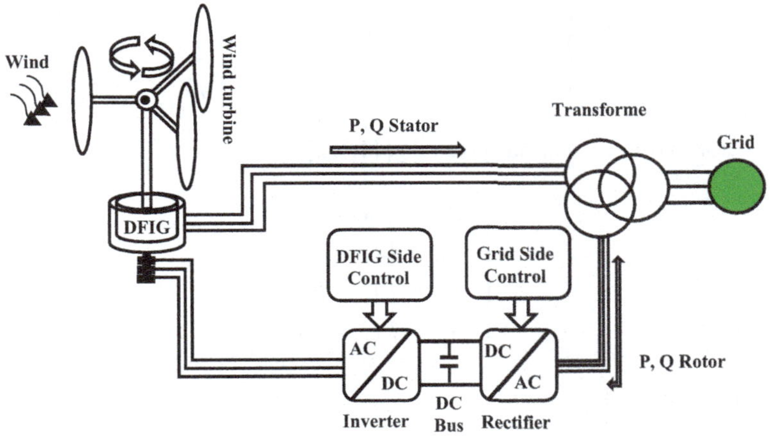

Fig. 1. Topology of DFIG system (VAWT).

The wind system converts the kinetic energy of the wind into electrical energy. This transformation depends on: air density, turbine cross section area and the wind speed. The wind power is given by [2, 7]:

$$P_e = \frac{1}{2}\rho A v^3 C_p(\lambda, \beta) \tag{1}$$

where; ρ (1.225 kg.m^{-3}) is the air density, $A = HD$. The turbine blade sweep area (m^2), v (m.s^{-1}) is the wind speed, $C_p(\lambda, \beta)$ is the power coefficient, λ Speed ratio, β the pitch angle.

The power coefficient is unique for each turbine, it is given as [7, 8]:

$$C_p(\lambda, \beta) = \frac{1}{2}\left(\frac{116}{\lambda_i} - 0.4\beta - 5\right)e^{-\left(\frac{21}{\lambda_i}\right)} \tag{2}$$

Such as the tip speed ratio is defined as [6, 7]:

$$\lambda = \frac{R_e \Omega_t}{v} \tag{3}$$

2.1 Modelling of the DFIG

A classical modeling of the DFIG in the Park reference frame is used. The voltage and flux equations of the DFIG are given as follows [9, 10]:

$$\begin{cases} v_{ds} = R_s i_{ds} + \frac{d\varphi_{ds}}{dt} - \omega_s \varphi_{qs} \\ v_{qs} = R_s i_{qs} + \frac{d\varphi_{qs}}{dt} + \omega_s \varphi_{ds} \\ v_{dr} = R_r i_{dr} + \frac{d\varphi_{dr}}{dt} - (\omega_s - \omega_r)\varphi_{qr} \\ v_{qr} = R_r i_{qr} + \frac{d\varphi_{qr}}{dt} - (\omega_s - \omega_r)\varphi_{dr} \end{cases} \tag{4}$$

The stator and rotor flux are given as [11, 12]:

$$\begin{cases} \varphi_{dr} = l_s i_{ds} + l_m i_{dr} \\ \varphi_{qs} = l_s i_{qs} + l_m i_{qr} \\ \varphi_{dr} = l_r i_{dr} + l_m i_{ds} \\ \varphi_{qr} = l_r i_{qr} + l_m i_{qs} \end{cases} \tag{5}$$

Mechanic equation:

$$\begin{cases} T_{em} = p(\varphi_{ds} i_{qs} - \varphi_{qs} i_{ds}) \\ T_g - T_{em} = J\frac{d\Omega_t}{dt} + f\Omega_t \end{cases} \tag{6}$$

The active and reactive stator power in the Park reference, are written as [13, 14]:

$$\begin{cases} P_s = v_{ds} i_{ds} + v_{qs} i_{qs} \\ Q_s = v_{qs} i_{ds} - v_{ds} i_{qs} \end{cases} \tag{7}$$

In order to eliminate the coupling between the powers we applied the FOC method of stator. The principle of this strategy consists in aligning of the stator vector flux φ_{ds} with the axis d allows getting constant electrical voltages and currents in permanent mode $\varphi_{ds} = \varphi_s$, $\varphi_{qs} = \frac{d\varphi_{qs}}{dt} = 0$ [7, 8]:

The equation of the electromagnetic torque can be simplified as the following form:

$$T_{em} = -p\left(\varphi_s \frac{l_m}{l_s} i_{qr}\right) \tag{8}$$

According to FOC, the equation systems (9) can be simplified as [9, 15, 19]:

$$\begin{cases} P_s = -v_s \dfrac{l_m}{l_s} i_{qr} \\ Q_s = -v_s \dfrac{l_m}{l_s} i_{dr} + \dfrac{v_s^2}{l_s.\omega_s} \end{cases} \tag{9}$$

The DFIG control stator powers is achieved through the control of the DFIG rotor currents. For that, a relationship between the rotor currents and the rotor voltages is established as [17–19]:

$$\begin{cases} v_{dr} = R_r i_{dr} + l_r \sigma \dfrac{di_{dr}}{dt} - \left(g\omega_s l_r \sigma i_{qr}\right) \\ v_{qr} = R_r i_{qr} + l_r \sigma \dfrac{di_{qr}}{dt} + g\omega_s l_r \sigma i_{dr} + g \dfrac{l_m v_s}{l_s} \end{cases} \tag{10}$$

with: $\sigma = \left(1 - \frac{l_m^2}{l_s . l_r}\right)$.

3 Control Strategy of DFIG

3.1 High Order Sliding Mode Control of the DFIG

Sliding mode control (SMC) is one of the most interesting nonlinear control approaches. Nevertheless, a few drawbacks arise in its practical implementation, such as chattering phenomenon and undesirable mechanical effort. In order to reduce the effects of these problems, high order sliding mode seems to be a very attractive solution [11, 12]. This method generalizes the essential sliding mode idea by acting on the higher order time derivatives of the sliding manifold, instead of influencing the first time derivative as it is the case in SMC, therefore reducing chattering and avoiding strong mechanical efforts while preserving SMC advantages [13, 14]. In order to ensure the stator active and reactive powers convergence to their references, a robust high-order sliding mode strategy is used [15].

Super-twisting sliding mode control is a viable alternative to the conventional first order sliding mode control for the systems of relative degree 1 in order to avoid chattering without affecting the tracking performance. The control objective is to drive the system trajectory to reach the sliding manifold $S = \dot{S} = 0$ in finite time [16, 17]. The sliding mode will exist only if the following condition is verified: $S\dot{S} < 0$.

3.2 Switching Function

The switching function applied in the sliding surface, $s(t)$ of first order SMC as shown in Eqs. (11) and (12). This traditional sliding surface, $s(t)$ relates to the tracking error, $e(t)$ and the first time derivative of tracking error $\dot{e}(t)$ [15]. The non linear surface is that of Slotine which is the simplest to apply this surface is written in the following form [12, 17–19]:

$$s(t) = \left(\lambda + \frac{d}{dt}\right)^{n-1} e(t) \tag{11}$$

$$e(t) = y(t) - r(t) \tag{12}$$

where; λ: a positive constant, while is n represents the order of the uncontrolled system. Both $y(t)$ and $r(t)$ indicate the output position and desired position respectively; $e(t)$: error of the variable to be adjusted; n: relative degree, equal to the number of times to derive the output to display the control [18, 19].

3.3 Control Laws

The control laws of ST-SMC consists of three parts, namely; the equivalent control, the continuous state function, and the discontinuous input with integrator as shown in Eqs. (22) and (23). We used a high order sliding mode controller (HOSMC) to control the stator active and reactive power of DFIG [16–19]. The sliding surfaces is given as follows:

$$\begin{cases} e_{P_s} = P_s^{ref} - P_s \\ e_{Qs} = Q_s^{ref} - Q_s \end{cases} \tag{13}$$

The derivative of the sliding surfaces is:

$$\begin{cases} \dot{e}_{P_s} = \dot{P}_s^{ref} - \dot{P}_s \\ \dot{e}_{Q_s} = \dot{Q}_s^{ref} - \dot{Q}_s \end{cases} \tag{14}$$

The sliding surface derivative e_{Ps}, e_{Qs} is calculated by substituting the active and reactive power expression (9) as follows [11, 12]:

$$\begin{cases} \dot{e}_{P_s} = \dot{P}_s^{ref} + v_s \frac{l_m}{l_s} i_{qr} \\ \dot{e}_{Q_s} = \dot{Q}_s^{ref} + v_s \frac{l_m}{l_s} \mathrm{i}_{dr} \end{cases} \tag{15}$$

We have;

$$\begin{cases} i_{dr} = \frac{1}{l_r\sigma}\left(v_{dr} - R_r i_{dr} - g\omega_s l_r \sigma i_{qr}\right) \\ i_{qr} = \frac{1}{l_r\sigma}\left(v_{qr} - R_r i_{qr} - g\omega_s l_r \sigma i_{dr} - g\frac{l_m v_s}{l_s}\right) \end{cases} \tag{16}$$

We replace the expression of the rotor current of the axis q and the axis d described by the Eq. (16) in the preceding equation we obtain:

$$\begin{cases} \dot{e}_{P_s} = \dot{P}_s^{ref} + v_s \frac{l_m}{l_s}\frac{1}{l_r\sigma}\left(v_{qr} - R_r i_{qr} - g\omega_s l_r \sigma i_{dr} - g\frac{l_m v_s}{l_s}\right) \\ \dot{e}_{Q_s} = \dot{Q}_s^{ref} + v_s \frac{l_m}{l_s}\frac{1}{l_r\sigma}\left(v_{dr} - R_r i_{dr} - g\omega_s l_r \sigma i_{qr}\right) \end{cases} \tag{17}$$

Derivative of Eq. (17) can get:

$$\begin{cases} \ddot{e}_{P_s} = \ddot{P}_s^{ref} + \frac{l_m v_s}{l_s l_r \sigma}\left(-R_r i_{qr} - g\omega_s l_r \sigma i_{dr}\right) + \left(\frac{l_m v_s}{l_s l_r \sigma}\right)\dot{v}_{qr} \\ \ddot{e}_{Q_s} = \ddot{Q}_s^{ref} + \frac{l_m v_s}{l_s l_r \sigma}\left(-R_r i_{dr} - g\omega_s l_r \sigma i_{qr}\right) + \left(\frac{l_m v_s}{l_s l_r \sigma}\right)\dot{v}_{dr} \end{cases} \tag{18}$$

There are several techniques of algorithms generating the convergence of S and $\dot{S}$ to zero. The most used in literature are twisting and super-twisting [15].

3.4 Super-Twisting Algorithm

This algorithm has been developed for the control of systems with relative degree equal to one with respect to the sliding surface. This law of command was proposed by Emelyanov in 1990. It was studied by Levant in [5]. Super-twisting does not use information about S this can be seen as a benefit. It is composed of two parts, a discontinuous part u_2 and a continuous part u_1 [16, 17]. The super twisting algorithm has been developed and analyzed for systems that can be written in the form (19) and satisfy the conditions given by the relations (20). The trajectories generated on the super-twisting algorithm are represented in the phase plan of the sliding variables shown in Fig. 2. The super twisting algorithm converges in a finite time [15, 16].

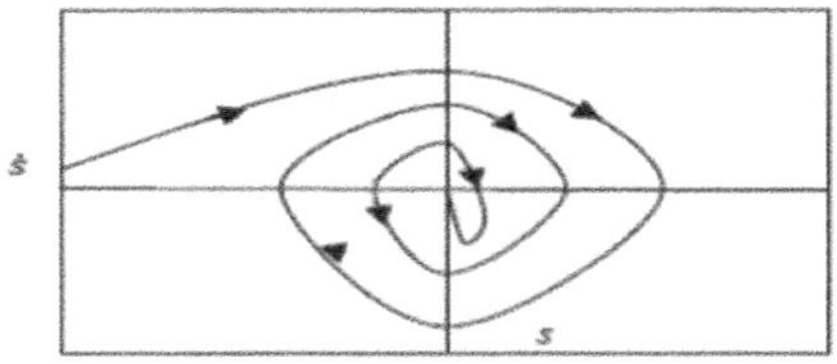

Fig. 2. Convergence of the Super Twisting algorithm in the plane.

In the case of a HOSM control, the control appears explicitly in $\ddot{S}$ can be written in the following form [16, 17]:

$$\ddot{S} = \rho(x,t) + \varphi(x,t)v \tag{19}$$

where; $\rho(x,t); \varphi(x,t)$ are uncertain functions; $\rho(x,t) = \frac{\delta}{\delta u}\dot{S}(x,t,u); \varphi(x,t) = \frac{\delta}{\delta t}\dot{S}(x, t,u) + \frac{\delta}{\delta x}\dot{S}(x,t,u)[A(x)x + B(x)u]$; v represents the variable that defines the vector u of the system, given by (19).

S is the sliding surface variable chosen to ensure the convergence in finite time to the n-order sliding set.

$$\begin{cases} 0 < K_m \leq |\varphi| \leq K_M \\ |\rho| \leq C_0 > 0 \end{cases} \tag{20}$$

where; K_m, K_M, C_0 are positive constants.

The proposed algorithm is described as follows [15]:

The super-twisting control $u(t)$ can be described as [12, 14]:

$$u(t) = u_1(t) + u_2(t) \tag{21}$$

$$\dot{u}_1 = \begin{cases} -u & |u| > 1 \\ -w.sign(S) & |u| \leq 1 \end{cases} \tag{22}$$

$$u_2 = \begin{cases} -\alpha|S_0|^{\rho}.sign(S) & |S| > S_0 \\ -\alpha|S|^{\rho}.sign(S) & |S| \leq S_0 \end{cases} \tag{23}$$

The sufficient condition to generate the convergence in finite time is:

$$\begin{cases} w > \dfrac{C_0}{K_m} \\ \alpha^2 \geq \dfrac{4C_0K_M(w + C_0)}{K_m^2K_m(w - C_0)} \\ 0 < \rho \leq 0.5 \end{cases} \tag{24}$$

For $\rho = 1$ the algorithm converges to the origin exponentially.
We can written $\dot{e}_{P_s}$, $\dot{e}_{Q_s}$ such as the form [15, 16]:

$$\dot{e} = G + \left(\frac{l_m v_s}{l_s l_r \sigma}\right) v_{qr} \tag{25}$$

$$\begin{cases} \dot{e}_{P_s} = G_{P_s} + \left(\frac{l_m v_s}{l_s l_r \sigma}\right) v_{qr} \\ \dot{e}_{Q_s} = G_{Q_s} + \left(\frac{l_m v_s}{l_s l_r \sigma}\right) v_{dr} \end{cases} \tag{26}$$

We define the functions G_{P_s} and G_{Q_s} as follows;

$$\begin{cases} G_{P_s} = \dot{P}_s^{ref} + \frac{l_m v_s}{l_s l_r \sigma}\left(-R_r i_{qr} - g\omega_s l_r \sigma i_{dr} - g\frac{l_m v_s}{l_s}\right) \\ G_{Q_s} = \dot{Q}_s^{ref} + \frac{l_m v_s}{l_s l_r \sigma}\left(-R_r i_{dr} - g\omega_s l_r \sigma i_{qr}\right) \end{cases} \tag{27}$$

The second derivative is given by:

$$\begin{cases} \ddot{e}_{P_s} = \dot{G}_{P_s} + \left(\frac{l_m v_s}{l_s l_r \sigma}\right) \dot{v}_{qr} \\ \ddot{e}_{Q_s} = \dot{G}_{Q_s} + \left(\frac{l_m v_s}{l_s l_r \sigma}\right) \dot{v}_{dr} \end{cases} \tag{28}$$

with;

$$\begin{cases} \dot{G}_{P_s} = \ddot{P}_s^{ref} + \frac{l_m v_s}{l_s l_r \sigma}\left(-R_r i_{qr} - g\omega_s l_r \sigma i_{dr}\right) \\ \dot{G}_{Q_s} = \ddot{Q}_s^{ref} + \frac{l_m v_s}{l_s l_r \sigma}\left(-R_r i_{dr} - g\omega_s l_r \sigma i_{qr}\right) \end{cases} \tag{29}$$

$\ddot{e}_{P_s}$, $\ddot{e}_{Q_s}$ written the same the form of (19) to appear the control vector u, then the super twisting algorithm can be applied. When we identify the previous relation by the relation (29), We apply the super-Twisting theorem, the second-order gliding mode control algorithm of the active and reactive stator power of the DFIG is given by:

$$\begin{cases} V_{qr} = u_{P_s} + \alpha_{P_s} |S(P_s)|^{\rho} sign(S(P_s)) \\ u_{P_s} = w_{P_s} sign(S(P_s)) \end{cases} \tag{30}$$

$$\begin{cases} V_{dr} = u_{Q_s} + \alpha_{Q_s} |S(Q_s)|^{\rho} sign(S(Q_s)) \\ u_{Q_s} = w_{Q_s} sign(S(Q_s)) \end{cases} \tag{31}$$

where; $w_{P_s} > \frac{C_{P_s}}{K_{P_{sm}}}$; $\alpha_{P_s}^2 \geq \frac{4C_{P_s}K_{P_sM}(w_{P_s}+C_{P_s})}{K_{P_{sm}}^2 K_{P_{sm}}(w_{P_s}+C_{P_s})}$; $0 < \rho \leq 1$; $w_{Q_s} > \frac{C_{Q_s}}{K_{Q_{sm}}}$; $\alpha_{Q_s}^2 \geq \frac{4C_{Q_s}K_{Q_sM}(w_{Q_s}+C_{Q_s})}{K_{Q_{sm}}^2 K_{Q_{sm}}(w_{Q_s}+C_{Q_s})}$; $0 < \rho \leq 1$

4 Simulation Results

Figure 3 is presented wind speed on (m/s) [±8 m/s] and Fig. 4 present the behavior of the stator current and electromagnetic torque of DFIG respectively. All these variables depend on the wind speed variation. Figure 5 illustrate the robust super twisting high order sliding mode controller for stator active and reactive powers and there references. Results simulations with super twisting high order sliding mode method they proved very interesting performances in terms of reference tracking (time response, overshoot), sensitivity to perturbation.

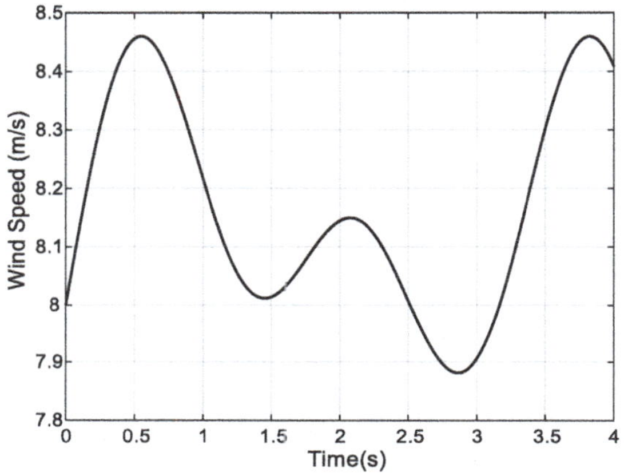

Fig. 3. Wind speed and rotor speed of DFIG.

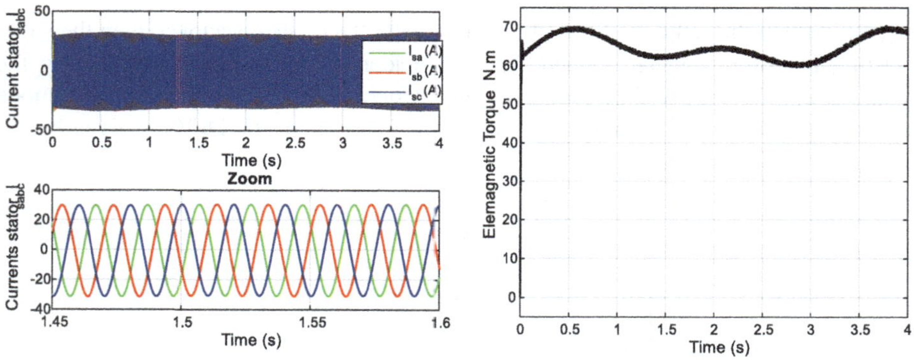

Fig. 4. Current stator and electromagnetic torque of DFIG.

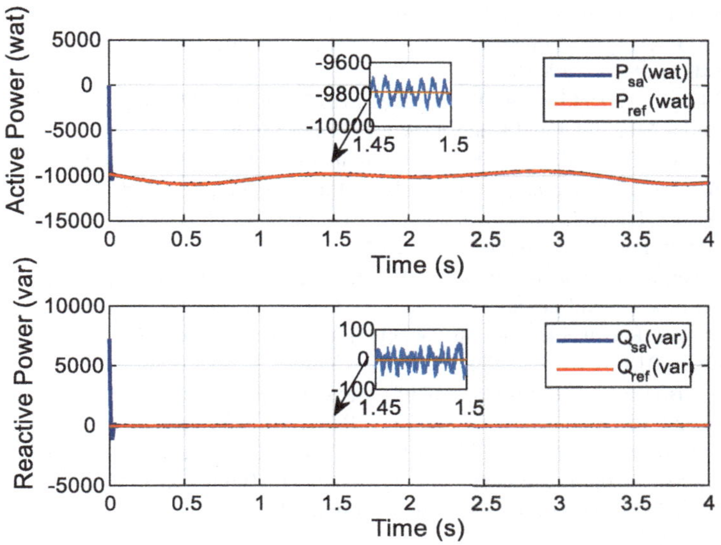

Fig. 5. Trajectory tracking of active reactive power.

5 Conclusions

The modeling, the control and the simulation of an vertical axis wind turbine (H-darrieus) with direct attack based on a doubly fed induction generator (DFIG) connected directly to the grid by the stator and fed by converter on the rotor side has been presented in this paper. Our objective was the implementation of a super twisting high order sliding mode method of active and reactive powers generated by the stator side of the DFIG, in order to ensure of the high performance and a better execution of the DFIG. ST-HOSMC is used to remove the chattering. Compared with the conventional sliding mode controller, the ST-HOSMC system results in robust control performance without chattering. The chattering free improved performance of the FSMC makes it superior to conventional SMC, and establishes its suitability for the system drive and with excellent ability of changes in the wind speed and better quality of the generated power when the speed is varying.

References

1. Saihi, L., Bouhenna, A., Chenafa, M., Mansouri, A.: A robust sensorless SMC of PMSM based on sliding mode observer and extended Kalman filter. In: 2015 4th International Conference on Electrical Engineering (ICEE), pp. 1–4. IEEE, December 2015
2. Bakou, Y., et al.: Design of robust control based on R^{∞} approach of DFIG for wind energy system. In: 2019 1st Global Power, Energy and Communication Conference (GPECOM), pp. 337–341. IEEE, June 2019

3. Ferroudji, F., Khelifi, C., Outtas, T.: Structural dynamics analysis of three-dimensional bi-axial sun-tracking system structure determined by numerical modal analysis. J. Sol. Energy Eng. **140**(3), 031004 (2018)
4. Anaya-Lara, O., Jenkins, N., Ekanayake, J.B., Cartwright, P., Hughes, M.: Wind Energy Generation: Modelling and Control. Wiley, Hoboken (2011)
5. Ardjoun, S.A.E.M., Abid, M.: Fuzzy sliding mode control applied to a doubly fed induction generator for wind turbines. Turk. J. Electr. Eng. Comput. Sci. **23**(6), 1673–1686 (2015)
6. Belounis, O., Labar, H.: Fuzzy sliding mode controller of DFIG for wind energy conversion. Int. J. Intell. Eng. Syst. **10**(3), 163–172 (2017)
7. Roummani, K., Koussa, K., Ferroudji, F., Meguellati, F., Bakkou, Y., Saihi, L.: A new study of direct-driven wind energy conversion system under variable wind speed. In: 2018 6th International Renewable and Sustainable Energy Conference (IRSEC), pp. 1–6. IEEE, December 2018
8. Saihi, L., Boutera, A.: Robust control of a variable-speed wind turbine with fixed pitch angle and strategy MPPT control associated on a PMSG. In: 2016 8th International Conference on Modelling, Identification and Control (ICMIC), pp. 326–331. IEEE, November 2016
9. Adjoudj, M., Abid, M., Aissaoui, A. G., Ramdani, Y., Bounoua, H.: Sliding mode control of a doubly fed induction generator for wind turbines. Rev. Roum. Sci. Techn.–Électrotechn. et Énerg **56**(1), 15–24 (2011)
10. Tria, F.Z., Srairi, K., Benchouia, M.T., Mahdad, B., Benbouzid, M.E.H.: An hybrid control based on fuzzy logic and a second order sliding mode for MPPT in wind energy conversion systems. Int. J. Electr. Eng. Inform. **8**(4), 711–726 (2016)
11. Belgacem, K., Mezouar, A., Massoum, A.: Sliding mode control of a doubly-fed induction generator for wind energy conversion. Int. J. Energy Eng. **3**(1), 30–36 (2013)
12. Zhu, R., Chen, Z., Wu, X., Liu, H.: High order sliding mode control of doubly-fed induction generator under unbalanced grid faults. In: IECON 2013–39th Annual Conference of the IEEE Industrial Electronics Society, pp. 1662–1667. IEEE, November 2013
13. Saihi, L., Boutera, A.: Robust sensorless sliding mode control of PMSM with MRAS and Luenberger extended observer. In: 2016 8th International Conference on Modelling, Identification and Control (ICMIC), pp. 174–179. IEEE, November 2016
14. Medjebouri, A., Mehennaoui, L.: Mode Glissant du Second Ordre Appliqué pour l'Asservissement de Position d'une Articulation Robotique Rigide **55**(3), 145–149 (2014)
15. Heng, C.T., Jamaludin, Z., Hashim, A.Y.B., Abdullah, L., Rafan, N.A.: Design of super twisting algorithm for chattering suppression in machine tools. Int. J. Control Autom. Syst. **15**(3), 1259–1266 (2017)
16. Bouyekni, A., Taleb, R., Boudjema, Z., Moumna, M.: Commande par mode glissant d'ordre 2 pour une capture maximale d'énergie d'une turbine éolienne. Revue des Energies Renouvelables **21**(1), 19–26 (2018)
17. Bounadja, E., Djahbar, A., Boudjema, Z.: Variable structure control of a doubly fed induction generator for wind energy conversion systems. Energy Procedia **50**, 999–1007 (2014)
18. Saihi, L., et al.: Hybrid control based on sliding mode fuzzy of DFIG power associated WECS. In: AIP Conference Proceedings, vol. 2123, no. 1, p. 030015. AIP Publishing (2019)
19. Saihi, L., Berbaoui, B., Glaoui, H., Djilali, L., Abdeldjalil, S.: Robust sliding mode H∞ controller of DFIG based on variable speed wind energy conversion system. Period. Polytech. Electr. Eng. Comput. Sci. (2019)

Estimation of Solar Power Output Using ANN Model: A Case Study of a 20-MW Solar PV Plan at Adrar, Algeria

K. Bouchouicha[1(✉)], N. Bailek[2], M. Bellaoui[1], and B. Oulimar[1]

[1] Unité de Recherche en Energies renouvelables en Milieu Saharien, UERMS, Centre de Développement des Energies Renouvelables, CDER, 01000 Adrar, Algeria
k.bouchouicha@gmail.com

[2] Materials and Energy Research Group, SERL, Department of Matter Sciences, Faculty of Sciences and Technology, University Center of Tamanrasset, 10034 Tamanrasset, Algeria

Abstract. In this paper, the Artificial Neural Networks (ANN) model have been exploited for estimation of the energy production of the 20-MW solar photovoltaic (PV) plant installed in Adrar in terms of energy yield, using a minimal knowledge of meteorological parameter include global solar radiation, relative humidity, wind speed, and ambient or air temperature. A comparative analysis has been carried out between the proposed artificial neural networks model and multiple linear regression (MLR) models in PV output estimation and performance is measured based on evaluation indexes, namely mean absolute percentage error, normalized mean absolute error, and normalized root mean square error respectively. Further, the comparison of the ANN model with MLR based models reveals the advantages and supremacy of the ANN model.

Keywords: ANN model · Multiple linear regression models · PV plant · Energy production estimation

1 Introduction

Several photovoltaic solar power plant projects have been implemented, and others are underway in Algeria as part of the national program renewable energies, which is expected to meet a large percentage of national electricity demand. The objective of this strategy is to produce a 40% of the electricity needs in 2030, of which about 22,000 MW [1, 2]. Seven solar power plants totaling a 53 MW capacity are also underway in the region of Adrar, to cover In Salah, Adrar and Timimoun. One of these stations is directly located in the Adrar city with a nominal power of 20000 kWc, equipped with 81,840 YL245P-29b PV modules. This station is locally monitored and supervised by SKTM (Sharikat Kahraba wa Takat Moutajadida) the representative of the electricity and renewable energy company, Algerian National Society for Electricity and Gas (SONELGAZ) group's subsidiary.

Numerous factors influence the generation of power in a solar photovoltaic system namely solar radiation, weather conditions (ambient temperature, wind) and location

M. Hatti (Ed.): ICAIRES 2019, LNNS 102, pp. 195–203, 2020.
https://doi.org/10.1007/978-3-030-37207-1_20

(altitude, installation angle,…). Because of the inconsistency in solar radiation and factors affecting environment such as meteorological parameters, the generation of power in a solar photovoltaic system is a stochastic process, which not only affects the stability of the solar system but the working capital and maintenance costs as well [3]. So, to advance the solar photovoltaic system stability, accurately estimation output of a photovoltaic solar installation is essential, especially for the large systems integrate into the electric grid. Which help producers to improve and to implement the economic strategies required [4]. This estimation can help manufacturer's device some operational strategies or policies in a way that can achieve better management.

Further, several previous types of research have worked towards the accuracy in estimating the power of a solar photovoltaic system, such as artificial neural networks (ANN), support vector machines (SVM), statistical methods [5–8]. ANN techniques for estimating irradiation have been shown to have greater accuracy than other techniques such as linear, nonlinear and fuzzy approaches by Graditi et al. [6, 7].

The goal of this paper is the estimation of the energy production of the 20-MW solar PV plant installed in Adrar in terms of energy yield, using a minimal knowledge of meteorological parameter include global solar radiation, relative humidity, wind speed, and ambient or air temperature. In this context, an artificial neural networks model have been developed using MATLAB. Meteorological parameters are prepared and simulations have been carried out by selecting the model based on evaluation indexes. The results obtained can be further exploited in estimating solar photovoltaic system power that employs the same photovoltaic module operated in the same climate zone;

2 Data and Method

2.1 Data

In this work, the 2 years of the data recorded have been obtained from the electricity and renewable energy company SKTM (Sharikat Kahraba wa Takat Moutajadida) a Sonelgaz (Algerian National Society for Electricity and Gas) group's subsidiary [9], the experimental validation is carried out by using real 15-minutes of Meteorological parameters include ambient or air temperature, relative humidity, wind speed, and air pressure. In addition of Global Horizontal Irradiance (GHI) and DC output powers, recorded from 01/01/2017 to 31/12/2018 of PV plant located in Saharian arid climate (Adrar, Algeria), with a nominal power of 20000 kWc, equipped with 81,840 YL245P-29b PV modules, the electrical specifications of the PV module are presented in Table 1. In Fig. 1, the solar energy potential map of the city of Adrar is presented [10]. As can be seen in the figure, this region is rich in solar potential. The site of measure is located at 27°55 N latitude and 0°19 W longitudes, the specifics information of the site are given in Table 2.

Table 1. General characteristics of the PV module at AM1.5G, 1000 W/m2 157 and 25°C.

PV Module	YL245P-29b PV Module
Cell technology	Polycristalline
Maximum Power P_{max} (W)	245
Short-circuit current I_{sc} (A)	8.63
Open-circuit voltage V_{oc} (V)	37.8
Current of Maximum power I_{PMax} (A)	8.11
Voltage of maximum power U_{PMax} (V)	30.2
Module area (mm)	1650 × 990

Table 2. Geographic and data records period of the studied station.

Station	Adrar
Latitude (°N)	27.91
Longitude (°E)	−0.32
Data series period	2017–2018
Mean Temperature (°C)	29.99
Mean Relative humidity (%)	20.32
Mean wind speed (m/s)	5.54
Mean Air pressure (hPa)	981.78
Mean GHI (W/m2)	609.46
Mean DC output powers (MWatt)	9.09

2.2 Model

This section describes the proposed methodology for estimating the output PV production of 20-MW solar PV, using multiple linear regression and the neural network model, The first set consists of regression modelling, the regression technique is briefly described in Massidda and Marrocu [8].

ANN model is designed in such a way that the variables at the output are calculated from variables at the input side by the composition of basic connections and functions [11, 12]. A general mathematic representation of an individual neuron within an ANN architecture is shown by Eq. (1) and The general configuration of multilayer artificial neural network of current research is depicted in Fig. 1.

$$y = \sum\nolimits_{j=1}^{n} w_{ij}x_{ij} + \theta_i \tag{1}$$

In this equation, the inputs x_{ij} to the neuron are multiplied by weights w_{ij} and summed up together with the constant and i neuron bias term θ_i, w_{ij} is the connection directed from j neuron to i neuron (at the hidden layer) (Fig. 2).

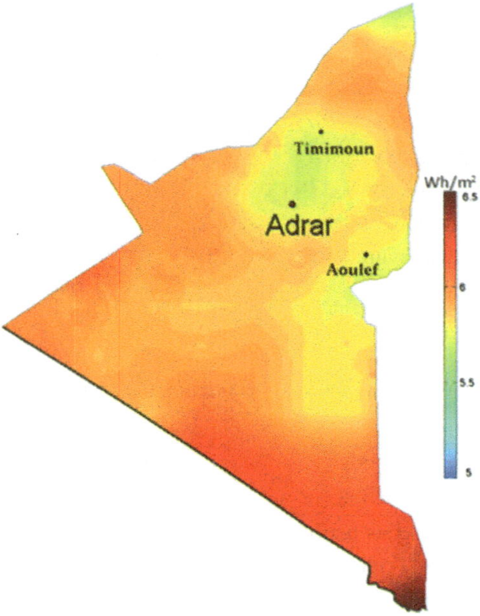

Fig. 1. The Annual average of the daily GHI, Adrar [10].

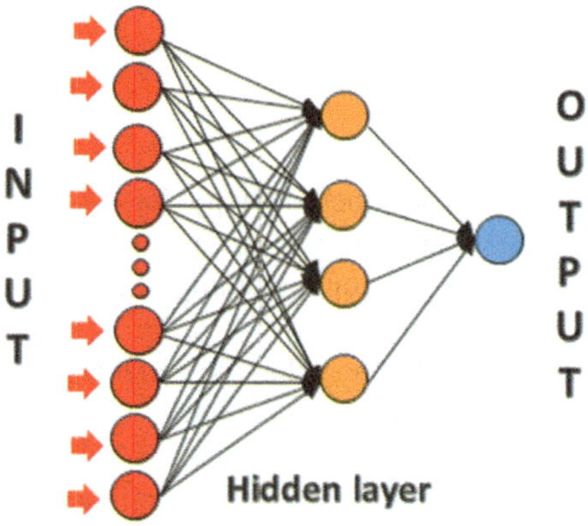

Fig. 2. ANN general configuration for prediction of DC output powers.

In this study, the ANN is a network of interconnected neurons arranged in multiple layers, including input layer of five inputs, one output layer. The ANN model have been designed and simulated with five input parameters which include ambient temperature, relative humidity, wind speed, air pressure and Global Horizontal Irradiance to forecast DC output powers. Levenberg-Marquardt training algorithms have been used whose description can be found in the MATLAB. The following can be briefly outlined for ANN Model.

The training and testing data used for the ANN proposed model are given in Table 2. Among the collected data, 80% of data was used for training. Testing was done for 20% of data (Table 3).

Table 3. The ANN developed model to predict DC output powers data.

Input		Output	
Tair	Air temperature	P_{dc_out}	DC output powers
Rh	Relative humidity		
Pair	Air pressure		
Ws	Wind speed		
GHII	Global Horizontal Irradiance		

Evaluation Indexes

For validation of the models, evaluation indexes have been used and they are defined as follows [13–15]:

$$MBE = \frac{1}{N}\sum_{i=1}^{N}\left(X_{Sim,i} - X_{Mes,i}\right) \tag{2}$$

$$MAE = \frac{1}{N}\sum_{i=1}^{N}\left|X_{Sim,i} - X_{Mes,i}\right| \tag{3}$$

$$RMSE = \sqrt{\frac{\sum_{i=1}^{N}\left(X_{Sim,i} - X_{Meas,i}\right)^2}{N}} \tag{4}$$

$$R = \frac{\sum_{i=1}^{N}\left(X_{Sim,i} - \overline{X}_{Sim}\right)\left(X_{Meas,i} - \overline{X}_{Meas}\right)}{\sqrt{\sum_{i=1}^{N}\left(X_{Sim,i} - \overline{X}_{Sim}\right)^2} \times \sqrt{\sum_{i=1}^{N}\left(X_{Meas,i} - \overline{X}_{Meas}\right)^2}} \tag{5}$$

mean bias error (MBE), mean absolute error (MAE), and root mean square error (RMSE).

where $X_{Meas/Est}$ is the hourly or daily measured/simulated from power output PV system values and N is the number of evaluated pair's data. Besides these scores, there are the corresponding normalized ones, the relative MBE (rMBE) and relative RMSE (rRMSE). The two normalized scores are computed as follows:

$$MBE(\%) = 100 \times \left(\frac{1}{N}\sum_{i=1}^{N}\left(\frac{X_{Est,i} - X_{Meas,i}}{X_{Meas,i}}\right)\right) \tag{6}$$

$$MAE(\%) = 100 \times \left(\frac{1}{N} \sum_{i=1}^{N} \left| \frac{X_{Est,i} - X_{Meas,i}}{X_{Meas,i}} \right| \right) \quad (7)$$

$$RMSE(\%) = \sqrt{\frac{1}{N} \sum_{i=1}^{N} \left(\frac{X_{Est,i} - X_{Meas,i}}{X_{Meas,i}} \right)^2} \quad (8)$$

3 Results and Discussion

e objective of this work is to estimate the energy produced by a PV plant. All variables are acquired every 15 min. In this way, we obtain the input data. Then, we select the variables that have some degree of direct correlation with the PV production using the Pearson coefficient criterion, the irradiation represents the higher linear correlation with the PV energy production, followed by temperature parameter.

To model that, a mathematical expression has been developed using multiple linear regression (MLR), to estimate the final output PV energy production. For this purpose, 70% of data values were used to develop the model, whereas the remaining data was used to validate the model.

Three regression models are selected as the most appropriate model using a forward selection algorithm:

$$M1 = f(GHI) = a * GHI + b \quad (9)$$

$$M2 = f(GHI, Tair) = a * GHI + b * Tair + c \quad (10)$$

$$M3 = f(GHI, Tair, Pair) = a * GHI + b * Tair + c * Pair \quad (11)$$

Table 4. Performance indices and optimal regression parameters for developed MLR models.

	Scores						Regression Coefficients		
Model	MBE		RMSE		R	*a*	*b*	*C*	*d*
	(W)	(%)	(W)	(%)					
M1	−320,69	−3,519	1730,39	18,987	0,936	14,89	167,94		
M2	−189,69	−2,081	1371,98	15,054	0,960	15,78	−97,42	2369,45	
M3	−183,87	−2,017	1347,63	14,787	0,961	15,52	−74,07	63,05	−60037,40

The results of developed models are provided in Table 4. The provided results include the validation mean estimated errors, as well as the regression coefficients corresponding.

In this work, based on the comparison of ANN model has been employed in estimating solar PV system power. During model development, a number of network configurations and training regimes were tested, in addition to varying combinations of input fields, in an effort to determine the optimal number of hidden neurons. The best ANN architecture is obtained with two hidden layers; the first contains 18 neurons, while the second one with 23 hidden neurons.

From Table 5, it is learnt that by employing the ANN model, the annually averaged mean absolute percentage error (MAE) is 0.007% which is obtained by comparing the measured data with forecasted data in a solar photovoltaic system employing 2017 and 2018. Also, the averaged normalized mean absolute error (NMAE) obtained is 0.092% and nRMSE is 0.109% respectively which is within the permissible error limits.

Table 5. Performance indices of ANN developed model.

	MBE		MAE		RMSE		R
	(W)	(%)	(W)	(%)	(W)	(%)	
2017	−11.913	−0.135	604.01	6.82	918.88	10.375	0.981
2018	6.596	0.071	662.79	7.17	969.15	10.48	0.98
All	0.235	−0.003	642.591	7.051	952.175	10.448	0.980

Further, the graphical representation showing a comparison of the measured data and estimated data obtained by employing the ANN model (Fig. 3).

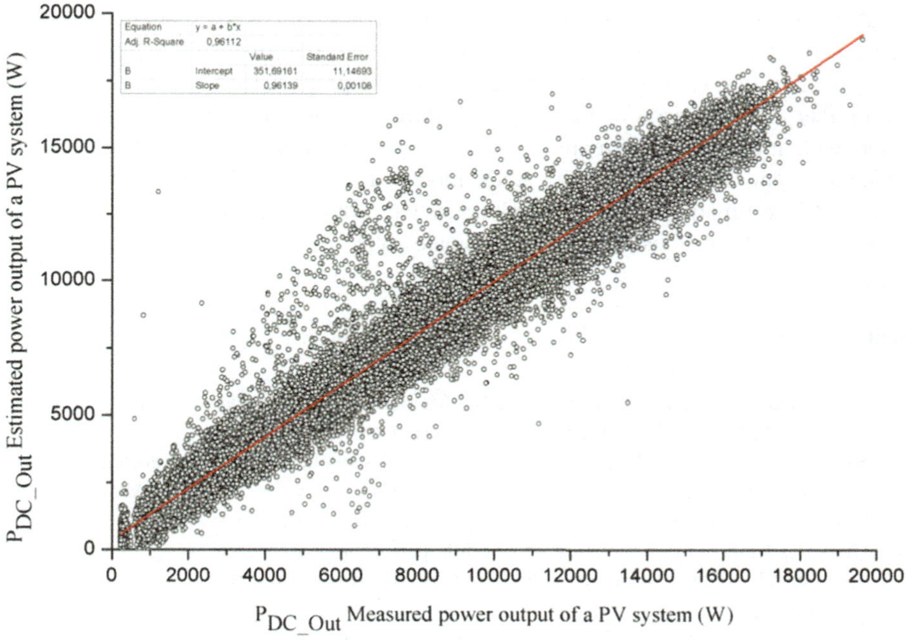

Fig. 3. Correlation between the experimental data and predicted values of the ANN model for prediction of the power output of a PV system.

A comparative analysis has been carried out between the ANN model and multiple linear regression models in PV output estimation and performance is measured based on evaluation indexes. The following can be briefly summarized as:

The mean percentage error obtained by employing MLR based model lies between -2.02% to -3.5% whereas by employing artificial neural network model, the mean percentage error varies between -0.351% to 0.07% which is comparatively less and far accurate than MLR models.

It is learned that among the different artificial neural network (ANN) model gave the best result with averaged MAE of 7.05%, and nRMSE of 10.44% respectively. From Table 5, it has been found that the root mean square errors (RMSE) is the lowest and coefficient of determination (R2) is the highest values in comparison the MLR results.

The results of the comparison confirm the accuracy and supremacy of the proposed ANN models.

4 Conclusion

In this work, Neural Networks models have been exploited for estimation of the energy production of the 20-MW solar PV plant installed in Adrar in terms of energy yield, using a minimal knowledge of meteorological parameter include global solar radiation, relative humidity, wind speed, and ambient or air temperature. Simulations have been carried out by selecting the model based on evaluation indexes.

A comparative analysis has been carried out between the ANN model and multiple linear regression models in PV output estimation and performance is measured based on evaluation indexes, to check for accuracy of the proposed ANN model.

Further, a comparison of the proposed artificial neural networks models has been made with multiple linear regression models based models and results reveals the advantages and supremacy of the ANN model. The results obtained can be further exploited in estimating solar photovoltaic system power that employs the same photovoltaic module operated in the same climate zone

References

1. Ghezlouna, A., Saidaneb, A., Oucher, N., Merabet, H.: Actual case of energy strategy in Algeria and Tunisia. Energy Procedia **74**, 1561–1570 (2015)
2. Sahnoune, F., Imessad, K., Bouakaz, D.M.: Energy consumption renewable energy development and environmental impact in Algeria - trend for 2030. In: AIP Conference Proceedings, vol. 1814, p. 020072 (2017). https://doi.org/10.1063/1.4976291
3. Huld, T., Amillo, A.: Estimating PV module performance over large geographical regions: the role of irradiance, air temperature, wind speed and solar spectrum. Energies **8**(6), 5159–5181 (2015)
4. Cole, W.J., Morton, D.P., Edgar, T.F.: Optimal electricity rate structures for peak demand reduction using economic model predictive control. J. Process Control 24(8), 1311–1317 (2014)

5. Antonanzas, J., Osorio, N., Escobar, R., Urraca, R., Martinez-de-Pison, F.J., Antonanzas-Torres, F.: Review of photovoltaic power forecasting. Sol. Energy **136**, 78–111 (2016)
6. Graditi, G., Ferlito, S., Adinolfi, G.: Comparison of photovoltaic plant power production prediction methods using a large measured dataset. Renew. Energy **90**, 513–519 (2016)
7. Moretón, R., Lorenzo, E., Pinto, A., Muñoz, J., Narvarte, L.: From broadband horizontal to effective in-plane irradiation: a review of modelling and derived uncertainty for PV yield prediction. Renew. Sustain. Energy Rev. **78**, 886–903 (2017)
8. Massidda, L., Marrocu, M.: Use of multilinear adaptive regression splines and numerical weather prediction to forecast the power output of a PV plant in Borkum. Germany. Solar Energy **146**, 141–149 (2017)
9. Ministry of Energy and Mines: Renewable Energy and Energy Efficiency Algerian Program. Sonelgaz Group, Algerian Ministry of Energy and Mines (MEM), Algeria ((2011)). www.mem-algeria.org/francais/uploads/enr/Programme_ENR_et_efficacite_energetique_en.pdf
10. Bouchouicha, K., Aoun, N., Bailek, N., Oulimar, B., Bellaoui, M.: Analyse du Potentiel solaire de la région d'Adrar. In: Conférence Internationale sur les Matériaux, le Patrimoine et l'Environnement en Zones Arides, CIMaPEZA 2019, 17–19 February 2019, Adrar (2019)
11. Elsheikh, A.H., Sharshir, S.W., Elaziz, M.A., Kabeel, A.E., Guilan, W., et al.: Modeling of solar energy systems using artificial neural network: a comprehensive review. Sol. Energy **180**, 622–639 (2019)
12. Bellaoui, M., Bouchouicha, K., Aoun, N., Oulimar, B., Babahadj A.: Daily time series estimation of global horizontal solar radiation from artificial neural networks. In: 1st International Conference of Computer Science and Renewable Energy, ICCSRE 2018, Ouarzazate, 22–24 November 2018, Morocco (2018)
13. Bouchouicha, K., Razagui, A., Bachari, N.I., Aoun, N.: Hourly global solar radiation estimation from MSG-SEVIRI images-case study: Algeria. World J. Eng. **13**(3), 266–274 (2016). https://doi.org/10.1108/wje-06-2016-036. ISSN 1708-5284
14. Bouchouicha, K., Bailek, N., Mahmoud, M.E.-S., Alonso, J.A., Slimani, A., Djaafari, A.: Estimation of Monthly Average Daily Global Solar Radiation Using Meteorological-Based Models in Adrar, Algeria. Appl. Solar Energy **54**(6), 448–455 (2018)
15. Bailek, N., Bouchouicha, K., Aoun, N., Mohamed, E.-S., Jamil, B., Mostafaeipour, A.: Optimized fixed tilt for incident solar energy maximization on flat surfaces located in the Algerian Big South. Sustain. Energy Technol. Assessments **28**, 96–102 (2018). https://doi.org/10.1016/j.seta.2018.06.002

Validation Modeles and Simulation of Global Horizontal Solar Flux as a Function of Sunshine Duality in Southern Algeria (Adrar)

I. Oulimar[1(✉)], A. Benatiallah[2], and K. Bouchouicha[1]

[1] Unité de Recherché en Energie Renouvelables en Milieu Saharien (URERMS), Centre de Développement des Énergies Renouvelables (CDER), Adrar, Algeria
oulimar.brahim@gmail.com

[2] Laboratoire des Energies Environnment et Système d'informations, Université d'Adrar, Adrar, Algeria

Abstract. This work is a study and modeling of global horizontal solar radiation according to duration of sunshine applied on south of Algeria Adrar area), using data of monthly sunshine duration and data extracted from the satellite images captured by the European MSG2 satellite. By this information's we have establish regression models giving the horizontal total as a function of insolation duration.

The performance of these models was evaluated and validated against real radiometric measurements by calculating the most used statistical scores. The results of this modeling of monthly global horizontal irradiations values are very encouraging and promising compared to the existing literature and practical values.

Keywords: Solar radiation · Modeling · Statistical scores · Fraction of sunshine · Clarity index

1 Introduction

Renewable Energy has become the first vector of sustainable economic development in many countries of the world. Recourse to development of this energy type is a strategic choice for most countries, in order to provide comprehensive and sustainable solutions to environmental challenges and issues related to preservation of fossil energy resources [1].

The needs and possibility of solar energy exploitation are directly related to climatic conditions. This source is intermittent in time and space, which poses many problems in sizing solar energy systems. These aspects require a perfect knowledge of solar potential before undertaking any action or valorization program. This goal can only be achieved if solar radiometric data is available continuously in space and time.

M. Hatti (Ed.): ICAIRES 2019, LNNS 102, pp. 204–211, 2020.
https://doi.org/10.1007/978-3-030-37207-1_21

In addition, it is not practical and economical to install a dense solar radiation measuring network instruments, such as a pyranometre network, over a large area in a big country like Algeria (2.400.000 km^2). But in order to be able for interpolate from a measurement database, it is necessary for the study region to have a network of measuring instruments for a number of microclimates.

Our region (City of Adrar) in particular the South of Algeria is characterized by a huge solar energy potential. The duration of insolation on almost the whole national territory exceeds the 2000 h annually and can reach the 3900 for the southern region. The energy received daily on a horizontal surface of 1 m^2 is of the order of 2500 $kWh/m^2/year$ [1].

The needs and the possibility of exploitation of solar energy are directly related to climatic conditions. This source is intermittent in time and space, which poses problems in sizing solar energy systems. These aspects require a perfect knowledge of the solar field before undertaking any action or valorization program. This goal can only be achieved if the radiometric data is available continuously in space and time.

The number of radiation measurement stations is very limited in this region. However, the duration of insolation and the state of the sky are available in the majority of weather stations. It is therefore interesting to use calculation methods to obtain the different components of the radiation from the available data.

For all these reasons, our study concerns the region of Adrar and aims to give a good estimate of the solar potential on a horizontal plane using data of the duration of insolation, we also used radiation data. solar station of the URER/MS measuring station as well as data evaluated and derived from the processing of satellite images [2].

2 Meteorological Parameters

2.1 Sunshine Duration

Depending on the weather conditions, the sky can be more or less covered with clouds during a day. These obscure the Sun, totally or partially, thus preventing radiation from reaching the ground directly. It is said cloudiness is more or less important depending on whether there are many or few clouds.

The effective time of sunshine or SS insolation is time during which direct solar radiation reaches the soil of the site in question during a day exceeds the threshold of 100 to 120 W/m^2. Direct radiation is radiation that reaches the earth's surface without having been deflected since it was emitted by the Sun.

The maximum duration of sunshine SS0 (potential insolation) (Fig. 1) for the site of Adrar corresponds to duration of day calculated by Eq. (1).

$$SS_0 = d = \frac{2}{15} arcos(-tan(L)tan(\delta)) \tag{1}$$

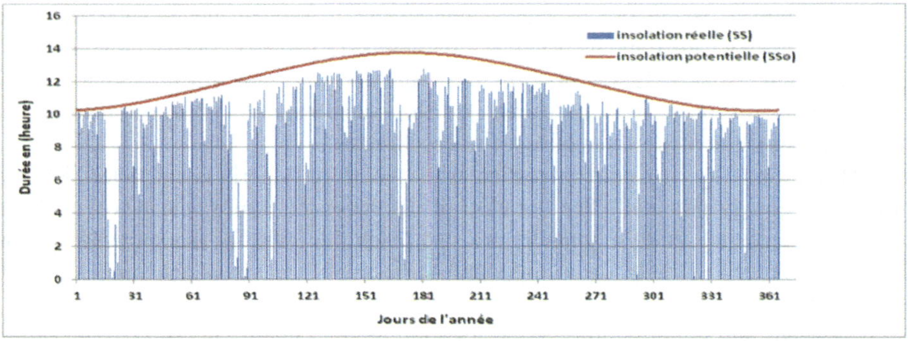

Fig. 1. Real and potential insolation in Adrar

2.2 Sunshine Rate

On a clear day, the ground receives direct solar radiation all duration of day, or precisely during the maximum duration of sunshine SS_0. The ratio of the effective duration to maximum duration of sunshine is give by [7]:

$$\rho = \frac{SS}{SS_0} \tag{2}$$

The average sunshine rate is around 80% for Adrar site and is presented in (Fig. 2), with a cumulative 3500 h per year.

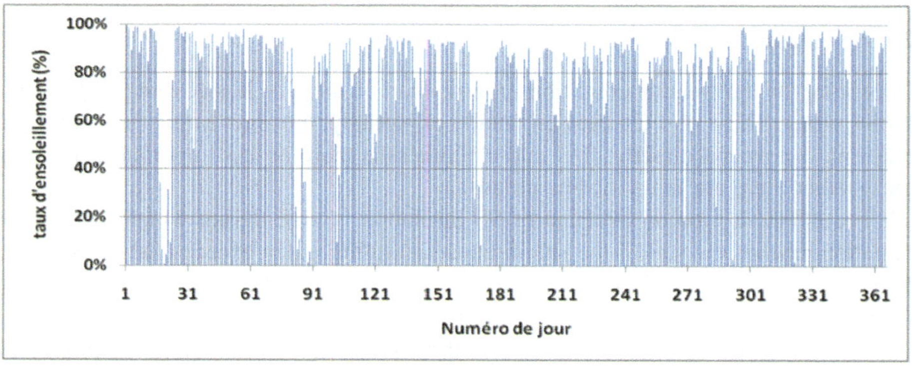

Fig. 2. Sunshine rate in Adrar.

3 Meteorological Data

The data of sunshine duration used in this work are taken from the services of the National Office of Meteorology (ONM), while solar radiation data are taken from a work research thesis, following an estimate of solar radiation using satellite images [2].

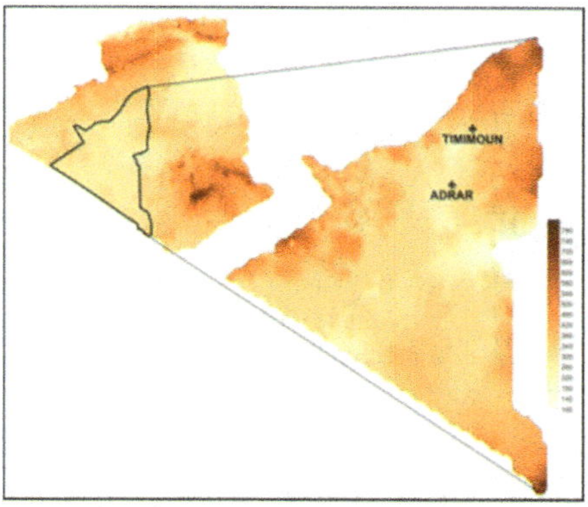

Fig. 3. Geographic map of Adrar

Table 1. Geographical coordinates of Adrar

Station	Adrar
Latitude (°)	27.88
Longitude (°)	−0.28
Altitude (°)	263

This work is done horizontal global radiation estimation from MSG2 satellite High-Resolution Visible (HRV) images. The evaluation was carried out against the data of five stations for the year 2010–2017. According to its results, the quality of estimation in general, give a good coherence with the measurements, expressing a high correlation, and the error relative mean quadratic varies between 4% and 12% for all stations (Table 1).

3.1 Modeling Parameters

The modeling in this study requires the use of two main parameters:

The fraction of sunshine (σ) that is given by formula (3):

$$\sigma = \frac{S}{S_0} \tag{3}$$

With S is the measured average monthly duration and S0 is the theoretical duration of the day (maximum duration).

The clarity index (Kt) calculated using formula (4):

$$K_t = \frac{H}{H_0} \tag{4}$$

With H is the estimated monthly average satellite irradiation and H0 is the monthly average of the maximum irradiation.

3.2 Models of Estimations the Horizental Solar Radiation

In this study we tested four types of regression model to express the relative global radiation (H/H0) as a function of relative insolation (S/S0).

The equations of these models are given in Table 2. With a, b, c, d are empirical constants. H and H0 are respectively the monthly averages of global horizontal ground

Table 2. Regression models based on sunshine duration.

No	Modeles	Sources	Équations
1	Linear	Angström [3], Prescott [3]	$\frac{H}{H_0} = a + b.\frac{S}{S_0}$
2	Quadratic	Akinoglu and Ecevit [4]	$\frac{H}{H_0} = a + b.\frac{S}{S_0} + c.\left(\frac{S}{S_0}\right)^2$
3	Cubic	Bahel et al. [2]	$\frac{H}{H_0} = a + b.\frac{S}{S_0} + c.\left(\frac{S}{S_0}\right)^2 + d\left(\frac{S}{S_0}\right)^3$
4	Logarithmic	Ampratwum and Dorvlo [5]	$\frac{H}{H_0} = a + b.\log\left(\frac{S}{S_0}\right)$

radiation and horizontal global radiation outside the atmosphere, they are expressed in Wh/m^2/Day, S and S0 are respectively the average effective sunshine duration of the day per month and the average maximum duration of sunshine (astronomical duration) of the day per month, S0 can be calculated by the following formula:

$$S_0 = \frac{2}{15}\omega_s \tag{5}$$

With ω_s the hour angle of the site.

The following table summarizes the different types of models to use in our study.

4 Simulation Methodology

The Matlab software was used to program the calculation of the coefficients of each station model, as well as the application of the most used statistical scores to assess the quality of the estimation model, the correlation coefficient (R, Eq. (8)), the average error (MBE: Mean Bias Error) (6) and the Root Mean Square Error (RMSE: Root Mean Square Error) (Eq. 7) [2].

The Matlab 'fittype' command was used to insert the models tested using as input data; the data of the fraction of sunshine and the index of clarity, at the output of each manipulation the coefficients of models are compute is to save.

After having found the coefficients of each model and for each station, we apply them on our measured data of insolation duration to find the calculated irradiation.

Third step, comparison of calculated data to those measured using scatterplot regression graphs, and calculation of statistical scores.

$$MBE(\%) = \frac{1}{N}\sum\nolimits_{i=1}^{N} e_i \tag{6}$$

$$RMSE(\%) = \left[\frac{1}{N}\sum\nolimits_{i=1}^{N} e_i^2\right]^{\frac{1}{2}} \tag{7}$$

$$R = \frac{\sum_{i=1}^{N}\left(Y_{ie} - \overline{Y}_{ie}\right)\left(Y_{im} - \overline{Y}_{im}\right)}{\sqrt{\sum_{i=1}^{N}\left(Y_{ie} - \overline{Y}_{ie}\right)^2} \times \sqrt{\sum_{i=1}^{N}\left(Y_{im} - \overline{Y}_{im}\right)^2}} \quad (8)$$

With: e_i is the relative difference between the measured and the estimated values, Y^{-} is the average irradiation per month,

Y_{im} and Y_{ie} are respectively the measured values and the estimated values of global radiation, it is expressed by [6]:

$$e_i = \frac{(Y_{im} - Y_{ie}) * 100}{Y_{im}} \quad (9)$$

5 Results and Discussion

We began with the statistical evaluation which presents coefficients values of regression models applied to the studied stations, as well as the results of the statistical scores applied to the results of these models in relation to the data measured on the stations concerned (Table 3).

Table 3. Coefficients and statistical scores of Adrar

	Coefficients				Statistical scores		
	A	b	C	d	MBE	RMSE	R
MODEL1	0,367	0,389			0,122	3,295	0,9908
MODEL2	1,314	−2,004	1,499		0,053	3,164	*0,9914*
MODEL3	**1,181**	**−1,498**	**0,859**	**0,268**	**0,053**	**3,162**	***0,9914***
MODEL4	0,747	0,304			0,142	3,351	0,9905

Another evaluation was made with respect to the experimental measurements, through a graphical comparison between the irradiation calculated by the eight models and that extracted from the satellite images. This phase gathers the results of the Adrar station (Figs. 4 and 5).

In this section we present the statistical evaluation results obtained for the four models compared to experimental data for global horizontal radiation per measurement site.

Adrar Site: Table 3 shows that all models give a good estimate of solar irradiance on a horizontal plane, with an MBE that varies between (0.05% and 0.14%) so all models slightly overestimate the monthly values of the horizontal solar irradiation, the RMSE varies between (3.16% and 3.35%), the correlation coefficient also gives good values with values greater than 0.99.

The best model for the Adrar site is the third model with R = 0.99 and RMSE = 3.16%.

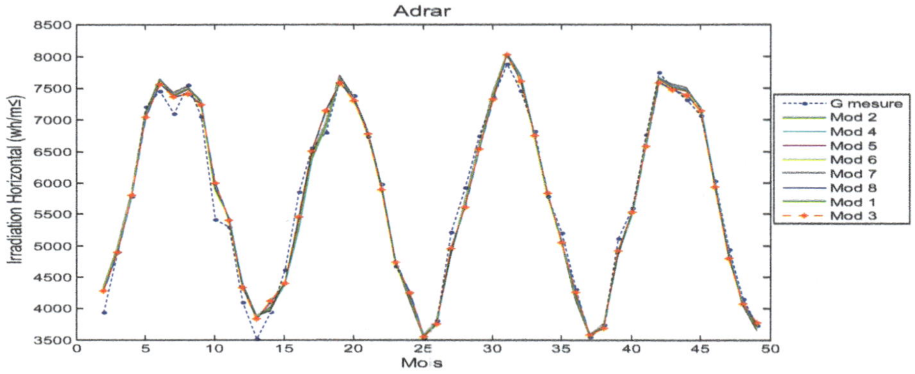

Fig. 4. Comparison between the irradiation calculated by models and estimated by satellite for Adrar.

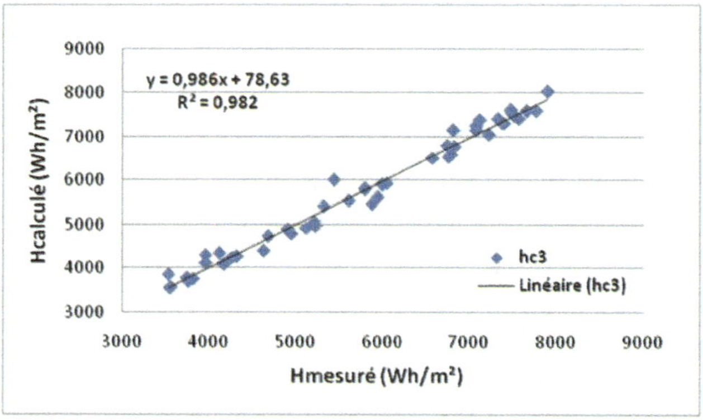

Fig. 5. Regression curve between data of best model and real data (Adrar)

6 Discussion and Interpretation

The evaluation is carried out by comparing the results of the models with the estimated satellite data, we evaluated the estimates made for 48 months for the Adrar sites through the most used scores for the evaluation of estimation models.

All models give good estimates of the monthly average of daily irradiation on a horizontal plane. With a slight overestimation compared to the actual measurement for the Adrar site (Fig. 4).

The mean error (MBE) for all models varies between (0.05 and 0.14%) and the RMSE gives very small values and varies around (3%).

For the correlation coefficients, all the models give a good correlation between the monthly values estimated by the models and the measured values ($R > 0.99$).

7 Conclusion

In this study, we have developed regression models giving total solar horizontal radiation as a function of insolation duration for Adrarsite important located and capital of department in a south Algeria. The results obtained by most of these models indicate the presence of a slight overestimation compared to actual real measurement for Adrar site.

The average error for all models varies between (0.05 and 0.14%) and the RMSE gives very small values and varies around (3%), for correlation coefficients all models also give a good correlation between the monthly values estimated by models and measured values ($R > 0.99$). The best model that gives good results for both sites is the polynomial type 3 (model 3).

Finally we can conclude that for all Saharan sites, with a climate similar to that of Adrar region, we can use duration of insolation which is very available on weather stations to estimate the horizontal solar irradiation with good accuracy, in the absence of measured data. Without recourse to direct measurement which is very expensive and requires a lot of human and material intervention.

Referances

1. December 2017. http://www.energy.gov.dz. http://www.energy.gov.dz/francais/index.php?page=potentiels
2. Bouchouicha, K.: Modelisation Multispectrale Des Images Satellitaire - Application: Quantification du bilan d'énergie Sol-Atmosphère. université d'oran, oran, these de doctorat en science (2017)
3. Şen, Z.: Solar Energy Fundamentals and Modeling Techniques. Springer, London (2008)
4. Akinoblu, B.G., Ecevit, A.: A further comparison and discussion of sunshine-based models to estimate global solar radiation. Energy **15**(10), 865–872 (1990)
5. Dorvlo, A.S.S., Ampratwum, D.B.: Estimation of solar radiation from the number. Appl. Energy **63**, 161–167 (1999)
6. El-Metwally, M.: Sunshine and global solar radiation estimation at different sites in Egypt. J. Atmos. Solar Terr. Phys. **67**, 1331–1342 (2005)
7. Sayigh, A.A.M.: Solar Energy. Solar Energy Engineering. Academic Press, Inc., Riadh (1977)

Direct and Indirect Nonlinear Control Power of a Doubly-Fed-Induction Generator for Wind Conversion System Under Disturbance Estimation

Bouiri Abdesselam(✉), Benoudjafar Cherif, and Boughazi Othmane

Department of Electrical Engineering, Faculty of Technology,
University Tahri Mohamed, Bechar SGRE Laboratory of Bechar University
Bechar, Bechar, Algeria
abdou.dwidi@gmail.com, benoudjafer@gmail.com,
othmaneaaa64@gmail.com

Abstract. It's difficult to control the powers in the wind energy conversion system (WECS) based on doubly-fed-induction generator (DFIG), because of its complex system, which is connected to several variables, among them: the first of which is wind and also, who cares more, sensitivity to disturbance caused by the parametric variations of the DFIG, which makes control ineffective, especially when it comes to linear control or when the control is not based on the effective estimation facing these variations. This paper presents nonlinear control strategy for statoric active and reactive powers of DFIG in WECS using backstepping controller(BSC) in two control method direct(BS-dir) and indirect (BS-ind): direct by a power sensor and indirect control by estimation electric current from power during parametric variations: and we choose the optimum control method in this case.

Keywords: Wind energy conversion system (WECS) · Doubly-fed induction generator (DFIG) · Nonlinear controller · Backstepping controller(BSC) · Method direct(BS-dir) · Indirect(BS-ind)

1 Introduction

For future energy problems, wind energy is a renewable resource and an alternative solution, due to the increasing concern about CO2 emission, which have emerged and developed with the development of power electronic and generators, for several advantages Among them variable speed of about ±30% around the synchronous speed which allows to better exploit the resources wind turbines, and the power converters are sized to use a fraction (25%–30%) of the total power produced by the generator to achieve full control of the system, Consequently, the reduction of losses in power electronics components [1, 2]. These advantages can only be realized by the using of doubly fed induction generator (DFIG) in wind energy conversion system (WECS), the Control of the DFIG is more difficult than the control of a standard induction machine [3], especially when it comes to non-linear control Various control methods have been

M. Hatti (Ed.): ICAIRES 2019, LNNS 102, pp. 212–219, 2020.
https://doi.org/10.1007/978-3-030-37207-1_22

developed in the literature [4, 5]. Where was using Backstepping approach has been introduced in 1990 by Krstic, Kanellakopoulos and Kokotovic [6, 7]. However they did not show the essential difference in their control method and where it can be used and needed.

This article presents direct and indirect backstepping control applied to control active and reactive powers of the rotor side converter RSC under disturbance estimation a result of parametric variations of DFIG this can be ensured by a comparative study between direct controller, and indirect backstepping control in, the analyzed results in the software MATLAB/SIMULINK.

The paper is organized as follows. Firstly we present the mathematical generator modeling of the doubly-fed-induction generator (DFIG), and then we will present the nonlinear synthesis control active and reactive powers by two control method using backstepping controller where we test their performance during parametric variations. Finally, we will confirm the effectiveness control by the obtained results (Figs. 1, 2 and 3).

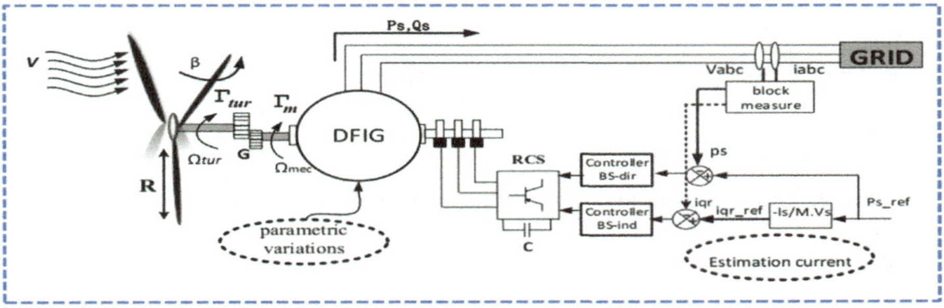

Fig. 1. Diagram control of the DFIG in wind energy conversion system

2 Modeling of the Control Strategy of DFIG

The mathematical models of three phases DFIG in the Park frame can be represented by the expressions of stator, rotor voltages and flux components as follows [4, 8, 9]:

$$\begin{cases} v_{ds} = R_s i_{ds} + \dfrac{d\varphi_{ds}}{dt} - w_s \varphi_{qs} \\ v_{qs} = R_s i_{qs} + \dfrac{d\varphi_{qs}}{dt} + w_s \varphi_{ds} \\ v_{dr} = R_r i_{dr} + \dfrac{d\varphi_{dr}}{dt} - w_g \varphi_{qr} \\ v_{qr} = R_r i_{qr} + \dfrac{d\varphi_{qr}}{dt} + w_g \varphi_{dr} \end{cases} \tag{1}$$

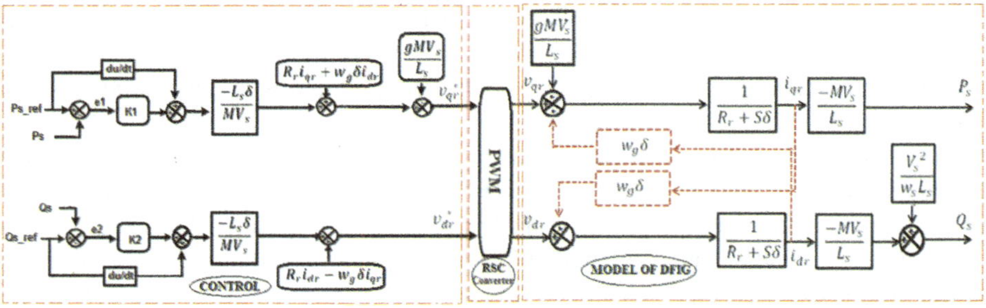

Fig. 2. Block diagram for control of the DFIG by backstepping direct method

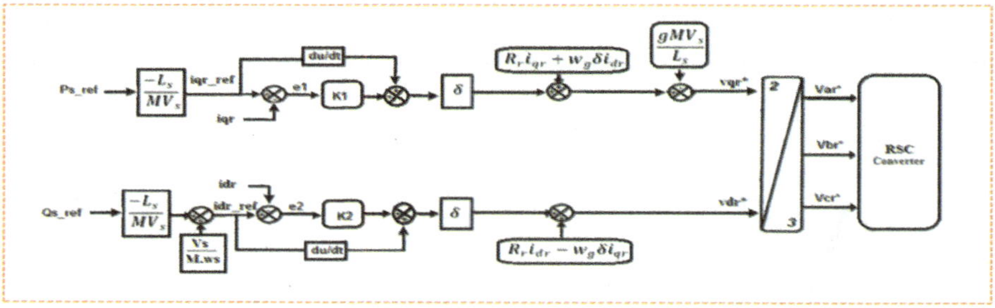

Fig. 3. Block diagram for control of the DFIG by backstepping indirect method

Where:

$$w_g = w_s - w_r \tag{2}$$

$$\begin{cases} \varphi_{ds} = L_s i_{ds} + M i_{dr} \\ \varphi_{qs} = L_s i_{qs} + M i_{qr} \\ \varphi_{dr} = L_r i_{dr} + M i_{ds} \\ \varphi_{qr} = L_r i_{qr} + M i_{qs} \end{cases} \tag{3}$$

The stator active and reactive powers are written:

$$\begin{cases} p_s = v_{ds} i_{ds} + v_{qs} i_{qs} \\ Q_s = v_{qs} i_{ds} - v_{ds} i_{qs} \end{cases} \tag{4}$$

3 The Strategy of the Non-linear Control of DFIG

To realize a statoric active and reactive power vector control, we choose a d-q reference-frame synchronized with the stator flux [9]. By using the oriented flux vector aligned with d-axis, we have: $\varphi_{ds} = \varphi_s$ and $\varphi_{qs} = 0$. By supposing that the electrical supply network is stable, having for simple voltage v_s, that led to a stator flux ϕ_s constant. In addition, if the per phase stator resistance is neglected, which is a realistic approximation for medium and high power machines used in wind energy conversion, we obtain: $v_{ds} = 0$, $v_{qs} = V_s$ and $\varphi_s = \frac{V_s}{W_s}$. Hence, the active and reactive powers (p_s, Q_s) exchanged between the stator of the DFIG and the grid can be written versus rotoric currents as:

$$\begin{cases} p_s = \frac{-MV_s}{L_s} i_{qr} \\ Q_s = \frac{-MV_s}{L_s} i_{dr} + \frac{V_s^2}{w_s L_s} \end{cases} \tag{5}$$

The rotoric voltages can be expressed by:

$$\begin{cases} v_{qr} = R_r i_{qr} + \delta \frac{di_{qr}}{dt} + w_g \delta . i_{dr} + \frac{gMV_s}{L_s} \\ v_{dr} = R_r i_{dr} + \delta \frac{di_{dr}}{dt} - w_g \delta . i_{qr} \end{cases} \tag{6}$$

Where $\delta = L_r - \frac{M^2}{L_s}$ is the leakage factor, the generator slip is, $g = (w_s - w_r)/w_r$.

3.1 Backstepping Controller Synthesis

The Backstepping approach has been introduced in 1990 by Krstic, Kanellakopoulos and Kokotovic [6, 7] and ensures overall stability control systems, usually multivariate and higher order. I is a method for synthesis of recursive class having triangular nonlinear systems, regardless of their order.

The backstepping control scheme is a nonlinear control method based on the Lyapunov theorem. The advantage of backstepping compared with other control methods lies in its design flexibility [10].

For our system, we first choose the Lyapunov V1 function in a quadratic form:

$$V_1 = \frac{1}{2} e_1^2 \tag{7}$$

Its derivative is given by:

$$\dot{V}_1 = e_1 . \dot{e}_1 \tag{8}$$

In order to ensure the stability of the subsystem, according to Lyapunov $\dot{V}_1$, it must be negative. For this we choose:

$$\dot{V}_1 = -k.e_1^2 < 0 \tag{9}$$

Where k is a constant parameter.

3.2 Direct Backstepping Control Method

$$\begin{cases} e_1 = p_{s\_ref} - p_s \Rightarrow \dot{e}_1 = \dot{P}_{s\_ref} - \dot{p}_s = -k_1 e_1 \\ e_2 = Q_{s\_ref} - Q_s \Rightarrow \dot{e}_2 = \dot{Q}_{s\_ref} - \dot{Q}_s = -k_2 e_2 \end{cases} \tag{10}$$

The errors e_1, e_2 represent respectively the error between the measured stator active power p_s and the active power reference p_{s_ref} that will be directly proportional to the rotor current to the axis q as well as the reactive power stator Q_s and the reactive power reference will be proportional to the rotor current to the axis d

Substituting Eq. (5) in (10) gives us:

$$\begin{cases} \dot{P}_{s\_ref} + \frac{MV_s}{L_s} \dot{i}_{qr} = -k_1 e_1 \\ \dot{Q}_{s\_ref} + \frac{MV_s}{L_s} \dot{i}_{dr} = -k_2 e_2 \end{cases} \tag{11}$$

$k_1, k_2 > 0$ Gain of active and reactive power design respectively

Substituting the derivatives of the currents $\dot{i}_{ds}$, $\dot{i}_{qs}$ from the equation of the voltage Eq. (6) in (11) the rotoric voltages by direct method can be expressed:

$$\begin{cases} v_{qr} = -\frac{L_s.\delta}{M.V_s} \left[\dot{P}_{s\_ref} + k_1.e_1 \right] + R_r i_{qr} + w_g \delta.i_{dr} + \frac{gMV_s}{L_s} \\ v_{dr} = -\frac{L_s.\delta}{M.V_s} \left[\dot{Q}_{s\_ref} + k_2.e_2 \right] + R_r i_{dr} - w_g \delta.i_{qr} \end{cases} \tag{12}$$

3.3 Indirect Backstepping Control Method

From the Eq. (6) can be rewritten as following:

$$\begin{cases} p_s = \frac{-MV_s}{L_s} i_{qr} \Rightarrow i_{qr\_ref} = \frac{-l_s}{M.V_s} P_{s\_ref} \Rightarrow \dot{i}_{qr\_ref} = \frac{-l_s}{M.V_s} \dot{P}_{s\_ref} \\ Q_s = \frac{-MV_s}{L_s} i_{dr} + \frac{V_s^2}{w_s L_s} \Rightarrow i_{dr\_ref} = \frac{-l_s}{M.V_s} Q_{s\_ref} + \frac{V_s}{M.w_s} \Rightarrow \dot{i}_{dr\_ref} = \frac{-l_s}{M.V_s} \dot{Q}_{s\_ref} \end{cases} \tag{13}$$

$$\begin{cases} e_1 = i_{qr\_ref} - i_{qr} \Rightarrow \dot{e}_1 = \dot{i}_{qr\_ref} - \dot{i}_{qr} = -k_1 e_1 \\ e_2 = i_{dr\_ref} - i_{dr} \Rightarrow \dot{e}_2 = \dot{i}_{dr\_ref} - \dot{i}_{dr} = -k_2 e_2 \end{cases} \tag{14}$$

Substituting the derivatives of the currents $\dot{i}_{ds}$, $\dot{i}_{qs}$ from the equation of the voltage Eq. (6) in (14) the rotoric voltages by indirect method can be expressed:

$$\begin{cases} v_{qr} = \delta\left[\dot{i}_{qr\_ref} + k_1.e_1\right] + R_r i_{qr} + w_g\delta.i_{dr} + \frac{gMV_s}{L_s} \\ v_{dr} = \delta\left[\dot{i}_{dr\_ref} + k_2.e_2\right] + R_r i_{dr} - w_g\delta.i_{qr} \end{cases} \tag{15}$$

4 Simulation Results

Both backstepping methods (direct and indirect) in the tracking test give the same results the figures are identical (Fig. 4), unless it is related to the parametric of the DFIG (M, ls) during saturation of magnetic (Fig. 5) circuit or the grid voltage (Vs) during unbalanced grid voltage (Fig. 6): this last figure illustrated the trios precedence phenomenon (t = 2 s) in addition to connecting of RSC by PWM (converter).

4.1 Reference Tracking Tests

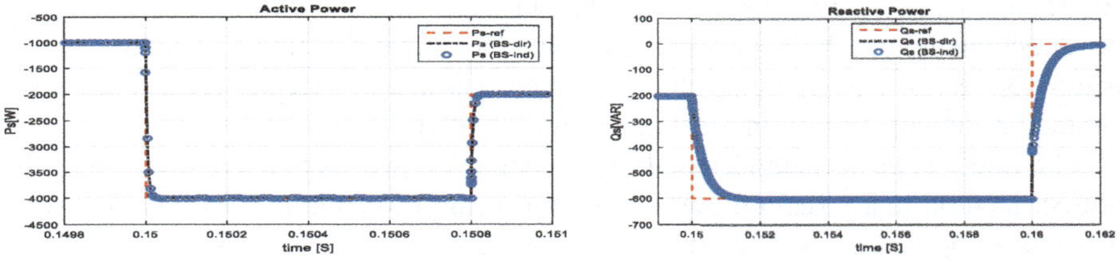

Fig. 4. Response to the active and reactive power without parametric variations or unbalanced grid voltage (reference tracking test).

4.2 Robustness Tests

In this test, one chooses the phenomenon of saturation of magnetic circuit to test the performance of duex method whose the value of the rotoric magnetizing inductance M are increased 10%.

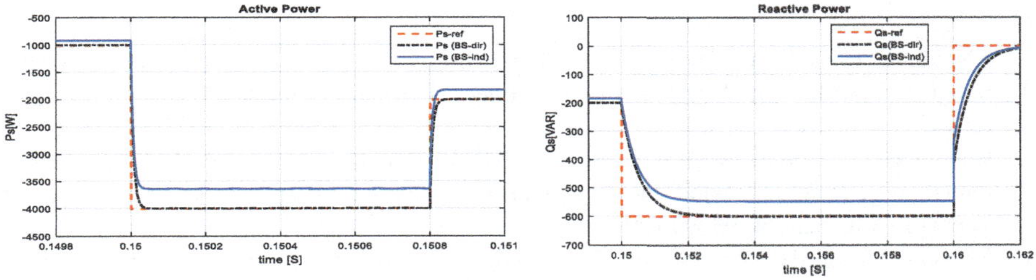

Fig. 5. Response to the active and reactive power with saturation of magnetic M are increased 10% (robustness test).

4.3 RSC Connection by PWM

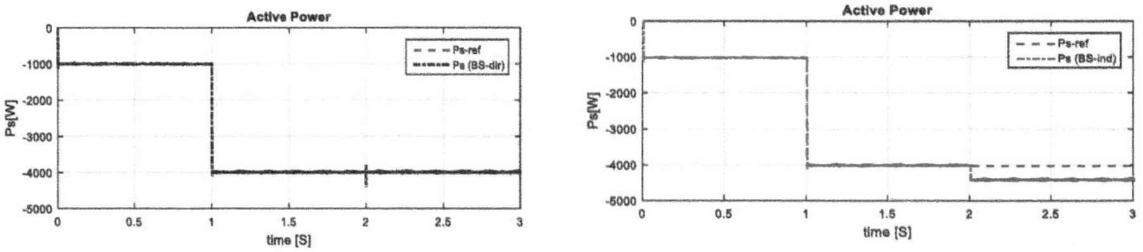

Fig. 6. Response to the active power with M and Ls and Vs are increased 10% (the influence of the converter - PWM)

5 Conclusion

Modeling and nonlinear control by two backstepping methods of the DFIG for the wind energy conversion system is dealt in this study, where we found that: the two methods (direct and indirect) are the same performance and if there was a change in (Rr, Rs, Lr, wr) they give the same results but when it is about change in (M, Vs, Ls) the direct method is high performance than the indirect method. And the indirect method is very sensitive to this change; the current is not correctly estimated because parametric incompatibility in machine with controller.

The direct control method based on the control power is efficient and robust than indirect method that based on control current from power through parametric estimation, this is shown clearly when the parameter is changed (M, Vs, Ls) or in the absence of the observer for these parameters or disrupted, In this case, the direct method is preferred.

Parameters of DFIG

Pn = 5 KW, Number of pole pairs p = 3, Nominal speed wr = 320 rad/s, Grid voltage Vs-fs = 220/380 v-50 H, Stator resistance Rs = 0.095 Ω, Rotor resistance Rr = 1.8 Ω, Stator inductance Ls = 0.094 H, Rotor inductance Lr = 0.088 H, Stator-rotor mutual inductance M = 0.082 H.

References

1. Chakib, R., Essadki, A., Cherkaoui, M.: Active disturbance rejection control for wind system based on a DFIG. Int. J. Electr. Comput. Energ. Electron. Commun. Eng. **8**, 1306–1315 (2014)
2. Boualouch, A., Frigui, A., Nasser, T., Essadki, A., Boukhriss, A.: Control of a doubly-fed induction generator for wind energy conversion systems by RSTController. Int. J. Emerg. Technol. Adv. Eng. **4**, 93–99 (2014)
3. Menso, S., Essadki, A., Nasser, T., Idrissi, B.B.: An efficient nonlinear backstepping controller approach of a wind power generation system based on a DFIG. Int. J. Renew. Energy Res. **7**(4), 1520–1528 (2017)

4. Loucif, M., Boumediene, A., Mechernene, A.: Backstepping control of double fed induction generator driven by wind turbine. In: 3rd International Conference on Systems and Control, Algiers, Algeria, 29–31 October 2013 (2013)
5. Boyette; A., Saadate, S., Poure, P.: Direct and indirect control of a Doubly Fed Induction Generator wind turbine including a storage unit. IEEE (2006). 1-4244-0136-4/06/2006
6. Krstic, M., Kanellakopoulos, I., Kokotovic, P.V.: Nonlinear design of adaptive Controllers for linear Systems. IEEE Trans. Autom. Control **39**, 738–752 (1994)
7. Krstic, M., Kanellakopoulos, I., Kokotovic, P.V.: Nonlinear and Adaptive Control Design. Wiley, New York (1995)
8. Hanzo, K.M.: Doubly Fed Induction Machine Modeling and Control for Wind Energy Generation, pp. 139–143. IEE Press, Piscataway (2011)
9. Ouamri, B., Ahmed, Z.F.: Comparative analysis of robust controller based on classical proportional-approach for power control of wind energy system. Rev. Roum. Sci. Techn.– Électrotechn. et Énerg **63**(2), 210–216 (2018)
10. Basri, M.A.M., Husain, A.R.: Robust chattering free backstepping sliding mode control strategy for autonomous quadrotor helicopter. Int. J. Mech. Mechatron. Eng. IJMME-IJENS **14**(03), 36–44 (2014)

Study and Implementation of Sun Tracker Design

Zakia Bouchebbat(✉), Nabil Mansouri, and Dalila Cherifi

Institute of Electrical and Electronic Engineering, University of Boumerdes, Boumerdes, Algeria

Abstract. The world is using up all the resources to meet the daily demands of energy, renewable energy solution has achieved a great step today to save the natural resources and solar technique is rapidly gaining its popularity as an important source of renewable energy. In this paper, we have constructed and studied the efficiency of a two axis solar tracker, which can track the sun throughout the day to obtain the maximum efficiency.

Keywords: Sun tracker · Photovoltaic

1 Introduction

The increasing demand for energy, the endless reduction in existing sources and the growing interest regarding environmental threats and abuse, have pushed humankind to explore new technologies for the generation of electrical energy using clean, renewable sources, such as solar energy, wind energy, etc. [1]. Among various non-conventional renewable energy sources, solar energy provides great potential for conversion into electric power, able to ensure an important part of the electrical energy needs of the planet. the conversion of solar irradiation into electrical energy describes one of the most promising and challenging energetic technologies, in improving development, being clean, silent and reliable, with very low maintenance expenses and minimal ecological impact. Solar energy is always available, practically never-ending, and requires no polluting residues or greenhouse gas emissions. Regarding the importance of investing in this advantageous source of energy, the efficiency improvement of the PV conversion equipment is one of the top priorities for many research groups all over the world. In the past, PV systems were fixed, the absorption of the light was limited, more energy can be extracted if the PV panel is installed on solar trackers [2, 3]. In this project one way of the two-axis solar tracker is studied, the solar tracker senses the direct solar radiations falling on photo-sensors as a feedback signal to ensure that the PV panel is tracking the sun all the time, it keeps the PV panel at a right angle to the sun's rays to get the maximum solar insolation. In more detail, a two-axis solar tracker system has been designed and implemented using four LDR's and two motors. The tracker's control algorithm has been implemented via Arduino micro-controller on a simple and cheap mechanical structure. The microcontroller implements the control circuit. The control circuit then positions the used motors to orient the solar panel optimally. The main objective of the project is to achieve maximum possible output

M. Hatti (Ed.): ICAIRES 2019, LNNS 102, pp. 220–227, 2020.
https://doi.org/10.1007/978-3-030-37207-1_23

from the solar panel at all times of day, hence the problem that is posed is the implementation of a system, which is capable of enhancing the production of power by 30–40% [4].

2 Photovoltaic Principles

The photovoltaic effect is the physical process of converting light into electricity. Solar cell is the elementary block for the PV panels; there are different kinds of solar cells according to the material made with, most cells are made from silicon and many other materials are also used such as GaAs (Gallium Arsenide), CdTe (Cadmium Telluride) and CIS (Copper Indium Diselenide CuInSe2) they vary from each other in terms manufacturing technology, light absorption, and energy conversion efficiencies (Fig. 1).

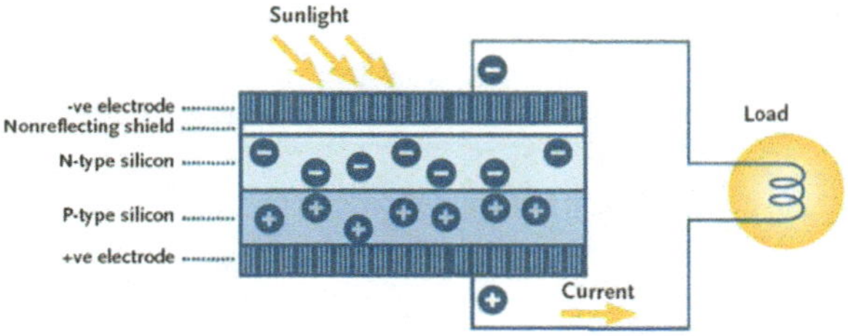

Fig. 1. Principle of operation for photovoltaic cell

3 Solar Trackers

A solar tracker is a device that orients solar photovoltaic panels towards the sun radiations since the sun's position in the sky varies each day according to the season and the time. Fix mount solar panels are limited in the energy absorption hence they have many drawbacks for the performance of a solar PV system. Solar trackers can solve the core problem by reducing the misaligned angle between sunlight's straight beams and the solar panels. There exists a wide variety of options to fulfill the functional requirements since the different solar power technologies direct us to various solutions with varying accuracy and a general rule of function. Among the various types, we can expose (Figs. 2 and 3):

3.1 Body Design of Solar Tracker

Since our project is based on a solar tracking two-axis system, we had to develop a very effective model, which can move the panel in a dual-axis. We constructed a structure of the model using solid works software since that would give a better visualization of what the model will look like (Fig. 4).

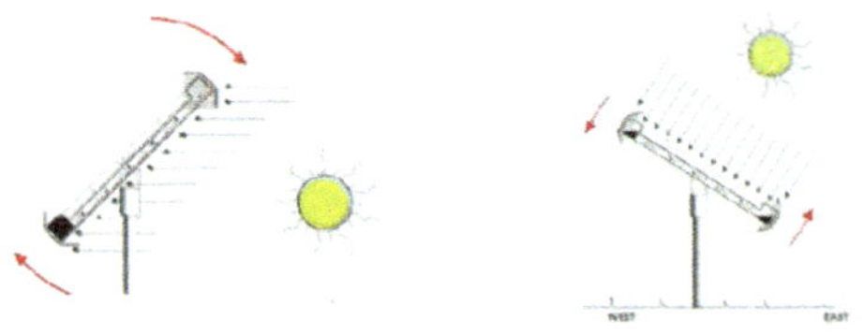

Fig. 2. Passive tracking system

Fig. 3. Single and dual axis tracking systems

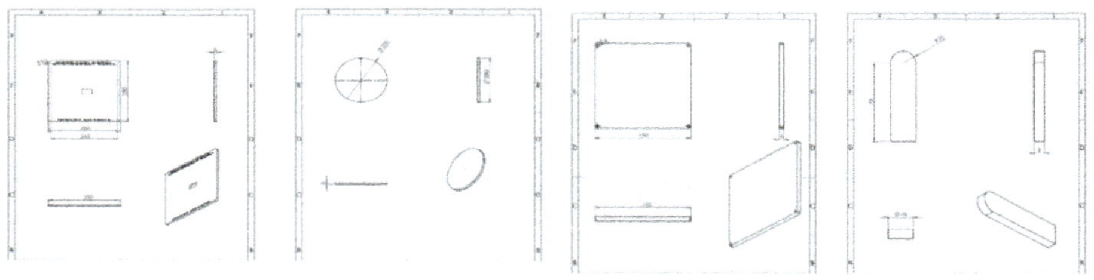

Fig. 4. Solid work structure design

The system relies on two different rotational movements in two different axes as shown in the following picture (Fig. 5):

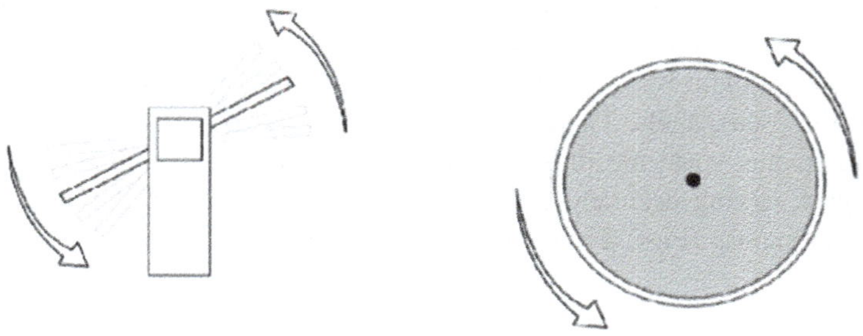

Fig. 5. Base and panel movement

The final structure is shown as follow (Fig. 6):

Fig. 6. Final structure of the solar tracker

3.2 Software and Hardware Development

The automatic solar tracker that we designed is a two-axis tracker, which will track the sun on both horizontal and vertical axis. To achieve this we had to build a prototype that consisted of many individual parts. Some of the key hardware that we have used are (Fig. 7):

- Arduino Uno microcontroller
- 4 Light Depending Resistors (LDR)
- 2 Servomotors
- 4 Resistors 220 Ω
- Wires
- Battery
- Switch

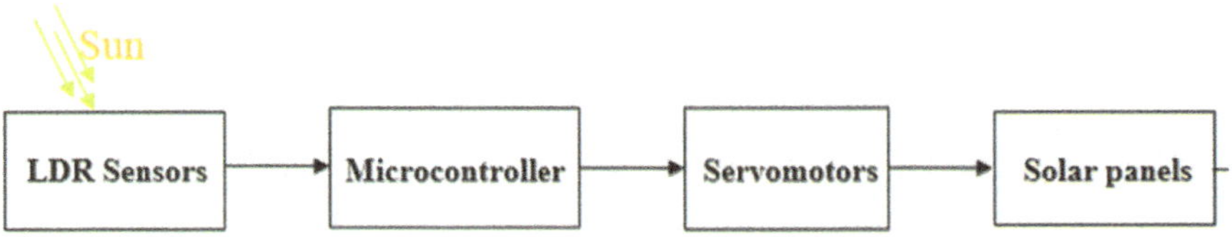

Fig. 7. Solar tracker block diagram

The figure below shows the hardware schematic diagram of the system, simulated using the Proteus software (Fig. 8):

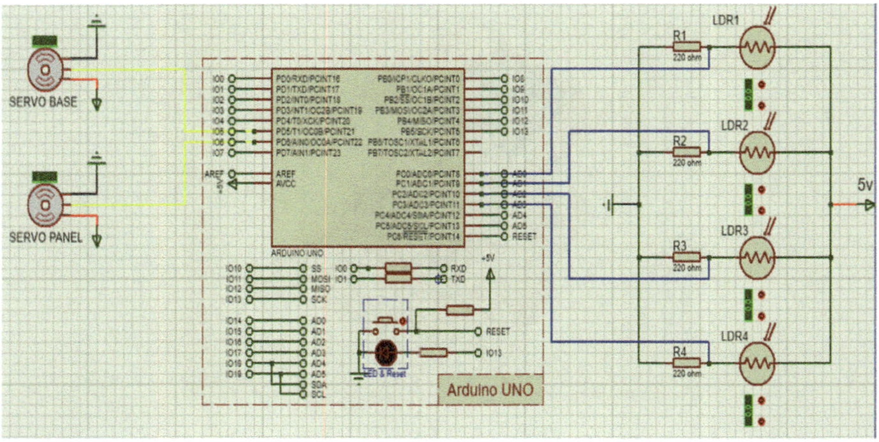

Fig. 8. Final circuit of the model

The readings of the four Light Dependent Resistors (LDRs) are taken as input by the micro-controller. The inputs are analog; they are converted to a digital value in the range between 0–1023. The average is computed hence, the direction is imposed on motors and the PV panel is now facing the direction with the greatest light intensity.

4 Experiments

The sole purpose of this project was to improve the efficiency of solar photovoltaic cell applications by embracing the automatic solar tracking system. During our project, we conducted various experiments where the results illustrated the possibility of improving efficiency through an automatic sun tracking system. Alongside this, we also conducted experiments to find out the characteristic curves of the solar panel (Fig. 9).

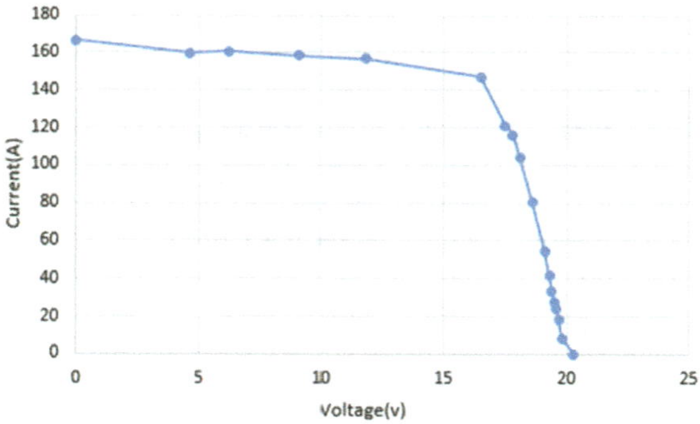

Fig. 9. Characteristics curve

In this analysis, we obtained the expected graphs. From the curve, we can find that the maximum voltage called also the open-circuit voltage that our panel can attain is 20.3 V, and the maximum current known as the short circuit current of the panel is 166.5 mA.

4.1 Power Improvement of Movable Solar Panel

This is the most important part of our project, It illustrates that the main objective is attained and the feasibility of our project. We took the whole day readings of the voltage and current we got the following graphs (Figs. 10 and 11):

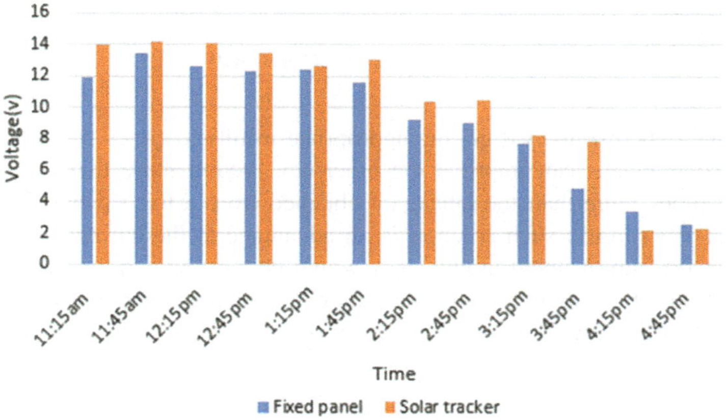

Fig. 10. Voltage vs. time results

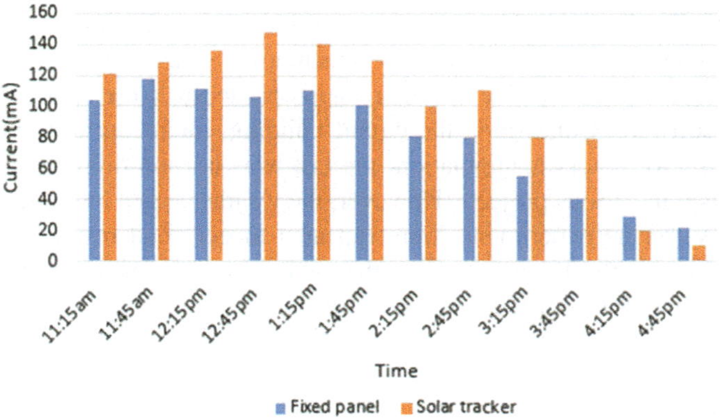

Fig. 11. Current vs. time results

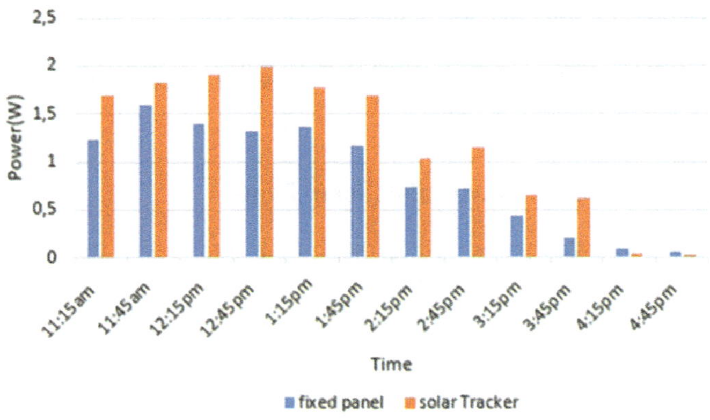

Fig. 12. Power vs. time results

From the experiments, we get the appropriate results to concretize that moveable arrays of solar panels can improve the efficiency around more than 40% which is almost 1.5 times more than the efficiency obtained from a fixed panel. Thus we can deduce that investing in such solution is an advantage at the current moment of the world since 1% improvement in efficiency would be worthy; and through our project, we reached the 40% efficiency with the automatic solar tracker that favorably is an appropriate way to harvest more solar energy (Fig. 12).

4.2 Discussion

The complexity of this project was very challenging. First, solar energy was something unfamiliar. How a solar panel works, the characteristics of the panel, the parameters that determine how the panel would behave. As a result, many studies had to be done before getting familiar with this form of renewable energy. Designing the model was not a big problem, using wood and some machines a simple structure was made for experiments, waiting to finalize the more appropriate structure that we made with SolidWorks software. Concerning the circuit building, microcontroller programming, sensors, and servo motors manipulation; the task was not too difficult since we practiced previously in various projects. However, the weather was a big issue for all of the experiments since the experiments were repeated to get good results. Besides, there were some issues with the circuitry like the calibration of sensors to eliminate errors and increase the stability of the system.

5 Conclusion

Because of the enormous scope of development, advantages and solutions that it can bring to our country, we choose solar energy as our subject of work. Thanks to this marvelous form of harvesting energy, many rural areas can now enjoy the blessing of electricity. Commercially, two-axis sun tracking is still rare even in countries where it is significantly studied. However, throughout our project with a dual-axis tracking system, we arrived at a 40% increase in efficiency compared to the fixed system and with more works; we believe that this figure can raise more. Besides, it is always wise to start early since an energy crisis is threatening the world. Even a 1% improvement in efficiency would save tons of fuels and ores in a year and that is not a small amount. Solar energy is unlimited, solar panels are easy to maintain and have a very long lifetime, which favors the use of it in our country. With a tracking system, solar energy can be harvested even more efficiently and probably two houses can be supplied when using a tracker with the electricity using the fixed panel that could only support one house. We hope that there will be more research about this technology and our country will move forward to implement a sun-tracking system to minimize the electricity crisis that is hitting us at the very moment.

References

1. Tudorache, T., Kreindler, L.: Design of a solar tracker system for PV PowerPlants. Acta Polytechnica Hungarica **7**(1), 23–39 (2010)
2. Kıvrak, S.: Passive Trackers. Afyon Kocatepe University, Renewable Energy Winter School Presentation (2012)
3. Fadil, S., Çapar, A.C., Çalar, K.: Two axis solar tracker design and implementation. In: 8th International Conference on Electrical and Electronics Engineering (ELECO), Bursa, Turkey, 28–30 November 2013
4. Otieno, O.R.: Solar tracker for solar panel. Project report, University of Nairobi, Kenya (2015)

Tuned PID by Genetic Algorithm for AGC with Different Wind Penetration

Soumia Kail(✉), Abdelkader Bekri, and Abdeldjebar Hazzab

Department of Electrical Engineering, CAOSEE Laboratory,
Tahri Mohammed University, Bechar, Algeria
kailsoumia@gmail.com, bekriabdkader@yahoo.fr,
a_hazzab@yahoo.fr

Abstract. In this paper, Automatic Generation Control (AGC) of two area non-reheat power system with the integration of wind power is presented. The influence of wind power on AGC is investigated with different wind penetration. In order to improve the frequency response in the presence of wind power, a Proportional Integral Derivative (PID) controller gains have been optimized using a Genetic Algorithm (GA). The simulation results confirm that GA is able to generate better dynamic performance.

Keywords: Automatic Generation Control (AGC) · Wind power · Proportional Integral Derivative (PID) · Genetic Algorithm (GA)

1 Introduction

Power system frequency regulation has been one of the important control problems in electric power system design and operation. Off-normal frequency can directly impact on power system operation and system reliability. Frequency deviation is a direct result of the imbalance between the electrical load and the power supplied by the connected generators.

A large frequency deviation affects power system operation, security, reliability, and efficiency and it can damage equipment, degrade load performance, cause the transmission lines to be overloaded and can interfere with system protection schemes, which leads to an unstable condition for the power system [1, 2].

Interconnected power systems regulate power flows and frequency using Automatic Generation Control (AGC). AGC is a feedback control system adjusting a generator output power to remain defined frequency [3]. AGC provides an effective mechanism for adjusting the generation to minimize frequency deviation and regulate tie-line power flows [2].

Frequency regulation has become more challenging under the conditions of low power system inertia due to the high wind energy penetration in the power system. Renewable energy sources like wind having new production and nonlinear control technology imply a new energy management system in modern power systems. This system is indispensable for stable operation of grid ensuring continuous adaptation of generation to demand [4]. Many research papers have reported in the literature on how a variable-speed wind turbine can participate effectively in system frequency regulations [5–8].

M. Hatti (Ed.): ICAIRES 2019, LNNS 102, pp. 228–235, 2020.
https://doi.org/10.1007/978-3-030-37207-1_24

The impact of wind power on the dynamic behavior of power system may cause a different system frequency response to a disturbance event. To solve AGC problems with the integration of wind power, many heuristic methods have been used such as Genetic Algorithm (GA), Particle Swarm Optimization (PSO), Bat Algorithm (BA), [9–11]. In this paper, Genetic Algorithm (GA) is used to tune PID controller gains for two area power system with wind power based-DFIG in both area to find the optimum gains to improve the frequency response. However, the conventional control techniques may not assure the desired performance due to the complexity and multi-variable conditions of the power system. The simulation results show that the PID controller tuned by GA gives better results over the conventional method.

2 An Overview of AGC Model of Interconnected Power System

The transfer function model of the investigated two-area interconnected non-reheat thermal power system with DFIG based wind turbine is shown in Fig. 1. The two-area power system associates two equal non-reheat thermal power generating units. Each unit has three major components, which are turbine, governor, and generator.

The turbine is represented by the transfer function:

$$G_T(s) = \frac{K_T}{1 + sT_T} \tag{1}$$

The transfer function of governor is:

$$G_G(s) = \frac{K_G}{1 + sT_G} \tag{2}$$

The generator and load is modeled by the transfer function:

$$G_P(s) = \frac{K_P}{1 + sT_P} \tag{3}$$

Where: $K_p = \frac{1}{D}, T_p = \frac{2H}{f_0 D}$.

Conventional AGC model is based upon tie-line bias control, where each area tends to reduce the Area Control Error (ACE) to zero. The area control error (ACE) consists of two components which are the power error and the frequency error and is given by

$$ACE_1 = \Delta P_{tie12} + \beta_1 \Delta f_1 \tag{4}$$

$$ACE_2 = \Delta P_{tie21} + \beta_2 \Delta f_2 \tag{5}$$

Where Δf_1 and Δf_2 are frequency deviations in area 1 and area 2, respectively.

β is frequency bias parameter.

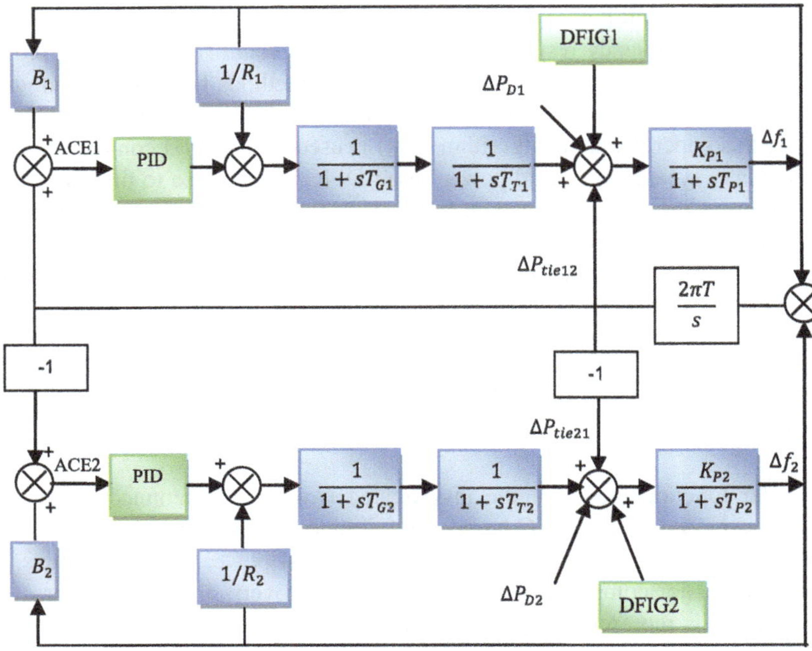

Fig. 1. Representation of two area system with DFIG-based wind turbine.

3 DFIG-Based Wind Turbine

During AGC, when wind generators participate in frequency control, the wind turbines prevent themselves from supplying their available power so as to maintain a reserve margin for frequency control. As the technology changes, the kinetic energy stored in the mechanical system of wind turbines can be extracted with the help of variable speed generators. The DFIG-based wind turbines are able to produce power with variable mechanical speed and extract the kinetic energy to aid the primary frequency control [5]. The model used for active power control with dynamic participation of DFIG-based wind turbine is shown in Fig. 2.

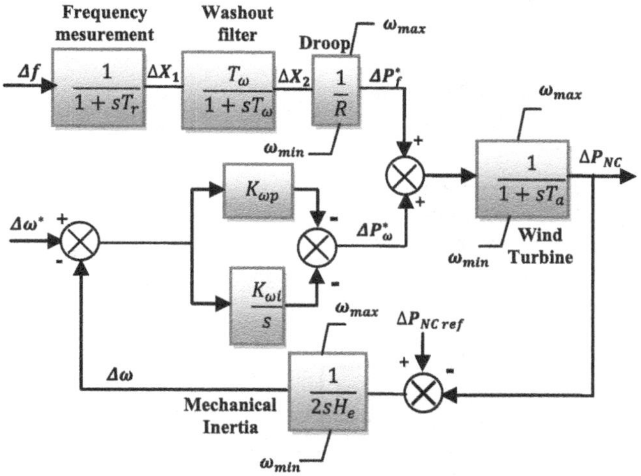

Fig. 2. DFIG-based wind turbine.

4 Proposed Methods

Proportional Integral Derivative (PID) controllers have been used form last seven decades by various electrical utilities. They play a major role in industrial process control. The PID controllers are used for minimizing the frequency deviations in single or multi area power systems employing AGC [12]. PID controller widely used because of simple design and its robust performance against a wide range of operating conditions. The mathematical description of PID as follow:

$$u(t) = K_P\ e(t) + K_I \int_0^t e(t).dt + K_D \frac{de(t)}{dt} \quad (6)$$

Where *e(t)* is the error value and *u(t)* is the control variable.

K_P, K_I and K_D, all non-negative, denote the coefficients for the proportional, integral, and derivative terms respectively. In this work, the PID parameters have been tuned by Ziegler–Nichols method and have been developed by genetic algorithm based minimisation approach.

Genetic Algorithms are stochastic optimization algorithms that were originally motivated by the mechanisms of natural selection and evolutionary genetics. A simple GA is an iterative procedure that maintains a constant size population N of candidate solutions. During each iteration step, or generation, three genetic operators (reproduction, crossover, and mutation) are performing to generate new populations (offsprings), and the chromosomes of these new populations are evaluated via the value of fitness which is related to some cost functions. On the basis of these genetic operators and the evaluations, the better new populations of candidate solution are formed [13] (Fig. 3).

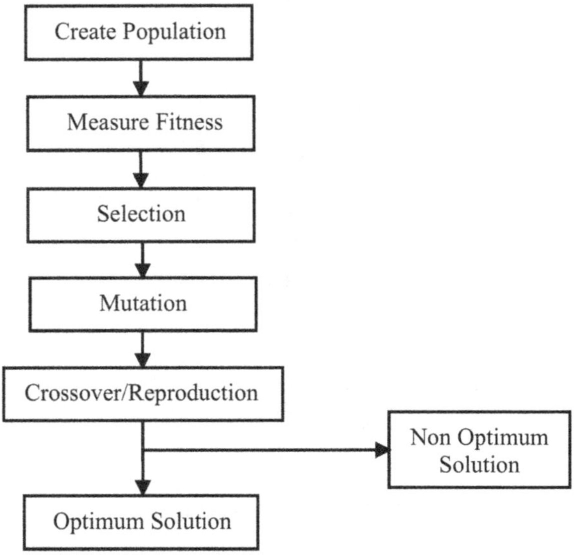

Fig. 3. Flow chart of GA.

5 Results and Discussions

A two-area non-reheat interconnected thermal power system as shown in Fig. 1 is considered. Each area has a rating of 2000 MW with a nominal load of 20 MW (0.01 p.u) which was applied at t = 0 s in area 1 and area 2. Performance of PID tuned by Ziegler–Nichols method and GA are plotted.

GA is used in a way to optimize PID gains. In this paper the performance of PID controller designed using the integral of squared-error (ISE) as fitness function, the ISE performance criterion formula is given by:

$$ISE = \int_{0}^{\infty} e^{2}(t)dt \tag{7}$$

Genetic algorithm parameters are taken as given below

The number of population = 50
The number of generation = 100
The probability of crossover is 0.8
The fitness scaling function is Rank.

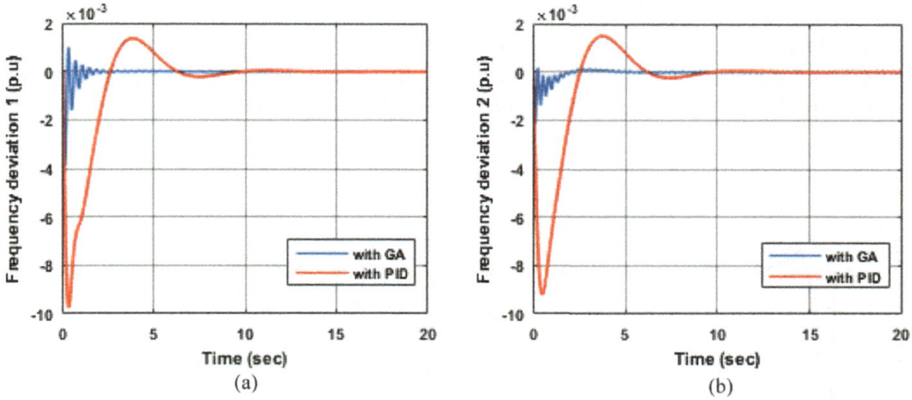

Fig. 4. Frequency deviation, (a) in area 1, (b) in area 2.

From the results shown in Fig. 4, it is observed that the GA gives better performance in term overshoot, undershoot and settling time. It is clear from Tabl. 1 below that the settling time of GA tuned PID is reduced compared to Ziegler–Nichols method and also compared to the other methods have proposed in some research papers [9–11]. This shows the efficacy of GA to obtain optimum values of PID controller gains (Table 1).

Table 1. Performance of PID tuned by Ziegler–Nichols method and by GA.

		Ziegler–Nichols	GA
df1	Settling time	9.90	4.645
	Peak overshoot	0.001283	0.001018
	Peak undershoot	−0.009557	−0.003804
df2	Settling time	9.90	6.265
	Peak overshoot	0.001388	0.0001629
	Peak undershoot	−0.009077	−0.002123

The different wind penetration levels of 5%, 10% and 20% are used to investigate the influence of wind power on AGC with 1% step load change in both areas. The penetration level of wind power can be increased by changing the droop setting (R) and system inertia constant (H). L_p % wind penetration means L_p % reduction in the system inertia, considering no frequency support from the DFIG. With the wind penetration level of L_p, the droop setting and the inertia constant without frequency support are respectively given by:

$$R_{L_P} = R/(1 - L_P) \tag{8}$$

$$H_{L_P} = H(1 - L_P) \tag{9}$$

Figure 5 shows frequency deviation responses of the two area for different wind power penetration, it is observed that the steady-state frequency deviation may increase as the wind power penetration level increases. But a high penetration may decline the dynamic performance of the system with increased settling time and higher steady-state errors.

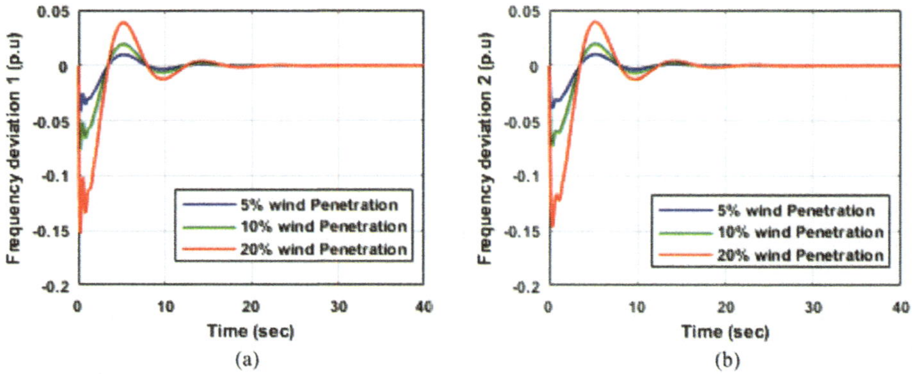

Fig. 5. Frequency response with different levels of wind penetration, (a) in area 1, (b) in area 2.

6 Conclusion

In this work, Automatic Generation Control (AGC) in two area power system with two non-reheat thermal power units has been presented. Each area is integrated with DFIG-based on wind turbine with 1% step change in load in both areas. PID controller parameters have tuned by Ziegler–Nichols method and by Genetic Algorithm (GA). The dynamic performance in terms of the frequency has been analyzed and compared. The simulation results show that the PID tuned its parameters via genetic algorithm has better result compared to the conventional method, less overshoot, small settling time and zero steady-state error. The effect of wind on AGC is investigated. By increasing the penetration level of the wind power, the inertia constant and permanent droop change. As the penetration of wind power increases, the frequency deviation increases.

References

1. Bevrani, H.: Robust power system frequency control (2014)
2. Bevrani, H., Hiyama, T.: Intelligent Automatic Generation Control. CRC Press, Boca Raton (2016)
3. Prakash, S., Sinha, S.K.: Four area load frequency control of interconnected hydro-thermal power system by intelligent PID control technique. In: Students Conference on Engineering and Systems, pp. 1–6. IEEE (2012)
4. Aziz, A., Oo, A.T., Stojcevski, A.: Frequency regulation capabilities in wind power plant. Sustain. Energy Technol. Assess. **1**(26), 47–76 (2018)

5. Ibraheem, Niazi, K.R., Sharma, G.: Study on dynamic participation of wind turbines in automatic generation control of power systems. Electr. Power Compon. Syst. **43**(1), 44–55 (2015)
6. Ahmadi, R., Sheikholeslami, A., Nabavi Niaki, A., Ranjbar, A.: Dynamic participation of doubly fed induction generators in multi control area load frequency control. Int. Trans. Electr. Energy Syst. **25**(7), 1130–1147 (2014)
7. Verma, Y.P., Kumar, A.: Dynamic contribution of variable-speed wind energy conversion system in system frequency regulation. Front. Energy **6**(2), 184–192 (2012)
8. Varzaneh, S.G., Mehrdad, A., Gharehpetian, G.B.: A new simplified model for assessment of power variation of DFIG-based wind farm participating in frequency control system. Electr. Power Syst. Res. **1**(148), 220–229 (2017)
9. Raja, S.K., Badathala, V.P.: LFC problem by using improved genetic algorithm tuning PID controller. Int. J. Pure Appl. Math. **120**(6), 7899–7908 (2018)
10. Charles, K., Urasaki, N., Senjyu, T., Elsayed Lotfy, M., Liu, L.: Robust load frequency control schemes in power system using optimized PID and model predictive controllers. Energies **11**(11), 3070 (2018)
11. Hussein, A.A.: Load frequency control for two-area multi-source interconnected power system using intelligent controllers. Tikrit J. Eng. Sci. **25**(1), 78–86 (2018)
12. Meena, S.K., Chanana, S.: Comparative study of load frequency control using PID and FGPI controller. In: 6th IEEE Power India International Conference (PIICON), pp. 1–6 (2014)
13. Pingkang, L., Hengjun, Z., Yuyun, L.: Genetic algorithm optimization for AGC of multi-area power systems. In: TENCOM 2002 Proceedings: IEEE Region 10 Conference on Computers, Communications, Control and Power Engineering, pp. 1818–1821 (2002)

A Robust Control Design for Minimizing Torque Ripple in PMSMS for Vehicular Propulsion

Aouadj Norediene(✉), Hartani Kada, and T. Mohammed Chikouche

Electrotechnical Engineering Laboratory,
University Tahar Moulay of Saida, Saida, Algeria
norediene.aouadj@gmail.com, kada_hartani@yahoo.fr,
tchikouche@yahoo.fr

Abstract. Permanent magnet synchronous motors are expected to be applied to propulsion systems of electric vehicles for their high power and torque density, high efficiency, large constant power operation region and cost-effectiveness. However, their large torque ripple is an obstacle in practical applications of PMSMs to vehicle propulsion. The torque ripple reduction of PMS machines for vehicular propulsion (traction applications) has also received great attention because large torque ripple can create uncomfortable vibrations and mechanical damage and cause vehicle noise, and even lead to vehicle instability. Here, our objective is to solve more particularly the problems caused by the torque ripple affecting the mechanical transmission of the electric traction chain. The purpose of this paper is to propose a new control technique of PMSMs for vehicular propulsion by decreasing the torque ripple and improving the dynamic performance of direct torque control using a new sliding mode backstepping control. In this proposed method, the control of the torque and flux, which is designed by the nonlinear backstepping control, replaces the hysteresis controllers in the conventional DTC and the sliding mode control is used as speed controller. The simulation results show that the proposed method can obviously reduce the torque and flux ripple, and can provide better speed tracking performance compared with the conventional DTC, so that the system has satisfactory dynamic and static performance.

Keywords: Sliding mode control · Backstepping control · Torque ripple · DTC · PMS in-wheel motor

1 Introduction

Traction motors used in electric vehicles (EVs), differently from motors used in the industrial applications, usually require frequent starts and stops, high rates of acceleration/deceleration, high torque and low-speed hill climbing, low torque and high-speed cruising. Moreover, the traction motors must have two important characteristics: a fast and robust torque response which is needed in a wide speed range to meet the instantaneous torque demand by the driver, and a low torque ripple which must not exceed $\pm 2\%$ in order to avoid uncomfortable mechanical vibrations and vehicle noise.

M. Hatti (Ed.): ICAIRES 2019, LNNS 102, pp. 236–245, 2020.
https://doi.org/10.1007/978-3-030-37207-1_25

Nowadays, permanent magnet synchronous motors (PMSM) have been extensively analyzed as feasible considerate for variable speed electric vehicle (EV) traction applications for their high power and torque density, high efficiency, large constant power operation region, robust mechanical construction and cost-effectiveness [1–3]. However, their large torque ripple is an obstacle in practical applications of PMSMs to vehicle propulsion. The torque ripple reduction of PMS machines for EV traction applications has also received great attention because large torque ripple can create uncomfortable vibrations and mechanical damage and even lead to vehicle instability [4]. For improving the dynamic performance of permanent magnet synchronous drives for electric vehicle propulsion, generally, vector control techniques is prefered. In recent years, direct torque control (DTC) scheme for PMSM drives has received enormous attention in industrial motor drive application due to its potential advantages and practically in the embedded systems (Electric Vehicle). Unfortunately, the major drawback in the direct torque control is high torque ripple, which is due to the presence of hysteresis controllers and the limited number of available voltage vector [3, 5, 6]. Recently, many approaches have been developed in order to obtain fast and robust torque response and to solve the problems caused by the torque ripple affecting the mechanical transmission of the electric traction chain [7–11]. The authors in this paper propose a new control technique of PMS in-wheel electric motors for vehicular propulsion to improve the dynamic performance of direct torque control (DTC) and decrease the torque ripple using a new sliding mode backstepping control. In this study, the sliding mode controller with exponential reaching law is designed.

The chattering phenomenon of SMC is basically a high frequency oscillation effect yielded by the switching inputs of the sliding mode control law. This unwanted effect deteriorates the system performance and could lead to the instability of the system. One way to avoid this effect is by designing appropriated control strategies such as the second order sliding mode control [12] or high order sliding mode control instead of implementing classical sliding mode controller strategies. Another way to solve this problem is by selecting appropriate sliding mode control laws that reduce this unwanted effect such as exponential reaching law [13, 14]. In order to eliminate the electrical speed sensor mounted on the rotor shaft of the PMSM to reduce the system hardware complexity and improve the reliability of the system, a sliding mode observer of speed motor, which was presented in [15], is used in this paper.

Backstepping control (SMB) is now regarded as one of the most robust control strategies in the power traction of the EV with PMS in-wheel electric motors because it offers the fast and accurate torque response with very low torque ripple and It can precisely track the speed trajectory and guarantee a high performance, even during the speed transients under load torque change which are similar to the EV-traction operations.

The following discussion is composed of three sections. Section 2 presents the dynamic mathematical model of the PMSM drive system. Section 3 presents the theoretical basis of the proposed control, and the sliding mode speed controller with exponential reaching law which can suppress the chattering and improve the reaching speed. Lyapunov stability theorem is employed to provide the stability analysis of the system. In Sect. 4, the simulation results provide the evidence of improvements of the proposed SMBC-DTC strategy by indicating a fast torque response and accurate speed tracking. Finally, some conclusions are given in Sect. 5.

2 Dynamic Mathematical Model of the PMSM

The dynamic model of the PMSM in the stationary $\alpha - \beta$ reference frame is given by the following equations [3]. The stator flux linkage is given by

$$\begin{cases} \dot{i_\alpha} = -\frac{R_s}{L_s} i_\alpha + \frac{u_\alpha}{L_S} + \frac{E_\alpha}{L_S} \\ \dot{i_\alpha} = -\frac{R_s}{L_s} i_\beta + \frac{u_\beta}{L_S} + \frac{E_\beta}{L_S} \\ \dot{\Omega_m} = \frac{p}{J}(T_e - T_L) - \frac{f}{J}\Omega_m \end{cases} \quad (1)$$

Where E_α, E_β are the back EMFs along $\alpha - \beta$ axis respectively and can be expressed as

$$\begin{cases} E_\alpha = \Omega_m \Phi_f(-sin\theta) \\ E_\beta = \Omega_m \Phi_f(cos\theta) \end{cases} \quad (2)$$

Stator flux equation:

$$\begin{cases} \Phi_{s\alpha} = L_s i_{s\alpha} + \Phi_f cos\theta \\ \Phi_{s\beta} = L_s i_{s\beta} + \Phi_f sin\theta \end{cases} \quad (3)$$

Stator voltage equation:

$$\begin{cases} v_{s\alpha} = R_s i_{s\alpha} + \frac{d\Phi_{s\alpha}}{dt} \\ v_{s\beta} = R_s i_{s\beta} + \frac{d\Phi_{s\beta}}{dt} \end{cases} \quad (4)$$

Torque equation:

$$T_{em} = \frac{3}{2} p\left(\Phi_{s\alpha} i_{s\beta} - \Phi_{s\beta} i_{s\alpha}\right) \quad (5)$$

The mechanical equation is expressed by:

$$J\frac{d\Omega}{dt} + f\Omega = T_{em} - T_r \quad (6)$$

Where:

$$\omega_m = p\Omega \quad (7)$$

3 Sliding Mode Backstepping DTC Approach

The configuration of the proposed sliding mode backstepping for PMSM drive system based on DTC is comprised of sliding mode speed control and flux and torque control, Fig. 1.

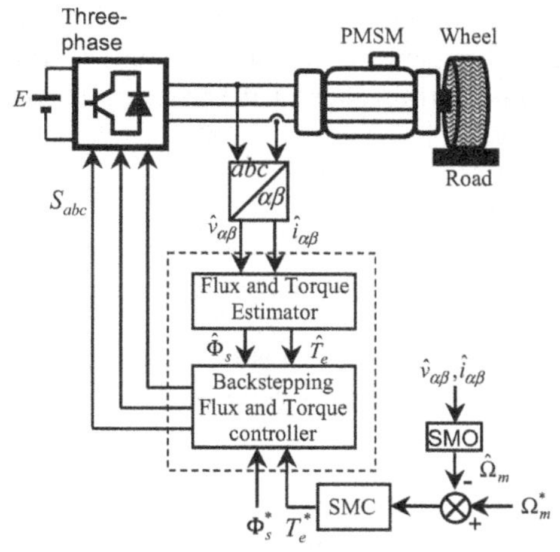

Fig. 1. Block diagram of the Sliding mode Backstepping DTC of PMSM for vehicular propulsion.

The proposed SMB-DTC scheme uses the error of stator flux and the error between the required reference electromagnetic torque (the output of SMC) and the estimated electromagnetic torque to generate the reference voltages u_α^*, and u_β^* which are used by a space vector modulation (SVM) to provide the inverter switching states and ensure the constant switching frequency. However, in conventional DTC scheme, the reference electromagnetic torque is generated by the standard PI based speed controller. Then, the hysteresis controllers are used to control torque and flux [2, 16].

3.1 Speed Controller Design

In order to improve the response of the system and mitigate the chattering effect, the reaching law method can ensure dynamic performance for the sliding mode speed controller [16, 17].

The expression of exponential reaching law is given by the following equation

$$\dot{s} = -\varepsilon\, sgn(s) - ks, \quad \varepsilon > 0, \quad k > 0 \tag{8}$$

When s > 0, the solution for Eq. (8) is given by

$$s(t) = -\tfrac{\varepsilon}{k} + \left(s_0 + \tfrac{\varepsilon}{k}\right)e^{-kt}, \quad s_0 = s(0) \tag{9}$$

When t is sufficiently large, the reaching speed will be faster than the exponential law.

When s(t) = 0, the time required to reach the sliding surface can be given by

$$t = \frac{1}{k}\left[ln\left(s_0 + \frac{\varepsilon}{k}\right) - ln\frac{\varepsilon}{k}\right] \tag{10}$$

The reaching speed can be regulating by adjusting the parameters ε and k directly. The advantages of the reaching law method of the sliding mode control system is that it can guarantee the dynamic performance of the reaching mode and restrain the chattering of the system, as well.

The system state variables is defined as

$$\begin{cases} x_1 = \omega^* - \omega \\ x_2 = \dot{x}_1 = \dot{\omega}^* - \dot{\omega} \end{cases} \tag{11}$$

Where ω^* is the motor speed reference, ω is the actual speed of the motor.

If the sliding surface of the system is defined as

$$s = cx_1 + x_2 \tag{12}$$

Where c must satisfy Hurwitz condition $c > 0$

Therefore, we have

$$\dot{s} = cx_2 + \ddot{\omega}^* + \frac{f}{J}\dot{\omega} - \frac{\dot{T}_e}{J} \tag{13}$$

According to the exponential reaching law and from Eqs. (8) and (13), we have

$$cx_2 + \ddot{\omega}^* + \frac{f}{J}\dot{\omega} - \frac{\dot{T}_e}{J} = -\varepsilon\, sgn(s) - ks$$

Then, we can get the sliding mode controller as

$$T_e^* = J\left[cx_2 + \ddot{\omega}^* + \frac{f}{J}\dot{\omega} + \varepsilon\, sgn(s) + ks\right] \tag{14}$$

Where T_e^* is the required reference torque.

It can be inferred from Eq. (14) that the integral term can act as a filter and attenuate the chattering effect in the sliding mode control.

We choose $V_1 = \frac{1}{2}s^2$ as the Lyapunouv function. The time derivative of function V_1 is as follows

$$\dot{V}_1 = s\dot{s} = s(-\varepsilon\, sgn(s) - ks) \tag{15}$$

The system is stable according to the Lyapunouv stability criterion. Therefore, the negative semi-effective of function V_1 can be guaranteed by an approximation choice of the parameters $\varepsilon > 0$ and $k > 0$, which results in the opposite signs for s and $\dot{s}$.

3.2 Backstepping Torque and Flux Control

The Back-stepping torque and flux controller are designed to achieve the satisfactory torque and flux tracking. Define the following torque and flux tracking errors [16]

$$\begin{aligned} e_T &= T_e^* - T_e \\ e_\Phi &= \Phi^* - \Phi_e \end{aligned} \tag{16}$$

Then, the derivative of torque tracking error dynamic can be obtained as follows

$$\dot{e}_T = J\left[cx_2 + \ddot{\omega}^* + \frac{f}{J}\dot{\omega} + \varepsilon\, sgn(s) + ks\right] - \frac{3}{2}p\begin{bmatrix} \Phi_\beta\left(\frac{R_s}{L_s}i_\alpha + \frac{E_\alpha}{L_s}\right) - \Phi_\alpha\left(\frac{R_s}{L_s}i_\beta - \frac{E_\beta}{L_s}\right) \\ +\left(i_\beta - \frac{\Phi_\beta}{L_s}\right)u_\alpha - \left(i_\alpha - \frac{\Phi_\alpha}{L_s}\right)u_\beta \end{bmatrix} \tag{17}$$

For stabilizing the flux torque components, the flux tracking error dynamics can be defined as

$$\dot{e}_\Phi = 2R_s\Phi_\alpha i_\alpha + 2R_s\Phi_\beta i_\beta - 2\Phi_\alpha u_\alpha - 2\Phi_\beta u_\beta \tag{18}$$

Define the Lyapunouv function V_2 for whole system as

$$V_2 = \frac{1}{2}\left(V_1^2 + e_T^2 + e_\Phi^2\right) \tag{19}$$

Take the derivative of the Lyapunouv function V_2, we can get

$$\dot{V}_2 = V_1\dot{V}_1 + e_T\dot{e}_T + e_\Phi\dot{e}_\Phi = V_1\dot{V}_1 - k_T e_T^2 - k_\Phi e_\Phi^2 \tag{20}$$

Where k_T and k_Φ are positive constants.

According to the above Eq. (20), the final control voltage outputs u_α^* and u_β^* as follows

$$\begin{aligned} u_\alpha^* &= \frac{1}{\Phi_\alpha i_\alpha + \Phi_\beta i_\beta - \Phi/L_s}\left\{\Phi_\beta\left[\left(J\left(cx_2 + \ddot{\omega}^* + \frac{f}{J}\dot{\omega} + \varepsilon\, sgn(s) + ks\right) + k_T e_T\right) \times \frac{2}{3p} + \Phi_\alpha\left(\frac{R_s}{L_s}i_\beta + \frac{E_\beta}{L_s}\right)\right.\right. \\ &\left.\left. -\Phi_\beta\left(\frac{R_s}{L_s}i_\alpha + \frac{E_\alpha}{L_s}\right)\right] + \left(i_\alpha - \frac{\Phi_\alpha}{L_s}\right) \times \frac{1}{2}\left(2R_s\Phi_\alpha i_\alpha + 2R_s\Phi_\beta i_\beta + k_\Phi e_\Phi\right)\right\} \\ u_\beta^* &= \frac{1}{\Phi/L_s - \Phi_\alpha i_\alpha + \Phi_\beta i_\beta}\left\{\Phi_\alpha\left[\left(J\left(cx_2 + \ddot{\omega}^* + \frac{f}{J}\dot{\omega} + \varepsilon\, sgn(s) + ks\right) + k_T e_T\right) \times \frac{2}{3p} + \Phi_\alpha\left(\frac{R_s}{L_s}i_\beta + \frac{E_\beta}{L_s}\right)\right.\right. \\ &\left.\left. -\Phi_\beta\left(\frac{R_s}{L_s}i_\alpha + \frac{E_\alpha}{L_s}\right)\right] + \left(i_\beta - \frac{\Phi_\beta}{L_s}\right) \times \frac{1}{2}\left(2R_s\Phi_\alpha i_\alpha + 2R_s\Phi_\beta i_\beta + k_\Phi e_\Phi\right)\right\} \end{aligned} \tag{21}$$

Substituting the final control voltage outputs (Eq. (21)) in the function $\dot{V}_2$, the time derivative of the Lyapunouv function is desired as

$$\dot{V}_2 = s(-\varepsilon\, sgn(s) - ks) - k_T e_T^2 - k_\Phi e_\Phi^2 \leq 0 \tag{22}$$

4 Simulation Results

In order to verify the effectiveness of the SMB-DTC proposed in this paper, a simulation model of sliding mode backstepping control for PMSM-DTC system is established in Matlab/Simulink. Table 1 gives the nominal parameters of the PMSM which used in the simulation.

Table 1. The specifications of motors

Symbol	Quantity	Value
R_s	Resistance	0,03 Ω
L_s	Inductance	0,2 mH
Φ_f	Permanent magnet flux	0,08 Wb
p	Pole pairs	4

The performance of the system using the proposed method is shown in Fig. 2. The motor is started with the reference speed Ωm* = 200 rad/s and the reference flux 0.08 Wb Then, the reference speed is increased to 300 rad/s. For testing the robustness of the controller, a stepwise load torque is changed from 40 to 60 Nm at t = 0.5 s. Figure 2 show the dynamic performance of the DTC and SMBC drives using IP antiwindup and SMC speed controllers respectively. The two controllers were tested in a wide range of motor speed. The results of Fig. 2 show that the startup of the motor, until it reaches the command speed value 200 rad/s, is made with 40 Nm initial load torque. When the motor runs at 200 rad/s/40 Nm steady state operation, a step speed command of 300 rad/s is given to the drive and the motor reaches again another operation point (300 rad/s/40 Nm). Finally, the controllers are tested to step load torque disturbance.

Simulation results shown in Fig. 2 provide the evidence of improvements of the proposed SMB-DTC strategy over the conventional DTC by indicating a fast torque response (Fig. 2(d)), and an accurate speed tracking performance. Figure 2(a) show the speed response. These responses are detailed in Fig. 2(a), (b) and (c). The actual speed converges with reference speed in very short time with a negligible overshoot and no steady state error. The sliding mode controller rejects the load disturbance rapidly and converges back the reference speed. A zoom details this result in Fig. 2(c).The proposed SMB-DTC method gives a better response of the stator flux linkage during transient-state. Figure 2(e) shows that the stator flux magnitude was constant at its nominal value 0,08Wb. Figures 2(d), (e) and (f) show that the torque and flux ripples are decreased. Figure 2(k), (l) show, in more detail, the comparison of the phase current in both control, during steps of speed command and load torque. It was observed, that the SMB-DTC has better performance than the conventional DTC using IP antiwindup

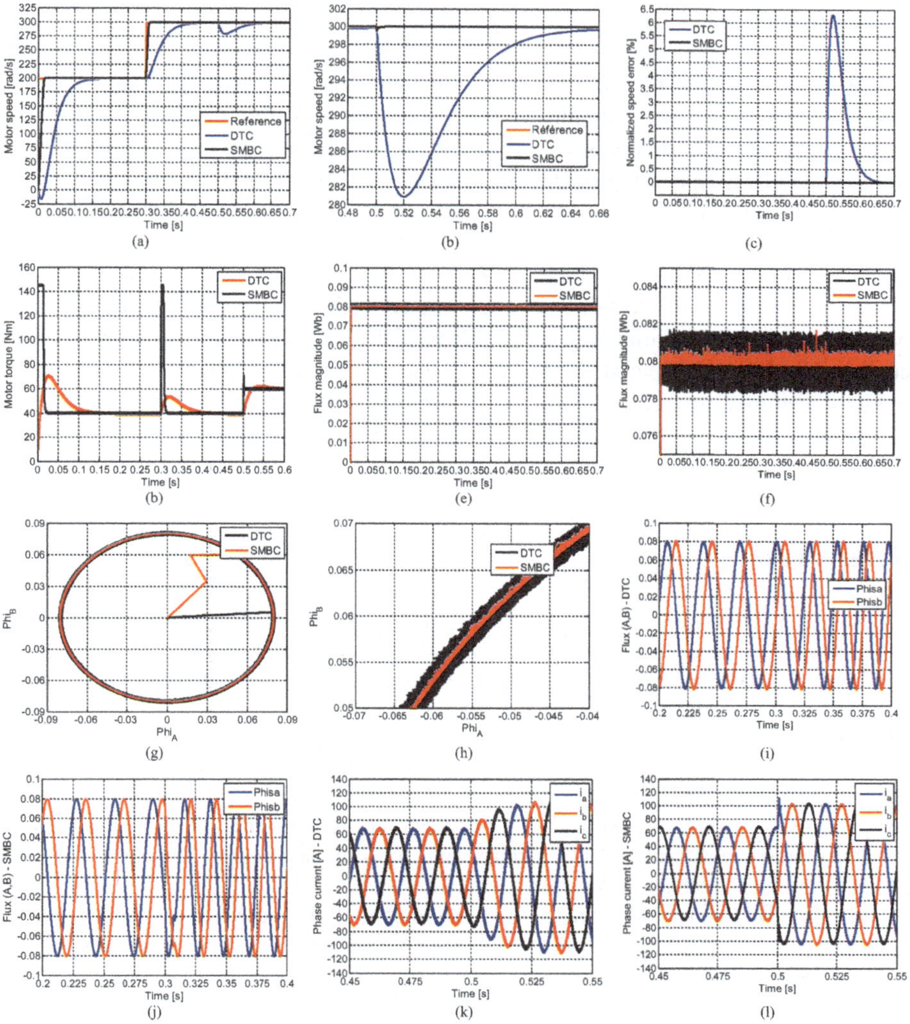

Fig. 2. Motor response with steps of speed command and load torque.

speed controller. Figure 2(k), (l) show that the stator current which increases when load disturbance is applied. Result shows that the torque and flux ripples are decreased in SMB-DTC and provide good speed tracking performance compared with the conventional DTC.

5 Conclusion

Based on the analysis of basic DTC of permanent magnet synchronous motor (PMSM), a new sliding mode backstepping control is proposed in this paper to improve the performance of DTC drives and to reduce torque and flux ripples. According to the speed equation, proper torque and flux controllers were recursively designed by the novel nonlinear backstepping approach, instead of the conventional hysteresis controller method. Lyapunov stability theorem is employed to provide the stability analysis of the system. In the proposed method, the control of the torque and flux which instead of the hysteresis controllers in the conventional DTC is designed by the nonlinear backstepping control. Simulation results from the conventional DTC and SMB-DTC are presented and compared. Result show that the proposed method can obviously reduce the torque and current ripple, can provide better speed tracking performance compared with the traditional DTC, so that the system has satisfactory dynamic and static performance.

References

1. Honda, Y., Nakamura, T., Higaki, T., Takeda, Y.: Motor design considerations and test results of an interior permanent magnet synchronous motor for electric vehicles. In: Conference Record of the 1997 IEEE Thirty-Second IAS Annual Meeting, Industry Applications Conference, IAS 1997, pp. 75–82 (1997)
2. Sekour, Mh, Hartani, K., Draou, A., Allali, A.: Sensorless fuzzy direct torque control for high performance electric vehicle with four in-wheel motors. J. Electr. Eng. Technol. **8**, 530–543 (2013)
3. Hartani, K., Miloud, Y., Miloudi, A.: Improved direct torque control of permanent magnet synchronous electrical vehicle motor with proportional-integral resistance estimator. J. Electr. Eng. Technol. **5**, 451–461 (2010)
4. Foo, G.H.B., Rahman, M.: Direct torque control of an IPM-synchronous motor drive at very low speed using a sliding-mode stator flux observer. IEEE Trans. Power Electron. **25**, 933–942 (2010)
5. Hartani, K., Bourahla, M., Miloud, Y., Sekour, M.: Electronic differential with direct torque fuzzy control for vehicle propulsion system. Turk. J. Electr. Eng. Comput. Sci. **17**, 21–38 (2009)
6. Hartani, K., Merah, A., Draou, A.: Stability enhancement of four-in-wheel motor-driven electric vehicles using an electric differential system. J. Power Electr. **15**, 1244–1255 (2015)
7. Garcia, C.E., Prett, D.M., Morari, M.: Model predictive control: theory and practice—a survey. Automatica **25**, 335–348 (1989)
8. Zhang, Y., Zhu, J.: Direct torque control of permanent magnet synchronous motor with reduced torque ripple and commutation frequency. IEEE Trans. Power Electron. **26**, 235–248 (2011)
9. Malla, S.G., Rao, M.H.L., Malla, J.M.R., Sabat, R.R., Dadi, J., Das, M.M.: SVM-DTC permanent magnet synchronous motor driven electric vehicle with bidirectional converter. In: International Multi-Conference on Automation, Computing, Communication, Control and Compressed Sensing (iMac4s), pp. 742–747 (2013)

10. Sekour, M., Hartani, K., Merah, A.: Electric vehicle longitudinal stability control based on a new multimachine nonlinear model predictive direct torque control. J. Adv. Transp. **2017**, 19 (2017)
11. Huaqiang, Z., Xinsheng, W., Pengfei, W.: Study on direct torque control algorithm based on space vector modulation. Electr. Mach. Control **16**, 13–18 (2012)
12. Bartolini, G., Ferrara, A., Usai, E.: Chattering avoidance by second-order sliding mode control. IEEE Trans. Autom. Control **43**, 241–246 (1998)
13. Zhu, Q., Zhang, P.: A novel reaching law based on integral sliding mode control of permanent magnet synchronous motors. In: 25th Chinese Control and Decision Conference (CCDC), pp. 4646–4651 (2013)
14. Guo, D., Dinavahi, V., Wu, Q., Wang, W.: Sliding mode high speed control of PMSM for electric vehicle based on flux-weakening control strategy. In: 36th Chinese Control Conference (CCC), pp. 3754–3758 (2017)
15. Hartani, K., Draou, A.: A new multimachine robust based anti-skid control system for high performance electric vehicle. J. Electri. Eng. Technol. **9**, 214–230 (2014)
16. Ning, B., Cheng, S., Qin, Y.: Direct torque control of PMSM using sliding mode backstepping control with extended state observer. J. Vib. Control **24**, 694–707 (2016)
17. Azar, A.T.: Advances and Applications in Sliding Mode Control systems. Springer, Cham (2015)

New Direct Power Control Based on Fuzzy Logic for Three-Phase PWM Rectifier

T. Mohammed Chikouche, K. Hartani, S. Bouzar(✉), and B. Bouarfa(✉)

Electrotechnical Engineering, Tahar Moulay University of Saida, BP-138, En-Nasr, Saida, Algeria
tchikouche@yahoo.fr, kada_hartani@yahoo.fr

Abstract. This work proposes a new control strategy of a sinusoidal current absorption AC/DC converter as an alternative to harmonic pollution problems. It is based on the instantaneous direct active and reactive power control of PWM rectifiers known as direct power control (DPC). Although DPC has been considered a powerful and robust control system for PWM rectifiers, high power ripples and variable switching frequency are the two most notable drawbacks of conventional DPC. To overcome these drawbacks, we propose a new strategy of the DPC based on the fuzzy technique. The proposed strategy must provide unity power factor operation with good DC bus voltage regulation and stability and low harmonic distortion of network currents. To confirm the effectiveness of the proposed DPC of two-level PWM rectifier, numerical simulations were carried out under the Matlab/Simulink environment. Better performances will be obtained in transient and permanent state for the appreciable adjustment of the active and reactive instantaneous powers and especially for the case of rates harmonic distortion of current.

Keywords: Direct power control · Fuzzy logic · Power ripple · Power quality · PWM rectifier

1 Introduction

In recent years, significant research has been conducted on the control strategies of three-phase PWM rectifiers. These strategies can be classified according to the use of current control loops or active and reactive power control loops. There are two common control strategies, a voltage oriented control (VOC) and a direct power control (DPC). The VOC can indirectly control the active and reactive input power by controlling the input current of the PWM rectifier. This provides good steady state performance and dynamic responses.

The DPC is another type of high performance control strategy for PWM rectifiers based on the instantaneous power theory first proposed in [1] and more clearly presented in [2, 3]. The basic idea of this control is to choose the best switching state of the power switches through a switching table with hysteresis comparators to maintain a perfectly sinusoidal current and also achieve a unit power factor [4]. Despite the DPC has a simple structure, fast response and robustness, it has two major drawbacks: high

M. Hatti (Ed.): ICAIRES 2019, LNNS 102, pp. 246–258, 2020.
https://doi.org/10.1007/978-3-030-37207-1_26

steady-state power ripples and variable switching frequency, that is caused by hysteresis controllers and the switching table.

In to overcome these disadvantages, various modified DPC configurations have been proposed. In the past, some work has been done to improve conventional DPC by proposing new switching tables [4–9].

In [4], a new proposed switching table allows an improvement of the performances of the control DPC, compared to the conventional table proposed in [8]. Another control strategy called DPC-SVM, proposed by [10] is used, which has the advantages of VOC and DPC [11]. Other nonlinear control methods have been introduced by researchers in recent years, such as sliding mode control [12] and predictive controls [13, 14], which offer a frequency of constant switching and fast dynamic response. Direct Power Predictive Control (P-DPC), combining the DPC strategy with a predictive control strategy [15], chooses a single voltage vector for the next control period to minimize the error between reference powers and the real powers. The optimal voltage vector is determined by a control scheme that minimizes the cost function [16]. The use of predictive DPC algorithms for the control of PWM rectifiers has been presented in [17] (Ant-06) for the first time. Afterwards, other researchers have tried to use prediction algorithms with different basic methods to improve them [18, 19]. However, in such a system, the switching frequency is variable and depends on the sampling frequency, the converter load and parameter variations. To avoid the above disadvantages, predictive control with DPC, which operates with a constant switching frequency, has been developed [20, 21]. Unfortunately, these control techniques require complex calculations. In order to achieve better performance of DPC, a new switching table is proposed in this paper, based on the analysis on the change of active and reactive power, to select the optimal rectifier voltage. The aim of this method is also to eliminate the harmonic current and therefore reduce the total harmonic distortion of the line current and improve the power factor.

This paper proposes an improved DPC for PWM rectifier using fuzzy logic control to overcome the drawbacks and limits of the classical switching table used in conventional DPC. This improved DPC is based on the analysis of instantaneous active and reactive power variations behaviors for different voltage vector in the twelve sectors, so that, we can have power ripple minimization dynamic response and robustness against external load disturbance [22, 23].

To better organize our task, we divided our work into three parts. Section 2 is devoted to the study and modeling of an AC-DC converter, PWM rectifier. In Sect. 3, the authors proposed a new strategy of DPC based on the fuzzy technique, that we name fuzzy DPC to improve the simulation results of the proposed switching table of the DPC control in [4]. In this section, the fuzzy control uses control rules based on the study of variations in the active and reactive powers, caused by the application of each of the control vectors during a complete period of the grid voltage. In Sect. 4, we will present the simulation results obtained from the fuzzy DPC of a two-level three-phase PWM rectifier. Finally, our work will end up with a conclusion.

2 Modeling of Three-Phase PWM Rectifier

2.1 Language Mode of Operation Rectifying/Regeneration

The main advantage of the PWM voltage rectifier over other sinusoidal current absorption converters is its ability to operate in both rectifying mode and regeneration mode [24]. The overall equivalent diagram of the three-phase PWM rectifier with sinusoidal current absorption, illustrating the two modes of its rectifying and regenerating operation, is shown in Fig. 1.

During these two modes of operation, the DC bus voltage is controllable by exchanging part of the power transited to charge or discharge the capacitor. Note that for ideal rectifier operation, the inverter must behave as a network-side voltage source and as a load-side power source. Respecting the equilibrium of powers requires the control of the active and reactive fundamental powers and the minimization of those due to the harmonics. In addition, the DC component of the DC bus voltage must be controllable regardless of the nature of the load connected to the output of the rectifier, linear or nonlinear, passive or active.

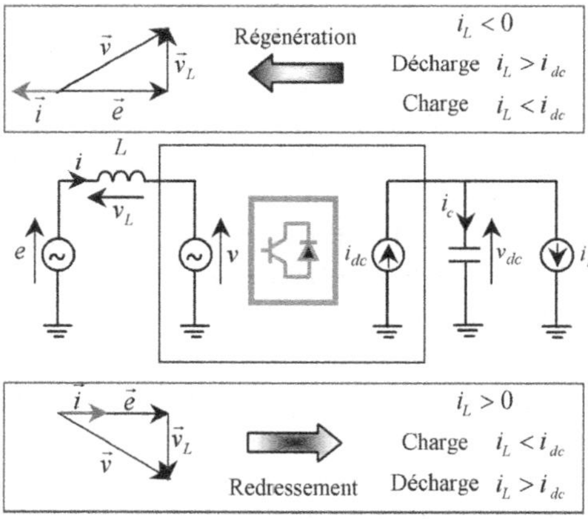

Fig. 1. Principle of the bidirectional transit of the active power in PWM rectifier and its equivalent circuit.

2.2 Modelling of the Power Circuit of PWM Rectifier

Widely described in the literature [4, 25, 26], the power part of the PWM rectifier comprises six power transistors with antiparallel diodes to ensure the bidirectional power conversion described before. Figure 2 shows the diagram of the three-phase voltage PWM rectifier on which our study is based.

In this study, we consider the ideal case of two-level three-phase PWM rectifier which is modelled by ideal switches with instantaneous commutation. From the diagram of Fig. 2, we can deduce the equations of the system [4, 25].

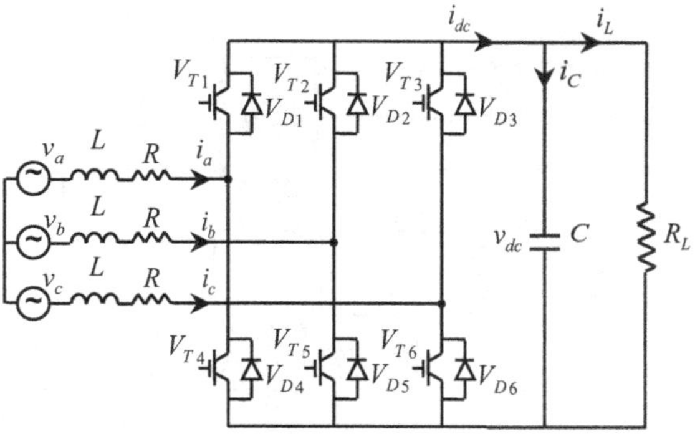

Fig. 2. Power scheme of PWM rectifier.

The model of the rectifier in ( ) frame is given by the following equation system:

$$\begin{cases} L\frac{di_a}{dt} = v_a - Ri_a + \frac{S_b+S_c-2S_a}{3}v_{dc} \\ L\frac{di_b}{dt} = v_b - Ri_b + \frac{S_a+S_c-2S_b}{3}v_{dc} \\ L\frac{di_c}{dt} = v_c - Ri_c + \frac{S_a+S_b-2S_c}{3}v_{dc} \\ C\frac{dv_{dc}}{dt} = S_a i_a + S_b i_b + S_c i_c - \frac{v_{dc}}{R_L} \end{cases} \tag{1}$$

With the source phase voltages are expressed as:

$$\begin{cases} v_a = V_m \sin(\omega t) \\ v_b = V_m \sin(\omega t - 2\pi/3) \\ v_c = V_m \sin(\omega t - 4\pi/3) \end{cases} \tag{2}$$

3 Principe of Fuzzy DPC

The direct power control strategy has been proposed by Noguchi [8]. The overall structure of the DPC applied to the three-phase PWM rectifier bridge is illustrated in Fig. 3. The principle of DPC was detailed by the authors in [4].

In this section, we present the modelling steps and the principle of fuzzy DPC applied to the three-phase PWM rectifier bridge, Fig. 4. The main feature of the proposed fuzzy DPC is the suppression of the hysteresis comparators and the switching table.

After estimating the two instantaneous active and reactive powers, in a method identical to that of the conventional DPC, we compared the estimated reactive power and its reference value. And in the same way, we compare the estimated active power with its control value which is generated by an IP regulator of the DC bus voltage. The active and reactive power errors obtained, as well as the sector of membership of the

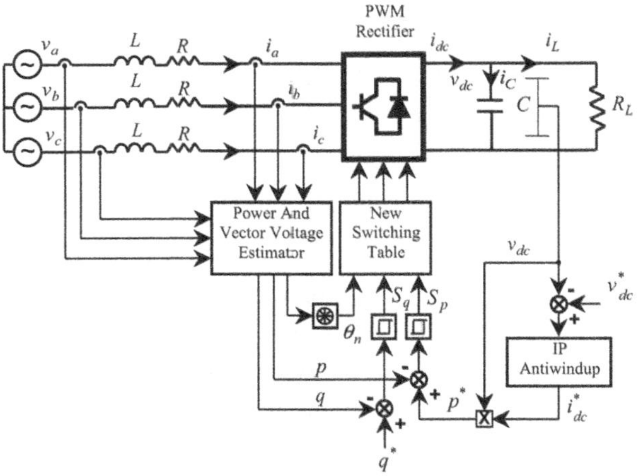

Fig. 3. The block diagram of the DPC control of the three-phase PWM rectifier.

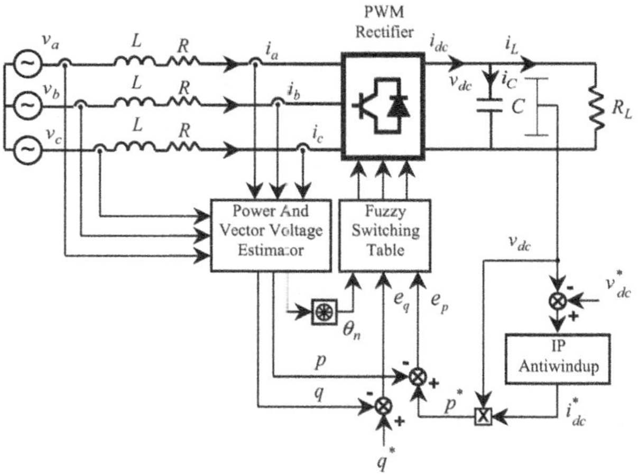

Fig. 4. Block diagram of the fuzzy DPC of a three-phase PWM rectifier.

voltage vector are used as inputs of a fuzzy controller, denoted FLC. This controller is responsible for generating the states, and for controlling the switches of the three-phase rectifier bridge and driving the active and reactive power to their reference values in an optimal manner [22].

The active and reactive power errors are expressed by the following relations:

$$\begin{aligned} e_p &= p^* - q \\ e_q &= q^* - q \end{aligned} \tag{3}$$

Where, the active and reactive instantaneous powers are defined as follows [27, 28]:

$$\begin{cases} p = e_\alpha i_\alpha + e_\beta i_\beta \\ q = e_\alpha i_\beta - e_\beta i_\alpha \end{cases} \quad (4)$$

Table 1 summarizes the different possibilities for determining the sector in which the voltage vector evolves, according to the signs of the two components e_α and e_β, and comparing e_α with ($e_{\alpha\max\_1} = 1/2V_s$, $e_{\alpha\max\_2} = \sqrt{3}/2V_s$) and e_β with ($e_{\beta\max\_1} = 1/2V_s$, $e_{\beta\max\_2} = \sqrt{3}/2V_s$).

Table 1. Location of the voltage vector position.

Sign of e_α	Sign of e_β	Quadrant	Boundaries of the sectors	Belonging sector
$e_\alpha > 0$	$e_\beta > 0$	1	$e_\beta < e_{\beta\max\_1}$	2
			$e_{\beta\max\_1} < e_\beta < e_{\beta\max\_2}$	3
			$e_\beta > e_{\beta\max\_2}$	4
$e_\alpha < 0$	$e_\beta > 0$	2	$e_\beta < e_{\beta\max\_1}$	7
			$e_{\beta\max\_1} < e_\beta < e_{\beta\max\_2}$	6
			$e_\beta < e_{\beta\max\_2}$	5
$e_\alpha < 0$	$e_\beta < 0$	3	$e_\beta > -e_{\beta\max\_1}$	8
			$-e_{\beta\max\_2} < e_\beta < -e_{\beta\max\_1}$	9
			$e_\beta < -e_{\beta\max\_2}$	10
$e_\alpha > 0$	$e_\beta < 0$	4	$e_\beta > -e_{\beta\max\_1}$	1
			$-e_{\beta\max\_2} < e_\beta < -e_{\beta\max\_1}$	12
			$e_\beta < -e_{\beta\max\_2}$	11

In this strategy, the fuzzy controller FLC is designed to have three fuzzy state variables to directly reach the power control. Active and reactive power errors (e_p, e_q) are multiplied by "scaling factors" to obtain the normalized magnitudes e_p^n and e_q^n using the trapezoidal and triangular membership functions, Fig. 6. These variables are used by the fuzzification block to be transformed into fuzzy values $\tilde{e}_p$ and $\tilde{e}_q$. These values are used with the position of the voltage vector (belonging sector) by the block of Mamdani-type fuzzy control rules to obtain directly the optimal switching state (S_a, S_b, S_c) of the switches constituting the PWM rectifier at means of fuzzy logical rules. These rules are designed to limit instantaneous active and reactive power errors simultaneously for each sampling time, as shown in Fig. 4.

The fuzzy the basic scheme of the proposed fuzzy controller is given in Fig. 5. To develop this controller the following steps are followed:

- Normalization of inputs variables by multiplying them by scaling factors.
- Fuzzification of the normalized variables.
- Selection of the appropriate output which must be between 0 and 1, taking as reference Table 2.

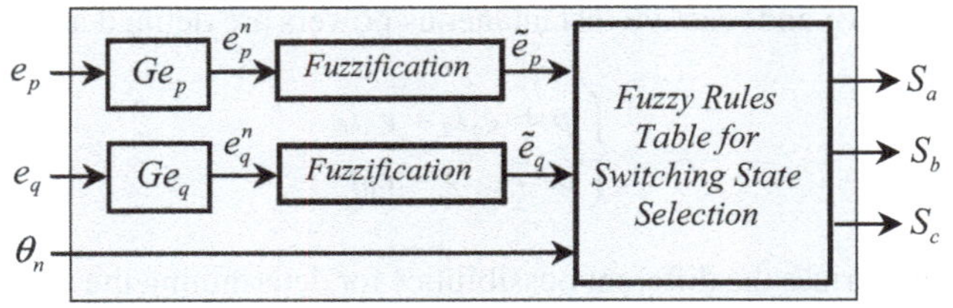

Fig. 5. Structure of the fuzzy controller for the estimation of the switching state of the three-phase PWM rectifier switches.

Controller's performance is based both on the shape of the membership functions and the fuzzy rules.

Figure 6 gives the membership functions for the two input variables of the controller (e_p and e_q) where each variable is subdivided into three fuzzy sets N, Z and P. The discourse universes used are (−0.4, 0.4) for active and reactive power errors. Trapezoidal and triangular membership functions have been chosen and distributed over the discourse universe as shown in Fig. 6. These functions remain symmetrical with respect to zero.

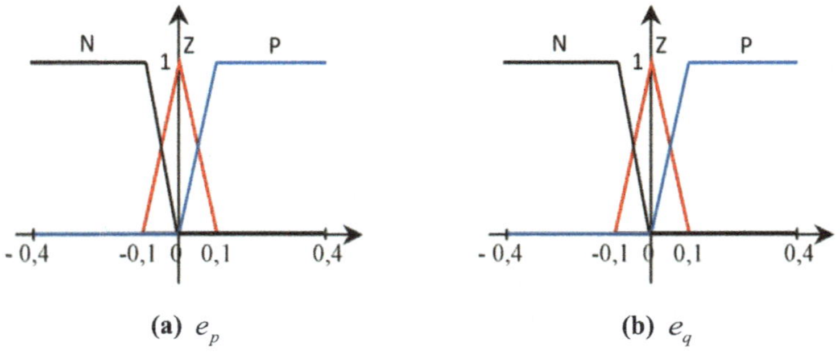

Fig. 6. Membership functions for the input variables of the fuzzy controller.

For more accuracy of the position θ_n of the source-side voltage vector, in the frame $\alpha - \beta$, the discourse universe of this fuzzy variable is divided into twelve fuzzy sets θ_1 to θ_n as shown in Fig. 7.

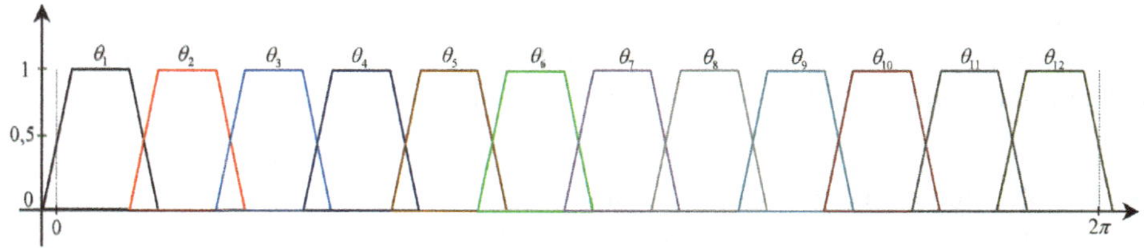

Fig. 7. The membership functions for the voltage vector position θ_n.

The output variable of the fuzzy controller is designed to have seven singleton subsets, one zero voltage vector, and six nonzero voltage vectors. The membership functions of the output voltage vectors are shown in Fig. 8.

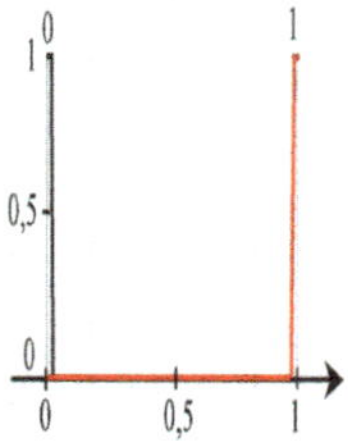

Fig. 8. The membership functions of the output switching states S_a, S_b, S_c.

We use the DPC control switching table defined in [4] for constructing the inference table of the fuzzy controller. The set of control rules is characterized by the input variables and the output variable i.e. control variable [22]. The fuzzy inference rules are described by the flowchart in Fig. 9. Each rule generates a large control action based on the input values of the variables. The total number of rules is 108 shown in Table 2. This table shows the voltage vector applied to the converter for a network voltage vector position. In our case, Mamdani (max-min) method is used as an inference method.

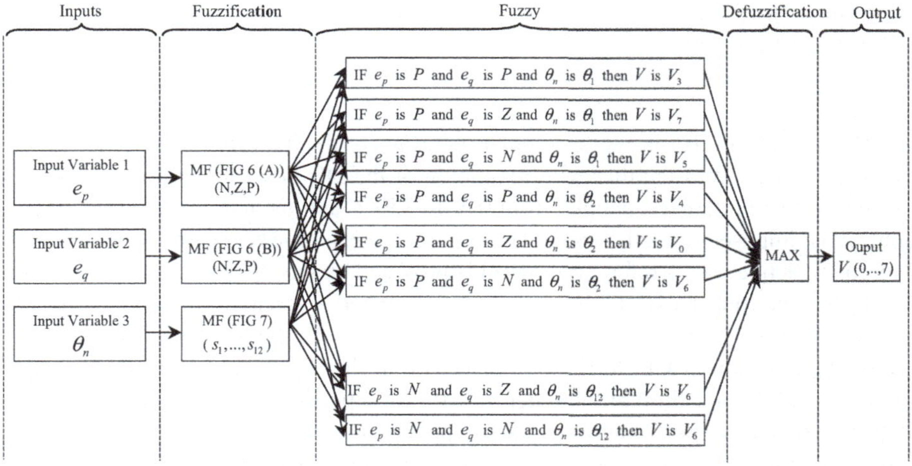

Fig. 9. Membership functions for the input variables of the fuzzy controller.

Table 2. Selection table of the output voltage vector.

e_p	e_q	θ_1	θ_2	θ_3	θ_4	θ_5	θ_6	θ_7	θ_8	θ_9	θ_{10}	θ_{11}	θ_{12}
P	P	V_3	V_4	V_4	V_5	V_5	V_6	V_6	V_1	V_1	V_2	V_2	V_3
	Z	V_7	V_0	V_7	V_0	V_7	V_0	V_7	V_0	V_7	V_0	V_7	V_0
	N	V_5	V_6	V_6	V_1	V_1	V_2	V_2	V_3	V_3	V_4	V_4	V_5
Z	P	V_1	V_2	V_2	V_3	V_3	V_4	V_4	V_5	V_5	V_6	V_6	V_1
	Z	V_7	V_0	V_7	V_0	V_7	V_0	V_7	V_0	V_7	V_0	V_7	V_0
	N	V_6	V_1	V_1	V_2	V_2	V_3	V_3	V_4	V_4	V_5	V_5	V_6
N	P	V_1	V_2	V_2	V_3	V_3	V_4	V_4	V_5	V_5	V_6	V_6	V_1
	Z	V_1	V_1	V_2	V_2	V_3	V_3	V_4	V_4	V_5	V_5	V_6	V_6
	N	V_6	V_1	V_1	V_2	V_2	V_3	V_3	V_4	V_4	V_5	V_5	V_6

4 Simulation Results

The control system was implemented in the environment of MATLAB/Simulink. And the parameters of the PWM rectifier in this paper are given in Table 3.

Table 3. Parameters of PWM rectifier.

Line resistance: $R =$ 0.3 Ω
Line inductance: $L = 37$ Mh
DC bus capacitor: $C_{dc} = 1100$ μF
Voltage source: $V = 125$ V
Line voltage frequency: $F = 50$ Hz
DC voltage: $V_{dc} = 350$ V

The set of Figs. 10 shows the simulation results of the fuzzy direct power control of the three-phase PWM rectifier with regulation of the DC bus voltage. In this case, the DC bus voltage reference stepped from 0 V to 350 V and the reactive power reference control is imposed directly equal to zero to have a power factor close to unity. Moreover, active power reference control is provided by the DC bus voltage regulation.

To test the robustness of the regulation, we applied a variation of the load (increase and decrease) between the instants $t = 0,2$ s and $t = 0,4$ s. Figure 10(1) shows the DC bus voltage response with load disturbance. From this figure and after the load step, there is small voltage drop of the order of 0.1% for a period of 0.02 s, and in the steady state the IP antiwindup controller keeps the DC bus voltage of the rectifier at the reference input value. There is therefore a satisfactory operation both in transient and steady states.

Figures 10(2) and (3) represent the currents source in the (α, β) and (a, b, c) frame respectively, corresponding to the considered operation which responds instantaneously to the load variation.

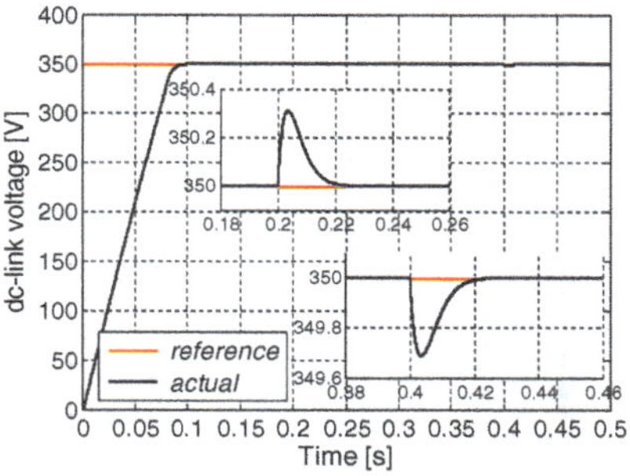

(1) DC bus voltage.

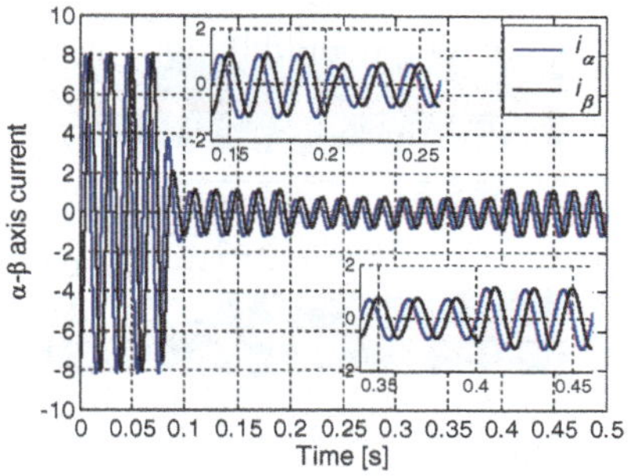

(2) i_α , i_β currents.

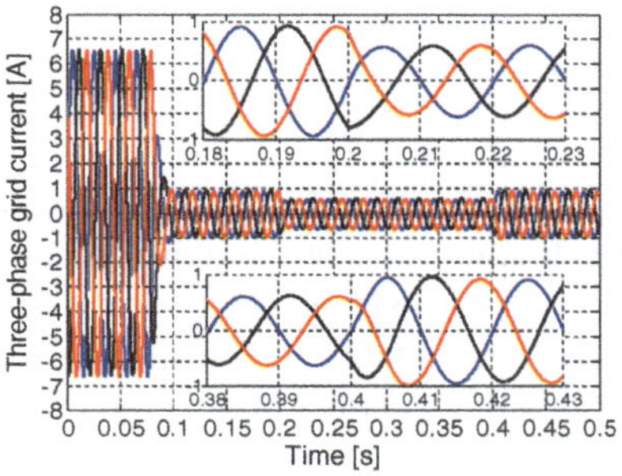

(3) i_a , i_b , i_c grid currents.

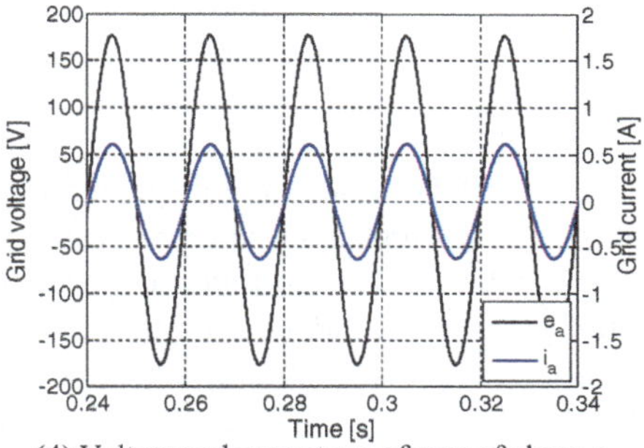

(4) Voltage and current waveforms of phase a .

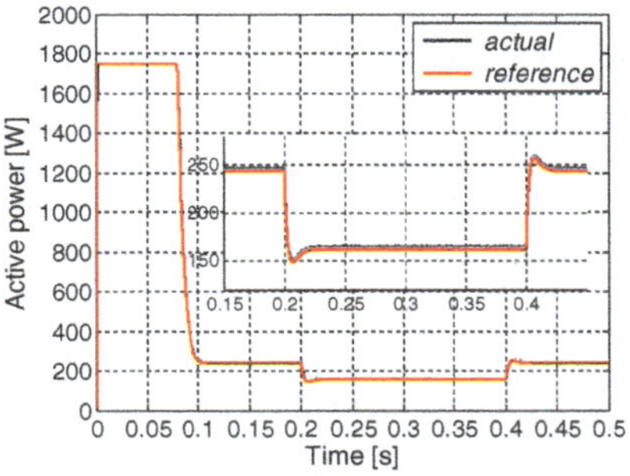

(5) Active power.

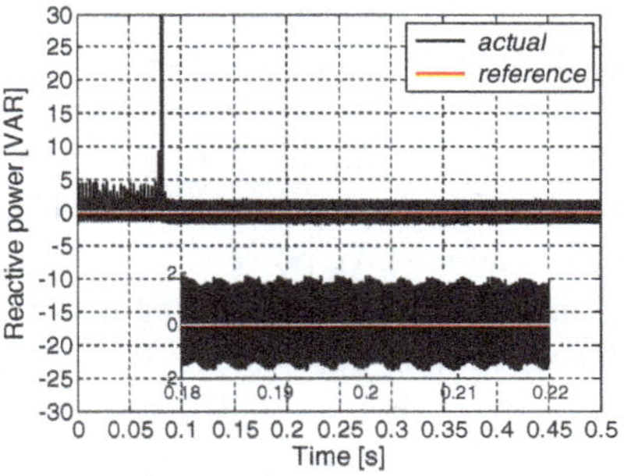

(6) Reactive power.

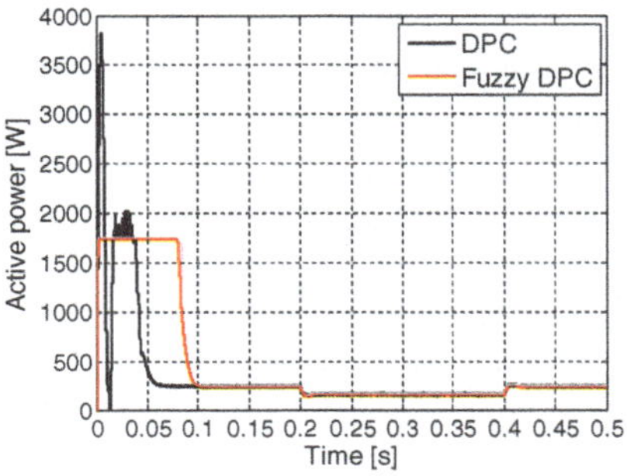

(7) Active power for DPC and Fuzzy DPC.

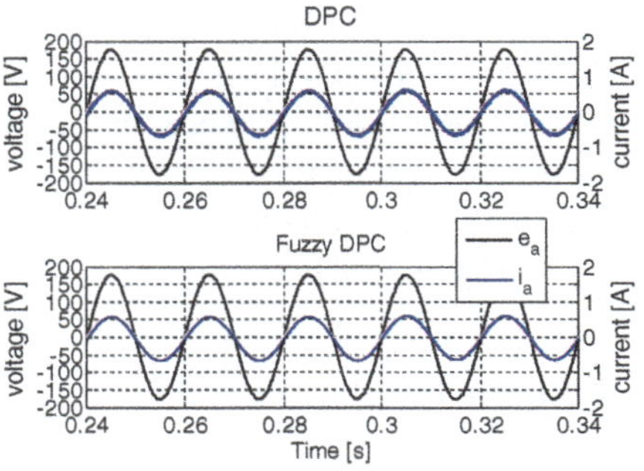

(8) Voltage and current waveforms of phase a for (a) DPC and (b) Fuzzy DPC.

Fig. 10. Simulation results.

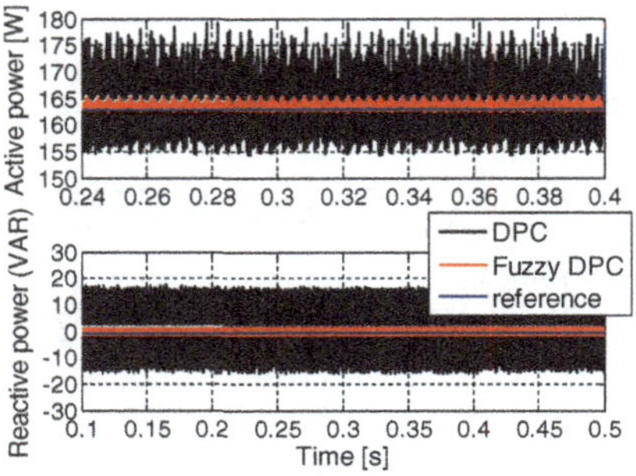

(9) Active power and reactive power.

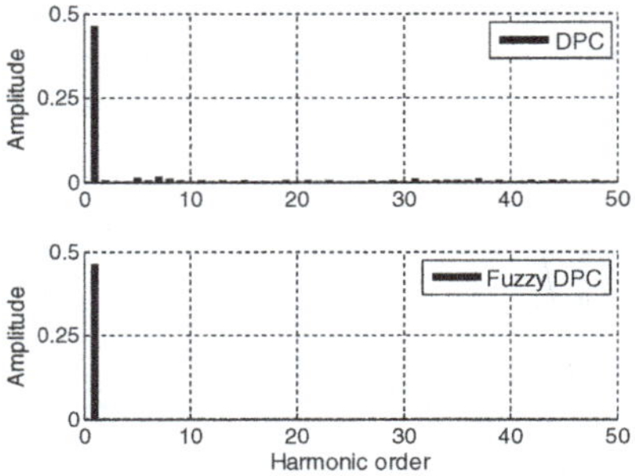

(10) Current harmonics.

Fig. 10. *(continued)*

Figure 10(4) shows the voltage and current of a phase power source. It can be seen that the line current is in phase with line voltage. It confirms that the power factor is unity and the source currents are nearly a sine wave, even in the transient state (THD = 0.31%). It can be seen that the system behaves suitably with respect to load variation and provides a unit power factor, and the PWM rectifier absorbs sinusoidal currents.

In Fig. 10(5), the active power reaches a maximum value in a transient state, then it drops almost instantaneously towards its limit value with a ripple of 1 W amplitude in a steady state. It can be seen from Fig. 10(5) and (6), that the active and reactive powers properly follow their references in spite of the load variation which confirms the robustness of the control. It is observed in Fig. 10(6) that the reactive power is not disturbed during the application of the load, which shows the good decoupling between the active and reactive powers. Consequently, the Fuzzy DPC ensures good control of active and reactive powers during all sectors.

Noting in Fig. 10(9) a significant attenuation of the ripples of the active and reactive powers. Figure 10(10) shows the harmonic spectrum of the response of the grid current i_a. It is noted that all the low render harmonics are well attenuated, which gives a rate of harmonic distortion (THD = 0.31%). This shows that the Fuzzy DPC is more robust than the conventional DPC.

Finally, it is clearly seen that both control methods can achieve fast and accurate tracking of the power reference, and decoupled control of active power and reactive power is achieved, and the operation of unity power factor is clearly seen. Therefore, the Fuzzy DPC presents much better the steady state performance in terms of power ripples and current harmonics.

5 Conclusions

In this article, we have proposed a new strategy of the DPC based on the fuzzy technique. This strategy provides unity power factor operation with good DC voltage regulation and stability and low harmonic distortion of grid currents. The simulation results show that better performances are obtained in steady state and transient for the appreciable adjustment of the active and reactive instantaneous powers and the DC bus voltage. Thus, the currents absorbed by the PWM rectifier have a sinusoidal wave, and the performances in steady state and transient are clearly better compared to the conventional DPC.

References

1. Takahashi, I., Noguchi, T.: A new quick-response and high-efficiency control strategy of an induction motor. IEEE Trans. Ind. Appl. **5**, 820–827 (1986)
2. Malinowski, M., Kazmierkowski, M.P.: Control of three-phase PWM rectifiers. In: Control in Power Electronics–Selected Problems (2002)
3. Malinowski, M., Kazmierkowski, M.P., Trzynadlowski, A.M.: A comparative study of control techniques for PWM rectifiers in AC adjustable speed drives. IEEE Trans. Power Electron. **18**, 1390–1396 (2003)
4. Mohammed-Chikouche, T., Hartani, K.: Direct power control of three-phase PWM rectifier based on new switching table. J. Eng. Sci. Technol. **13**, 1751–1763 (2018)
5. Baktash, A., Vahedi, A., Masoum, M.: Improved switching table for direct power control of three-phase PWM rectifier. In: Power Engineering Conference, AUPEC 2007, Australasian Universities, pp. 1–5 (2007)
6. Bouafia, A., Gaubert, J.-P., Krim, F.: Analysis and design of new switching table for direct power control of three-phase PWM rectifier. In: Power Electronics and Motion Control Conference, EPE-PEMC, pp. 703–709 (2008)
7. Alonso-Martínez, J., Carrasco, J.E.-G., Arnaltes, S.: Table-based direct power control: a critical review for microgrid applications. IEEE Trans. Power Electron. **25**, 2949–2961 (2010)
8. Noguchi, T., Tomiki, H., Kondo, S., Takahashi, I.: Direct power control of PWM converter without power-source voltage sensors. IEEE Trans. Ind. Appl. **34**, 473–479 (1998)
9. Gong, B., Wang, K., Zhang, J., You, J., Luo, Y., Wenyi, Z.: Advanced switching table for direct power control of a three-phase PWM rectifier. In: Transportation Electrification Asia-Pacific (ITEC Asia-Pacific), IEEE Conference and Expo, pp. 1–5 (2014)
10. Malinowski, M., Jasinski, M., Kazmierkowski, M.P.: Simple direct power control of three-phase PWM rectifier using space-vector modulation (DPC-SVM). IEEE Trans. Ind. Electron. **51**, 447–454 (2004)
11. Maghamizadeh, M., Fathi, H.: Virtual flux based direct power control of a three-phase rectifier connected to an LCL filter with sensorless active damping. In: Power Electronics and Drive Systems Technologies Conference (PEDSTC), pp. 476–481 (2016)
12. Lu, W., Li, C., Xu, C.: Sliding mode control of a shunt hybrid active power filter based on the inverse system method. Int. J. Electr. Power Energy Syst. **57**, 39–48 (2014)
13. Zhang, Y., Peng, Y., Qu, C.: Model predictive control and direct power control for PWM rectifiers with active power ripple minimization. IEEE Trans. Ind. Appl. **52**, 4909–4918 (2016)
14. Zhang, Y., Peng, Y., Yang, H.: Performance improvement of two-vectors-based model predictive control of PWM rectifier. IEEE Trans. Power Electron. **31**(8), 6016–6030 (2016)
15. Song, Z., Chen, W., Xia, C.: Predictive direct power control for three-phase grid-connected converters without sector information and voltage vector selection. IEEE Trans. Power Electron. **29**, 5518–5531 (2014)

16. Cho, T.Y., Lee, K.-B.: Virtual-flux-based predictive direct power control of three-phase PWM rectifiers with fast dynamic response. IEEE Trans. Power Electron. **31**, 3348–3359 (2016)
17. Antoniewicz, P., Kazmierkowski, M.: Predictive direct power control of three-phase boost rectifier. Bull. Pol. Acad. Sci. Tech. Sci. **54** (2006)
18. Larrinaga, S.A., Vidal, M.A.R., Oyarbide, E., Apraiz, J.R.T.: Predictive control strategy for DC/AC converters based on direct power control. IEEE Trans. Ind. Electron. **54**, 1261–1271 (2007)
19. Antoniewicz, P., Kazmierkowski, M.P.: Virtual-flux-based predictive direct power control of AC/DC converters with online inductance estimation. IEEE Trans. Ind. Electron. **55**, 4381–4390 (2008)
20. Amirkafi, M., Moghani, J.S., Khoshsaadat, A.: Performance improvement of predictive direct power control of a PWM rectifier based on virtual flux under unbalanced grid voltage. In: Power Electronics, Drives Systems and Technologies Conference (PEDSTC), pp. 265–270 (2018)
21. Mehreganfar, M., Saeedinia, M.H., Davari, S.A., Khaburi, D.A.: Direct power control of AFE rectifier by line voltage sensorless predictive technique and MRAS inductance estimator. In: Power Electronics, Drives Systems and Technologies Conference (PEDSTC), pp. 247–252 (2018)
22. Lamterkati, J., Khaffalah, M., Ouboubker, L., El-Afia, A.: Fuzzy logic based improved direct power control of three-phase PWM rectifier. In: Proceedings of the International Conference in Electrical and Information Technologies (ICEIT), Tangiers, Morocco, pp. 125–130 (2016)
23. Baktash, A., Vahedi, A., Masoum, M.A.S.: New switching table for improved direct power control of three-phase PWM rectifier. Aust. J. Electr. Electron. Eng. **5**, 161–167 (2009)
24. Rodríguez, J.R., Dixon, J.W., Espinoza, J.R., Pontt, J., Lezana, P.: PWM regenerative rectifiers: state of the art. IEEE Trans. Ind. Electron. **52**, 5–22 (2005)
25. Hartani, K., Miloud, Y.: Control strategy for three phase voltage source PWM rectifier based on the space vector modulation. Adv. Electr. Comput. Eng. **10**(3), 61–65 (2010)
26. Blasko, V., Kaura, V.: A new mathematical model and control of a three-phase AC-DC voltage source converter. IEEE Trans. Power Electron. **12**(1), 116–123 (1997)
27. Herrera, R.S., Salmerón, P., Kim, H.: Instantaneous reactive power theory applied to active power filter compensation: different approaches, assessment, and experimental results. IEEE Trans. Ind. Electron. **55**, 184–196 (2008)
28. Superti-Furga, G., Todeschini, G.: Discussion on instantaneous strategies for control of active filters. IEEE Trans. Power Electron. **23**, 1945–1955 (2008)

Advanced Lateral Control of Electric Vehicle Based on Fuzzy Front Steering System

Aouadj Norediene(✉), Hartani Kada, and Merah Abdelkader

Electrotechnical Engineering Laboratory,
University Tahar Moulay of Saida, Saïda, Algeria
norediene.aouadj@gmail.com, kada_hartani@yahoo.fr,
merah_aek@yahoo.fr

Abstract. This paper presents a robust control method for stabilizing vehicle lateral motion of electric vehicle equipped with active front steering system. There are two main objectives, the first is to improve electric vehicle lateral handling performance by using a fuzzy logic technique for controlling the yaw rate, and the second is to solve the problems caused by the torque ripple affecting the mechanical transmission of the electric traction chain, by using a new sliding mode backstepping control which is based on direct torque control. In this proposed method, the control of the torque and flux, which is designed by the nonlinear backstepping control, replaces the hysteresis controllers in the conventional DTC. The sliding mode control which is used as speed controller. The simulation results show that the proposed fuzzy AFS control can stabilize electric vehicle lateral motion and enhance lateral handling performance. Simultaneously, the new sliding mode backstepping control can obviously reduce the torque ripple, and can provide better speed tracking performance.

Keywords: Active front steering · Electric vehicle · Backstepping control · Sliding mode controller · Fuzzy logic controller

1 Introduction

The electric vehicles (EVs) have been receiving considerable attention as a solution for energy and environmental problems. as Moreover, EVs have many advantages over internal combustion engine vehicles owing to the use of electric motors and inverters in EV drivetrains [1]. With the superior control performance of electric motors compared to ICEVs, electric vehicles could not only be clean, but also able to achieve higher levels of safety and handling [2, 3]. Traction motors used in EVs, differently from motors used in the industrial applications, usually require frequent starts and stops, high rates of acceleration/deceleration, high torque and low-speed hill climbing, low torque and high-speed cruising. Moreover, the traction motors must have two important characteristics: a fast and robust torque response which is needed in a wide speed range to meet the instantaneous torque demand by the driver, and a low torque ripple which must not exceed $\pm 2\%$ in order to avoid uncomfortable mechanical vibrations and vehicle noise. The control of an electric vehicle is carried out by the control of its traction chain, which is based by controlling the in-wheel motors. In a few years ago,

M. Hatti (Ed.): ICAIRES 2019, LNNS 102, pp. 259–271, 2020.
https://doi.org/10.1007/978-3-030-37207-1_27

direct torque control (DTC) scheme for PMSM drives has received enormous attention in industrial motor drive application due to its potential advantages and practically in the embedded systems (Electric Vehicle). Unfortunately, the major drawback in the direct torque control is high torque ripple, which is due to the presence of hysteresis controllers and the limited number of available voltage vector.

Recently, many approaches have been developed in order to obtain fast and robust torque response and to solve the problems caused by the torque ripple affecting the mechanical transmission of the electric traction chain [4–7]. The authors in this paper propose a new control technique of PMS in-wheel electric motors for vehicular propulsion to improve the dynamic performance of direct torque control (DTC) and decrease the torque ripple using a new sliding mode backstepping control. In this study, the sliding mode controller with exponential reaching law is designed. The chattering phenomenon of SMC is basically a high frequency oscillation effect yielded by the switching inputs of the sliding mode control law. This unwanted effect deteriorates the system performance and could lead to the instability of the system. One way to avoid this effect is by designing appropriated control strategies such as the second order sliding mode control [8] or high order sliding mode control instead of implementing classical sliding mode controller strategies. Another way to solve this problem is by selecting appropriate sliding mode control laws that reduce this unwanted effect such as exponential reaching law [9]. In order to eliminate the electrical speed sensor mounted on the rotor shaft of the PMSM to reduce the system hardware complexity and improve the reliability of the system, a sliding mode observer of speed motor, which was presented in [10], is used in this paper.

Several studies on yaw-rate control in the field of vehicle dynamics and control have been reported for stabilizing vehicles' cornering motions. For ICEVs without a torque-distribution mechanism and EVs without in-wheel motors, active front and rear systems are used for controlling the yaw rate [9]. In contrast, for EVs with all-wheel independent drive systems, such as in-wheel motors, yaw-rate control methods have been studied based on the active yaw moment generated owing to the torque difference between the left and right motors [11, 12].

The following discussion is composed of seven sections. Section 2 presents the theoretical basis of the SMBC-DTC control for the PMSM drive system, and the sliding mode speed controller with exponential reaching law which can suppress the chattering and improve the reaching speed. Lyapunov stability theorem is employed to provide the stability analysis of the system. The objective of Sect. 3 is the development of an Active Safety System (AFS) using the fuzzy logic control by adding a corrective steering angle to driver maneuvers, which gives a better roadhandling. Section 4 presents the simulation results via Matlab/Simulink which show the effectiveness of the proposed AFS system in lateral stability for EV under double lane change and various road surface conditions. Finally, some conclusions are given in Sect. 5.

2 Sliding Mode Backstepping DTC Approach

The configuration of the proposed sliding mode backstepping for PMSM drive system based on DTC is comprised of sliding mode speed control and flux and torque control, Fig. 1 [13, 14]. The dynamic model of the PMSM in the stationary $\alpha - \beta$ reference frame is given in [4]. The proposed SMB-DTC scheme uses the error of stator flux and the error between the required reference electromagnetic torque (the output of SMC) and the estimated electromagnetic torque to generate the reference voltages u_{α}^{*}, and u_{β}^{*} which are used by a space vector modulation (SVM) to provide the inverter switching states and ensure the constant switching frequency. However, in conventional DTC scheme, the reference electromagnetic torque is generated by the standard PI based speed controller. Then, the hysteresis controllers are used to control torque and flux [5].

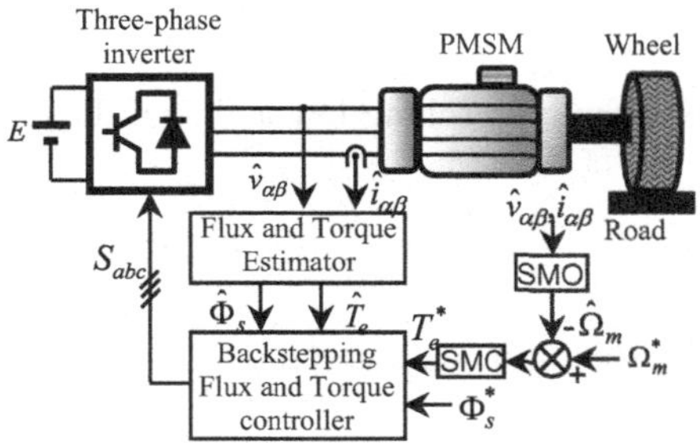

Fig. 1. Block diagram of the sliding mode backstepping DTC of PMSM for vehicular propulsion.

2.1 Speed Controller Design

In order to improve the response of the system and mitigate the chattering effect, the reaching law method can ensure dynamic performance for the sliding mode speed controller [15]. The exponential reaching law is given by [16].

$$\dot{s} = -\varepsilon\, sgn(s) - ks, \quad \varepsilon > 0, \quad k > 0 \tag{1}$$

When $s > 0$, the solution for Eq. (1) is given by

$$s(t) = -\frac{\varepsilon}{k} + \left(s_0 + \frac{\varepsilon}{k}\right)e^{-kt}, \quad s_0 = s(0) \tag{2}$$

The reaching speed can be regulating by adjusting the parameters ε and k directly. The advantages of the reaching law method of the sliding mode control system is that it can guarantee the dynamic performance of the reaching mode and restrain the chattering of the system, as well [16].

The system state variables is defined as [16]

$$\begin{cases} x_1 = \omega^* - \omega \\ x_2 = \dot{x}_1 = \dot{\omega}^* - \dot{\omega} \end{cases} \tag{3}$$

Where ω^* is the motor speed reference, ω is the actual speed of the motor. If the sliding surface of the system is defined as

$$\mathrm{s} = \mathrm{c}\mathrm{x}_1 + \mathrm{x}_2. \tag{4}$$

Where c must satisfy Hurwitz condition $c > 0$
Therefore, we have

$$\dot{s} = cx_2 + \ddot{\omega}^* + \frac{f}{J}\dot{\omega} - \frac{\dot{T}_e}{J} \tag{5}$$

According to the exponential reaching law and from Eqs. (1) and (5), we have

$$cx_2 + \ddot{\omega}^* + \frac{f}{J}\dot{\omega} - \frac{\dot{T}_e}{J} = -\varepsilon\, sgn(s) - ks$$

Then, we can get the sliding mode controller as

$$T_e^* = J\left[cx_2 + \ddot{\omega}^* + \frac{f}{J}\dot{\omega} + \varepsilon\, sgn(s) + ks\right] \tag{6}$$

Where T_e^* is the required reference torque.

It can be inferred from Eq. (6) that the integral term can act as a filter and attenuate the chattering effect in the sliding mode control.

We choose $V_1 = \frac{1}{2}s^2$ as the Lyapunouv function. The time derivative of function V_1 is as follows

$$\dot{V}_1 = s\dot{s} = s(-\varepsilon\, sgn(s) - ks) \tag{7}$$

The system is stable according to the Lyapunouv stability criterion. Therefore, the negative semi-effective of function V_1 can be guaranteed by an approximation choice of the parameters $\varepsilon > 0$ and $k > 0$, which results in the opposite signs for s and $\dot{s}$ [16].

2.2 Backstepping Torque and Flux Control

The Back-stepping torque and flux controller are designed to achieve the satisfactory torque and flux tracking. Define the following torque and flux tracking errors [16]

$$\begin{aligned} e_T &= T_e^* - T_e \\ e_\Phi &= \Phi^* - \Phi_e \end{aligned} \tag{8}$$

Then, the derivative of torque tracking error dynamic can be obtained as follows

$$\dot{e}_T = J\left[cx_2 + \ddot{\omega}^* + \frac{f}{J}\dot{\omega} + \varepsilon\, sgn(s) + ks\right] - \frac{3}{2}p\left[\begin{array}{c} \Phi_\beta\left(\frac{R_s}{L_s}i_\alpha + \frac{E_\alpha}{L_s}\right) - \Phi_\alpha\left(\frac{R_s}{L_s}i_\beta - \frac{E_\beta}{L_s}\right) \\ + \left(i_\beta - \frac{\Phi_\beta}{L_s}\right)u_\alpha - \left(i_\alpha - \frac{\Phi_\alpha}{L_s}\right)u_\beta \end{array}\right] \tag{9}$$

For stabilizing the flux torque components, the flux tracking error dynamics can be defined as

$$\dot{e}_\Phi = 2R_s\Phi_\alpha i_\alpha + 2R_s\Phi_\beta i_\beta - 2\Phi_\alpha u_\alpha - 2\Phi_\beta u_\beta \tag{10}$$

Define the Lyapunouv function V_2 for whole system as

$$V_2 = \frac{1}{2}\left(V_1^2 + e_T^2 + e_\Phi^2\right) \tag{11}$$

Take the derivative of the Lyapunouv function V_2, we can get

$$\dot{V}_2 = V_1\dot{V}_1 + e_T\dot{e}_T + e_\Phi\dot{e}_\Phi = V_1\dot{V}_1 - k_T e_T^2 - k_\Phi e_\Phi^2 \tag{12}$$

Where k_T and k_Φ are positive constants.

According to the above Eq. (12), the final control voltage outputs u_α^* and u_β^* as follows

$$u_\alpha^* = \frac{1}{\Phi_\alpha i_\alpha + \Phi_\beta i_\beta - \Phi/L_s}\left\{\Phi_\beta\left[\left(J\left(cx_2 + \ddot{\omega}^* + \frac{f}{J}\dot{\omega} + \varepsilon\, sgn(s) + ks\right) + k_T e_T\right) \times \frac{2}{3p} + \Phi_\alpha\left(\frac{R_s}{L_s}i_\beta + \frac{E_\beta}{L_s}\right) - \Phi_\beta\left(\frac{R_s}{L_s}i_\alpha + \frac{E_\alpha}{L_s}\right)\right] + \left(i_\alpha - \frac{\Phi_\alpha}{L_s}\right) \times \frac{1}{2}\left(2R_s\Phi_\alpha i_\alpha + 2R_s\Phi_\beta i_\beta + k_\Phi e_\Phi\right)\right\}$$

$$u_\beta^* = \frac{1}{\Phi/L_s - \Phi_\alpha i_\alpha + \Phi_\beta i_\beta}\left\{\Phi_\alpha\left[\left(J\left(cx_2 + \ddot{\omega}^* + \frac{f}{J}\dot{\omega} + \varepsilon\, sgn(s) + ks\right) + k_T e_T\right) \times \frac{2}{3p} + \Phi_\alpha\left(\frac{R_s}{L_s}i_\beta + \frac{E_\beta}{L_s}\right) - \Phi_\beta\left(\frac{R_s}{L_s}i_\alpha + \frac{E_\alpha}{L_s}\right)\right] - \left(i_\beta - \frac{\Phi_\beta}{L_s}\right) \times \frac{1}{2}\left(2R_s\Phi_\alpha i_\alpha + 2R_s\Phi_\beta i_\beta + k_\Phi e_\Phi\right)\right\} \tag{13}$$

Substituting the final control voltage outputs (Eq. (13)) in the function $\dot{V}_2$, the time derivative of the Lyapunouv function is desired as

$$\dot{V}_2 = s(-\varepsilon\, sgn(s) - ks) - k_T e_T^2 - k_\Phi e_\Phi^2 \leq 0 \tag{14}$$

3 Proposed Fuzzy Active Front Steering

AFS is an anti-skid program that detects cornering adhesion losses and acts immediately by adding a corrective steering angle, which improves road handling and prevents accidents caused by loss of vehicle control. The basic principle of active steering is to provide a steering correction with respect to the angle of rotation of the steering wheel. This shift is created by an electric motor interposed between the steering wheel and the rack, Fig. 2. For better roadhanlding and to avoid accidents due to loss of the vehicle control, active steering (AFS) uses a corrective steering angle to facilitate the vehicle's rotational movement round its vertical axis. Figure 3 shows a schematic diagram of AFS system. The control system processor (AFS Algorithm) receives signals from the various embedded sensors. Based on sensor signals and observation of status information, the processor calculates the corrective steering angle that allows for better roadhandling.

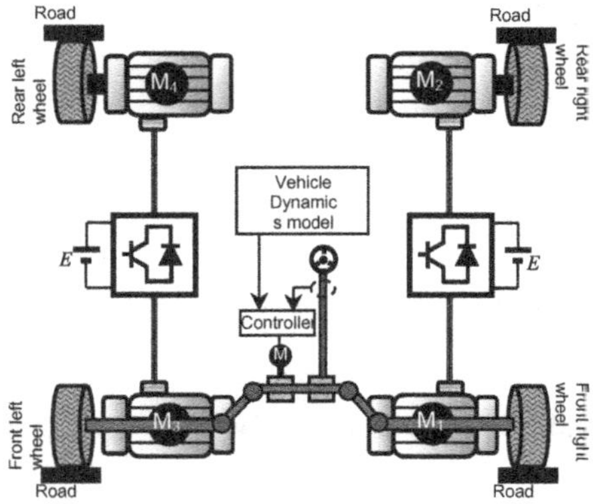

Fig. 2. Block diagram of the AFS based on fuzzy logic control.

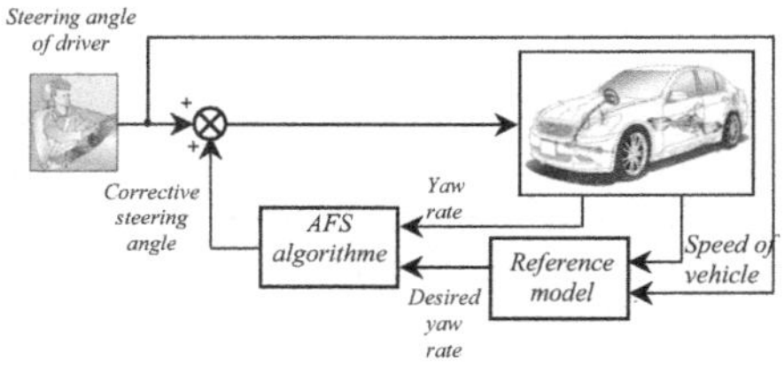

Fig. 3. System scheme of AFS for EV.

3.1 Control Variables

In general, the existing AFS methods use the yaw rate (r) and/or vehicle sideslip angle (β) as control variables. The yaw rate plays a crucial role in the dynamic control of the vehicle. Furthermore, the desired yaw rate derived from the vehicle bicycle model [17] is a function of the steering angle of the front wheels. The desired yaw rate can be interpreted as the desired vehicle response whish by the driver. For these reasons, the yaw rate is selected as one of the main control variables. When the vehicle sideslip angle increases to high values, the yaw moment generated by the tire side forces generally decreases [18]. But when the vehicle sideslip angle is large enough, the generated yaw moment becomes negligible and can hardly be increased by changing the steering angle. Thus, the vehicle tends to lose its stability. In addition, a small vehicle sideslip angle implies a heading consistency of the vehicle, which gives the driver a great opportunity to control the vehicle while cornering [19]. For the above reasons, the vehicle sideslip angle is closely related to the stability of the vehicle and the possibility of vehicle control, for this it should be chosen as a control variable. Moreover, the yaw rate is most related to vehicle handling, whereas the vehicle sideslip angle is mainly related to vehicle stability. These two states variables are not independent, but they are intrinsically linked by the vehicle dynamics (see the equations of motion vehicles).

3.2 The Desired Variables

The driver will have a large possibility to control the vehicle in cornering if the vehicle sideslip angle is very small. For this possibility, we have to impose the desired vehicle sideslip angle close to zero. The vehicle sideslip angle and the vehicle yaw rate can be derived from the vehicle bicycle model which will be developed from the non-linear vehicle model using simplifying assumptions.

3.3 The Bicycle Model

To simplify designing the controller, the widely used single track model shown in Fig. 4 is chosen in this study.

For lateral efforts:

$$F_{yi}(\alpha_i) = F_z.D_i.\sin[C_i.\arctan(B_i.\alpha_i - E_i.(B_i.\alpha_i - \arctan(B_i.\alpha_i)))] \tag{15}$$

The resulting lateral forces are obtained by linearization of Eq. (15) for small angles α_i (i.e. $\alpha_i = 0$):

$$F_{yi} \approx \alpha_i.(D_iC_iB_i) \tag{16}$$

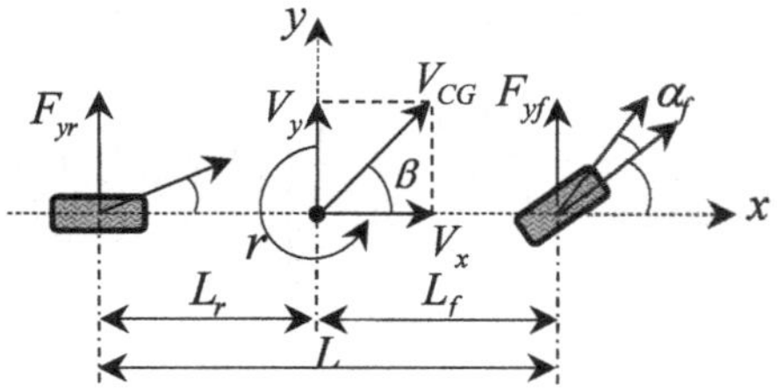

Fig. 4. Single track vehicle model (i.e. Bicycle model).

The front and rear cornering stiffness can be expressed by:

$$C_{yi} = D_iC_iB_i \tag{17}$$

Then the lateral tire forces can be further approximated as follows:

$$\begin{aligned} F_{yf} &= 2C_{yf}\alpha_f \\ F_{yr} &= 2C_{yr}\alpha_r \end{aligned} \tag{18}$$

The tire sideslip angles for the front and rear axles for vehicle equipped with the AFS can be expressed as follows:

$$\begin{aligned} \alpha_f &= \delta_f - \arctan\left(\frac{v_y + L_f r}{v_x}\right) \approx \delta_f - \frac{v_y + L_f r}{v_x} \\ \alpha_r &= -\arctan\left(\frac{v_y + L_r r}{v_x}\right) \approx -\frac{v_y + L_r r}{v_x} \end{aligned} \tag{19}$$

Where δ_f is the steering angle of the front wheels, and $r = \psi^{\cdot}$ is the yaw rate.

The vehicle lateral dynamics equations that include lateral and yaw motions can be written as Eqs. (20) and (21):

$$\dot{v}_y = -2\frac{C_{yf} + C_{yr}}{M_v v_x} v_y + \left(2\frac{-C_{yf}L_f + C_{yr}L_r}{v_x} - v_x\right) + \frac{2C_{yf}}{M_v}\delta_f + \frac{1}{M_v}F_w \tag{20}$$

$$\dot{r} = \left(2\frac{-C_{yf}L_f + C_{yr}L_r}{J_v v_x}\right) v_y - \left(2\frac{C_{yf}L_f^2 + C_{yr}L_r^2}{J_x v_x}\right) r + \frac{2C_{yf}L_f}{J_v}\delta_f + \frac{L_w}{J_v}F_w \tag{21}$$

By choosing β and r as state variables, the bicycle model can be written as:

$$\dot{x} = Ax + bu + wF_w, x = \begin{bmatrix} v_y & r \end{bmatrix}^T, u = \delta_f$$

$$\begin{bmatrix} \dot{\beta} \\ \dot{r} \end{bmatrix} = \begin{bmatrix} -2\frac{C_{yf} + C_{yr}}{M_v v_x} & 2\frac{-C_{yf}L_f + C_{yr}L_r}{M_v v_x^2} - 1 \\ 2\frac{-C_{yf}L_f + C_{yr}L_r}{J_v} & -2\frac{C_{yf}L_f^2 - C_{yr}L_r^2}{J_v v_x} \end{bmatrix} \begin{bmatrix} \beta \\ r \end{bmatrix} + \begin{bmatrix} \frac{2C_{yf}}{M_v v_x} \\ \frac{2C_{yf}L_f}{J_v} \end{bmatrix} \delta_f + \begin{bmatrix} \frac{1}{M_v v_x} \\ \frac{l_w}{J_v} \end{bmatrix} \tag{22}$$

3.4 The Reference Model

In designing the controller to improve vehicle maneuverability, it is important for the yaw rate response to track the targeted value. The reference model for yaw rate rd and front wheel angle $\boldsymbol{\delta_f}$ are represented by the first order delay system as shown below [20, 21]:

$$r_d(s) = \frac{k_r}{1 + \tau s}\delta_f(s) \tag{23}$$

$$k_r = \frac{v_x}{L_f + M_v L_f L_r v_x^2 / 2L_f(L_f + L_r)C_{yr}}, \tau = \frac{J_v v_x}{M_v L_r v_x^2 + 2C_{yf}L_f(L_f + L_r)}$$

Here, k_r is stability factor, τ is the desired time constant.

3.5 Fuzzy Logic AFS Controller Design

In this work, to improve lateral stability, lateral control and handling performance of the electric vehicle, the main control objective is to track the desired yaw rate and vehicle sideslip angle. To ensure this, the fuzzy logic control is proposed by using the yaw rate error $\boldsymbol{e_r} = \boldsymbol{r} - \boldsymbol{r_d}$ and the vehicle sideslip angle error $\boldsymbol{e_r} = \boldsymbol{\beta} - \boldsymbol{\beta_d}$ as inputs and the corrective steering angle $\boldsymbol{\Delta\delta_f}$ as output. The Fig. 5 shows the designed controller block diagram. The goal of this proposed control is to implement a controller having a good yaw rate tracking, path keeping and disturbance rejection to improve yaw dynamics. The controller generates the corrective steering angle $\boldsymbol{\Delta\delta_f}$ according to the input current fuzzy and the fuzzy rules. The membership functions for the two input variables Δ_1 and Δ_2, and the output variable $\boldsymbol{\Delta\delta_f}$ are shown in Fig. 6 and the basic rules of the fuzzy controller can be summarized in Table 1.

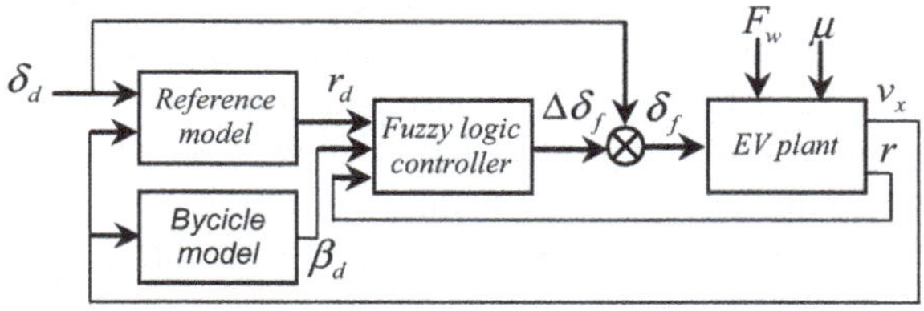

Fig. 5. Block diagram of AFS with fuzzy logic controller.

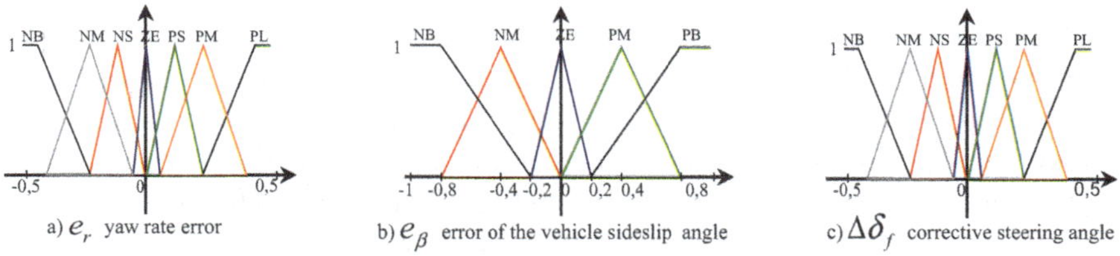

Fig. 6. Membership functions for fuzzy inputs and output variables.

Table 1. Fuzzy logic rules.

$\Delta\delta_f$	$e_\beta = \beta_d - \beta$				
$e_r = r - r_d$	NB	NM	ZE	PM	PB
NB	NB	NB	NM	NM	NS
NM	NB	NM	NM	NS	NS
NS	NM	NM	NM	NS	NS
ZE	ZE	ZE	ZE	ZE	ZE
PS	PS	PS	PM	PM	PM
PM	PS	PS	PM	PM	PB
PB	PS	PM	PM	PB	PB

4 Simulation Results

In this section, simulations results via Matlab/Simulink will show the improved handling and stability of electric vehicle under double lane change maneuvers with the changes of cornering stiffness on various road conditions with different adhesion coefficient. The main objective of double lane change maneuver under various road conditions is to verify the effectiveness of the integrated control for vehicle handling stability and safety performance. The vehicle's specifications are summarized in Table 2. Simulations start with a vehicle speed of 36 km/h. Then, a front steering angle reference as double lane change test is applied, where in the first phase (t = 0 ÷ 15) s on a dry (non-slippery $\mu = 0.9$) road and in the second phase (t = 15 ÷ 25 s) on a slippery road which selects the wet road and the friction coefficient is set to be 0.3, while keeping the longitudinal velocity constant $V_x = 36\,\text{km/h}$. The simulation results including lateral and vertical vehicle dynamics behaviors are shown in Fig. 7, which shows the variable range of the longitudinal velocity that changes from 0 km/h to 36 km/h. Figure 7(b) shows the steering wheel angle, where we notice the Fuzzy AFS controller subtracts or adds a smooth, oscillation-free corrective steering angle, throughout this maneuver. Consequently, the corrective steering angle generated by the fuzzy AFS controller reduces vehicle lateral acceleration to ensure vehicle comfort, Fig. 7(b). Figure 7(c) shows the vehicle yaw rate for double lane change maneuver, where it follows its desired trajectory despite the various road conditions that characterize the passage of the vehicle from dry road to slippery road. It can be seen clearly in Fig. 7(d) that the sideslip angle of the vehicle is kept in allowance range, approximately zero, which increases stability during double lane change, and therefore the vehicle follows the reference path. Figure 7(e) shows the longitudinal velocity of the vehicle v_x when the vehicle reaches constant speed at 5 s which is allowed from acceleration of the vehicle mass. From Fig. 7(f), we can immediately recognize that the lateral velocity v_y depends slowly on the driver's steering command. We can clearly see that this speed occurs only during cornering and it vanishes when the vehicle is travelling on a straight road. We notice that in Fig. 7(j) there is a difference between the driving torques in the cornering which is the same on dry and slippery roads. As shown in Fig. 7(k), the traction forces applied to the driving wheels have the same behavior as the driving torques during cornering. The simulation results show that double lane change test at speed 36 km/h under various road conditions has a considerable effect on the handling stability of vehicle. The proposed fuzzy AFS system can keep the vehicle stable, and the path tracking ability is improved, under this condition.

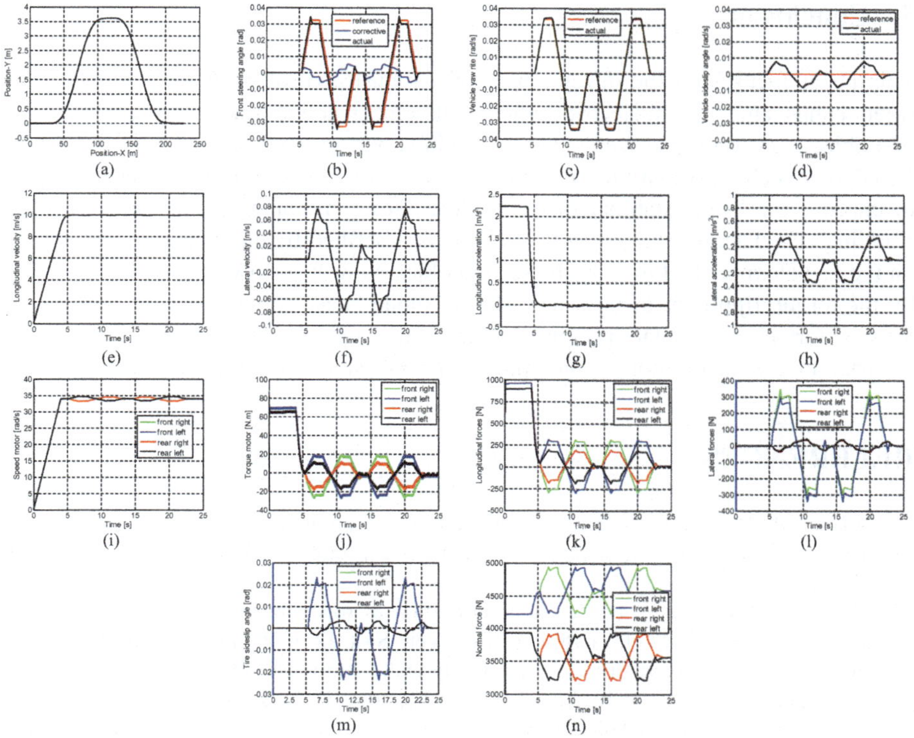

Fig. 7. The simulation results for the trajectory, steering angle, yaw rate, sideslip angle, longitudinal and lateral velocity, longitudinal and lateral acceleration, speed and torque of motors, traction and lateral forces, sideslip of the wheel and the normal forces in the double lane turn maneuver.

Table 2. The specifications of the vehicle used in simulation.

Symbol	Quantity	Value
M_v	Vehicle mass	1562 kg
J_v	Vehicle inertia	2630 kg.m^2
J_ω	Wheel inertia	1,284 kg.m^2
L_f	Distance from the gravity center to front axle	1,104 m
L_r	Distance from the gravity center to rear axle	1,421 m
h_{cg}	Height gravity center of the vehicle	0,5 m
S_f	Frontal area of vehicle	2,04 m^2
ρ	Air density	1,2 kg.m^{-3}
C_{px}	Drag coefficient	0,25
C_{rr}	Rolling resistance coefficient	0,01
C_f	Longitudinal stiffness of each tire lateral	37407 N/rad
C_r	Lateral stiffness of each tire lateral	51918 N/rad
R_ω	Wheel radius	0,294 m

5 Conclusion

This paper focuses on the development of a robust control method for stability and safety enhancement of an electric vehicle with a fuzzy active front steering system. The first objective was to improve the lateral handling performance of the vehicle by using a fuzzy logic technique that provides a corrective steering angle to the driver maneuvers to improve the lateral stability of the vehicle during critical situations, cornering and skidding. The lateral behavior of the vehicle is described by the dynamic model based on the bicycle model. However, the second objective was to solve the problems caused by the torque ripple by using a new sliding mode backstepping control. The various tests performed in simulation highlight the robust control method which can stabilize electric vehicle lateral motion and enhance lateral handling performance.

REFERENCES

1. Sekour, M., Hartani, K., Merah, A.: Electric vehicle longitudinal stability control based on a new multimachine nonlinear model predictive direct torque control. J. Adv. Transp. **2017** (2017)
2. Ma, C., Xu, M., Wang, H.: Dynamic emulation of road/tyre longitudinal interaction for developing electric vehicle control systems. Veh. Syst. Dyn. **49**, 433–447 (2011)
3. Hori, Y.: Future vehicle driven by electricity and control-research on four-wheel-motored UOT Electric March II. IEEE Trans. Industr. Electron. **51**, 954–962 (2004)
4. Hartani, K., Miloud, Y., Miloudi, A.: Improved direct torque control of permanent magnet synchronous electrical vehicle motor with proportional-integral resistance estimator. J. Electr. Eng. Technol. **5**, 451–461 (2010)
5. Sekour, M.H., Hartani, K., Draou, A., Allali, A.: Sensorless fuzzy direct torque control for high performance electric vehicle with four in-wheel motors. J. Electr. Eng. Technol. **8**, 530–543 (2013)
6. Foo, G.H.B., Rahman, M.: Direct torque control of an IPM-synchronous motor drive at very low speed using a sliding-mode stator flux observer. IEEE Trans. Power Electron. **25**, 933–942 (2010)
7. Zhang, Y., Zhu, J.: Direct torque control of permanent magnet synchronous motor with reduced torque ripple and commutation frequency. IEEE Trans. Power Electron. **26**, 235–248 (2011)
8. Bartolini, G., Ferrara, A., Usai, E.: Chattering avoidance by second-order sliding mode control. IEEE Trans. Autom. Control **43**, 241–246 (1998)
9. Imine, H., Fridman, L.M., Madani, T.: Steering control for rollover avoidance of heavy vehicles. IEEE Trans. Veh. Technol. **61**, 3499–3509 (2012)
10. Hartani, K., Draou, A.: A new multimachine robust based anti-skid control system for high performance electric vehicle. J. Electr. Eng. Technol. **9**, 214–230 (2014)
11. Kang, J., Yoo, J., Yi, K.: Driving control algorithm for maneuverability, lateral stability, and rollover prevention of 4WD electric vehicles with independently driven front and rear wheels. IEEE Trans. Veh. Technol. **60**, 2987–3001 (2011)
12. Yim, S., Choi, J., Yi, K.: Coordinated control of hybrid 4WD vehicles for enhanced maneuverability and lateral stability. IEEE Trans. Veh. Technol. **61**, 1946–1950 (2012)

13. Xu, Y., Lei, Y., Sha, D.: Backstepping direct torque control of permanent magnet synchronous motor with RLS parameter identification. In: 2014 17th International Conference on Electrical Machines and Systems (ICEMS), pp. 573–578 (2014)
14. Sun, H., Cui, X., Tang, C.: Back-stepping adaptive SVM direct torque control of SPMSM drive system. Indonesian J. Electr. Eng. Comput. Sci. **12**, 6587–6593 (2014)
15. Liu, J., Wang, X.: Advanced sliding mode control for mechanical systems design, analysis and MATLAB simulation. In: Advanced Sliding Mode Control for Mechanical Systems. Springer (2011)
16. Ning, B., Cheng, S., Qin, Y.: Direct torque control of PMSM using sliding mode backstepping control with extended state observer. J. Vib. Control **24**, 694–707 (2018)
17. Nakhaie-Jazar, G., Naghshineh-Poor, A., Aghabaik-Lavassani, M.: Time and energy optimal control by a new way based on central difference approximation of equation of motion with application to robot control. In: Second IEEE Conference on Control Applications, pp. 377–382 (1993)
18. Shibahata, Y., Shimada, K., Tomari, T.: Improvement of vehicle maneuverability by direct yaw moment control. Veh. Syst. Dyn. **22**, 465–481 (1993)
19. Fu, C., Hoseinnezhad, R., Jazar, R., Bab-Hadiashar, A., Watkins, S.: Electronic differential design for vehicle side-slip control. In: 2012 International Conference on Control, Automation and Information Sciences (ICCAIS), pp. 306–310 (2012)
20. Ariff, M.M., Zamzuri, H., Idris, N.N., Mazlan, S., Nordin, M.: Direct yaw moment control of independent-wheel-drive electric vehicle (IWD-EV) via composite nonlinear feedback controller. In: 2014 First International Conference on Systems Informatics, Modelling and Simulation, pp. 112–117 (2014)
21. Nagai, M., Yamanaka, S., Hirano, Y.: Integrated control of active rear wheel steering and yaw moment control using braking forces. JSME Int. J. Ser. C Mech. Syst. Mach. Elem. Manuf. **42**, 301–308 (1999)

Smart and Resilient Cities

Thermal Comfort in Southern Algeria: Some Useful Investigation and Case Study

B. Hebbal[1(✉)], Y. Marif[1], M. M. Belhadj[1], Y. Chiba[2], and M. Zerrouki[3]

[1] Faculty of Mathematics and Matter Sciences, LENREZA Laboratory, University of Ouargla, Ouargla, Algeria
hebbalbelkhir@gmail.com, yacine.marif@yahoo.fr, mhmu_belhadj@yahoo.com

[2] Faculty of Technology, University of Medea, Medea, Algeria
chibayounes@gmail.com

[3] Faculty of Applied Sciences, University of Ouargla, Ouargla, Algeria
moussa.zerrouki@yahoo.fr

Abstract. This paper aims to discussing some studies that have proposed in order to create an acceptable thermal comfort and improve people's quality of life in Algerian Saharan contemporary buildings. On the other hand and in climate conditions of Ouargla town situated in northern east of Algerian Sahara, the impact of insulation in the cooling load of one room with air condition of 26 °C was studied. The investigation finds that the energy demand for summer cooling was lower in the case of insulated exterior walls; the peak cooling load was in the region of 1.2 kW.

Keywords: Algerian south · Passive cooling · Simulation · Experimental

1 Introduction

The Algerian south is of a desert climate with vast area, summer season characterized by high global solar radiation intensity with a maximum of 1000 W/m^2 and very high air-temperature exceeding 45 °C. A review of ancient architecture in the south of Algeria revealed that important technologies have been used for several years. For example, buildings in the historic villages or Ksour were erected adjacent to each other in order to reduce the surfaces exposed to the sun with massive external walls, inner courtyard and small windows. Furthermore, the Ksour were surrounded by the palm groves in order to taking the advantage of the freshness of the palm. By use of these technologies, people were able to live in comfort without any electrical air conditioning system.

Nowadays, new materials have been introduced in construction without a proper study of their suitability for the climate in hot regions. This has resulted in extensive recourse to conventional air conditioning to guarantee human comfort. It is estimated that about 60% of the total building power consumption is consumed by air conditioning equipment [1]. According to Ghedamsi et al. [2] study, the energy consumption can be increased to 179.78 TWh in 2040 in Algeria. The energy efficient housing standard is rapidly spreading across world, it becomes necessary for Algeria to exploit this standard for facing the important increase in the electricity consumption.

M. Hatti (Ed.): ICAIRES 2019, LNNS 102, pp. 275–283, 2020.
https://doi.org/10.1007/978-3-030-37207-1_28

2 Overview of Some Passive Strategies

Algerian south regions are a part of the huge northern African Sahara and they characterized by an arid continental climate. Numerous researches are recorded in these regions about the improving of building thermal performance by introducing many passive cooling designs. The following sections summarizes some studies carried out at four cities: Ouargla, Ghardaïa, Bechar and Adrar.

2.1 Studies of Ouargla

To increase interior comfort in Ouargla typical house, Bouchahm et al. [3] studied the use of 6.50 m wind tower for passive evaporative cooling system (interior cross section is 0.70 × 0.75 m^2). The adjacent wall is ventilated to minimize heat gain from solar radiation. At the bottom of the tower is a pool filled up with water (Fig. 1a). The site measurements in July month are not encouraging, the inside air temperature is higher than 40 °C with an increase of humidity obtained by the humidification procedure (Fig. 1b). However, the simulation results confirm the advantage of the application of this strategy. Inside temperature can be significantly reduced (28.3–32.3 °C) using a higher height of wetted column (5.5–4.5 m) and a smaller size of the conduits partition inside the tower (0.14–0.116 m) by increasing their number.

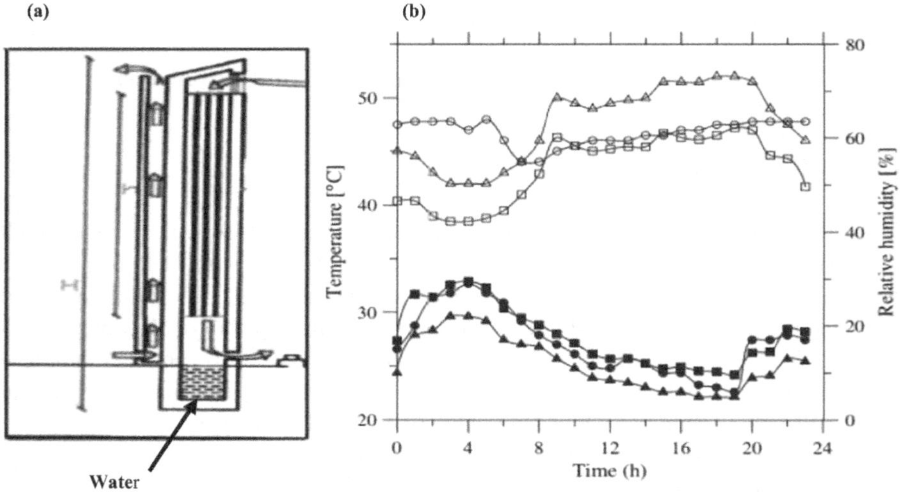

Fig. 1. (a) Wind tower configuration (b) Hourly measured and calculated temperatures and humidity [3]. Temperature: Δ: T_{ext} measured, □: T_{int} calculated, O: T_{int} measured. Relative humidity: ▲: H_{ext} measured, ■: H_{int} calculated, ●: H_{int} measured.

Another approach was introduced by Saifi et al. [4] based on the use of vertical green wall in order to increase interior thermal comfort. Three test cells of volume

1 × 1.2 × 0.8 m^3 have been constructed in Ouargla university (cell without green wall and cell covered with two types of plants Jasmine and Aristolochia) using bricks, cement in the outside and inside wall surface, the roof is constructed by concrete slab and cement in the inside surface (Fig. 2a). The solar radiation was generated using four lamps. The results showed that, the wall vegetation offer an acceptable level of indoor comfort regardless to the harsh outdoor conditions of Ouargla city in summer season (Fig. 2b).

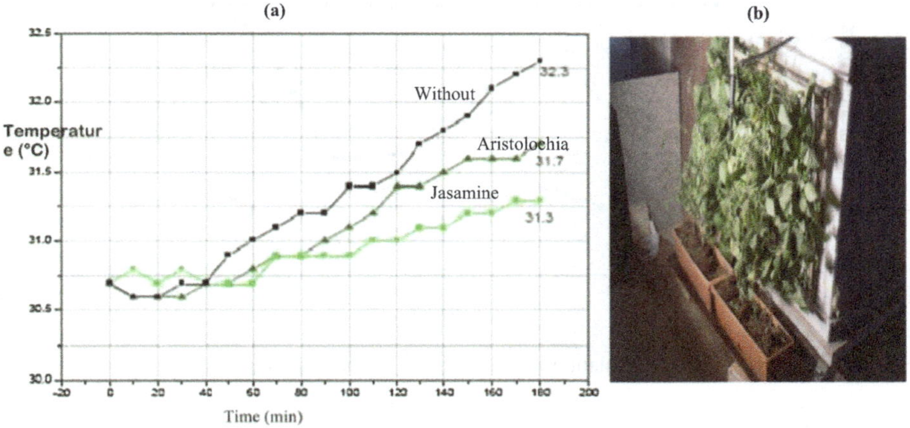

Fig. 2. (a) Wall internal surface temperature; (b) Test cell with green wall [4]

The use of phase change materials (PCM) can be also limit energy consumption in buildings. A promising simulation results were achieved by Ghedamsi et al. [5] using PCM, the optimal concentration of 15% for all wall orientations presents a significant reduction of temperature fluctuations on the interior surface compared with the wall without PCMs (Fig. 3a and b)

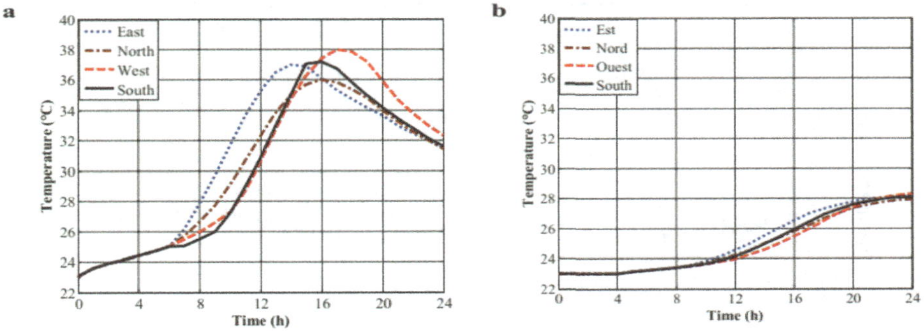

Fig. 3. Wall internal surface temperature for all orientation: (a) without (b) with PCMs (15%) [5].

2.2 Studies of Ghardaïa

Bekkouche et al. [6] affirmed in their paper conducted on a non-air-conditioned building in Ghardaïa that the integration of interior thermal insulation with 8 cm thick insulation layer of polystyrene and the use of shading process such as eaves of roof in hot seasons can decrease sensitively the building inside air temperature (Fig. 4).

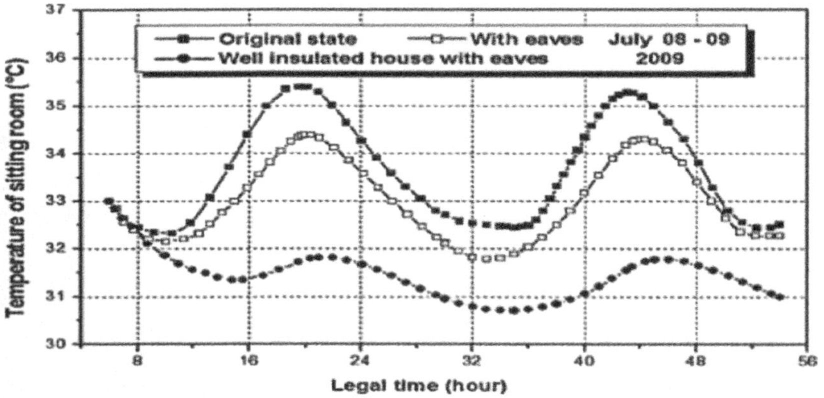

Fig. 4. Thermal insulation and eave effect in the building temperature [6]

In another work [7], they proved that owing to their high thermal resistance, the hollow brick is thermally more effective than the stone and cinderblock. A new configuration of the sun-facing wall has been proposed with the use of two layers of hollow brick arranged horizontally, an air gap of 1 cm and 6 cm layer of polystyrene. The simulation findings showed that the new configuration significantly reduces fluctuations of internal temperatures (Fig. 5).

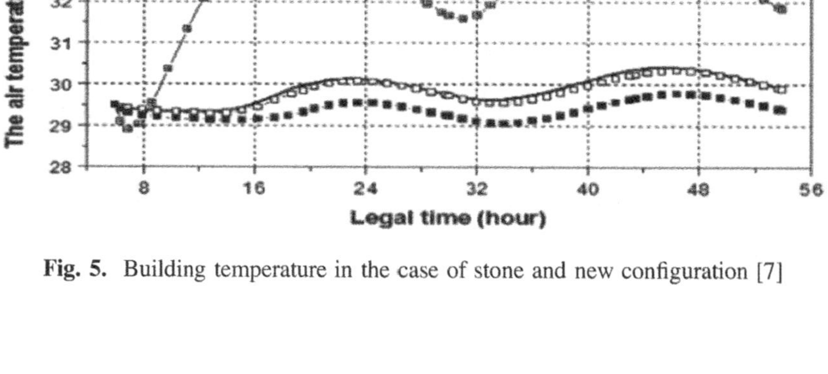

Fig. 5. Building temperature in the case of stone and new configuration [7]

2.3 Studies of Adrar

Under the climatic conditions of Adrar region new hybrid passive cooling system consisting of an earth-to air heat exchanger (EAHE) coupled with a wind tower was simulated by Benhammou et al. [8]. The results show that a system with 5.1 m tower height, 0.57 m^2 cross section area and 7 m EAHE pipe length can significantly promote natural ventilation and offer an acceptable level of indoor temperature inferior to 32 °C (Fig. 6). The authors also examined the configuration proposed by Bouchahm et al. [3] (wind tower with wet surface); it is observed that the room air temperature is much higher than the temperature of the wind tower coupled to the EAHE. In another paper [9], the same authors found that the proposed system will be more effective if the building is properly insulated.

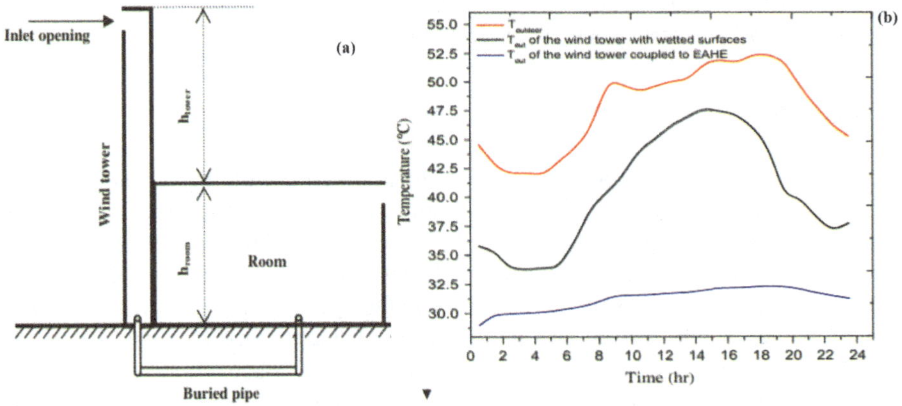

Fig. 6. (a) The earth-to-air heat exchanger coupled to a wind tower (b) Impact of EAHE in the room temperature [8]

2.4 Study of Bechar

Using TRNSYS-COMIS software, the effect of a horizontal opening "skifa" in a typical traditional house thermal comfort was studied by Fezzioui et al. [10] (Fig. 7a). Numerical studies have shown a stronger effect of the horizontal opening dimension in the house outside air flow. For this reason, it allows good air circulation in the house. This increase in air circulation results in an decrease of the interior house temperature (Fig. 7b).

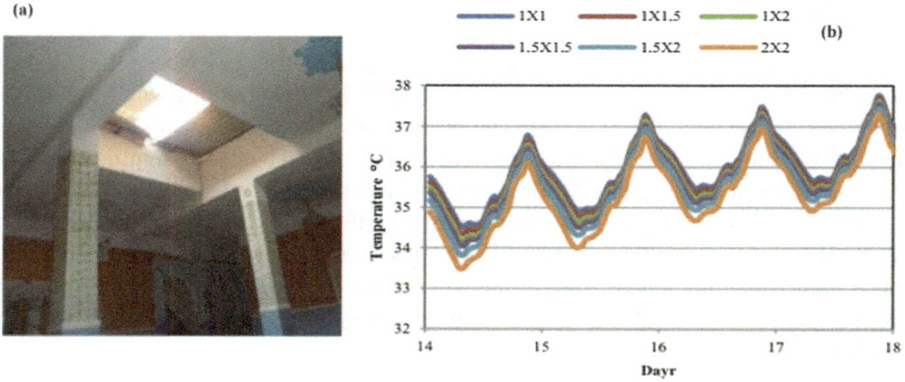

Fig. 7. (a) The traditional house with horizontal opening in the ceiling (b) Temperature for different sizes of horizontal opening from the 14 to 17 July [10]

3 Case Study

The use of conventional cooling system with improving building envelope thermal performance is necessary to maintain the thermal comfort in the south of Algeria. The mean objective of this section is to assess the thermal performance of an insulated room equipped with an air conditioning system. The test building (Fig. 8a) is a single family home of four persons. Construction materials in Ouargla region were limited to bricks, cement in the outside surface and plaster in the inside surface of walls, the roof is constructed by concrete slab and plaster in the inside surface. A conditioned room (Room1) with 26 °C room air temperature is a partitioned space that can treat as a single load. Heat within a room envelope originates especially via external and internal methods. The cooling load can be calculated as [11]:

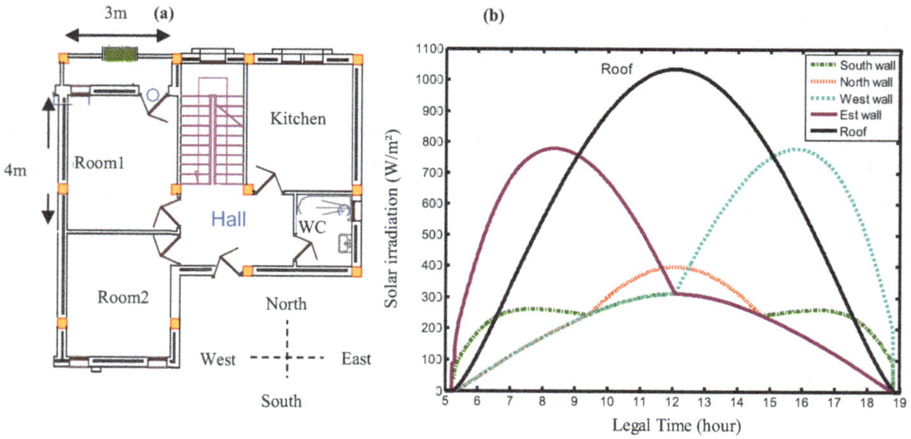

Fig. 8. (a) The model of test building (b) Incident solar radiation on walls and roof for typical day of July

$$Q_{cooling} = Q_{external} + Q_{internal} \quad (1)$$

The external cooling load can be calculated by using Eq. (2). Where A_{wall} is the wall area, h_{int} is the internal heat transfer and the room air temperature T_r is considered to be equal to 26 °C.

$$Q_{external} = h_{int}(T_{int} - T_r)A_{wall} \quad (2)$$

The interior surface temperatures of the exterior walls T_{int} can be calculated using the governing equation of heat conduction through this wall. The heat conduction transfer is supposed to be unidirectional and the conductivity is assumed to be constant for each material constituting the wall. In order to solve the heat conduction equation, a finite difference method with energy balance approach is adopted. The equations for all nodes can be written in tri-diagonal matrix and vector matrix. The system was solved by means of Tri-Diagonal Matrix Algorithm (TDMA) [12]. The internal cooling load can be calculated as follow

$$Q_{internal} = Q_{cond} + Q_{occupant} + Q_{lighting} + Q_{equipment} \quad (3)$$

The heat generated by lighting, persons and appliances can be found from appropriate tables in the CNERIB document in the case of 26 °C room temperature [13]. $Q_{lighting}$(fluorescents lighting) = 45 W, $Q_{occupant}$(four person) = 228 W and $Q_{equipment}$(Television) = 150 W. For simplification reasons the latent cooling load, loads caused by infiltration of the air, heat transmitted from internal walls and the cooling load due to sun radiation transmitted from windows are ignored. The incident global solar radiation was estimated using Algerian solar atlas numerical model [14]. Figure 8b shows the instantaneous variation of global solar radiation on the roof and walls of different orientation for representative day of July, it is evident that the roof received the maximum of solar radiation in the day.

For sizing the air conditioning system, the cooling load of the room1 can be calculated without and with 5 cm-thick internal insulation layer of polystyrene in

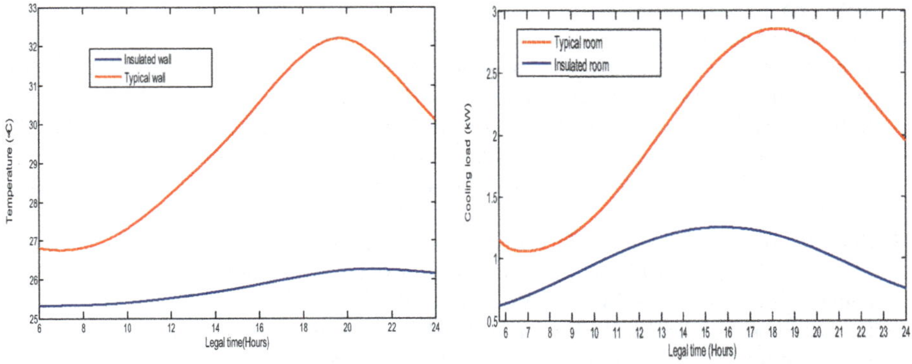

Fig. 9. West wall interior surface temperatures and Room cooling load for typical day of July

external walls (wall facing west, north and roof). West's facing wall internal surface temperature and room1 cooling load for the design day of July are represented in Fig. 9. From which it could be observed that the presence of insulation significantly reduces fluctuations of internal surface temperatures. The maximum internal surface west facing wall temperature decreased to 32 °C at 26.3 °C. Thus providing a higher level of indoor comfort, the peak cooling load decreased to 2.8 kW at 1.2 kW.

4 Conclusions

The contribution is divided into two parts. The first one summarizes the recent findings associated to the use of passive techniques to increase interior comfort in the desert of Algeria, after reviewing it is observed that the desired comfort temperature has not reached using these passive techniques. The purpose of the second part is to simulate the use of external wall insulation in order to reduce the cooling load of a typical one room in Ouargla town. Based on the simulation results obtained for typical days in summer period, it could be concluded that a difference in the energy demand of 1.6 kW is obtained between the ordinary and insulate room.

References

1. Ministry of Energy and Mines: Renewable energy and energy efficiency program (2018). http://www.mem-algeria.orgS
2. Ghedamsi, R., Settou, N., Gouareh, A., Khamouli, A., Saifi, N., Recioui, B., Dokkar, B.: Modeling and forecasting energy consumption for residential buildings in Algeria using bottom-up approach. Energy Build. **121**, 309–317 (2016)
3. Bouchahm, Y., Bourbia, F., Belhamri, A.: Performance analysis and improvement of the use of wind tower in hot dry climate. Renew. Energy **36**, 898–906 (2011)
4. Saifi, N., Settou, N., Necib, H., Damene, D.: Experimental study of thermal performance and the contribution of plant-covered walls to the thermal behavior of building. Energy Procedia **36**, 995–1001 (2013). ARECE13
5. Ghedamsi, R., Settou, N., Saifi, N., Dokkar, B.: Contribution on buildings design with low consumption of energy incorporated PCMs. Energy Procedia **50**, 322–332 (2014). TMREES14
6. Bekkouchen, S.M.A., Benouaz, T., Yaiche, M.R., Cherier, M.K., Hamdani, M., Chellali, F.: Introduction to control of solar gain and internal temperatures by thermal insulation, proper orientation and eaves. Energy Build. **43**, 2414–2421 (2011)
7. Bekkouchen, S.M.A., Benouaz, T., Cherier, M.K., Hamdani, M., Benamrane, N., Yaiche, M. R.: Thermal resistances of local building materials and their effect upon the interior temperature case of a building located in Ghardaïa region. Constr. Build. Mater. **52**, 59–70 (2014)
8. Benhammou, M., Draoui, B., Zerrouki, M., Marif, Y.: Performance analysis of an earth-to air heat exchanger assisted by a wind tower for passive cooling of buildings in arid and hot climate. Energy Convers. Manage. **91**, 1–11 (2015)
9. Benhammou, M., Draoui, B., Hamouda, M.: Improvement of the summer cooling induced by an earth-to-air heat exchanger integrated in a residential building under hot and arid climate. Appl. Energy **208**, 428–445 (2017)

10. Fezzioui, N., Benyamine, M., Draoui, B., Roulet, C.A.: The traditional house with horizontal opening: a trend towards zero energy house in the hot, dry climates. Energy Procedia **96**, 934–944 (2016)
11. Hatamipour, M.S., Mahiyar, H., Taheri, M.: Evaluation of existing cooling systems for reducing cooling power consumption. Energy Build. **39**, 105–112 (2007)
12. Marif, Y., Benhammou, M., Zerrouki, M., Belhadj, M.M.: Thermal performance of the outside and inside wall insulation in the existing building in the south of Algeria. ISESCO J. Sci. Technol. **9**(16), 53–59 (2013)
13. Document Technique Réglementaire (DTR): C 3-4, Règles de calcul des apports calorifiques des bâtiments: Climatisation. Ministère de l'habitat et de l'urbanisme, CNERIB (1998)
14. .Capderou M.: Theoretical and experimental models. In: Solar atlas of Algeria (in French). Tome 1, vol 1. Algerian University Publications Office (1987)

A Simple Design of Printed Antenna with DGS Structure for UWB/SWB Applications

Tarek Messatfa(✉), Fouad Chebbara, Belhedri Abdelkarim, and Annou Abderrahim

Department of Electronic and Telecommunications, Electrical Engineering Laboratory (LAGE), Université Kasdi Merbah Ouargla, 30000 Ouargla, Algeria
messatfa.tarek@univ-ouargla.dz

Abstract. In this paper a circular microstrip patch antenna with Defected Ground Structure (DGS) has been designed and simulated for Ultra and Super Wide Band (UWB/SWB) applications by using the Computer Simulation Technology (CST). The aim of this work is to increase the bandwidth of an antenna by using (DGS). The total size of antenna is 25×30 mm^2. This proposed antenna covers the frequency range of UWB and SWB for S11 < −10 dB was from 2.77 GHz to 30 GHz, this proposed antenna which gives a useful structure for modern wireless communication systems include point to point communication such as WVB (Wireless Video Broadcast), Satellite Communication and Radar Applications, WLAN applications "IEEE802.11a" in (5.12–5.825 GHz) and WiMAX system in (3.4–3.7 GHz) and for short range communication such as Biomedical applications.

Keywords: Defected ground structure (DGS) · Ultra wideband (UWB) · Super wideband (SWB) · Microstrip patch antenna · WLAN and WiMAX

1 Introduction

In the recent years, the wireless technology is one of the main areas of research in the world of communication systems and the development in communication systems requires the development of low cost, minimal weight, low profile antennas that are capable of maintaining high performance over a wide range of frequencies.

The performance and advantages of microstrip patch antennas make them the perfect choice for communication systems. The most of the proposed designs for patch antennas suffer from narrow bandwidths, and a number of techniques have been proposed and investigated to render them suitable for UWB communication applications by enhancing their bandwidth [1]. Ultra-wideband (UWB) antennas are one of the most important elements for UWB systems. With the release of the 3.1–10.6 GHz band, applications for shortrange and high-bandwidth handheld devices are primary research areas in UWB systems [2]. The one of these techniques is defected ground structure or DGS, where the ground plane metal of a microstrip circuit is intentionally modified to enhance performances.

DGS is realized by etching off a simple shape in the ground plane, depending on the shape and dimensions of the defect, the shielded current distribution in the ground

M. Hatti (Ed.): ICAIRES 2019, LNNS 102, pp. 284–291, 2020.
https://doi.org/10.1007/978-3-030-37207-1_29

plane is disturbed, resulting a controlled excitation and propagation of the electromagnetic waves through the substrate layer. The shape of the defect may be changed from the simple shape to the complicated shape for the better performance [3]. When considering patch antenna design for ultra-wideband applications, defected ground structures present a novel technique for modifying the characteristics of the microwave device [4].

In this paper, a circular patch antenna is proposed, by apply the technique of DGS on the ground plane of antenna to enhance its bandwidth. The size of the proposed antenna in [5] is bigger than the size of the proposed structure in this paper. The simulation results have demonstrated that the antenna has a bandwidth of 166% (2.77–30 GHz).

2 Proposed Antenna Design

The design of the proposed antenna is shown in Fig. 1. It is a circular patch antenna with patch radius R = 9.5 mm which is designed on a fr4 substrate (permittivity 4.3 and loss tangent 0.025) of size w = 25 mm by l = 30 mm and a thickness of 1.6 mm.

On the bottom side of the substrate, a defected ground plane with dimensions 25 mm * 10 mm (Wg × Lg). This antenna is fed by a 50 Ω microstrip line with size Wf = 3.05 mm by Lf = 10.82 mm. A semi-circular shape with a radius Rc has been etched from the upper middle of the ground plane, and two quarters of a circle, with a radius Re has been also etched from the upper edges of the ground plane, as it is shown in Fig. 1.

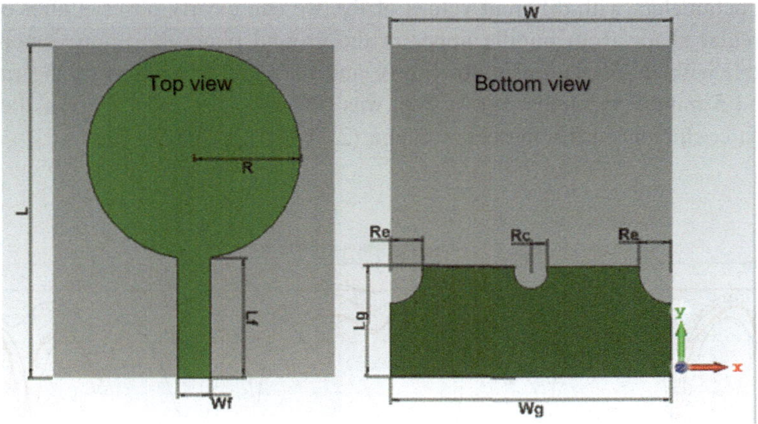

Fig. 1. Design of the proposed antenna with DGS

The radius R of the circular patch can be obtained by [6]

$$R = \frac{F}{\sqrt{\left\{1 + \frac{2h}{F\pi\varepsilon_r\left[ln\left(\frac{\pi F}{2h}\right) + 1.7726\right]}\right\}}} \tag{1}$$

Where,

$$F = \frac{8.791 \times 10^9}{f_r\sqrt{\varepsilon_r}} \tag{2}$$

Where,

R = patch radius, f_r = operating frequency, h = thickness of substrate, ε_r = dielectric permittivity of the substrate.

3 Simulated Results and Discussion

The proposed antenna was simulated using the software *CST Microwave Studio* from 2.77 to 30 GHz.

3.1 Effect of the Semicircular Slot with Radius Rc

First we present the effect on the reflection coefficient S11 due the variation of the radius (Rc) of the semi-circular on middle upper in the ground plane. Figure 2 shows the S11 parameters with different values of Rc, we can clearly observe after etching a semi-circular shape from middle upper in the ground plane the improvement in the bandwidth with the variation of Rc (mm), and between the values of the radius Rc (1 mm, 1.5 mm, 2 mm) the best value was 1.5 mm, because it gives the lowest reflection coefficient in the frequency range (2.77 GHz to 30 GHz).

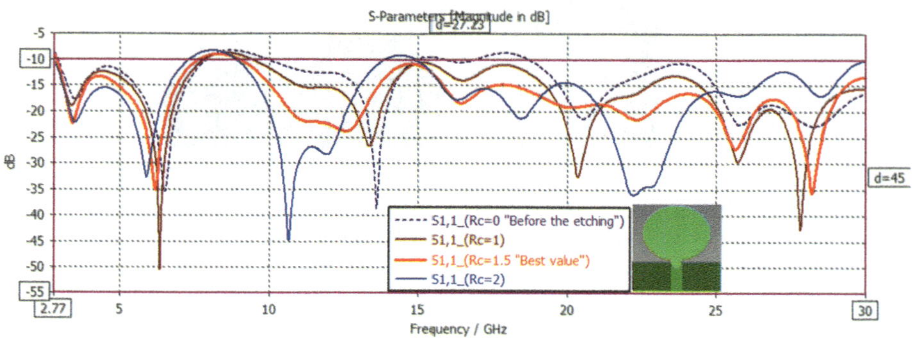

Fig. 2. The effect of the radius Rc variation on return losses due to the semicircular slot in the ground plane of proposed antenna

3.2 Effect of the Two Quarters of a Circle with Radius Re

The reflection coefficient results for the different values of radius Re of the two quarters of a circle, etched from the edges in the ground plane is shown in Fig. 3 Three values of radius Re (2 mm, 2.5 mm, and 2.85 mm) are simulated and we chose the value (2.85 mm), as it gives the best return losses which is lower than −10 dB in the entire frequency range.

We can notice that after the last changes in the ground plane, we got a bandwidth that covers the entire frequency range from fmin = 2.77 GHz to fmax = 30 GHz. The antenna bandwidth where the return loss is lower than −10 dB occupies the band of frequency from 3.1 GHz to 10.6 GHz, so it works well in UWB applications, and also in SWB applications.

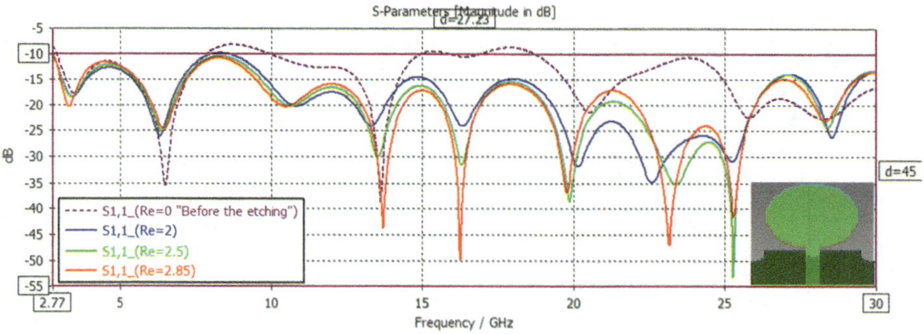

Fig. 3. The effect of radius Re variation on both the reflection coefficient due to the two quarters of a circle in the edges of the ground plane.

3.3 VSWR

Figure 4 shows the simulated voltage standing wave ratio (VSWR) of the proposed antenna with DGS. We can be seen that the simulated results give a VSWR $\leq$ 2 for 2.77 GHz–30 GHz, therefore the results are acceptable and more adapted in the band of 2.77 GHz to 30 GHz. Because the VSWR is less than 2 on this band.

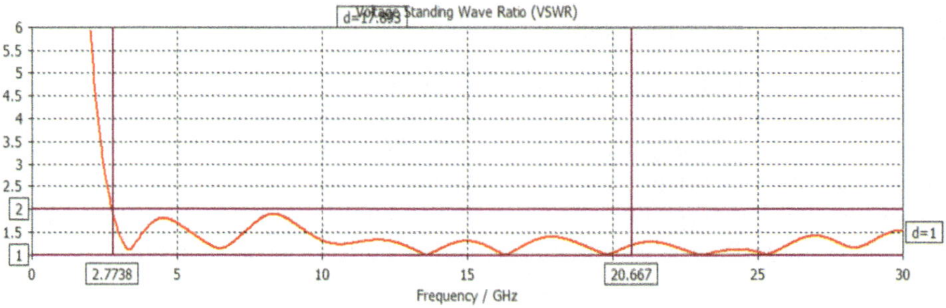

Fig. 4. The frequency (GHz) versus VSWR of the proposed antenna with DGS

3.4 The Radiation Patterns

Figure 5 show the radiation patterns of the proposed antenna with DGS on E-plane and H-plane at different frequencies (3.1 GHz, 5 GHz and 9 GHz). We can be noted the E and H plane radiation patterns of the proposed antenna are symmetric and almost omnidirectional. On the other hand we notice that radiation pattern is comparable to that of a dipole antenna because the three Figures have dumbbell shape. Furthermore the radiation pattern is stable over the band defined by FCC.

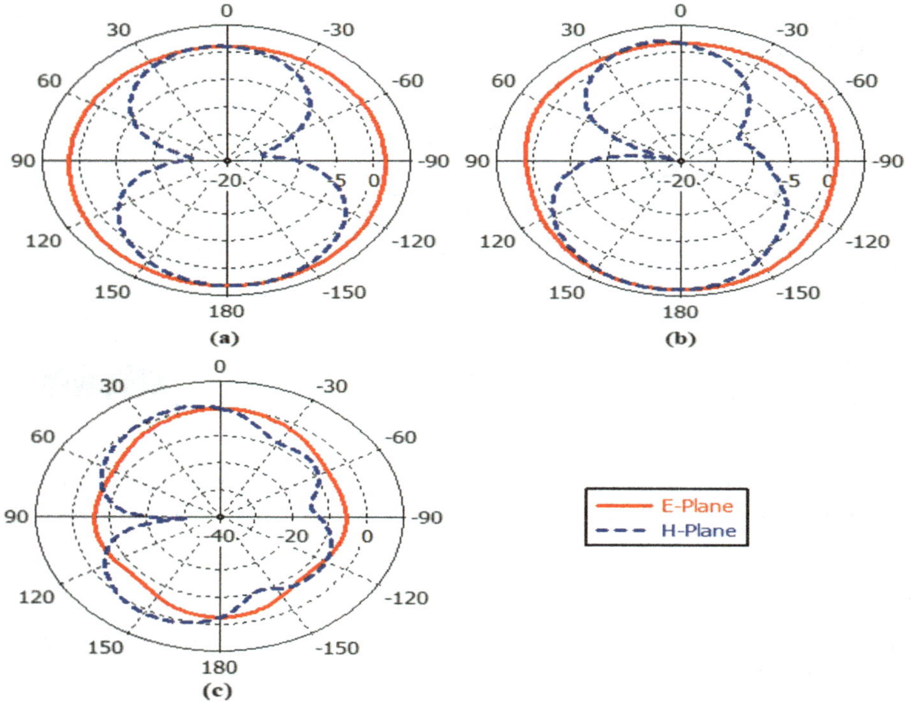

Fig. 5. The Simulated radiation pattern for the proposed antenna with DGS on E-plane and H-plane at: (a) f = 3.1 GHz, (b) f = 5 GHz and (c) f = 9 GHz

3.5 Surface Current Distribution

Figure 6 shows the current distributions on patch and ground where the currents are mainly concentrated on the edges of patch and DGS ground. Hence, the radiation of the patch is more along the edges which lead to effective radiation. The DGS in the ground plane increases the current path which in turn increases the electrical length of the microstrip line. The current distribution was found to be minimum at the non- feeding port and at the DGS. The Maximum surface current is localized at feeding port.

Fig. 6. Surface current of antenna with DGS at f = 5.5 GHz

3.6 Efficiency

The variation of efficiency (%) versus frequency of the proposed antenna with DGS is shown in Fig. 7. The efficiency of this antenna is inversely proportional with frequency. The antenna has a good efficiency for frequencies less than 12 GHz ($\eta(\%) > 80\%$), an acceptable efficiency for frequencies between 12 GHz to 20 GHz ($60\% > \eta\ (\%) > 80\%$), but is shows a weak efficiency for frequencies over 20 GHz (efficiency < 60%).

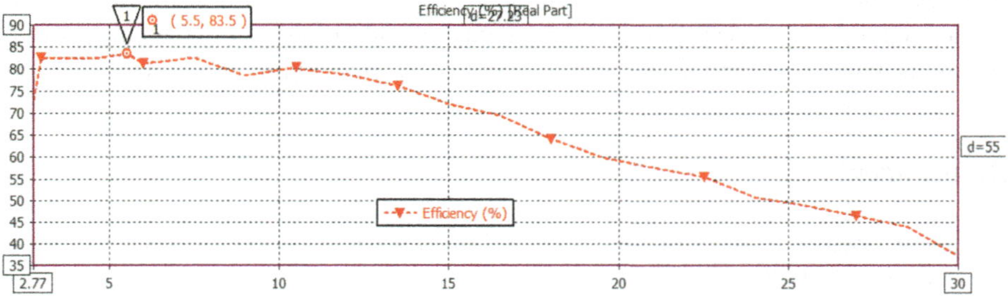

Fig. 7. Efficiency (%) versus frequency (GHz) plot of proposed antenna with DGS

4 Comparison with Other Research Work

Table 1. Comparison of the main parameters between this antenna and other antennas

Antennas	Parameters			
	Dimensions (mm^2)	Bandwidth (GHz)	Min S_{11} (dB)	Complexity
This Work	25 × 30	27.23 (2.77–30)	-55.57	Low
[5]	35 × 77	17.36 (1.44–12)	−34.92	Very high
[7]	35 × 30	8.83 (3.14–11.92)	−47.63	High
[8]	26 × 30	11.5 (3–14.5)	−43.5	High
[9]	55 × 56	13.6 (1.25–14.86)	−30.08	Low

From the Comparative Table (Table 1) between this work and other works in [5, 7–9] we can conclude that our structure of antenna with DGS has:

The widest bandwidth, the smaller size, the Less complicated in design and the best value of $S_{11.}$

5 Conclusion

A circular patch antenna with defected ground structure has been designed and simulated. The antenna design is very simple, compact and it operates with a very good behavior from 2.77 GHz to 30 GHz with the return loss ($S11 < -10$), giving an extremely wide band allowing to be used in many applications, it is suitable and operates well at UWB and SWB communication applications and useful for modern wireless communication systems, because it covers both the short and the long range of communications frequencies and, making possible its integration into portable devices.

References

1. Garg, R., Bhartia, P., Bahl, I.J., Ittipiboon, A.: Microstrip Antenna Design Handbook. Artech house, Norwood (2001)
2. F. S. P. T. Force: Report of the spectrum efficiency working group (2002)
3. Arya, A.K., Kartikeyan, M.V., Patnaik, A.: Defected ground structure in the perspective of microstrip antennas: a review. Frequenz **64**(5–6), 79–84 (2010)
4. Breed, G.: An introduction to defected ground structures in microstrip circuits. High Freq. Electron. **7**, 50–54 (2008)
5. Chen, K.R., Row, J.S.: A compact monopole antenna for super wideband applications. IEEE Antennas Wirel. Propag. Lett. **10**, 488–491 (2011)
6. Balanis, C.A.: Antenna Theory: Analysis and Design. Wiley, New York (2016)
7. Vijayalakshmi, S., Jothilakshmi, P.: Design and implementation of wideband antenna using defected ground structure (DGS) for ultrawide band application. Int. J. Innovative Res. Comput. Commun. Eng. **5**, 178–183 (2017)

8. Elajoumi, S., Tajmouati, A., Errkik, A., Mediavilla Sánchez, Á., Latrach, M.: Microstrip rectangular monopole antennas with defected ground for UWB applications. Int. J. Electr. Comput. Eng. (IJECE) **7**(4), 2027–2035 (2017)
9. Anusudha, K., Karmugil, M.: Design of circular microstip patch antenna for ultra wide band applications. In: 2016 International Conference on Control, Instrumentation, Communication and Computational Technologies (ICCICCT), pp. 304–308. IEEE, December 2016

Enhancing Perceived Quality and Comfort Optimization Through a Parametric CAD/Simulation-Based Process

Hamdaoui Abd El Djalil(✉) and Adad Mohamed Cherif

Department of Architecture, Larbi Ben Mhidi University, Oum El Bouaghi, Algeria
hamdaouihere@gmail.com, adad_cherif30@yahoo.fr

Abstract. Computer aided parametric design opens the door to a set of solutions in regard to modeling and design process. This study aims to investigate the collaboration of parametric design and simulation-based tools to generate a parametric model dotted with a perceived quality and an optimized comfort. The study introduces the perceived quality notion as one of the most important factors to consider when designing a sustainable model, and seeks to highlight the connections between this notion, the current parametric design and the simulation-based environment. A responsive process will integrate these concepts in a proposed model where variations of parameters will be tested and applied to bypass design related problems. By automatically generating many outcomes, forms and shapes using algorithms and configurable parametric systems based on inserted data, the model will be susceptible to modification via a list of parameters and commands and the overall workflow will be more efficient. Parametric design as an approach allows for a flexible design exploration and easy access to modeling and simulation data, providing multiple possibilities to improve constructions quality and to analyze the impact of the chosen design process on the comfort levels of structures especially in arid areas where intelligent environmental solutions are essential.

Keywords: Perceived quality · Parametric CAD · Sustainable design · Simulation · Comfort optimization

1 Introduction

Parametric design is one of the fields where Computer-Aided Design (CAD) is used to visualize and design digital models in architecture and other domains, this design approach gives the designer a wide range of options and the ability to generate many solutions based on algorithms and scripting, also known as algorithmic design. In association with simulation applications, it is possible to reach for a better understanding of the design nature and how it can be improved in different aspects like visual and perceived quality, performance and comfort. This collaborative platform allows analyzing and assessing the performance of the digitally created model and to adapt solutions that are more suitable. Designers can profit from this emerging CAD technology, specifically in the pre-design phase, thanks to the precise form manipulation,

M. Hatti (Ed.): ICAIRES 2019, LNNS 102, pp. 292–304, 2020.
https://doi.org/10.1007/978-3-030-37207-1_30

the versatile process and flexible commands that supports creativity and innovation. Generative Components, Grasshopper, Digital Project, are examples of parametric software that provides more freedom to designers, so they can concentrate on the innovative side of design production. It is important to understand the relationship between parametric design and simulation tools and the related practices from design exploration to quality enhancement and comfort optimization, in order to come out with a sustainable design that has a positive impact on the environment while incorporating new technologies. London City Hall by Norman Foster and The Heydar Aliyev Center by Zaha Hadid are a depiction of contemporary architecture and how parametrization is used in a dynamic way with algorithmic design. Parametric CAD tools must not only be restricted to design 3D models but also need to achieve a higher objective such as boosting the design quality and optimize other related aspects in collaboration with integrated or external simulation tools to combine the aesthetics with functionality.

2 Perceived Quality Notion

Before explaining the perceived quality notion, it is important to understand what perception is, Perception is the ability to organize sensations into useful information, it is the process that allows detecting and interpreting the environment. In the same light, perceived quality is an intangible parameter that is evaluated in relation to a variety of preferences and tendencies, according to Aaker (Aaker, 1991) this notion articulates around seven distinct dimensions (Fig. 1).

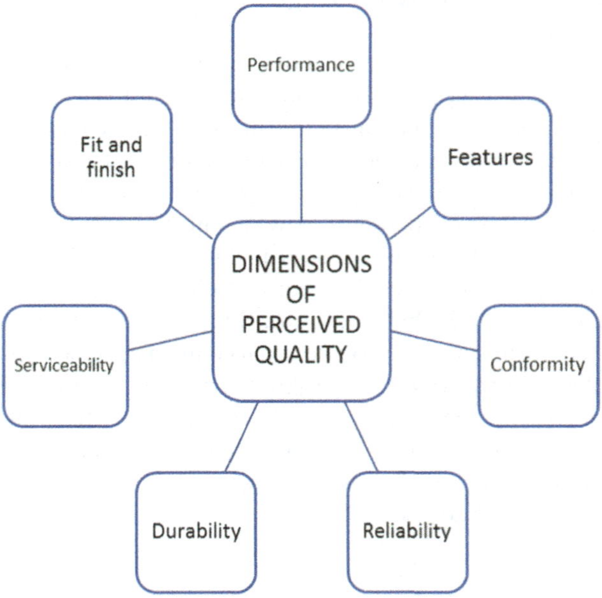

Fig. 1. Dimensions of perceived quality according to Aaker.

Architecture is an art of representation based on visual and mental perception that refers to images (Harries, 1997). Thus, the perceived architectural quality concerns all senses, but in this study, this concept is divided to a set of selected dimensions that are merged in the architectural model according to the relationship with parametric design and computer simulation applications and their contribution in this specific field (Fig. 2).

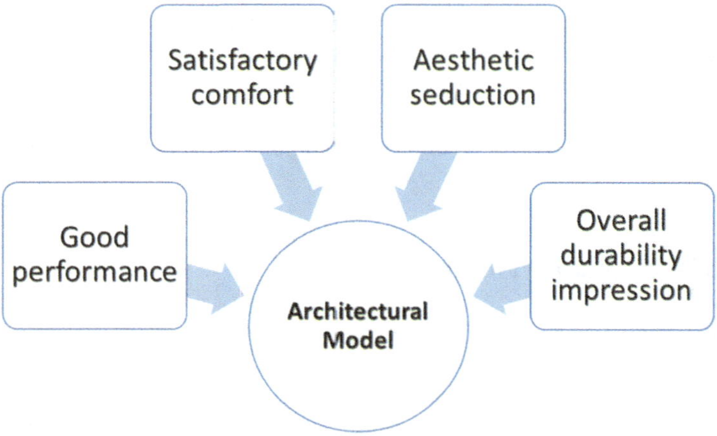

Fig. 2. Selected dimensions that are incorporated in the architectural model.

3 A Glimpse of Parametric Design

Parametric design offers three-dimensional designing tools that allow generating many variations of forms and shapes based on parameters, algorithms and scripting. Branko Kolarevic describes parametric design in this manner: "In parametric design, it is the parameters of a particular design that are declared, not its shape. By assigning different values to the parameters, different objects or configurations can be created. Equations can be used to describe the relationships between objects, thus defining an associative geometry - the constituent geometry that is mutually linked. That way, interdependencies between objects can be established, and object's behavior defined as transformations" (Kolarevic, 2003). In this context, Burry defines the parametric design by saying that "the ability to define, determine and reconfigure geometrical relationships is of particular value" (Burry, 1999). From this perspective, a parametric system comprises various interacting components (Fig. 3), inside this system, the geometry generation is controlled by a list of rules and parameters that are linked to each other, if a particular element of a geometry has its properties modified, all other related parts will be updated accordingly (Woodbury, 2010).

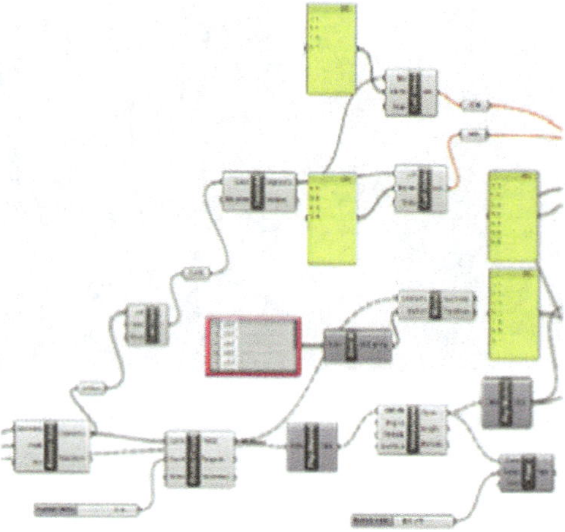

Fig. 3. Components in a parametric system.

A parametric system consists of interconnected parameters; each one of these parameters holds a value that defines the geometric transformation (Jabi, 2013), the object's behavior and other model design details (Figs. 4 and 5).

Fig. 4. Lightwall Cirie, Turin, Italy ecoLogicStudio (http://www.ecologicstudio.com/v2/project.php?idcat=3&idsubcat=4&idproj=10).

On a larger scale, parametric system can generate complex structures using algorithms and scripting (Fig. 6), allowing a fluid assimilation of advanced technologies in the construction field such as automatic mechanical designs, computationally created robotic structures and contemporary curtain walls (Scott, 2009).

Parametric design offers advanced level of modeling configuration thanks to the versatility of parametric CAD that gives the designer the potential to manipulate different aspects of his model with high efficiency and a lower amount of effort (Fig. 7) compared to traditional CAD (Forbes & Ahmed, 2010).

Fig. 5. The programmed wall, ETH Zurich (http://gramaziokohler.arch.ethz.ch/web/e/lehre/81.html).

Fig. 6. The Heydar Aliyev Center by Zaha Hadid (Copyright by Iwan Baan, https://www.archdaily.com/448774/heydar-aliyev-center-zaha-hadid-architects)

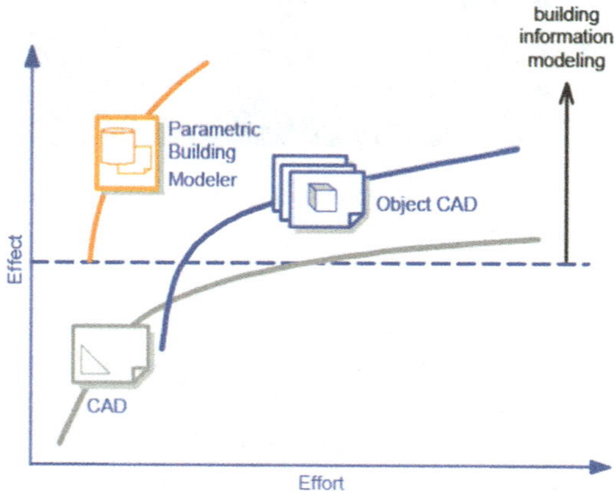

Fig. 7. The effect/effort ratio of parametric modeler (Autodesk, 2003)

4 Simulation-Based Engineering Science (SBES)

Simulation-Based Engineering Science is an new field based on mathematics and complex algorithms for the purpose of producing accurate simulations of physical complex behaviors in many domains like engineering, multiscale modeling, optimization, big data and data-driven simulations. In this context, Bryan Crutchfield illustrates that "AM lets you prototype the part, test the part, select the best design and move into production, all in a fraction of the time compared to traditional methods" (Waterman, 2017).

Simulation software and 3D CAD modeling tools like nTopology, ANSYS, Autodesk Within, allow for a variety of operations to be achieved such as finite element analysis, thermal simulation, offering many solutions for generative design and model optimization (Fig. 8).

Fig. 8. A simulation of distortion in an additively manufactured metal bracket, simulated with Simufact (Waterman, 2017, January. Optimize your Additive Manufacturing Know-How. Digital Engineering, vol. 22, p. 20).

5 Method

This research follows a mixed methods approach based on data analysis from previous works in the parametric design field to find solutions through defining perceived quality related concepts and comfort optimization aspects in order to establish the guidelines for the implementation of these notions in architectural design development. While it is a complex process to effectively, categorize and classify intangible concepts and integrate them in a coherent design strategy as controllable dimensions, a procedure of data collection and analysis (simulation-based) is used to understand multiple perspectives of design quality in relation to advanced modeling technologies (Fig. 9).

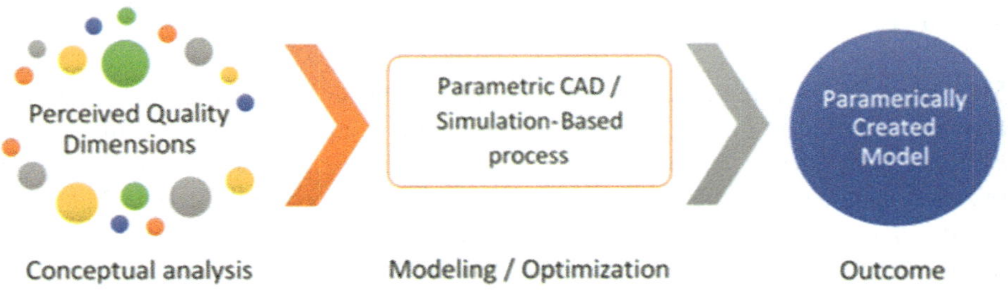

Fig. 9. A graphic demonstrating the research method.

6 Results

A parametric system (Fig. 10) is used to generate a model that demonstrates a controllable-curved structure dotted with a mechanism that responds to the sun's position. The curved roof consists of a collection of circular openings that cover amorphous silicon solar cells with photovoltaic and thermal hybrid solar collectors to generate electricity and thermal energy. Amorphous silicon cells also called thin-film silicon solar cells absorb more light and can be applied on curved surfaces because of its flexibility.

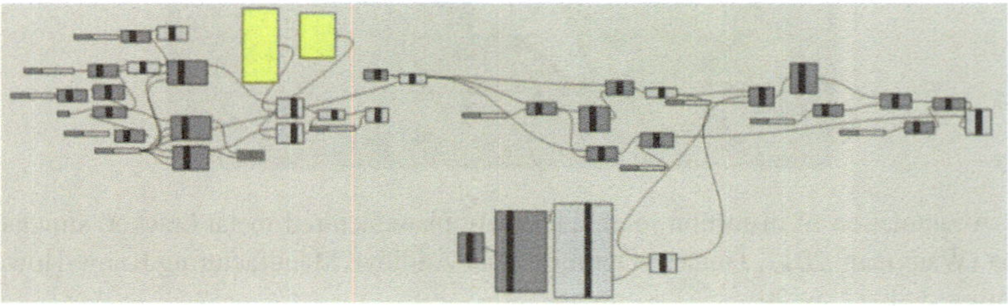

Fig. 10. Parametric system that generates the model.

The model is composed of three parts, a base, a column grid supporting system, and an automated roof (Fig. 11), the column grid is connected to the roof via a data matching algorithm where a component provides a list of data to choose from. This list allows to connect each point with its destined target through a line component without breaking the connection if the geometry changes. If the necessary information is provided, multiple columns can be manipulated by one component. A cull pattern component is used to specify the points that need to be connected, a list of data that contains values (True/False), determines the connection pattern, points with (True) value are connected while others with (False) value are omitted (Fig. 12).

The circular openings on the curved roof are reconfigurable to change form and orientation according to the position of the sun through a point attractor, this method is used to apply multiple changes on a geometry such as rotation, scaling and re-orientation, these changes are bound to a set of conditions and commands related to the

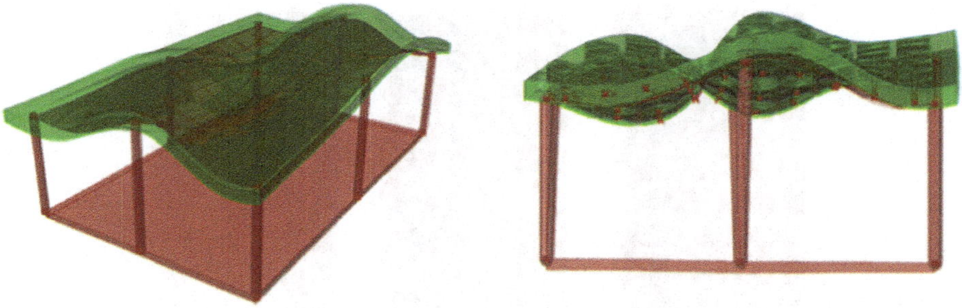

Fig. 11. Adjustable curved roof with control points.

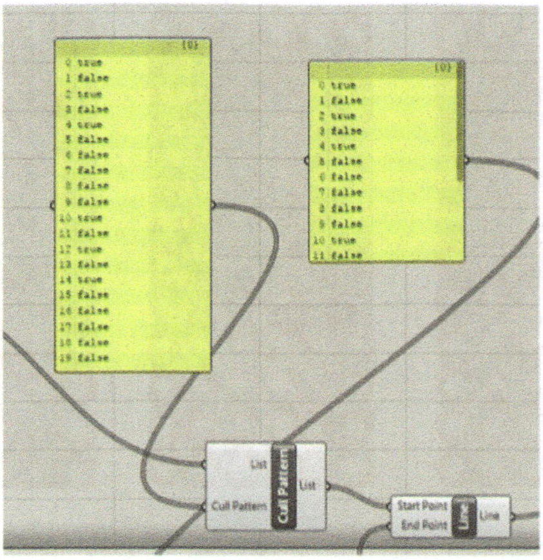

Fig. 12. A list of data to determine the connection pattern.

designer and the parametric environment. Along with attractor point, a morphing process is integrated in the design; this function allows to optimize the geometrical transformation by deforming and bending entities to panelize the curved surface easily. Thus, this combination generates many optimized and adjustable solutions (Fig. 13).

In order to enhance the performance and optimize the design, a Ladybug extension component is connected to the parametric system to import the selected weather data file (Fig. 14) for the purpose of setting up and running a simulation to analyze the orientation of openings in relation to the sun's position.

A weather data file (EPW) of the city of Bechar is inserted into the system to identify the sun path, location, latitude, temperature through each day of the year and humidity. The algorithm calculates the position and the distance between the attractor point, which is the sun, and the circular openings that respond dynamically by changing its position according to sun path (Fig. 15).

Ladybug simulates several solutions in regard to the model orientation to determine the best possible choice that allows the roof to profit from the maximum solar energy while protecting the interior from the high exposition to the sun (Fig. 16), the

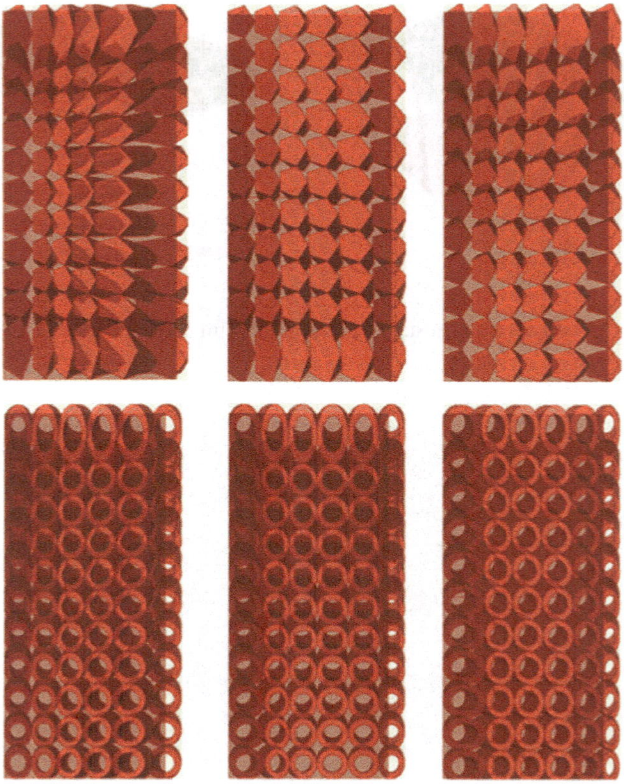

Fig. 13. The diverse outcomes generated from the parametric system.

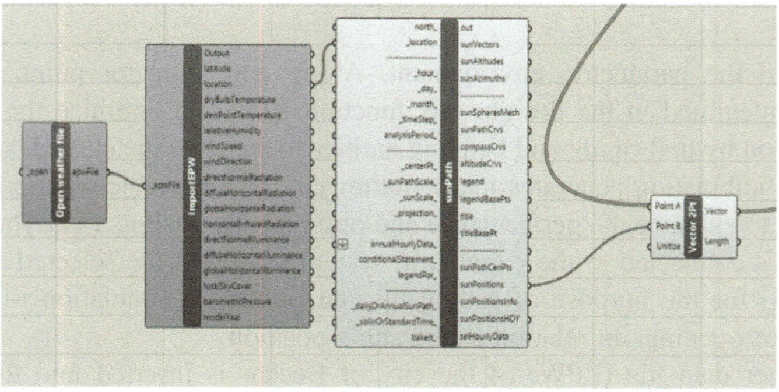

Fig. 14. Importing weather data file to the parametric system.

calculations are based on the weather data of the city of Bechar provided by the Ladybug extension.

The geometry of the curved roof and the calculated orientation affects the energetic performance of the model through absorbing the solar energy and converting it to

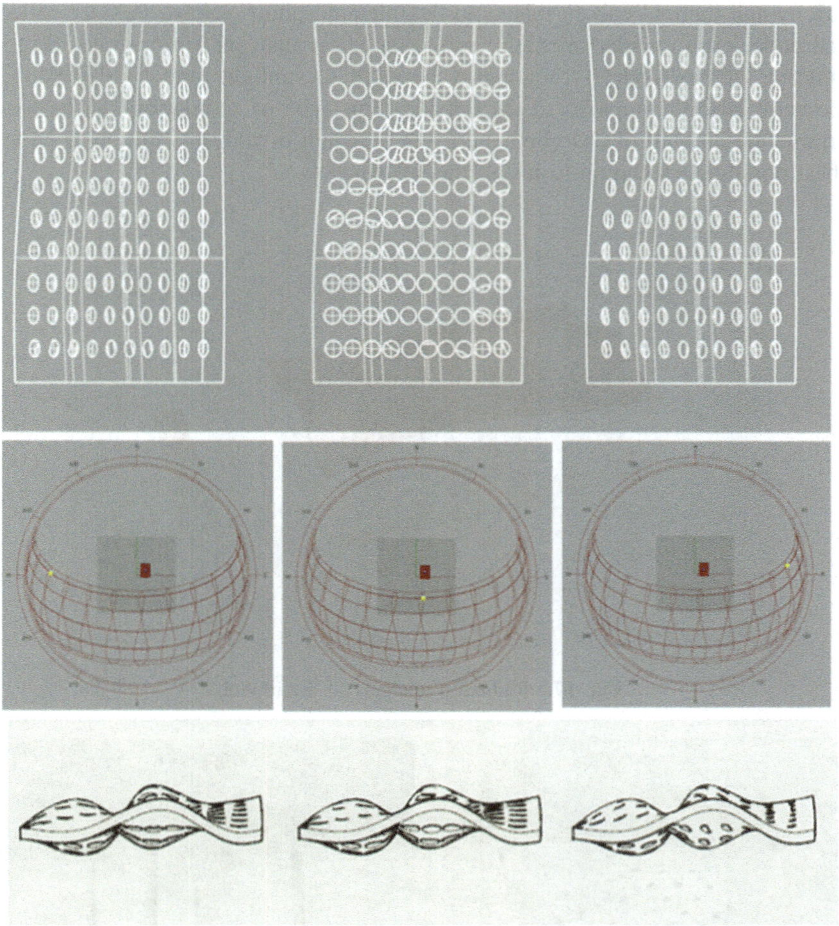

Fig. 15. Openings change their orientation according to the sun's position.

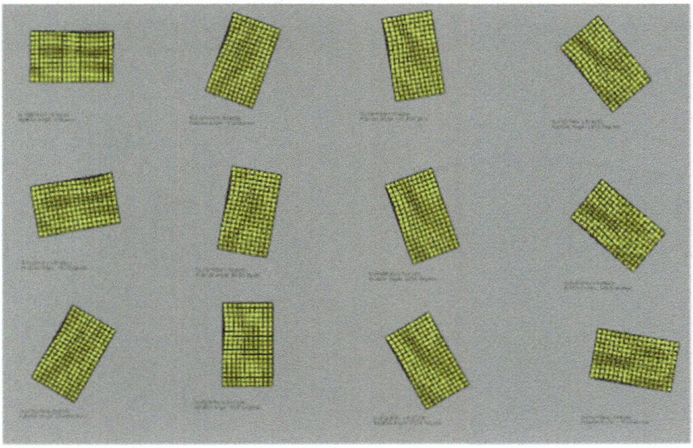

Fig. 16. Testing different positions to choose the best orientation.

electricity and thermal energy thanks to flexible amorphous cells with solar collectors placed under each circular opening. As the radiation analysis demonstrates (Fig. 17), the interior of the model is not highly exposed to sunlight; and by extension, the temperature is low compared to the outside. This type of simulation can be used to determine the amount of absorbed energy in a certain period, in addition to choosing the best possible orientation to install the solar panels.

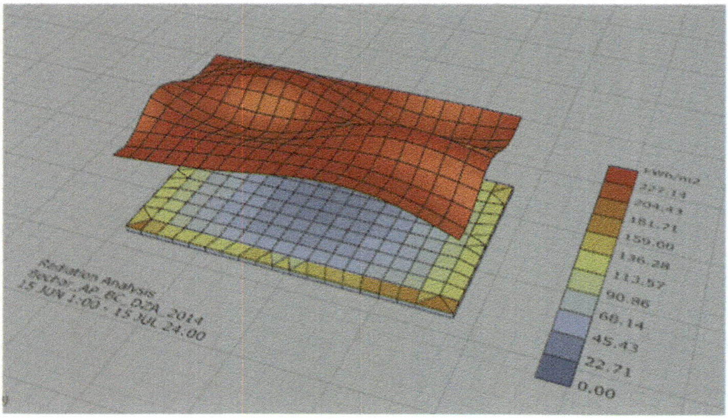

Fig. 17. Radiation analysis of the model.

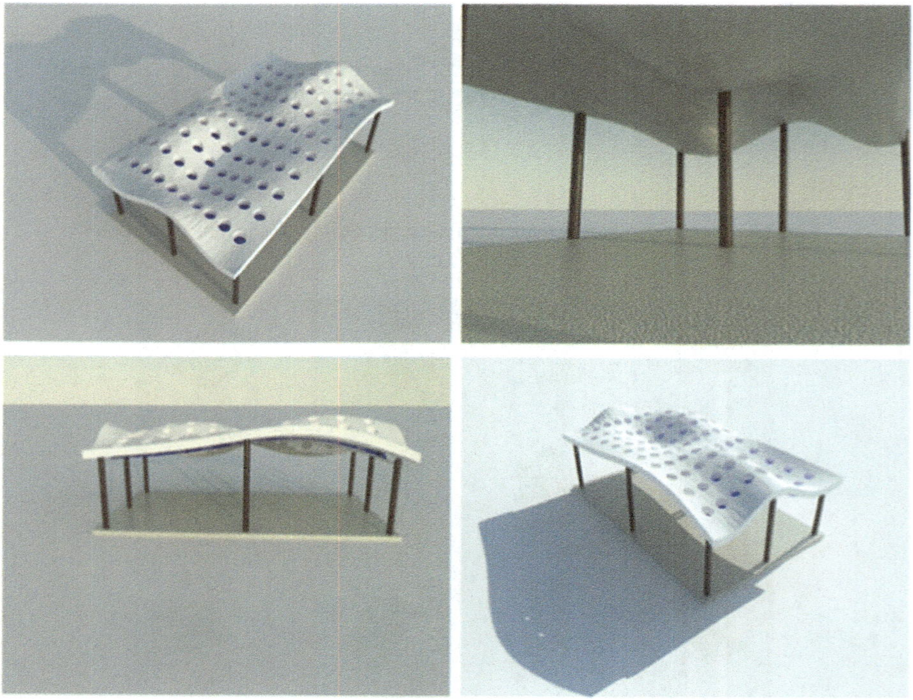

Fig. 18. The final result realized parametrically.

To ensure a better thermal insulation, Fiberglass that consists of fine glass fibers, is used as an insulating material, a loose-fill or blanket is placed in the lower layer of the curved roof. This insulator offers a high resistance to heat and fire and a low thermal conductivity and while it is efficient, it requires a careful installation by experts. Rendering gives the final appearance to an architectural model with a sustainable design (Fig. 18) that combines the perceived quality dimensions such as visual aesthetic, comfort, performance and the application of a parametric modeling/simulation process.

7 Conclusion

Parametric CAD as a modeling tool gives the designer the potential to create a model susceptible to adjustment, which interacts responsively with different agents of the parametric environment, allowing for a flexible form exploration, a shorter time of realization and a better overall quality. The integration of the simulation-based process in the design experience provides the necessary support for optimizing the performance, based on real data transferred efficiently between the system components.

The dynamic parametric CAD/simulation combination that consists of intertwined elements, can produce diverse propositions and test multiple solutions for the sake of assimilating the perceived quality dimensions and comfort factors in the architectural design; from the pre-design phase where the concept model is being created, to the optimization phase where it is tested, until the last stage where the result is rendered in the form of a 3D model.

The incorporation of digital tools in the design process introduces new methods to solve problems related to architectural design in arid areas and permits to build a better understanding of how the sustainable design should be in this environment.

References

Aaker, D.: Managing Brand Equity. Free Press Simon & Schuster Inc., New York (1991)

Benjamis, D.: Beyond Efficiency. Digital Workflows in Architecture. The MIT Press, Cambridge (2012)

Brian, J.: Design Computing: An Overview of an Emergent Field. Routledge, Abingdon (2016)

Burry, M., Perella, S.P. (eds.) AD Profile 141: hypersurface architecture II. Academic Editions, London (1999)

Daniel, D.: Modelled on Software Engineering: Flexible Parametric Models in the Practice of Architecture. Routledge, Abingdon (2013)

Forbes, L., Ahmed, S.: Modern Construction: Lean Project Delivery and Integrated Practices (Industrial Innovation Series), 1st edn. CRC Press, Boca Raton (2010)

Harries, K.: The Ethical Function of Architecture. MIT Press, Cambridge (1997)

Jabi, W.: Parametric Design for Architecture. Laurence King Publishing, London (2013)

Jillian, W., Heike, R.: Landscape Architecture and Digital Technology: Re-conceptualising Design and Making. Routledge, Abingdon (2016)

Kolarevic, B.: Architecture in the Digital Age: Design and Manufacturing. Spon Press, London (2003)

Leatherbarrow, D.: "Sitting in the City, or The Body in the World", in Body and Building. MIT Press, Cambridge (2001)
Marble, S.: Digital Workflows in Architecture. Walter de gruyter, Berlin (2013)
Scheurer, F.: Digital Craftsmanship: From Thinking to Modeling to Building. Birhauser, Basel (2012)
Schumache, P.: Parametricism - a new global style for architecture and urban design. AD Archit. Des. Digit. Cities **79**, 14–23 (2009)
Scott, M.: Contemporary Curtain Wall Architecture. Princeton Architectural Press, New York (2009)
Tuba, K., Benachir, M.: Distributed Intelligence in Design. Wiley, Hoboken (2011)
Waterman, P.J.: Optimize your additive manufacturing know-how. Digital Eng. **22**, 20 (2017)
Woodbury, R.: Elements of Parametric Design. Routledge, Abingdon (2010)

Compact CPW-Fed Ultrawideband Circular Shape-Slot Antenna

Abderrahim Annou[1,2(✉)], Souad Berhab[1,3], Fouad Chbara[1,2], and Tarek Messatfa[1,2]

[1] Department of Electronic and Telecommunications, Université Kasdi Merbah Ouargla, 30000 Ouargla, Algeria
annou.abderrahim@univ-ouargla.dz, berhab_souad@yahoo.com

[2] Laboratory of Electrical Engineering (LAGE), Université Kasdi Merbah Ouargla, 30000 Ouargla, Algeria

[3] STIC Laboratory, Faculty of Technology, University of Tlemcen, Tlemcen, Algeria

Abstract. A compact CPW-fed ultra-wideband (UWB) circular–shape slot antenna is presented. The proposed antenna comprises a circular-shaped slot in the centre of the radiating surface. The antenna is fabricated onto an inexpensive FR4 substrate with an overall dimension of only 18 × 23 mm^2. The suggested antenna is analyzed using full-wave electromagnetic solvers, CST Microwave Studio (MWS) and Ansoft HFSS, to validate the obtained simulated results. Moreover, a comprehensive parametric study has been applied to understand the effect of some parameters on the antenna's performance. The obtained results show that the proposed antenna achieves good impedance matching, constant gain and stable radiation patterns over the operating bandwidth of 3.1–10.6 GHz (109%) that covers the entire UWB. The stable radiation pattern with a maximum gain of 3.58 dBi and a directivity of 80%, making the proposed antenna very suitable for using in UWB applications.

Keywords: Ultrawideband (UWB) · Coplanar waveguide (CPW)-fed · Partial ground plane · Slot antenna

1 Introduction

The UWB technology has received considerable research attention for the large variety of applications in the S-band and X-band covered by the UWB frequency range [1], especially for WLAN and WiMAX wireless applications, and for its usability in the industrial and civil domain.

A low-cost miniature antenna with a broad impedance bandwidth and a stable radiation pattern along with linear phase and low manufacturing cost [2] is an indispensable element of the UWB system.

The CPW technique provides a simple stricture of a single conductor layer above a dielectric substrate, which help reduce the cost of the antenna fabrication In addition, the CPW provides the benefits of good impedance matching, omnidirectional patterns,

M. Hatti (Ed.): ICAIRES 2019, LNNS 102, pp. 305–312, 2020.
https://doi.org/10.1007/978-3-030-37207-1_31

minimum surface wave, other advantages of CPW technique are enlarging the antenna bandwidth [3] and miniaturizing the antenna size [4]. Different shapes of the CPW technique are reported, i.e. bow-tie [5], wide rectangular slot [6], and monopoles [7, 8].

In his study, a planar CPW-fed circular-shaped slot antenna with a simple design and a compact dimension of only 18 × 23 mm^2, significantly smaller than the UWB antennas reported in [9–11], proposed for UWB application.

By etching circular-shaped slot from the radiating surface and using CPW-fed, the proposed antenna achieved a wide bandwidth ranging from 3.15 to 10.7. Further, a comprehensive parametric study on the structure is performed to understand the effect of various dimensions of the main parameters. The suggested antenna is simulated with commercially available package CST software, which is based on the FIT method (Finite Integral Technique). The paper is organized as follows; Sect. 2 describes the antenna design process, the Sect. 3 presents the simulated results and the parametric study and Sect. 4 concludes the paper.

2 Antenna Design

To fulfill the requirement for UWB communications, the design antenna evolution has involved three main steps.

The design evolution starts firstly with a wideband elliptic design of CPW-fed elliptic patch antenna, as shown in Fig. 1a. The first proposed design achieves the −10-dB impedance bandwidth ranging from 3.2 to 9.6 GHz. Secondly, and by etching a circular-shape slot from the elliptic patch, as illustrated in Fig. 1b, wider bandwidth has performed. Finally, with the aim to make the proposed antenna suitable for UWB applications, the ground plan is shaped of two identical semi-ecliptics positioned on the right and left feed line sides, as shown in Fig. 1c.

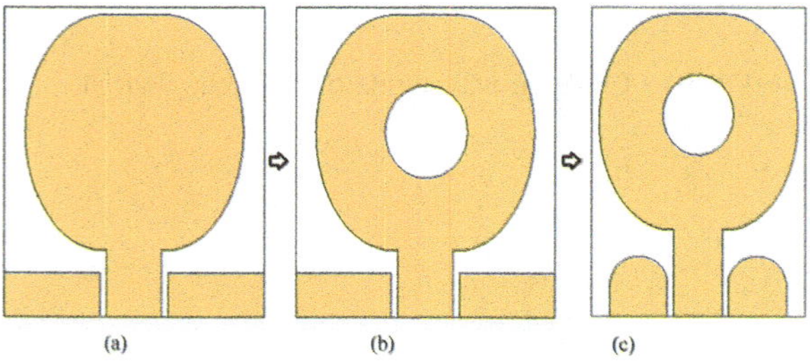

Fig. 1. The proposed antenna design evolution.

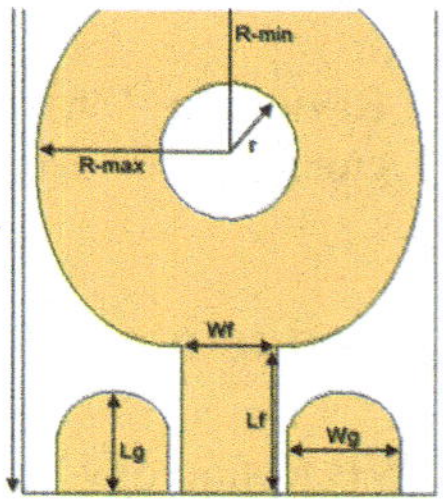

Fig. 2. The proposed antenna geometry.

The geometry and configuration of the final proposed antenna is shown in Fig. 2.

The antenna consists of slotted elliptical patch, defined by a maximum radius R-max, a minimum radius R-min and a slot radius r. The radiating element is printed on the opposite side of an inexpensive FR4 substrate of thickness 1.6 mm, with relative permittivity 4.3 and loss tangent 0.02. The ground plane is chosen to be semi-ellipse conductors positioned on the right and left sides of the feed line. A 50 CPW feed line, having a metal microstrip of width Wf and a gap of distance g, is used to excite the proposed antenna. The optimized parameters of the proposed antenna are the following: R_max = 4.2 mm, R_min = 4 mm, r = 2 mm, wf = 2.5 mm, the gage between the feed line and ground plan g = 0.15 mm, the ground plan parameters are Wg = 4.9 and Lg = 4.5 mm, which represent the width and high off the ground plane respectively.

The overall size of the proposed antenna is only 18 × 23 mm^2, which can be considered as one of very compact UWB slot antennas.

3 Simulated Results and Parametric Study

Of all the investigated design parameters, some of them have a very noticeable effect in determining the, performance of the antenna [12]. In the simulation, only one parameter is varied each time, whereas the others are kept constant.

3.1 Effect of Changing Elliptical Patch Parameters (R-max, R-min)

Usually, a printed planar elliptic patch providing better impedance match in a wider frequency range and has similar effect of bevelling the radiating element [13, 14]. Figures 3 and 4 show the effect of elliptic patch radius. Parametric analysis on the elliptic patch radius is done in the frequency band of 2 to 12 GHz to yield a wider band.

It can be seen that the elliptic patch radius has an effect on the impedance matching bandwidth. The –10 dB impedance bandwidth is obtained from 3.2 to 9.6 GHz, for the optimum values of R-max, R-min of 4.2 mm, 4 mm, respectively.

3.2 Effect of Changing Patch Radius (R)

The wide-slot antenna is well known to have wide impedance bandwidth, though its operating bandwidth is limited due to the degradation of the radiation patterns at higher frequencies [14]. Figure 5 shows the effect of circular-shape slot radius r, in the frequency band of 2 to 12 GHz. As observed, the −10 dB impedance bandwidth is practically the same for the r radius values.

3.3 Effect of Changing (Wg)

The ground plan is shaped of two semi-elliptic of the size of Wg × Lg, emplaced on the right and left of the feed line sides. Figure 6 and 7 show the effect Wg and Lg on the improvement of the bandwidth, and matching impedance bandwidth (under −10 dB) overall the UWB frequency range.

As shown in Fig. 5, the antenna bandwidth is insensitive to the variation of Wg from 15 mm to 18 mm and a little effect on return loss amplitude has recorded in frequency band [3–4 GHz].

Figure 6 presents the variation effect of Lg on the impedance matching bandwidth. As illustrated, the parameter variation Lg has a clearly effected the impedance matching under −10 dB. The good characteristic of the return loss and the bandwidth are obtained when Lg is 4.5 mm.

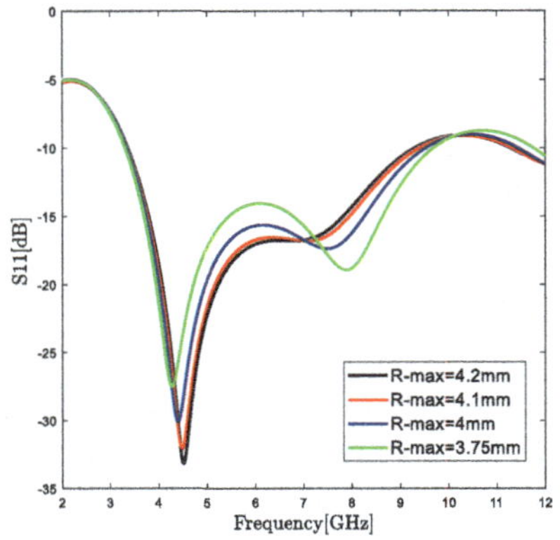

Fig. 3. Variation of patch radius R-max on the return loss response

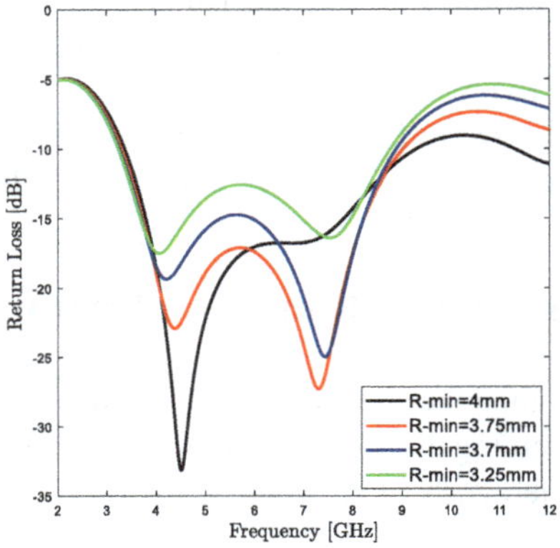

Fig. 4. Variation of patch radius R-min on the return loss response.

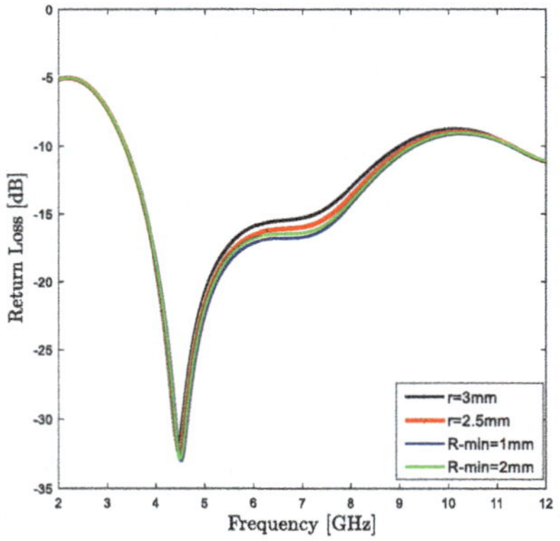

Fig. 5. Variation of patch radius r on the return loss response.

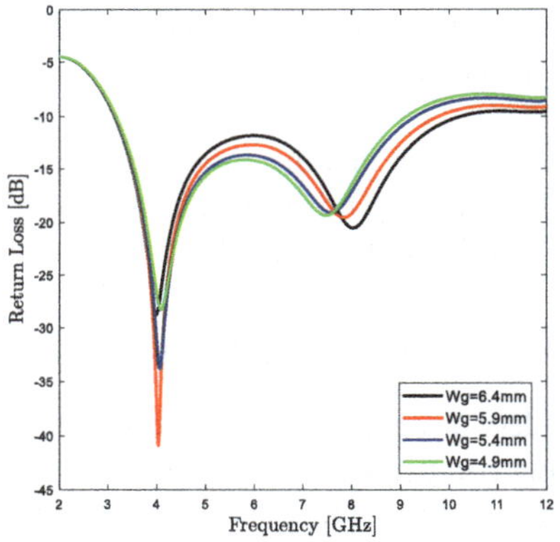

Fig. 6. Variation of Wg on the return loss response.

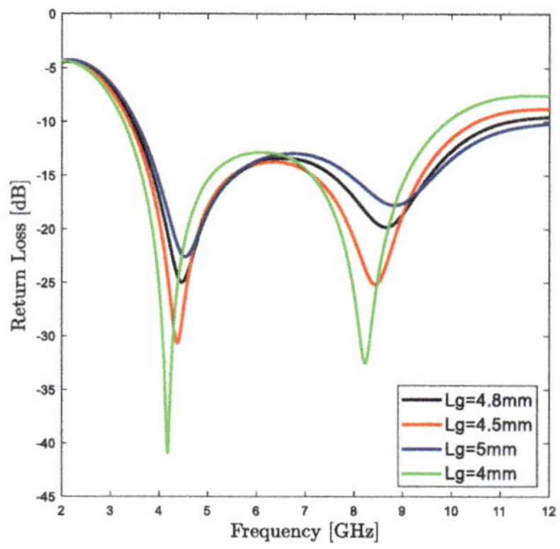

Fig. 7. Variation of Lg on the return loss response.

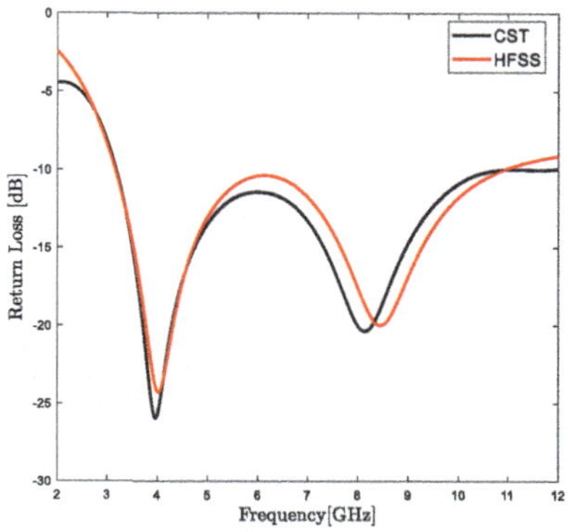

Fig. 8. Simulated return losses of the proposed antenna.

Figure 8 presents the comparison between CST and HFSS return losses in the frequency band [2−12 GHz].

As shown in Fig. 8, a good agreement is observed. The proposed antenna exhibits a wideband performance from 3.1 to 10.6 GHz (109%). Two dominant nulls are observed in S11 characteristics, by emplacing two semi ecliptics in the right and the left of the feed line. Despite being compact in size than the antenna proposed in [9, 13, 15, 16], the antenna covers the entire UWB frequency band.

The simulated peak gain of the proposed antenna is shown in Fig. 9. The proposed antenna has an average peak gain of 3 dBi. The maximum peak gain is 3.58 dBi at 8.5 GHz, where the aperture angle of the radiation patterns become narrower. The radiation efficiency of the proposed antenna is shown in Fig. 10. The proposed antenna achieves a maximum radiation efficiency of 80%.

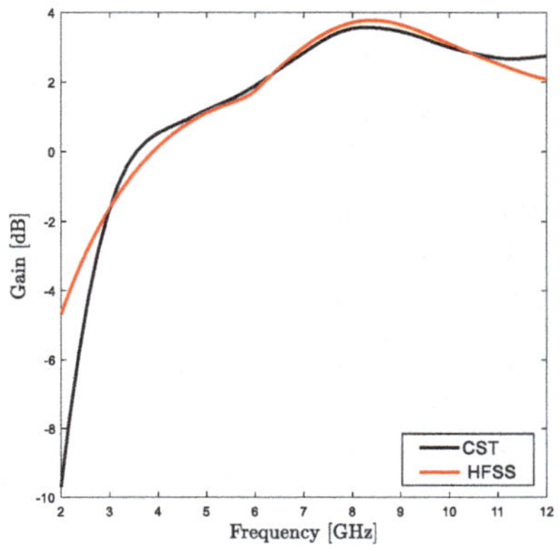

Fig. 9. Simulated peak antenna gain.

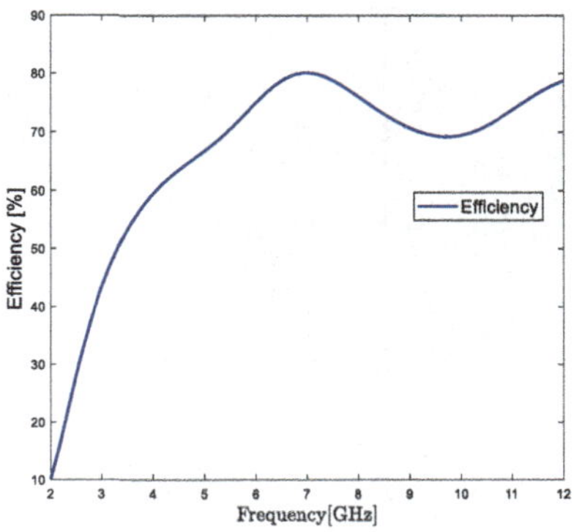

Fig. 10. Simulated radiation efficiency.

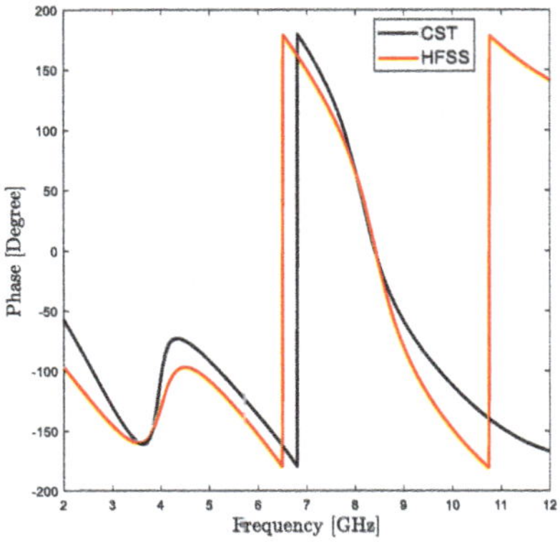

Fig. 11. Simulated phase of the input impedance.

Figures 12 shows the radiation pattern of the proposed antenna, simulated at the frequencies 4 GHz, 6 GHz and 8 GHz.

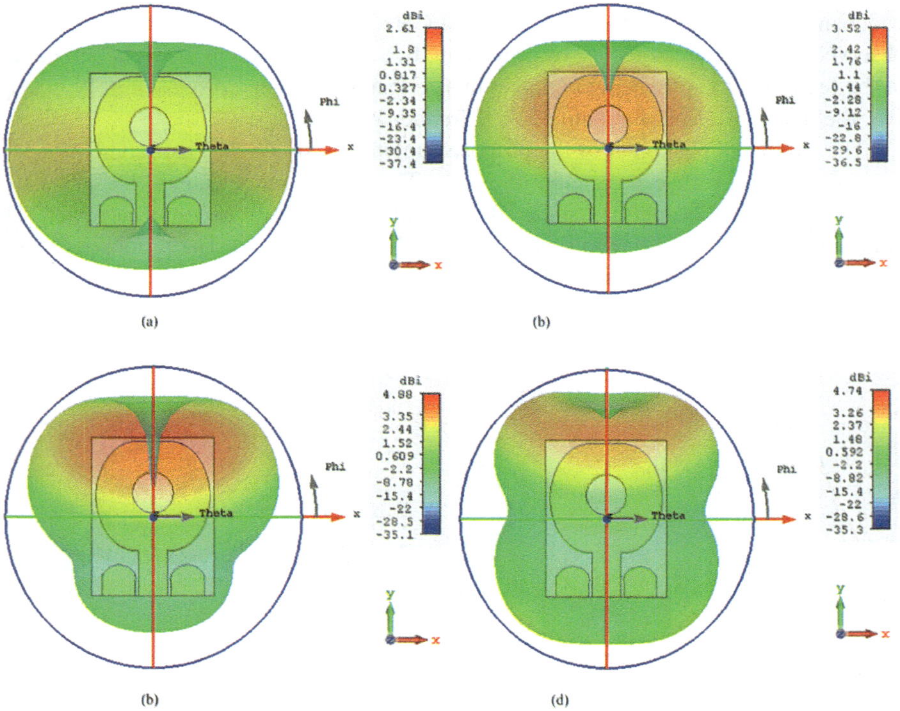

Fig. 12. Simulated radiation patterns of the proposed antenna at different frequencies. (a) 3.1 GHz. (b) 6 GHz. (c) 8 GHz. (d) 10.6 GHz.

It is observed that the radiation pattern is stable over the whole frequency bandwidth of the antenna. The diagram of the E-plan is bidirectional, due to the absence of a ground plan reflector on the bottom of the antenna. The H plan present a single lobe for the low frequencies [3–4 GHz], and two principal lobes at frequencies [5–7 GHz], and three lobs at the frequencies over 7 GHz, Fig. 11 show that the maximum of gain is focused toward the Z-axis, which present the direction of propagation.

In the table below, a comparative analysis is provided with respect to existing UWB antennas. The comparison is done by taking into consideration of overall dimension, bandwidth and return loss amplitudes of all the already reported papers with respect to the proposed UWB antenna.

Parameters	Dimension (mm^2)	Bandwidth (GHz)	S_{11} (dB)
This Work	18 × 23	(3.1–10.6)	−27.5
[9]	35 × 30	(3.14–11.92)	−47.63
[13]	25 × 20	(2.51–16.48)	−34.92
[15]	26 × 30	(3–14.5)	−43.5
[16]	55 × 56	(1.25–14.86)	−30.08

4 Conclusion

A compact CPW-fed ultra wideband (UWB) circular–shape slot antenna has proposed. The ultra-wideband property for the proposed antenna is achieved by using the elliptic patch and positioned semi-elliptic conductors on the right and left of the feed line. The proposed antenna presents a low profile and overall size of 18 × 23 mm^2. The obtained results confirm that the antenna achieves good impedance matching, constant gain, linear phase, stable radiation patterns over the UWB application. The proposed antenna is compared with existing works, in terms of dimension, bandwidth and return loss amplitudes. Hence, the proposed antenna fulfills the requirements of UWB applications, which selected as an excellent candidate for ultrawideband.

References

1. Federal Communications Commission: First report and order. Revision of part 15 of the commission's rules regarding ultra-wideband transmission systems FCC 02-48. Federal Communications Commission, Washington, DC (2002)
2. Sarkar, D., Srivastava, K.V., Saurav, K.: A compact microstrip-fed triple band-notched UWB monopole antenna. IEEE Antennas Wirel. Propag. Lett. **13**, 396–399 (2014)
3. Mitra, D., Das, D., Bhadra Chaudhuri, S.R.: Bandwidth enhancement of microstrip line and CPW-fed asymmetrical slot antennas. Prog. Electromagn. Res. Lett. **32**, 69–79 (2012)
4. Shameena, V.A., Mridula, S., Pradeep, A., Jacob, S., Lindo, A.O., Mohanan, P.: A compact CPW fed slot antenna for ultra wide band applications. AEU Int. J. Electron. Commun. **66** (3), 189–194 (2012)

5. Sadek, S., Katbay, Z.: Ultra wideband CPW bow-tie antenna. In: 2009 International Conference on Electromagnetics in Advanced Applications, Torino, pp. 261–263 (2009)
6. Sharma, P., Jha, S.K., Bhattacharya, P.P.: A new CPW-fed patch antenna for UWB applications. J. Telecommun. Inf. Technol. **2017**, 75–78 (2017)
7. Jose, L.A., Atulbhai, P.J., Dwivedi, R.P.: CPW ultra-wideband tunable notched antenna. In: 2017 International Conference on Nextgen Electronic Technologies: Silicon to Software (ICNETS2), Chennai (2017)
8. Balaji, M., Vivek, R., Joseph, K.O.: CPW feed circular monopole antenna for UWB applications with notch characteristics. In: 2015 IEEE International Conference on Electrical, Computer and Communication Technologies (ICECCT), Coimbatore, p. 14 (2015)
9. Tiwari, R.N., Singh, P., Kanaujia, B.K.: Small-size scarecrow-shaped CPW and microstrip-line-fed UWB antennas. J. Comput. Electron. **17**, 1047 (2018)
10. Zhang, X.-M., Ma, J., Li, C.-X., Ma, A.-S., Wang, Q., Shao, M.-X.: New planar monopole UWB antenna with quad notched bands. Prog. Electromagn. Res. Lett. **81**, 39–44 (2019)
11. Chakraborty, M., Pal, S., Chottoraj, N.: Realization of high performance compact CPW-fed planar UWB antenna using higher order asymmetry for practical applications. Microwave Opt. Technol. Lett. **58**(2), 2515–2519 (2016)
12. Berhab, S., Abri, M., Fellah, B.: Proposed of multi-band inset fed stairs bow-tie antenna. In: Conference on Electrical Engineering CEE (2019)
13. Huang, C.-Y., Hsia, W.-C.: Planar elliptical antenna for ultra-wideband communications. Electron. Lett. **41**(6), 296–297 (2005)
14. Antonino-Daviu, E., Cabedo-Fabres, M., Ferrando-Bataller, M., Valero-Nogueira, A.: Wideband double-fed planar monopole antennas. Electron. Lett. **39**, 1635–1636 (2003)
15. Elajoumi, S., Tajmouati, A., Errkik, A., Sanchez, A., Latrach, M.: Microstrip rectangular monopole antennas with defected ground for UWB applications. Int. J. Electr. Comput. Eng. **7**(4), 2027–2035 (2017)
16. Karmugil, M., Anusudha, K., Karmugil, M., Anusudha, K.: Analysis of circular patch antenna with slot and DGS for UWB applications, vol. 10, no. 37, pp. 157–167 (2017)

Efficient Management of Channel Bonding in the Current IEEE 802.11ac Standard

Fadhila Halfaoui, Mohand Yazid(✉), and Louiza Bouallouche-Medjkoune

Research Unit LaMOS (Modeling and Optimization of Systems), Faculty of Exact Sciences, University of Bejaia, 06000 Bejaia, Algeria
hal.fadila@gmail.com, yazid.mohand@gmail.com, Louiza_medjkoune@yahoo.fr

Abstract. IEEE 802.11ac is a Very High Throughput (VHT) Wireless Local Area Network (WLAN) introduced to reach 7 Gbps over 5 GHz bandwidth. This data rate is achieved through several enhancements to the physical (PHY) layer and medium access control (MAC) sub-layer. Channel bonding is one of the new technologies introduced in IEEE 802.11ac that increases channel bandwidth by combining multiple 20 MHz channels. To access the wide channel, IEEE 802.11ac specified a multichannel MAC procedure, known as: Dynamic Multichannel Access (DMA) that operates under the Enhanced Distributed Channel Access (EDCA) operating rules. In this paper, we demonstrate that: (i) the basic rules of EDCA function causes the access equity problem between the data streams, (ii) the DMA procedure is not effectively managing the bandwidth allocated to the IEEE 802.11ac WLAN. This is why, our purpose in this work is proposing a new MAC procedure called EDMA (Enhanced Dynamic Multichannel Access) that enhances the existing DMA procedure, in order to efficiently use the radio resources while distributing them equitably on different data streams. The simulation results show that the new EDMA procedure provides satisfactory results (in terms of throughput) by comparing it with the DMA procedure.

Keywords: IEEE 802.11ac WLAN · Channel bonding · Multichannel access procedures · Simulation and performance analysis

1 Introduction

The IEEE 802.11 standard is one of the most promising technologies to provide ubiquitous networking access. The working groups have always strived to improve this wireless technology through creating new amendments to the base of IEEE 802.11 standard. Since its beginning in 1997, the network throughput and coverage were increased by every successive standard amendment, starting from IEEE 802.11b through IEEE 802.11a and IEEE 802.11g until to IEEE 802.11n/ac [1]. The data rate has gone from megabits per second to gigabits per second, which was achieved by the cable technology not long ago [2]. IEEE 802.11ac is another amendment which defines VHT (Very High Throughput) networks and data rates up to 7 Gbps [3]. The IEEE 802.11ac amendment improves the achieved Throughput coverage and QoS (Quality of Service)

M. Hatti (Ed.): ICAIRES 2019, LNNS 102, pp. 313–321, 2020.
https://doi.org/10.1007/978-3-030-37207-1_32

capabilities, compared to previous generations, by introducing improvements and new features in the PHY and MAC layers. In the PHY layer, higher Modulation and Coding rates (256 QAM: Quadrature Amplitude Modulation), wider bandwidth channels (up to 160 MHz) at 5 GHz and eight spatial streams are used to enable higher spectral efficiency [4]. IEEE 802.11ac adopted downlink Multi-user Multiple-Input Multiple-Output (MU-MIMO) which allows multiple frames to be simultaneously transmitted from the access point (AP) to different receivers via multiple spatial streams [5].

In the MAC layer, IEEE 802.11ac includes many of the improvements first introduced with IEEE 802.11e and IEEE 802.11n [6]. Frame aggregation is a feature of the IEEE 802.11n and IEEE 802.11ac that increases throughput by sending two or more consecutive data frames in a single transmission, followed by a single acknowledgment frame, denoted Block Ack (BA) [3], in order to reduce transmission overheads in the PHY and MAC layers. IEEE 802.11ac also introduces other novel features, e.g., the TXOP Sharing, [2] and channel bonding. The TXOP Sharing mechanism extended the existing TXOP Transmission Opportunity to send simultaneously down-link multiple data streams to multiple receivers [4]. TXOP is period during which a particular station can transmit several frames without contention. The significant increase of the data rates offered by 802.11ac is explained also by the use of channel bonding techniques [7]. The 802.11 standard has defined Distributed Coordination Function (DCF), which has been the basic method of medium access from the beginning of the IEEE 802.11 standard [1]. DCF has been designed for the unique support of transfer of data and the web, and it was not initially foreseen that the DCF could be used by QoS demanding applications such as voice and video. For supporting QoS, DCF has been enhanced to EDCA. EDCA is a distributed channel access function enabling prioritization defined initially in IEEE 802.11e and it is the access method used by IEEE 802.11n/ac for its benefits [8].

The rest of this paper is organized as follows. In Sect. 2, we briefly introduce the basic principle of Channel Bonding technique and EDCA operating rules. The Enhanced Dynamic access is presented in Sect. 3. In Sect. 4, the performance evaluation by simulation is conducted. Finally, the paper is concluded in Sect. 5.

2 Background

2.1 Channel Bonding Technique

Channel bonding technique allows the use of 80 and 160 MHz wide channel bandwidths by respectively bonding four and eight 20 MHz channels. 20, 40 and 80 MHz channels are defined as mandatory supports, and 160 MHz Channel as optional support [9]. Efficiently utilizing the 80/160 MHz wide channel is challenging due to legacy 802.11a and 802.11n [10] stations operating in 20 or 40 MHz wide channels. 802.11ac networks are expected to have less chance to operate in its full 80 MHz bandwidth but will have to efficiently share the 80/160 MHz channel with the legacy 802.11a/n networks [10]. In this paper we consider a channel of 80 MHz. Channels consisting of 40 MHz or wider always require a primary 20 MHz channel. To access the wide channel IEEE 802.11n defined a 20/40 MHz static and dynamic access rules which are extended to the IEEE 802.11ac 80 MHz channel. The 802.11ac station checks if the primary channel is idle for an AIFS plus a Backoff (BO) counter time and if the secondary channels are also idle for a PIFS (Point coordination function Inter Frame Spacing) duration immediately preceding the expiration of the Backoff counter. If one of the secondary channels are not idle, the station can reattempt to access the 80 MHz channel (according to the static access) by restarting the channel access attempt with a new Backoff counter chosen from its current Contention Window (CW) [10], or (according to the dynamic access) transmit data over a narrower channel 20 MHz (primary channel) or 40 MHz (primary channel and one secondary channel).

2.2 EDCA Function

The EDCA function proposed in IEEE 802.11e standard is a channel access scheme providing QoS supporting in WLAN by providing prioritized differentiation of traffics. EDCA clarifies four queues with different priorities, instead of a unique traffic queue in DCF [8]. These four queues are mapped to four Access Categories (AC), VO (Voice), VI (Video), BE (BestEffort) and BK (Background). The four access categories are characterized by their proprietary contention parameters such as contention window range $(CWmax/CWmin)$ and Arbitrated Inter Frame Space (AIFS) [8] and the Transmission Opportunity Limit (TXOPLimit). The access category with higher priority has more opportunities to access the medium. In virtual contention, the activated ACs inside the same node start their Backoff procedures at the same time to compete the channel. If the Backoff counters of two or more ACs within the same node decrease to zero at the same time slot, the virtual collision happens. The virtual collision management (VCM) makes sure that the AC with the highest priority have the right to access the channel, and the other ACs involved in the virtual collision will double their CW values within the CW range as a real collision occurs [8].

3 Enhanced Dynamic Multichannel Access Procedure

As seen in Subsect. 2.2, when an internal collision occurs, the access category with highest priority wins the access to the channel and the other access categories with lower priorities double their contention windows and reattempt channel access with another backoff time counter. This leads to reduce the chance to access the channel for the following access categories, namely: BestEffort and Background. The Virtual Collision Management (VCM) improves the throughput for AC with higher priority, and at the same time it causes the access equity problem for ACs with lower priorities [8]. The access equity problem of channel resource allocation among the different access categories happens both in one node and among multiple nodes. The dynamic multichannel access procedure allows ACs to transmit data over narrow channels (20 MHz or 40 MHz). For sending over a 40 MHz channel, the primary channel should be free during an AIFS duration plus a backoff time counter and the first secondary channel should be free during a PIFS duration preceding the expiration of the backoff time counter. In this situation, the second secondary channel can be busy and the third channel free or the opposite. In this case, a channel of 20 MHz remained unused. For sending over a 20 MHz channel, the primary channel should be free for an AIFS duration plus one unit of the backoff time counter, under the assumption that at least one secondary channel is busy. In this situation, one or two channels of 20 MHz or 40 MHz remained unused.

To efficiently utilize these available channels, we propose an enhanced version of the existing dynamic multichannel access procedure to improve bandwidth utilization and avoid the access equity problem between the priority and non-priority streams. In the following, we describe step by step the operating functioning of both dynamic and enhanced dynamic multichannel access procedures by presenting their algorithms. Table 1 summarizes all the parameters that will be used in this section.

Table 1. PHY and MAC parameters of DMA and EDMA procedures

Parameter	Description
AC[h]	Access category : h ∈ {VO, VI, BE, BK}
PCH , SCHs	Primary channel , Secondary channels
ADJ_SCH	Number of adjacent 20 MHz secondary channels
N_ADJ_SCH	Number of not adjacent 20 MHz secondary channels
CW , CWmin[h]	Current Contention Window, Minimum contention window of an AC[h]
TXOP, TXOPLimit[h]	Remaining time or current TXOP time, TXOP length of an AC[h]
i , j	Number of retransmissions, Index of the current A-MPDU
BO , PIFS, SIFS	Backoff counter, PIFS space time, SIFS Short IFS space time
AIFS[h]	Arbitration IFS space time before Multichannel Access of an AC[h]
AIFS_h , A _MPDU[j, h]	Current AIFS[h], A_MPDU frame numbered j of an AC[h]
$T_{A\text{-}MPDU[j,h]}$	$T_{A\text{-}MPDU[j,h]}$ frame transmission time
BA[j, h] , $T_{BA[j,h]}$	Block ACK frame for an A _MPDU[j, h], BA[j, h] frame transmission time
m , m′	Maximum retransmissions attempts, Minimum retransmissions attempts
T_r	The time required to send an A_MPDU and receive its acknowledgment

Algorithm 1: DMA algorithm

```
input : A-MPDUs of AC[h] arrive, m=m′, CW
=CWmin[h], TXOP=TXOPLimit[h],i=0, j=1;
AIFS_h← AIFS[h];

1.BO←Random [0, CW ];
   2.If (BO >= PIFS) then
            Repeat
                 sense PCH;
                 if (PCH is sensed idle)
                 AIF(S_h ←AIFS_h-1;
                 else
                 AIFS_h←AIFS[h];
            Until(AIFS_h=0);

            Repeat
              sense PCH for a slot time;
              if (PCH is sensed idle)
               BO ←BO-1;
              else
              Goto 2:
            Until(BO=PIFS);

            Repeat
            Sense PCH for a slot time;
             Sense SCH and compute ADJ_SCH
             idle
```

```
      for a slot time;
          if (PCH is sensed idle)
              BO←BO-1;
           else
              Goto2:
      Until(BO=0);


3.Transmit_A_MPDU(ADJ_SCH, h,i,j,TXOP);

   Else
   BO←PIFS-BO;
   Repeat
     sense PCH for (AIFS[h]-  (BO)) ;
   until (PCH is sensed idle);

   Repeat
   Sense PCH for a slot time;
   Sense and compute ADJ_CH idle for  slot
   time;
          if
            PCH is sensed idle then
            PIFS=PIFS-1

          else Go to (2);
    Until(PIFS=0)

   GOTO3.
```

Transmit_A_MPDU(N_ADJ_CH,h,i,j,TXOP) is a function that will be executed by BE or BK Access categories. It allows ACs with low propriety to send simultaneously over non adjacent channel not used by the access categories with height propriety. Initially, Access with lower BO transmits. After AC_BE and AC_BK sent their data in turn until the end of the TXOP assigned to the high priority category.

- Before starting each transmission, a virtual 802.11ac station (AC[h]) generates a random BO counter in number of time slots taken in its minimal contention window (CWmin[h]).
- The AC[h] must listen the primary channel during the AIFS[h] if the BO is greater than the value of PIFS and during AIFS[h]-(PIFS-BO) otherwise.
- If the primary channel is sensed idle for this duration, the AC[h] sense the primary channel for BO duration. Before the BO expiry of a PIFS value it listen the secondary channels.
- In the case of dynamic access, the AC [h] will transmit the A-MPDU on the primary channel and adjacent secondary channels sensed free.
- In the case of the enhanced dynamic access, the AC [h] will transmit the A-MPDU on the primary channel and adjacent secondary channels sensed free, and the non-adjacent secondary channels sensed free will be allocated for low priority CAs to transmit in parallel with high priority CAs.
- After receiving a B-Ack the AC[h] can transmit another A-MPDU if its TXOP is not exceeded.
- If the B-Ack is not received, the AC[h] retransmits the packet after doubling its contention window.
- At each transmission failure (i), the AC[h] doubles its contention window to the minimum retransmission number (m′). The packet is destroyed after a maximum retransmission number (m) and the contention window is reset to the minimum value (CWmin).

Algorithm 2: EDMA algorithm

```
input : A-MPDUs of AC[h] arrive, m=m′, CW =
CWmin[h], TXOP=TXOPLimit[h],i=0, j=1;
AIFS_h← AIFS[h],counter←0;
1.BO←Random [0, CW ];
2.If (BO >= PIFS) then
       Repeat
              sense PCH;
               if (PCH is sensed idle)
                    AIFS_h ←AIFS_h-1;
               else
                   AIFS_h←AIFS[h];
          Until(AIFS_h=0);
          Repeat
                sense PCH for a slot time;
                    if (PCH is sensed idle)
                        BO ←BO-1;
                    else
                    Goto 2:
          Until(BO=PIFS);

          Repeat
               Sense PCH for a slot time;
        Sense SCH and compute      ADJ_SCH and
             N_ADJ_SCH idle for
             a slot time;
                          if (PCH is sensed
idle)
                             BO ←BO-1;
                    else
                    Goto 2:
      Until(BO=0);

       3. If(N_ADJ_SCH=0)

Transmit_A_MPDU(ADJ_SCH, h,i,j,TXOP);
 Else
 Transmit_A _MPDU(ADJ_SCH, h,i,j,TXOP);

Transmit_A_MPDU(N_ADJ_CH,h,i,j ,TXOP);

Else
             BO←PIFS-BO;
```

```
     Repeat
        Sense PCH for (AIFS[h]-(BO)) ;
     until (PCH is sensed idle);
        Repeat
         Sense PCH for a slot time;
     Sense and compute ADJ_CH and N_ADJ_SCH
    idle for slot time;
         PIFS=PIFS-1;
               if
                PCH is sensed idle then
                PIFS=PIFS-1;
                 else Go to 2;
             Until(PIFS=0)
              Goto  3;

Void Transmit _A_MPDU (CH,h,i,j,TXOP);

  Tr  ← TA-MPDU[j,h] /(CH +1)+SIFS+TBA[j,h]);

 Repeat
 Transmit A _ MPDU[j, h] over (CH+1) channels;
  Wait for SIFS ;
      if   (BA[j, h] is received) then
           TXOP ← TXOP- Tr    ;
             j ← j + 1 ;
            Wait for SIFS ;
   else
            i ← i + 1 ;
            if ( i < m) then
              if ( i < m′) then
                   CW ← CWmin[h] * 2i ;
            Go to 1;
   else
                      Destruct;

    Until (TXOP < (SIFS + Tr ));
```

4 Simulation and Performance Analysis

In evaluating the performance of dynamic access and our proposal we have taken an interest the overall throughput of the network. Throughput is exprimed in terms of Mbps (Mega Bits Per Second). It refers to the total number of A-MPDUs transmitted by an ACs and that are successfully received during the simulation time. To carry out our simulation, we have implemented an 802.11ac PHY/MAC simulator with C language on Linux using multithreading programming. Each calculated data is the average of 5 simulations performed using the same parameters. The results are presented with a 95% confidence interval. Table 2 denotes the 802.11ac PHY and MAC parameters used for simulations as described in the 802.11ac standard [11].

Table 2. Dynamic Multi channel Access, Enhanced Dynamic Multi channel Access, PHY and MAC parameters.

Parameters	Numerical values
Bandwidth (BW) , Number of Channels (NC) for 20/40/80 MHz	20/40/80MHz , 1/2/4
Number of Spatial Streams (NSS) , Propagation time, Slot time	8NSS, 1 μs , 9 μs
Maximum length of MSDU, MPDU length , Number cf MPDUs	11 414 bytes, 5500 bytes,16
MPDU header length, Sub frame header length, PHY header time	36 bytes, 4 bytes, 68 μs
Non High-Throughput (non-HT)/BW = 20	52 Mbps
High-Throughput (HT)/BW = 20 MHz	624 Mbps
Block ACK (BA) length, Block ACK (BA) transmission time	40 bytes , 6 μs
SIFS , PIFS	16 μs, 25 μs
AIFS space of [BK, BE, VI, VO]	[79, 43, 34, 34]
Minimal Contention Window CWmin of [BK, BE, VI, VO]	[15, 15, 7, 3]
Maximal Contention Window CWmax of [BK, BE, VI, VO]	[1023, 1023, 15,7]
TXOP of [BK, BE, VI, VO]	[1, 1, variable, (TXOP-VI/2)]

In Fig. 1a we analyze the overall throughput versus the busy rate of the secondary channels. The overall throughput of two access methods decrease with the increase of the busy rate of the secondary channels witch decreases according to the 802.11a/ac

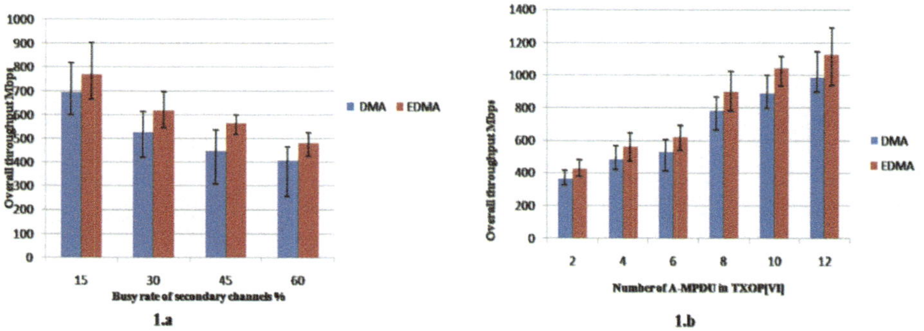

Fig. 1. **a.** Overall throughput versus busy rate of secondary channels **b.** Overall throughput versus TXOP[VI] length.

stations operating on the secondary channels of the 80 MHz wide channel. This is justified that the ACs transmit on narrow channels (20 MHz, 40 MHz). Transmission on narrow channels, does not allow to send a large data during a defined time which decreases the overall throughput. We note that our proposal offers the best overall throughput because low priority CAs have the chance to send their data on non-adjacent free channels not used by high priority ACs, which greatly improves the total throughput of the network.

Figure 1b illustrate the overall throughput versus number A-MPDUs to be transmitted during TXOP[VI]. Throughput increases by increasing the number of A-MPDUs sent for a TXOP [VI] time for the dynamic access and the enhanced dynamic access. When the number of A-MPDU sent is small, the ACs are forced to redo the contention for the access to the channels that increases the overhead communications related to EDCA. By increasing the number of A-MPDUs sent during a TXOP, the different overheads are reduced which allows to send more data and thus improve the total throughput of the network This figure also shows that enhanced dynamic access offers better throughput compared to dynamic access. The overall throughput improvement of the network is due to the improvement of throughput of low priority ACs. Our proposal allows low priority ACs to use channels not used by high priority ACs to send data simultaneously.

Figure 2a illustrates the overall throughput versus number of MPDUs in A-MPDUs. The length of the A-MPDU is an important parameter which really affects the achievable throughput. For a number under or equal 24 MPDUs, the throughputs of the two methods progressively and logically increase. From a number of 32 MPDUs, we see an increase with an average rate for dynamic access method and the enhanced dynamic access. This is justified that the big length of A-MPDUs consume a lot of time for transmission in a narrow channels. The enhanced dynamic access improves the overall throughput of the network for the same reasons explained in the previous two paragraphs.

Figure 2b shows the variation of throughput of the network according to the number of stations over 80 MHz channel bandwidths. The throughput decrease for the two methods with the increasing of the number of station in the network because the

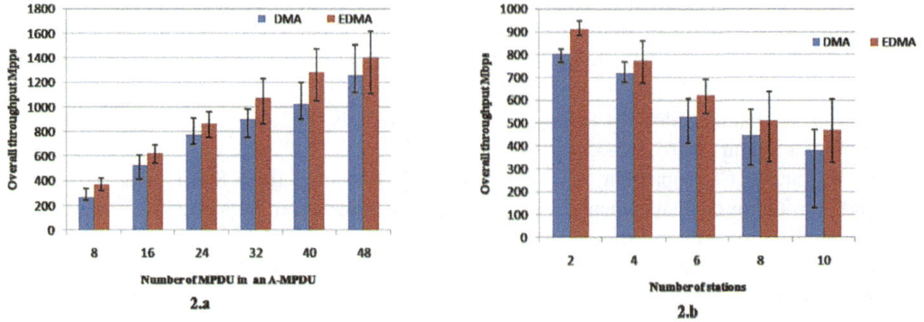

Fig. 2. **a.** Overall throughput versus number of MPDUs in an A-MPDU. **b.** Overall throughput versus number of stations.

use of the bandwidth is degraded and the probability of collision increases. We notice that, the throughput of the network in a case of the use of an improved dynamic access is the best. Thus, it shows once again that it uses the available bandwidth efficiently.

5 Conclusion

In this paper we proposed an improvement of the dynamic access to the multi channels where we elaborated the algorithm of the two methods. The dynamic access in its initial version does not allow using efficiently the wide channels since free non-adjacent channels remains unused during the period of contention. The proposed improvement is made to improve the throughput of low priority ACs that suffers from starvation since the EDCA favors high priority CAs to access free channels. To effectively use wide channels, EDMA allow high priority ACs to affect free non-adjacent channels not used for low priority ACs to send their data simultaneously. Through the performance evaluation via simulations by using multithreading programming, we have analyzed the performance of the DMA and the EDMA depending on comparative studies according to busy rate of secondary channels, number of MPDU in an A-MPDU, TXOP[VI] length and network density parameters. The results showed that EDMA method outperforms DMA regardless of the specification of the 802.11ac network. This is thanks to the improvement of low priority ACs throughput.

References

1. Dolińska, I.: The EDCA implementation in NS-3 network simulator. Zesz. Nauk. Uczel. Vistula **59**(2), 19–29 (2018). Informatyka
2. Liao, R., Bellalta, B., Barcelo, J., Valls, V., Oliver, M.: Performance analysis of IEEE 802.11 ac wireless backhaul networks in saturated conditions. EURASIP J. Wirel. Commun. Netw. **2013**(1), 226 (2013)
3. Daldoul, Y., Meddour, D.E., Ksentini, A.: IEEE 802.11 n/ac data rates under power constraints. In: 2018 IEEE International Conference on Communications (ICC), pp. 1–6. IEEE, May 2018
4. Sharon, O., Alpert, Y.: Coupled IEEE 802.11 ac and TCP goodput improvement using aggregation and reverse direction. arXiv preprint: arXiv:1803.10148 (2018)
5. Yazid, M., Ksentini, A., Bouallouche-Medjkoune, L., Aïssani, D.: Performance analysis of the TXOP sharing mechanism in the VHT IEEE 802.11 ac WLANs. IEEE Commun. Lett. **18**(9), 1599–1602 (2014)
6. Verma, L., Fakharzadeh, M., Choi, S.: Wifi on steroids: 802.11AC and 802.11AD. IEEE Wirel. Commun. **20**(6), 30–35 (2013)
7. Karmakar, R., Chattopadhyay, S., Chakraborty, S.: Impact of IEEE 802.11 n/ac PHY/MAC high throughput enhancements on transport and application protocols A survey. IEEE Commun. Surv. Tutor. **19**(4), 2050–2091 (2017)
8. Chen, F., Li, B., Yang, M., Yan, Z., Zuo, X.: Fairness oriented MAC protocol for the next generation wlan. In: 2017 IEEE Wireless Communications and Networking Conference (WCNC), pp. 1–6. IEEE (2017)

9. Mammeri, S., Yazid, M., Bouallouche-Medjkoune, L., Mazouz, A.: Performance study and enhancement of multichannel access methods in the future generation VHT WLAN. Future Gener. Comput. Syst. **79**, 543–557 (2018)
10. Park, M.: IEEE 802.11 ac: Dynamic bandwidth channel access. In: 2011 IEEE International Conference on Communications (ICC), pp. 1–5. IEEE, June 2011
11. 802.11AC/D7.0: Wireless LAN Medium Access Control (MAC) and Physical Layer (PHY) Specifications: Enhancements for Very High Throughput for Operation in Bands Below 6 GHz (2013)

Remote Control of Several Solenoid Valves for Irrigation System, via GSM (SMS) and Web Page Controller

A. Benbatouche[1,2(✉)], B. Kadri[1,2], and N. Touati[1,2]

[1] Laboratory Information Processing and Telecommunications, Université Tahri Mohamed Bechar, P.Box: 417 route de Kenadsa, Bechar, Algeria
hamid.bat.91@gmail.com, kadri_bf@yahoo.fr

[2] Smart Networks and Renewable Energies Laboratory, Université Tahri Mohamed Bechar, P.Box: 417 route de Kenadsa, Bechar, Algeria

Abstract. The work presented in this paper is a realization of remote control of the solenoid valves with an intelligent system, it is possible to control it via a GSM link by SMS, or by a graphical interface via an internet connection. This system can control several solenoid valves constitute an automated irrigation network controlled remotely. The advantage of this achievement lies in the considerable reduction in the cost of the solenoid valve and the consumption of electrical energy. This system gives us the advantage of saving energy because the consumption will be at the moment of opening or closing the solenoid valves. It is an easy system anyone can use it.

Keywords: GSM · Solenoid valves · Intelligent system · Automated irrigation · Graphical interface

1 Introduction

Water is one of the essential services and basic elements for the earth and human life, in our day we record its scarcity due to the growth of the population [1]. The field of agriculture is the largest consumer of the largest quantities of water through the use of old irrigation systems.

So, this domain has two used irrigation systems. First classical system, that rely on the displacement of the human to control its irrigation network and do it manually which is tiring and time consuming to manage large areas agriculture, and on the other hand, can also be a waste of water. The second control system, it's easy to control, and it's easier to manage the information irrigation network. This clever method doesn't help you wasting water to rely on it with the old methods and also less tiring. Precision intelligent irrigation provides the amount of water needed by the plants, also, it increases the efficiency of agriculture [1, 2].

In our days, the use of internet and cellphones (using 2G, 3G, 4G network) are essential, for example to consult an email or an information through web, to make a call or send a message (SMS) is guaranteed by a simple terminal such as mobile or tablet, using the GSM link. That's why we based our study in development of an intelligent

M. Hatti (Ed.): ICAIRES 2019, LNNS 102, pp. 322–328, 2020.
https://doi.org/10.1007/978-3-030-37207-1_33

irrigation system that will be controlled by simple means that they are accessible to everyone. Our system can be controlled by two methods, are easy to implement in irrigation networks, and does not take time in management of large agricultural areas. The first method that is easiest is to control the solenoid valve network by sending SMS, the system will decrypt the code sent, then it will generate a system command using this code received. This method has been chosen due to the coverage of the GSM 2G network by everyone in the world and it takes a few seconds for the system to react, the system also generates an SMS that will be sent to the controller (system manager) to inform him, the all mains state of solenoid valves (open or close). The second method requires an internet connection in two places, which means that the internet service is must be in these places where the system is implemented, and the same thing for the control side, which is the holder or the controller of this system. A simple graphical interface has been set up in the system for an easiest management, this interface guarantees the control of each solenoid valve separately, and even its current state in real time, also the latest SMS that the system has received and an SMS that generate it and send to the manager by the system.

2 System Objective

This system is developed for improving and solving some problems in controlling large agricultural areas, and easiest their management, here some objectives of this system:

- Realize a complete irrigation network.
- Automation of irrigation system.
- Remote system control from anywhere.
- The destitution of human intervention.
- Smart irrigation and precision farming.
- Water conservation.

3 Equipments and Materials Used

The system is developed with the help of simple electronic components that are available in local markets (Algerian markets) and even the global markets. The materials are chosen for their price, efficiency and also their quality of operation. We will sit a brief definition to each material.

3.1 Arduino MEGA2560

In the realization and design of the prototype, we use an Arduino MEGA2560 card. This board is one of the most popular Arduino boards where the ATmega2650 microcontroller is integrated, this board has a 16 MHz clock and 54 pin Input/Output including 14 pin PWM, 16 analog and 4 link UART [3] (Fig. 1).

Fig. 1. Carte Arduino MEGA2560

3.2 Shield GSM SIM900A

To ensure the remote management of the irrigation system, it is essential to equip our system with a module GSM, The Sim900a GSM shield is a class B GPRS 900/1800 MHz dual-band module. The modem is powered by a selectable interface voltage, which allows with 5 V and 3.3 V connection without need level conversion. The baud rate is configurable 9600-115200 bits/s [2, 3]. This module supports two modes, text mode and PDU mode, in our case the data is sent in text format. This module is linked with the microcontroller of the Arduino using the URATs connexion (Tx, Rx) [4, 5] (Fig. 2).

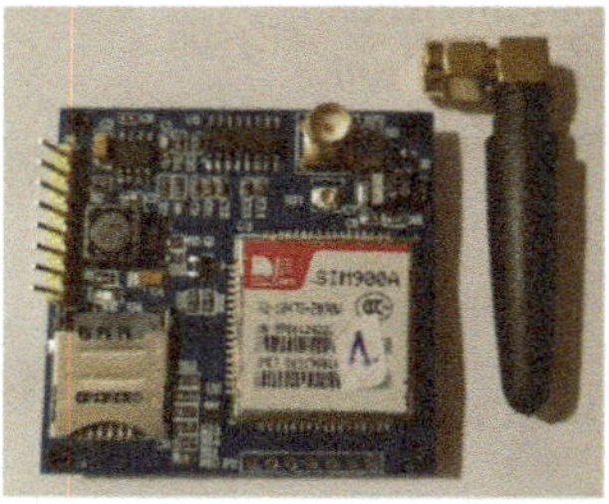

Fig. 2. Module GSM SIM900A

3.3 ESP8266 Wireless Module

ESP8266 is a SoC designed by Espressif Which is based on a 32 bit RISC processor with Tensilica Xtensa LX106 processor. It has features such as built-in Wi-Fi (802.11 b/g/n), General I/O (GPIO), Integrated Circuit (I^2C), Analog/Digital Conversion, Serial Peripheral interface (SPI), UART (universal asynchronous Receiver/Transmitter) [4, 5] (Fig. 3).

Fig. 3. ESP8266 Module

4 Electrical Diagram and Principle of Operation

The irrigation system that we have designed as an example consists of two solenoid valves controlled remotely via a GSM network, and via a web interface. Figure 4 reveals its electrical diagram.

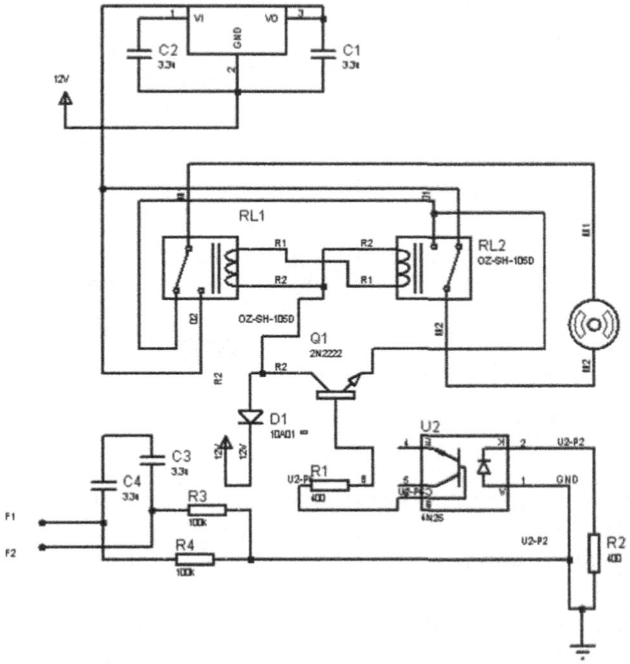

Fig. 4. Electrical diagram of the control board

The opening or closing command of the solenoid valves is performed by inverting the power supply at the terminals of the electric motor by switching two relays (RL1 and RL2) mounted in opposition. If one relay is switched to +12 V the other will be

0 V and so the opposite. The relays are switched by transistor Q1. The optocoupler U2 provides galvanic isolation between the digital control and the power unit. The detection of the opening or closing limits is obtained by reading the signals from the different limit sensors (F1 and F2).

The control of the desired watering cycle is ensured by sending of a simple message in text mode (SMS), the farmer can control or interrogate the state of his system any time and from any place, by sending a message containing the verification code or the control code. After each operation, an SMS will be sent to the farmer to inform him that the operation has been validated.

The web graphical interface is presenting in the system to better facilitate the control and the management of the irrigation cycle wanted by the farmer. Figure 5 shows the functionality provided by this interface, to access to this control interface, an internet connection is required which is the opposite with SMS control.

Fig. 5. Web control interface programmed in Arduino

In the web interface we find several informations, the first part we find the text message (SMS) received or sent by the system, for example the message (#a1) is the control message of the solenoid valve (a) which is sent by the farmer, and the display message (Electrovanne 01 ouvert) is the message sent to the user to confirm that the system reacted using the control message was received. The second part we find the button that gives us the possibility to control the solenoid valves that are installed in our system, after the execution of any control operation we can see and check the status of each solenoid valve, which they help to manage our system well, even if the user has forgotten the last command task that made.

An emergency button has been added to stop all (turn off all system solenoid valves) in the event of a problem or system malfunction.

5 Organigrame

See Fig. 6.

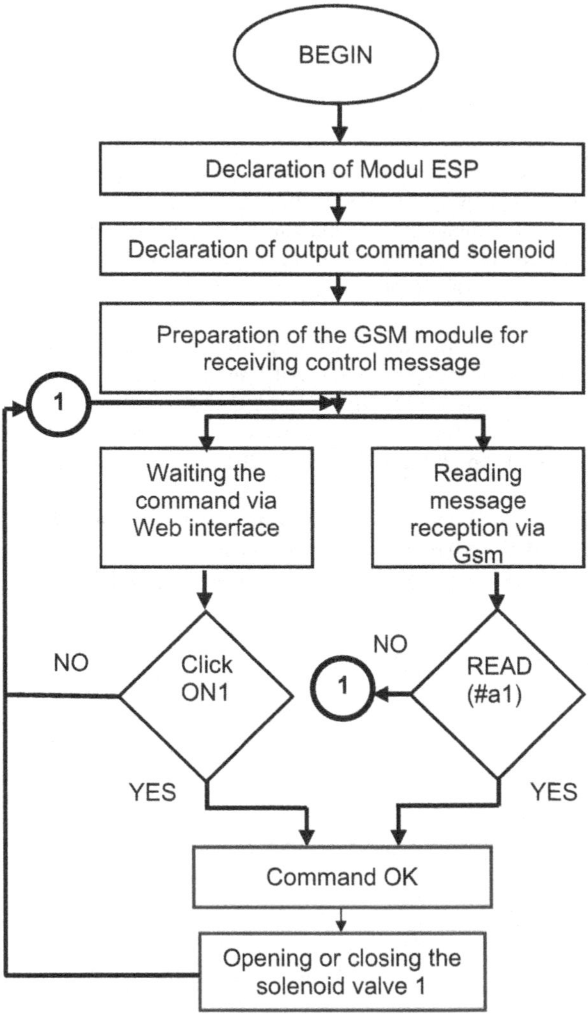

Fig. 6. Organigram de of Operation

6 Conclusion

We have just presented in this article the design and the realization of a remote control with a low-cost intelligent irrigation network based on the development of a simple electronic card allowing the remote control via the GSM network and via a graphical interface of irrigation system (electrovalves connect). The interesting thing of this realization is lying in the considerable reduction in terms of the energy consumed during the opening or closing of the solenoid valves, in fact the energy consumption is detected only when closing or opening the solenoid valves. which is not the case in conventional irrigation systems.

References

1. Pernapati, K.: IoT based low cost smart irrigation system. In: 2nd International Conference on Inventive Communication and Computational Technologies, pp. 1312–1315. IEEE (2018). https://doi.org/10.1109/ICICCT.2018.8473292
2. Newlin Rajkumar, M., Abinaya, S., Venkatesa Kumar, V.: Intelligent irrigation system—an IOT based approach. In: International Conference on Innovations in Green Energy and Healthcare Technologies (ICIGEHT 2017), pp. 1–5. IEEE (2017). https://doi.org/10.1109/IGEHT.2017.8094057
3. Sinulingga, E.P.: Electrical appliances control prototype by using GSM module and Arduino. In: 4th International Conference on Industrial Engineering and Applications, pp. 355–358. IEEE (2017). https://doi.org/10.1109/IEA.2017.7939237
4. Shukla, R., Somani, S.B., Shete, V.V.: International Conference on Inventive Computation Technologies (ICICT), pp. 1982–1993. IEEE (2017). https://doi.org/10.1109/INVENTIVE.2016.7823277
5. Saha, S., Majumdar, A.: Data centre temperature monitoring with ESP8266 based Wireless Sensor Network and cloud based dashboard with real time alert system. In: 2017 Devices for Integrated Circuit (DevIC), pp. 307–310. IEEE (2017). https://doi.org/10.1109/DEVIC.2017.807395

Looking over the Horizon 2030: Efficiency of Renewable Energy Base Plants in Algeria Using Fuzzy Goal Programming

Samir Ghouali[1,2(✉)], Mohammed Seghir Guellil[3,4], and Mostefa Belmokaddem[4]

[1] Faculty of Sciences and Technology, Mustapha Stambouli University, 29000 Mascara, Algeria
s.ghouali@univ-mascara.dz
[2] Faculty of Engineering Science, STIC Laboratory, University Abou Bekr Belkaid Tlemcen, 13000 Chetouane, Tlemcen, Algeria
[3] Faculty of Economics, Business and Management Sciences, MCLDL Laboratory, University of Mascara, 29000 Mascara, Algeria
m.guellil@univ-mascara.dz
[4] Faculty of Economics, Business and Management Sciences, POLDEVA Laboratory, University of Tlemcen, 13000 Tlemcen, Algeria
belmo_mus@yahoo.fr

Abstract. Renewable energy shapes have been broadly utilized in the previous decades, featuring a Green move in energy production. A genuine explanation for this swing to Renewable energy generation is internationals directives, which set the worldwide objectives for energy production from inexhaustible sources, greenhouse gas emissions and increase in energy efficiency. The ability expansion-planning problem of the renewable energy industry implies some important decisions concerning the optimal mix of different plant types, locations where each plant should be built, and capacity extension decisions over the planning horizon for each plant.

The aim of this paper is to analyse the relationship between the type of renewable energy by combining the geographical, climatic and ecological criteria, in the Algerian framework, using a multi criteria analysis, precisely Fuzzy Goal Programming model, based on a multi-source, multi-sink network, in order to determine the optimal number of renewable energy plants for electric generation in the Algerian territory.

Keywords: Algeria · Multi criteria analysis · Fuzzy Goal Programming · Renewable energy · Sustainable development

1 Introduction

The Economic and Energy Development present today comes from non-Renewable Energy such as fossil fuels exhaustible energies, nuclear energy etc, these two energies are the most used by man, and they no longer have the coast. Why? Because they are not eternal and polluting. However, their biggest disadvantage is that these energies are

M. Hatti (Ed.): ICAIRES 2019, LNNS 102, pp. 329–337, 2020.
https://doi.org/10.1007/978-3-030-37207-1_34

too polluting, the combustion of these energies is very much responsible for global warming. Nowadays, this energy issue is in accord, as we need such as travel, the use of means.

Over time, these need only increase. However, the awareness of the risks involved in the use of these resources has led to the emergence and development of new energy sources. It has become inevitable that less polluting ones must replace fossil fuels.

The issue of green energy is well aligned with the ambitious program of Renewable Energy development and energy efficiency or the latter is based on a strategy focused on the development of inexhaustible resources such as photovoltaic solar energy, Solar thermal energy, Wind energy, Hydraulic energy, Biomass…

The issue of Renewable Energies is nowadays at the heart of energy and economic policy debates. The energy transition is a major challenge for emerging countries today: if Renewable Energy plays a major role in their development, their development must be efficient and controlled.

Our work is on this theme and from this; we will dedicate our study so that we can find ways to link green energy with the most strategic implementation of energy bases on the ground.

Knowing that there is a very important link or one that goes in the same direction between the different Renewable Energy sources and their territorial implementation, it is therefore a question of trying to answer the following: What type of Renewable Energy is it best suited for Algeria and which refers to an analysis based on ecological, economic, geographical and climatic criteria?

To answer this problem and to the hypothesis we will proceed to the application of the technique called Goal Programming and more precisely the technique of Fuzzy Goal Programming which is a multi-criteria analysis applied in this precise field.

2 Literature Review

Mezher et al. (1998) taking into account the problems of energy resources, the author of the research gave another vision of the strategy for the implementation of energy bases; it is seen from two points. From an economic and environmental point of view while including different objectives such as: efficiency, cost, job creation and the case of Lebanon is examined to illustrate the usefulness of the proposed GP technique.

Bal and Chabot (2001) they talk about any energy resulting directly or indirectly from solar radiation, characterized as inexhaustible energies, but it must not be denied that they are limited to a given place and time. Speaking about the development of Renewable Energy in Europe, they highlighted the adoption of a European directive on electricity produced from Renewable sources in order to minimize production costs as much as possible while producing even more. In the French context, it will be a question of increasing from 15% Renewable Energy electricity to 21% from 2010.

Kim et al. (2008), the purpose of this research is to determine the locations and sizes of distributed generations (DGs) for reducing losses and improving the voltage profile in distribution systems. The strategic placement of DGs can help reduce power losses and improve the profile of the supply voltage. The author of the research to address the problem of multi-objective DG placement integrating the voltage characteristics of each individual charging component adopts Fuzzy Goal Programming. The initial objective of functions and constraints is transformed into a multi-objective function with fuzzy sets by the Fuzzy Goal Programming method; the multi-objective function transformed with fuzzy sets presents imprecise natures for the criterion of reducing losses and improving the voltage profile, as well as the number of total capacity.

San Cristóbal (2012), apply a solution for the problem of planning the expansion of Renewable Energies involves decisions that involve an optimal mix of different types of energy on different Spanish lands, where each base type should be planted according to the different criteria imposed. In this research, Jose R et al. applies a development of the Goal Programming method based on networks and sources, with the specific aim of implementing five renewable energies for power generation in five different autonomous regions, the aim of this model is to combine the relationship between the economy, energy and socio-environmental to define and examine the different objectives that must be implemented. As each base type can be planted in each location, the purpose of this method is to maximize the number of locations that correspond to the geographical location. In other words, minimize the number of deviations from objectives.

Scala et al. (2013) extended the Goal Programming based method for a solution to a problem of optimization of the multi-carrier energy base management that was formalized by a non-linear method of multi-criteria analysis theory, the goal is to provide an efficient solution for the distribution of energy bases according to the criteria of energy demand and price.

Chang (2015), this author proposes a multi-objective programming model to address the problem of planning for capacity expansion in the Renewable Energy industry. This model involves decisions about the optimal combination of different types of implantations. According to the different criteria. Different types of bases should be located in appropriate locations to minimize total deviations from predefined targets for energy generated, investment costs, emissions avoided, jobs created, and operating and maintenance costs. The proposed method is superior to the model proposed by Ramon and Cristobal, as it can avoid underestimating the level of aspiration and expanding the potentially possible region and obtaining results closer to actual conditions.

Hanane (2015) highlighted the Algerian potential in Renewable Energies. Through the chapters of her thesis she first started by explicitly defining the types of energy, in the second chapter she succeeded in clearly identifying the Algerian government's strategy towards these energies in the context of development through the energy development and efficiency program in 2011, which subsequently highlighted the importance of diversifying the implantation of Renewable Energies throughout the territory.

Jayaraman et al. (2015), the sustainable growth and development of countries requires a strategic vision and a plan aiming growing energy needs while preserving and protecting the natural environment. Studies that couple energy and the environment often have conflicting objectives that require the explicit use of quantitative models to develop appropriate solutions for policy planning and analysis. Multi-purpose programming provides an analytical framework for solving multi-objective problems that simultaneously satisfy the decision-maker's preferences. The author of this research highlighted the development of a polynomial programming model to study the effects of electricity consumption and greenhouse gas emissions on the United Arab Emirates economy. The results provide valuable insights into the opportunities for improvement and quantitative justification of the investments required and efforts to implement sustainable development plans.

Ali et al. (2017), concretizes the notion that Renewable sources can provide a clean and intelligent solution to the increased demand, so photovoltaics and wind turbines are caught up in this research as much as Distributed Generation. The location and sizing of the DG have significantly affected the system losses. In this research, the author uses the Ant-Lion Optimization Algorithm (ALOA) method, a method that is proposed for optimal location and sizing of DG-based Renewable sources for different distribution systems.

Jayaraman et al. (2017) in this research shed light on sustainable development including Renewable Energy; they introduced the weighted GP to achieve the objectives taking into account the criteria: economic, electricity consumption, gas emissions.

Amine et al. (2018) They propose an efficient method, called multi-segment fuzzy goal programming (MS-FGP), which addresses decision-making problems with high levels of uncertainty. The results show that the proposed methodology can assist decision makers in determining the most sustainable renewable energy source portfolio for electricity generation under uncertain conditions and in imprecise environments.

3 Our Application

In the following study, reference will be made to the multi-criteria analysis in relation to energies for an optimized and feasible solution, involving four different aspects and perspectives in the same study and trying to achieve an optimal result to simultaneously satisfy the four criteria of the model. Table 1 presents the values of the objectives highlighted by the state to be expected before 2030.

Table 1. Objectives to be achieved

Goal before 2030	Power/MW	Investissements/MRD	$MTCo_2$/an	Job
Value	28700	148	193000	570000
Tolerance	1000	12	200	5000
Weights	0,3	0,15	0,25	0,3

Table 2 shows each energy in relation to its location, we used five types of Renewable Energy as follows: (1) Solar photovoltaic, (2) Solar thermal, (3) Wind turbine, (4) Biomass, (5) Geothermal, distributed over five strategic locations: (6) Illizi, (7) Laghouat, (8) Adrar, (9) Algiers, (10) Mascara.

Table 2. Geographical location of energies

Energies	Places
1 Solar photovoltaic	6 Illizi, Laghouat
2 Solar thermal	7 Laghouat
3 Wind turbine	8 Adrar
4 Biomass	9 Alger
5 Geothermal	10 Mascara

Table 3 represents each type of Renewable Energy according to four criteria: power in MW; the investment devoted to the project; the carbon dioxide emission avoided in millions of tonnes and the jobs to be expected.

Table 3. Renewable energies for electricity generation

Alternative	Power/Mwh	Investissement/Mrd	MT Co_2/an	Jobs
Solar photovoltaic	13575	35	92300	310000
Solar thermal	2000	35	13600	45000
Wind turbine	5010	28	33600	115000
Biomass	1000	13,2	2980	25000
Geothermal	15	8,8	6470	340

Table 4. Stations for electricity generation

Variable	Type of energies	Stations	Power Mw/station	Investissement/station	Co_2 emissions/station	Jobs/station
X1	Solar photovoltaic	14	970	2,5	6593	22143
X2	Solar thermal	6	333	5,8	2267	7500
X3	Wind turbine	7	716	4	4800	16429
X4	Biomass	2	500	6,6	1490	12500
X5	Geothermal	1	15	8,8	6470	340

4 Model Formulation

With reference to the data presented in Table 4 and the objectives highlighted in Table 1, we can construct four objective functions as follows:

4.1 Electricity Production in MW

$$970 * X_1 + 333 * X_2 + 716 * X_3 + 500 * X_4 + 15 * X_5 \quad (1)$$

4.2 Investment/Billion

$$2.5 * X_1 + 5.8 * X_2 + 4 * X_3 + 6.6 * X_4 + 8.8 * X_5 \quad (2)$$

4.3 CO_2 Emissions/Million Tons/Year

$$6593 * X_1 + 2267 * X_2 + 4800 * X_3 + 1490 * X_4 + 6470 * X_5 \quad (3)$$

4.4 The Number of Jobs Created via Each Energy

$$22143 * X_1 + 7500 * X_2 + 16429 * X_3 + 12500 * X_4 + 340 * X_5 \quad (4)$$

The formulation of this Fuzzy Goal Programming Model can be expressed as follows (Table 5):

Table 5. Goals

Goal 1: $28,7.10^3$	Goal 2: 148.10^9	Goal 3: 193.10^3	Goal 4: 570.10^3

$$Max\, z = 0.3 * \mu_1 + 0.15 * \mu_2 + 0.25 * \mu_3 + 0.3 * \mu_4 \quad (5)$$

St

$$970 * X_1 + 333 * X_2 + 716 * X_3 + 500 * X_4 + 15 * X_5 + n_1 \geq 28700 \quad (6)$$

$$\mu_1 + \frac{1}{1000} * n_1 = 1 \quad (7)$$

$$2.5 * X_1 + 5.8 * X_2 + 4 * X_3 + 6.6 * X_4 + 8.8 * X_5 - p_2 \leq 148 \quad (8)$$

$$\mu_2 + \frac{1}{12} * p_2 = 1 \quad (9)$$

$$6593 * X_1 + 2267 * X_2 + 4800 * X_3 + 1490 * X_4 + 6470 * X_5 - p_3 \leq 193000 \quad (10)$$

$$\mu_3 + \frac{1}{200} * p_3 = 1 \quad (11)$$

$$22143 * X_1 + 7500 * X_2 + 16429 * X_3 + 12500 * X_4 + 340 * X_5 + n_4 \geq 570000 \quad (12)$$

$$\mu_4 + \frac{1}{5000} * n_4 = 1 \tag{13}$$

$$X_1 \geq 14, \quad X_2 \geq 6, \quad X_3 \geq 7, \quad X_4 \geq 2, \quad X_5 \geq 1, \quad \mu_i, n_i, p_i \geq 0 \quad i = 1 \text{ to } 5$$

5 Solution

Solving the above problem using LINGO lead to the optimal solution set, as follows (Table 6):

Table 6. LINGO output solution of the Yaghoobi and Tamiz Model (2008)

Deviations	Value	The degree of membership functions	Value	Variable	Value
n_1	0.000000	μ_1	1.000000	X_1	20
p_2	1.713863	μ_2	0.857235	X_2	6
p_3	0.000000	μ_3	1.000000	X_3	7
n_4	0.000000	μ_4	1.000000	X_4	4
				X_5	2
Global optimal solution found					

It can be seen from the optimal solution that the all Fuzzy Goals are almost completely satisfied. $\boldsymbol{n}_1 = \boldsymbol{p}_1 = \boldsymbol{n}_4 = 0$, $\boldsymbol{p}_2 = 1.713863$ and $\boldsymbol{\mu}_1 = \boldsymbol{\mu}_3 = \boldsymbol{\mu}_4 = 1$; $\boldsymbol{\mu}_2 \approx 0.86$ all these indicate that the problem solution is logically satisfying 100% for the first, third and the fourth objective, except the second objective is satisfying at 86%, with a good allocation between Efficiency of Renewable Energy Base Plants In Algeria Referring to Ecological, Economic, Geographical & Climatic Criteria.

5.1 Analysis

Analytical models involving multiple objectives such as energy, the environment or employment play a more than important role in policy planning and development.

In the previous model, we developed a Fuzzy Goal Programming system proposed by Yaghoobi and Tamiz (2008). The latter includes an optimal distribution of electricity capacity to satisfy consumption, investment, reduction of greenhouse gas emissions and preservation of labor growth before 2030 for Algeria.

The model provides a mathematical justification for the changes needed to be made before 2030, whether in terms of satisfying consumption patterns, portfolio or the importance of using clean energy sources for a more secure future.

5.2 Comments on the Above Wording

- The objective function includes either positive or negative deviations of each criterion from its related objective.
- The constraints show that there is linear relationship between the level of achievement of each criterion, the objectives and the deviations. Moreover, whenever the deviations are small; it differs in terms of the objectives achievement degree.
- The remaining constraints ($X_1 \geq 14$) require that the optimal solution must preserve at least the current number, which is a real assumption.
- The Output shows the presence of a zero deviation in n_1, p_3 and n_4 and small non-zero deviation in p_2 the value found is 1.713863u, we will exceed objective number 2 by a value of 1.71 billion which is represented by p_2.

This analytical approach, we will notice that to simultaneously satisfy all four-program criteria with a quantitative optimization of the energy base implementations the results of the variables must be as follows:

- 20 photovoltaic solar energy stations,
- 6 solar thermal energy stations,
- 7 wind energy stations,
- 4 bioenergy (biomass) stations,
- 2 geothermal station.

Replacing the n_i, p_i, μ_i and X_i by their values found via the LINGO software in each objective function, we obtain the values we want to achieve, i.e. 28700 MkWh for the first objective, 148 $*$ 10 $*$ 9 for the second and so on until we reach 570000 positions for the last Goal.

6 Conclusion

Environmental decisions are regularly complex and multi-dimensional and include changes and especially instability. In this paper, we have built a Fuzzy Goal Programming model that coordinates between the different criteria to simultaneously achieve the objectives. Energy consumption, reduction of GHG emissions (greenhouse gases) as well as the use of energy and the Algerian investment portfolio. The model provides a quantitative and mathematical justification for additional investments and an increase in emissions to reach the perfect compromise that will satisfy energy demand before 2030.

Our study focuses on green or Renewable Energies, which is why in our research we have linked the energy issue with different ecological and socio-economic aspects while introducing the Algerian state to this. The announced exhaustion of fossil fuels and its instability have pushed Algeria to plunge into an air of clean energy and to promote new models of sustainable development. To reduce greenhouse gas emissions and respect environmental practices, particularly the creation of jobs in the long term, our research highlights several factors and criteria and, at the end, we conclude our research with a practical case study, the aim of which was to demonstrate through a Fuzzy Goal Programming model how to optimize the performance of each energy base on its geographical location in relation to the criteria imposed.

References

Ali, E.S., Elazim, S.M.A., Abdelaziz, A.Y.: Ant Lion Optimization Algorithm for optimal location and sizing of renewable distributed generations. Renew. Energy **101**, 1311–1324 (2017). https://doi.org/10.1016/j.renene.2016.09.023

Atmania, H.: La Strategie D'implantation Des Energies Renouvelables En Algérie: Cas De La Photovoltaïque. Memory of Magister, University of Oran 2 -Mohamed Ben Ahmed (2015)

Bal, J.-L., Chabot, B.: Comptes Rendus de l'Académie des Sciences, Series IIA - Earth and Planetary Science, pp. 333, 827 (2001)

Chang, C.T.: Multi-choice goal programming model for the optimal location of renewable energy facilities. Renew. Sustain. Energy Rev. **41**, 379–389 (2015). https://doi.org/10.1016/j.rser.2014.08.055

Hocine, A., Kouaissah, N., Bettahar, S., Benbouziane, M.: Optimizing renewable energy portfolios under uncertainty: a multi-segment fuzzy goal programming approach. Renew. Energy (2018). https://doi.org/10.1016/j.renene.2018.06.013

Jayaraman, R., Colapinto, C., La Torre, D., Malik, T.: A weighted goal programming model for planning sustainable development applied to gulf cooperation council countries. Appl. Energy**185**, 1931–1939 (2017). https://doi.org/10.1016/j.apenergy.2016.04.065

Jayaraman, R., La Torre, D., Malik, T., Pearson, Y.E.: A polynomial goal programming model with application to energy consumption and emissions in United Arab Emirates. In: 2015 International Conference on Industrial Engineering and Operations Management (IEOM) (2015). https://doi.org/10.1109/ieom.2015.7093869

Kim, K.H., Song, K.B., Joo, S.K., Lee, Y.J., Kim, J.O.: Multiobjective distributed generation placement using fuzzy goal programming with genetic algorithm. Eur. Trans. Electr. Power **18**(3), 217–230 (2008). https://doi.org/10.1002/etep.226

Mezher, T., Chedid, R., Zahabi, W.: Energy resource allocation using multi-objective goal programming: the case of Lebanon. Appl. Energy **61**(4), 175–192 (1998)

San Cristóbal, J.R.: A goal programming model for the optimal mix and location of renewable energy plants in the north of Spain. Renew. Sustain. Energy Rev. **16**(7), 4461–4464 (2012). https://doi.org/10.1016/j.rser.2012.04.039

Scala, A., Allmann, S., Mirabella, R., Haring, M.A., Schuurink, R.C.: Green leaf volatiles: a plant's multifunctional weapon against herbivores and pathogens. Int. J. Mol. Sci. **14**, 17781–17811 (2013)

Yaghoobi, M.A., Jones, D.F., Tamiz, M.: Weighted additive models for solving fuzzy goal programming problems. Asia Pac. J. Ope. Res. **25**(5), 715–733 (2008). https://doi.org/10.1142/s0217595908001973

Search and Substitution of Web Services Operations: Composition and Matching Techniques

Rekkal Sara[1,2(✉)], Rekkal Kahina[1,2], and Amrane Bakhta[1,2]

[1] Computer Science Department, Faculty of Sciences, University of Oran 1 Ahmed Ben Bella, Oran, Algeria
rekkal_sarah@yahoo.fr

[2] Electrical Engineering Department, Faculty of Sciences, University of Tahri Mohammed, Bechar, Algeria

Abstract. Measuring the similarity between web services operations is a very effective solution for both the research and the substitution. But, unfortunately, cannot often be the case. It is therefore recommended to use other means such as a composition. The composition consists of grouping the limited functionalities of the operations (respectively web services) in order to obtain more complex ones, in general, responding to a complex request. In this article, we focus on both the composition (e.g. compound operations meeting the substitution or the user's needs) and on the matching technique (similar operations considering relations between both inputs and outputs parameters) to ensure the research and the substitution. The proposed approach is validated by an experimental study, conducted on real web services, belonging to different domains.

Keywords: Composition · Matching technique · Research · Substitution · Similarity · Web services

1 Introduction

Nowadays, many companies publish their features on the internet; this new generation of applications is known as Web Services. A web service is a computer program accessible via a network to meet a given need. They are developed by providers who described them in WSDLs (Web Services Description Language) files and published them in UDDIs (Universal Description, Discovery, and Integration) directories for customers who discover, select and use them. A web service includes one or more operations, each of them, taken alone, has a limited functionality. The manipulation process begins when the client expresses its need precisely in the form of a query querying the UDDI directory. Faced with this need, many web services are returned: the customer can then choose the one that best meets his needs and invokes it. This process seems simple, but as this technology is not yet mature, there are still many problems for which solutions are needed, among them:

1- The increase in the number of web services causes a consummation of research time and a deficiency in the selection.

M. Hatti (Ed.): ICAIRES 2019, LNNS 102, pp. 338–347, 2020.
https://doi.org/10.1007/978-3-030-37207-1_35

2- The volatile environment in which operate the web services causes a malfunction during their uses. This situation is inadmissible and therefore requires the use of the substitution which consists of replacing the failed operation (respectively the web service) by other that meets the same need. These problems have been much discussed in several works, among them:

The authors of [1] proposed to select the pairs of operations whose degree of similarity greater than or equal 0.7, and the degree of similarity between their messages (inputs and outputs) are greater than or equal 0.75. These substitution conditions are valid only when the client seeks to replace an operation with another that takes the same inputs and produces the same outputs.

In [2], the authors have worked on ontology similarity which is related to the mapping ontology, however, the measurement of similarity of ontologies and concepts is not an easy task and required specific features in semantic web service descriptions.

The authors of [3, 4] proposed to annotate Web services manually with additional semantic information, and then use these annotations to compose the services. Annotating a collection of web services manually is a very difficult and unworkable task.

Our previous work done in [5] consists of forming clusters grouping similar operations using the K_means algorithm. This approach responds perfectly to:

1- The research: once the searched operation is detected. The research process stops. This operation will be returned as well as the similar ones from the same cluster.
2- The substitution: the failed operation will be substituted by other operations of the same group. With the exception that there are, sometimes, clusters with unique operations, they will be returned alone, or when they fail, we are forced to adopt other means such as composition to cover their substitution. This document is, therefore, the continuation of an existing work. We attempt to reinforce the preceding work by a matching technique and by composition to ensure more and more the research and the substitution. We are trying to reform the clusters while including:
    1- Similar operations considering both inputs and outputs parameters,
    2- Compound operations leading the same task,
    3- Functionally equivalent operations.

The rest of this paper is organized as follows: Sect. 2 presents our basic modules. Section 3 describes the new approach. Section 4 is reserved for the implementation and results of our experiments, and in the end, a conclusion to close the paper.

2 Basic Modules

Since this work is a continuation of an existing work; we, briefly, present its basic modules.

2.1 Syntactic and Semantic Methods

To measure the degree of similarity between the descriptions extracted from the WSDLs files, the recourse to methods of automatic processing of the natural language is obligatory because it is difficult, even impossible to access their source codes.

In [5] two types of analysis were considered:

1- Syntactic analysis based on the similarity of the structures of the chains to be compared. Several methods exist of which we chose to use Jaro-Winkler because it is the best-known algorithm and according to [6] it is the most powerful and fastest measure.
2- Semantic analysis based on the resemblance of the semantic level. Several methods exist, of which Wu-Palmer was chosen because it has the advantage of being simpler to implement and gives better results according to [6].

2.2 Hungarian Maximum Matching

This algorithm is used to solve the problems of assignment, problems that can be summarized as follows: considering a costs matrix, it is necessary to choose a single element by line and by column so as to make their sum minimum or maximum.

2.3 Similarity Description

Any operation takes a set of inputs parameters and produces a set of outputs parameters. Generally, when two operations to be compared belong to the same field of study, we can ignore the relation that links their inputs (intersection, difference, equality or inclusion), and be limited to the relation that links their respective outputs. Hence the idea of grouping and substituting operations functionally similar (same results) in the same field of study.

2.4 Similarity Process

In our work conducted in [5], the similarity study between operations goes through several steps, the main ones being:

1- Accessing into WSDLs files and extracting the necessary descriptions for the evaluation of the similarity such as operations identifiers', outputs messages identifiers and the outputs parameters identifiers and their associated types.
2- The constitution of clusters using the K-means algorithm where the distance (Mi, Oj) must be greater than or equal a Threshold, this is its assignment condition such as the threshold equal 0.7, Mi: is the center of the cluster (the first operation affected to the cluster) and Oj is the new operation to affect. The distance (Mi, Oj) is the degree of similarity between the center and this new operation.

3 Proposed Approach

3.1 Objectives and Research Issues

As we mentioned previously, our goal is to reinforce the preceding work by a matching technique and by a composition to ensure more the research and the substitution. The main points to reinforce are:

Table 1. Similarity rules.

Inputs relations	Outputs relations	Decisions
Inputs1 $\{= / \subset\}$ Inputs2	Outs1 = Outs2	O1 Similar to O2
Inputs1 $\{= / \subset\}$ Inputs2	Outs1 $\subset$ Outs2	O1 retrained to participate in a Composition process
Inputs1 $\{= / \subset\}$ Inputs2	Outs1 $\supset$ Outs2	O1 Similar with excess with O2
Inputs1 $\{= / \subset\}$ Inputs2	Outs1 $\cap$ Outs2	O1 retrained to participate in a Composition process
Inputs1 $\{\neq, \supset, \cap\}$ Inputs2	Outs1 $\{\subset / =\}$ Outs2	O1 is functionally equivalent to O2

1- Ensuring the substitution during a call failure of a unique operation (which alone forms a cluster).
2- Ensuring the substitution of operations producing the same outputs (same function) but also taking the same inputs. Especially when:
   a. A client tends to substitute an operation by another, taking and producing the same things.
   b. The failed operations participate in a composition process. It must be replaced by another, taking and producing the same things.
3- The inclusion relation is taken into account, e.g., O7 includes O3, so O7 can substitute O3, even if it produces undesired additional outputs. Our previous work does not consider this relationship because of the degree of similarity that may be below the threshold.
4- When the client refuses to use another operation, even if it belongs to the same cluster, because of some non-functional criteria (e.g., price), we try to offer him other choices (e.g. compound operations) that satisfy his needs.
5- Matching technique clearly visualize the existing relationships between parameters (Equality (=), Difference ($\neq$), Intersection ($\cap$), Inclusion ($\subset, \supset$)). We even find that it is more significant than scores of similarity between them.
6- In this work, we try to satisfy the customer by proposing all the operations that can satisfy his needs (all results are returned) without imposing one or that he feels compelled to use one. We offer him several operations and he has the decision to choose the best.

3.2 Matching Technique and Composition Process

3.2.1 Description

For any compared operations:

1- We ignore the non-functional criteria.
2- Let O1 and O2 two operations extracted from two different Web services WS1 and WS2. Let {Inputs1, Outs1}: the Inputs and the Outputs parameters of O1 and {Inputs2, Outs2}: respectively the Inputs and the Outputs parameters of O2.

Four relations may exist between two sets of parameters even inputs or outputs: { Inputs1, Inputs2} and {Outs1, Outs2}: Equality, Inclusion, Difference, Intersection.

Let O2 the failed operation. O1 is similar to O2 according to some rules summarized in Table 1.

For the Outputs, we did not consider the differences $\{\neq\}$: two different outputs mean two different operations, so one cannot substitute the other.

3.2.2 Reassemble the Similar Operations

The grouping of operations, according to the previous rules described in Table 1, will be done as follows:

a. Similar operations will be grouped into a single cluster named e.g. G.
b. Similar operations with excess will be grouped into a single cluster named e.g. G1.
c. Similar by composition will be grouped into a single cluster named e.g. G2.
d. Operation similar functionally will be grouped in a single group named e.g. G3.

Given these entire groups meet the same need, with some differences between them, they will be grouped together on the same cluster, Fig. 1. They will be returned, if needed, for research or for substitution as mentioned previously.

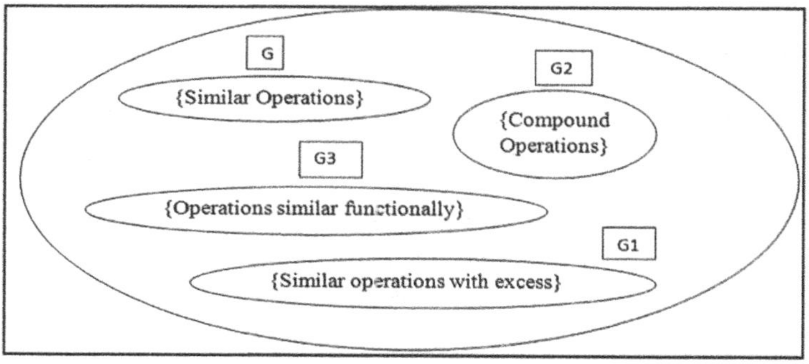

Fig. 1. The grouping of similar operations.

3.2.3 Matching Module

Let T a matrix such that:

1. Columns refer to the parameters of an operation: e.g. O2.
2. Lines refer to the parameters of another operation: e.g. O1.
3. ***the matrix is filled with values obtained from the following formula*** **:** $\boldsymbol{Sim_{Parameter}()}$ **:** is the main function. It calculates the similarity between the parameters. The similarity measurements are obtained by the following formula:

$$Sim_{Parameter}(Parameter\ i, Parameter\ j) = [Identifiers_{Similarity}() + Types_{Similarity}()]/2$$

Where:

1. $\boldsymbol{Identifiers_{Similarity}()}$ **:** is the function that measures the similarity between the identifiers. It flows the some tasks described in [5].

2. $\underline{\boldsymbol{Types}_{\boldsymbol{Similarity}}()}$ *:* is the function that measures the similarity between the types of parameters. We had used the proposed Table by [7] and [8] described also in [5].

After the filling of the matrix of similarity between parameters "T", the Hungarian method is applied, not to obtain a maximum average, but to identify the best scores and consequently the similar parameters.

Two similar parameters must have a similarity score greater than or equal to 0.7: Threshold. At the base of all of this, we determine the relationship that may link two parameters $\{\neq, \cap, \supset, \subset, =\}$ as described above.

3.2.4 Composition Process

Let Table 2 a matrix such that:

- Lines refer to competing operations (those having $\{\supset, \cap\}$ as a relation between theirs outputs parameters as motioned in Table 1).
- Columns refer to the output parameters of the failed operation to be replaced or to study the similarity with it.

Table 2. Similarity study between the parameters of the competing operations and those of the failed operation.

	Parameter1	Parameter2	Parameter3	Parameter4
O1	Parameter_{1i}	/	/	/
O2	/	$\text{Parameter}_{2i'}$	$\text{Parameter}_{2j'}$	$\text{Parameter}_{2k'}$
O3	$\text{Parameter}_{1i''}$	/	/	/

The cells of the matrix refer to the parameter i of the operation i, considered better to substitute the parameter j (Fig. 2). Such that:

$$\text{Sim}_{\text{Score}} \geq \text{Threshold}$$

And

$$Sim_{Score} = Max\left\{ \sum_{l=1}^{n} Sim_{parameter}(Parameter\ l, \quad Parameter\ j) \right\}$$

Where:

$\mathbf{j} = \{1, 2, 3 \ldots m\}$: $\mathbf{m}$ = total number of parameters included in the failed operation to be replaced e.g., $\mathbf{n}$ = total number of parameters included in an operation_i.

- $\text{Sim}_{\text{parameter}}$ (): the function that measures the similarity between the parameters. It is defined above.

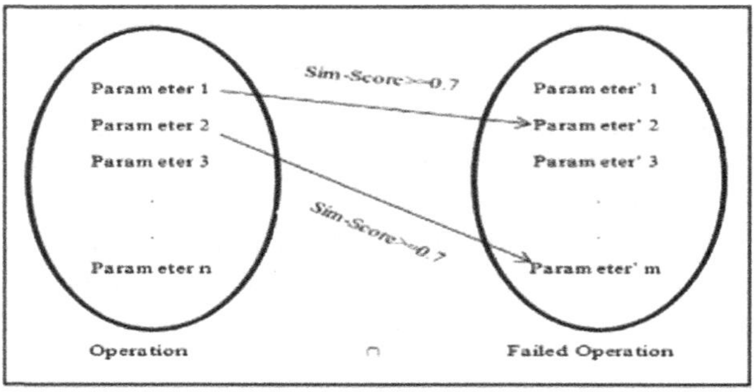

Fig. 2. Example of a matching module between two operations linked by intersection relation.

3.2.4.1 Parameters' Choice and Orchestration

After determining the best correspondence between parameters, Fig. 3. We, now pass to the last step which consists in determining among these competitors, the best for the substitution. At the base of this choice, we compose the selected operations by specifying which parameter must be invoked and at which order (orchestration).

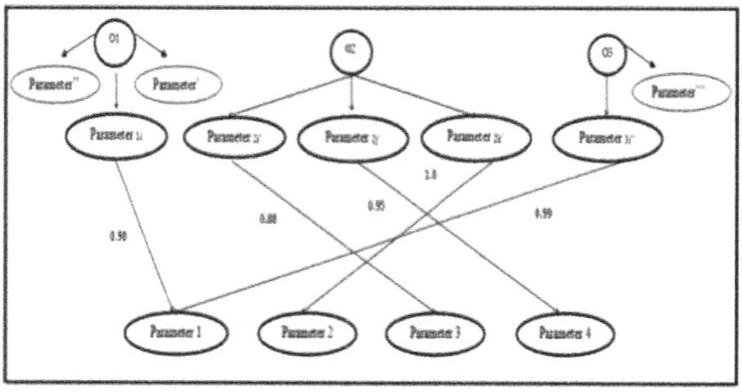

Fig. 3. Matching module between different parameters.

A graph can be represented by a matrix (Table 2) and vice versa. Moving from one line to another (arc to another) allows us to choose the best matching score, Fig. 4.

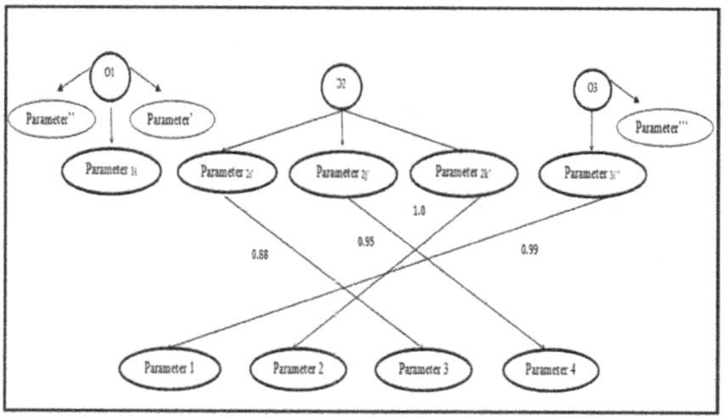

Fig. 4. Parameters' choice.

Let the similarity degree between (Parameter1i, Parameter1) = 0.90 and (Parameter3i, parameter1) = 0.99. The best parameter to substitute Parameter1 is Parameter3i, because of the degree of the similarity which equals to 0.99 and much higher than 0.90.

In this example, the best composition is: O3 {Parameter 3i" → Parameter 1}, O2 {Parameter 2k' → Parameter 2, Parameter 2i' → Parameter 3, Parameter 2j' → Parameter 4}.

4 Experiment Results

4.1 WSDLs Files Used and Implementation

The experiment has been carried out on real web services belonging to different fields: communications, transport, finance, weather… The approach has been applied on an Intel processor machine (I3-3110M CPU2.40 GHZ) with 4 GB RAM and Windows 07 as the operating system.

4.2 Experimental Stud

The tool was tested with samples of real WSDLs files, as mentioned before. This experience tries to: select Some WSDLs Files, extract web services operations, choose an operation to be the failed one. The tool returned all operations that are:

- Similar considering both inputs and outputs parameters.
- Compound operations leading the same task.
- Functionally equivalent operations.

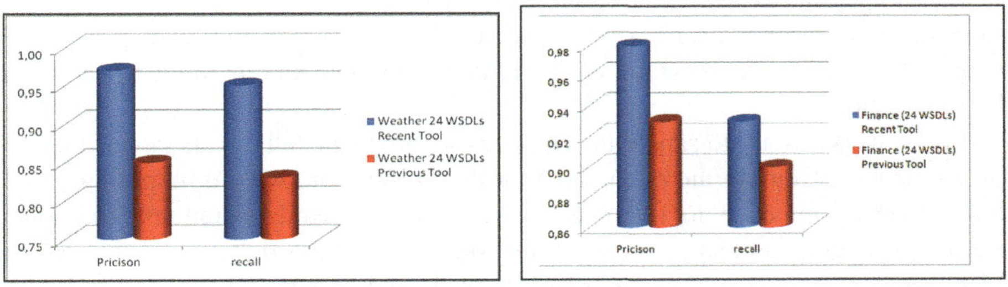

Fig. 5. Results of precision and recall for weather and finance WSDLs files.

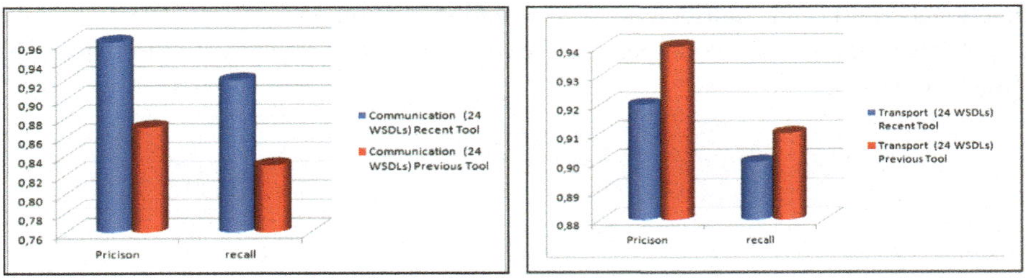

Fig. 6. Results of precision and recall for communication and transport WSDLs files.

4.3 Results' Discussion

4.3.1 Human Evaluation

As mentioned previously, the approach was tested and experimented on real web services (communications, transport, finance, weather…). We had run several tests to obtain more consistent results according to human evaluation and according to our previous work done on [5], so we had performed a manual assessment of the similarity of the operations on the same sample, two measures were calculated: Precision & Recall; Figs. 5 and 6 show the obtained results:

4.3.2 Results Discussion

4.3.2.1 The Previous System

It has been noticed that the results returned by the first system are included in those identified manually by the human being and those returned by this new system. What seemed logical to us, because it takes into account that the outputs parameters.

4.3.2.2 The New System

The returned results are very relevant and very close to the group manually identified. Except that some operations were not returned, this means that the system had failed to identify all operations due to the vocabulary used.

It is important to clarify that the tool returns the competitors' operations.

5 Conclusion

Studying the similarity between web services operations is a key solution to many problems such as research, substitution and composition.

It is important to clarify at this point that there is no formal definition to the substitution; it depends on the specific purpose in the user's mind. Our goal, in this work, is to reinforce the previous work and to ensure more and more the research and the substitution.

In this work, we try to satisfy the customer by proposing all the operations that can satisfy his needs (during the research or for the substitution) without imposing one or that he feels compelled to use one. We offer him several operations (compound operations, operations that are functionally similar (focusing only on outputs parameters), similar operations (focusing on both inputs and outputs parameters), and those similar with excess) and he has the decision to choose the best suits his needs.

As future work and in order to produce a complete tool, the consideration of non-functional criteria can contribute to the orientation of choice for a given client. It would also be interesting as a second perspective to compare our results with those given by similar tools.

References

1. Tibermacine, O., Tibermacine, C., Cherif, F.: A practical approach to the measurement of similarity between WSDL-based web services. In: Proceedings of the Frensh-Speaking Conference on Software Architecture (CAL 2014), France (2014)
2. Weinstein, P., Birmingham, W.: Comparing concepts in differentiated ontologies. In: Proceedings of KAW 1999 (1999)
3. DAML-S Coalition, Ankolekar, A., et al.: DAML-S: web service description for the semantic web. In: ISWC (2002)
4. Poalucci, M., Kawmura, T., Payne, T., Sycara, K.: Semantic matching of web services capabilities. In: Proceedings of International Semantic Web Conference (ISWC) (2002)
5. Sara, R., Fatima, A., Lakhdar, L.: A new approach for grouping similar operations extracted from WSDLs files using k-means algorithm. Int. J. Adv. Comput. Sci. Appl. **8**(12), 84–91 (2017)
6. Boutahar, J., Rachad, T., El houssaini, S.: A new efficient matching method for web services substitution. J. Comput. Sci. Issues **11**(2) (2014)
7. Plebniurbe, P., Pernin, B.: URBE: web service retrieval based on similarity evaluation. IEEE Trans. Knowl. Data Eng. **21**, 1629–1642 (2009)
8. Stroulia, E., Wang, Y.: Structural and semantic matching for assessing web service similarity. Int. J. Coop. Inf. Syst. **14**, 407–437 (2008)

Matrix Product Calculation in Real Grid Environment Under the Middleware Unicore

M. Meddeber(✉), A. Moussadek, and N. Hocine

Faculty of Exact Science, University of Mascara, Mascara, Algeria
m.meddeber@yahoo.fr, mo_sidahmed77@hotmail.com,
hocine10bac2011@gmail.com

Abstract. Grid computing implies cooperation and sharing resources belonging to distributed machines. Users can send their tasks on remote computing resources instead of executing them locally. Subsequently, Tasks scheduling is an important problem in a grid.

In this work, we propose to realise a real Grid computing under the Middleware Unicore. We have developed a python application that computes matrix product to experiment our computing Grid. User's applications are divided on several Tasks and then scheduled on the Grid to be executed. After that the Grid recovers the results and sends them to the user. We notice that when increasing the machines number, response time decreases significantly.

Keywords: Grid computing · Middlewares · Unicore · Matrix product · Resources

1 Introduction

A computational Grid is a hardware and software infrastructure that provides consistent pervasive and inexpensive access to high end computational capacity. An ideal grid environment should provide access to all the available resources seamlessly and fairly [1]. To build a grid, software called middleware is used. Middleware unifies access to heterogeneous computing resources. It is placed between the existing operating systems and the user. It hides from the host the various systems installed and provides a set of routines (controls and programming library) independent of the system. Among the middleware that exists Globus [2], GLite [3], Condor [4], Unicore [5].

In the literature there is a lot of works that exploit the grid. The realization of a grid dedicated to the execution of parallel applications MPI: It presents the deployment of the Globus middleware by taking into account different constraints of the participating sites [6], UNICORE 7 - Middleware services for distributed and federated computing [7]. Design and implementation of the Condor-UNICORE bridge [8]. The Computation Grid: Introduction and Generalities [9].

The objective of this work is to study a scheduling of tasks in a real grid based on the middleware unicore. We have exploited the resources of the installed grid, this exploitation is carried out via mechanisms offered by unicore in order to cover all the needs in terms of computing resources and make calculations of matrix products.

M. Hatti (Ed.): ICAIRES 2019, LNNS 102, pp. 348–355, 2020.
https://doi.org/10.1007/978-3-030-37207-1_36

This work is organized as follows: The second section describes grids computing and their layered architecture. A detailed explanation of the UNICORE middleware is given in the third section. The fourth section presents an overview of the realized grid and different matrix product tests. Finally, we conclude the paper and give some perspectives.

2 The Grid Computing

A grid is a collection of machines, sometimes referred to as nodes, resources, members, donors, clients, hosts, engines, and many other such terms. They all contribute any combination of resources to the grid as a whole. Some resources may be used by all users of the grid, while others may have specific restrictions [10].

2.1 Grid Architecture

A Grid computing consists of four layers [11]:

(1) Application Layer: includes different types of applications: scientific, technical, management, finances, portals, it is the layer of the Grid users.
(2) Middleware layer: This is the brain of the grid, defined as a set of functions allowing resources (servers, memories, networks, etc.) to participate to a unified grid.
(3) Service-Protocol Layer: Constitute of Different services and protocols of Authentication, authorization, encryption, communication, etc.
(4) Network layer: Ensuring connectivity of resources on the grid, it is the network infrastructure of the grid (Material level).

Figure 1 illustrates the different layers of the Grid.

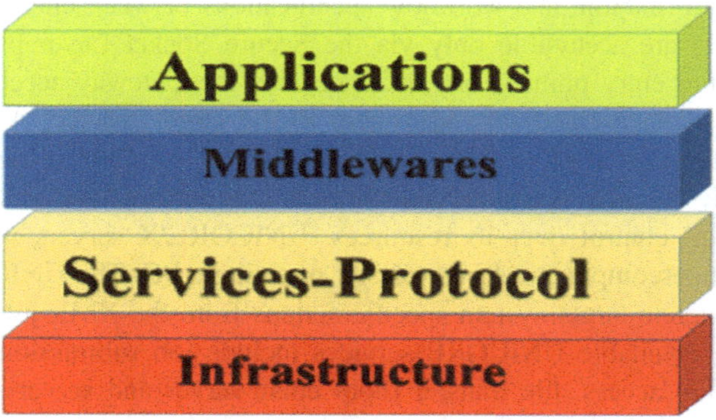

Fig. 1. Grid architecture

2.2 Tasks Scheduling

Scheduling is a process of allocating tasks onto available resources in time. Such process has to respect constraints given by the tasks and the Grid. Usually one or more optimization criterion is used to make scheduling decisions.

Applications may be broken down into any number of individual jobs. Those, in turn, can be further broken down into sub jobs. Jobs are programs that are executed at an appropriate point on the grid. They may compute something, execute one or more system commands, move or collect data, or operate machinery. A grid application that is organized as a collection of jobs is usually designed to have these jobs execute in parallel on different machines in the grid [12].

Finally, the results of all of the jobs must be collected and appropriately assembled to produce the ultimate output/result for the application.

3 UNICORE Middleware

UNICORE [13] is the acronym of "Uniform Interface to Computing Resources". UNICORE is a middleware dedicated to the grid, Developed mainly by Fujitsu Laboratory Europe in Germany [14]. The main purpose of the system is to simplify access to computing resources, which can be distributed onto several different places. UNICORE makes distributed computing resources and data available in a transparent manner and secured on intranet and internet networks.

UNICORE consists of three layers: a user, server, and target system tier [15].

(1) The user is represented by various clients. The primary clients are the UNICORE Rich Client, a graphical user interface based on the Eclipse framework, and the UNICORE commandline client (UCC). The clients use SOAP Web services to communicate with the server tier. XML documents are used to transmit platform and site independent descriptions of computational and data related tasks, resource information, and workflow specifications between client and server.

(2) The servers are accessible only via the Secure Socket Layer protocol. As the single secure entry point to a UNICORE site, the Gateway accepts and authenticates all requests, and forwards them to the target service. A further server, UNICORE/X, is used to access a particular set of Grid resources at a site. UNICORE supports many different system architectures and ensures that organization full control over its resources. UNICORE/X servers may be used to access a supercomputer, a Linux cluster or a single PC. The UNICORE/X server creates concrete target system specific actions from the XML job received from the client. Available UNICORE services include job submission and job management, file access, file transfer (both client-server and server-server), storage operations (mkdir, ls, etc.), and workflow submission and management.

(3) The target system tier consists of the Target System Interface (TSI), which directly interfaces with the underlying local operating system and resource management system.

4 Experimentations and Results

4.1 Scenario Description

We present a scenario for using Unicore grid multi-sites with five machines in a local network, each machine in a different site:

- First, we install in the 1st machine the services: UNICORE/X, XUUDB, Registry, WORKFLOW, and Gateway. This site is called SITE-A.
- In the other machines, we just install the service UNICORE/X, under the names (SITE-B to SITE-E).
- The multi-site installation allows several UNICORE installations to work together, and the users can use the resources of more than one site.

In Unicore there is a basic scheduler based on the Round-Robin algorithm that submits tasks to the available sites corresponding to the needs on resources. For advanced resource management, we install the SLURM scheduler (Simple Linux Utility for Resource Management) [16].

SLURM offers more functions in tasks scheduling such as execution and tasks tracking, fault tolerance, parallelism. After the installation of our Grid, we connect another machine as client. As illustrated on the Fig. 2.

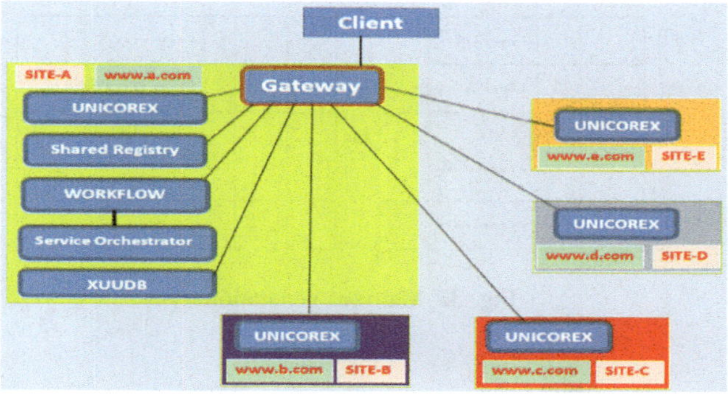

Fig. 2. General architecture of Unicore

Characteristics of the used machines and the details are shown in the following Table 1:

The user in client machine executes an application in java and python, which runs randomly two matrixes with the same size (Fig. 3).

Our application will divide the work into independent tasks for the purpose of distributing them on the grid. So we generate scripts according to the dimension n x n chooses beforehand (for example if the dimension is 100×100 we generate 100 scripts) (Fig. 4).

Table 1. The installation details of the grid

Machine	Characteristics	Operating system	Installed services	Site names
Pc	Processor: Intel (R) Core (TM) i7, memory:12288 Mo, Hard disk: 750 Go	Fedora 29	UNICORE/X, XUUDB, Registry, WORKFLOW, Gateway	Site-A
Pc	Processor: Intel (R) Core (TM) i7, memory: 8192 Mo, Hard disk: 500 Go	Fedora 29	UNICORE/X	Site-B
Pc	Processor: Intel (R) Core (TM) i5, memory: 6144 Mo, Hard disk: 1 TB	Ubuntu 18.04	UNICORE/X	Site-C
Pc	Processor: Intel (R) Core (TM) i5, memory: 6144 Mo, Hard disk: 128 Go(SSD)	Ubuntu 18.04	UNICORE/X	Site-D
Pc	Processor: Intel (R) Pentium (R) 4, memory: 4096 Mo, Hard disk: 500 Go	Fedora 29	UNICORE/X	Site-E

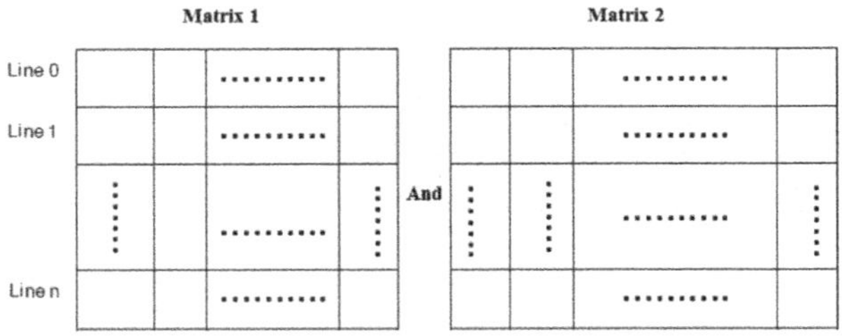

Fig. 3. Matrixes generation

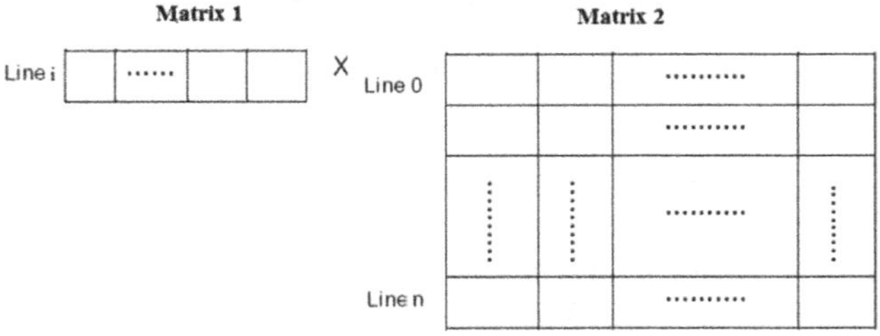

Fig. 4. Deviding work

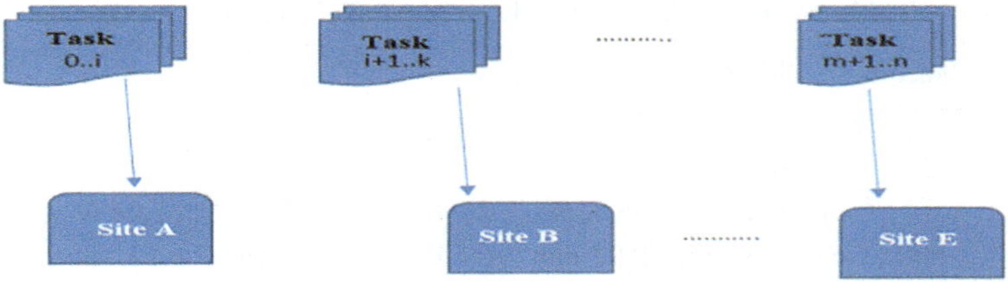

Fig. 5. Tasks scheduling

After tasks creation, they are sent to the nodes of the grid fairly, Fig. 5 shows the operation:

After the execution, result is returned to the user.

In what follows, we will present the results obtained for several execution scenarios. Table 2 present results obtained by a Parallel execution scenario and the elapsed response time, we fix at each time the matrixes size, we vary the Grid nodes number and we measure the response time:

We notice that when we increase the number of machines in the grid the response time decreases.

Table 2. Response time in parallel execution mode.

Machines number		Matrixes size				
		One	Two	Three	Four	Five
100	Time	0,48 min	0,37 min	0,30 min	0,27 min	0,20 min
	Gain	**0%**	**13%**	**37,5%**	**43,75%**	**58,33%**
500	Time	5,65 min	2,85 min	1,97 min	1,53 min	1,22 min
	Gain	**0%**	**49,5%**	**65,15%**	**73%**	**78,6%**
1000	Time	33,98 min	17,08 min	11,47 min	8,70 min	6,97 min
	Gain	**0%**	**49,75%**	**64,25%**	**74,4%**	**79,5%**
2000	Time	256,77 min	126,38 min	83,58 min	63,35 min	50,67 min
	Gain	**0%**	**50,79%**	**67,45%**	**75,33%**	**80,27%**
8000	Time	16693,83 min	8465,03 min	5643,03 min	5000,85 min	3983,67 min
	Gain	**0%**	**49.3%**	**66.20%**	**70,05%**	**76,14%**

The second test is to run the same scenario but on an Advanced Parallel execution mode. Table 3 illustrates in detail the response time results.

The advanced parallel mode consists in distributing the tasks on the different machines of the Grid (the same thing as the first mode). In the same machine, we exploit all the processors. That is to say that we will have tasks running in parallel inside each multiprocessor machine.

Thanks to the advanced parallelism, we noticed that each time we increase the machines number, the response time decrease significantly but when we introduced the fourth machine in the calculation we found that the response time has stopped

Table 3. Response time in advanced parallel execution mode.

Machines number		Matrixes size				
		One	Two	Three	Four	Five
100	Time	0,27 min	0,22 min	0,20 min	0,18 min	0,17 min
	Gain	**0%**	**18,52%**	**26%**	**33,33%**	**37,04%**
500	Time	1,67 min	0,92 min	0,63 min	0,72 min	0,53 min
	Gain	**0%**	**45%**	**62,28%**	**56,88%**	**68,26%**
1000	Time	8,60 min	4,33 min	3,00 min	4.03 min	3,47 min
	Gain	**0%**	**49,65%**	**65,11%**	**53,13%**	**59,65%**
2000	Time	4.85 min	31,58 min	20,55 min	39,95 min	31,97 min
	Gain	**0%**	**51,3%**	**68,31%**	**38,39%**	**50,70%**

decreasing because of the performance of the fourth machine which slowed down the response time Gains (Fig. 6). To solve this problem it is necessary to introduce a load balancing mechanism.

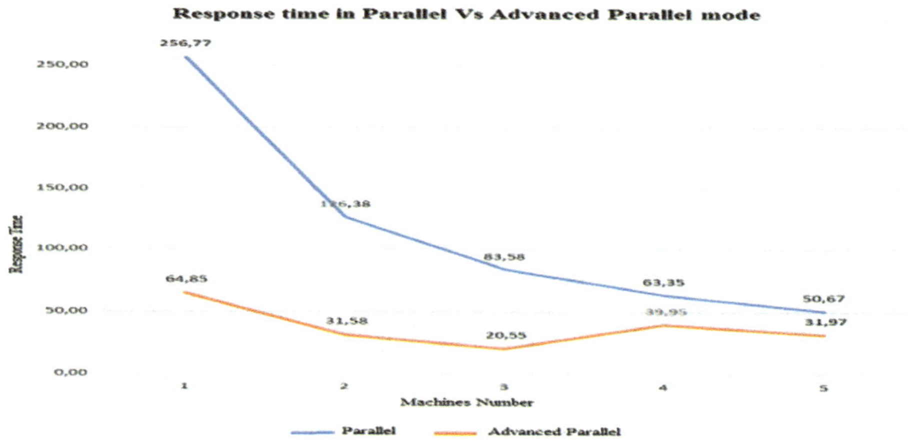

Fig. 6. Response time gains in parallel vs advanced parallel mode.

5 Conclusion

During this work, we designed and implemented Grid software architecture. This architecture is based on the middleware UNICORE. It is thus composed of five machines.

We have developed a python application that computes matrix product to experiment our computing Grid. We started a scheduling so that the tasks are distributed to the nodes in a fair way. We used the results to make graphical comparison between the execution of the application for different number of nodes as well as according to two modes of execution (parallel, Advanced) on the same samples of data (matrixes).

As a future works, it is desirable to exploit the resources of this grid by developing applications for image processing because it is a very important research area.

We wish also to integrate a load balancing and fault tolerance mechanisms to achieve effective results.

References

1. Salehi, M.A., Deldari, H., Mokarram Dorri, B.: Balancing load in a computational grid applying adaptive, intelligent colonies of ants. Informatica **32**, 327–335 (2008)
2. Globus toolkit. http://toolkit.globus.org/toolkit/. Accessed 30 Mar 2019
3. GLite - Lightweight Middleware for Grid Computing. http://grid-deployment.web.cern.ch/grid-deployment/glite-web/. Accessed 30 Mar 2019
4. HTCondor - Home. https://research.cs.wisc.edu/htcondor/. Accessed 30 Mar 2019
5. UNICORE—Distributed computing and data resources. https://www.unicore.eu/. Accessed 30 Mar 2019
6. Romaric, D.: Une grille de calcul pour la recherche. Lille novembre (2003)
7. Benedyczak, K., Schuller, B., Petrova El Sayed, M., Rybicki, J., Grunzke, R.: Unicore 7—middleware services for distributed and federated computing. In: International Conference on High Performance Computing and Simulation, pp. 613–620. IEEE (2016)
8. Nakada, H., Yamada, M., Itou, Y., Nakano,Y., Matsuoka, S., Frey, J.: Design and implementation of condor-unicore bridge. In: Eighth International Conference on High-Performance Computing in Asia-Pacific Region (HPCASIA 2005). IEEE (2005)
9. Parashar, M., Lee, C.A.: Grid computing: introduction and overview. Proc. IEEE **93**(3), 479–484 (2005). Special issue on grid computing
10. Amir, A., Armstrong, J., Berstis, V., Bieberstein, N., Bing-Wo, R., Ferreira, L., Hernandez, O., Kendzierski, M., Magowan, J., Murakawa, R., Neukoetter, A., Takagi, M.: Introduction to Grid Computing with Globus. IBM Redbook, 1st October 2004
11. Foster, I., Kesselman, C., Tuecke, S.: The anatomy of the grid: enabling scalable virtual organizations. Int. J. High Perform. Comput. Appl. **15**(3), 200–222 (2001)
12. Magouls, F.: Grid Computing Theory and Technologies (2010)
13. https://www.unicore.eu/. Accessed July 2019
14. https://www.fujitsu.com/uk/about/local/corporate/subsidiaries/fle/aboutus/. Accessed July 2019
15. Streit, A., Bala, P., Beck-Ratzka, A., Benedyczak, K., Bergmann, S., Breu, R., Daivandy, J. M., Demuth, B., Eifer, A., Giesler, A., et al.: Unicore 6—recent and future advancements. Ann. Telecommun. **65**(11–12), 757–762 (2010)
16. SchedMD—Slurm Support and Development. https://www.schedmd.com/. Accessed 13 Mar 2019

Resources Allocation in Cloud Computing: A Survey

Karima Saidi[1(✉)], Ouassila Hioual[2,3], and Abderrahim Siam[1]

[1] ICOSI Laboratory, Abbes Laghrour University, Khenchela 40004, Algeria
kariming2008@gmail.com, siamabderrahim@gmail.com
[2] Abbes Laghrour University, Khenchela 40004, Algeria
ouassila.hioual@gmail.com
[3] LIRE Laboratory, Constantine2, Constantine, Algeria

Abstract. Cloud Computing is a model in which resources (Computing, Networking, Storage…) are consumed on any utility such as computing power, storage space, servers and applications. With the growing of the Cloud resources demand and the user's number, the need of the quality of service and the resources allocation become a crucial challenge. This paper presents the challenges of resource allocation and some recent contributions to minimize them, hence the need for paper reviews and surveys to identify the area of our research.

Keywords: Cloud computing · Resource allocation · Quality of Service QoS · Deep learning · MCDA

1 Introduction

Various definitions about the Cloud Computing have been given in the literature. In general, we can say that Cloud Computing is a popular trend that use the technology, it attempts to provide easy and inexpensive access to computing resources. In addition, Cloud Computing is a computing model based on an internet. Cloud virtualization technology plays a very important role, allowing resources to be shared, by allocating virtual machines on demand instead of renting physical machines. The importance of virtualization lies in the fact that it allows almost complete isolation between customers who will actually have the illusion that they have just rented a dedicated physical machine (Yazir et al. 2010). There are four types of virtualization by (Akintoye and Bagula 2017): full virtualization, para-virtualization, native virtualization and operating system virtualization. The main issues related to using the Cloud are resource allocation and task scheduling. The last one can be expressed as the allocation of different types of work using existing resources (Manvi and Shyam 2014), and the allocation of resources as the assignment of tasks to virtual machines and the placement of VMs on physical machines PMs.

The IT resource allocated is based on a Service Level Agreement (SLA) (Alhamad et al. 2010) which is a service level agreement between the cloud customer and the service provider. The latter expresses in detail the quality of service (QoS) such as reliability, response time and throughput. These are performance parameters to be

M. Hatti (Ed.): ICAIRES 2019, LNNS 102, pp. 356–364, 2020.
https://doi.org/10.1007/978-3-030-37207-1_37

respected by the service provider, so resource allocation is the effective allocation and planning of resources to achieve the quality of service performance objectives identified by the SLA (Chana and Singh 2014).

In this work, we try to synthesize a set of contributions that focus on solving the problem of resource allocation in the field of Cloud Computing. The rest of the document is structured as follows: In Sect. 2, some basic concepts and challenges of resource allocation in Cloud Computing are introduced. Section 3 will classify the methods used in resource allocation. Section 4 presents recent contributions in this area with orientation of future research. We're going to conclude the paper with a conclusion.

2 Concepts and Challenges

2.1 Some Basic Concepts in RA

To fully understand the principle of resource allocation in cloud computing, we must first introduce what it really means to "allocate resources efficiently and dynamically". These terms are closely defined by a set of parameters (criteria). As shown in Fig. 1, efficiency and dynamism are keywords that include many useful requirements for resource allocation. We present the following main parameters:

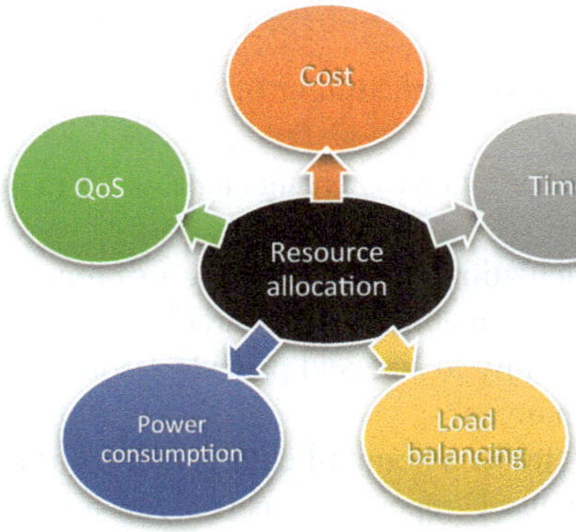

Fig. 1. The parameters of RA

From the perspective of cloud users

- Cost: the allocation of resources must be carried out at a lower cost
- Response time: the allocation of resources must be carried out in a minimum time

For the cloud provider

- Energy consumption
- Load balancing
- The execution times
- QoS quality of service

2.2 The Challenges of Research in RA

Existing problems related to research in the area of Cloud resource allocation that have not been fully resolved according to (Alnajdi et al. 2016) include (see Fig. 2):

Fig. 2. The different challenges of the resource allocation area

- ***Scheduling***: means the parallel scheduling of tasks in case of increasing demands on resources
- ***VM migration***: it is the user's need to change a provider in order to ensure the high storage of his data
- ***Availability***: in long calculations, it is necessary to propose techniques that automatically deal with the unavailability of resources
- ***Elasticity***: indicates the ability to size and provide resources quickly, i.e. the ability to allocate resources dynamically.
- ***Energy efficiency:*** refers to the amount of carbon emissions released by data centers due to the various and huge IT operations performed in them.
- ***Control***: the need to build a control mechanism that allows resources to be allocated on remote servers.

3 Classification of Methods Used in RA

We can classify the resource allocation methods into two different categories, namely:

(a) The reactive methods that make up the techniques that monitor the effectiveness of resource allocation before deciding on the implementation of an action.
(b) The proactive methods that integrate the prediction of future resource requirements. There are methods based on one model and others without a model.

4 Comparison and Orientation of Future Research

4.1 Comparison

There are many publications in recent years in which the authors discuss some of the challenges of resource allocation. This research presents different algorithms and techniques for resource allocation that have been proposed in a Cloud environment. We present some new contributions between 2016 and 2019 in order to have a clearer vision in our work.

In order to have an analysis of these different research studies based on different parameters, we propose a comparative table (see Table 1).

Firstly, the authors Akintoye and Bagula (2017), have proposed solutions to solve two problems in the Cloud/Fog environment: the first problem is the assignment of tasks to virtual machines. In this case, they use the *HABBP (Hungarian Algorithm Based Binding Policy)* algorithm as a heuristic solution to the problem of linear programming, this allocation solution is one by one where each VM performs only one task and each task (cloudlet) must be assigned to a single VM to minimize the total cost. The 2nd problem is the placement of virtual machines. In this case, the authors used a genetic algorithm called *GABVMP (Genetic Algorithm Based Virtual Machine Placement)* to solve and optimize this problem.

Liu et al. (2017), proposed a hierarchical framework to solve the resource allocation problem at a global level on virtual machines and power management, and at a local level on local servers. This work is based on deep learning (and more precisely Deep Reinforcement Learning DRL). Decision-making is done automatically from the reduction of the state and action spaces for the allocation of global resources. On the other hand, the workload predictor is responsible for providing future workload forecasts to facilitate the local operation of the *LSTM (long short-term memory)* power management algorithm.

The authors Gawali and Shinde (2018), have proposed a heuristic algorithm that efficiently schedules tasks and allocates resources in the cloud. The authors have combined the modified analytical hierarchy process (MAHP), BATS bandwidth aware divisible scheduling, BAR optimization, longest expected processing time preemption (LEPT) and division and conquest methods to perform task and resource planning. The MAHP process allows scientific tasks to be prioritized, and the combination of BATS + BAR optimization methods allows resources to be allocated according to bandwidth constraints and cloud resource load. In addition, LEPT preemption was used to give the status of the virtual machine, and a modified divided methodology to conquer was proposed to aggregate the results after task preemption.

Kumar et al. (2017), have proposed an algorithm based on the combination of two algorithms: the first is the Teaching learning-based optimization algorithm (TLBO) and the second is the Grey Wolves optimization algorithm (GW). The proposed algorithm works more efficiently, compared to others, by balancing time and costs. Theoretically, the proposed algorithm works much better than the other two for task scheduling.

Malekloo et al. (2018), have proposed a multi-objective approach to managing resources in the cloud. It balanced energy consumption with the system's ability to meet QoS quality of service and SLA contract requirements. This strategy is performed

Table 1. Compared recent contributions in RA based on different parameters

Papers	Algorithms Or Approachs	Contributions	Great Challenge	Classify the method	Parameters				
					Cost	Time	Load Balancing	Energy consumption	QoS
Akintoye, S. B. and Bagula, A (2017)	(HABBP) heuristic solution to the linear programing problem Genetic Algorithm Based Virtual Machine Placement (GABVMP)	Used the simulator CloudSim	- Parallel planning of the tasks - Migration of VM to PM	Reactive	✓ □	✓ □	✓□	✗□	✓□
Liu, N., Li, Z., Xu, J., Xu, Z., Lin, S., Qiu, Q., Tang, G. and Wang, Y (2017)	Deep Reinforcement Learning (DRL),	Used a Global Resource Allocation Framework based on Deep Neural Networks (DNN) and A deep Q-learning framework	- Energy efficiency - Elasticity	Proactive	✗ □	✗ □	✓□	✓□	✓□
Gawali, M. B. and Shinde, S. K (2018)	- MAHP modified analytical hierarchy process - BATS bandwidth aware divisible scheduling - LEPT longest expected processing time preemption	-Cloud Simulator in the experimental phase Uses real Cyber-shake and Epigenomics scientific work-flows as core tasks for the system	Task scheduling	Reactive	✓ □	✓ □	✗□	✗□	✓□
Kumar, P., Yadav, P. S., Bhutani, K., Arora, N., Jain, D. and Dabas, B (2017)	Teaching learning-based optimization algorithm (TLBO) and grey wolves (GW)	Balance the time and the costs	- Tasks scheduling - Elasticity	Reactive	✓ □	✓ □	✓□	✗□	✓□
Malekloo, M. H., Kara, N. and El Barachi, M (2018)	MACO multi-objective Ant Colony Optimization	Two types of algorithms virtual machine placement and virtual machine consolidation algorithms	- Energy efficiency - VM migration	Reactive	✓ □	✗ □	✗□	✓□	✓□
Gilesh, M. P., Kumar, S. D. and Jacob, L (2018)	Greedy and meta-heuristic algorithm	Virtual datacenter (VDCE) embedding in DCC on the least cost migration using NetworkX library and fnss	- VM migration - Availibility	Reactive	✓ □	✓ □	✗□	✗□	✓□
Wang, W. Jiang, Y. and Wu, W (2017)	Auction mechanism negociation	Allocate virtual machines to PM with a minimum of energy cost	Energy efficiency	Reactive	✓ □	✓ □	✗□	✓□	✓□
Lin, J. Dai, Y. Chen, X. and Wu, Y (2017)	Machine learning Genetic Algorithm	Software self-adaptation technology e.g. MAPE-K and control loop	Reduce the cost and minimize the resp time	Proactive Reactive	✓ □	✓ □	✗□	✗□	✓□
Jyoti, A. and Shrimali, M (2019)	Multi-agent Deep Learning (MADRL-DRA) Dynamic Optimal Service Load-Aware Service Broker (DOLASB)	Based on load balancing and Service Broker strategy Use the CloudSim simulator	- Elasticity - Tasks scheduling - Energy efficiency	Reactive Proactive	✓ □	✓ □	✓□	✓□	✓□
Alsadie, D. Tari, Z. Alzahrani, E.J. and Zomaya, A.Y (2018)	K-means clustring technique	Determine the appropriate VM type based on Google Cloud traces	Energy efficiency	Proactive	✗ □	✗ □	✓□	✓□	✓□

using two types of algorithms: VM placement and VM consolidation algorithms that allow optimal solutions to be found by reducing the total energy consumption of the data center, minimizing the number of active PMs to shut down unused servers, and, in addition, reducing the number of VM migrations.

The authors Gilesh et al. (2018), have proposed a model to address the problem of finding the most cost-effective set of virtual machine migrations in order to minimize the cost of migration and optimize the integration of a virtual data center. The idea was to integrate a set of virtual machine migrations, while respecting cost reduction, into a new virtual data center created in all Cloud data centers. The authors used the Greedy and metaheuristic algorithm to solve this problem.

The authors Wang et al. (2017) have developed an approach to allocating energy-efficient resources by allocating virtual machines to PMs in order to minimize energy costs. This approach is based on two steps: the first is the auction-based allocation of virtual computers and the second was the negotiation-based consolidation of virtual computers using the multiagent system.

The proposed work by Lin et al. (2017), reduces the workload (with minimal response time and cost) associated with system maintenance and configuration. The authors used machine learning to build an effective knowledge model and applied software self-adaptation technology, which is a capability that allows a software system to adjust to respond to frequent changes from external environments. In addition, the authors applied the genetic algorithm to solve the problem of allocating computing resources, which ensures that the solution can be to optimize the configuration of resources. In this work, the virtual machine pool is responsible for managing the number of virtual machines, creating and closing virtual machines at the right time.

The authors Jyoti and Shrimali (2019), considered load balancing and the Service Broker strategy as two main areas to solve the problem of resource allocation. First, a multi-agent deep learning model (MADRL-DRA) was used. In this model, the local user agent (LUA) is used to predict the environmental activities of the user task and allocate the task to the virtual machine (VM) based on priority. Then, a load balancing (LB) is performed in the VM, which increases the flow rate and reduces the response time of the resource allocation task. Secondly, DOLASB (Dynamic Optimal Service Load-Aware Service Broker) is used in the GUA (Global User Agent) to plan the task and provide services to users based on the cloud brokers available to minimize the costs of cloud customers and at the same time generate a profit for Cloud Service Broker CSB. Finally, the authors proposed the BD-MIP algorithm that provides an optimal solution to the problems of optimizing multi-service configuration, virtual machine allocation and CSB.

The important contribution of Alsadie et al. (2018) is to design an approach to find virtual machines of the appropriate size to optimize resource utilization, thereby reducing energy waste in data centers. The authors used the K-means Clustering technique. Energy efficiency was a major challenge in the proposed approach, it allowed fewer VM instances to be used and proved to be able to reduce the number of rejected tasks.

4.2 Orientation of Future Research

The allocation of resources in Cloud Computing is one of the most relevant issues to be addressed. In this research, we try to guide our future work by citing some challenges in this area.

After a thorough analysis of the existing work, several gaps and disadvantages were identified. For (Akintoye and Bagula 2017), the proposed allocation model does not address the case where several tasks are assigned to a single virtual machine, nor does the placement of virtual machines study load balancing in each physical machine. On the other hand, (Jyoti and Shrimali 2019), the proposed model is based on load balancing and service broker to solve the challenges of scheduling tasks in parallel. In addition, addressing the reduction of energy waste has proven to be a major challenge for cloud researchers to optimize the use of resources according to real needs and to balance the workload in a suitable way. For example, the authors (Liu et al. 2017), (Alsadie et al. 2018) proposed models to determine the type of VM appropriate to the IT resource requirements of the task group. Among the parameters that have been taken into consideration in (Lin et al. 2017) the prediction of response times using machine learning. Other parameters were treated to increase the efficiency of the allocation and other areas were introduced, such as (Liu et al. 2017), (Jyoti and Shrimali 2019) introduced deep reinforcement learning (DRL) and obtained more precise results.

On the other hand, among the approaches that transfer VMs to PMs in the same Cloud considering the cost of migration, we can cite the works of (Wang et al. 2017), (Malekloo et al. 2018). There are other research studies by (Jyoti and Shrimali 2019), (Gawali and Shinde 2018) that have focused on solving some challenges such as parallel task scheduling.

(Kumar et al. 2017) and (Pradhan et al. 2016) proposed changes to resource allocation algorithms by satisfying client requests and reducing wait times. (Pradhan et al. 2016) has modified the oldest algorithm, the Round Robin, which is a simple step in obtaining an optimal planning model. In addition, several meta heuristic algorithms have been combined, as well as several mechanisms and techniques are used (see Fig. 3) according to the table above to allocate resources efficiently and meet users' expectations.

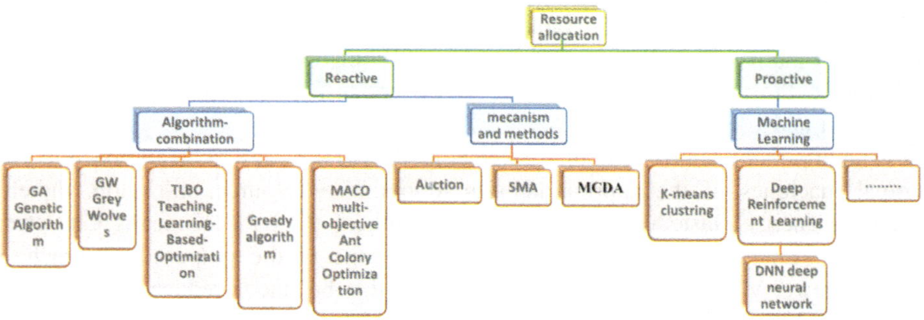

Fig. 3. The taxonomy of resource allocation methods

Our work is based on the combination of MCDA multi-criteria decision support methods as (Gawali and Shinde 2018) but our role is to help a decision-maker to select one of several alternatives based on decision criteria, and in the field of machine learning more specifically preference learning or possibly deep learning.

We choose the field of artificial intelligence because it has given almost zero chances of errors and failure rates, and it also allows for high accuracy in the allocation of resources in Cloud Computing (Madni et al. 2017). Currently, methods of learning and automatic preference prediction are among the most recent research areas in disciplines such as machine learning. The sub-domains of machine learning are used because they allow the rapid and automatic creation of models capable of analyzing large and complex data and obtaining faster and more accurate results, even on a very large scale (Jyoti and Shrimali 2019). The objective of using the two sub-domains is to build a decision model based on preferences (Jyoti and Shrimali 2019).

We can classify our problem as classification problems that determine them through supervised learning in which the entry and exit spaces are clearly distinguished from each other. In this type of problem, input instances are mapped to preference models (Kotsiantis et al. 2007).

5 Conclusion

The number of articles published in recent years which are based on the cloud has been enormous, especially those that address the issue of resources allocation. And many more are being revised and published due to the increasing number of resource requests in the Cloud environment. With the large number of articles, it is difficult for new researchers in the field to identify potential areas for discussion. This study aimed to address this challenge.

References

Yazir, Y.O., et al.: Dynamic resource allocation in computing clouds using distributed multiple criteria decision analysis. In: Proceedings of the 3rd International Conference on Cloud Computing IEEE, pp. 91–98, July 2010

Akintoye, S.B., Bagula, A.: Optimization of virtual resources allocation in cloud computing environment. In: Proceedings of the Africon IEEE, pp. 873–880, September 2017

Manvi, S.S., Shyam, G.K.: Resource management for Infrastructure as a Service (IaaS) in cloud computing: A survey. J. Network Comput. Appl. IEEE, (41), 424–440 (2014)

Alhamad, M., Dillon, T., Chang, E.: Conceptual SLA framework for cloud computing, In: Proceedings of the 4th International Conference on Digital Ecosystems and Technologies IEEE, pp. 606–610, October 2010

Chana, I., Singh, S.: Quality of service and service level agreements for cloud environments: issues and challenges. In: Mahmood, Z. (ed.) Cloud Computing. Computer Communications and Networks, pp. 51–72. Springer, Cham (2014). https://doi.org/10.1007/978-3-319-10530-7_3

Alnajdi, S., Dogan, M., Al-Qahtani, E.: Asurvey on resources allocation in Cloud computing. Int. J. Cloud Comput. Serv. Archit. IJCCSA **5**(6), 1–11 (2016)

Liu, N., et al.: A hierarchical framework of cloud resource allocation and power management using deep reinforcement learning. In: Proceedings of the 37th International Conference on Distributed Computing Systems ICDCS IEEE, pp. 372–382, June 2017

Gawali, M.B., Shinde, S.K.: Task scheduling and resource allocation in cloud computing using a heuristic approach. J. Cloud Comput. **1**(7), 1–16 (2018)

Kumar, P., Yadav, P.S., Bhutani, K., Arora, N., Jain, D., Dabas, B.: Allocating resource dynamically in cloud computing, In: Proceedings of the International Conference on Infocom Technologies and Unmanned Systems (Trends and Future Directions) ICTUS IEEE, pp. 249–254, December 2017

Malekloo, M.H., Kara, N., El Barachi, M.: An energy efficient and SLA compliant approach for resource allocation and consolidation in cloud computing environments. Sust. Comput. Inf. Syst. **17**, 9–24 (2018)

Gilesh, M.P., Kumar, S.D., Jacob, L.: Bounding the cost of virtual machine migrations for resource allocation in cloud data centers. In: Proceedings of the 33rd Annual ACM Symposium on Applied Computing, pp. 201–206, April 2018

Wang, W., Jiang, Y., Wu, W.: Multiagent-based resource allocation for energy minimization in cloud computing systems. Trans. Syst. Man Cybern. Syst. IEEE **2**(47), 205–220 (2017)

Lin, J., Dai, Y. Chen, X, Wu, Y.: Resource allocation of cloud application through machine learning: a case study. In: Proceedings of the International Conference on Green Informatics ICGI IEEE, pp. 263–268, August 2017

Jyoti, A., Shrimali, M.: Dynamic provisioning of resources based on load balancing and service broker policy in cloud computing. Cluster Comput., 1–19 (2019)

Alsadie, D., Tari, Z., Alzahrani, E.J., Zomaya, A.Y.: Dynamic resource allocation for an energy efficient VM architecture for cloud computing. In: Proceedings of the Australasian Computer Science Week Multiconference on ACSW 2018 ACM, Brisband, Queensland, Australia, pp. 1–8, January 2018

Pradhan, P., Behera, P.K., Ray, B.N.B.: Modified Round Robin Algorithm for Resource Allocation in Cloud Computing. In: Proceedings of the International Conference on Computational Modeling and Security: Procedia Computer Science, no. 85, pp. 878–890 (2016)

Madni, S.H.H., Latiff, M.S.A., Coulibaly, Y., Abdulhamid, S.M.: September, Recent advancements in resource allocation techniques for cloud computing environment: a systematic review. Cluster Comput. **3**(20), 2489–2533 (2017)

Kotsiantis, S.B., Zaharakis, I., Pintelas, P.: Supervised machine learning: a review of classification techniques. Emerging Artif. Intell. Appl. Comput. Eng. **160**, 3–24 (2007)

The Role of Solar PV Energy in the Arabic Traditional Tent for Raising the Quality of Tourism Services in Taghit City

M. Haidas[1(✉)], A. Dahbi[2], and O. Abdelkhalek[1]

[1] Smart Grid and Renewable Energies Laboratory,
Tahri Mohammed University of Bechar, Béchar, Algeria
haidasmed08@gmail.com, othmane_elec@yahoo.fr
[2] Unité de recherche en énergies renouvelables en milieu saharien,
URERMS, Centre de développement des énergies renouvelables, CDER,
Bouzaréah, Adrar, Algeria
Dahbi_j@yahoo.fr

Abstract. Taghit is one of the famous towns in desert of Algeria, it is situated exactly in Bechar, where tourists frequent it from all over the world. This city is rich with many monuments and attractions, as well as customs and traditions, where their residents offering good services for tourists. However they face many obstacles, such as electricity lack in some touristic areas, which decreases the quality of some services. In order to improve the quality and diversification of tourism services, we propose solutions by introducing PV solar energy into the traditional Arabian tents, aiming to be developed in diversify tourism space. Furthermore, these solutions create jobs and reduce electricity costs. These guests will stay in the traditional Arab tents, in hotels or in rental houses. In order to guarantee a good reception and satisfaction, we propose an accommodation optimization using genetic algorithm.

Keywords: PV solar · Arabic traditional tent · Taghit · Guests · Genetic algorithms

1 Introduction

The objective of this paper is to determinate the number of traditional Arabic tents supplied by PV solar energy, hotels and rental houses by using the genetic algorithms, from one hand. On the other hand the demonstration of the renewable energies importance, especially solar energy, this is to raise the level of tourism services in many ways. That is way we have devoted our study to the Arabic traditional tent because of its social, cultural and tourist impact. Therefore, we have proposed some axis of study in this domain in order to stimulate the national economy and get a strong investment basing on solving of the following problems: Limited use of this tent, electricity lack, lack of hotels, lot of tourists congestion in a particular area at peak times and lack of service facilities. Note that each PV solar energy system is normalized and dimensioned according to the number of tents and the requested loads [1–5]. The Arabic traditional tent represented a home, but now it's rare. Therefore, it is presented and

M. Hatti (Ed.): ICAIRES 2019, LNNS 102, pp. 365–371, 2020.
https://doi.org/10.1007/978-3-030-37207-1_38

exploited as a cultural heritage, generally in tourism investment. With renewable energies, this tent gives more solutions for economic investment exploitation.

2 The Role of the PV Solar Energy in the Arabic Traditional Tent

A. Economic Energy

With PV solar system, we obtain an economic energy and more independence (Fig. 1).

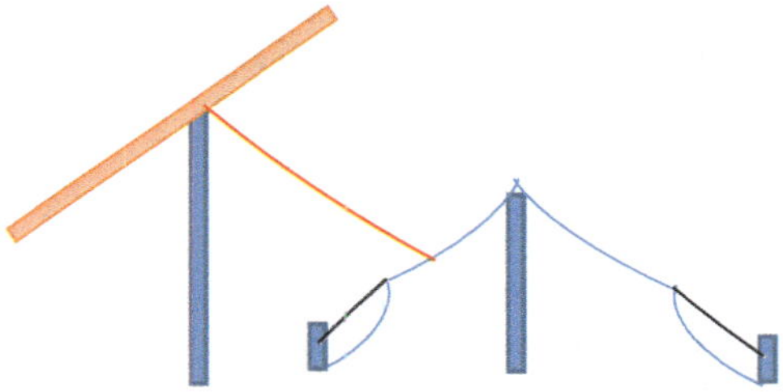

Fig. 1. PV solar energy for Arabic traditional tent

B. More Activities in Arabic Traditional Tent with PV Solar System

In this case, with the autonomous electricity, we can use television, refrigerator, lamp, tele-operation and others. Therefore, this tent will be a full home where we find all necessaries needs without forgetting toilet and shower.

C. Arabic Traditional Tent with PV Solar System as Supermarket

By installing of the renewable energy system, this removes big limits due to autonomous electricity. So, new roles have been given to this tent. Thanks to this system the tent can be transformed to the supermarket, where tourists will be able to shop and buy their needs, like foods, clothes and all things requiring storage in refrigerator.

D. Arabic Traditional Tent with PV Solar System as Hotel

The big problem which disturbs the tourists is the accommodation when Taghit undergoes a rent crisis, especially during some peak month period, because of hotels and rental houses filling. For these raisons, we proposed a magnificent solution, which is building hotels from Arabic traditional tent with PV solar energy (Fig. 2).

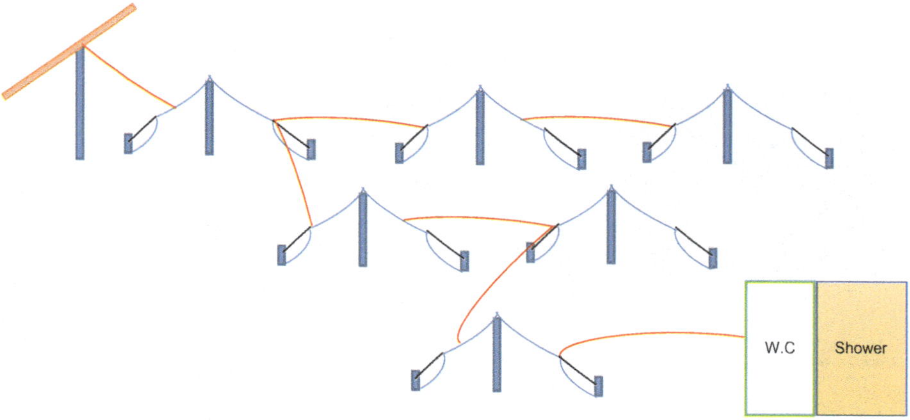

Fig. 2. PV solar hotel made by Arabic traditional tents

E. Good Dispatching and Exploitation in all Interesting Zones

Another advantage of this proposed system allows displacing the Arabic traditional tents in any importance area in Taghit for good dispatching and more exploitation for all interesting zones, moreover, it gives comfort and blessing to tourists without hassle. The proposed places for installing these tents with PV solar system are shown on Figs. 3 and 4. In order to determinate the best dispatching of PV solar systems in Taghit area, we have used artificial intelligence algorithms, as like genetic algorithm and particle swarm optimization [6, 7].

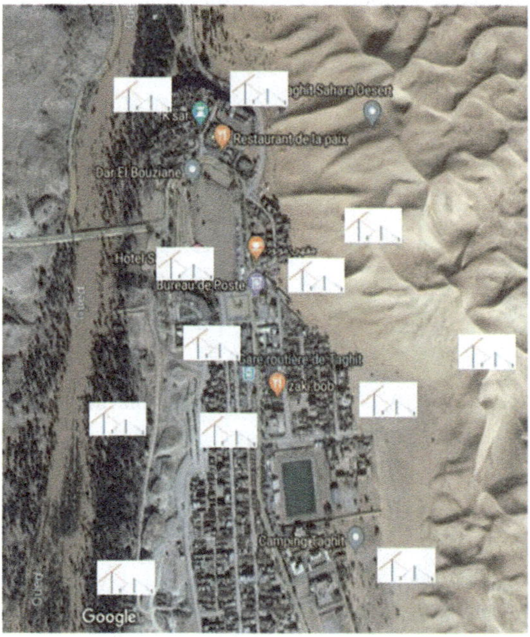

Fig. 3. The proposed zones for installing PV solar system with Arabic traditional tent in Taghit [8]

Fig. 4. The proposed zones for installing PV solar system with Arabic traditional tent in south of Taghit [9, 10]

A. The Impact of "The Third International Conference on Artificial Intelligence in Renewable Energetic Systems" on the Exploitation of Arabic Traditional Tents Equipped with Solar PV System

In this conference, we can exploit the renewable energy in the Arabic traditional tent even at night by integrating storage battery system, for example after finishing works in the first day, desert theater show can be organized (Fig. 5), or introducing an applied fieldwork, which contributes in dissemination, publicity and awareness about this renewable energy.

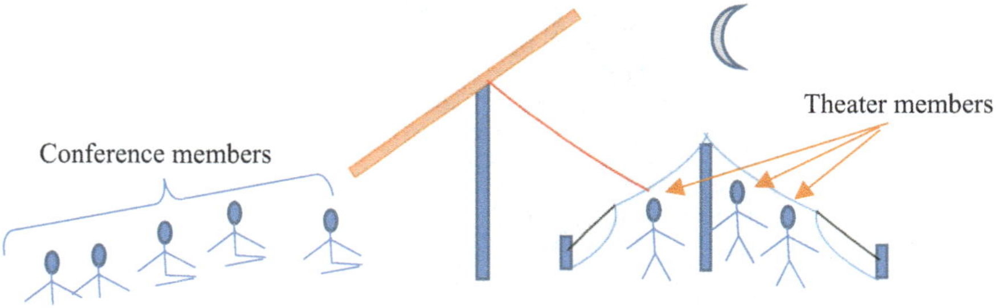

Fig. 5. Applied fieldwork for PV solar energy with Arabic traditional tent in international conference (At night)

3 Sufficiency Optimization of Traditional Arabic Tents Supplied by Solar Energy for Taghit Guests Using Genetic Algorithm

In this case, we have proposed a new model as shown on the first equation; so, the goal is to determinate sufficiency numbers of traditional Arabic tents supplied by solar energy, hotels and rental houses for Taghit guests especially tourists. All proposed data in this study are illustrated in Table 1.

Table 1. Data of proposed mode

Data	Value
Number of tourists	$N_{Tourists} = 3100$
Number of persons in one tent	$n_{PTent} = 10$
Number of persons in one hotel	$n_{PHotel} = 200$
Number of persons in one rental house	$n_{PHouse} = 200$

Our proposed model is presented by the objective function $F(N_{Tent}, N_{Hotel}, N_{House})$ as below:

$$F(N_{Tent}, N_{Hotel}, N_{House}) = N_{Tourists} - (n_{PTent}N_{Tent} + n_{PHotel}N_{Hotel} + n_{PHouse}N_{House}) \tag{1}$$

Where:

N_{Tent} is the number of traditional Arabic tents.
N_{Hotel} is the number of hotels.
N_{House} is the number of rental houses.

The constraints are given as below:

$$-N_{Tent} + N_{House} \leq 0 \tag{2}$$

$$N_{Tent} + N_{House} \leq 200 \tag{3}$$

$$N_{Hotel} + N_{House} \leq 0 \tag{4}$$

$$N_{Tent} + N_{Hotel} \leq 150 \tag{5}$$

$$\begin{cases} 1 \leq N_{Tent} \leq 150 \\ 1 \leq N_{Hotel} \leq 4 \\ 1 \leq N_{House} \leq 150 \end{cases} \tag{6}$$

The solution of this problem is needed to find the best target variables, so, an introducing of an artificial intelligence by using genetic algorithms [6, 7, 11] as illustrated in Fig. 6 have given an optimization solution for the sited objective function, where the best results are shown on Table 2.

```
clear all
clc
i=0;
while true
A=[-1 0 1;1 0 1 ;0 1 -1;1 1 0];
b=[0;200;0;150];
ub=[150;4;150];
lb=[1;1;1];
Tourists=@(x)(-(10*x(1)-200*x(2)+15*x(3))+3100);
[x,feval]=ga(Tourists,3,A,b,[],[],lb,ub)
i=i+1
if abs(feval)<2
   x
   feval
   i
   break
end
end
```

Fig. 6. The proposed source code for genetic algorithm optimization of the proposed objective function

Table 2. Data of proposed model

Number of iteration i	Number of traditional Arabic tents	Number of hotels	Number of rental houses
I = 571	$N_{Tent} = \mathbf{140.0658}$	$N_{Hotel} = \mathbf{4.000}$	$N_{House} = \mathbf{59.9352}$

4 Conclusion

In this work, we have given some interesting axis to prove the importance of introducing the PV solar energy system in Arabic traditional tent for raising the quality of tourism services in Taghit city. Finally, this conference about artificial intelligence in renewable energetic systems will be a first step to encourage Taghit's residents to invest in renewable energy. An optimization in sufficiency of traditional Arabic tents supplied by solar energy, hotels and rental houses for Taghit guests by genetic algorithms had given the best values.

References

1. Sumathi, S., Ashok, L., Surekha, P.: Solar PV and Wind Energy Conversion Systems An Introduction to Theory, Modeling with MATLAB/SIMULINK, and the Role of Soft Computing Techniques. Springer, Switzerland (2015)
2. Reinders, A., Verlinden, P., van Sark, W., Freundlich, A.: Photovoltaic Solar Energy: From Fundamentals to Applications. Wiley, Hoboken (2017)
3. Solanki, C.S.: Solar Photovoltaic Technology and Systems – A Manual for Technicians, Trainers and Engineers. PHI Learning Private Limited, Delhi (2013)
4. Hatti, M.: Renewable Energy for Smart and Sustainable Cities: Artificial Intelligence in Renewable Energetic Systems. Springer, Switzerland (2019)
5. Bhattacharyya, S.C., Palit, D.: Mini-Grids for Rural Electrification of Developing Countries: Analysis and Case Studies from South Asia. Springer, Switzerland (2014)
6. Solanki, C.S.: Introduction to Genetic Algorithms. Springer, Heidelberg (2008)
7. Messac, A.: Optimization in Practice with MATLAB. Cambridge University Press, Cambridge (2015)
8. https://www.google.com/maps/@30.9191869,-2.033254,1339m/data=!3m1!1e3. Accessed 21 July 2019
9. https://www.google.com/maps/@30.8027752,-2.0552052,1340m/data=!3m1!1e3. Accessed 21 July 2019
10. https://ar.wikipedia.org/wiki/%D8%AA%D8%A7%D8%BA%D9%8A%D8%AB. Accessed 21 July 2019
11. Michalewicz, Z.: Genetic Algorithms + Data Structures = Evolution Programs. Springer, Heidelberg (1996)

Self-management of Autonomous Agents Dedicated to Cognitive Radio Networks

M. Z. Baba-Ahmed[1(✉)], S. Tahraoui[1], A. Sedjelmaci[1], M. Bouregaa[2], and M. A. Rabah[3]

[1] Department of Electronic, University of Hassiba Ben Bouali, Laboratory LTT of Tlemcen, Chlef, Algeria
zaki.babaahmed@gmail.com, s.tahraoui@univ-chlef.dz, sedjelmaci_amina@yahoo.fr

[2] Department of Electrotechnic, University of Mustapha Stambouli, Laboratory LTT of Tlemcen, Mascara, Algeria
bouregaa.abdou@yahoo.fr

[3] Center of Development of Satellites (ASAL), Oran, Algeria
arabah6@gmail.com

Abstract. An autonomous network is a network that is able to self-manage and deliver a service based on the resources of its nodes, it follows the concept of autonomous computing, its goal is the creation of self-management networks to support the growing complexity of internet and allow the expansion of networks beyond their current sizes. In the field of networks, cognitive radio could be considered as an intelligent agent able to adapt to its operational context, it also offers a balanced solution to the problems of spectrum congestion. The concept of cognitive radio is based on the dynamic use of any available and detectable frequency band of the radio spectrum for communications between networks of two categories, namely primaries, which have controlled and prioritized access to spectrum and the secondary ones say cognitive. In this paper, we are interested in the dynamic and intelligent management of radio resources as part of a cognitive radio network using mechanisms based on autonomous learning techniques. We have proposed a new approach for the efficient management and sharing of radio spectrum.

Keywords: Autonomous network · Cognitive radio · Learning · Modeling · Adaptation

1 Introduction

Cognitive radio (CR) has been used frequently to refer to a system that is aware of its environment and uses this information [1]. CR was formally presented in 1998 by Joseph Mitola at a seminar at the Stockholm Royal Institute of Technology, later published in an article by Mitola and Maguire Jr. [2]. The principle of cognitive radio included in the IEEE 802.22 and IEEE 802.16h standards [3], Sometimes it is considered more closely as a system with great frequency agility to explore the opportunities that may exist in the spectrum of frequencies. This allows for good spectrum management by occupying or exploiting unoccupied radio spectrum bands and, of

M. Hatti (Ed.): ICAIRES 2019, LNNS 102, pp. 372–380, 2020.
https://doi.org/10.1007/978-3-030-37207-1_39

course, improving spectrum management. That does it through the software restricted radio SDR (Software Defined Radio). The SDR is a radio system capable of adapting to any frequency band and receiving any modulation using the same hardware [4]. The CR system requires four main functions that allow it to use spectrum opportunistically [5]. These functions consist of the main steps of the CR terminal for spectrum management. These are: spectrum detection, spectrum decision, spectrum sharing and spectrum mobility [6]. Autonomy in networks has brought about a real change in IT systems and the realization of autonomous systems has revolutionized the world. Even if people could somehow find enough qualified people, the complexity exceeds the human capacity to manage it. As computing evolves, connections, dependencies, and interactions with overlapping applications require administrative decision-making and faster responses than anything man can offer. Based on the concept of autonomous computing but in the domain of networks, autonomous networks are generated by four complementary and self-managed aspects, namely self-optimization, self-configuration, self-protection and self-management. healing. These four elements form the basis of our study on cognitive systems [7]. We focused on two characteristics of cognitive systems, namely self-management and quality of service of certain types of transmission. Our objective is therefore to satisfy this type of application for autonomous networks based on the RC [8]. In this article, we propose a new approach in the cooperative and competitive multi-agent system for autonomous agents of cognitive radio based on self-management agents of different types of transmission ensuring the good functioning of cognitive systems. We will see in section two, the learning base proposed autonomous cognitive radio based on allocation of spectrum for autonomous cognitive radio networks so to see the importance of self-management in cognitive radio. Then we will see in section three the results of simulations of some types of applications for cognitive radio with the Java Agent Development Framework (JADE) platform.

2 Proposed Approach

2.1 Target

The wide deployment of digital systems and the complex management of modern technology both require the need for autonomy which has become a necessity in all application areas. Networking is benefiting more and more from functionalities that allow computer systems to get closer and closer to the human brain networks. several works in the field of dynamic spectrum allocation for autonomy-based cognitive radio [9] and QoS [10] are intended for the study of the hardware layer of the Open Systems Interconnection (OSI) model, as part of present work and continuity with our previous work [11], we are interested in a new approach to autonomy for upper layers (OSI model), which is presented as a complement to what has already been done in an earlier article [11], in order to show a new aspect of cognitive radio (CR), while remaining closely related to this CR area. In addition, this approach enables to address different points of autonomous management of a cognitive system by improving the negotiation between users and also the learning operation of the selected decisions.

In view of what has been reported in the article [11], the secondary user, also known as a cognitive radio user, SUi tries to communicate at any cost with a primary user PUj having a license on the radio spectrum (i and j are two different integer numbers), but sometimes it is difficult to ensure channel availability. The Learning Base for Autonomous Agent (LBAA) allows for a better channel allocation by introducing autonomous mechanisms that are capable of adapting to any radio environment.

2.2 Principle

In Fig. 1, it is assumed that we are in a competitive and cooperative environment. Each secondary user (SU) detects a part of the spectrum with a set of primary users (PUs). The secondary users (SUs) proceed with the negotiation in order to share the spectrum with one of the primary users (this step is detailed in the article [9]). The secondary users (SUs) compete with each other to share a specific number of channels with the corresponding primary user (PU):

- The primary users (PUs) displayed in red cannot satisfy the requests of the corresponding secondary users (SUs) but its information will be stored in LBAA. The primary users (PUs) displayed in yellow have received no requests from the secondary users (SUs); The primary users (PUs) displayed in green share the spectrum with the requesting SU, as shown in Fig. 1.

However, each secondary user is going to cooperate with the other secondary users (SUs) through a Learning Base for Autonomous Agent (LBAA). This database allows detecting all the secondary agents or users (SUs) that are in need and to collect all information about them in order to help them choose the best primary users (PUs) that are compatible with their requests.

In its display table, the Learning Base for Autonomous Agent (LBAA) stores all the information on all secondary and primary users (SUs and PUs), in an autonomous manner (the number of channels requested by the SUs, the number of channels that can be allocated by the PUs, and the failure or acceptance of negotiation in case of risk of interference). If the secondary user (SU) shares a spectrum with a primary user (PU), it will be removed from the learning base and the information collected on the PUs will be recorded and suggested to other SUs in need. Now, if the secondary user (SU) is not satisfied in the area it has detected, the results found will be communicated to other SUs in need of PUs through the Learning Base for Autonomous Agent (LBAA); this SU will then be redirected to other PUs which can be suggested by the same autonomous base.

Figure 1 enables us to have more insight about the autonomy of this learning base. It helps us see and understand the suggestions proposed to the new SUs as well as to those which have still not found a corresponding PU (like the SU4 in the given example). It can also be noted that the primary users PU2,5,6 and PU13 suggested to the secondary user SU4,5,6,7 and SU8 are in red or yellow (without even being consulted by these secondary users). This has become possible thanks to the information provided by LBAA.

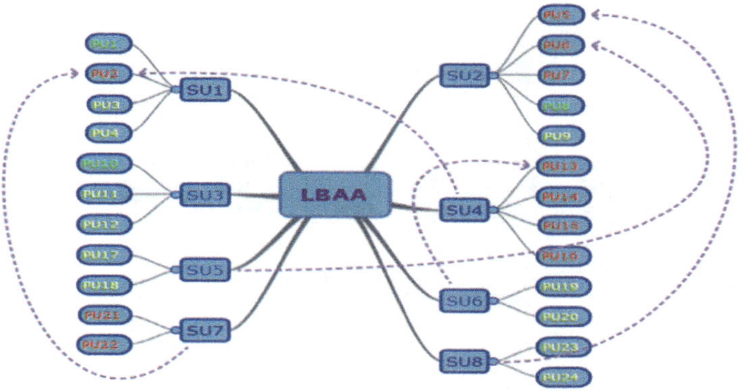

Fig. 1. LBAA approach.

2.3 Architecture of LBAA Approach

To better understand our approach, we tried to introduce an architecture based on real parameters. The first step is based on the secondary user (SU) who seeks to share the spectrum with the primary user (PU) either through direct negotiation (see article [11]), or by including the learning base of an autonomous agent (our approach). We were inspired by [12] in developing a real architecture capable of being self-managed in a radio environment. Figure 3 illustrates three high levels of standalone elements, such as a network manager, like an operator or server, which resides in the network areas or is present in the service layers. In the cognition level, different secondary users communicate with the LBAA which suggests the PUs that are compatible with their criteria. In the same context, the SUs will in turn communicate with the PUs through the primary providers that are present in the network level. These hierarchical, or rather structured, levels allow us to understand the functioning of our approach despite the autonomy of its individuals. During the autonomous management phase, the LBAA carries out the collection of information, the filtering and reporting of the data collected from the element to be managed using the sensors of the radio environment. These data are derived from the resources exchanged between the secondary user (SUi) and the Learning Base for Autonomous Agent (LBAA). The LBAA also monitors and analyzes the models, if necessary, and teaches the element the knowledge-acquisition. Also, this LBAA may include requests for suggestions to other SUs in need, in future situations. The planning step provides a mechanism structure for the actions to be taken as needed; the main objective is to facilitate communication between cognitive users. The execution of one part of the autonomous management phase gives the architecture a harmonious control of the orders in progress. The autonomous manager makes it possible to know if the commands have completed the required actions and also to update its knowledge base.

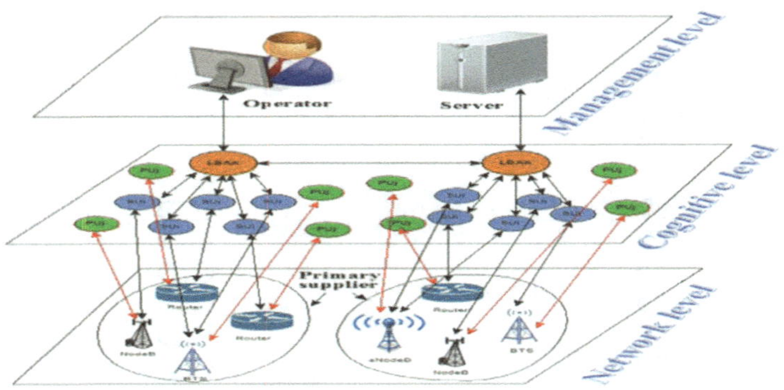

Fig. 2. Architecture of autonomous manager reacting with a cognitive environment.

3 Simulation Results of Our Contribution

The new approach was simulated in the same JADE platform [13] To validate the effectiveness of our proposed LBAA approach, JADE is a software Framework to develop agent applications in compliance with the FIPA specifications [14] for inter-operable intelligent multi-agent systems. With 24 agents for the primary users (PUs), 8 agents for the secondary users (SUs) and one agent for the LBAA database. Figure 3 shows the average of the convergence times of the eight SUs for the ten attempts to communicate with the different PUs. SU1 has the highest average convergence time, exceeding 273 ms, and this may degrade and eventually harm its quality of service (QoS). The convergence time is also high for SU2, with more than 254 ms. This is also true for SU3 and SU4 whose convergence times exceed 230 ms. Also, the convergence time of the second four SUs is much smaller than that of the first four ones thanks to the suggestion from the LBAA base. Here, the convergence time is between 215 and 220 ms for SU7 and SU8. This time is considerably smaller than those mentioned above. The convergence time of SU6 is less than 185 ms; SU5 has an optimized time of 138.9 ms which represents the best average compared to those of all the other SUs.

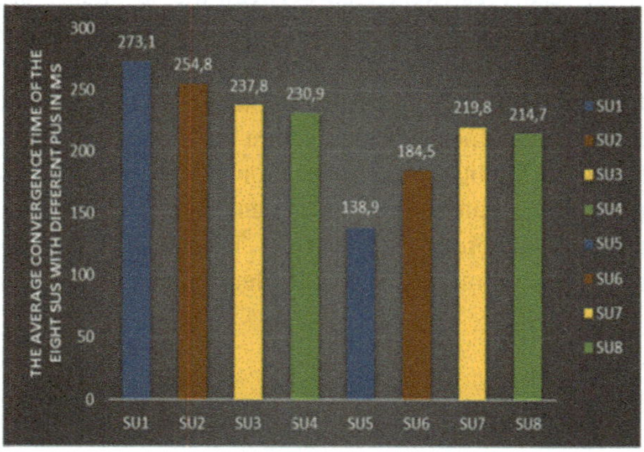

Fig. 3. The average convergence time between the eight SUs and their PUs for ten attempts communication.

On the other hand, Fig. 4 shows the average success rate of spectrum sharing for the 8 SUs with different PUs. For the first four SUs, the average success rate is 85% for SU1, which has the smallest success rate but still with a good result; it is 88% for SU2 and SU4, and 90% for SU3, which attained the best result among the 4 SUs. For the second four SUs, SU7 has a success rate of 85%, a rather mitigated result mainly because there is no suggestion from the LBAA. On the other hand, the success rate for SU8 is the same as that reached for the best of the first four SUs, namely 90%. Also, the success rate for SU6, which is 95%, is very close to the ideal spectrum sharing. Finally, SU5, which has the best success rate of 100%, makes the secondary user flexible to all suggestions and negotiations.

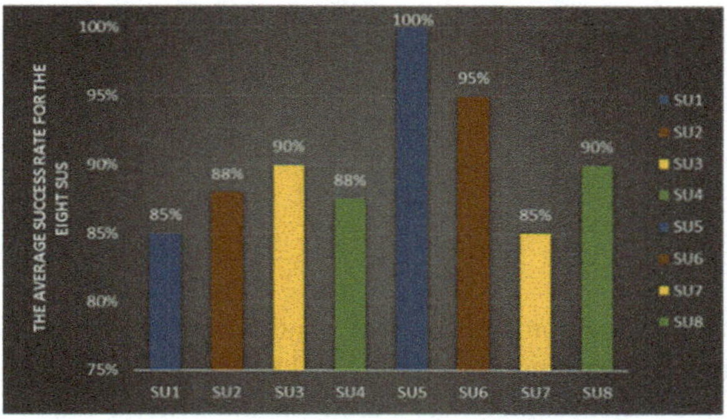

Fig. 4. The average success rate of spectrum sharing between the eight SUs and their PUs for ten attempts communication.

3.1 Interpretation of the Simulation Results of Our Contribution

The results found for the 10 communication attempts allows us to estimate the precious time that the secondary users (SUs) can gain while seeking the adequate PUs, and with a quality that meets their radio spectrum allocation requirements. Moreover, most of the SUs using the LBAA base have a success rate that is nearly complete (100%). The use of this database is not compulsory but rather suggestive. All decisions made by the agents, whether SUs, PUs or LBAA, are programmed autonomously in order to interact dynamically and in a flexible way with their environments.

3.2 Comparing the Simulation Results from Different Contributions

The QoS requirements of video conferencing are described in Cisco networks. Cisco is a network that was originally created for the city of San Francisco in the United States of America. There are two main types of video traffic, namely the interactive video (video conferencing) and video streaming (both unicast and multicast). Our study focused on interactive video, with the other three types of transmission cited before. When provisioning for interactive video (videoconferencing), the following guidelines are recommended:

The loss rate should not exceed 1%,
The one-way delay or latency (delay time for transmission) should not exceed 150 ms,
The jitter should not be more than 30 ms [15].

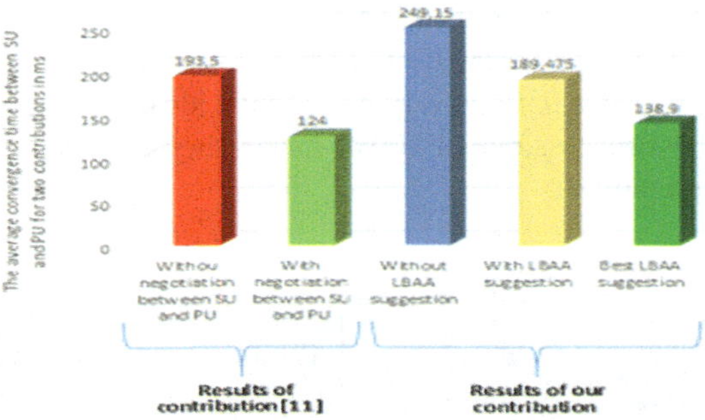

Fig. 5. Comparison between two contributions of the convergence time of SU and PU communication.

Figure 5 illustrates the comparison between different results of simulations of the two contributions (the reference contribution [11] and the new contribution) for the transmission of the video conference, based on the criteria required for the video conferencing quality of service. It was decided to continue on the same track and compare the two contributions on this same type of transmission. The average convergence time for the first contribution is 124 ms in the presence of negotiation between the secondary user SU and the corresponding primary user PU. This value is much smaller than that obtained in the case where there is no negotiation; here the average convergence time is equal to 193.5 ms. Compared to the second contribution, it was found that the average convergence time for the first four SUs without LBAA suggestion is 249.15 ms. This time is too high compared to that of the contribution proposed by [11]. On the other hand, the average convergence time for the second four SUs with the suggestion from the LBAA is much smaller than that of the first four ones, but this time remains a little worse than that of the contribution suggested by [11], with 124 ms. The secondary user SU5 presents the best result in the presence of a suggestion from the LBAA, with an average convergence time equal to 138.9 ms, which is less than the time of 150 ms required for the QoS in the literature [15] and quite close to the value found by [11], which remains the best result obtained for the convergence time.

Figure 6 compares the average success rates found in the two contributions regarding the QoS of the video conference. The results reached with our approach are much better than those reported in reference [11]. The average success rate exceeds 91% when there is negotiation between the secondary user (SU) and the primary user (PU), whereas in the case without negotiation, this rate is only 68.75%. The results of the approach developed in this article are much better, with a success rate of 93% for the second four SUs in the presence of the LBAA suggestion. This rate is higher than that of the first four SUs (without LBAA suggestion) and that of the contribution proposed in reference [11]. Finally, the secondary user SU5 has the best average success rate of spectrum sharing with 100% success rate, a value that ensures the QoS required in the literature [15], where the loss rate must be less than 1%.

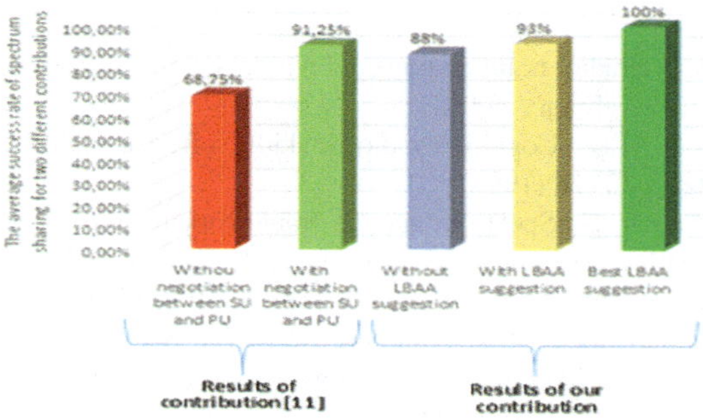

Fig. 6. Comparison between two contributions of the success rate of spectrum sharing between SU and PU

3.3 Interpretation of Results from the Comparison of the Two Contributions

The results of this contribution are satisfactory. They have addressed issues related to spectrum management. The results obtained have a direct positive impact on the quality of service (QoS). The response time, error rate and jitter in our contribution are in accordance with the QoS requirements of video conferencing. Our results are better than those proposed in the article [11]. Autonomy in decision making allows cognitive systems to be less dependent on administrators, with good management of the radio spectrum.

4 Conclusion

This article addressed the aspect of autonomy in cognitive radio systems. Multi-agent systems were used to study the behavior of these agents and to analyze the interactions between them, using a self-management system allowing cognitive users to better share the radio spectrum with minimal interference between the various primary and secondary users. Several architectures and descriptions have been put forward in this article to better understand the aspect of this new approach of the LBAA and to better appreciate the usefulness of this autonomous learning base.

This new contribution allowed us to develop an algorithm that illustrates the communication of secondary users with the autonomous learning base and with the primary users. Subsequently, the results of the learning base simulation were found to be conclusive, with a significant gain for the cognitive radio. The system relies on autonomous management and learning; it is capable of manipulating the spectrum in an opportunistic and reliable manner. Finally, the results of our new approach were compared with those of the contribution provided in [11] in order to better apprehend the quality of service of video conferencing.

References

1. Haykin, S.: Cognitive radio: brain empowered wireless communications. IEEE J. Sel. Areas Commun. **23**(2), 1–9 February 2005
2. Mitola, J., Maguire, G.Q.: Cognitive radio: making software radios more personal. IEEE Pers. Commun. **6**(4), 13–18 (1999)
3. Kasbekar, G.S., Sarkar, S.: Spectrum auction framework for access allocation in cognitive radio networks. IEEE/ACM Trans. Networking (TON) **18**(6), 1841–1854 (2010)
4. Benmammar, B., Amraoui, A., Krief, F.: A survey on dynamic spectrum access techniques in cognitive radio networks. Int. J. Commun. Netw. Inf. Secur. **5**(2), 68 (2013)
5. Akyildiz, I.F., et al.: NeXt generation/dynamic spectrum access/cognitive radio wireless networks: a survey. Comput. Netw. **50**(13), 2127–2159 (2006)
6. Benmammar, B., Benmouna, Y., Amraoui, A., et al.: A parallel implementation on a multi-core architecture of a dynamic programming algorithm applied in cognitive radio ad hoc networks. Int. J. Commun. Netw. Inf. Secur. **9**(2), 196 (2017)
7. Strassner, J., De Souza, J.N., van der Meer, S., et al.: The design of a new policy model to support ontology-driven reasoning for autonomic networking. J. Netw. Syst. Manag. **17**(1–2), 5–32 (2009)
8. Bargaoui, H., et al.: Hybrid QoS based routing protocol for inter and intra wireless mesh infrastructure communications. Wirel. Netw. **22**(7), 2111–2130 (2016)
9. Zubow, A., Döring, M., Wolisz, A.: Distributed spectrum allocation for autonomous cognitive radio networks. In: Proceedings of 20th European Wireless Conference European Wireless 2014, VDE, pp. 1–7 (2014)
10. Raiss El-Fenni, M., El-Azouzi, R., El-Kamili, M., et al.: Dynamic spectrum allocation based on cognitive radio for QoS support. In: Proceedings of the 13th ACM International Conference on Modeling, Analysis, and Simulation of Wireless and Mobile Systems. ACM, pp. 351–354 (2010)
11. Baba-Ahmed, M.Z., Benmammar, B., Bendimerad, F.T.: Spectrum allocation for autonomous cognitive radio networks. Int. J. Adv. Comput. Technol. **7**(2), 48 (2015)
12. Mihailovic, A., Nguengang, G., Kousaridas, A.: An Approach for Designing Cognitive Self-Managed Future Internet, the European Commission Seventh Framework Program ICT-2008–224344 through the Self-NET Project (2010). https://www.ict-selfnet.eu
13. JAVA Agent Development Framework. Open source platform for peer-to-peer agent based applications. http://jade.tilab.com/
14. Foundation for Intelligent Physical Agents. Specifications (1997). http://www.fipa.org
15. Szigeti, T., Hattingh, C.: End-to-End QoS Network Design: Quality of Service in LANs, WANs, and VPNs. Published Nov 9, 2004 by Cisco Press. Part of the Networking Technology series

Crown Planar Antenna Element for KA Band Satellite Applications

M. A. Rabah[1(✉)], M. Bekhti[1], M. Debbal[2], and Y. Benabdelleh[3]

[1] Satellite Development Center, Algerian Space Agency,
Bir El Djir, 31130 Oran, Algeria
marabah@cds.asal.dz

[2] Belhadj Bouchaib University Center, Ain Temouchent, Algeria
debbal.mohammed@gmail.com

[3] Operations Center for Telecommunications Systems (CEST)/ASAL,
14 Rue Omar Aissaoui El Hammadia, Bouzareah, Alger, Algeria
BenabdellahYoucef@gmail.com

Abstract. This article presents a new design of planar satellite antenna for Ka-band applications, where the effectiveness of the triangular planar antenna shape is investigated to get Crown shape; This analysis will be carried out by the simulation of planar antenna element with CST Microwave Studio ®, where numerical simulation with 5.35 dBi of antenna gain at 28.32 GHz and return loss less than −10 dB at all a frequency band between [27–29.5 GHz], can be exploited to develop the next generation of small communication satellite.

Keywords: Planar antenna · Ka-band intelligence · Satellite · Gain

1 Introduction

The development of telecommunications in recent years requires the realization of equipment increasingly efficient and less expensive. This evolution appears in many telecommunications systems. In the microwave domain, previously the circuits were heavy and their design was impressive but in recent years circuits responding to electrical performance increasingly ambitious were realized.

The patch antennas are a device that is very developed and also usable. Due to the ease of manufacture and creation of the new models.

Our goal is to create a new design of a satellite antenna that radiates in the Ka-band in the best way. In order to achieve satisfactory simulation results, we are directed to available commercial software which is CST MWS, which is a specialized tool for three-dimensional electromagnetic simulation of high frequency component and microwave applications.

To meet these requirements; several techniques have been developed which have proved their efficiency for planar antennas (Balanis 2012; Garg et al. 2001). Where we can found among the various methods (Yahya and Debidni 2011; Kil et al. 2007) to increase the gain is to put a patch element in an array configuration that can be a useful technique to improve the characteristics of planar antennas.

M. Hatti (Ed.): ICAIRES 2019, LNNS 102, pp. 381–386, 2020.
https://doi.org/10.1007/978-3-030-37207-1_40

This article focuses on the design of a new design of antenna element for Ka-band satellite applications; where we aim to reduce the weight and ensure the rigidity of the structure. A parametric study by a variation on substrates heights for new antenna structures for Ka-band applications where we present graph comparison of the reflection coefficient, the VSWR and the gain of this polygonal patch antenna.

2 Antenna Design Procedure

Triangular patch antenna shapes are currently considered very promising, especially for the design of compact antennas. It's designed to improve the performance of planar antennas such as bandwidth, multi-frequency resonance, and gain.

Basing on the theoretical analysis in (Chan 2002; Lin et al. 2005) proposed antenna has a regular crown shape with microstrip fed substrate permittivity of 4.3 will be presented in Fig. 1; and the antenna dimension are given in Table 1.

Table 1. Polygonal antenna parameters

Parameters	Dimensions (mm)
Patch length (Lg)	4.95
Patch width (Wg)	8.58
Substrate length (Ls)	20
Substrate width (Ws)	10
Microstrip feed length (Lt)	14
Microstrip feed width (Wt)	1

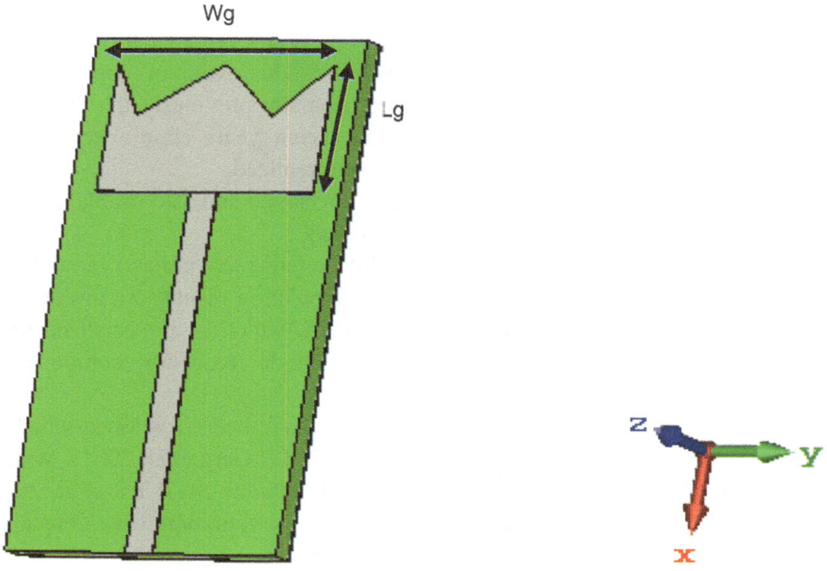

Fig. 1. Polygonal antenna with micro-strip feed.

3 Parametric Study

In this section, a parametric variation analysis of the substrate heights with commercial software of the proposed antenna is presented. This study aims to exploit the radiation characteristics of this antenna to operate in Ka band.

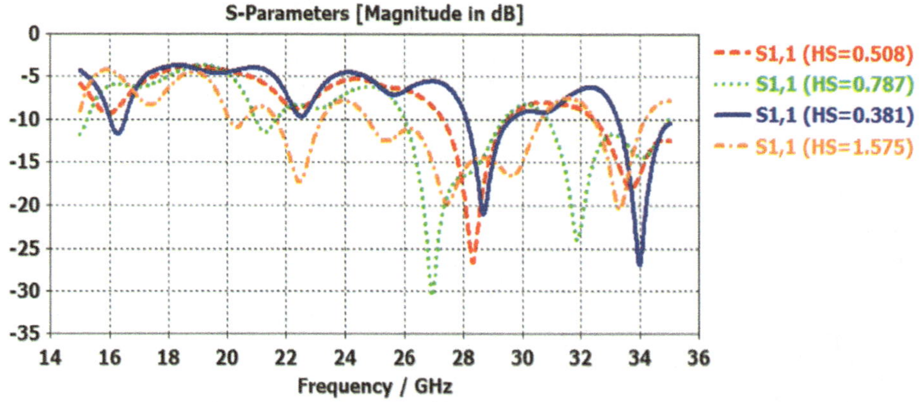

Fig. 2. The comparison results of the input return loss (S11).

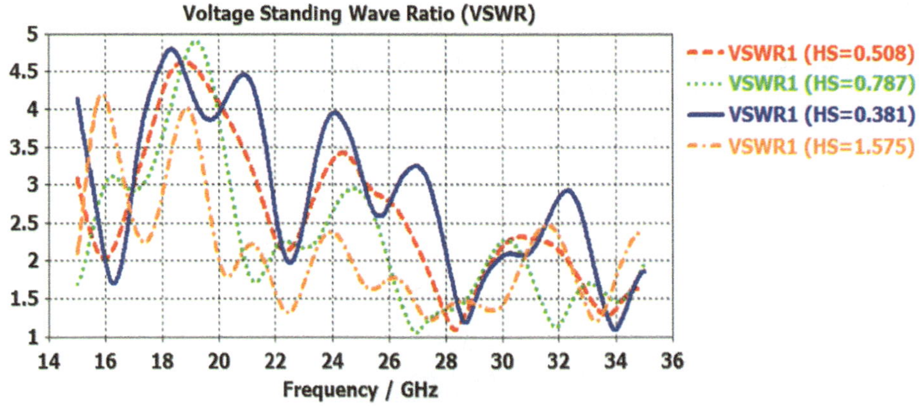

Fig. 3. The comparison results of the VSWR

The analysis results of the substrate height variation are given by the comparison in the return losses (Fig. 2), ensuring the adaptation conditions; and also given by the graph of comparison for VSWR values (Fig. 3).

It is well shown that for a height of 0.508 mm the return loss value is minimum, it is −26.58 dB for a resonance frequency of 28.32 GHz and this can also show in the value of the corresponding VSWR where it is 1.1 also less than 1.5.

Table 2 summarizes the values of the main characteristics of the proposed antenna for a frequency of 28.32 GHz. And it is clear that the value of the gain of the planar antenna is greater than 3 dBi for the different values of the substrates.

Table 2. Comparison of return loss and gain for different substrates heights at 28.32 GHz.

Substrate thickness (mm)	Return loss (dB)	Gain (dBi)	VSWR
1.575	−14.67	3.82	1.45
0.787	−15.92	4.14	1.38
0.508	−26.58	5.35	1.1
0.381	−12.92	4.76	1.58

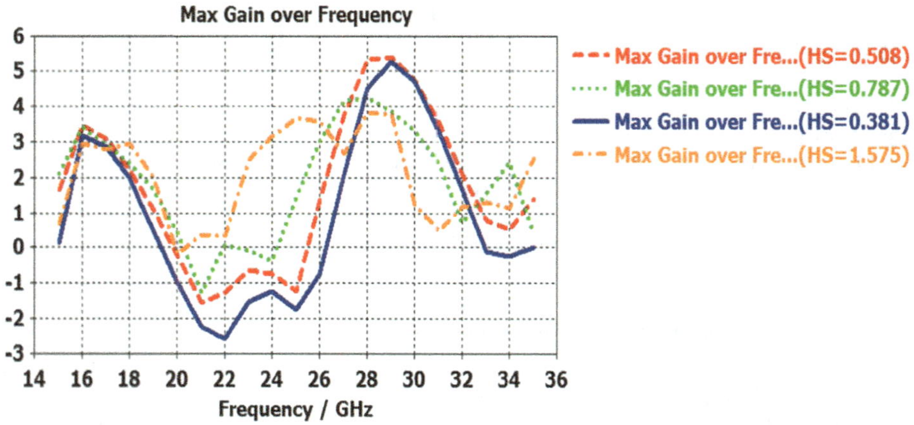

Fig. 4. Antenna gain variation comparison.

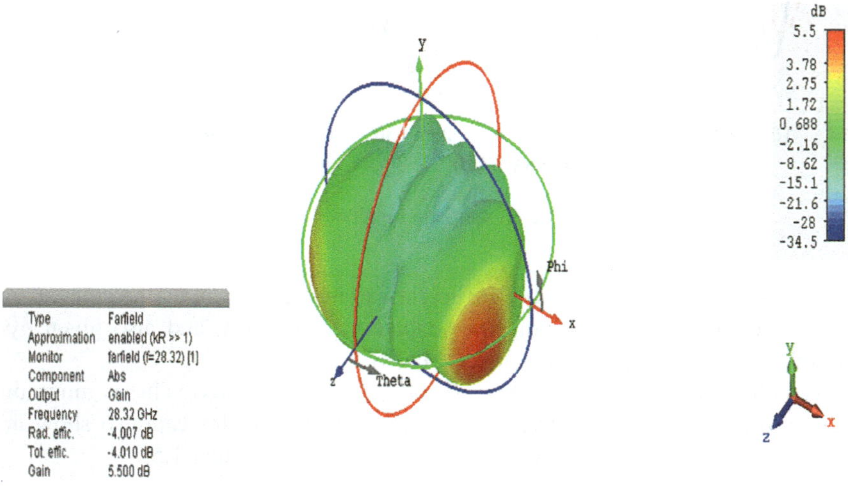

Fig. 5. 3D antenna pattern for resonance frequency of 28.32 GHz and h = 0.508 mm.

Although the return loss is −31.18 for h = 1.575 mm but the gain is low compared with h = 0.787 mm where the gain is 7.1 dBi that we bring to use this height since the value of the return loss is less than −10 dB and the VSWR value is acceptable for resonant frequency 24.804 GHz.

Figure 6 shows the variation of gain versus frequency when it is reached more than 5 dBi for a frequency range between 27.97 and 29.54 GHz (Fig. 4).

For the same dimension of the antenna cited in Table 1 and for the height h = 0.508 mm, Fig. 5 illustrates the 3D radiation pattern in resonant frequency of 28.32 GHz and Fig. 6 shows a 2D plot of the radiation pattern to describe the energy distribution of the proposed antenna (Fig. 5).

It is clear in 2D the radiation pattern of the proposed polygon antenna around 28.32 GHz for variation in the Phi angle; the energy is concentrated in a direction of an aperture of 51.8 at −3 dB.

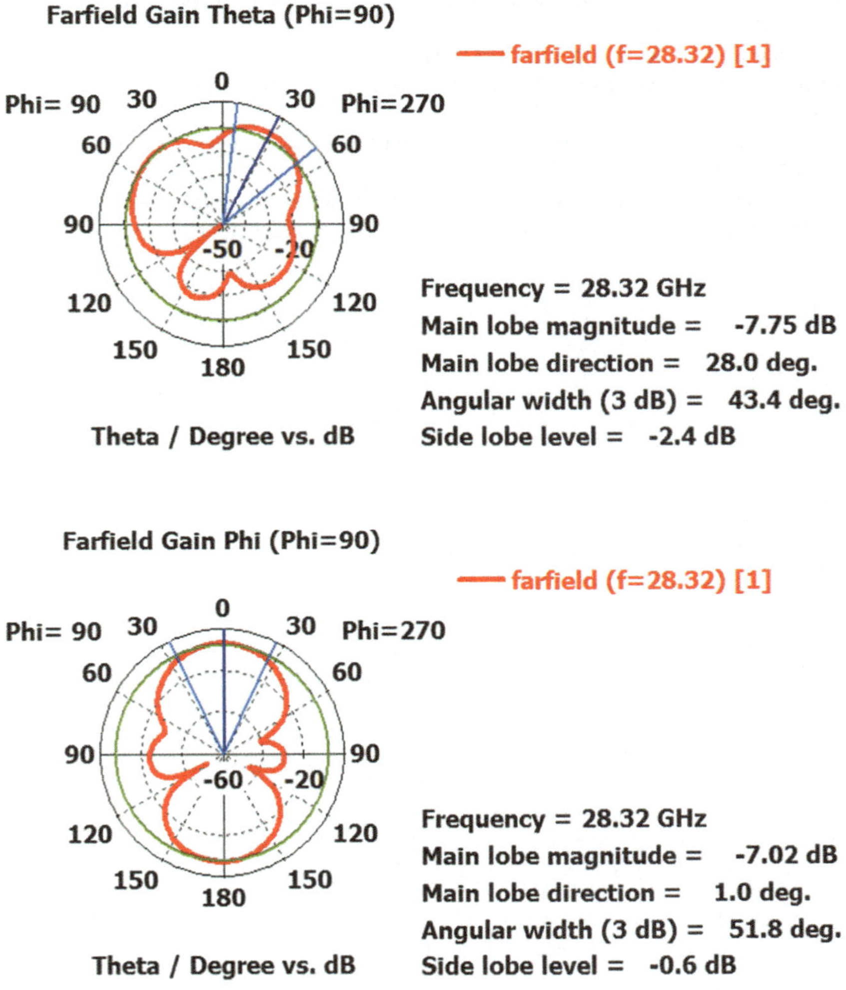

Fig. 6. 2D pattern: (a) for Theta, (b) for Phi

4 Conclusion

The parametric study presented in this analysis for a new design of a patch antenna for a Ka band applications, shows us that the characteristics of the antenna are improved for the smaller thickness, which present the efficiency to use a this antenna to meet specific needs at these substrate heights.

References

Balanis, C.A.: Antenna Theory Analysis and Design. Wiley, New York (2012)

Garg, R., Bhartia, P., Bahl, I., Ittipiboon, A.: Microstrip Antenna Design Handbook. Artech House Publishers (2001)

Yahya, R., Debidni, T.A.: Compact CPW-fed antenna with ultra-wide band and dual polarization. In: 11th Mediterranean Microwave Symposium, MMS, Hammamet, Tunisia, pp. 110–112 (2011)

Kil, J.H., Kwan, J.K., Kim, Y.J., Lee, H.M.: High gain antenna using parasitic shorted annular patch structure. In: Asia-Pacific Microwave Conference, APMC, Bangkok, Thailand, pp. 1–4 (2007)

Chan, H.M.: Microstrip-fed dual frequency printed triangular monopole. Electron. Lett. **38**(6), 619–620 (2002)

Lin, C.-C., Kan, Y.-C., Kuo, L.-C., Chuang, H.-R.: A planar triangular monopole antenna for UWB communication. IEEE Microw. Wireless Compon. Lett. **15**(10), 624–626 (2005)

An Approach Based on (Tasks-VMs) Classification and MCDA for Dynamic Load Balancing in the CloudIoT

S. Benabbes(✉) and S. M. Hemam

ICOSI, Abbes Laghrour University, Khenchela, Algeria
benabbes.sofiane@gmail.com, sofiane.hemam@gmail.com

Abstract. Cloud IoT is a new paradigm that has emerged as a result of the integration of Cloud Computing and the Internet of Things. It provides a set of intelligent services and applications that can strongly influence on our daily lives. In the CloudIoT, physical sensors are responsible for detecting and transmitting data to the cloud in order to be treated and stored. In this context, the amount of data to be processed through the CloudIoT is in increasing usually and, in this situation, the load balancing mechanism is needed in order to distribute the captured data between the different CloudIoT resources. In this paper, we propose an approach that allows to balance the load between the different virtual machines of a CloudIoT. The proposed approach is composed of four essential components: The Classifier which assigns Tasks to its corresponding class and assigns the Virtual Machines (VM) to its corresponding class too, the Dispatcher which sends the tasks to the VM classes, the Local Balancer whose role is to balance the tasks between different VMs belonging to the same class used the spooling method for duplicate same VM, finally, the Manager component or general balancer that is responsible to balance the load between VMs classes with the test of the maximum cloud allowance bar to release VM. The results obtained through the proposed the case study show that our approach allows effectively the load balancing between VM.

Keywords: Cloud computing · CloudIoT · Internet of Things · Load balancing · Worst-Case Execution Time · Machine learning

1 Introduction

Cloud Computing is a computer model based on the internet. According to NIST "National Institute of Standards and Technology", the Cloud computing is a model for enabling ubiquitous, convenient, on-demand network access to a shared pool of configurable computing resources that can be rapidly provisioned and released with minimal management effort or service provider interaction. These resources are, for example, networks, servers, storage space, applications and services. They can be provisioned quickly with a minimum management effort and an interaction with service provider (Mell and Grance 2009). On the other hand, the Internet of Things (IoT) are based on intelligent nodes (things), self-configured, interconnected in a dynamic and global network infrastructure (Botta et al. 2015). It is generally characterized by small

M. Hatti (Ed.): ICAIRES 2019, LNNS 102, pp. 387–396, 2020.
https://doi.org/10.1007/978-3-030-37207-1_41

objects of real-world, widely distributed, with limited storage and processing capacity, which implies problems about the reliability, performance, security, and confidentiality (Botta et al. 2015). In contrast to the Internet of Things, cloud computing has unlimited capacity in terms of storage, power and processing, it is a much more mature technology. From these two technologies (Cloud-Computing and Internet of Things) appears a new paradigm in which Cloud Computing and the Internet of Things are merged. This new paradigm is called CloudIoT. In CloudIoT, the amount of data to be stored and processed is daily increasing, which requires a mechanism that balances the storage and processing load between the different VMs of the CloudIoT. Load balancing is considered to be one of the key issues, it is necessary to distribute the workload between multiple nodes to ensure that no nodes are overloaded (Sidhu and Kinger 2013). The algorithms that use the principle of first-come-first-serve have their limit in load balancing, since they do not achieve the goal of load balancing between dynamic nodes or between VMs in Cloud. Thus, some VMs will be overloaded while others will be underloaded (Chien and Loc 2016).

In this paper, we propose a model that allows load balancing between the different VMs of a CloudIoT. The proposed model is composed of four essential components: The Classifier which is responsible to constitute the five classes of VMs; from the fastest to the lowest class, then constitute the same number of Tasks classes, by using one of the unsupervised classification techniques,. The Dispatcher, which calculates at first the distances between the received VM and the centers of the x classes created, then it sends the VM to the nearest class according to the distances previously calculated, then sends the tasks of class to VM of the class corresponding. The local balancer whose role is to balance the load between the different VMs, belonging to the same class, by taking into consideration the Worst Case Execution Time of a task and its size. Finally, the general balancer component, which is responsible for balancing the load between classes by migrating tasks, from an overloaded class to a less loaded class of VM. The results obtained through the proposed case study show that our model actually allows load balancing between all VMs.

Our paper is structured as follows: after the introduction, the Sect. 2 will be devoted to the related work. In Sect. 3 we first present the general architecture of the proposed model, then the functionality of each component will be explained in Sect. 4. As for Sect. 5, it will be devoted to the case study in order to validate the functionality of the various proposed components, and we will conclude this paper by a conclusion in Sect. 6.

2 Related Work

Load balancing is an important technique for improving system performance. User requests must be evenly distributed between machines or nodes in order to obtain a fast response (Cardellini et al. 1999). The goal of load balancing is to increase resource utilization, reduce response time, and overload of some nodes. Load balancing can be achieved by different approaches with different degrees of efficiency. We can distinguish between two types of load balancing: static and dynamic. These approaches depend on how a load balancing algorithm is based to make its decision, that is, it is based on the current state of the system or not (Alakeel 2010). Static load balancing can be defined by a situation in which a computational load is partitioned between nodes by an algorithm executed before the execution of the program, i.e. the assignment of a set requests to a set of resources; and in the dynamic approach, the load balancing algorithm is based on the current state of the system, so that requests are moved dynamically from an overloaded node to an underloaded node, depending on the availability of resources system (Olejnik et al. 2009). There are two categories of dynamic load balancing algorithms: centralized and distributed algorithms. In the distributed dynamic load balancing algorithm, all nodes in the system must execute the algorithm and interact with each other to perform load balancing. The interaction between the nodes can be cooperative or uncooperative. Interaction between nodes in distributed dynamic load balancing algorithms generates more messages than centralized messages. However, it is a reality that even with the greatest complexity of execution; dynamic algorithms have the potential to offer better performance than static algorithms (Olejnik et al. 2009). Several work related to load balancing in Cloud Computing has been proposed in recent years. In this paper, we quote the most recent works. For example, the authors Chien and Loc (2016), the authors proposed a load balancing algorithm that improves the performance of the cloud environment based on the method of estimating the end-of-service time. They managed to improve the user's service time and response time. The problem of energy consumption was addressed by (Yakhchi et al. 2015). Thus, they explain that energy consumption has become a major challenge in cloud computing infrastructures. For this, they proposed a new method of load balancing, called ICAMMT. The latter is used to manage energy consumption in Cloud Computing data centers.

To satisfy the needs of users, the authors Kapoor and Dabas (2015) proposed an algorithm that aims to maximize user satisfaction, minimizing the response time of tasks and improving resource utilization through fair allocation of cloud resources. The authors Fahim et al. (2016) proposed a new load balancing design. Their model is based on two main phases: the first is the pre-estimation of the execution time of the instructions of a task, based on the methods that calculate the worst execution time. As for the second phase, it consists of scheduling tasks according to classification levels in order to allocate tasks to VMs with the guarantee of equality in the distribution of the load, and decrease the degree of load unbalance taking into account the characteristics of the tasks before their allocations. Al-Rayis and Kurdi (2013) explain that, fundamentally, load balancers can be deployed based on several different architectures. The centralized load balancing architecture, which relies on only one central load balancer, maintains the entire system in a balanced state by selecting the cloud resource that should support the work received. In the hierarchical load balancer architecture, a primary load balancer (parent) receives all job requests and then distributes them to other connected load balancers (children) where each load balancer in the tree can use a different algorithm. The authors Hemam et al. (2017) proposed a new approach based on the idea of cloning a cloud service on one or more nodes, when the number of user requests will be important at a given moment. They proposed an algorithm that averages the load of the nodes in order to select less loaded nodes.

Unlike this work, our job is to balance the load at the level of each class in a decentralized way, since the local balancers are independent of each other, which allow a gain in response time on one side, and they allow you to select the most appropriate VM for scheduling the received task on another side. As for the central balancer, it allows to establish a contact between local balancer components of the overloaded and the underloaded classes in order to migrate tasks from one class to another.

3 Proposed Architecture

This section describes the proposed architecture (see Fig. 1). The latter consists of two types of load balancer: the local balancers that are at the level of each class of VMs, and a general balancer. In addition to these two components, our architecture contains the Classifier component, whose role is to classify Tasks and classify VMs using the K-means method, and the dispatcher component, which allows to send the tasks of the users to the adequated classes. The classes are: Very Fast (VF), Fast (F), Medium (M), Slow (S), and Very Slow (VS).

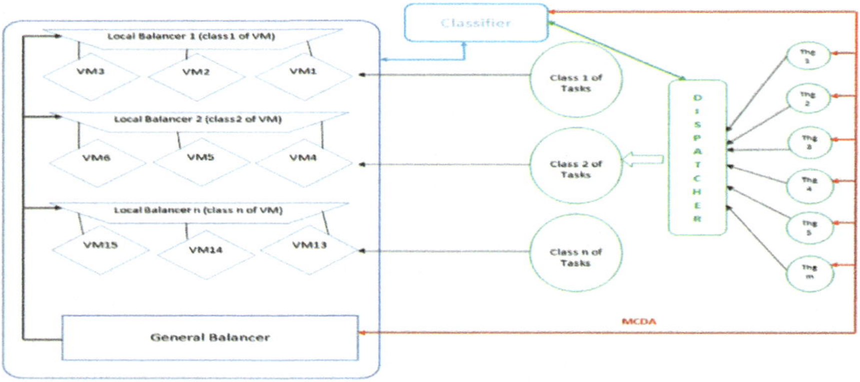

Fig. 1. Proposed architecture.

4 Functionality of the Proposed Model

As shown in Fig. 2, the components of the proposed model interact with each other to ensure a reasonable load between the different VMs, thereby reducing response and service times.

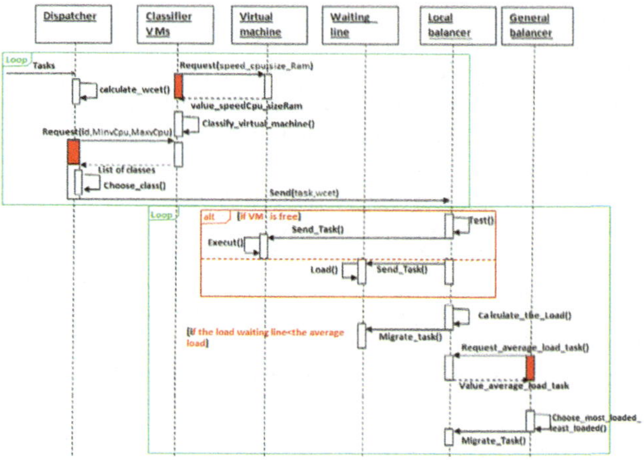

Fig. 2. Sequence diagram- model.

4.1 Classifier

The classifier applies the K-means method. It classifies all the VMs and Tasks in into five classes (VF, F, M, S, VS). VMs are characterized by two criteria (CPU Speed and RAM Size) and Tasks are characterized by (WCET and size). The method first assigns the first five VMs to the five classes respectively and then, at each iteration, calculates the distance between the VMs and the gravity center of the class.

$$D_{xy} = \sqrt{(Xcpu - Ycpu)^2 + (Xram - Yram)^2} \tag{1}$$

Where D_{xy} is the distance between two nodes.

$$\left.\begin{aligned} Ci : CPUi &= \sum\nolimits_{j=1}^{n} VMi(cpu)/n \\ Ci : RAMi &= \sum\nolimits_{j=1}^{n} VMi(ram)/n \end{aligned}\right\} \tag{2}$$

Where C_i (CPU, RAM) is the center.

The Classifier assigns the VM to the class where the distance is the minimum, until the classes structure will be the same. The classes are named VF, F, M, S, VS, and each of them is characterized by its gravity center Ci.

4.2 Dispatcher

The dispatcher stars by calculating the worst execution time of a task based on the work of (Fahim et al. 2016), and then based on this calculated time, the size of the task, and according to the gravity center Ci of each class, it selects the most suitable class for the task.

4.3 Local Balancer

It receives the task, issued by the dispatcher, with its worst execution time to assign it to the VM that suits it. At each arrival or departure of a task, this balancer assures the balancing of the local load of its class. For this, it calculates the load of each VM (VM_L) according to formula (3).

$$VM_L = Vp * \sum_{j=1}^{m} Tj \tag{3}$$

Where Vp is the speed of the processor, Tj is the worst execution time of the task and n is the number of VMs.

Then it calculates the load average of the class according to formula (4).

$$Avg_c = \sum VM_L / n \tag{4}$$

After that, it selects the overloaded VMs, those with their load is above the average, to migrate some tasks to the least loaded VMs. The choice of the task to be migrated is based on two essential criteria which are: the size of the task and its worst execution time. To minimize the cost of migration, the local balancer should choose the task with the smallest possible size but the worst execution time is the highest. To select a task that meets both criteria, the TOPSIS method is used (Yezza 2015).

4.4 General Balancer

In the same way as the local balancer, the general balancer calculates the class load average in order to migrate the tasks from the most loaded classes to the least loaded classes. In this context, we can distinguish two cases. The first case is when the receiving class is inferior to the sending class, in which case the candidate task must be chosen from those with the higher worst execution time. In the contrary case, it chooses the task with the lowest worst execution time.

5 Case Study

In this part we will present a detailed case study to clarify the functioning of our approach.

5.1 Scenario 1: Classification of VMs

In this scenario, we consider that the system is composed of 20 VMs where each VM has its own characteristics, as shown in Table 1:

Table 1. List of virtual machines.

VM ID	VM name	CPU speed	RAM size	VM ID	VM name	CPU speed	RAM size
0001	VM1	5.0	4	0011	VM11	1.8	2
0002	VM2	6.0	5	0012	VM12	2.2	4
0003	VM3	3.2	2	0013	VM13	2.1	3
0004	VM4	3.5	3	0014	VM14	2.0	5
0005	VM5	4.2	3	0015	VM15	6.1	5
0006	VM6	2.5	4	0016	VM16	3.3	6
0007	VM7	6.2	3	0017	VM17	4.8	2
0008	VM8	4.1	4	0018	VM18	4.0	7
0009	VM9	4.0	4	0019	VM19	2.9	3
0010	VM10	4.0	3	0020	VM20	3.1	8

The K-means algorithm is based on the calculation of the distances between the VMs (Formula 1), and the centers of gravity of the classes Ci (Formula 2). After the execution of this algorithm we obtain five classes of VMs. These classes of VMs are shown in Table 2. As mentioned above, the classes are named respectively, Very Fast, Fast, Medium, Slow, Very Slow, and each class is characterized by its gravity center (Ci).

Table 2. Classification of virtual machines in 5 classes (K-means).

Class1 (VF) C1	Class2 (F) C2	Class3 (M) C3	Classe (S) C4	Classe (VS) C5
VM1, VM2, VM7, VM15	VM16, VM18, VM20	VM3, VM11, VM13, VM19	VM6, VM12, VM14	VM4, VM5, VM8, VM9, VM10, VM17

5.2 Scenario 2: Assigning Tasks to Classes

At the arrival of the task, the dispatcher will calculate the worst execution time (WET) by executing the WCET algorithm (Fahim et al. 2016). Let's see Table 3 of the tasks below. This table represents all the tasks that are characterized by their size and their WET.

Table 3. List of tasks with their WET.

Task	Size (MI)	WCET	Task	Size (MI)	WCET
T1	105	4.2	T11	52	2.2
T2	127	5.1	T12	410	8.1
T3	210	6.4	T13	119	4.8
T4	240	6.6	T14	221	6.4
T5	158	5.9	T15	265	7
T6	721	12.3	T16	220	6.4
T7	320	7.3	T17	147	5.7
T8	410	8.1	T18	152	5.8
T9	110	4.6	T19	149	5.7
T10	58	2.4	T20	36	1.1

After calculating the WET, the dispatcher sends the tasks to the local balancers according to the gravity center of classes and to the characteristics of the tasks (WET, size). Table 4 below shows the result of distribution of tasks on the different classes:

Table 4. Results of task distributions.

Class1 (VF) C1	Class2 (F) C2	Class3 (M) C3	Class4 (S) C4	Class5 (VS) C5
T2, T3, T4, T5, T6, T7, T8, T12, T14, T15, T16, T17, T18, T19	T1, T9, T13		T10, T11	T20

5.3 Scenario 3: Assigning Tasks to the VMs of a Class

When the task arrives at the class with its previously calculated WET, the local balancer assigns the task to its corresponding VM based on the formula (4). So to minimize response and service times, the local balancer must consider the average load of the class. Table 5 shows how tasks are distributed by the local balancer between the VMs of VF class.

Table 5. The state of the tasks of the VF class.

Class	VM	Execution	Waiting line						
VF	VM 1	T2							
	VM 2	T3	T17	T19	T20				
	VM 7	T4	T6	T7	T8	T12	T14	T15	T16
	VM 15	T5							

5.4 Scenario 4: Intra-class Tasks Migration

The local balancer of the VF class starts by calculating the loads of each VM (VM1, VM2, VM7, VM15) belonging to its class according to Formula 3, as shown in Fig. 3 (histograms in blue), and then it calculates the average load of the VF class according to formula 4. To balance the load between the VMs, the local balancer component must migrate the tasks from VM7 to the other VMs (from the most loaded to the least loaded). To select the tasks to be migrated, the component uses the TOPSIS method (Yezza 2015) to choose the tasks with the worst execution time on one side and the smallest possible size to minimize the transfer time. In this scenario, and by applying all the steps of the TOPSIS method, the local balancer will select the T8 task to migrate it to the VM1, and so one until the VMs be balanced.

5.5 Scenario 5: Inter-class Migration of Tasks

Periodically, the general balancer sends a request to the local balancers asking them the average load of their class. So, local balancers communicate the average load of their class to the general balancer in order to balance the load classes. To do this, it migrates tasks from overloaded classes to the least loaded classes, while taking into account the class performance when selecting tasks. In the same way as the local balancer, the general balancer calculates the average load of classes and after that it establish a contact between local balancers of the selected classes in order to migrate tasks from the class C1 to the other classes, and finally an equilibrium state between classes is obtained as indicated by the green histograms of Fig. 4.

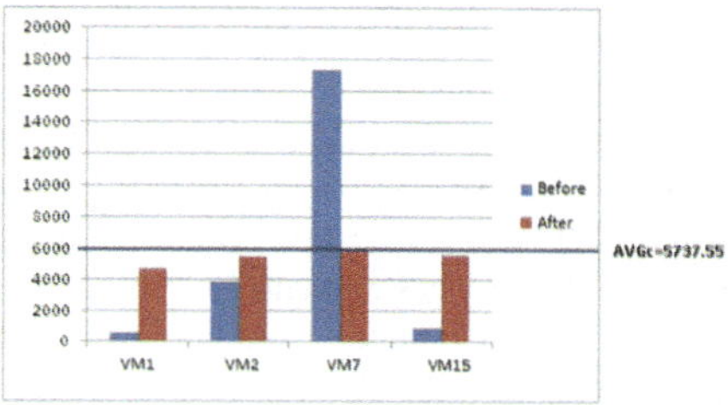

Fig. 3. Intra-class migration.

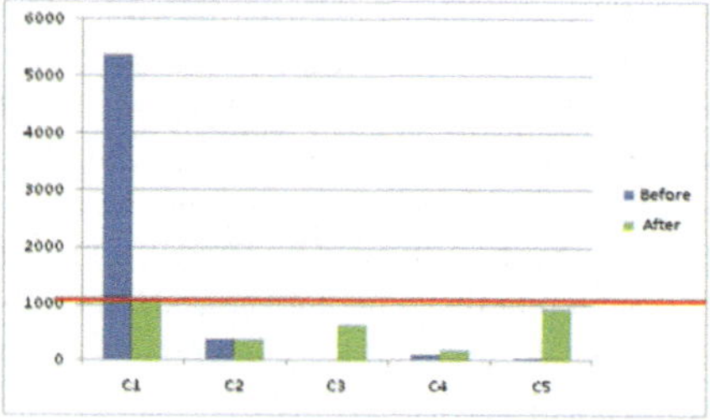

Fig. 4. Inter-classes load balancing.

6 Conclusion

In this paper, we discussed the problem of load balancing in the CloudIoT. For this, we have proposed an approach based on the characteristics of a task: worst execution time and its size, on one side, and on the characteristics of the VMs (CPU speed and memory size) of the other side. The proposed approach is composed by three main phases. In the first phase, the classifier component classifies the VMs into five classes. In the second phase, the dispatcher component estimates the worst execution time of the task in order to route it to the class whose gravity center is closest to him. As for the third phase, it consists in balancing the load, intra-class and inter-class by using the local-balancer and central-balancer components respectively, based on the average load, in order to migrate the tasks of the overloaded VMs to the least loaded VMs (respectively classes overloaded to the least loaded classes).

References

Botta, A., Donato, W., Persico, V., Pescapé, A.: Integration of cloud computing and internet of things: a survey. Future Gener. Comput. Syst. **56**, 684–700 (2015)

Yezza, A.: La méthode TOPSIS expliquée pas à pas, Sopra Steria Group (2015)

Sidhu, A.K., Kinger, S.: Analysis of load balancing techniques in cloud computing. Int. J. Comput. Technol. **4**(2), 737–741 (2013). ISSN 2277-3061

Alakeel, A.M.: A guide to dynamic load balancing in distributed computer systems. Int. J. Comput. Sci. Netw. Secur. **10**(6), 153–160 (2010)

Al-Rayis, E., Kurdi, H.: Performance analysis of load balancing architectures in cloud computing. In: IEEE European Modeling Symposium, pp. 520–524 (2013). ISBN 978-1-4799-2577-3

Chien, N.K., Loc, H.D.: Load balancing algorithm based on estimating finish time of services in cloud computing. In: 18th International Conference on Advanced Communication Technology (ICACT), pp. 228–233 (2016). ISBN 978-89-968650-6-3

Mell, P., Grance, T.: The NIST definition of cloud computing. In: The NIST Definition of Cloud Computing, National Institute of Standards and Technology, p. 2 (2009)

Olejnik, R., Alshabani, I., Toursel, B., Laskowski, E., Tudruj, M.: Load balancing metrics for the SOAJA framework. Scalable Comput. Pract. Exp. **10**(4), 419–428 (2009)

Kapoor, S., Dabas, C.: Cluster based load balancing in cloud computing. In: Proceedings of the Conference 2015 Eighth International Conference on Contemporary Computing (IC3), pp. 76–81. IEEE, India, 20–22 August 2015

Yakhchi, S., Ghafari, S.M., Yakhchi, M., Fazeli, M., Patooghy, A.: ICA-MMT: A load balancing method in cloud computing environment. IEEE, Sousse (2015)

Hemam, S.M., Hioual, O., Hioual, O.: Load balancing between nodes in a volunteer cloud computing by taking into consideration the number of cloud services replicas. In: Proceedings of the Conference 3rd International Conference of Cloud Computing Technologies and Applications (CloudTech), vol. 1, pp. 1–8. IEEE, Morocco, 24–26 October 2017

Cardellini, V., Colajanni, M., Yu, P.S.: Dynamic load balancing on web-server systems. IEEE Internet Comput. **3**(3), 28–39 (1999)

Fahim, Y., Benlahmar, E.H., Elhoussine, L., Eddaoui, A.: Une nouvelle conception d'équilibrage de charge dans le cloud computing. In: Proceedings of the conference 4ème Journée sur les Technologies d'Information et de Modélisation TIM 2016, pp. 167–172, Morocco, 02 June 2016

A Novel Communication Mode for Energy-Efficient Based Chain in Wireless Sensor Networks

Mohammed Kaddi[1(✉)], Khelifa Benahmed[2], and Mohammed Omari[1]

[1] LDDI Laboratory, Mathematics and Computer Science Department, University of Adrar, Adrar, Algeria
kadimohamed1983@yahoo.fr, omarialgeria@gmail.com

[2] Mathematics and Computer Science Department, Faculty of ES, University of Bechar, Bechar, Algeria
benahmed_khelifa@yahoo.fr

Abstract. In this paper, we propose energy efficient chain based routing protocol for WSNs (Wireless Sensor Networks) called CBRP (Chain- Based Routing Protocol). The CBRP protocol mainly divides a WSN into a number of chains using greedy algorithm as in PEGASIS protocol (Power Efficient GAthering in Sensor Information Systems). All chain leader nodes form also a chain using a new objective function. Simulation results have shown that our proposed CBRP excels regarding energy consumption in WSNs, and resulting in an extension of lifetime for the network compared with PEGASIS and IEEPB (Improved Energy Efficient PEGASIS Based). We simulated the proposed technique using MATLAB environment.

Keywords: WSN · Routing · Energy consumption · Lifetime · PEGASIS · IEEPB

1 Introduction

The development of mobile networks and wireless networks opens a new era in the telecommunications field. The recent evolution of the wireless communication means has allowed the manipulation of information through portable computing units with very particular characteristics (low storage capacity, autonomy energy source, limited power, etc.) that access the network through a wireless communication interface.

To make this communication more attractive, flexible and with less infrastructure, a new generation of networks called WSN has emerged. These sensors incorporate wireless sensing and communication devices in a single circuit, at a reduced size, and at a reasonable cost.

The WSNs represent a particular class of ad hoc networks. They are one of ten emerging technologies that will change the world. They are composed of a large number of materially small sensors, and usually placed near the monitored objects. These sensors are able to harvest, process and route environmental data from the monitored area in an autonomous manner to collection stations called base station (BS).

M. Hatti (Ed.): ICAIRES 2019, LNNS 102, pp. 397–407, 2020.
https://doi.org/10.1007/978-3-030-37207-1_42

The limits imposed are the limitation of processing capacity, storage and especially energy because they are usually powered by small batteries, which are usually y irreplaceable. As a result, energy is the most valuable resource of a sensor network because it has a direct impact on sensor life, or even the entire network.

Several research projects have emerged to optimize node energy consumption through the use of innovative conservation techniques to improve network performance, including maximizing its lifetime.

In the hierarchical structure there are cluster-based approach and chain–based approach. In this latter type, the node is organized in one or more chains. In this type there are some of routing protocols such as: LEACH (Low-energy adaptive clustering hierarchy) [1], PEGASIS [2] and IEEPB [3].

In this paper, in order to extend a WSNs lifetime, we present energy efficient chain based routing protocol for WSNs called CBRP (Chain- Based Routing Protocol). This proposed protocol is actually an improvement of the three protocols: LEACH, PEGASIS and IEEPB, where it uses another mechanism of chain formation and a new mode of data transmission. The data transmission is done by multi-hop mode using also a chain between the chosen leaders.

The rest of this paper is organized as follows. In Sect. 2, we present the structure of WSNs. In Sect. 3, the related work is discussed. Section 4 explains our proposed protocol. In Sect. 5, we give a performance evaluation of the proposed method via simulations and compare the results with PEGASIS and IEEPB protocols. Finally, Sect. 6 concludes the paper and future work is pointed out.

2 Structure of WSNs

Sensor networks have a wide variety of applications. A WSN is made up of nodes which are wireless sensor units, router and BS (Fig. 1).

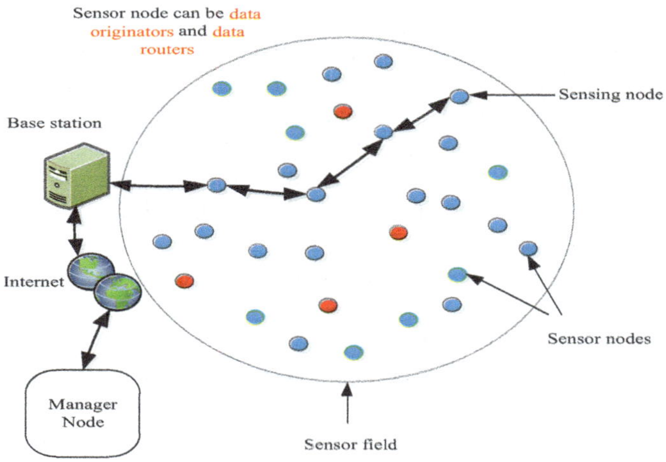

Fig. 1. WSN communications architecture.

2.1 Sensor Nodes

A sensor is a small electronic device equipped with a limited energy source capable of detecting and measuring a physical value (event, information) if they are within their perception range. Sensor nodes consist of a processing unit, a sensing unit, a power supply unit and a communication unit. And, depending on the specificity of the application, the sensor nodes may also include additional components, such as a location finding system to determine their location (Global Positioning System "GPS", for example), a mobilizer to change their location or configuration (for example, the orientation of the antenna). Figure 2 shows the block diagram of sensor node.

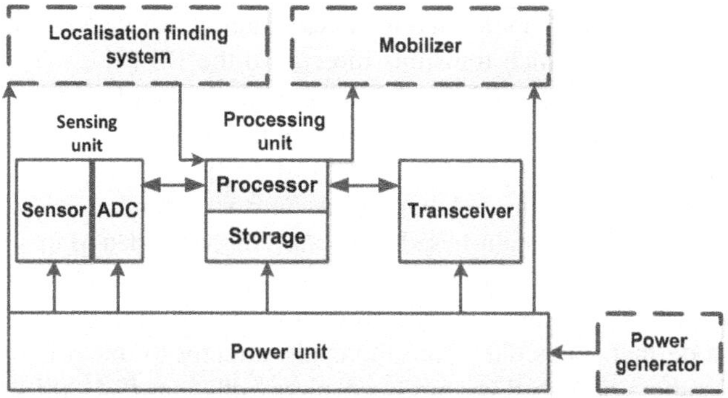

Fig. 2. Sensor node structure.

2.2 Base Station (Sink)

A sink/base station is the type of sensor node which possesses large memory, high power and it is the entity, where information is required. The BS can be a part of the WSN field or it could be the node outside the field of sensors. It can also a gateway to other sensor nodes in the WSN field.

2.3 Router

Router is a microprocessor-controlled device connected between two or more data lines from different networks which makes connection possible between two or more different network present at same or different geographical locations. It also forwards data packets between computer networks. A Router is used to connect different networks; it extracts the packet destination, selects the best destination path and forwards the packet to the next device on selective path. The router determines the destination of the packet by reading the address information in the packet, when a data packet comes in one of the lines. Using the information in the routing table, the packet is directed to the next network.

3 Related Work

Routing algorithms and data-gathering raises an important topic in WSNs due to the power limitation of sensor node [4]. There are two categories for the existing routing/data-gathering algorithms: hierarchy algorithms and non-hierarchy algorithms. In non-hierarchy algorithms, sensor nodes have the same role and collaborate to perform the sensing task and multi-hop communication. On the other hands, in the hierarchical algorithms, the network divided into several logical groups called clusters within a fixed area and nodes perform different tasks. The hierarchical algorithms include cluster based, tree-based and chain based routing protocols. In this paper, we focus on chain based routing protocols. The key in chain-based routing, however, is to form chains among the nodes so that each node will receive and transmit only to one pre-determined one-hop neighbor. Data is thereby aggregated through the chain until it reaches the chain leader, which transmits directly to the BS [5].

In recent years, many types of search have explored chain-based routing in WSNs. Various protocols have been proposed to improve the performance and extend the life of wireless sensor networks.

PEGASIS (Power-Efficient GAthering in Sensor Information Systems) [2] an extension of LEACH. It is a chain-based protocol. The basic idea of this protocol is that the nodes will be organized in a chain, so that the nodes transmit and communicate only with their nearest neighbors. Gathered data moves from node to node, are fused, and the only node that sends data from the chain directly to the BS is the leader.

In [6], the authors proposed an improved energy efficient PEGASIS based protocol (PEGASIS-E). PEGASIS-E uses average distance among the sensor nodes as the criteria for chaining, thereby providing better performance in terms of energy dissipation and amount of information sent to the sink. The simulation results obtained show that PEGASIS-E gives an increase in the network lifetime as compared to PEGASIS.

The authors of [7] proposed an Energy-Efficient Chain-Based routing protocol (EECB) that is an improvement over PEGASIS. EECB uses distances between nodes and the base station and remaining energy levels of nodes to decide which node will be the leader that takes charge of transmitting data to the base station.

In [8], the authors proposed a Chain Oriented Sensor Network (COSEN) protocol, which is a two-layer protocol where a number of chains are formed in order to cover the whole region. COSEN reduce the latency associated with the original PEGASIS protocol by forming multiple smaller chains instead of a single longer chain. At first several small, fixed-length chains constructed using the same distance-based greedy heuristic of the original PEGASIS protocol and a chain leader is selected in each chain based on the highest remaining energy. Leader nodes are selected for a certain number of rounds. These chain leaders again create a higher-level chain and among those leaders, one is chosen depending on the distance to the sink and remaining energy level; to accumulate all data from the network and send that data to the sink at each round whereas other nodes only communicate with its neighbor in the chain.

Energy Efficient Chain Based Sensor Network (ECBSN) [9] overcomes several problems of PEGASIS. ECBSN overcomes the problem of excessive delay; instead of one long chain in the network, numbers of short chains are formed. In ECBSN, every first node in a chain becomes a leader node. Thus, it ignores the suitable proportion of nodes energy and distance between node and BS, which optimize the leader selection according to the various application environments. ECBSN protocol has certain deficiencies like the non-optimal selection of leader nodes in rounds, aggregation, and transmission of data by head nodes that leads to unbalanced energy consumption.

An improved ECBSN protocol called IECBSN is proposed in [10]. IECBSN adopts a new method of selection of leader nodes based on selection value (SV) parameter, which outperforms ECBSN.

EAPHRN (Energy Aware PEGASIS Based Hierarchical Routing) [11] is a chain based routing protocol in which node does not connect to the next closest node but connects to a random node that is not far than the Distance Threshold (DT). EAPHRN is divided into two phases, Chain Setup phase, and Leader election phase.

A Chain-Based Routing Protocol to Maximize the Lifetime of Wireless Sensor Networks (CHAIN-BASED) proposed in [12]. CHAIN-BASED is based on constructing multiple chains in the direction of the BS. The first node of each chain sends data to the closest node in the same chain. This latter collects aggregates and transmits data to the next closest node. This process repeats until reaching the last node, which aggregates and transmits data directly to the BS.

The work in [13] is similar to our work, the authors compare five hierarchal cluster-based routing protocols and introduce the mobility of BS in the five protocols and compare their performance. The main difference between our work and [4] is that our work provides analytical simulations in terms of network lifetime and average energy consumption for different chain based routing protocols.

Energy Efficient PEGASIS Based (EEPB) [14] is an enhanced PEGASIS algorithm in WSN. In PEGASIS distance algorithm is used to form the data chain and it can result in communication distance between two sensors being too long. Thus, the sensors die early as they consume a lot of energy for the transmission of data. In the chaining process, the average distance of the chain (known as “threshold distance”) is considered by a node. If the distance from the closest node and the upstream node is greater than thresh distance, the closest node is the “far node”. If the closest node joins the chain, it becomes “long chain”. EB-PEGASIS avoids this phenomena using distance threshold. It not only does the energy saving on threshold but also balances energy consumption of all sensor nodes.

Improved Energy Efficient PEGASIS Based (IEEPB) [15] protocol, overcomes the deficiencies of EEPB. When EEPB builds a chain, the threshold adopted is uncertain and complex to be determined and thus forms “long chain”. It also selects the leader and ignores the node energy and the distance between the BS and node that optimizes the selection of leader. Based on this, IEEPB compares the distance between two nodes twice and finds the shortest path to link two adjacent nodes. The chain construction is simplified such that formation of “long chain” is avoided. Also while selecting the leader, IEEPB considers the node’s energy, distance between the BS and the node, normalizes these two factors and assigns different weight co-efficient to them. Finally the node with the minimum weight becomes the leader. IEEPB has higher energy efficiency and hence longer network lifetime.

The objective of the algorithm proposed in [16] is to improve the PEGASIS protocol by proposing an algorithm named EPEGASIS (Enhanced PEGASIS). This algorithm belongs to the chain-based routing algorithms. In EPEGASIS, each node uses an optimal communication distance to choose the relay node among its neighbors for data transmission. In order to avoid excessive energy consumption for certain nodes with a particular location, Wang et al. [16] have defined a protection mechanism based on the average residual energy of its neighbors. Mobile BS technology is used to collect data for a balanced energy consumption in different regions.

4 The Proposed CBRP Protocol

The CBRP is chain based method for hierarchical network architecture. It is divided into three steps: the chains creation step, the inter-leader path construction step and data transmission step.

4.1 Chains Creation Step

In this step, the formation of the chains will be determined using a greedy algorithm, where the nodes communicate only with the nearest neighbor nodes (Fig. 3).

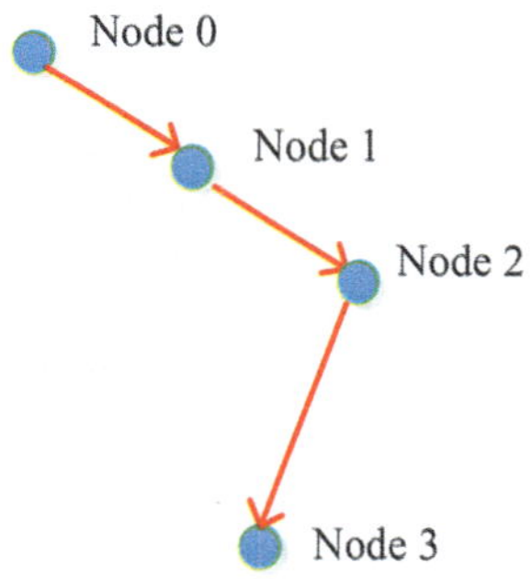

Fig. 3. Chain Construction using greedy Algortihm.

The main idea of CBRP is that nodes are formed into a several chains. Firstly, the BS broadcasts the whole network a hello message to obtain basic network information such as identifier, position, and energy level of all nodes. Secondly, the BS selects all nodes that have an energetic level greater than or equal to the residual energy average of entire network as leader node. Finally, each leader broadcasts the join_req, then the ordinary nodes join the nearest leader to form the chains based on Euclidean distance. The latter is used by ordinary nodes to calculate the distance between them and the leaders.

$$distance = \sqrt{(xi - xj)^2 + (yi - xj)^2} \quad (1)$$

Where x_i and y_i are the coordinates of ordinary node i and x_j and y_j are the coordinates of leader j.

In each chain, we run the following operations:

- the leader set the node which is farthest from it as end node, the latter joins the chain first and is named as node 1.
- Each node gets information of distance between itself and surrounding nodes which have not joined the chain, it finds the nearest node names it as *j*; where *j* represent the *j*-th node that has joined.
- Repeat the same operations to connect node $j + 1, j + 2$….and continue this till all sensor nodes have joined the chain.

4.2 Inter-chain Path Construction Step

In order to send collected data from leaders to BS, we choose inter-chain path according to the distance between the leader nodes, the distance between each leader node and the BS and residual energy of leaders as follows:

- Calculate distance between leader node and BS ($Dist_{leader\_bs}$) using the Euclidean distance.
- Determine residual energy of leader (RE_{leader}).
- Calculate weight of each leader by using the following objective function (OF):

$$\mathrm{OF} = w * RE_{leader}(1 - w) * Dist_{leader\_bs} \tag{2}$$

Where w is used to control the contribution of weight of RE_{leader} and $Dist_{leader\_bs}$

- Compare the different objective function of each leader and the leader with minimum value is selected as the leader of inter_chain path in this round (Fig. 4).

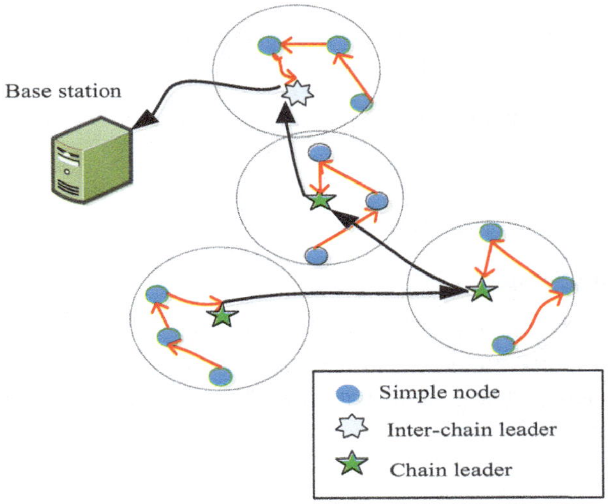

Fig. 4. Intra-chain and inter-chain communication.

4.3 Data Transmission Step

Once the chains creation step and inter-chain path construction step are accomplished, the data transmission step will take place as the last phase. In each chain formed, the transmission of data from ordinary nodes to its leader is carried out as follows:

- Each node transmits its collected data to the adjacent node.
- The adjacent node then fuses its data with the data of previous node and sends it further.
- Once the data reaches leader, it sends data of all nodes to next leader.
- One round of communication terminates when the BS receives all the data captured by all the whole network live nodes from last leader.

5 Simulation Results

5.1 Radio Model

In the simulation part, we use the same model of propagation of the radio signal introduced in [5] for the energy consumption, which represents a first-order radio model.

To transmit a packet of L bit, at a distance d the transmitter consumes:

The energy consumption for the transmission of a packet of L bit at a distance d is given by the following equation:

$$E_{TX}(L,d) = \begin{cases} L * E_{elec} + L * \varepsilon_{fs} * d^2, \cdots d < d_o \\ L * E_{elec} + L * \varepsilon_{mp} * d^4, \cdots d \geq d_o \end{cases} \tag{3}$$

Where

- E_{elec} is the amount of energy consumed to send or receive 1 bit.
- ε_{fs} and ε_{mp} d are the energy for signal amplification (Free space and multiple path).

The energy consumption for the reception a message of L bits is:

$$E_{TX}(L) = L * E_{elec} \tag{4}$$

5.2 Simulation Settings

We chose the MATLAB language as a simulation environment to test the performance of our proposed CBRP protocol, and we compared it with the PEGASIS and IEEPB protocols. The next Table 1 summarizes all the parameters taken to simulate our proposed CBRP protocol.

Table 1. Simulation parameters.

Parameters	Values
Area of simulation	$100 \times 100\ m^2$
Number of nodes	100
Coordinate of the BS	(50, 50)
Initial energy (E_0)	0.5 J/node
Transmitter Electronics (E_{TX})	50 nJ/bit
Receiver Electronics (E_{TX})	50 nJ/bit
Size of data packet	1000 bits
ε_{fs}	10 pJ/bit/m^2
ε_{mp}	0.0013 pJ/bit/m^4
Energy of Data Aggregation (E_{DA})	50 nJ/bit
Value of w	0.5

5.3 Results of Simulation

The results obtained are shown in following figures:

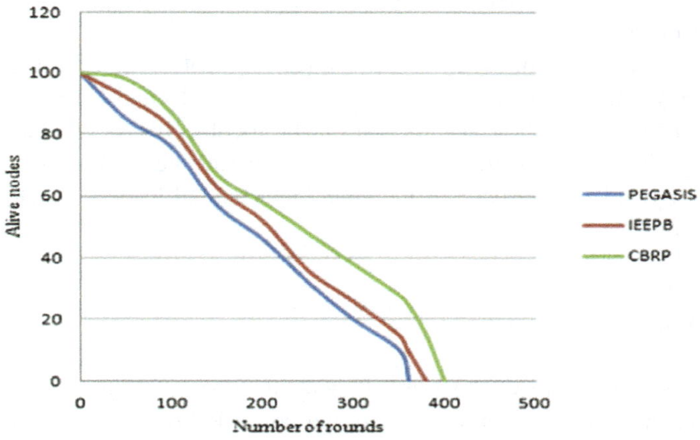

Fig. 5. Number of live nodes vs. number of rounds.

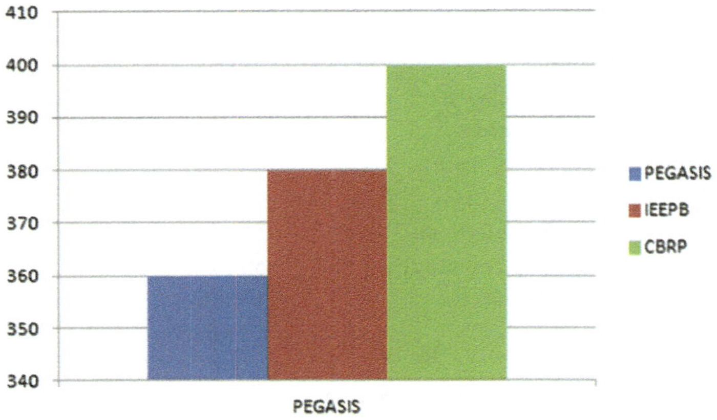

Fig. 6. Network lifetime.

5.4 Results Analyses

Figure 5 shows the number of live nodes as a function of the number of rounds in the three protocols: CBRP, PEGASIS and IEEPB. We can see that the number of live nodes is higher in each round for our CBRP protocol compared to the IEEPB and PEGASIS protocols.

Figure 6 shows network lifetime of this three protocols. It can be seen that CBRP has longer network lifetime as compared to both PEGASIS and IEEPB protocols.

We can deduce from these results that our protocol CBRP is efficient in terms of saving energy in the WSN, this is due to the gathering network nodes into several groups (chains) and the presence of the chain connects all leaders which reduces the transmissions to the BS.

6 Conclusion

In this paper, we offered a novel energy efficient chain based protocol in WSN. Results of simulation show the efficiency of our proposed CBRP protocol compared with the two others protocols based chain: PEGASIS and IEEPB protocols because in our CBRP protocol there are several chains instead of one, also there is a chain connects all the group in order to transmit the data to the BS. As a future work, we will compare our CBRP protocol with other protocols and we will test it in mobile networks.

References

1. Heinzelman, W., Chandrakasan, A., Balakrishnan, H.: Energy-efficient communication protocol for wireless microsensor networks. In: Proceedings of the 33rd Hawaii International Conference on System Sciences (HICSS 2000), January 2004
2. Lindsey, S., Raghavendra, C.S.: PEGASIS: power efficient gathering in sensor information systems. In: Proceedings of the IEEE Aerospace Conference, Big Sky, Montana, March 2002

3. Dhawan, H., Waraich, S.: A comparative study on LEACH routing protocol and its variants in wireless sensor networks: a survey. IJCA **95** (2014)
4. Manap, Z., Ali, B.M., Ng, C.K., Noordin, N.K., Sali, A.: A review on hierarchical routing protocols for wireless sensor networks. Wirel. Personal Commun. **72**(2), 1077–1104 (2014)
5. Öberg, K.: System design choices in smart autonomous networked irrigation systems. Master of Science thesis Stockholm, Sweden (2014)
6. Nehra, V., Sharma, A.K.: PEGASIS-E: power-efficient gathering in sensor information system extended. Global J. Comput. Sci. Technol. Netw. Web Secur. **13**(15) (2013)
7. Yu, Y., Yichang, S.: An energy-efficient chain-based routing protocol in wireless sensor network. In: Computer Application and System Modelling (ICCASM), vol. 11, pp. 486–489 (2010)
8. Abassum, N., Mamun, Q.E.K., Haque, A.A., AUrano, Y.: A chain oriented data collection protocol for energy-aware and delay-constrained WSN. African J. Inf. Commun. Technol. **2** (3), 11 (2006)
9. Mahajan, S., Malhotra, J.: Enhanced chain technique based data collection sensor network. Int. J. Comput. Sci. **8**(3), 83–87 (2011)
10. Mahajan, S., Malhotra, J., Sharma, S.: Improved enhanced chain based energy efficient wireless sensor network. Wirel. Sens. Netw. **5**, 84–89 (2013)
11. Alhasan, H., Qatawneh, M., Sleit, A., Almobaideen, W.: EAPHRN: energy-aware PEGASIS-based hierarchal routing protocol for wireless sensor networks. J. Am. Sci. **7**, 753–758 (2011)
12. Guyennet, H., Hadjila, M., Feham, M.: A chain-based routing protocol to maximize the lifetime of wireless sensor networks. Wirel. Sens. Netw. **5**(5), 116–120 (2013)
13. Javaid, N., Ain, Q., Khan, M.A., Javaid, A., Khan, Z.A., Qasim, U.: On sink mobility trajectory in clustering routing protocols in WSNs. J. Basic Appl. Sci. Res. (JBASR) (2013). arXiv:1304.0984 [cs.NI]
14. Devasena, A., Sowmya, B.: A study of power and energy efficient clustering protocols in wireless sensor networks. Int. J. Adv. Res. Comput. Sci. Manag. Stud. **1**(6) (2013)
15. Feng, S., Qi, B., Liangrui, T.: An improved energy-efficient PEGASIS-based protocol in wireless sensor networks. In: IEEE Eighth International Conference on Fuzzy Systems and Knowledge Discovery, pp. 2230–2233 (2011)
16. Wang, J., Yu, G., Yin, X., et al.: An enhanced PEGASIS algorithm with mobile sink support for wireless sensor networks. Wirel. Commun. Mobile Comput. **2018**, 9 (2018). https://doi.org/10.1155/2018/9472075. Article ID 9472075

Smart Grid, Energy Control and Management

Static Behavior of a PV/Wind Hybrid System Structure

F. Ferroudji[1,2(✉)], L. Saihi[1], and K. Roummani[1]

[1] Unité de Recherche en Energies Renouvelables en Milieu Saharien, URERMS, Centre de Développement des Energies Renouvelables, CDER, 01000 Adrar, Algeria
fferroudji@yahoo.fr

[2] Laboratoire de Mécanique des Structures et Matériaux, Université Batna 2, Batna, Algeria

Abstract. Wind and solar power play an important role in supplying cities with renewable energy. The main objective of this study is to access the structural strength performance of a photovoltaic (PV)/Wind hybrid system for applications in isolated area. Therefore, this study requires two steps. First part involves modelling of the three-dimensional (3D) assembly of PV-Wind hybrid structure using Computer Aided Design (CAD) software, namely, SOLIDWORKS (version 2016) and the second part involves in static response analysis subjected to the critical wind (Eurocode 1991-1-4 standard) and gravitational loads by using finite element (FE) method (SOLIDWORKS Simulation). The results from this analysis identify that the static strength and stability are qualified (no structural failure).

Keywords: PV/wind hybrid system · Static analysis · Finite element analysis · Renewable energy

1 Introduction

The renewable energy systems are stand-alone, provide electric current independently of electric distribution network. They supply the dwellings, the farms, the villages and the commercial applications like public telecommunications and the hertzian relays. The connection of the system to the electrical communication, enables us to decrease the invoice of electricity and the addition of a storage batteries system, can ensure autonomy of several days and provides auxiliary current during a storm or a power cut. The most common electric energy sources for these systems are small aero generators and photovoltaic solar panels (PV) which convert the solar radiation directly into electricity or indirectly the wind into electrical energy. PV/Wind hybrid energy systems are usually less costly and more reliable than the systems that use a single source of energy. These systems do not exhaust the worldwide resources limited out of fossil fuels, do not pollute and do not heat the atmosphere and do not produce dangerous waste [1–4].

In recent decades, Computer-Aided-Design (CAD) technology and engineering software packages (SolidWorks, Ansys, Abaqus, etc.) were used to solve efficiently

M. Hatti (Ed.): ICAIRES 2019, LNNS 102, pp. 411–416, 2020.
https://doi.org/10.1007/978-3-030-37207-1_43

various complex problems. Designing the required three-dimensional (3D) models along with application of numerical methods (finite elements (FE), finished volumes, etc.) is used for improving and accelerating the design process before the investment in prototyping with various shapes [5, 6]. PV/Wind hybrid system structures naturally are located and worked in outdoor environment. However, it must resist the forces of nature (winds, snow, sand and seismic) and gravitational load. Analyzing the external load acting on the structure is first and important for the static analysis. The structural static response is to guarantee the strength, stiffness and stability of the structure [7, 8].

A PV-wind hybrid system is investigated in this present study. Firstly, the three-dimensional (3D) full-scale model for the structure was realized using Computer Aided Design (CAD) software, namely, SolidWorks (version 2016). For second step, numerical static analysis of the structure by the finite element (FE) method (SolidWorks Simulation) is used to determine the von Mises stresses, deflections and factor of security caused by loads including the critical wind of 36 m/s (130 km/h) and gravity loadings.

2 PV-Wind Structure Description

The structure PV-wind hybrid system (as shown in Fig. 1a) is made up of three principal parts (sub-structures): (i) the photovoltaic part (Fig. 1b) consists of an electrical equipment box and two PV panels with a capacity of 100 W each (thirty six polycrystalline silicon solar cells). Typical dimensions of a 100 W PV panel are presented in Table 1. (ii) The wind part (Fig. 1c) is made up by a small horizontal axis wind turbine 500 W (model Whisper 100) with its rotor (2.1 m diameter) and fixed part constituting the circular tower of 1.5 m height and diameter of 70 mm. (iii) The structure support is a circular tower of diameter 100 mm and 7 m of foundation. The 3D commercial software SOLIDWORKS 2016 [9] was adopted for the modeling of the whole structural design of the PV-wind hybrid system (over 60 parts modeled).

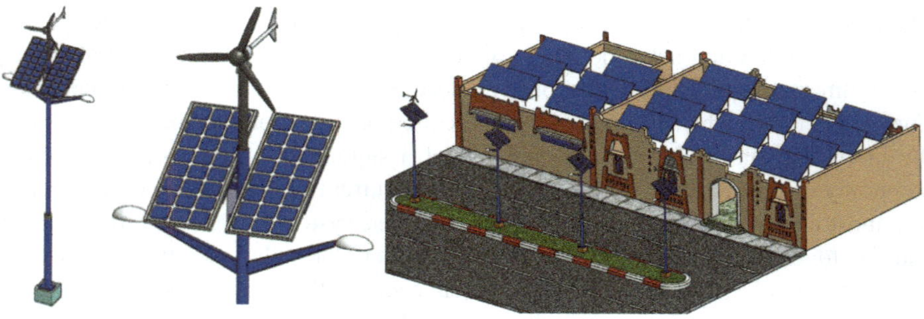
(a) (b)

Fig. 1. CAD model of a PV-Wind hybrid kit: (a) complete kit, (b) both parts PV and Wind turbine.

Table 1. Typical dimensions of a 100 W PV panel (STP080S-12/12Bb)

Parameter	Value
Power output	100 W
Length	1195 mm
Width	540 mm
Depth	30 mm
Weight	8 kg

3 Static Structural Analysis

The numerical simulations for static strength of the PV-W structure were established with FE analysis software SolidWorks Simulation. The FE analysis is used in the following five steps: (1) the modelling of the geometric model, (2) the definition of material properties, (3) the definition of boundary conditions, (4) Divided into grid, and (5) the results [10].

The PV-W structure designed in SOLIDWORKS can be directly used. There is no need to converse. 3D geometry and FE model for PV-W structure are shown in Fig. 1a and 1b. The mesh PV-Wind structure was generated by using solid 3-D tetrahedral elements. The structure is consisted of 125,557 nodes and 91,232 units for a total of 373,404 degree of freedom (DOF). The global size of the unite (54.632 to 273.16) mm. The simplified structure and meshed of the model are shown in Fig. 2. According to the working conditions, the definitions of material parameters of each component the PV-W structure are selecting the Table 2. These materials (plain carbon steel, galvanized steel and aluminium alloy) were selected from SOLIDWORKS Simulation library.

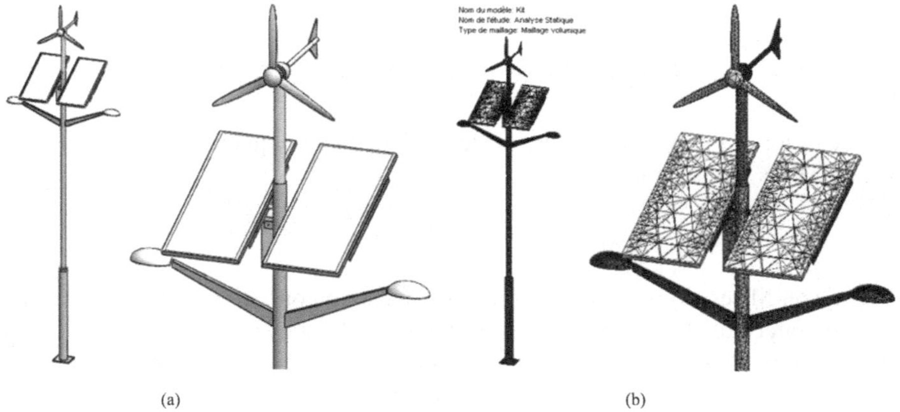

Fig. 2. Model of the simplified kit; (b) Meshed model of PV-Wind hybrid kit.

The objective of static analysis is to guarantee the force, stiffness and stability of the structure [10]. Analyzing the external load acting on the structure is first and important

Table 2. Material parameter.

Material	Plain Carbon Steel	Galvanized steel	Aluminium alloy
Elastic modulus [GPa]	210	200	69
Poisson ration [—]	0.28	0.29	0.33
Yield strength [MPa]	220.59	203.94	27.57
Density [kg.m^{-3}]	7800	7870	2700

for the static analysis. The loads of the PV-W hybrid system mainly include wind load, snow load and seismic load, besides its gravity. In this study, the gravity and wind load were mainly considered. This study presents the analysis of the structure's performance under maximum wind load critical (130 km/h) [11, 12].

$$q_p = \frac{1}{2}.\rho.v_p^2 \tag{1}$$

Where q_p the velocity is pressure, v_p (36 m/s) is the peak wind velocity $[m/s]$, and ρ is the air density $(1.25\mathrm{kg/m^3})$.

4 Resultants and Discussions

In this study was used for simulate the static comportment of PV-wind hybrid system, Von Mises stress criterion. SOLIDWORKS Simulation software found all stresses components σ_X, σ_Y, σ_Z in X, Y and Z direction on each point and based on rupture theory of ductile materials. The results are conforming to Von Mises-Hencky theory equation [13]:

$$\sigma_{\mathrm{Von}} = \left\{\left[(\sigma_X - \sigma_Y)^2 + (\sigma_Y - \sigma_Z)^2 + (\sigma_X - \sigma_Z)^2\right]/2\right\}^{1/2} \tag{2}$$

The factor of safety at a location is calculated by below equation using the values of the von Mises stress limit [14]:

$$S_F = \sigma_{limit}/\sigma_{vonMises} \tag{3}$$

The von Mises stress contour, the displacement contour, deformation contour and factor of safety of the static analysis are respectively shown in Figs. 3, 4, 5 and 6. Figure 3 show that the maximum stress is occurred in the contact area between the circular mast and the tube of the support of PV panels, whose value is 34.6 MPa. This value is lower than the yield elastic limit acceptable for material used. As shown in Fig. 4, the maximum displacement of the PV-W structure is on the level of the free end of PV panel, whose size is 0.909 mm representing approximately 0.076% of length of the PV panel. This value is very weak compared to the structure size. The strain contour of the PV-W structure is seen in Fig. 5 were the maximum occurres in the contact area between the circular mast and the tube of the support of PV panels. Its value is equal to 1.06×10^{-4}, which respect the elastic hypothesis ($\varepsilon \ll 1$). Figure 5 show that the minimum value of factor of safety is 6.02. It occurres in the contact area between the circular mast and the tube of the support of PV panels. This value indicates that the PV-W structure is not overloaded and that the static design is sure since the factor value is larger than 1.

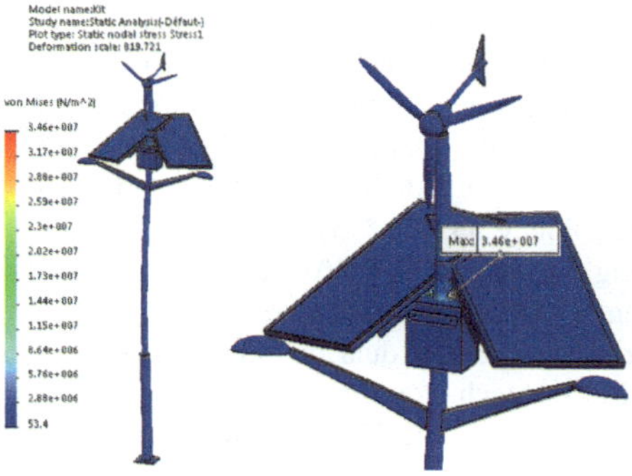

Fig. 3. von Mises stress contour

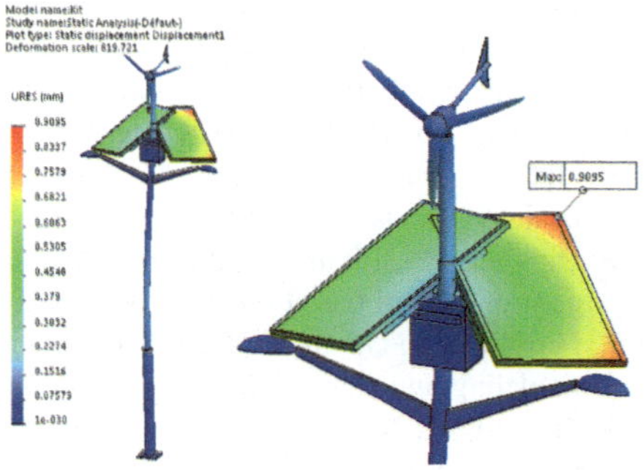

Fig. 4. Displacement contour

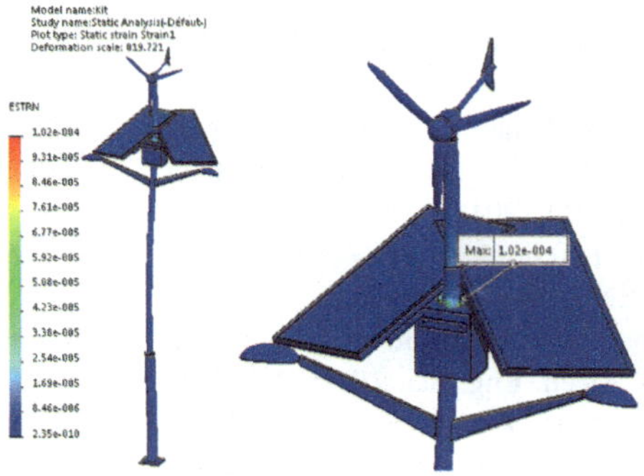

Fig. 5. Deformation contour

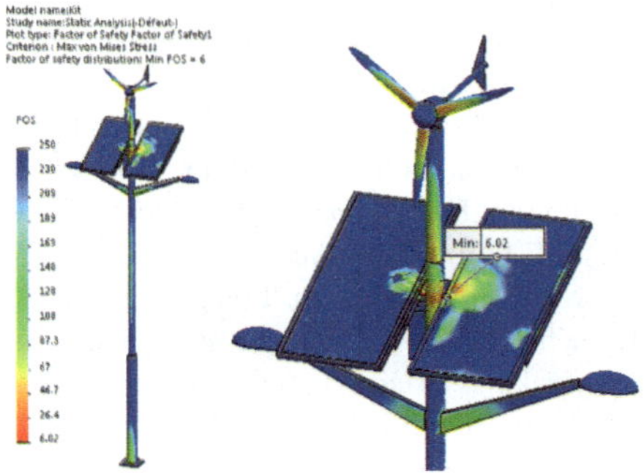

Fig. 6. Factor of safety contour

5 Conclusion

In this study, the structural static response of the PV-Wind hybrid kit was investigated using FE analysis subjected to its gravity and the critical wind loadings (Eurocode 1991-1-4 standard). SOLIDWORKS software was used in order to creation of a 3D model for the PV-W structure (all components and all assemblies). Static analysis responses was obtained by using SOLIDWORKS Simulation, in order to identify stiffness, resistance and reliability of the PV-W structure. Based on static analysis results, it can be concluded that the structure of the PV-W will not fail because the maximum stresse for the more fragile material of the structure is lower than the yield strength of the material (with factor of safety is higher than 6) and related maximum displacement meet material deflection. Future work and investigation will base on the evaluation of structural dynamic and fatigue analyses.

References

1. Ekren, O., Ekren, B.Y.: Size optimization of a PV/wind hybrid energy conversion system with battery storage using simulated annealing. Appl. Energy **87**, 592–598 (2010)
2. Khelifi, C., Ferroudji, F., Ouali, M.: Analytical modeling and optimization of a solar chimney power plant. Int. J. Eng. Res. Africa **25**, 78–88 (2016)
3. Diaf, S., Diaf, D., Belhamel, M., Haddadi, M., Louche, A.: Methodology for optimal sizing of autonomous hybrid PV/wind system. Energy Policy **35**, 5708–5718 (2007)
4. Bingbing, L., Lijun, Z., Chunzhe, L., Hongbing, N.: Design of PV/Wind hybrid generation control system based on PLC. Mat. Sci. Forum **770**, 348–351 (2014)
5. Ferroudji, F., Outtas, T., Khelifi, C., Mansouri, R.: Large-scale dual axis sun tracking system modeling and static analysis by FEM. Int. J. Mech. Mechatron. Eng. **14**(04), 92–97 (2014)
6. Stefano, B., Naoufel, W., Albertodel, B., Pietro, P.: Matching 3D face scans using interest points and local histogram descriptors. J. Comput. Graph. **37**, 509–525 (2013)
7. Petrova, R.V.: Introduction to Static Analysis Using SolidWorks Simulation. CRC Press, Boca Raton (2014)
8. Wang, S.Z.: Structural Mechanics and Finite-Element Method. Press of Haerbin Industry University, Harbin (2003)
9. SolidWorks: SolidWorks Corporation, 300 Baker Avenue, Concord, MA 01742 (2016) http://www.solidworks.com/
10. Ferroudji, F., Outtas, T., Khelifi, C.: Structural dynamics analysis of three-dimensional biaxial sun-tracking system structure determined by numerical modal analysis. J. Solar Energy Eng. Trans. ASME **140**(3), 031004–031011 (2018)
11. EN: Eurocode 1: Actions on Structures, General Actions—Part 1–4: Wind Actions, British Standards Institution, London, Standard No. EN 1991-1-4 (2004)
12. Khelifi, C., Ferroudji, F.: Stress and fatigue analyses under wind loading of the dual axis sun tracking system via finite element analysis. J. Mech. Eng. Sci. **10**(2), 2008–2015 (2016)
13. Guan, X., Lili, S., Xing, W., Yongjun, T., Yiguang, Z.: Research on the model and finite element analysis of vehicle chassis detection system based on COSMOSWorks, In: Proceedings IEEE Symposium Computer Society (ICCCS 09), pp. 127–130 (2009)
14. Ferroudji, F.: Conception et optimisation en dynamique forcée d'un nouveau prototype de système de suiveur solaire à deux axes, PhD Thesis, University of Baatna 2, Algeria (2015)

Tasks Scheduling and Consistency Management in Mono-Masters Grid Environment

M. Meddeber(✉) and H. Hamadouche

Faculty of Exact Science, University of Mascara, Mascara, Algeria
m.meddeber@yahoo.fr

Abstract. Grid computing consists of geographically distributed computing and storage resources that are used in large scale scientific applications. Job scheduling and data replication are two well-known techniques to boost the performance of Data Grid. There has been extensive research on integrating both techniques to further improve performance in Data Grid. However, most of the current works are heuristic based without performance guarantees.

Our goals are: (a) Integrating data replication and job scheduling into one framework to minimize the total job execution time in the Grid, (b) Achieving good management of the replica consistency.

Keywords: Grid computing · Scheduling tasks · Data replication · Consistency management

1 Introduction

The Grid concept was firstly introduced by Foster and Kesselman who defined it as a distributed and decentralised computer environment composed of large number of heterogeneous resources which are managed by different owners such as companies, universities or another business or scientific organisations. Such resources may be of different type such as computational machines, data storage nodes, databases or scientific tools, etc. These resources are interconnected by high speed communication network. Most of these resources are built from relatively cheap components like regular PCs. Also other common hardware or existing infrastructure such as Internet is usually used. It is important to notice that the Grid environment is changing dynamically in time mainly due to load of the system, available network bandwidth and also due to other events such as resource or network failure or recovery [1].

Maintaining data consistency is one of the most important tasks in data design and management. In a centralized system this task is more or less easily accomplished because the data is located in a specific location, and updates are performed locally without competition, which facilitates data management. But in large-scale distributed systems such as grid computing, consistency management becomes difficult because the data is distributed over geographically remote clusters, the data is fragmented on these clusters, in addition some of them can be replicated on several clusters. The clusters are interconnected to each other forming an interconnection network and

M. Hatti (Ed.): ICAIRES 2019, LNNS 102, pp. 417–424, 2020.
https://doi.org/10.1007/978-3-030-37207-1_44

communicate with each other via messages, whatever the chosen architecture, updates are carried out by remote requests, and read/write requests are concurrent, so the time to propagation of updates, failure resistance and concurrence management are taken into account consideration, it is therefore necessary to find good ways to maintain data consistency during their updates.

This work consists in defining the scheduling of dependent tasks assignment and data consistency management for grid computing. We consider that the grid is a set of mono-master clusters. Our goals are: (a) Integrating data replication and job scheduling into one framework to minimize the total job execution time in the Grid, (b) Achieving good management of the replica consistency.

The rest of this paper is organized as follows. I. Theoretical concepts are illustrated in Sect. 2. The proposed strategy and some simulation results are explained in Sect. 3. Finally, Sect. 4 concludes the paper and gives some perspectives.

2 Theoretical Concepts

2.1 Grid Computing

Ian Foster suggests a grid checklist to identify a real grid system. The suggestion can be concluded into three points [2]:

1. A grid integrates and coordinates resources and users that live within different control domains. Current internet technologies address communication and information exchange among computers but do not provide integrated approaches to the coordinated use of resources at multiple sites for computation.
2. A grid is built from multi-purpose protocols and interfaces that address such fundamental issues as authentication, authorization, resource discovery, and resource access. In a large-scale grid environment, each resource is integrated from multiple institutions, each with their own policies and mechanisms. Thus it is important that these protocols and interfaces should be standard and open. Otherwise, we are dealing with an application specific system.
3. A grid allows its constituent resources to be used in a coordinated fashion to deliver various qualities of service, relating for example to response time, throughput, availability, and security, and/or co-allocation of multiple resource types to meet complex user demands, so that the utility of the combined system is significantly greater than that of the sum of its parts.

2.2 Tasks Scheduling

Scheduling is a process of allocating tasks onto available resources in time. Such process has to respect constraints given by the tasks and the Grid. Usually one or more optimization criterion is used to make scheduling decisions.

First we have to distinguish between local and global scheduling. Local scheduling is used to assign the task to the CPU(s) of the single machine. Global scheduling is the problem of deciding where to execute the job selection of a suitable machine(s) is the key issue in this case.

We can also distinguish between static and dynamic scheduling. In the static case we assume that all information about jobs, resources, Grid topology and other constraints are known before the start of the scheduling process and stay same in time. Such situation is not realistic in the Grid environment which is highly dynamic. Jobs and resources are appearing and disappearing in time, errors often occurs, only uncertain and imprecise information about the topology, tasks or resources is available. If the scheduling process has to deal with such dynamic behavior of the system we are talking about dynamic scheduling [3].

2.3 Data Replication

Data replication is introduced into data grids as a method for optimizing data access [4]. Data replicas can be considered as a cache of data. Identical copies (replica) of data are created and distributed to various storage resource sites. Users or applications can access the nearest replica instead of looking for the original data and transferring them to where they are needed; therefore, the time dependence on data access latency is reduced. The responsibilities of a data replication management service (RMS) are: (a) Create a replica for an entire dataset or part of a dataset, (b) Manage data replicas such as add, delete and modify the replica files, (c) Register a new replica into RMS, (d) Catalog the registered replicas so that users can query and access them, (e) Select the optimal replica according to the requirements of users or applications to best adapt their execution, (f) Assure consistency among replicas coming from the same dataset, automatically updating replicas once the original dataset is changed.

2.4 Consistency Management

In the case where a file is modified, these changes must be propagated to all existing replicas. Since a general update propagation scheme requires expensive distributed locking mechanisms in order to ensure that only one of the replicas is modified at any given time, and that no access occurs until the modification is propagated to all the replicas, some of the following restrictions might be imposed on replica modifications [5]:

1. Only a master copy is allowed to be modified. Before accessing a secondary copy an update with the master can occur thus removing the need for distributed locking. Of course, distributed locking is required among multiple masters.
2. In versioning systems, a modification of a replica results in a new version. Version information is stored in the metadata catalogue. A read access might request a certain version or, as default value, the most recent version. In the latter case, the consistency service needs to determine the most recent version, which is then provided. Depending on the VO policies, new versions might automatically be installed at all places where replicas of the old version exist. Alternatively, new versions might only be installed upon request. Efficient versioning can be achieved by specifying master copies as the only replicas from which versioning can be done.

3 Proposed Strategy

In this section we describe our strategy that we applied for solving the Grid scheduling data consistency management problem. We discuss the most important characteristics of our solution and then we define the goals of our research. Finally a description of the Grid scheduling simulator we have developed to evaluate our techniques and some results are given.

3.1 Presentation

In our study we model a Grid as a collection of n clusters with different computational facilities. We assume that in the grid there is a central scheduler, to whom every Cluster Manager (CM) connects and the grid clients send their dependent tasks to the central scheduler. The central scheduler is responsible for scheduling tasks among CMs. A cluster is a set of worker nodes that includes one computing element and one storage element [6]. Computing element executes tasks and Storage element contains data files and/or their replicas (see Fig. 1).

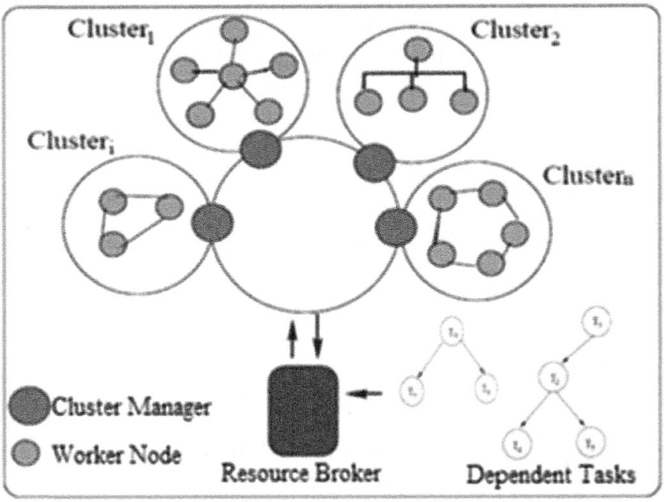

Fig. 1. Grid model

A cluster Manager is responsible for scheduling tasks among Worker Nodes. A cluster Manager is also responsible for load balancing in his cluster and between grid clusters.

Data must be managed effectively and efficiently. Multiple copies of data are stored at different geographical locations, making access to data faster and more reliable for all users. Efficient network performance is required to schedule and dispatch the number of jobs required by the end user.

We assume that at each cluster we find a Master. The master is responsible for executing writing tasks (a writing task can modify the replica contents) and keeping the set of replicas of a file consistent.

Reading tasks can be executed in any worker node in the cluster.

If a Replica is modified, the master informs the cluster manager to inform all clusters manager of the grid.

The following sequence diagram (Fig. 2) highlight the exchange of messages between actors in a chronological way, as the evolution of time reads from top to bottom.

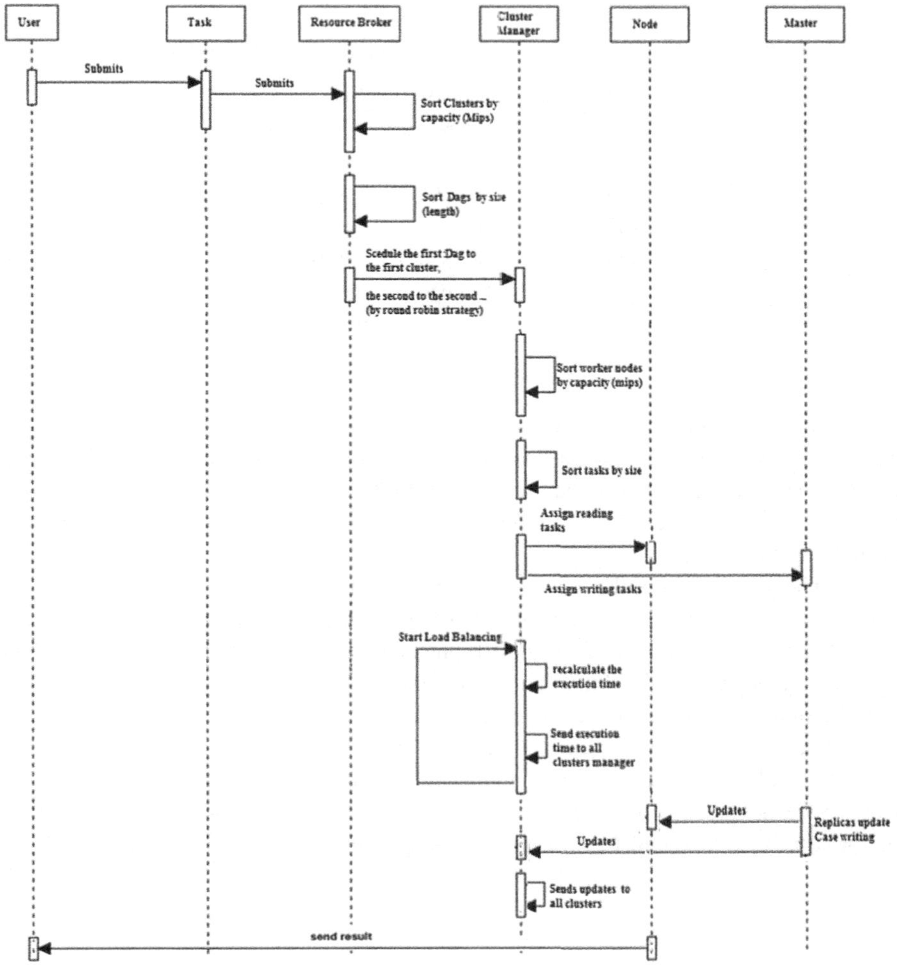

Fig. 2. Sequence diagram showing interaction between actors.

3.2 Implementation and Results Simulation

We developed our application on a machine running under the Windows system (Windows 7), with the following features: (a) Core i3 CPU 2.67 Ghz. (b) Memory capacity 8 GB of RAM.

We implement our approach in our simulator created in java netbeans and have realized some experimentations concerning response time and number of conflict.

(1) In this first series of experiments, we measured the response time. Performance comparison in a Grid with 3 sites, 25 machines, number of data files is 10, the number of tasks change between (100–600).

Simulation results were performed with the parameters described in the Table 1.

Table 1. Comparative response time between MonoMaster & MultiMaster environment

Nbr of tasks	Response time/sec	
	MultiMaster	MonoMaster
100	26,586	25,586
200	53,69	53,69
300	40,877	40,877
400	26,355	93,882
500	21,892	91,563
600	102,864	194,238

Compared to multimaster environment [7], the results of monomaster environment are lower. The average response time of tasks increases significantly when increasing the tasks number. This is due because of the tasks dependencies; a writing task must wait to be executed by a master and therefore all dependent tasks must wait even if they are reading tasks.

(2) The count of conflicts is a very important metric in the evaluation of our strategy; it allows us to evaluate the effectiveness of the adopted consistency management approach. Number of conflicts represents the number of non-updated replicas.

Performance comparison in a Grid with 3 sites, 25 machines, and number of tasks is 100, the number of data files change between (5 to 25).

Simulation results are given in Fig. 3.

Compared to multimaster environment, the results of monomaster environment are better since it reduces considerably the number of conflicts. This is due because in a monomaster environment, the consistency is guaranteed inside the cluster.

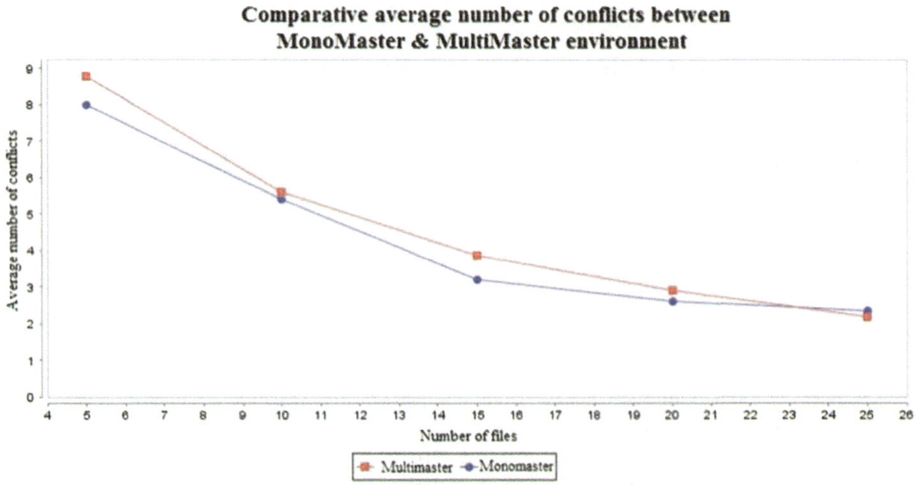

Fig. 3. Comparative average number of conflicts between MonoMaster & MultiMaster environment

4 Conclusion

Grid computing is an emerging technology that presents new challenges. The most important of these are the management of the large-scale aspect, the heterogeneity and the dynamic nature of the sites.

The research community interested in this technology addresses standard problems of distributed systems such as task scheduling, load balancing, file replication, and data consistency, trying to propose solutions that can take into account, as much as possible, the intrinsic characteristics of grids.

The replica transfer scheduling technique is exclusively proposed for minimizing the cost of transferring the data from one computing site to another and Consistency service performs.

This work consists in defining the scheduling of dependent tasks assignment and data consistency management for grid computing. We consider that the grid is a set of mono-master clusters. Our goals are: (a) Integrating data replication and job scheduling into one framework to minimize the total job execution time in the Grid, (b) Achieving good management of the replica consistency.

The software product produced for our module (simulator) meets the main functionalities of the initially defined specification phase. The experimental results obtained are very satisfactory and meet our expectations in a MonoMaster environment.

For a continuation of our work, several perspectives can be considered:

(a). Simulation model of the real Grid environment allowing: Network simulation, Resource failure and recovery, Specific machine parameters (RAM, HDD, architecture specific features).
(b). Preparing collection of the real workload traces including: Job descriptions, Grid environment descriptions, Evaluation of proposed techniques using collected data sets.

References

1. Klusacek, D.: Scheduling in Grid Environment. *PhD thesis, Masaryk University, Brno*, 2008)
2. Magoulès, F.: Fundamentals of Grid Computing: Theory, Algorithms and Technologies. CRC Press, Chapman and Hall book, Boca Raton (2010). ISBN 978-1-4398-0367-7
3. Casavant, T., Kuhl, J.: A taxonomy of scheduling in general-purpose distributed computing systems. IEEE Trans. Software Eng. **14**(2), 141–154 (1988)
4. Dllmann, D., Hoschek, W., Martinez, J., Segal, B.: Models for replica synchronisation and consistency in a data grid (2001). https://fr.calameo.com/read/0003272658b9de2da6964. Accessed Jun 2019
5. Guy, L., Kunszt, P., Laure, E., Stockinger, H., Stockinger k.: Replica Management in Data Grids, presented at Global Grid Forum 5
6. Meddeber, M., Yagoubi, B.: Tasks assignment for grid computing. Int. J. Web Grid Serv. **7** (4), 427–443 (2011)
7. Meddeber, M., Yagoubi, B.: Dependent tasks assignment and data consistency management for grid computing, Accepted in the International Journal of Multiagent and Grid Systems (2019)

Feasibility Analysis of a Solar PV Grid-Connected System Using PVsyt Software Tools

T. Touahri(✉), S. Laribi, R. Maouedj, and T. Ghaitaoui

Unité de Recherche en Energies Renouvelables en Milieu Saharien, URERMS Centre de Développement des Energies Renouvelables, CDER, 01000 Adrar, Algeria
touahri01@gmail.com

Abstract. The Renewable energy sources have become an effective and practical option for integration with fossil fuel power plants, as well as reducing the use of traditional fuels that pollute the environment. This paper presents a exploration of smart grid-connected photovoltaic installations in the southwest of Algeria. This power plant is located in the area of Kabertene Adrar with a capacity of 3 MW and has been under operation since 2015.

The analysis and design of the system is based on the simulation with PVsyst software. The design of 3 MW grid-connected PV system consists of 12244 PV panels of 245 W each and 6 inverters of 500 kW each. The results of the evaluation show that investment in photovoltaic technology is very promising at this site, as the amount of power injected into the grid during the year is 5980 MWh

Keywords: Renewable energy · PV system · Grid-connected · Design · Power

1 Introduction

Energy efficiency will play an important role in the national energy context, characterized by strong growth in consumption driven in particular by the domestic sector with the construction of new housing, the construction of public utility infrastructure and the revival of the industry [1]. Algeria is committed to the path of renewable energies to provide global and sustainable solutions to environmental challenges and the problems of conservation of fossil energy resources by launching an ambitious program for the development of renewable energies. Renewable Energy adopted by the Government in February 2011 and revised in May 2015 [2].

The national renewable energy park currently consists of 24 plants with an aggregate capacity of 354.3 megawatts (MW). This renewable energy park consists of 23 photovoltaic (PV) plants with an overall capacity of 344.1 MW and one for wind energy of 10.2 MW [3].

The Grid connected photovoltaic system is one of the photovoltaic technologies where electricity generated by solar panels is connected to the grid. This system consists of a number of solar panels and one or several units of inverters, power

M. Hatti (Ed.): ICAIRES 2019, LNNS 102, pp. 425–433, 2020.
https://doi.org/10.1007/978-3-030-37207-1_45

conditioning unit and the equipment of grid connection [4]. PVsyst is a simulation program for PV systems, the PVsyst software was developed by the University of Geneva. The software is used for feasibility studies, sizing and simulation of photovoltaic systems. Where the user chooses the location, loads and database of the various components and then the program calculates the size of the system [5].

In this work, present analyzed and evaluation of the performance of the solar power station connected to the network with a capacity of 3 MW In the area Kabertane of Adrar state. The evaluation is based on design of grid connected PV system which employing the PVsyst simulation software to carry out the technical conception and performance analysis.

2 Power Plant Details

2.1 Location Information

The photovoltaic station is located in the Kabertene in the north-east of Adrar state in latitude 28°27′04.8″N, longitude 0°02′49.8″W, the annual ambient air temperature is varies between 14.66 °C and 39.09 °C. The Satellite view of the PV plant is shown in Fig. 1 and the PV plant system of Kabertene is shown in Fig. 2.

Fig. 1. Satellite view of the PV plant Kabertene.

2.2 System Description

The grid-connected solar PV power systems comprise of:

- PV modules, connected in series and parallel, depending on the size of the photovoltaic solar array, to generate direct current power from solar intercepting the sun [6].
- Maximum power point tracking (MPPT), ensuring that photovoltaic solar modules generate DC power at the best power output at any given time during the sunshine hours [7].

Fig. 2. Satellite view of the PV plant Kabertene.

- DC/AC adapter connected to the network, ensuring that AC power that has been safely converted and converted to the grid of facilities is fed whenever the grid is available [8].
- The equipment of grid connection such as DC/AC breakers fuses, etc.

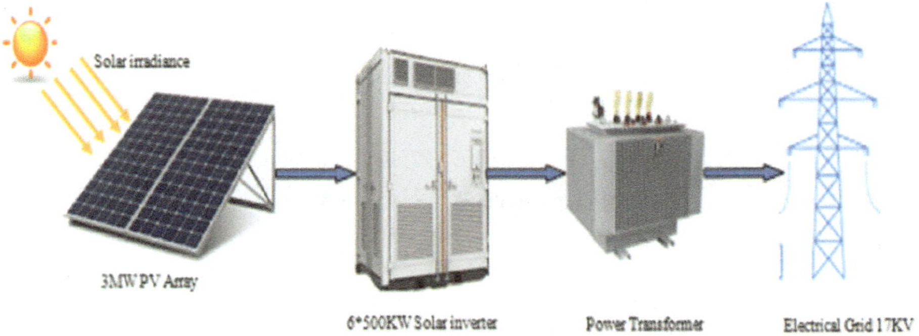

Fig. 3. Schematic of the grid connected PV system.

The Fig. 3 shows the components of the grid connected solar PV plant with her respective specification.

The photovoltaic plant of the Kabertene consists of 6 independent segments each one consisting of 2040 solar panels type poly crystalline silicon YL245P-29b. Each unit has 60 solar cells. Table 1 shows the important characteristics of the solar panel YL245P-29b. All panels solar are free from any shades effect and are fixed with a tilt angle of 30°, facing south at an azimuth angle of 0°.

Table 1. PV module specifications.

PV modules	Specification
Type	Silicon Polycrystalline
Maximum power (Pmax)	245 W
Peak Efficiency	15%
Maximum power voltage (Vpm)	30.2 V
Maximum power current (Ipm)	8.11 A
Open circuit voltage (Voc)	37.8 V
Short circuit current (Isc)	8.63 A
Nominal operating cell temperature (NOCT)	46 °C
Power Tolerances	0%/+2%
Number of Cells	60
Maximum System Voltage	1000 V
Module area	1.63 m^2
Weight	19.1 kg

Table 2. Sungrow SG500MX inverter specifications.

PV modules	Specification
Model	SUNGROW
Type	SG500MX
Input (DC)	
Isc	1344 A
Maximum PV input current	1120 A
Maximum DC voltage	1000 V
Voltage (VMPPmin)	500 V
Voltage (VMPPmax)	850 V
Output (AC)	
Nominal AC power	500 KW
Nominal output voltage	3–315 V
Output frequency	50 Hz
Maximum output current	1008 A
Power factor	−0.9–0.9
Operating temperature	−30 °C to + 55 °C

The panels are connected between them in series to form strings; each string consists of 22 panels. The strings are connected in parallel at the array junction box. Each 92 strings is connected with an inverter SUNGROW model SG500MX. The inverter has an efficiency of 95.6%. The specifications are given in Table 2.

3 Methodology

3.1 PVsyst Software

PVsyst is the most developed instrument used in the modeling and analysis of photovoltaic solar systems for differents applications (grid-connected, standalone, water pumping). These tools allow the user to optimize most of the parameters required for the solar PV plant, and also allow the user to use shadow effects as well. A huge database is installed with the tool related to the PV system components that are currently available on the market. It is also possible to perform a detailed economic analysis [9].

PVsyst is a simulation and design software. It is designed to provide easy-to-use ways to develop projects. PVsyst contains a large base of meteorological data for several locations in all countries of the world. You can also manually add measured data to sites that are not in the software. The results are presented in the form of a complete report that includes graphs and tables [10]. This software tool consists of photovoltaic modules, charge controller, inverters, and grid interface [11].

3.2 The Simulation Steps of PVsyst

In this project, the simulation is used PVsyst to analysis- the system and evaluate the performance of the grid connected PV system. The Simulation steps of PVsyst are as follows [12]:

- Initially you must create a project in the project design list.
- During the creation of the project, the geographical location and meteorological data are determined for the specific location.
- The appropriate data relating to the factory design is determined. This includes the direction of the photovoltaic modules, the area available to install the photovoltaic plant or the energy required by load, the type of photovoltaic modules and the type of inverter to be used that constitutes the PVsyst database [9].
- After entering the necessary data, the simulation is performed by adding successive variables.

4 Result and Discussion

In this section we analyzed the simulation results of a proposed photovoltaic system from the simulation model of 3 MW PV system plants by PVsyst software tools, the results obtained are mainly produced energy, specific production, performance ratio and arrow losses were obtained. The obtained results are analyzed for assessing the performance grid-connected solar power station.

The balances and main results of Grid connected PV system are shown in Table 3. System balances consists of the horizontal global irradiation, horizontal diffuse irradiation, ambient temperature, global incident in collector plane, effective global collector for IAM and shadings, effective energy at the output of the array, the energy injected into the grid, and performance ratio.

Table 3. Balances and main results of 3 MW grid connected PV system.

	Globhor kWh/m²	DiffHor kWh/m²	T Amb °C	GlobInc kWh/m²	GlobEff kWh/m²	E Array kWh	E Grid kWh	PR
January	141.4	26.20	14.66	204.6	200.3	549132	533917	0.864
February	147.8	35.86	17.63	190.1	186.1	499959	485565	0.846
March	197.5	47.47	22.88	224.6	219.4	569226	552922	0.815
April	211.0	62.37	26.29	212.7	206.5	531643	516642	0.805
May	229.3	73.50	32.25	210.2	203.3	511757	496848	0.783
June	232.5	67.08	36.04	204.1	197.3	485922	471003	0.764
July	237.3	68.23	39.06	212.1	205.2	495414	480725	0.751
August	217.8	73.69	38.09	211.3	205.0	497012	482848	0.757
September	187.9	62.45	34.07	202.9	197.5	488107	474148	0.774
October	163.5	55.02	28.95	198.4	193.8	493798	479559	0.801
November	141.7	30.31	20.41	197.7	193.5	514338	500115	0.838
December	129.5	27.93	15.88	193.6	189.3	519626	505343	0.865
Year	2237.1	630.11	27.24	2462.4	2397.1	6155933	5979636	0.805

Globhor global horizontal irradiation
GlobEff Effective Global
E Aval Availabie Solar Enregy
EUnused Unused Enregy
E Miss Missing Enregy
E user Enregy supplied to the user
E Load Earegy need of the user
SolFrac Solar Fraction (E user/ E Load)

Figure 4 shows the simulation of normalized productions such as collection losses ($Lc = 1.16$ KWh), system losses ($Ls = 0.16$ KWh) and produced useful energy ($Yf = 5.43$ KWh) per installed kWp/day.

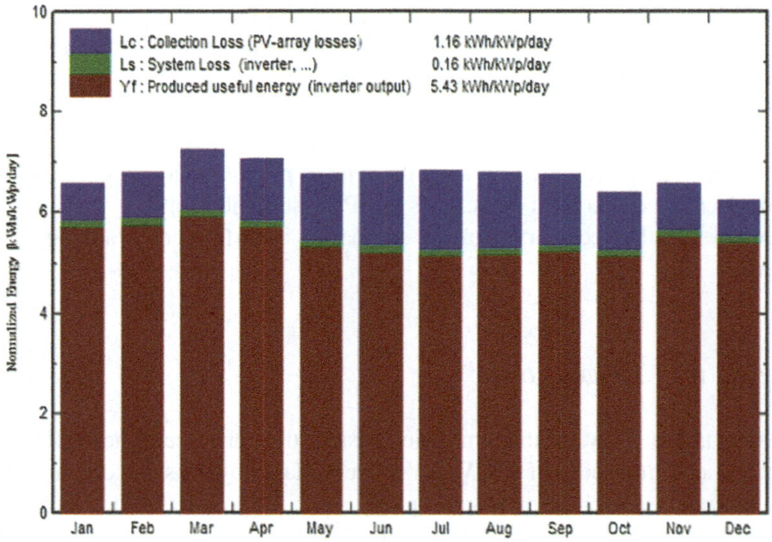

Fig. 4. Normalized energy productions of the grid-connected PV system

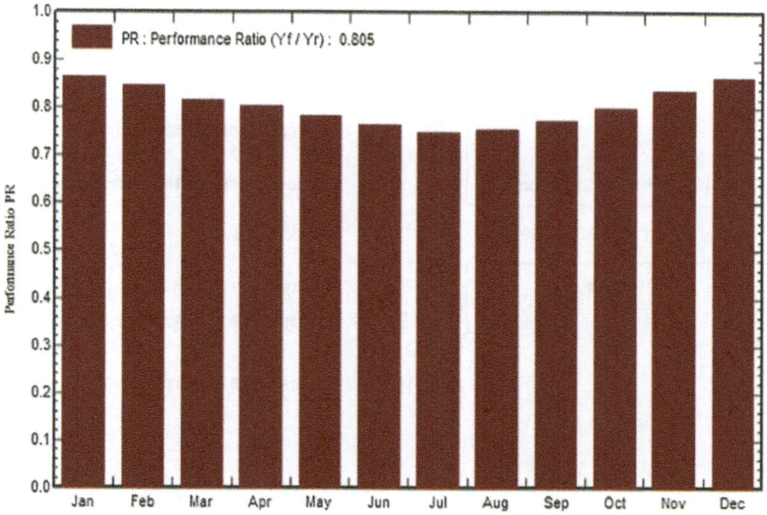

Fig. 5. Monthly performance ratio PR over the monitored period.

The energy that is injected into the grid is not the same energy produced by PV array, Because there is a loss of energy produced from convert DC energy produced into AC energy to feed the grid, the amount of energy injects 5979636 kWh into grid annually. The PV plant injected more energy to the grid in the month of March 552922 kWh. The lowest amount of energy that is injected into grid is 471003 kWh which is in the month of June.

The performance ratio usually ranges from 60% to 80% [13]. The annual average performance ratio (PR) is 0.805 as shown in Fig. 5, in December; the highest performance was 86.5%, while the lowest performance was 75.1% in July.

Figure 6 represents detailed of the loss diagram over the whole year in the proposed grid connected PV system. The horizontal global irradiance is 2237 KW/m^2 and the effective irradiance on collectors is 2397 KWh/m^2. The largest losses is the array losses around 15.9% and the inverter loss reach 2.8%. The amount energy injected into grid is 5980 MWh where the efficiency of the system is 15.4%.

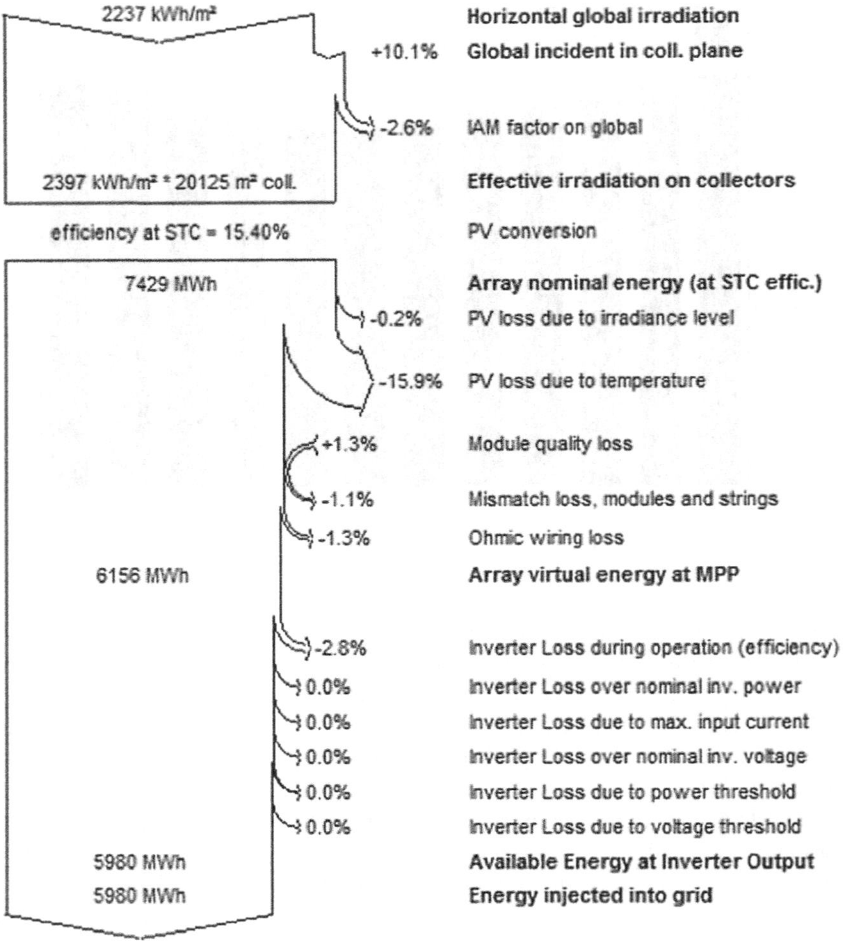

Fig. 6. Loss diagram of the system over the monitored period.

5 Conclusion

This study presents a feasibility analysis of a solar 3 MW PV grid-connected system using PVsyt software tools in Kabertene (Adrar), Algeria. From the simulation analysis the following results are obtained:

- The energy that injected into the grid is 5980 MWh/year.
- The maximum energy injected into the grid is 552,9 MWh in the month of March, and the least energy is 471 MWh in the month of June.
- The average performance ratio (PR) of grid connected PV system is 80.50%.

Through these results we find that the area of Kabertene has great potential for the use of solar energy, which contributes to the national plan to intensify renewable energy and reduce the dependence on fossil fuels, which causes the emission of greenhouse gases and global warming.

References

1. Ministry of Energy National Energy Efficiency Program. http://www.energy.gov.dz/francais/uploads/2016/Programme-National/Programme-National-Efficacite-Energetique.pdf
2. Ministry of Enregy National New and Renewable Energies Program. http://www.energy.gov.dz/francais/uploads/2016/Programme-National/Programme-National-Energies-Nouvelles-Renouvelables.pdf
3. Renewable energies Project 343 MWp, June 2019. http://www.sktm.dz/?page=article&id=56
4. Normazlina, M.I., Chee Wei, T., Yatim, A.H.M.: A techno-economic assessment of grid connected photovoltaic system for hospital building in Malaysia. In: IOP Conference Series: Materials Science and Engineering, vol. 217, pp. 2016–2029 (2017)
5. Jaydeep, V.R.: Performance evaluation of grid-connected solar photovoltaic plant using PVSYST software. J. Emerg. Technol. Innovative Res. (JETIR) **2**(2), 372–378 (2015)
6. Chandrakant, D., Deepak, P., Anshul, A., Akash, K., Sudhakar, S., Murali Manohar, K., Amit Bhimte, S.R.: Performance simulation of grid-connected rooftop solar PV system for small households: a case study of Ujjain, India Energy Reports 4, pp. 546–553 (2018)
7. Manju, S., Sagar, N.: Progressing towards the development of sustainable energy: a critical review on the current status, applications, developmental barriers and prospects of solar photovoltaic systems in India. Renew. Sustain. Energy Rev. **70**, 298–313 (2017)
8. Laib, I., Hamidat, A., Haddadi, M., Ramzan, N., Olabi, A.G.: Study and simulation of the energy performances of a grid-connected PV system supplying a residential house in North of Algeria. Energy **152**, 445–454 (2018)
9. Nallapaneni, M.K.: Simulation tools for technical sizing and analysis of solar PV systems. In: 6th World Conference on Applied Sciences, Engineering & Technology, UMPO, Indonesia, pp. 26–27 (2017)
10. Priya, Y., Nitin, K., Chandel, S.S.: Simulation and performance analysis of a lkWp PHOTOVOLTAIC SYSTEM USING PVsyst. In: International Conference on Computation of Power, Energy, Information and Communication, pp. 358–363. IEEE (2015)
11. Nallapaneni, M., Rohit, K.M., Ruth, R.P., Mobi, M.: Performance analysis of 100 kWp grid connected Si-poly photovoltaic system using PVsyst simulation tool. 1st International Conference on Power Engineering, Computing and Control, Energy Procedia, vol. 117, 180–189, March 2017
12. PVsyst Photovoltaic Software. http://www.pvsyst.com/en/
13. Florin, A., Nebi, C., Vjollca, K.: Design optimization and simulation of the photovoltaic systems on buildings in southeast Europe. Int. J. Adv. Eng. Technol. **1**, 58–68 (2011)

Numerical Investigation of Thermal Regulation Improvement of Curved PV Panel Using PCM

M. L. Benlekkam[1,2(✉)] and D. Nehari[2]

[1] Department of Mechanical Engineering, University Center of Ahmed Ibn Yahya Alwancharissi, Tissemsilt, Algeria
mohamed_benlekkam@yahoo.fr

[2] Laboratory of Smart Structure, University Center of Belhadj Bouchaib, Ain Temouchent, Algeria
mohamed_benlekkam@yahoo.fr, nehari_dr@yahoo.fr

Abstract. The elevation of photovoltaic (PV) cells temperature reduce its electrical conversion efficiency (above to characterization temperature usually 25 °C). The use of a phase change material (PCM) layer linked to a PV cells panel can decrease its temperature rise. This paper based on a numerical study of the effect of the tilt angle of curved PV panel with 100 mm of curvature radius (Technology so-called PV-mirror) on its thermal performance. For this raison a 2D model of heat transfer conjugated with a solid-liquid phase change was developed and validated with an experimental data for two PV/PCM systems with and without fins, where a good agreement was obtained. The Results reveal that the tilt angle (more than 45°) allows a good thermal regulation, a noticeable decrease in PV cell temperature more than 20 °C compared with a single curved PV which maintain its efficiency et 14%. In addition the heat transfer was dominated by conduction for tilt angle less than 45°, which increase the front temperature of PV/PCM system. However, the convection dominates in the other system, where the cold liquid PCM flow absorbs more heat, which decreases the PV cells temperature.

Keywords: PV cell · Latent heat · Phase change · PCM · PV-mirror

1 Introduction

The photovoltaic cell temperature elevation reduces its electrical efficiency. In fact, only *15–20%* of the sunlight converted to electricity. While the rest of sunlight dissipated into heat [1], that causes the increase of the PV operating temperature. As a result, the electrical conversion efficiency dropped by *0.5% per 1* °C; and affected the solar cell lifetime Norton et al. [2]. Therefore, maintaining the temperature of PV cells at adequate level can improve their efficiency. Several researches try to decrease the PV cells temperature rise. Overall, the use of a thin layer of Phase Change Material (hereafter PCM) behind PV panel has been an effective solution widely used for passive cooling of PV cells [3–7], which can absorb a large quantity of heat during its melting process at constant temperature. The experimental and numerical studies of Huang et al. [8–13] was conducted under the thermal regulation of building integrated

M. Hatti (Ed.): ICAIRES 2019, LNNS 102, pp. 434–441, 2020.
https://doi.org/10.1007/978-3-030-37207-1_46

photovoltaic panel BIPV by using PCM. Starting by the first research [12] which, they have succeed to maintain the temperature of the PV cells under *40* °C for *80* min, with a PCM RT25 that have a melt temperature of *32* °C, then considering the low thermal conductivity of RT25 *0.18* W m^{-1} K^{-1}, an internal fins in Aluminum used to transport the thermal load into the PCM in order to reduce the temperature rise on PV. In other hand Huang et al. [8–11] investigated the use of inner fins in BIPV to enhance the heat transfer in PCM losing the excess heat in PCM. They have studied the effect of internal fins spacing on BIPV thermal regulation and PCM behavior, they found that the use of fins improve thermal distribution of PV/PCM system and reduce its temperature significantly. Recently, a novel technology of solar photovoltaic panel was introduced in world of solar energy, so-called PV-mirror, which combined by two types of panel: a plate one placed at the focal point of reflector mirror and the second (curved PV) behind the curved mirror [14], or several PV panels arranged rear the mirror to take the curvature form [15]. This system made with special mirror that concentrates the large part of the sunlight into a flat PV panel positioned at the reflector's focal point. On the other hand, the rest part of the sunlight is converted to electric power via curved one. Nehari et al. [16] studied the impact of the fins length [5, 10, 15, 20, 25, 30, 35 mm] on the heat transfer improvement through the PCM layer. Results confirmed that the internal fins reduce significantly the increase of the temperature compared to the case configuration without inner fins. Furthermore, the configurations *of L = 25, 30 and 35* mm allow better cooling of the PV panel. Benlekkam et al. [17] numerically evaluated the impact of fins tilt angle and its orientation on the thermal cooling of PV/PCM system. They found that the best results were obtained with diverged configuration that maintained the temperature about *34* °C for *170* min. In addition the fins tilt angle with diverged configuration less than *30°* allows a good thermal cooling rate, and maintained the PV cell efficiency at average value of *14%* for more than *3* h compared with PV system without PCM. Moreover the fins tilt angle *25°* provides the best performances. Furthermore the PCM "RT25" was more useful for thermal regulation of PV/PCM system under medium and low insolations intensities for the same thermal operating conditions. Elarga et al. [18] developed a physical model to study the incorporation of PCM layer on a double-facade of PV panel for three different climates, they have found that this technique can improve the conversion of solar energy into electricity independently of the climate. Also they concluded that the use of PCM can increase the cooling of the building more than *20%* per month. An experimental study was carried out by Nikolaos et al. [19] to evaluate the efficacy of PCM to mitigate the influence of temperature on photovoltaic module performance. The PV/PCM system was tested under Mediterranean climatic conditions. They found that the chosen PCM "RT27" succeeded in reducing the PV module temperature by an average of 11 °C compared to a system without PCM, furthermore the conversion efficiency of the PV module increases by about *8.6%*. Since the phase change materials are pure and combined, Farouk et al. [20] studied their influence on the performance improvement of the PV modules. They found that the use of pure PCM can reduce the PV temperature by more than 6.5 °C by an average of *2.7* °C, and increase the electrical efficiency by an average of 3%. On the other hand, the combined PCM can reduce the PV temperature by an average of *5.6* °C and improves the efficiency of the PV module by about *5.8%*. The present study carried out to evaluate the effect of tilt angle of curved PV panel in the range of *15°* to *90°* in an interval of *15°* on its thermal regulation and the heat transfer mode inside the PCM.

2 Numerical Model and Validation

The studied domain of curved PV/PCM considered in the present paper is illustrated in Fig. 1-a, as shown the PV/PCM system is composed by a curved PV panel attached on a curved container filled by PCM. The PCM used is Rubitherm RT25 [17, 21] and its thermophysical properties are assumed to be constant during the melting process. Natural convection was adopted over the PV panel and the back wall of the container of PCM. The PCM is in perfect contact with the PV panel and the back wall. The incident energy I_T is absorbed and dissipated into heat inside the PV/PCM system. In our case, we use the same initial and thermal boundary condition considered by Huang et al. [12] which are:

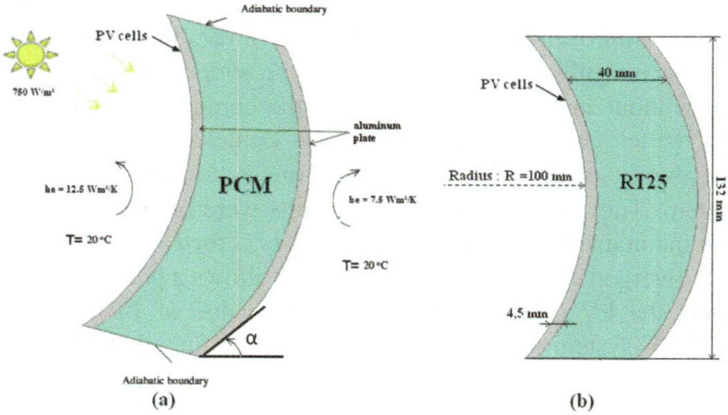

Fig. 1. Schematic diagram of the curved PV/PCM system: (a) Heat transfer and boundary conditions; (b) Geometry of the system (dimensions in mm).

(i) the initial temperature of the system PV/PCM is T_{PV};
(ii) the front and rear surfaces of the system have respectively the values h_1 and h_2;
(iii) concerning the top and bottom boundaries, the adiabatic conditions are used.

The 2D unsteady governing equations of energy and momentum heat transfer are solved by using the implicit finite volume method with the ANSYS FLUENT 17. In addition, the Boussineq approximation was adopted to take account the change in density of the PCM in liquid phase as a function of temperature.

For the numerical solution, the pressure-velocity coupling is accounted for by using SIMPLE algorithm [22], whereas the Quick scheme was adopted for convective discrimination. The grid size adopted in present simulation was 48 × 132 and a variable time step with a minimum value of 0.01 s are used for all the simulations, and the convergence was confirmed at each time step, with the convergence criterion of 10^{-6} for all variables.

3 Validation

All the simulations were conducted using the heat transfer coefficients on the front and rear surfaces are respectively *12.5* W m^{-2} K^{-1} and *7.5* W m^{-2} K^{-1} and the insolation was *750* W/m^{-2}.

The present model Benlekkam et al. [17, 21] was validated successfully with numerical and experimental data of [12] for two PV/PCM system with and without fins. Our validation was performed with a same initial and boundary conditions, material properties and geometry, then we compared our results with the predicted and experimental temperature evolution of front and rear surface of PV/PCM system with fins Fig. 2-b, and for the PV/PCM system without fins Fig. 2-c In addition we compared the isothermal contours of temperature in Fig. 2-a, where a good agreement was obtained.

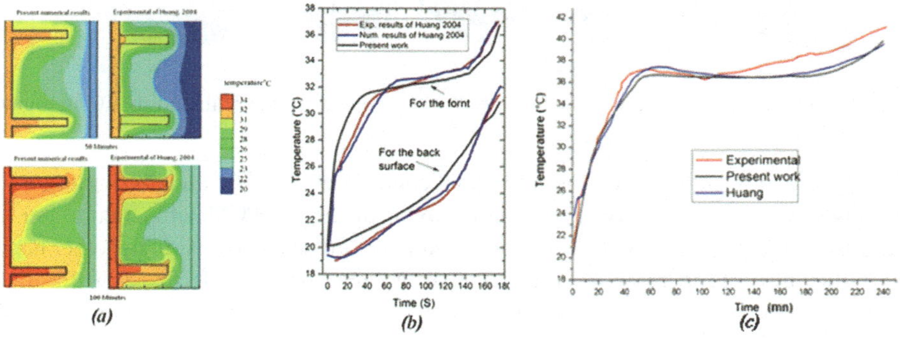

Fig. 2. Model validation with Huang et al. [12] (PV/PCM system with fins)

4 Results and Discussion

For all performed simulations the tilt angle was varied from the vertical (α) to *15*° in an interval of *15*°. All numerical simulations were performed along *200* min until the PCM will be completely melted. Initially the PCM starts to melt close to the front face, where the heat transfer was dominated by conduction but once enough PCM is melted, natural convection in the liquid melt starts to dominate.

Figure 3-a and b shows the distribution of the isothermal contours of the curved PV/PCM system at the *50th* and *100th* min. It is clear that the temperature increase to reaches its mean value of *39* °C and *40* °C for the PV/PCM system with tilt angle *45*° to *90*° at *50* and *100* min respectively. While rising to its maximum value *46* °C and *68* °C for tilt angle of 15°. This behavior due to the heat transfer mode domination.

From the inclination 0° to 45°, we observed the increase of the melted PCM quantity which allows the domination of convection. Therefore, it can maintain the temperature of the PV due to the convective flow of the cold PCM. On the other hand, the inclination lower than 45° reduces the heat transfer via convection until the case of pure conduction for the horizontal system.

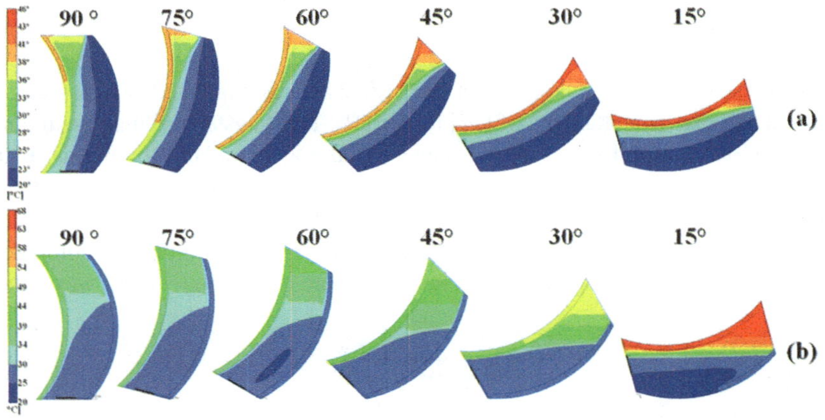

Fig. 3. Predicted isothermal contours of PV/PCM system at: (a) the 50th and (b) 100th minute

The Fig. 4-a and b, shows the isothermal contours at the *150*th and *200*th min for all tilt angle, we can see always the same behavior of the PV/PCM temperature and an increase of the melted PCM volume compared to the precedent cases. In the other hand, we notice also the increase of convection phenomena with the decrease of tilt angle until *45*°. However, for the lower ones than *45*°, we can confirm the domination of conduction phenomena as mentioned in the precedent time of simulation. We can observe also a kind of stratification of the temperature due to the strong domination of the conduction.

When the tilt angle of curved PV/PCM system decreases, the velocity of melted PCM decreases, which reduces the convective heat transfer inside the PCM enclosure. Therefore, reduces the heat transfer from the PV to the PCM (cases lower than *45*°). Besides, there is no convection for horizontal panel and the heat transfer by convection is the maximum for a vertical tilt angle.

The Fig. 5-a shows the temperature evolution along the time for all tilt angle investigated. It is clear that the temperature increase similarly for all cases up to *30* min due to the conduction heat transfer in the solid PCM. Then the PCM starts to melt and the liquid PCM quantity increase absorbing more heat from the PV panel as latent heat, therefore maintain the system temperature. As we explained above the tilt angle higher than *45°* allows a good thermal regulation due to the natural convection in liquid PCM, as shown in Fig. 6-b we can see that the thermal enhancement it's about *20* °C for vertical PV/PCM system compared with the PV panel without PCM. However the lower tilt angle than 45° reduce the velocity in the melted PCM to allow the domination of the pure conduction as shown in Fig. 5-a for the PV panel tilted by *15*°.

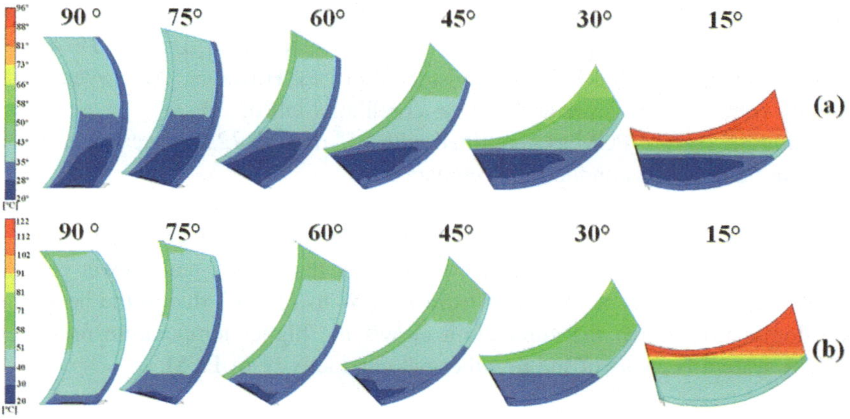

Fig. 4. Predicted isothermal contours of PV/PCM system at: (a) the 150th and (b) 200th minute

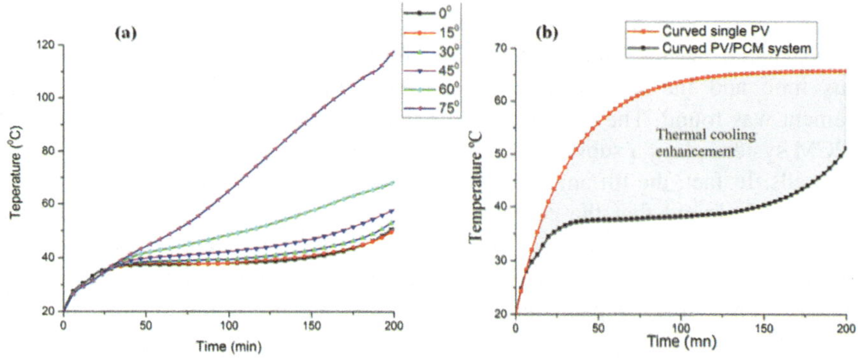

Fig. 5. PV panel temperature evolution: (a) The effect of tilt angle on PV cell temperature. (b) Thermal cooling enhancement.

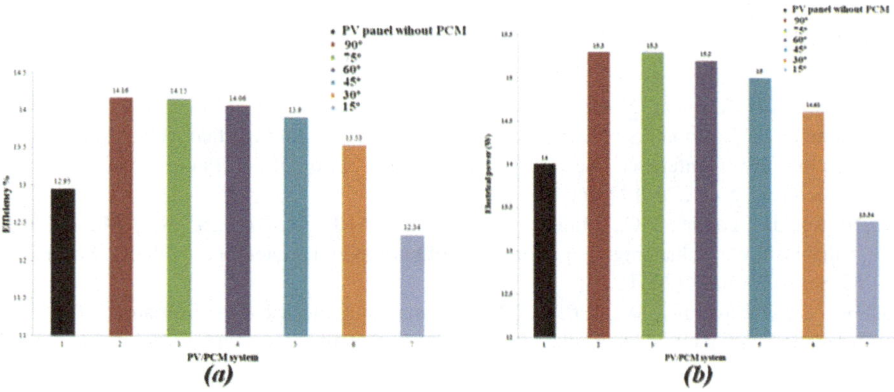

Fig. 6. (a) Solar cells electrical efficiency. (b) Electrical output power of PV cells.

From the Fig. 6-a, it's clear that the use of PCM can maintain the efficiency of PV cells at an average value greater than *14%* more than 3 h; a drop of *1%* from the maximum efficiency of the PV cells considered (15%). However, the curved PV cells efficiency without PCM layer at the same conditions fall of *2%*.

In addition, the PV/PCM system tilted with *15°* drop to *12.3%* because of the high temperature due to the conduction domination.

The electrical output power is presented in Fig. 6-b, it's clear that the use of PCM can improve the power generation of PV cells due to the passive cooling that maintain the PV cell temperature at lower values. We can see that the PV cells power generation decrease with respect to the tilt angle from *15.3* W to *13,3* W, due to the heat transfer mode domination which the conduction allows the higher temperature presented in PV/PCM system because of the low thermal conductivity of PCM.

5 Conclusion

In the current work, a numerical model was presented to investigate the effect of the tilt angle of curved PV/PCM system. The developed numerical model was compared with numerical and experimental data of literature for both: Average temperature of PV cells versus time and the isothermal contours for 50th and *100th* min, where a good agreement was found. The obtained results indicate that the heat transfer mode in the PV/PCM system plays a substantial role for its thermal regulation and the cooling rate of PV cells. In fact, the tilt angle of PV/PCM system controls the heat transfer inside the PCM. We found that the convection dominate for a tilt angle higher than *45°*. Whereas, the small tilt angle lower than *45°* allows the domination of the conduction where the natural convection completely disappears. In addition, the results reveal that the temperature of the PV panel increases with the decrease of the inclination, and the inclinations higher than *45°* can maintain the curved PV panel temperature under *50* °C for *170* min comparing with other tilted by *15°*, where its temperature increased up to *102* °C. Moreover, the vertical PV/PCM system improves the conversion efficiency and the power generation by *2%* and *1.3* W respectively for more than three hours.

References

1. Emery, K.B., Caiyem, J., Dunlavy, Y., Field, D., Kroposki, H., Moriarty, B., Ottoson, T., Rummel, L., Strand, S.: Temperature dependence of photovoltaic cells, modules and systems. In: Conference Record of the Twenty Fifth IEEE Photovoltaic Specialists Conference, pp. 1275–1278 (1996)
2. Norton, B., Eames, P.C., Mallick, T.K., Huang, M.J., McCormack, S.J., Mondol, J.D., Yohanis, Y.G.: Enhancing the performance of building integrated photovoltaics. Sol. Energy **85**(8), 1629–1664 (2011)
3. Ho, C., Tanuwijava, A., Lai, C.-M.: Thermal and electrical performance of a BIPV integrated with a microencapsulated phase change material layer. Energy Build. **50**, 331–338 (2012)

4. Hasan, A., McCormack, S.J., Huang, M.J., Norton, B.: Characterization of phase change materials for thermal control of photovoltaics using differential scanning calorimetry and temperature history method. Energy Convers. Manag. **81**, 322–329 (2014)
5. Atkin, P., Farid, M.M.: Improving the efficiency of photovoltaic cells using PCM infused graphite and aluminium fins. Sol. Energy **114**, 217–228 (2015)
6. Ma, T., Yang, H., Zhang, Y., Lu, L., Wang, X.: Using phase change materials in photovoltaic systems for thermal regulation and electrical efficiency improvement: a review and outlook. Renew. Sustain. Energy Rev. **43**, 1273–1284 (2015)
7. Browne, M.C., Norton, B., McCormack, S.J.: Heat retention of a photovoltaic/thermal collector with PCM. Sol. Energy **2**(133), 533–548 (2016)
8. Huang, M.J.: Two phase change material with different closed shape fins in building integrated photovoltaic system temperature regulation. In: World Renewable Energy Congress-Sweden, pp. 2938–2945. Linköping University Electronic Press, Sweden (2011)
9. Huang, M.J., Eames, P.C., Norton, B.: Thermal regulation of building-integrated photovoltaics using phase change materials. Int. J. Heat Mass Transf. **47**(12), 2715–2733 (2004)
10. Huang, M.J., Eames, P.C., Norton, B.: Phase change materials for limiting temperature rise in building integrated photovoltaics. Sol. Energy **80**(9), 1121–1130 (2006)
11. Huang, M.J., Eames, P.C., Norton, B., Hewitt, N.J.: Natural convection in an internally finned phase change material heat sink for the thermal management of photovoltaics. Sol. Energy Mater. Sol. Cells **95**(7), 1598–1603 (2011)
12. Huang, M.J., Eames, P.C., Norton, B., Hewitt, N.J.: The effect of phase change material crystalline segregation on the building integrated photovoltaic system thermal performance. In: World Renewable Energy Congress (WRECX), pp. 1338–1343 (2008)
13. Huang, M.J.: The effect of using two PCMs on the thermal regulation performance of BIPV systems. Sol. Energy Mater. Sol. Cells **95**(3), 957–963 (2011)
14. Holman, Z., Yu, Z.J.: Solar and alternative energy high-efficiency solar power with integrated storage (2015)
15. Zhengshan, J.Y., Fisher, K.C., Wheelwright, B.M., Angel, R.P., Holman, Z.C.: PVMirror: a new concept for tandem solar cells and hybrid solar converters. IEEE J. Photovolt. **5**(6), 1791–1799 (2015)
16. Nehari, T., Benlakam, M., Nehari, D.: Effect of the fins length for the passive cooling of the photovoltaic panels. Periodica Polytech. Eng. Mech. Eng. **60**(2), 89–95 (2016)
17. Benlekkam, M.L., Nehari, D., Madani, H.I.: The thermal impact of the fin tilt angle and its orientation on performance of PV cell using PCM. Int. J. Heat Technol. **36**(3), 919–926 (2018)
18. Elarga, H., Goia, F., Zarrella, A., Dal Monte, A., Benini, E.: Thermal and electrical performance of an integrated PV-PCM system in double skin façades: a numerical study. Sol. Energy **136**, 112–124 (2016)
19. Savvakis ,N., Tsoutsos, T.: Phase change materials in photovoltaics: the assessment of system performance in the present mediterranean climate conditions (2016)
20. Hachem, F., Abdulhay, B., Ramadan, M., El Hage, H., El Rab, M.G., Khaled, M.: Improving the performance of photovoltaic cells using pure and combined phase change materials–experiments and transient energy balance. Renew. Energy **107**, 567–575 (2017)
21. Benlekkam, M.L., Nehari, D., Madani, H.Y.: Numerical performances study of curved photovoltaic panel integrated with phase change material. Mech. Mech. Eng. **22**(4), 1439–1451 (2018)
22. Patankar, S.: Numerical Heat Transfer and Fluid Flow. CRC Press, Boca Raton (1980)

Prediction of Energy Storage Capacitor Values Based on Neural Networks. (Case of a Planar Capacitor)

B. Mimene[1(✉)], Y. Chiba[2], A. Tlemçani[3], and B. Kehileche[2]

[1] Laboratory of Advanced Electronic Systems (LSEA), University of Medea, Medea, Algeria
mimen.bakhti@gmail.com
[2] University of Medea, Medea, Algeria
[3] Laboratory of Research in Electrical Engineering and Automation, University of Medea, Médéa, Algeria

Abstract. Energy storage is a very important operation in continuously operating systems, such as telecommunications systems, embedded systems and power systems. Energy storage can be performed by various means such as batteries and super capacitors. In our work, we used neural networks to determine the capacitance values C of the planar capacitors as a function of the relative permittivity εr, the distance d and the dimensioning (Width and Length) of the capacitor plates and as a function of the maximum desired charge Qmax. The results of simulation will be better and more satisfying if the databases are richer and good.

Keywords: Energy storage · Capacitor · Neural networks · Activation function · Normalized square error

1 Introduction

The methods of implementing electronic circuits with more optimal performance are based primarily on their technical characteristics and experience during the experimental or operating phases to predict future performance of the various components of the circuits in question.

New technologies bring to the design of products, methods and information that make it possible to do better than before and above all, to predict the future behavior of a material or a component. The results of these studies can be used to identify critical components and undesired failures, or to estimate failure probabilities and associated downtime.

The universal predictive aspect of neural networks and their ability to adapt to the desired behavior allow to expand their uses for system identification and control.

Artificial neural networks are one of the artificial intelligence approaches whose development is done through the methods by which man always tries to imitate nature and to reproduce his own modes of reasoning and behavior.

M. Hatti (Ed.): ICAIRES 2019, LNNS 102, pp. 442–449, 2020.
https://doi.org/10.1007/978-3-030-37207-1_47

In our study, we used gradient-retro-propagating multilayer neural networks to predict the capacitance values from appropriate parameters from a good choice of theoretical or experimental database.

2 Description of the Planar Capacitor [1–4]

A planar capacitor consists of two metal plates of surface A = La.Lo (La, Lo: are respectively the width and the length of the two plates). Between these two plates of distance d there is a dielectric of permittivity $\varepsilon = \varepsilon_0.\varepsilon_r$ (with $\varepsilon_0 = 8{,}854.10^{-12}$ F/m the permittivity of the vacuum and ε_r the variable relative permittivity) as shown in Fig. 1.

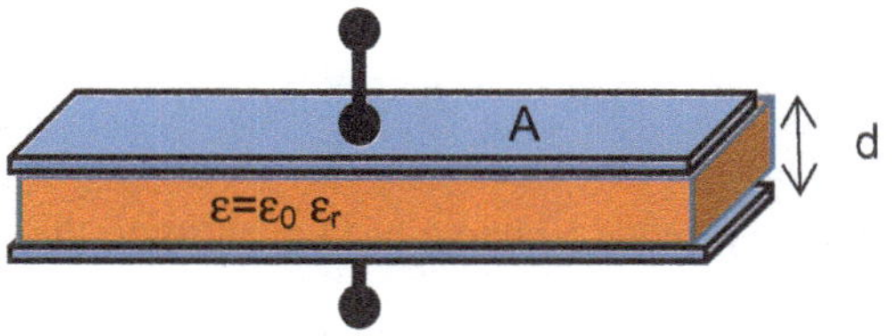

Fig. 1. Classic planar capacitor

The capacity of the capacitor is given by

$$C = \varepsilon_0.\varepsilon_r.\frac{A}{d} = \varepsilon_0.\varepsilon_r.\frac{L_a L_O}{d} \quad (1)$$

3 Energy Storage in a Capacitor [3–6]

A capacitor is an energy storage device. When we connect a battery to both plates of a capacitor, it charges. The potential difference gradually increases between the two plates and the battery has had to do more work to provide the same amount of charge due to the continuous increase in the potential difference as shown in Fig. 2.

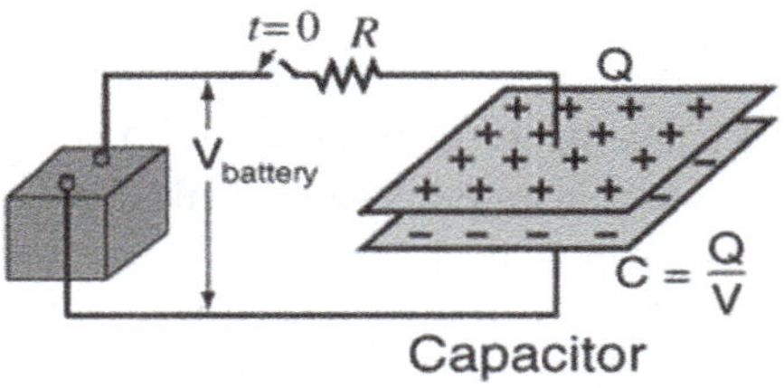

Fig. 2. The capacitor charge circuit

Storing energy on the capacitor involves doing the work to transport the charge from one capacitor plate to the other against the electrical forces. As the load accumulates in the charging process, each successive charging element requires more work to force it onto the positive plate. The sum of these constantly changing quantities requires an integral.

If a driver's capacitance is C, it is initially unloaded and the potential difference between its plates is V when connected to a battery. If q is the load on the plate at that moment, then q = CV.

We know that W = Vq, that is, the work done is equivalent to the product of potential V and load Q.

Therefore, if the battery delivers the infinitely small charge load dq to the capacitor at constant potential V, then

$$dW = Vdq = \frac{q}{C}dq \tag{2}$$

The total work done to provide a charge of Q quantity to the capacitor is given by

$$W = \int_{0}^{Q} \frac{q}{C}dq = \frac{1}{2}\frac{Q^2}{C} \tag{3}$$

Therefore, the energy stored in a capacitor is

$$E = \frac{1}{2}\frac{Q^2}{C} \tag{4}$$

By replacing Q = CV, we get

$$E = \frac{1}{2}CV^2 \tag{5}$$

So, if the capacity of a capacitor is charged to 10 F up to a potential of 10 V, the energy stored in it is 5000 J.

4 Mathematical Model of the Artificial Neural Network [4, 7–10]

Artificial neural networks are highly connected networks of elementary processors operating in parallel. Each elementary processor calculates a single output based on the information it receives. Any hierarchical network structure is obviously a network.

The mathematical model of an artificial neuron is shown in Fig. 3. A neuron essentially consists of an integrator that performs the weighted sum of its inputs. The result ***n*** of this sum is then transformed by a transfer function ***f*** which produces the output of the neuron. Following the notation presented in the previous section, the

R entries of the neurons correspond to the vector P = [P_1 P_2...... P_R] while W = [W_{11} W_{12}... W_{1R}] represents the vector of the neuron weights. The output ***n*** of the integrator is given by the following equation:

$$n = \sum_{j=1}^{R} W_{1,j}P_j - b$$

The structure of the multilayer networks consists of an input layer, a hidden layer that can consist of several sub-layers, and another output layer as shown in Fig. 3.

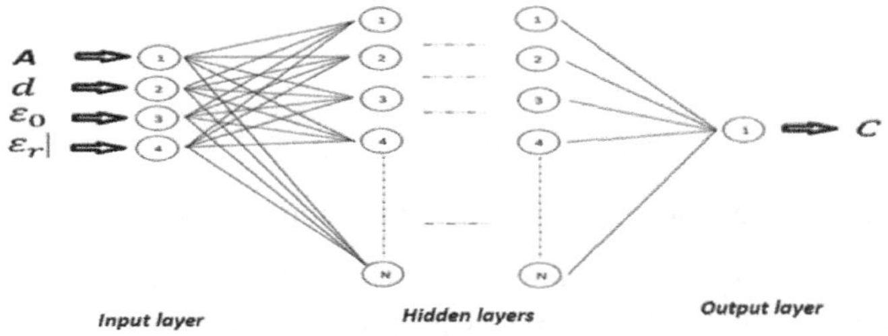

Fig. 3. Back propagation neural network

In our case we used a hidden layer of 10 neurons as shown in Fig. 4.

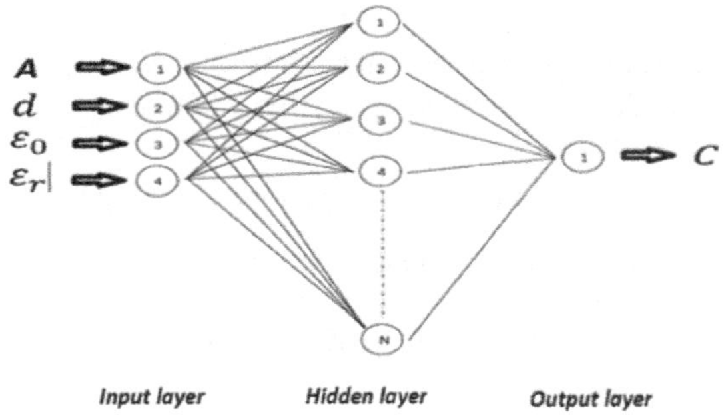

Fig. 4. Back propagation neural network of one hidden layer

The activation function chosen is the sigmoid function

$$f(x) = \frac{1}{1 + \boldsymbol{Exp}(x)}$$

The database used is related to the different types of insulation as shown in the following Table 1:

Table 1. Permittivity and calculated capacitance of used insulators database.

Insulation	ε_r: Permittivity (pF/m)	Capacitance (pF)	Insulation	ε_r: Permittivity (pF/m)	Capacitance (pF)
Dry air	1	$26{,}562 * 10^{-12}$	Polyethylene	2,25	$59{,}76 * 10^{-12}$
Bakelite	5	$132{,}81 * 10^{-12}$	Polyprpylène	2,2	$58{,}43 * 10^{-12}$
Rubber	4	$106{,}24 * 10^{-12}$	Polystyrene	2,4	$63{,}74 * 10^{-12}$
Silicone rubber	4,2	$11{,}56 * 10^{-12}$	Polycarbonate	2,9	$77{,}02 * 10^{-12}$
Cardboard	4	$106{,}24 * 10^{-12}$	Porcelain	5	$132{,}81 * 10^{-12}$
Mica	6	$159{,}36 * 10^{-12}$	Pressboard	3	$79{,}68 * 10^{-12}$
Paper	2	$53{,}12 * 10^{-12}$	Steatite	5,8	$154{,}04 * 10^{-12}$
Baked paper	5	$132{,}81 * 10^{-12}$	Styroflex	2,5	$66{,}4 * 10^{-12}$
Paraffin	2,2	$58{,}43 * 10^{-12}$	Teflon	2,1	$55{,}77 * 10^{-12}$
PVC	5	$132{,}81 * 10^{-12}$	Glass	5	$132{,}81 * 10^{-12}$
Plexiglass	3,3	$87{,}64 * 10^{-12}$	Stratifé glass-epoxy	5	$132{,}81 * 10^{-12}$
Polyster	3,3	$87{,}64 * 10^{-12}$			

The stages of the gradient retro-propagation algorithm are:

1. Initialization of weights to random values of small quantities.
2. Selection of a learning matrix [A, d, ε_(0), ε_r] of the learning base
3. Calculation of the output s_i of each Neuron
4. Propagation of layer-by-layer outputs
5. Evaluation of the error between the output calculated by the network and the desired output (s-d).
6. Backward propagation of the error to the back of all neurons from the output to the input

   - If $\boldsymbol{i}$ is an output neuron then $\delta_i = f'(s_i)(s_i - d_i)$
   - If $\boldsymbol{i}$ is a hidden neuron then $\delta_i = f'(s_i)\sum_k (W_{ki}.s_k))$

   k: neuron between the current layer and the output layer

7. Adjustment of weights by the gradient procedure

$$\boldsymbol{W_{ij}(t+1) = W_{ij}(t) + \mu.\delta_i.s_i}$$

μ: No learning

8. As long as the error is too important, go back to step 2 (following example).
9. Save the weights and end.

5 Simulation Results

The software MATLAB realized the simulation of our algorithm based on neural networks to prevent the values of the capacities. The algorithm consists of two main parts, the learning part and the test part.

In the learning part we used a part of the database to determine the weighting coefficients (weights) by setting the iteration number by determining the mean squared error between the desired value and the calculated value of capacitance.

The variation of the mean squared error between the desired outputs and calculated by the neural network as a function of the number of iterations represented in the following figure Fig. 5.

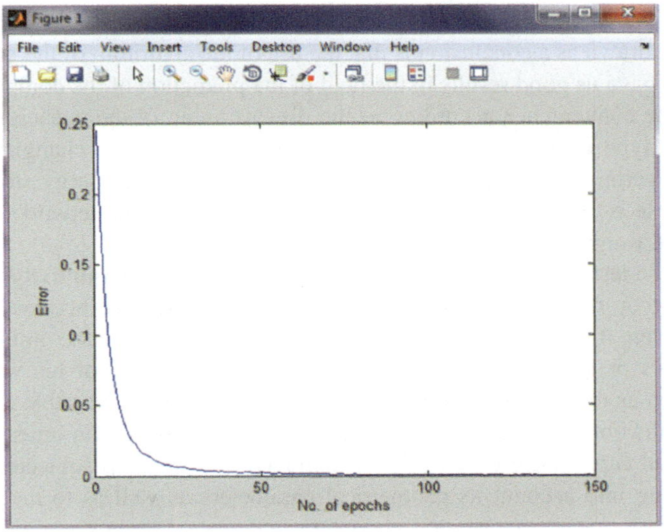

Fig. 5. Evolution curve of NSE according to the number of iterations

In the Test part, to determine the desired values, we used the weights determined by learning. Where the results are very satisfactory and show a very good convergence with a low quadratic error. The following Table 2 presents some examples of tests:

Table 2. Desired and calculated capacitance of some test insulators.

Area A $[m^2]$	Distance d [m]	Permittivity of the void ε_0 [pF/m]	Relative permittivity ε_r	Desired capacitance C_d[pF]	Calculated capacitance by NN C_{out} [pF]	Normalised square error
0.03	0.01	8.85	1.5	39.825	39.815	$5.021\ 10^{-4}$
0.03	0.01	8.85	2.2	58.430	58.420	$3.423\ 10^{-4}$
0.03	0.01	8.85	2.8	74.34	74.33	2.69010^{-4}
0.03	0.01	8.85	5.2	138.06	138.05	1.44910^{-4}

From these results it can be seen that the normalized quadratic error becomes rather weak, the higher the permittivity. So we can extend this study to the case study of energy storage in the capacitor based on a well-chosen database, so that we can do the predetermination stored energy following the load Q, the supply voltage V and the dimensioning of the capacitor.

6 Conclusion

What can be drawn as a conclusion is that the use of the algorithm of the gradient retro-propagation gave us good results in the field of the prediction of the computations with an acceptable convergence according to the dimensioning of the capacitors.

For this type of study, a very rich database is needed by changing the sizing parameters, permittivity and power parameters in the case of energy storage. So the learning phase is very important for this study based on neural networks by carefully choosing the weight coefficients and the activation function.

During the test, the choice to take slight variations on the permittivity parameter of the insulator or the dielectric is the final goal of the subject because in fact this variation, even if it is taken randomly, it actually represents the influence of the characteristics or external phenomena applied to the capacitor of his working environment such as temperature, humidity, pressure or even if it is variable capacitor use.

The results obtained were quite reliable and quite satisfactory in order to be able to predict a final capacitance of a capacitor deployed in a medium subjected to external stresses taking into account its geometrical parameters as well as to the construction characteristics.

References

1. Sarjeant, W.J., Ziirnheld, J., MacDougall, F.W.: Capacitors. IEEE Trans. Plasma Sci. **26**(5), 1368–1392 (1998)
2. Mouriès, G.: Condensateurs utilisés en électronique de puissance", Techniques de l'ingénieur, traité de génie électrique, D3280. Éditions T.I. (Techniques de l'Ingénieur) saint denis France, Mars 1989
3. Hantouche, C.: Condensateurs de puissance", Techniques de l'ingénieur, traité de génie électrique, D 4710. Éditions T.I. (Techniques de l'Ingénieur). Saint Denis France, Juin 1996
4. Martinez-Vega, J.: Matériaux diélectriques pour le génie électrique 2 – Applications spécifiques, Hermes Editions (2007)
5. Buller, S.: Impedance-based simulation models for energy storage devices in advanced automotive power systems. Doctor thesis. ISEA, RWTH Aachen University (2003)
6. Siami, S., Joubert, C., Glauze, C.: High frequency model for power electronics capacitors. IEEE Trans. Power Electron. **16**(2), 157–166 (2001)
7. Chan, C.C., Lo, E.W.C., Shen, W.X.: The available capacity computation model based on artificial neural network for lead-acid batteries in electric vehicles. J. Power Sourc. **87**, 201–204 (2000)
8. Hornik, K., Stinchcombe, M., White, H.: Multilayer feed forward networks are universal approximators. Neural Netw. **2**, 359–366 (1989)

9. Hornik, K., Stinchcombe, M., White, H., Auer, P.: Degree of approximation results for feed forward networks approximating unknown mappings and their derivatives. Neural Comput. **6**, 1262–1275 (1994)
10. Forgez, C., Lemaire-Semail, B., Clenet, S. : Réseaux de neurones appliqués à la modélisation de systèmes électrotechniques" (Neural networks applied to electrical engineering modelling). Revue International de Génie Electrique **2**(3), 305–334 (1999)

Secure Cluster Head Election Approach Based on Trust Management in Wireless Sensor Networks

Ahmed Saidi(✉), Khelifa Benahmed, and Nouredine Seddiki

Department of Mathematics and Computer Science, Faculty of Exact Sciences, Tahri Mohammed University of Bechar, Bechar, Algeria
ahmedduc@gmail.com

Abstract. Wireless sensor networks (WSNs) are interesting technologies that play an essential role in many recent applications such as internet of things, VANET, smart grids, and activities recognition. Lately, many trust management models based on clustering architecture are proposed. Most of these models are based on the assumption of pre-establishment of clustering architecture without addressing the problem of selecting the trusty cluster head (CH). Hence, it is necessary to select the most trustworthy node as CH before applying any security models. In this paper, we are focusing on the problem of selecting the most trustworthy node as CH by incorporating the trust management technique. We inject two types of malicious nodes, the first type provides wrong information in the election process in order to become a CH and the second type sends faulty recommendations to support the former type to achieve its goal. Our model shows resilience against 30% of malicious nodes injected in the network of a total 1000 nodes. The results of the simulation indicate that our model prevents the malicious nodes to be elected as CH.

Keywords: Wireless sensor networks · Security · Trust management · Direct trust · Indirect trust · Cluster head

1 Introduction

Recent innovation made in the fields of WSNs communication technologies and noticeable success of theoretical research over past decades has allowed the advent of wireless sensor networks. These networks consist of a large number of inexpensive sensors nodes (SN) deployed in a hostile area. Each sensor cooperatively monitors physical or environmental phenomena, for instance, vibration, sound, temperature, air, and motions [1]. Considering the advantages of WSN such as low cost, simplicity, and rapidity of deployment make this type of network ideal for various applications including battlefield monitoring, environment monitoring, and habitat monitoring. However, the sensors nodes are deployed in an unprotected area and they are not equipped with tamper-resistant hardware [2], they also have restricted resources like energy, memory and computational power which make them vulnerable to different types of attack. To overcome these attacks, many researchers have used different monitoring architecture. Cluster-based architecture is the most effective technique that

M. Hatti (Ed.): ICAIRES 2019, LNNS 102, pp. 450–461, 2020.
https://doi.org/10.1007/978-3-030-37207-1_48

is useful to conserve energy and reduces the routing overhead [3]. Clustering is the operation of grouping nodes into clusters each cluster managed by one cluster head (CH). Many researchers in the literature addressing the problem of electing the safe CH. Due to the fact that the CH plays a vital role in forwarding the data came from the lower level of the network (sensor node) toward the upper level of the network (Base station). Malicious nodes often try to increase their reputation in order to be elected as CH. If these nodes are authenticated with valid keys and legitimated identities, they are considered legitimate from a cryptographic point of view but they are malicious locating to their behaviors. In this regard, trust management technique is beneficial. Recently, it is proposed as a security solution that is able to deal with malicious, selfish and compromised nodes. But only a few works considered the trust factor when selecting the trusted node as CH. In this research work, we propose a secure CH election and misbehavior detection approach based on trust management technique composed of two phases. The first phase governed by the BS, allows the most trustworthy node to become a CH. After the stabilization of the clusters, the second phase starts. In this phase, each CH gathers the trust information from its members and uses it, in order to elect a surrogate CH before its energy depleted. Our objective is to make sure to select the most trustworthy node as CH.

2 Related Works

Research on trust management systems has received much attention newly, and it involves surveyed works on various operations such as secure CH election, secure routing, secure data aggregation, secure localization, and misbehavior detection. In this section, we only review a number of trust schemes for secure CH election and misbehavior detection related to our work. Clustering is an effective architecture that improves network control and scalability. Especially in the network that consists of thousands of nodes [1]. However, very few trust management schemes have designed for cluster-based architecture. The ideas proposed in the works [4, 5] consist of electing the safe CH through sharing a message that contains the weight of the node and its identity. The weight of the node is a combination of various metrics such as the connectivity degree, energy and mobility. Each node compares its weight with its neighbors and the node with the highest weight will be the cluster head. The authors proposed in [6] a secure clustering algorithm using swarm intelligent technique. The system creates a nominated list of the candidate nodes and the honey bee mating algorithm is used to elect the safe CH. In [7], authors proposed a monitoring scheme for cluster-based WSN. The proposed model classifies the trust into two layers: Intra-cluster and inter-cluster. In the case of intra-cluster evaluation, they evaluate the direct trust between the cluster members (CMs). In the case of inter-cluster evaluation, they evaluate the trust between the CMs and the CH. Using the aging factor and reward factor the trust raises very slowly when there are successful interactions between the nodes and drop quickly in the case of unsuccessful interactions between them. In order to minimize the memory overhead, the authors proposed in, [8] a trust management scheme for clustered WSN, the model suggests to evaluate the trust of a group of nodes rather than a single node. In [9] authors proposed a trust-based framework for cluster-based WSN. In their approach, each node records the behavior of its neighbors by operating in promiscuous mode and checks if the packet was successfully delivered to its destination. Then, the information recorded is sent to the CH for the next round election. Our research work is different from the other works above discussed in this section, as it introduces the trust factor as the key metric to promote a safe choice of CH and makes sure that it never be a malicious node. Then a multilevel trust evaluation mechanism is applied for each level of the network. Our objective is to eliminate the malicious nodes which try to increase their reputation in order to be elected as CH. Unlike the other works, we do not support the feedback sent from multiple cluster members. For trust evaluation, we use just the CH and the BS as reference nodes, especially in the case when there are no prior interactions between the nodes. We do that in order to reduce the communication overhead and to prevent the false recommendations provided by malicious nodes.

3 Our Approach

Our goal is to develop a lightweight trust management scheme that ensures the selection of trusty CH and eliminates the malicious nodes. Therefore, our model is based on the following assumptions:

- Nodes are trustworthy in the first stage of deployment.
- Nodes are able to localize themself the exact details on the localization technique is out of scope in this paper.
- All nodes are homogeneous and randomly deployed in the network.
- Each cluster monitored only by one CH.
- Each node interacts with its one-hop neighbors directly and with other nodes using multi-hop packets forwarding method.
- Each node has a unique ID registered at the BS.

A. Network Assumption
B. Metrics Used Including Trust

1. **The Weight of the Node**
   The weight of the node is a combination of various metrics, for instance, Connectivity, Distance, Energy, and the Trust factor. It is computed by a weighted mean function as it is described in Eq. (1) [4]:

$$weight = \alpha 1 * Ci + \alpha 2 * Di + \alpha 3 * Ei \tag{1}$$

   where $\alpha 1$; $\alpha 2$; $\alpha 3$ are the weight coefficients of each parameter.
2. **Connectivity Degree**
   In Eq. (2), Ci is the connectivity degree of the node represented by their neighbors. The neighbors of the node are a set of nodes that fall in its communication range [4]. Where *txrange* is the communication range of the node and dis(i,j) is the distance between the node i and j.

$$Ci = \bigcup_{j \in V, j \neq i} \{j | dis(i,j) < Txrange\} \tag{2}$$

3. **Distance**
   In our trust model we elect the node with the nearest members as CH [4] Eq. (3) represents the metric used:

$$Di = \sum_{j \in N(i)} \{\mathrm{j} | dis(i,j)\} \tag{3}$$

4. **Energy**
   Our objective is to select the node with higher remaining energy as CH. In this paper, we use the energy model based on [10]. The equations below describe our energy model:

$$ETX(k,d) = \begin{cases} K * Eelec(k,d) + K * Efs * d2 \ \text{(a)} \\ K * Eelec(k,d) + K * Eamp * d4 \ \text{(b)} \end{cases} \tag{4}$$

$$ERX(k) = K * Eelec \tag{5}$$

   Etx is the transmission energy consumption. The Eq. (4a) used in case of free space, Eq. (4b) used in case of multichannel. *K* the number of bits used in the message. *D* is the distance between the nodes, *Efs* is the free space energy and *Eamp* is the amplified energy.
5. **The Trust Factor**
   Our model calculates the trust value in two levels as described in Fig. 1:

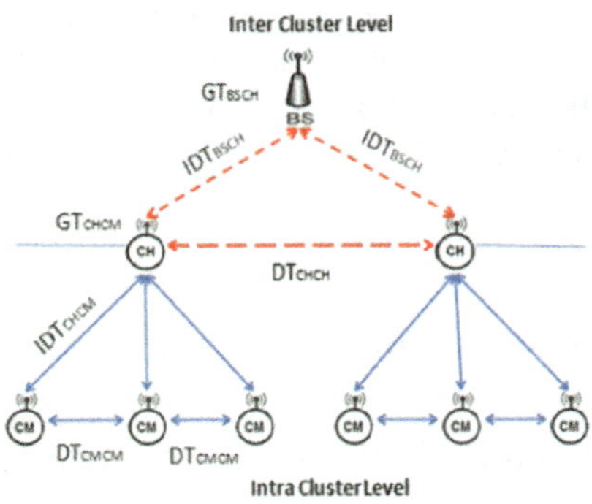

Fig. 1. Trust calculation.

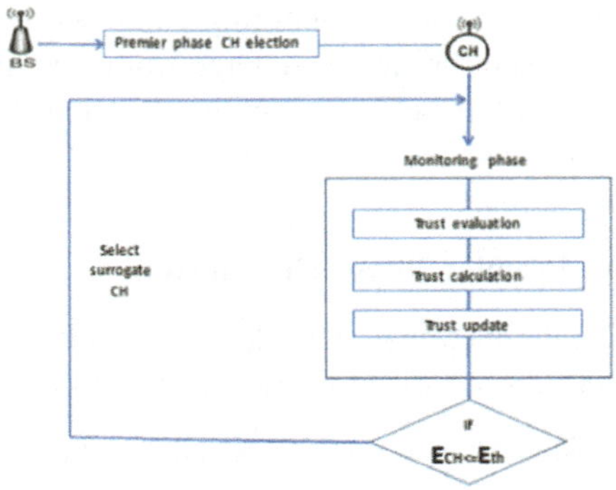

Fig. 2. Working principle.

5.1 Intra Cluster Evaluation (CH to CM)

DTCMCM: represents the trust value evaluated between the cluster members if there are previous interactions between them, otherwise they use the recommendations sent by their CH.
IDTCHCM: represents the trust value related to the direct evaluation between the CH and the cluster member.
GTCHCM: the CH calculates the final trust value of each CM using its direct evaluation (*IDTCHCM*) and the feedback sent from other nodes in the cluster (*DTCMCM*).

5.2 Inter Cluster Evaluation (BS to CH)

DTCHCH: represents the result of direct evaluation between the cluster heads.
IDTBSCH: represents the trust value related to direct evaluation between the BS and the cluster head.
GTBSCH: same as the intra-cluster evaluation the BS computes the final trust of the CH using its direct evaluation (*IDTBSCH*) and the feedback sent from other CHs (*DTCHCH*). In our model, all the trust values are in the range of [0,1] where 0 means no trust, 0.5 uncertain and 1 complete trust. As inspired by related works [7, 8] we use a type unsigned integer to optimize the memory required for saving the trust value rather than a real number.

4 Working Principle

As described in Fig. 2, our approach is composed of two phases. The first phase governed by the BS is responsible for premier phase CH election. After the CH was elected it sent a message (Start monitoring) to its cluster members and the monitoring phase starts. The monitoring phase is governed by The CH. In this phase, the CH uses the trust information gathered from the cluster members to choose the most trustworthy node as CH before its energy depleted. If the BS accepts this node, so it will become the CH for the next round. The new CH elects another CH using the same mechanism and so on.

A. Premier Phase CH Election

The Premier phase CH election algorithm is based on the ideas proposed in [4] with the modification made for our approach.

Step 1: each node Si broadcasts hello message periodically to discover its one-hop neighbors.
Step 2: each node Si calculates its metrics as described in Eqs. (2), (3) and (5).
Step 3: each node Si counts its weight according to Eq. (1) join the weight with its identity id and stored in list Lpi and broadcasts the list Lpi to all its one-hop neighbors.

Step 4: the node with the highest weight sends an advertisement to the BS. The BS multiple the trust value of each node with its weight and calculates the new weight of the node using Eq. (6):

$$Newweighti = weighti * trusti \tag{6}$$

The reason for multiple the trust value with the weight of the node to make sure that is a trusted node and avoid the malicious nodes that sent an amplified value of their weight.

Step 5: After the election of CH, each CH sends an advertisement to its neighbors.
Step 6: After the stabilization of each cluster, the CH broadcast TDMA schedule in order to avoid collision between the nodes.
Step 7: each CH sends a message "start monitoring" to its members and the monitoring phase starts.
Step 8: before the energy of CH is depleted, it broadcasts a message for new election round. In this case, each node sends its weight to its CH. the CH multiple the weight of the node with its trust value using Eq. (6).
Step 9: The CH chooses the node with the highest weight and highest trust value. Then, it sends its identity to the BS. If the BS accepts the chosen node, it will become the CH for the next round.

B. **Monitoring Phase**

The monitoring phase is executed in three levels as explained below:

1. **CM Level**

Since the interactions in the network take place in the round, each CM monitors all its one-hop neighbors during a specific time. For example, the CM x monitors the behavior of the CM y during a specific time, if the CM y shows good behavior, the CM x considered as a successful outcome from the CM y. Otherwise, if the CM y shows a bad behavior, for example, it drops the packets sent by x or does not forward them to their destination, so the CM x considered as an unsuccessful outcome from the CM y as follows:

$$Txy = \begin{cases} S = 1 & \textit{if a successful outcome} \\ U = -1 & \textit{if an unsuccessful outcome} \end{cases}$$

Therefore, the trust value is based on the number of successful and unsuccessful outcomes between the cluster members.

In our model, we used the beta distribution to simulate the interactions between the nodes. The beta distribution is a type of continuous probability distribution defined in the range [0, 1] denoted by alpha and beta described by the gamma function as Eq. (7):

$$P(x) = \frac{\Gamma(\alpha+\beta)}{\Gamma(\alpha)+\Gamma(\beta)} x^{\alpha-1}(1-x)^{\beta-1}\, \forall 0 \le x \le 1, a \ge 0, b \ge 0 \tag{7}$$

We chose this model because of its simplicity and flexibility, for a complete description of beta distribution and the beta reputation systems the reader can refer to the following references [11]. The parameters α and β model the aspects of trust between the cluster members. The final trust value is calculated using the expectation of the reputation function as described below:

$$DTCMiCMj = E(beta(S+1, U+1)) = \frac{S+1}{S+U+2} \quad (8)$$

If there is no possibility for direct evaluation between the cluster members, the evaluator node uses the recommendation sent by the CH. We note here that the CH is the only reference node that is responsible for sending recommendations. After each CM evaluates all its neighbors in the cluster and sends their trust values to the CH, the CH records all the trust values in one matrix of n × n dimension (Matrix A). We note that n represents the number of nodes in the cluster.

$$An, n = \begin{pmatrix} DTCM1CM1 & \cdots & DTCM1CMn \\ \vdots & \ddots & \vdots \\ DTCMnCM1 & \cdots & DTCMnCMn \end{pmatrix} \quad (9)$$

We assume that the trust value of the node to itself is equal to 1.

$$DTCMiCMi = 1 \ \forall i = 1..n$$

2. **CH Level**

In intra-cluster communication, the CH monitors all its cluster members. If the cluster member shows a good behavior the CH considered as a successful outcome from CM. However, if CM shows a bad behavior then the CH considered as unsuccessful outcomes from CM. using Eq. (10), the CH computes the trust of its members as follows:

$$DT\mathbf{CHCM} = E(beta(S_{CM}+1, U_{CM}+1)) = \frac{S_{CM}+1}{S_{CM}+U_{CM}+2} \quad (10)$$

When CH evaluates all its CMs, we get a set of trust values represented as a matrix of dimension (1,n), where n represents the number of nodes in the cluster. Then, the CH calculates the final trust value of its cluster members by multiple its own matrix (Matrix B) (11) with the matrix gathered from the CM level (Matrix A) (9) as described in the Eq. (12).

$$B1, n = (\,IDTCH1CM1 \quad \ldots \quad IDTCH1CMn\,) \quad (11)$$

$$GT1, n = B1, n \times An, n \quad (12)$$

We mention here, that in order to calculate the final trust of CM we take into consideration the feedback sent from trustworthy members only. They are chosen based on their direct evaluation with the CH as described in Algorithm 1. The example below describes our final trust calculation method at the CH level.

Example 1
X, Y, Z are three nodes in the cluster when all those nodes evaluate each other and sent their trust values to the CH. The CH joined all the trust values in one matrix and multiples with its own matrix for a final trust calculation.

$$\begin{bmatrix} GTCHCMx \\ GTCHCMy \\ GTCHCMz \end{bmatrix} = [IDTCHCMx \;\; IDTCHCMy \;\; IDTCHCMz] \times \begin{bmatrix} DTCMxCMx & DTCMxCMy & DTCMxCMz \\ DTCMyCMx & DTCMyCMy & DTCMyCMz \\ DTCMzCMx & DTCMzCMy & DTCMzCMz \end{bmatrix}$$

In this example, we count the final trust calculation of x, y, and z respectively.

$$GTCHCMx = IDTCHCMx \times DTCMxCMx + IDTCHCMy \times DTCMyCMx + IDTCHCMz \times DTCMzCMx$$
$$GTCHCMy = IDTCHCMx \times DTCMxCMy + IDTCHCMy \times DTCMyCMy + IDTCHCMz \times DTCMzCMy$$
$$GTCHCMz = IDTCHCMx \times DTCMxCMz + IDTCHCMy \times DTCMyCMz + IDTCHCMz \times DTCMzCMz$$

In the case of final trust calculation of Y, we choose the feedback from trustworthy nodes as described in Algorithm 1.

Algorithm 1 Recommendation CH to CM

```
if IDT_{CHxCMx} >= 0.6 then
                                        take recommendation from CMx
    DT_{CMxCMy} ← 1.
else if 0.5 < IDT_{CHxCMx} < 0.6 then
                                        take recommendation from CMx
    DT_{CMxCMy} ← 0.5.
else
                                        do not take recommendation from CMx
    DT_{CMxCMy} ← 0.
end if=0
```

3. **BS Level**

In inter-cluster communication, each CH monitors directly all other CHs in its one hop away. If there is no possibility for direct evaluation, then the CH uses the recommendation value (indirect trust) sent from the BS, as the only entity that is authorized for this operation. The recommendation values are the results of direct evaluation between the BS and the CHs, they are maintained by the BS in a matrix of (1,m) dimension (Matrix C) (13), where m represents the number of CHs in the network.

$$C1, m = (\, IDTBSCH1 \; IDTBSCH2 \quad \ldots \quad IDTBSCHm \,) \tag{13}$$

The BS gathers the matrix (Matrix D) (14) composed of the trust values sent by all the CHs.

$$Dm, m = \begin{pmatrix} DTCH1CH1 & \cdots & DTCH1CHm \\ \vdots & \ddots & \vdots \\ DTCHmCH1 & \cdots & DTCHmCHm \end{pmatrix} \tag{14}$$

Further, the BS calculates the final trust of each CH by multiplying the Matrix C with Matrix D. Same as an intra-cluster evaluation method, the BS takes into consideration the recommendations sent from the trustworthy CHs only as described in the Algorithm 2. For example, in order to calculate the final trust value of CHy, if the CHx sends its recommendation value about the CHy. The BS takes its recommendation into consideration based on its trust degree as described in Algorithm 2.

Algorithm 2 Recommendation BS to CH

if IDT_{BSCHx} >= 0.6 **then**
take recommendation from CHx
$DT_{CHxCHy} \leftarrow 1.$
else if $0.5 < IDT_{BSCHx} < 0.6$ **then**
take recommendation from CHx
$DT_{CHxCHy} \leftarrow 0.5.$
else
do not take recommendation from CHx
$DT_{CHxCHy} \leftarrow 0.$
end if=0

C. **Calculation of Surrogate CH**

When the energy of CH goes below the threshold, each CH gathers the weight of its cluster members. In order to elect the surrogate CH, the actual CH multiples the matrix of final trust (calculated at the monitoring phase) with the vector contains the weight of each CMs. Suppose we have a cluster composed of four nodes the CH, CM1, CM2, and CM3 with the following final trust values:

$$\textbf{\textit{GTCM1}} = 1 \; \textit{is a trusty node} \quad \textbf{\textit{GTCM2}} = 0.5 \; \textit{is an uncertain node} \quad \textbf{\textit{GTCM3}} = 0 \; \textit{is a malicious node}$$

Their weight values calculated by Eq. (1).

$$\textbf{\textit{weightCM1}} = 5 \; \textbf{\textit{weightCM2}} = 10 \; \textbf{\textit{weightCM3}} = 30$$

Now the CH calculates the new weight of each CM using Eq. (6):

$$\begin{aligned} \textbf{\textit{New\_weightCM1}} &= \textbf{\textit{weightCM1}} \times \textbf{\textit{GTCM1}} \rightarrow 5 \times 1 = 5 \\ \textbf{\textit{New\_weightCM2}} &= \textbf{\textit{weightCM2}} \times \textbf{\textit{GTCM2}} \rightarrow 10 \times 0.5 = 5 \\ \textbf{\textit{New\_weightCM3}} &= \textbf{\textit{weightCM3}} \times \textbf{\textit{GTCM3}} \rightarrow 30 \times 0 = 0 \end{aligned}$$

Here the CH takes the CM1 as the next CH because it's trustworthy compared to the CM2, also we observe that even if CM3 has the higher weight it is not selected as CH.

5 Performance Evaluation and Testing Criteria

In this section, we present the evaluation of our approach using MATLAB as a simulation tool. The Table 1 presents the parameters used to simulate our system. We evaluate the performance of our model considering two types of malicious nodes. The first type is malicious nodes that send an amplified weight value in order to be elected as CH. The second type of malicious nodes is the nodes that always send a false recommendation. Furthermore, we measure the effect of the final trust value of the trustworthy node in the presence of malicious nodes. Our goal is to measure the resistance of our model against bad-mouthing attack. Moreover, we justify how our schema resists the malicious nodes that lie in their weight and how we keep their weight minimum in order to avoid getting elected as a CH.

A. **Final Trust Calculation**

Figures 3, 4, 5 and 6 illustrate the final trust calculation at the CH level of two cluster members. The first is trusty and the other is malicious. In each Figure, we vary the number of malicious nodes from 0% up to 30% our goal is to measure how our model overcomes the faulty recommendations sent by malicious nodes. In Fig. 4, the final trust is stable because no malicious nodes are in the network. However, the final trust value starts fluctuating in the presence of 15% of malicious nodes but it is in the trusty range <0,5. In the case of 30% of malicious nodes are in the network and even the final trust is fluctuating but it is still not decreased below 0,5. When the percentage of malicious nodes exceeds the 30% the CH miss the accurate trust due to the huge number of malicious nodes.

Table 1. .

Parameter	Value
Connectivity	75 m
Number of nodes	1000
Dimension of area	800×800 m^2
Energy	5000 J
Number of rounds	From 300 up to 500 rounds
Number of malicious nodes	From 15% up to 50%

B. **Weight Calculation**

Since the weight of the nodes and the trust metrics are related to each other, we measure the weight of the node with and without the trust value. Same as the final trust calculation, we vary the number of malicious nodes that lie in their weight in order to be elected as CH. In Fig. 7, we observe that the average weight of malicious nodes follows the number of malicious nodes in the network. In the case of 30% up to 50% of malicious nodes, the weight of untrustworthy nodes exceeds the weight of trustworthy nodes. However, when we use the trust value, the weight of malicious nodes never exceed the weight of trustworthy nodes, even if there is 50% of malicious nodes existed in the networks. In this case, we ensure the selection of trustworthy node as CH.

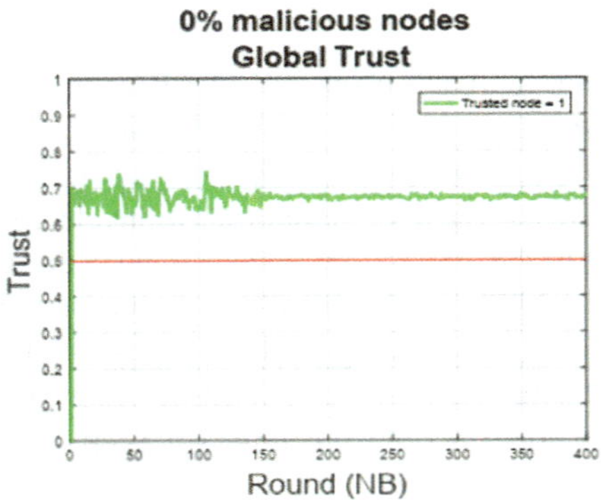

Fig. 3. Global trust calculation with 0% malicious

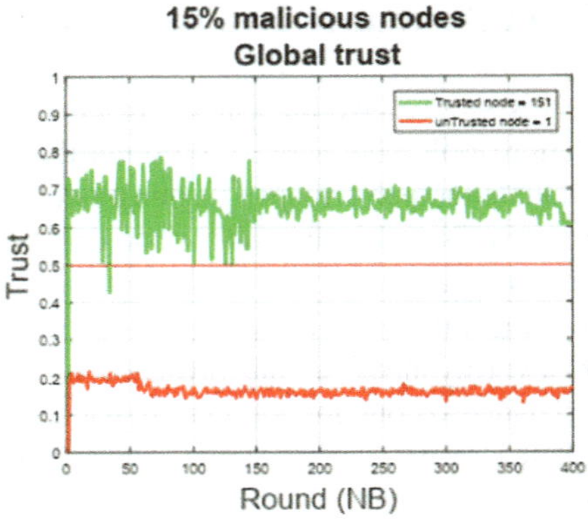

Fig. 4. Global trust calculation with 15% malicious

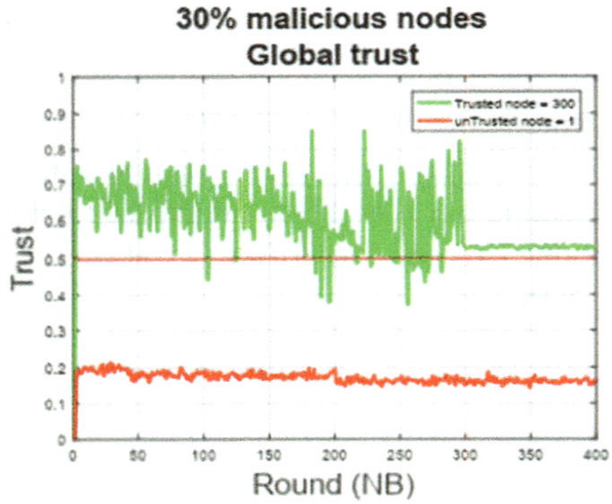

Fig. 5. Global trust calculation with 30% malicious

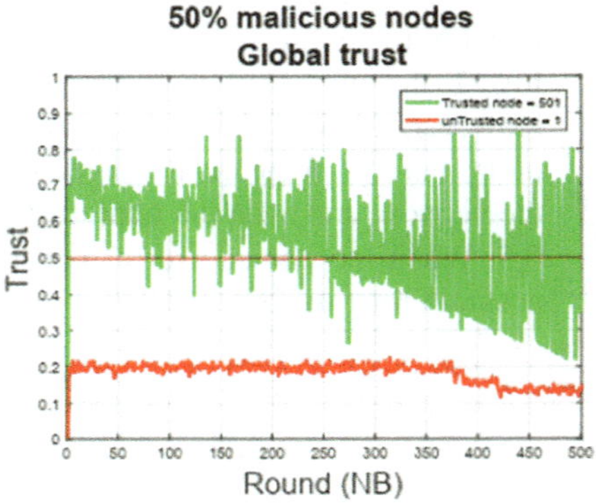

Fig. 6. Global trust calculation with 50% malicious

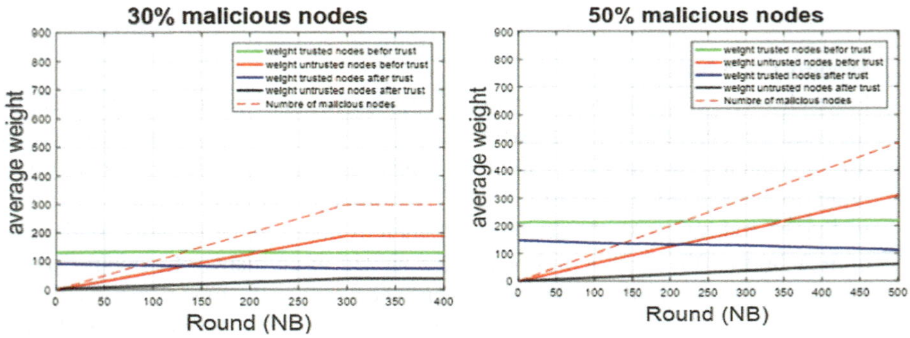

Fig. 7. Weight calculation before and after trust with 30% and 50% malicious nodes

6 Conclusion

In this research work, we propose a clustering architecture based on trust management that ensures the selection of trusty CHs. We introduce the trust factor as the key value for the election process. Furthermore, we prevent the malicious nodes that lie in their weight and send a larger value to be selected as CH by keeping their weight minimum compared to the trustworthy nodes. Moreover, the effect of false recommendations is minimized when the final trust value is assigned to the node and maintained only by the CH rather than the cluster members. In the future, we plan to extend this work by proposing a strategy to optimize the communication and the memory overhead of trust models. We want also to study the performance of our model against other types of attacks.

References

1. Alsaedi, N., Hashim, F., Sali, A., Rokhani, F.Z.: Detecting sybil attacks in clustered wireless sensor networks based on energy trust system (ETS). Comput. Commun. **110**, 75–82 (2017)
2. Ishmanov, F., Kim, S.W., Nam, S.Y.: A secure trust establishment scheme for wireless sensor networks. Sensors **14**, 1877–1897 (2014)
3. Rao, P., Jana, P., Banka, H.: A particle swarm optimization based energy efficient cluster head selection algorithm for wireless sensor networks. Wirel. Netw. **23**, 2005–2020 (2017)
4. Benahmed, K., Merabti, M., Haffaf, H.: Distributed monitoring for misbehavior detection in wireless sensor networks. Secur. Commun. Netw. **6**, 388–400 (2013)
5. Rehman, E., Sher, M., Naqvi, S.H., Badar Khan, K., Ullah, K.: Energy efficient secure trust based clustering algorithm for mobile wireless sensor network. J. Comput. Netw. Commun. **2017**, 1–9 (2017)
6. Sahoo, R., Sardar, A., Singh, M., Ray, S., Sarkar, S.: A bio inspired and trust based approach for clustering in WSN. Nat. Comput. **15**, 423–434 (2016)
7. Singh, M., Sardar, A.R., Majumder, K., Sarkar, S.: A lightweight trust mechanism and overhead analysis for clustered WSN. IETE J. Res. **63**, 297–308 (2017)
8. Shaikh, R.A., Jameel, H., d'Auriol, B.J., Lee, H., Lee, S., Song, Y.: Group-based trust management scheme for clustered wireless sensor networks. IEEE Trans. Parallel Distrib. Syst. **20**, 1698–1712 (2009)
9. Crosby, G.V., Pissinou, N., Gadze, J.: A framework for trust-based cluster head election in wireless sensor networks. In: IEEE Workshop for Dependability and Security in Sensor Networks and Systems, 24–28 April 2006; Columbia,Maryland, pp. 13–22. IEEE, New York (2006)
10. Heinzelman, B., Chandrakasan, P., Balakrishnan, H.: An application-specific protocol architecture for wireless microsensor networks. IEEE Trans. Wirel. Commun. **1**, 660–670 (2002)
11. Fang, W., Zhang, C., Shi, Z., Zhao, Q., Shan, L.: Beta-based trust and reputation evaluation system for wireless sensor networks. J. Netw. Comput. Appl. **59**, 88–94 (2016)

The Fire Risk in Green Building Caused by Photovoltaic Installations

Hadj Miloua[1,2](✉)

[1] Department of Mechanical Engineering, Faculty Technology, Djillali Llabes University, BP 89 Cité Arbi Ben Mhidi, Sidi Bel Abbes, Algeria
miloua_hadj@yahoo.fr

[2] LMSS, Djillali Llabes University, BP 89 Cité Arbi Ben Mhidi, Sidi Bel Abbes, Algeria

Abstract. This paper investigate the fire safety risks associated with green construction used photovoltaic panel to deliver electricity. The improper installation without adequate scheduled maintenance or the undetected arc faults caused by the deterioration of cables, connectors, conductors, and other system components caused by long-time weathering and aging effect favourite severed fire hazard to residential, commercial and wild-urban interface area. The aim of study is to investigate numerically the fire hazards and risks associated with green building technologies, materials and sustainable building practices to identify and reduce any potential risks to occupants and fire-fighters.

Keywords: Fire · Green building · Forest · Arc faults · Photovoltaic system

1 Introduction

Integration of renewable energy in order to gradually substitute traditional fossil fuels, especially solar energy is growing faster than ever before. Solar energy supplies over than 1.3% of worldwide power, and it is predicted to the largest electricity source by 2050 with about 11% of global power consumption [1]. For example the Algerian national objectives for renewable energy have been updated and elevated [2]. Although the global objective remains at 22 GW of installed power capacity for the period 2015–2030, the objectives according to the public statement made by the Ministry of Energy, have been elevated from 3 GW to 4.5 GW, expected to be achieved by 2020. This should provide 27% of Algeria's total electricity production instead of 20% as initially planned. Grid connected photovoltaic system: as abbreviated to PV system, it converts daylight into electricity, which used internally or fed in the public grid. In opposition, solar thermal systems convert daylight into heat. PV systems principally consist of the following components:

1. PV generator (one or more interconnected PV modules);
2. Generator connection box (optional) brings separate strings together and may include, for example, surge protection devices, string fuses;
3. Cabling on the DC side;

The original version of this chapter was revised: The author's name has been corrected to "Hadj Miloua". The correction to this chapter is available at https://doi.org/10.1007/978-3-030-37207-1_76

M. Hatti (Ed.): ICAIRES 2019, LNNS 102, pp. 462–469, 2020.
https://doi.org/10.1007/978-3-030-37207-1_49

4. Switch disconnectors, required for maintenance on the inverter and is often integrated in the inverter;
5. Inverter, formed direct current (DC) and DC voltage to grid compliant alternating current (AC) and AC voltage;
6. Cabling on the AC side;
7. Sub distribution;
8. Feed meter and/or counters for their own consumption (Fig. 1)

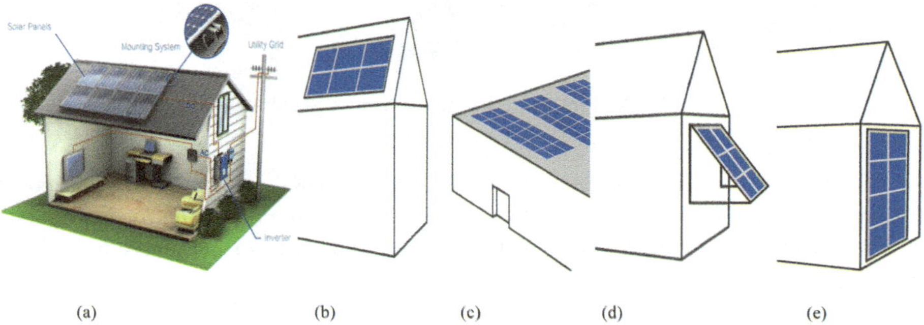

Fig. 1. (a) Basic design of a photovoltaic system, (b–c) Examples of building integrated PV systems and (d–e) additive PV system respectively.

Usually, PV systems are safe without risk to people or property. But PV system has a dark side, a hazards can be caused by installation and operation errors. Profoundly, they are exposed to a number of external threats to both their structure and function such as: Wind, Snow ice, Rodents, Theft (Fig. 2).

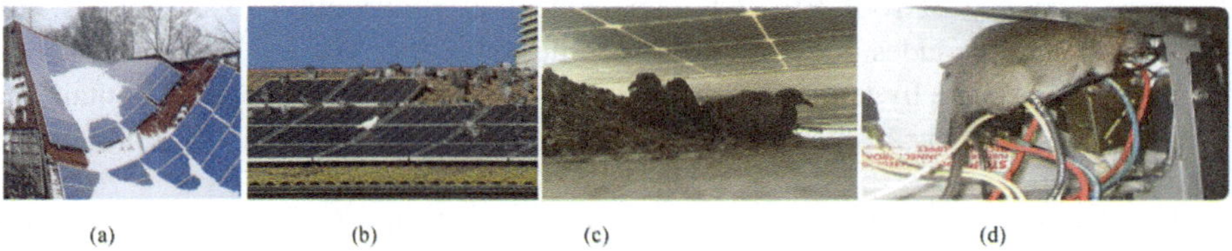

Fig. 2. (a) A roof with PV modules collapsed under snow load, (b–d) rodents and destroyed PV modules

Lightning and overvoltage 'OV' caused a fire: it is essential to retain information that during the daylight, PV panels and cables are energized with high DC voltages. With inappropriate precautions, these high DC voltages can present electrocution risks to installers, maintenance personnel, and firefighters. The safety criterion is an important consideration during installation, maintenance, and in the event of a fire. The increasing amount of (PV) systems and DC voltage level has a high potential of creating DC arc faults [3–9] (Fig. 3).

Fig. 3. Fire caused by photovoltaic systems.

There are two main considerations for fire safety of PV systems: Fires caused by PV apparatus and fires caused by external sources. It can be noted that rooftop panels incorrectly placed and improperly insulated cables and switches can make firefighting more difficult and dangerous. Also, the load of the panels means that burning roofs will collapse faster [10].

2 Results and Discussion

The computational fluid dynamics (CFD) technique is used to predict the fire dynamics in virtual green building used PV panel by a code named "the Fire Dynamic Simulator" (FDS). Important aspects of fire dynamics such as smoke propagation and temperature distribution were investigated. Before the simulation description it can be focus to some point in PV systems that affect fire spread along a roof PV array and their surrounding:

1 Some category of roof PV will offer some fuel contribution to a roof PV fire because they includes combustibles and reflected metal: The top surface of a rigid PV panel is made by tempered glass. But, the bottom of the panel may contain the polyester-based encapsulates and back sheets (this combustible used to protect the PV circuitry);
2 The flame propagates under the photovoltaic panel, the flame is redirected much closer to the roof surface and almost parallel to it. This increases the heat flow by thermal conduction on the roof surface, and spreads to green building which usually builds from woods;
3 Building integrated PV systems and surrounding by vegetation.
4 PV rooftop fires have been caused by electrical arcs that occurs near the combiner box, where many wires from PV panels are linked [11–18] where there is great voltage, before the current is converted from DC to AC at the inverter, and where the roof assembly could ignite and result in fire spread under the PV panels. Cheerfully, there have been some advances made by manufacturers during the past few years with regard to the electrical components that can reduce the potential for

ignition. Some PV panels have micro-inverters on each PV panel, which convert voltage from DC to AC. This can be expensive, but it reduces the probability of ignition [19].

5 In event where a PV system was not the cause of the fire, the PV system may still have an impact by limiting firefighter access in operations [20] (Figs. 4 and 5).

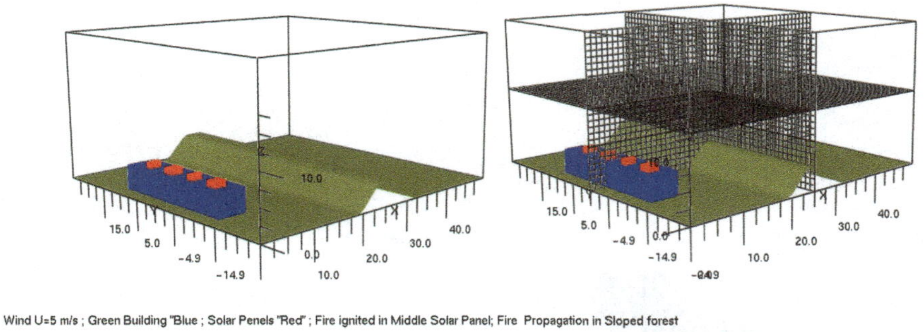

Fig. 4. Geometry model involve terrain vegetation, green building equiped with PV panel and grid location.

The fire shape inclined under a wind effect and propagated from ignition source to the grass vegetation and can caused a damage to wild urban interface and foresty (Fig. 6).

Here a flow velocity were presented to visualized the effect of building and their interaction with wind condition. We can see that house can play the role of turbulence promoter which can change the flow from stream line of laminar flow to the turbulent flow with vortex downstream building and can raise time of hot smoke and contribute to the heat transfer by convection and radiation and rise Rate of spread of fire (Fig. 7).

A numerical investigation to evaluate a radiation flux KW/m^2 with time to visualized the heat transfer by radiation from pool fire to the surrounding represented here by grass vegetation. The fire propagation depend directly to radiation flux apparition. Also, the imposed wind condition contribute to direction of the fire and radiation propagation (Fig. 8).

A smoke generated from fire after paralyze and combustion of solid materials such as PV panel, building constitutes and grass vegetation their temperature can be equivalent to 220 °C it is above many inflammation temperature of surrounding materials. A hottest region can be visualized downstream building which can favorite more fire propagation to the vegetation.

Heat release rate and masse loss rate of burner (cf. Fig. 9) are the most relevant parameters to quantify the fire degree. HRR can generated from fire can be divided to three heat transfer modes: thermal conduction, convection and radiation. The most dangerous is radiation it can be transferred without a contact from pool fire of PV panel to grass vegetation and exposed surface of vegetation due to slope hill.

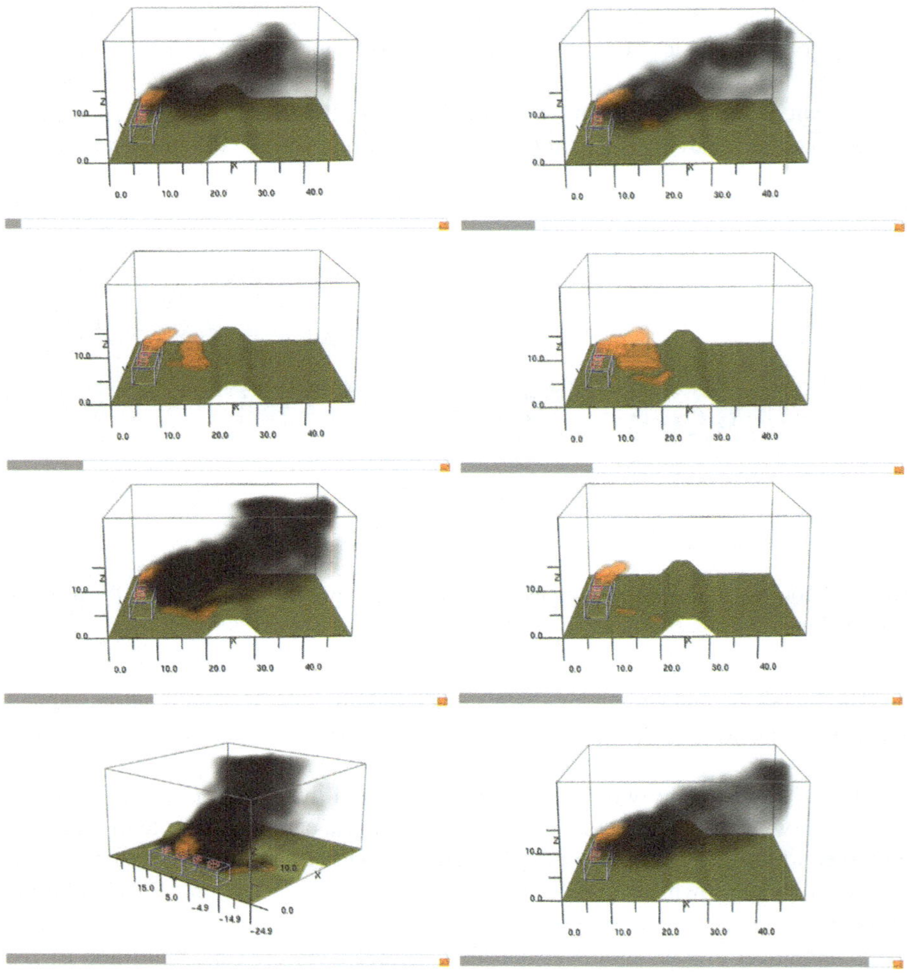

Fig. 5. PV rooftop fire in green building caused a forest fire under a wind condition.

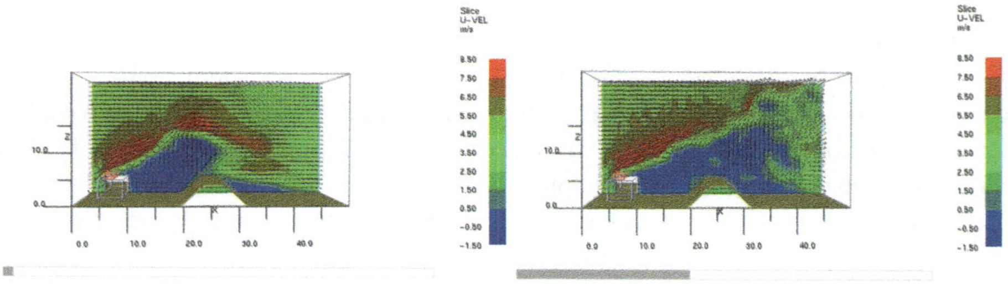

Fig. 6. Flow velocity under wind condition, building obstruction, fire smoke propagation.

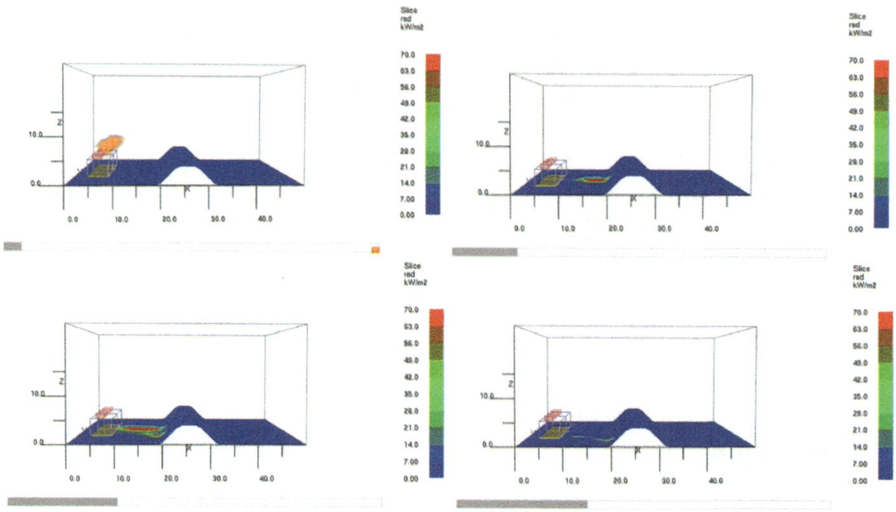

Fig. 7. Radiation flux during a fire propagation from PV to the grass vegetation

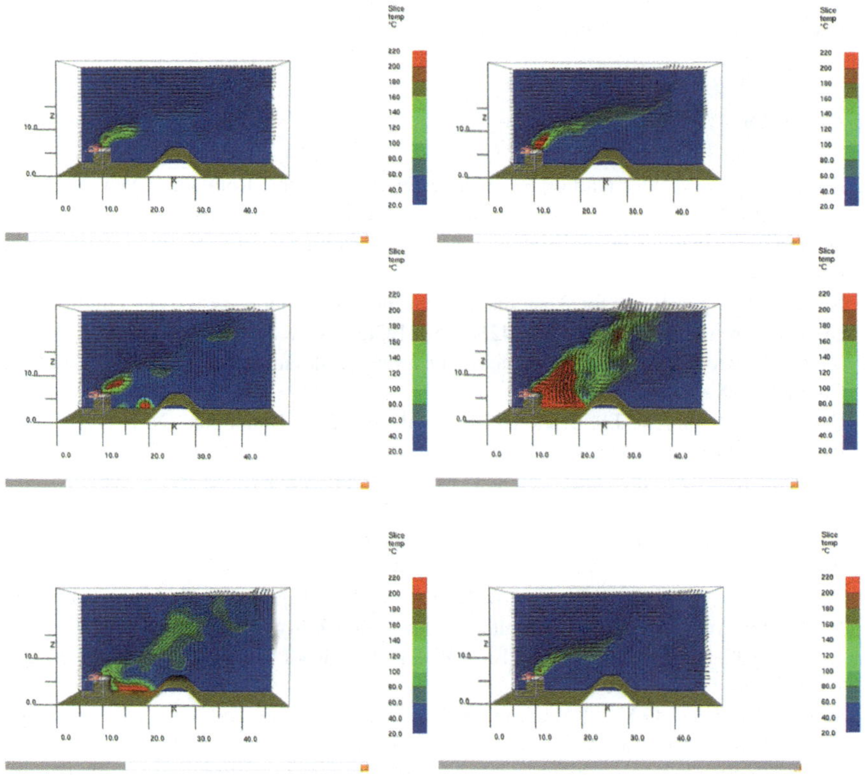

Fig. 8. Smoke and fire temperature

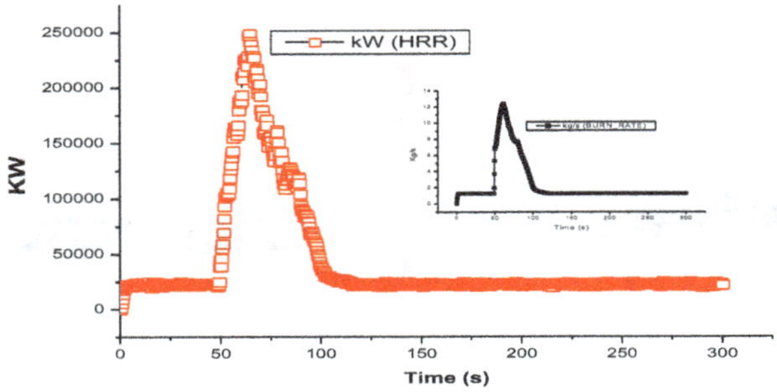

Fig. 9. Heat release rate HRR (KW) and Burning rate of PV panel and fuel bed (Kg/s) versus time.

3 Conclusion

Solar panel emergencies are not going away. Especially because the trend has been continuing to rise. Solar installations in the Algeria have doubled over the past year and are expected to continue. While the benefits of green energy are shining through to single family households and corporations, driving this type of energy use, like any use of technology, it creates safety implications that deserve our attention.

References

1. Quaschning, V.: Understanding Renewable Energy Systems. Routledge, New York (2016)
2. Saiah, S.B.D., Stambouli, A.B.: Prospective analysis for a long-term optimal energy mix planning in Algeria: towards high electricity generation security in 2062. Renew. Sustain. Energy Rev. **73**, 26–43 (2017)
3. Numerical: Biggest year for Australia's renewable energy industry (2017). http://numerical.co.in/numerons/collection/58a1400426cc65d01e2f3b1a. Accessed 10 Sept 2017
4. Smith, D.: Arc flash hazards on photovoltaic arrays. Colorado State University, Fort Collins, Colorado, USA (2013)
5. Earley, M.W., Sargent, J.S.: National Electrical Code 2011 Handbook. (NFPA) National Fire Protection Association, Quincy (2010)
6. Klement, K.: DC arc flash studied for solar photovoltaic systems: challenges and recommendations. IEEE Trans. Ind. Appl. **51**, 4239–4244 (2015)
7. Zhu, L., Ji, S., Liu, Y.: Generation and developing process of low voltage series DC arc. IEEE Trans. Plasma Sci. **42**, 2718–2719 (2014)
8. Zgonena, T., Ji, L., Dini, D.: Photovoltaic DC arc-fault circuit protection and UL subject 1699B. In: Proceedings of Photovoltaic Module Reliability Workshop (2011)
9. ASTM E108: Standard Test Methods for Fire Tests of Roof Coverings. ASTM International, West Conshohocken, PA (2011)
10. Mellit, A., Tina, G.M., Kalogirou, S.A.: Fault detection and diagnosis methods for photovoltaic systems: a review. Renew. Sustain. Energy Rev. **91**, 1–17 (2018)
11. UL 790: Standard for Standard Test Methods for Fire Tests of Roof Coverings, Underwriters' Laboratories, Northbrook, IL (2004)

12. Backstrom, B., Tabaddor, M.: Effect of Rack Mounted Photovoltaic Modules on the Fire Classification Rating of Roofing Assemblies. Underwriters' Laboratories, Northbrook, IL (2010)
13. Backstrom, B., Tabaddor, M.: Effect of Rack Mounted Photovoltaic Modules on the Flammability of Roofing Assemblies – Demonstration of Mitigation Concepts. Underwriters' Laboratories, Northbrook, IL (2010)
14. Backstrom, B., Sloan, D.: Effect of Rack Mounted Photovoltaic Modules on the Fire Classification Rating of Roofing Assemblies - Phase 2. Underwriters' Laboratories, Northbrook, IL (2012)
15. Backstrom, B., Sloan, D.: Characterization of Photovoltaic Materials – Critical Flux for Ignition/Propagation - Phase 3. Underwriters' Laboratories, Northbrook, IL (2012)
16. Backstrom, B., Sloan, D.: Report of Experiments of Minimum Gap and Flashing for Rack Mounted Photovoltaic Modules - Phase 4. Underwriters' Laboratories, Northbrook, IL (2012)
17. Backstrom, B., Sloan, D.: Considerations of Module Position on Roof Deck During Spread of Flame Tests - Phase 5. Underwriters' Laboratories, Northbrook, IL (2012)
18. Backstrom, B., Sloan, D.: "Validation of 42" PV Module Setback on Low Slope Roof Experiments - Project 7. Underwriters' Laboratories, Northbrook, IL (2012)
19. Hathout, A.: A comprehensive PV systems installation guide and designing a roof-based PV system as a demonstration system for troubleshooting errors in PV systems installations (2018)
20. Council: Fujisawa SST. Photovoltaics and Firefighters' Operations: Best Practices in Selected Countries (2017)

Black-box Accident Detection and Location System Based on the Raspberry Pi

Ibrahim Kadri[1,2(✉)], Boufeldja Kadri[1,2], Mohamed Beladgham[1,2], and Dahmane Oussama[1,2]

[1] Department of Electrical Engineering, Tahri Mohammed University-Bechar/lab: TIT, Mohamed Boudiaf District, 08000 Bechar, Algeria
hayamotoll@gmail.com, beladgham.tlm@gmail.com, oussamadahmane@gmail.com, Kadri_bf@yahoo.fr
[2] Department of Electrical Engineering, Tahri Mohammed University-Bechar/lab: CGRE, Mohamed Boudiaf District, 08000 Bechar, Algeria

Abstract. Nowadays, traffic accidents are one of the main reasons of fatalities in the world, many lives can be saved by reducing the time between the occurrence of an accident and the arrival of first help at its location. In research literature, numerous accident detection and notification systems were proposed to solve this problem. Including built-in unit, black-box, and mobile phone-based solutions. Our research has been devoted to design a black-box system that integrates Global System for Mobile Communication (GSM) and Global Positioning System (GPS) technologies to locate and send a Minimum Set of Data (MSD) containing important information about the accident collected using different sensors such as the accident type, longitude and latitude, time and speed of the vehicle to the emergency services (Hospitals, civil protection, and police). This system is developed based on the Raspberry Pi 3 in order to provide both low cost and high system performances.

Keywords: Accident detection · GSM · GPS · MSD · Raspberry Pi 3

1 Introduction

In today's world, Transportation has submitted great development over years. Unfortunately, this was a major reason of increasing the rate of fatal traffic accident [1]. Every year, nearly 1.2 million people die in road crashes. Besides, 20–50 million people are either severely injured or disabled [2].

As it was proven by previous researches, the traffic accident fatality is strongly depending on the response time of the emergency services: the time lapse taken by the intervention units to get to the accident site [3].

Automated car accident detection, or eCall systems, can reduce the response time of the emergency responders. In result, many lives could be saved, and the seriousness of the injuries can be decreased [3].

In order to provide rapid intervention services, we propose an Accident Detection and Location System (ADLS) that can send a Minimum Set of Data (MSD) from the vehicle to the emergency services [4]. This MSD contains the following information: -

M. Hatti (Ed.): ICAIRES 2019, LNNS 102, pp. 470–477, 2020.
https://doi.org/10.1007/978-3-030-37207-1_50

Accident type - Accident location (Longitude and latitude) - Time of the accident – Speed. The proposed system utilizes the GPS technology to calculate the exact longitude and latitude coordinates of the vehicle and send them through the Global System for Mobile communication (GSM) network to the emergency services. It also uses the Raspberry Pi to manage the acceleration and many other information collected using different types of sensors.

2 Related Works

eCall is an emergency call which allows to send an MSD to the Public Safety Answering Point (PSAP), it can be operated either manually by the occupant of the vehicle via the call button or automatically following a serious collision [5].

2.1 Built-in Units

Automobile manufacturers are currently working to equip their newer cars with an integrated (Built-in unit) eCall systems. The main advantage of such type is the possibility of accessing to the status of any in-vehicle sensor to determine the occurrence of an accident, such as accelerometers and airbag deployment monitors [3].

2.2 Black-Box Solutions

Unfortunately, most cars in countries where the average age of vehicles is much higher do not have automatic accident detection and notification systems (White et al. 2011). The perfect solution in this case is the use of 'Black-box' eCall solutions, standalone or deployable devices, which have personal, dedicated hardware for sensing, location and communication functions [3].

2.3 Mobile Phone Based Solution

With the widespread use of mobile communications devices, almost every passenger in a car owns a smartphone. This includes the main technical components, which are required by an eCall system (Accelerometer, GPS and GSM communication) [3]. Smartphone-based accident detection applications have both advantages and disadvantages relative to Built-in unit systems, e.g. they are not directly connected to the vehicle, and provide valuable data for accident analysis, including pictures and videos. However, they are not precise compared to other solutions because they can generate false positives [6].

3 Proposed System

The presented black-box system in this paper can reduce the response time by generating an automatic alerting message, indicating the location of the vehicle using a GPS device. This message will be sent to the emergency services, which will decide on the follow-up to be given to the alert. Since the time factor is essential in this type of emergencies, this process will save valuable time.

The proposed system architecture is depicted in Fig. 1. In the event of a traffic accident, an accident detection algorithm (ADA) will manage different data sources to detect the accident and give the order to the Raspberry Pi, which in turn will generate and send an alerting message to the emergency services using the GSM module containing the car's accident information.

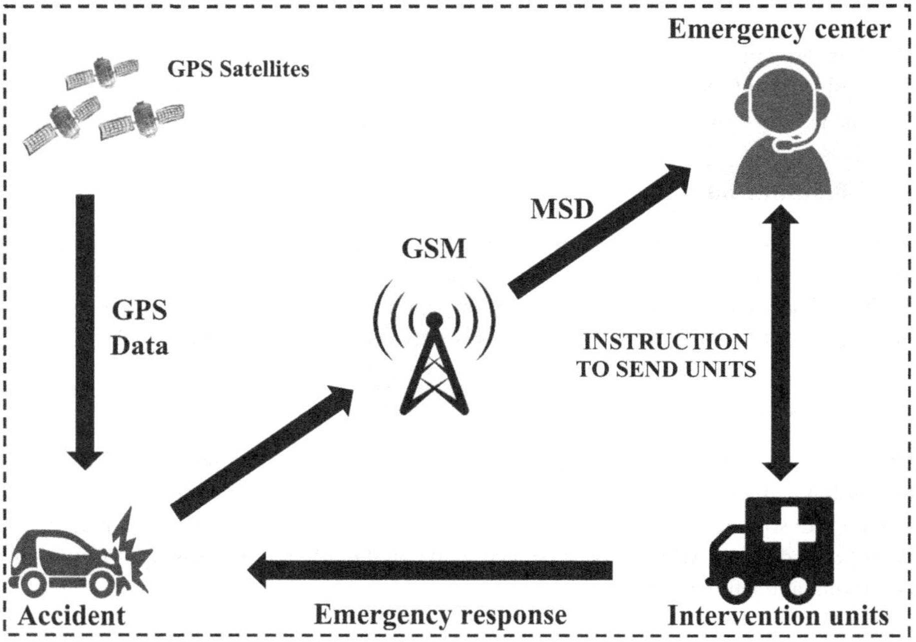

Fig. 1. Diagram of the proposed system.

4 Hardware Architecture

In order to provide and implement the eCall mechanism, an ADLS hardware architecture has been proposed and developed as shown in Fig. 2. The system is connected to the vehicle via the OBD-II communication. A Raspberry Pi 3 is utilized as a main system controller equipped with 1.2 Ghz of processing frequency to manage the different system's devices, the NEO6M V2 GPS module is used for locating the accident, where the responsible device of the mobile communication is a SIM900A GSM module.

The system is built as a mobile unit, this allows for possibility of displacement and maintenance.

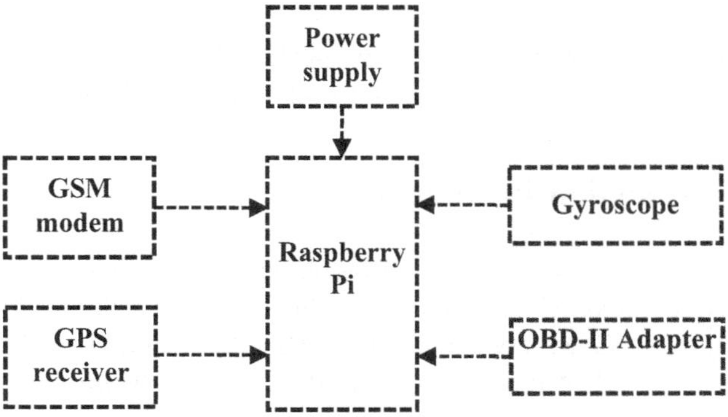

Fig. 2. System hardware architecture.

5 GPS for Accident Detection Systems

The Global Positioning System (GPS) is a satellite-based navigational system designed and operated by the US department of Defense for military and civilian use [7]. The GPS module determines the location of the car accident by comparing the time signals it receives from a number of satellites and triangulating on the known positions of each satellite. GPS can provide the accurate location of the accident place [8].

The GPS receiver provides data through Universal Asynchronous Receiver Transmitter (UART) depending on the NMEA(National Marine Electronics Association) standards, NMEA specify the communication by several sentences which are independent from each other [8].

The $GPRMC sentence from the NMEA 0183 communication protocol is widely used to transmit data in GPS devices. The time, latitude, longitude and speed can be extracted from the $GPRMC sentence. The format of the $GPRMC sentence is presented in Fig. 3 [9, 10].

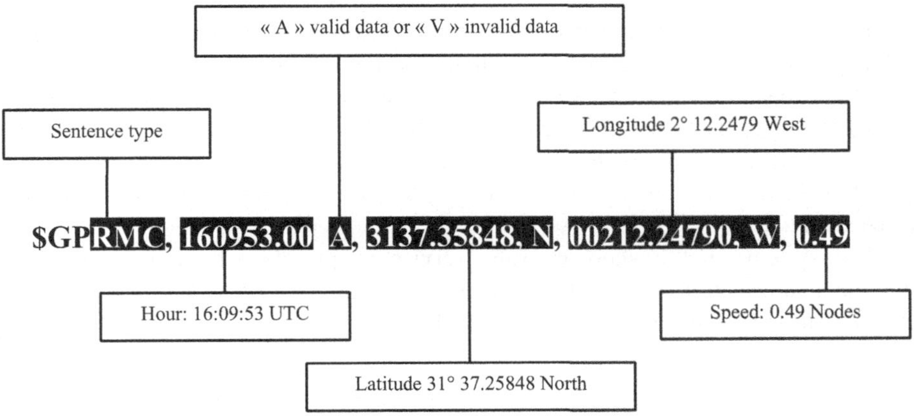

Fig. 3. The $GPRMC sentence.

The GPS data requires conversion before being useful, the latitude and longitude must be converted from the initial format provided by the $GPRMC sentence (Degrees, minutes, and seconds: ddmm.ssss) to decimal coordinates (dd.mmssss).

6 Data Transmission Procedure

The accident detection system must send the accident information in real-time to the emergency services in order to provide fast intervention, a Short Message Service (SMS) is sent through GSM modem which contains the longitude and latitude, occurrence time, speed, accident type and the accident severity based on the deceleration value.

The raspberry Pi can control the GSM modem via AT commands. At first, the echo is disabled using the 'ATE0'command. The 'AT+CMGF=1' is used to set up the text mode (SMS) and the 'AT+CMGS' command is utilized to send the SMS to the preselected recipient number.

7 Accident Detection Algorithm

In this paper, the vehicle accident detection phase is based on the use of different sensors depending on the type of the accident. The shock sensor is used to detect the vehicle collisions, the system has also gyroscope and flame sensor in case of vehicle overturn or burn respectively. The accident location phase will then be activated in order to obtain the exact position and time of the accident using the GPS technology, this information will be sent to the emergency services as well as the accident type through the GSM communication. The flowchart of the proposed algorithm for accident detection system is represented in Fig. 4.

8 Result and Discussion

A robust accident detection and location system has been developed and presented in this paper, an accident detection algorithm has been implemented and tested in order to detect the vehicle accident in its different types and send its location obtained through the GPS system using the GSM communication. Fig. 5 shows the test of the shock sensor using the system prototype by applying strong vibration, when Fig. 6 Shows the test of the flame sensor. Figs. 7 and 8 show the test of the vehicle overturn right and left respectively.

The system send a short message service to preselected number containing important information such as the accident type, time and the exact location of the accident, the message format is depicted in Fig. 9.

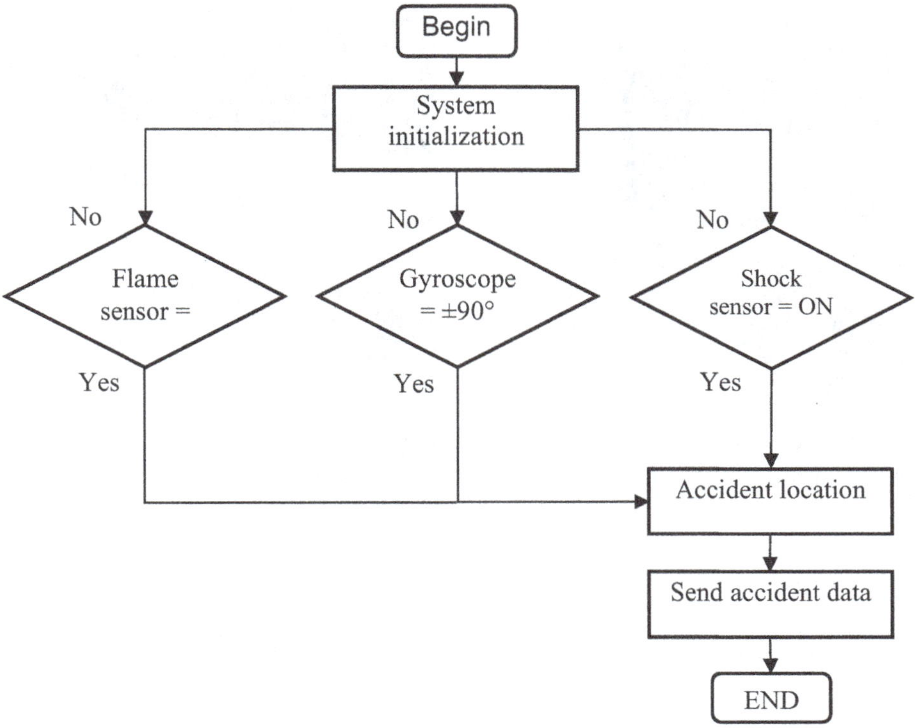

Fig. 4. Organigram of the accident detection algorithm.

Fig. 5. Test of the vehicle collision detection system.

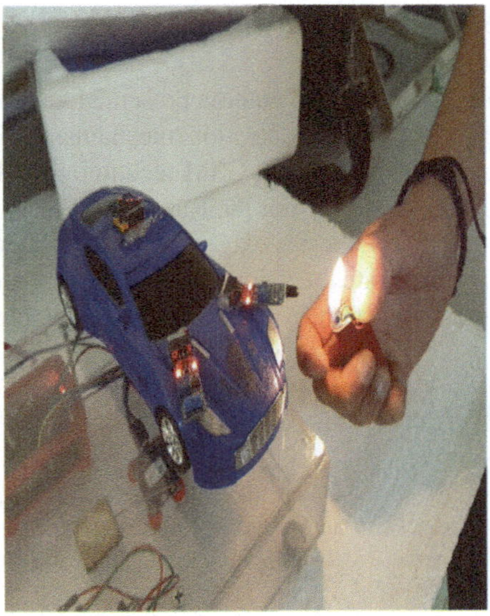

Fig. 6. Test of the vehicle fire detection system.

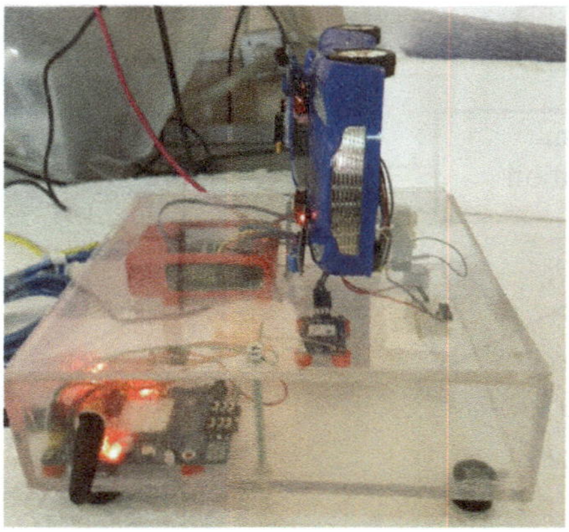

Fig. 7. Test of the vehicle overturn detection system (right).

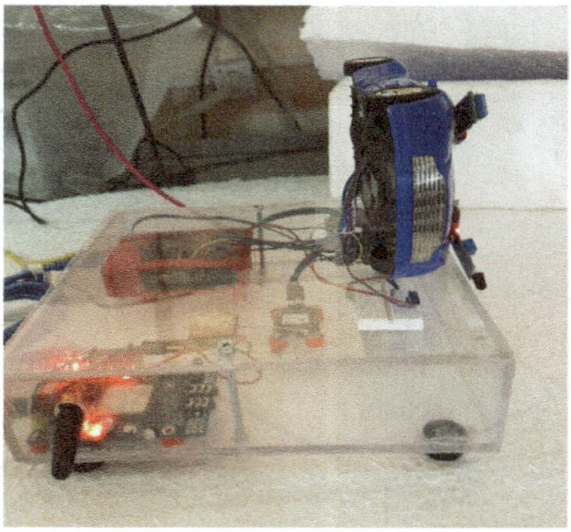

Fig. 8. Test of the vehicle overturn detection system (left).

Fig. 9. The SMS format.

9 Conclusion

In this article, the authors presented an Accident Detection and Location System, which provides robust detection mechanism of road accidents using fast reporting method using both GPS and GSM technologies. The main contribution in our system centered on the idea of using the Raspberry Pi as a main system controler. The response time of the system doesn't exceed 20 s which is a good result. In future work, we are planning to ameliorate the system in order to add more accurate accident detection procedure and test the system by simulating real vehicle accidents.

References

1. Meena, A., Iyer, S., Nimje, M., Jogjekar, S., Jagtap, S., Rahman, M.: Automatic accident detection and reporting framework for two wheelers. In: IEEE International Conference on Advanced Communication Control and Computing Technologies (ICACCCT), pp. 962–967 (2014)
2. World Health Organization, GLOPAL STATUS REPORT ON ROAD SAFETY, Geneva (2009)

3. Bonyár, A., Géczy, A., Krammer, O., Sántha, H., Illés, B., Kámán, J.: A review on current eCall systems for autonomous car accident detection. In: 40th International Spring Seminar on Electronics Technology (ISSE), pp. 1–8 (2017)
4. Megalingam, R.K., Nair, R.N., Prakhya, S.M.: Wireless vehicular accident detection and reporting system. In: 2010 International Conference on Mechanical and Electrical Technology (ICMET 2010), pp. 636–640 (2010)
5. Kremonas, P., Pâris, J.: Everything you wanted to ask, but did not know how…. In: The European Emergency Number Association (EENA) (2015)
6. White, J., Thompson, C., Turner, H., Dougherty, B., Schmidt, D.C.: WreckWatch : Automatic Traffic Accident Detection and Notification with Smartphones. In: Springer Science+Business Media, LLC, 285–303 (2011). https://doi.org/10.1007/s11036-011-0304-8
7. Drosos, B.K., Malesios, Ch.: Measuring the accuracy and precision of the garmin gps positioning in forested areas: a case study in taxiarchis - vrastama university forest. J. Environ. Sci. Eng. **1**(4), 566–576 (2012)
8. Amin, S., Bin, M., Reaz, I., et al.: Low cost GPS/IMU integrated accident detection and location system. Indian J. Sci. Technol. **9**(10), 1–9 (2016). https://doi.org/10.17485/ijst/2016/v9i10/80221
9. El-Rabban, A.: Introduction to GPS: The Global Positioning System, 2nd Ed, Artech House, Norwood (2009)
10. Motte, H., Wyffels, J., De Strycker, L., Goemaere, J.: Evaluating gps data in indoor environments. Adv. Electric. Comput. Eng. **11**(3), 25–28 (2011)

IoT-Based Smart Photovoltaic Arrays for Remote Sensing and Fault Identification

A. Hamied[1], A. Boubidi[1], N. Rouibah[2], W. Chine[1], and A. Mellit[1(✉)]

[1] Electronics Department, Renewable Energy Laboratory, Jijel University, Jijel, Algeria
adelmellit2013@gmail.com

[2] Electronics and Computer Sciences, USTHB, Algiers, Algeria

Abstract. In this paper a smart remote sensing prototype for fault detection and identification of photovoltaic arrays is presented. A simple fault detection and identification algorithm has been incorporated into the designed prototype. The designed system detect automatically the fault, and then identify the origin of some investigated faults (such as open circuit, short circuit, dust accumulation and shedding effect). The prototype has been tested experimentally at the Renewable Energy Laboratory of Jijel University, Algeria. The experimental results showed the capability of the prototype to remote sensing, detecting and faults identification. The status of the system can be checked online via a website using Internet of Things (IoT) and the user could be also notified by a SMS.

Keywords: Photovoltaic · Monitoring · Faults detection and identification · Remote sensing · IoT

1 Introduction

With reference to IEA, the overall capacity of photovoltaic (PV) installations around the world is about 500 GW [1]. Millions of PV plants were installed worldwide, so in order to avoid losses due to some faults, the status of the systems should be checked and monitored online. Some equipment are recently commercialized, however, most of them are mainly used to protect and isolate the system from faults [2], and they are not able to identify the type of faults, even to predict the nature or the origin of the faults.

Recently, some works have been started to develop and incorporate intelligent fault detection and diagnosis methods to detect and diagnosis of PV plants [3]. However, designing a sophisticated equipment that can detect, localize and identify different faults remain a challenging issue. Remote sensing of PV plants paly a very important role in supervising and monitoring of the plants, in addition, remote sensing and monitoring abilities can provide the information in advance when any abnormal happen on the PV plant without any additional efforts, particularly in plants located in inaccessible areas (e.g. remote Saharan areas in Algeria).

Internet of Things (IoT) has attracted enormous attention from different sectors including renewable energy applications. It enables communication and data sharing among a wide variety of devices, by using the Internet or others networks, without

M. Hatti (Ed.): ICAIRES 2019, LNNS 102, pp. 478–486, 2020.
https://doi.org/10.1007/978-3-030-37207-1_51

human interaction. Furthermore, the IoT can help users in checking and identifying malfunctioning of the systems in real time (online) [4]. The main role of the application of IoT in PV systems is to make it possible to access and check the status of the plant remotely via Internet.

Recently, there are some attempts about remote sensing, monitoring, fault detection and diagnosis of PV plants based on IoT. For example in [5], the first real-time remote monitoring of PV system was developed based on a low cost ZigBee wireless network, the system allows to control the power point. In [6], another application of remote sensing of solar power generation system using a ZigBee network. According to the authors the designed remote monitoring system can help users by decision-making reference to the safe operation and daily maintenance as well as management of PV power generation. An effective implementation of an intelligent remote MS for solar PV Power Conditioning Units (PCU) is presented. The system can be installed in solar PV-PCU to provide support for management and maintenance operations, is proposed in [7]. The most application of IoT in photovoltaic array are reported in [8] including monitoring, control, supervisor and remote sensing. A smart monitoring device has been designed in [9], the device is integrated with a PV module and can transmit data to IoT central for energy management and control of a PV array. An on-line IoT based remote monitoring system for PV arrays is proposed in [10], the authors also proposed a fault diagnosis approach based on extreme learning machine. In [11] the authors designed a low cost IoT based embedded photovoltaic monitoring system using a GPRS module.

In our previous works we have designed an IoT based prototype for monitoring current and voltage data of a PV array, as well as environmental data (such as air temperature and in-plane solar irradiance) [12], and then we have designed a low–cost prototype for monitoring data and fault detection of a stand-alone PV system [13]. The main contribution in this work is to implement and verify experimentally a new fault detection and identification procedure, as well as improving the low-cost prototype presented in [12, 13]. The investigated faults are short circuit, open circuit, dust accumulation, and shading effects on a PV array.

The paper is organized as follows: Sect. 2 presents a description of the investigated PV array, the experimental design, including procedure development and prototype designing. Results and discussion are reported in Sect. 3.

2 Methodology and Materials

The investigated PV array is installed at the RELab of Jijel University (Algeria). It consists of 18 PV modules, 12 storage batteries, two regulators and one inverter. More details about the system can be found in our last papers [12, 13]. The experiment has been carried out at the RELab, the investigated system consists of 3 PV modules connected in parallel, PV module specifications are reported in [13]. Figure 1a shows the facility test and Fig. 1b depicts the designed prototype, it consists of an acquisition board using an Arduino Mega 2560 microcontroller, a 16 $\times$ 4 LCD display to show the system status and monitored data in real time, a CDN package in MatLab environment has been used instead of using the ESP-01 Wifi [13], for data transmission to a host PC.

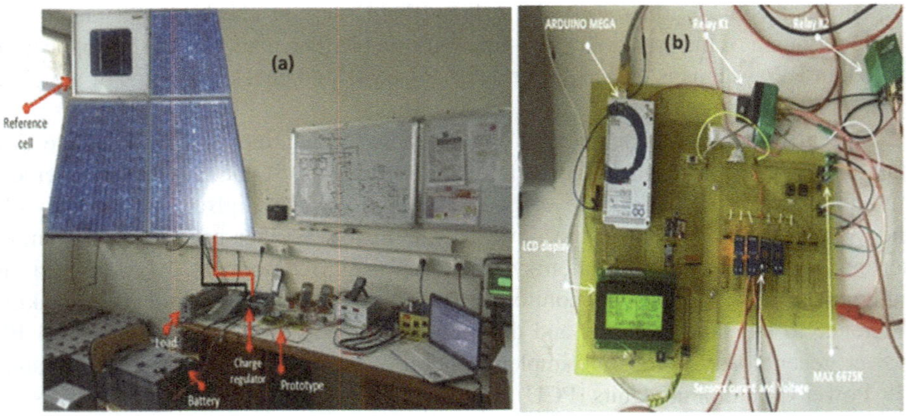

Fig. 1. (a) Test facility at RELab., Jijel University (b) Photo of the designed prototype

The employed components are reported in the following Table 1. The total cost of the prototype is about 72 euros.

Table 1. The used components, specifications and cost.

Component	Specification	Cost (Euro)
Current sensor	ACS712 (Imax = 30 A)	11
Voltage sensor	DC < 25 V	5
Temperature sensor	Max6623 Type K	10
Reference solar cell	Isofoton	15
Arduino mega2560 board	ATMEGA256016AU1738	23
Relay	MAXTOR (30 A–12 V)	5
LCD	LCD 16 × 4	6
		72 Euros

In the literature, numerous transmission techniques were developed for monitoring purpose, most typically the Ethernet or Zigbee wireless technology, etc. The well-known one diode model is used to estimate the produced output power. A detection and identification algorithm has been developed using MatLab and implemented into the Arduino Mega board for an experimental verification. The procedure of the process is shown in Fig. 2, it can be mainly divided into three parts:

- *Acquisition*
  It consists of collecting some physical parameters such as cell temperature, PV voltage, PV current, and in-plane solar irradiance (See Fig. 2 *Acquisition part*).
- *Detection*
  The detection part aims to detect the presence of any fault in the PV array. To do this, a simple comparison between measured (P_{max_mes}) and estimated (P_{max_cal})

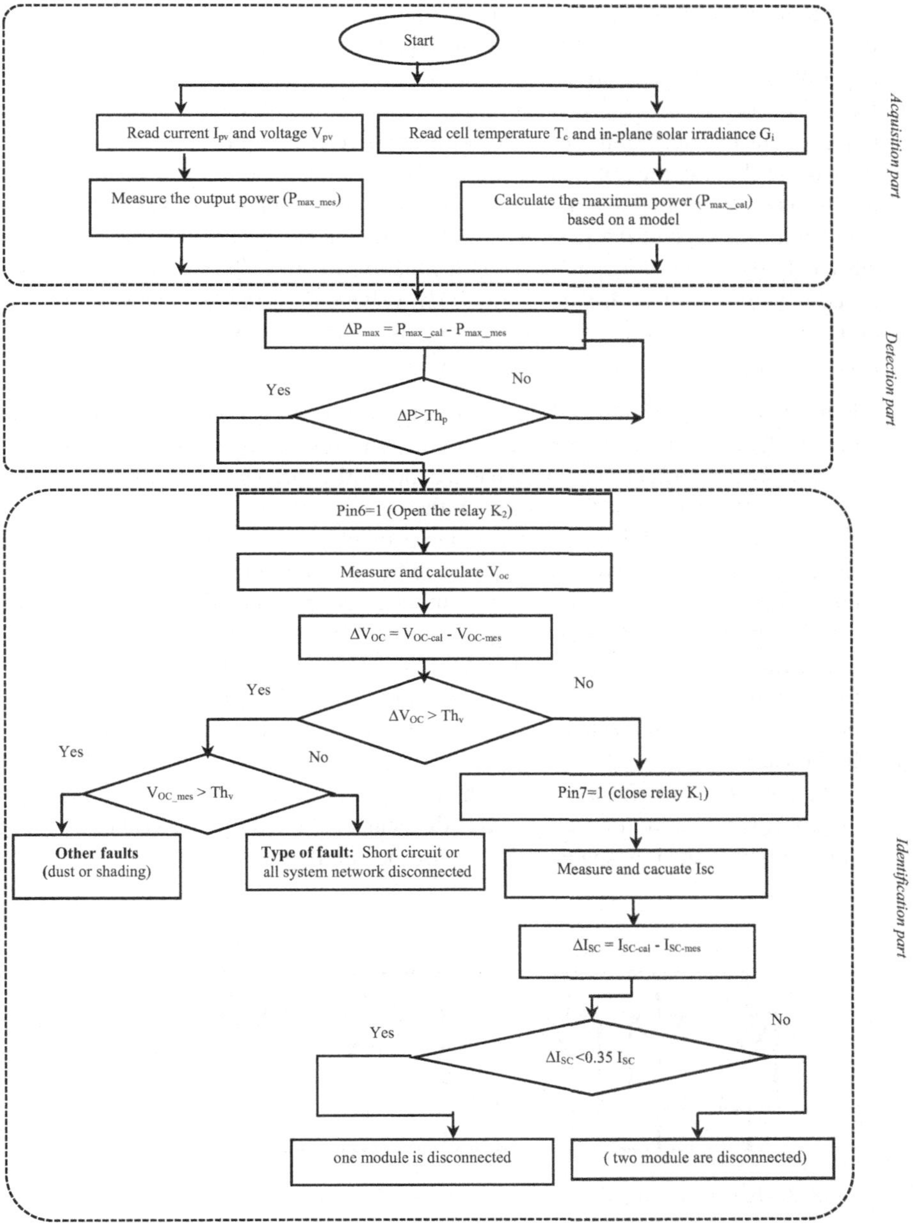

Fig. 2. Flowchart the designed procedure (detection and identification)

maximum powers under the same operating conditions, $\Delta P_{max} = P_{max\_cal} - P_{max\_mes}$ (Fig. 2, *Detection part*). A threshold Th_p has been defined after several experiments (Th_p = 0.25 W).

- *Identification*

  Once a fault is detected, the next step consist to determine the type of this fault. In the case of the presence of fault $\Delta P_{max} > Th_p$, so a decrease in the measured power is observed, it means that a decrease in the amount of voltage (V_{pv}) or current (I_{pv}). To identify the type of the fault, we are based on the analysis of the variation of the current I_{pv}(t) or the voltage V_{pv}(t) as a function of time. Thus, the identification process begin by measuring the open circuit voltage (V_{oc_mes}), so this isolation was done by opening relay K_2 (See Fig. 3). We start by comparing this quantity $\Delta V_{OC} = V_{OC\text{-}cal} - V_{OC\text{-}mes}$. There are two cases:

(1) In the case of the deviation ($\Delta V_{oc} < Th_v$), the threshold Th_v ($Th_v \approx 0.5$ V), it means that the fault don't affect the voltage, so we can exclude three types of faults (short circuit, shading effect and disconnected network), because the voltage in these types would be zero ≈ 0 V, except in the case of shading effect, less decrease in V_{OC}. As the PV modules are connected in parallel, one PV module can produce the same output voltage. So there is a possibility that one or more PV modules are disconnected. Thus, to identify the number disconnected PV modules, the short circuit current should be measured (I_{SC}_mes) and compared with the calculated one (I_{SC}_cal), and compute the difference ($\Delta I_{SC} = I_{SC}$_cal − I_{SC}-mes). The system isolation is done by closing relay K_1 (See Fig. 3). If ΔI_{sc} is less than 0.35 Isc_cal, one module is disconnected, otherwise two PV modules are disconnected.

(2) In the case of the deviation ($\Delta V_{oc} > Th_v$), it is necessary to re-examine the losses in voltage with respect to the threshold Th_v. If $V_{oc\_mes} < Th_v$ we conclude that there is a short circuit or our system is disconnected totally, otherwise other faults are happened, such as shading effect, dust accumulation, etc.

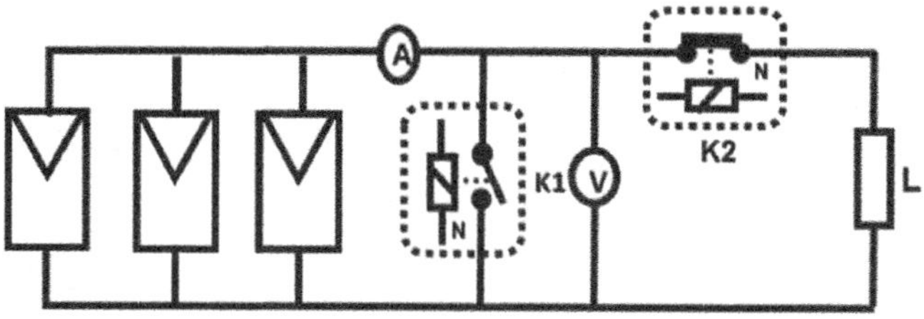

Fig. 3. Electrical circuit for measuring V_{OC} and I_{SC} using two relies K_1 and K_2

3 Results and Discussions

The experiments have been carried out on 7th July 2019. After connecting to the server, the data from the system have been sent to the network via IoT. The results can be displayed online on the designed webpage. So, the investigated faults are shown in Fig. 4, open circuit, dust accumulation, short circuit and shading effect.

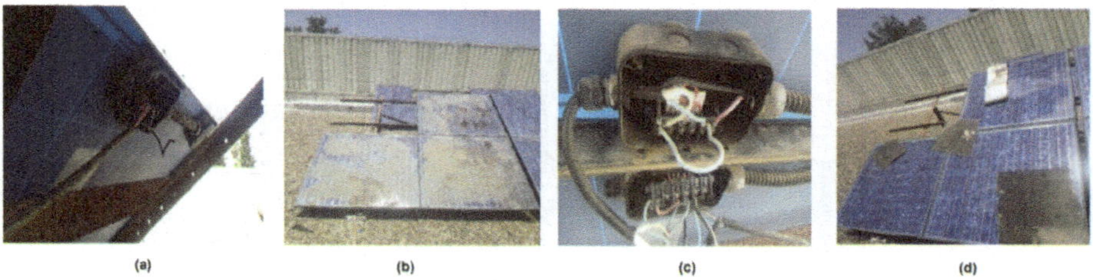

Fig. 4. The investigated faults: (a) open circuit, (b) dust accumulation, (c) short circuit and (d) shading effect.

To check the effectiveness of the procedure an experimental test has been carried out. Figure 5 shows the current variation *I(t)* during this PV modules test. As can be seen for Fig. 4, the variation of $I_{SC} \approx 19$ A, (it means three PV modules are connected, normal operation), in the case of $I_{SC} \approx 13$ A (one PV module is disconnected) and in the case of $I_{SC} \approx 6.2$ A (two PV modules are disconnected). This result demonstrate the effectiveness of the procedure.

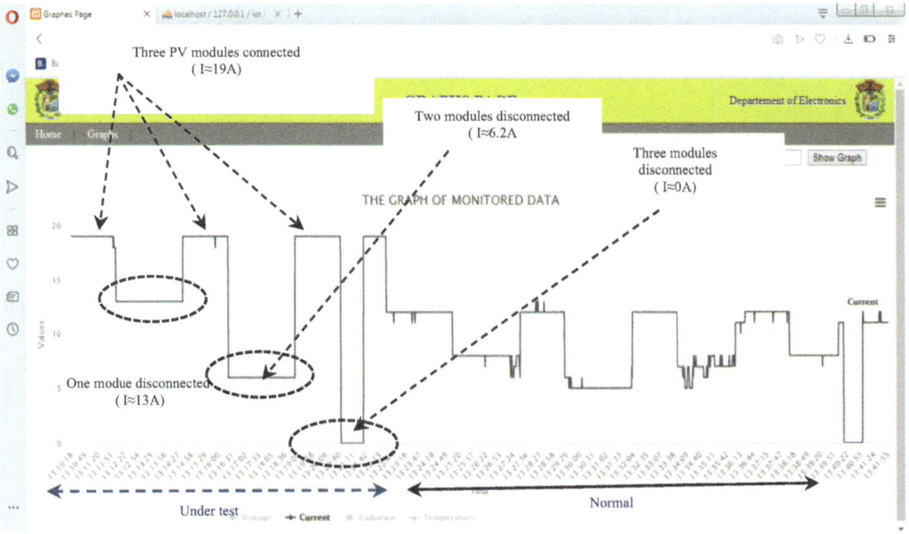

Fig. 5. Current evolution during the PV modules disconnection (online representation using IoT).

Figure 6 displays the evolution of the monitored data (current, voltage, cell temperature and in-plane solar irradiance) via IoT for a period of 3 h (7 July 2019). At the period 12 h 9 min–12 h 38 min the system work correctly (normal case), the measured current and voltage are 12.2 A and 11 V respectively, the cell temperature is 46° and the in-plane solar irradiance is 860 W/m^2. At 12 h 38 min 15 s the system detect automatically a fault, and then the identification procedure start by measuring the open

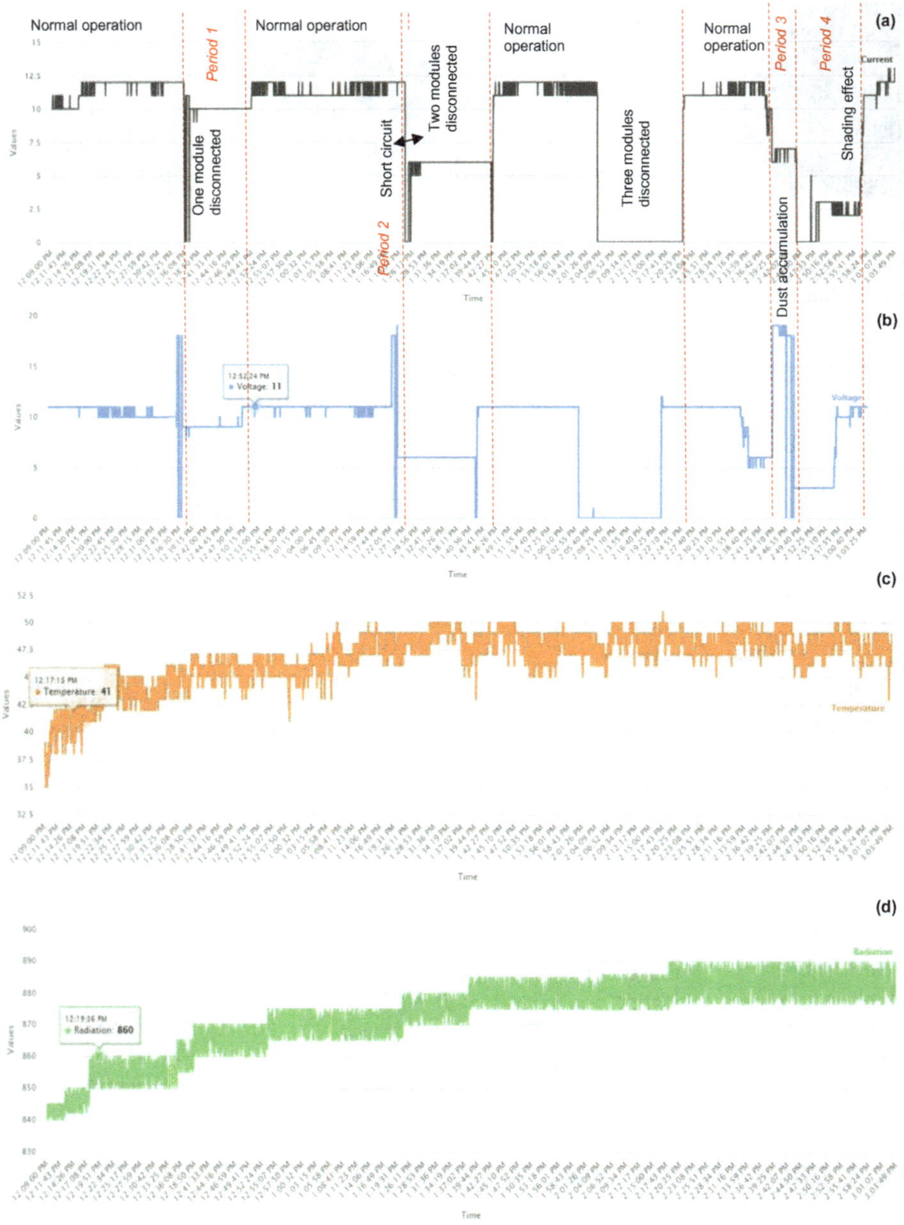

Fig. 6. Online evolution of the measured data (3 h test duration): (a) voltage, (b) current, (c) cell temperature and (d) in-plane solar irradiance. Period 1: open circuit of one PV module, period 2: short circuit, period 3: dust accumulation on the PV array and period 4: shading effect on the PV array.

circuit voltage (V_{OC} = 18 V), and at 12 h 38 min 30 s the I_{SC} is also measured (I_{SC} = 11.5A), then the system notify the user about the fault and display the type of the error online (in this case the system display this message: *One PV module is disconnected*), the duration of the fault is about 14 min. From 12 h 52 min to 13 h 26 min the system work correctly without any fault (normal case). The period from 13 h 26 min to 13 h 42 min a decrease in the current is observed and the system automatically indicate and display the type of the fault which corresponds to '*two modules are disconnected*'. The system work correctly without any fault from 13 h 42 to 14 h 4 min. While during the period from 14 h 4 min to 14 h 20 min the system is disconnected completely. At 13 h 26 min we have a short circuit, and the maximum current is about 12.2 A and voltage is 0 V. In the period from 14 h 39 min to 14 h 44 min the observed current is decreased (I_{SC} = 10A) as well as the voltage (V_{OC} = 11 V), which correspond to the '*dust accumulation*' on the PV array, the duration of the fault is about 5 min. The period from 14 h 52 min to 15 h 2 min, corresponding to the '*shading effect*', a decrease in the measured current (I_{SC} = 5A) and small decrease in the voltage (V_{OC} = 17A) is observed, the duration of the fault is about 10 min.

4 Conclusion

In this paper a smart photovoltaic remote sensing system for fault detection and identification has been developed. The designed prototype was verified experimentally at RELab of Jijel University, and shown the capability to identify the type of some investigated faults (e.g. short circuit, open circuit, dust accumulation and shading effect). The recorded data have been transmitted to the Internet via IoT technology. The stored data are available to the users with web-based interface, in which users can browse all recorded data in real-time and check the status of the system.

The designed prototype is more suitable for isolated areas (PV systems installed in Saharan regions, case of south of Algeria, as well as for domestic applications). The prototype enable users to monitor and check the installation remotely from anywhere and at any time. The major drawback of the identification algorithm is that is not able to identify faults that have the same symptoms or signatures, (e.g. different types of shading, dust accumulation, etc.).

This work could be further improved and extended for fault diagnosis of photovoltaic systems by using more sophisticated algorithms that are able to distinguish between faults that have the same signature and make a clear decision. Also, verification of the method for a large size PV array and faults localization.

References

1. IEA: Snapshot of global photovoltaic markets. Accessed March 2018
2. Daliento, S., Chouder, A., Guerriero, P., Pavan, A.M., Mellit, A., Moeini, R., Tricoli, P.: Monitoring, diagnosis, and power forecasting for photovoltaic fields: a review. Int. J. Photoenergy **2017**, 13 (2017)

3. Mellit, A., Tina, G.M., Kalogirou, S.A.: Fault detection and diagnosis methods for photovoltaic systems: a review. Renew. Sustain. Energy Rev. **91**, 1–17 (2018)
4. Barsocchi, P., Cassara, P., Mavilia, F., Pellegrini, D.: Sensing a city's state of health: structural monitoring system by Internet-of-Things wireless sensing devices. IEEE Consum. Electron. Mag. **7**, 22–31 (2018)
5. Xiaoli, X., Daoe, Q.: Remote monitoring and control of photovoltaic system using wireless sensor network. In: International Conference on Electric Information and Control Engineering, pp. 633–638 (2011)
6. Xu, X.L., Wang, H.: Construction of solar PV power generation remote monitoring system in the architecture of Internet of Things. In: Advanced Materials Research, vol. 347, pp. 178–182 (2012)
7. Adhya, S., Saha, D., Das, A., Jana, J., Saha, H.: An IoT based smart solar photovoltaic remote monitoring and control unit. In: 2nd International Conference on Control, Instrumentation, Energy and Communication (CIEC), pp. 432–436 (2018)
8. Kumar, N.M., Atluri, K., Palaparthi, S.: Internet of Things (IoT) in photovoltaic systems. In: National Power Engineering Conference (NPEC), pp. 1–4 (2018)
9. Spanias, A.S.: Solar energy management as an Internet of Things (IoT) application. In: 8th International Conference on Information, Intelligence, Systems & Applications (IISA), pp. 1–4 (2017)
10. Li, Y.F., Lin, P.J., Zhou, H.F., Chen, Z.C., Wu, L.J., Cheng, S.Y., Su, F.P.: On-line monitoring system of PV array based on Internet of Things technology. In: IOP Conference Series: Earth and Environmental Science, vol. 93, no. 1, p. 012078 (2017)
11. Kekre, A., Gawre, S.K.: Solar photovoltaic remote monitoring system using IOT. In: International Conference on Recent Innovations in Signal processing and Embedded Systems (RISE), pp. 619–623 (2017)
12. Hamied, A., Mellit, A., Zoulid, M.A., Birouk, R.: IoT-based experimental prototype for monitoring of photovoltaic arrays. In: 2nd International Conference on Applied Smart Systems (ICASS), pp. 1–5 (2018)
13. Mellit, A., Hamied, A., Pavan, A.M., Lughi, V.: A low-cost monitoring and fault detection system for stand-alone photovoltaic systems using IoT technique. In: ELECTRIMACS 2019 – Salerno, Italy, 21–23 May 2019 (2019)

Simulation of a Stand-Alone Mini-Central Photovoltaic System Designed for Farms

Benlaria Ismail(✉), Belhadj Mohammed, Othmane Abdelkhalek, Bendjellouli Zakaria, and Chakar Abdeselem

Laboratory Smart Grid and Renewable Energy (SGRE), University of Bechar, PO BOX 417, 08000 Bechar, Algeria
utmb@univ-bechar.dz

Abstract. With the high cost of electricity that continues to increase year on year, the renewable energy has attracted a lot of attention all over the world in the recent times due to the growing energy demand and increased prices of fossil fuels. The solar energy is the most prominent among all the renewable sources. Many farms are under a lot of pressure to maintain profitability. Through an attractive power purchase agreement, the mini photovoltaic centrals can offer for farms a free solar PV system allowing them to economize on energy at the lowest cost and over the long term. In this paper, we are interested in studying and simulating the various elements of a mini-central photovoltaic consisting of solar panels and DC-DC inverter with the MPPT control, DC-AC inverter and battery, to cover the needs, of the required electrical energy, of small farms.

Keywords: Photovoltaic central · DC-DC inverter · DC-AC inverter · MPPT · Small farms

1 Introduction

Given that one-day fossil fuels will end, a need arises to find alternative fuels. Renewable energy is considered as an alternative to fossil fuels and nowadays it attracts much attention. Among renewable energy sources, solar is the most important because it is available in all parts of the world. In one hour, the earth receives enough energy from the sun to meet its energy needs for nearly a year [1]. Also, this energy source is used in various industries including agriculture and it can be used in cultivating crops in the farthest corners of the world [2]. Photovoltaic is the direct conversion of sunlight to electricity. It is an attractive alternative to conventional sources of electricity for many reasons: it is safe, silent, and non-polluting [3].

Solar farming uses power generated from solar energy to operate agricultural or farming tools. It is simple, cost effective, reliable and long lasting. Most common agricultural tools such as watering systems, sprayers, etc., work on battery power and fuel oil. In solar farming, the battery power is replaced with solar power, so that the usage of electricity from grid-power and non-renewable sources can be reduced. Photovoltaic agriculture, the combination of photovoltaic power generation and agricultural activities, is a natural response to supply the green and sustainable electricity for agriculture. There are several main application modes of photovoltaic agriculture

M. Hatti (Ed.): ICAIRES 2019, LNNS 102, pp. 487–495, 2020.
https://doi.org/10.1007/978-3-030-37207-1_52

such as photovoltaic agricultural greenhouse, photovoltaic breeding, photovoltaic wastewater purification, photovoltaic water pumping and rural solar power station.

Nowadays, many farmers have started implementing solar technologies on their farms. Solar PV energy can be used in a wide range of applications, when it comes to the farming sector. The most common use of such technologies is that of implementing them on the roofs of the barns/buildings in order to generate electricity, and for water pumping. All this for powerfully promote the development environmental agriculture and increase economic benefits of farmers, Therefore, photovoltaic agriculture provides new opportunity for photovoltaic industry, thus not only to solve the dilemma of overcapacity for photovoltaic industry effectively, but also to accelerate the development of modern agriculture.

Our study is about modeling and simulation of the components of a mini-central photovoltaic to produce enough electricity for four farms, to reduce the costs, and expenses of consuming electricity for the farmer.

2 Description of the Mini-Central PV System

The station feeds four adjacent farms by electricity, which is used for irrigation especially and for lighting and some other uses as described in Fig. 1. The mini-central PV plant is based on an autonomous or stand-alone system, which does not include any extra energy source; the stand-alone system usually includes accumulators. In some cases, for example, water pumping, or the ventilation of a greenhouse, we need electric power during sunny periods only. Autonomous systems are practical in isolated places, where there is no conventional electricity grid [4].

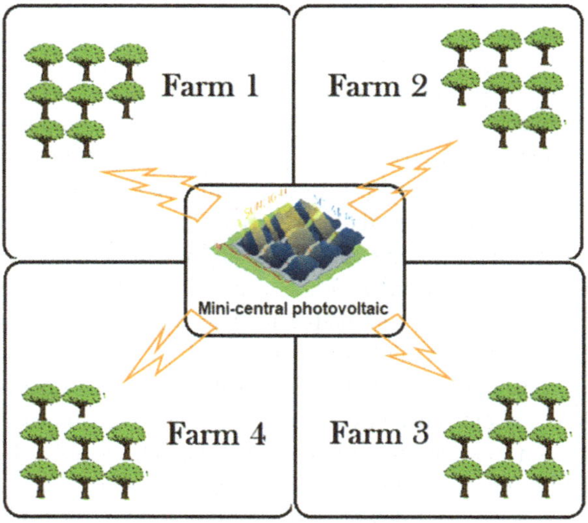

Fig. 1. Schematic location of the PV central with the four farms

3 Stand-Alone Solar PV System

The stand-alone electricity generation systems using PV technology has come up as a major and favored way to harness the solar energy due to its multi-dimensional advantages such as energy independence, safety, security, easier and timely installation, long-term back-up in case of storage system and power whenever and wherever you needed [6]. Therefore, the stand-alone solar PV system is an ultimate, convenient and self-sufficient alternative to provide electricity for remote locations where grid extension is practically unviable.

A stand-alone system based upon solar power comprises of a PV panels array to collect solar energy, a charge controller as a control unit, a battery as a storage device and an inverter for DC/AC conversion for AC loads [6].

The stand-alone system is the subject of this study of the mini-PV plant; it generally includes four (4) main elements [5]: One or more PV modules, the regulation system, one or more batteries, and the inverter. They are connected as in Fig. 2.

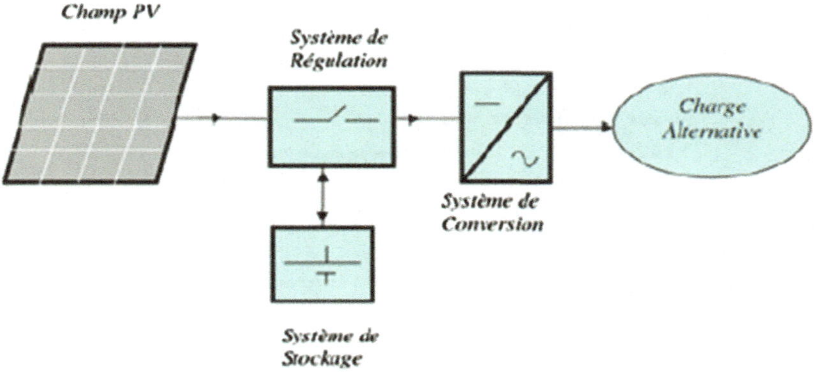

Fig. 2. Simplified diagram of an autonomous photovoltaic installation

The system of the mini-central PV consists of Photovoltaic array that transfers the solar energy to electrical energy, connected to a DC-DC converter controlled by an MPPT control to extract the maximum of power energy from the photovoltaic panels, this energy is stored in a bank of battery in the case of absence of sunshine, then also connected with a DC-AC inverter three-phase for the supply of AC loads (pumps and LED).

Most of the electricity consumption is concentrated in the morning period and is suitable for our system, however; in the evening period the batterys works alone. The battery should be large enough to store sufficient energy to operate the appliances at night and cloudy days [7].

The electrical energy requirement of the four farms is used for four pumps of 1.5 KW, they work five hours per day, and LEDs of 20 W for lighting are work for eight hours at night.

After the sizing of our system, the total energy consumption is estimated at around 12 KW, so the appliances should be powered by at least 40 modules of 300 Wp PV module to get about 12 KW of power, and the battery should be rated 48 V 1000 Ah for 3-day autonomy.

4 Simulation of System

The mini-central PV system consists of 300 W polycrystalline solar module of type "SunPower SPR-300E". The placement of the PV modules in series or in parallel is dependent on our need for voltage and the PV module Voc. It comprises 10 parallel strings with 4 series connected modules per string to get a power of 12 KW and a voltage of 250 V, Fig. 3.

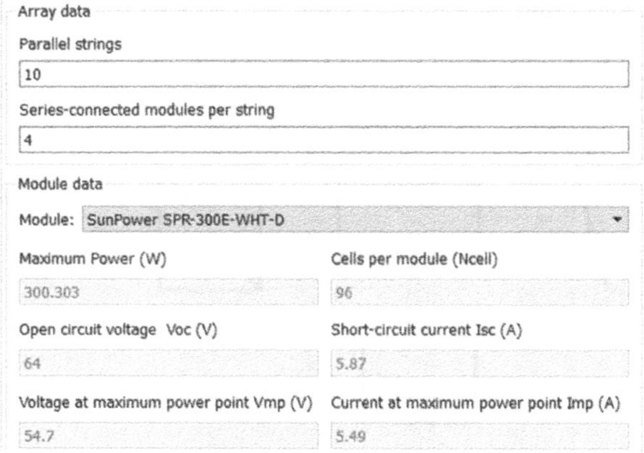

Fig. 3. PV Array parameters.

Using the Simulink model given in the Fig. 4, we plotted the (I_V) and (P-V) characteristics of the PV Array (irradiation 1000 w/m^2, Temperature 25°) as shown in Fig. 5.

The simulation (P_V) and (I_V) characteristic of the PV array shows that the maximum power of the field is around 12 KW, and the open circuit voltage is around 250 V as it is shown in Fig. 5.

The system's components are modeled in Matlab/Simulink software environment. Matlab/Simulink is selected, due to its reusability, extendibility, and flexibility in such systems. As stated above, the PV system consists 40 of 300 W solar module. It comprises 10 parallel strings with 4 series connected modules per string, a schematic of the Simulink model is shown in Fig. 6.

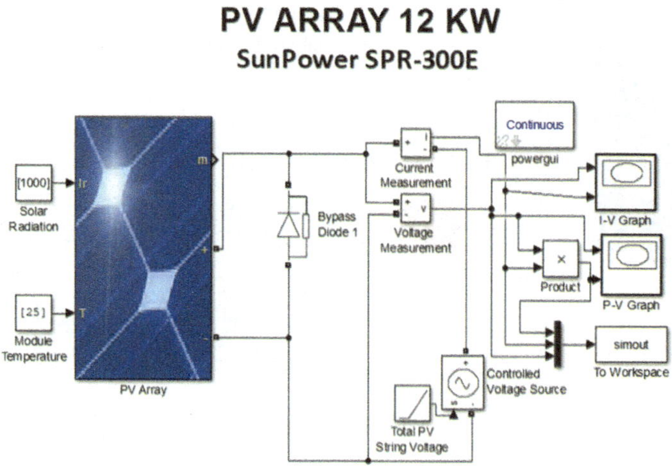

Fig. 4. Simulation of the PV Array

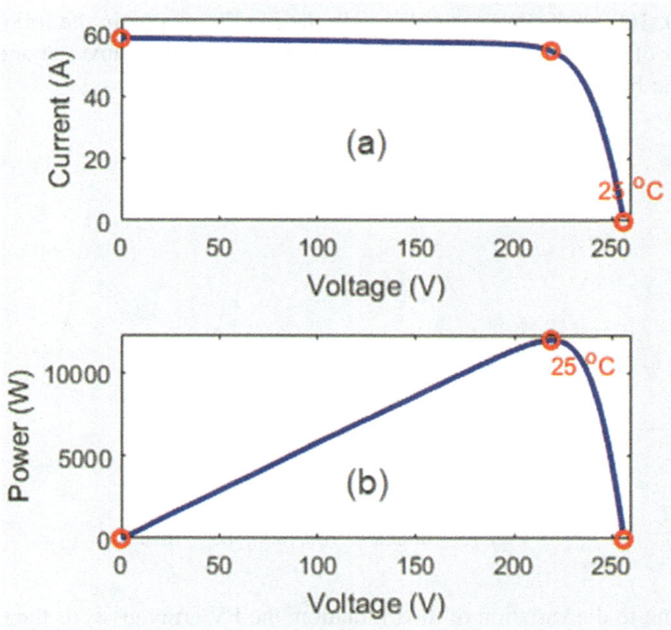

Fig. 5. (a) PV array I_V characteristics (irradiation 1000 w/m^2, Temperature 25°) (b) PV array P_V characteristics (irradiation 1000 w/m^2, Temperature 25°).

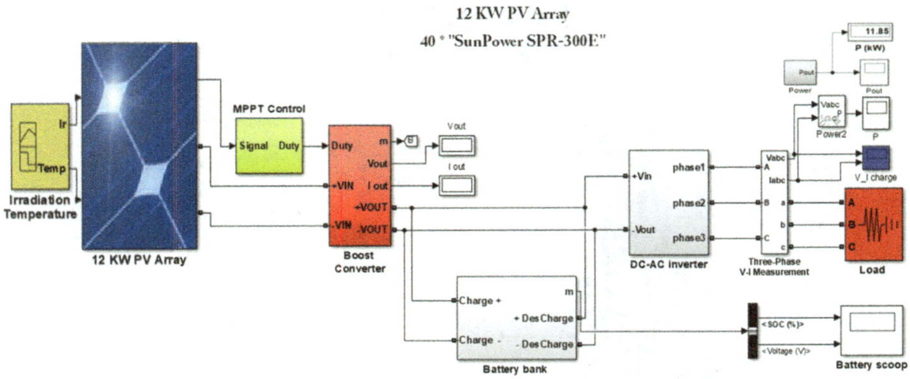

Fig. 6. Schematic Simulink model of the mini-central PV.

5 Results and Discussions

In this simulation for various system components and for variation scenario of irradiation (700, 1000, 800 W/m^2) as shown in the Fig. 7, we obtain the following results of the power of the PV array and the voltage at the output of the inverter and the state of charge of the battery.

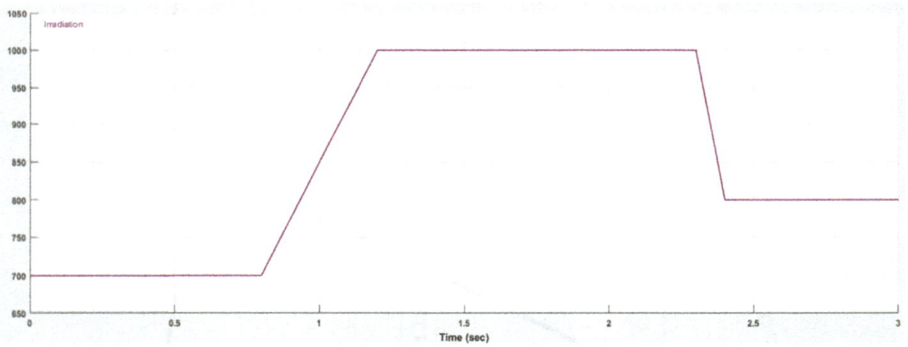

Fig. 7. Irradiation 700,1000,800 W/m^2.

According to the variation of the irradiation, the PV array gives us the maximum of power thanks to the role of the MPPT used. In the morning, the irradiation increases until it arrives at 1000 W/m^2 Fig. 7, the PV array gives a power around 12 KW Fig. 8.

According to the Fig. 8. The estimated energy generation by PV array energy system is about 12 KW. This means that the system could supply energy enough for the load at the time of operation of pumps used in irrigation.

When the PV array supply, the electrical energy comes from the PV array and the average energy storage in the battery will be rise as shown in Fig. 10.

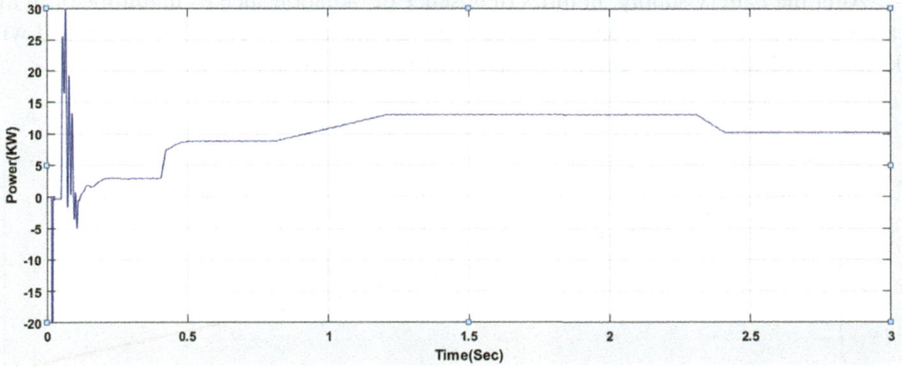

Fig. 8. POWER of the PV array.

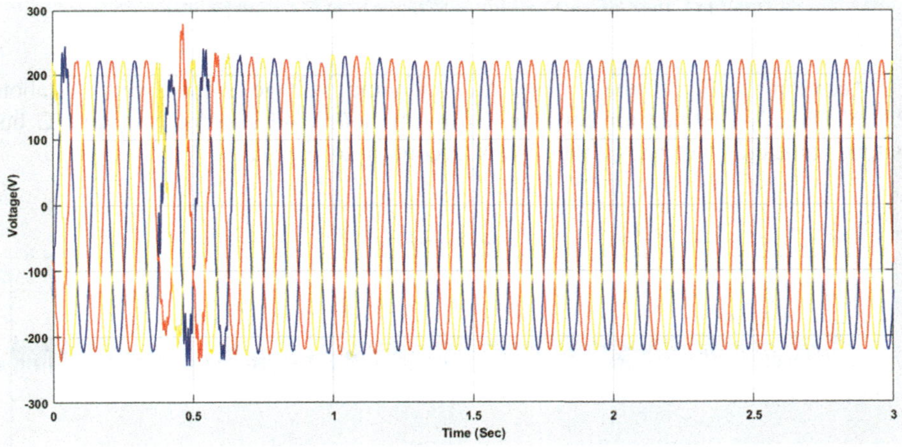

Fig. 9. Inverter voltage.

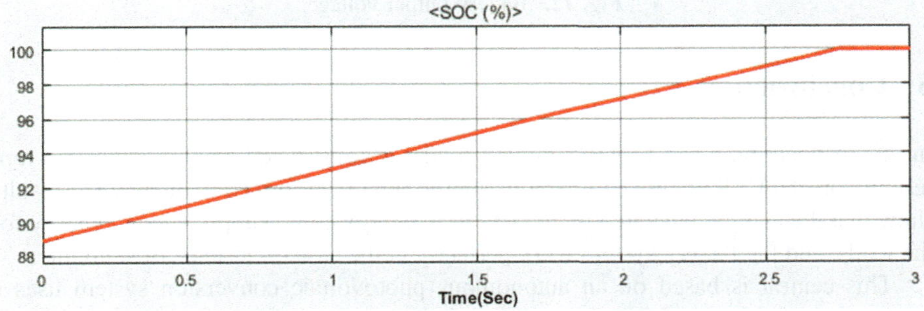

Fig. 10. Percentage battery charging in presence of irradiation

After the battery supply, in times of absence of radiation such as nighttime or in the case of cloudy days, the electrical energy to load the level of energy accumulated will decrease and the average energy storage will reduced as shown in Fig. 11.

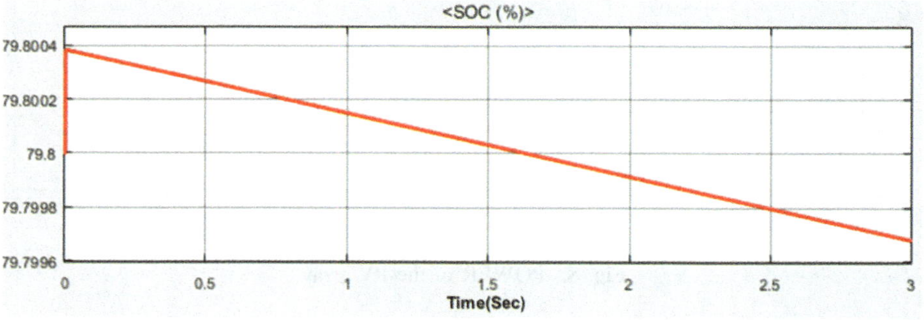

Fig. 11. Percentage battery discharging in absence of irradiation

According to Fig. 9. the average AC bus output voltage of the inverter is about 230 V Directly destined for consumption by the load, and also the average DC bus output voltage of battery is about 230 V according to Fig. 12.

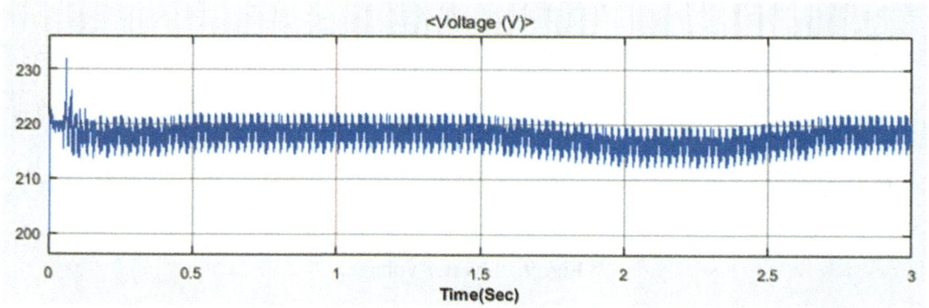

Fig. 12. Battery output voltage.

6 Conclusion

In this manuscript, a MATLAB Simulink model is constructed making a detailed representation of a 12 KW mini photovoltaic central and his various components. The results show that the energy generated by the PV array energy could supply enough energy for the loads, and the battery system work perfectly in the case of absence of sunlight.

This central is based on an autonomous photovoltaic conversion system uses a photovoltaic field with DC DC and DC AC converters and a battery bank to supply alternating loads, this construction is dedicated for use in farms located in isolated sites, his objective is to reduce the costs, and expenses of consuming electricity for the farmer.

References

1. Messenger, R.A., Ventre, J.: Photovoltaic Systems Engineering. CRC Press, Boca Raton (2003)
2. Torshizi, M.V., Mighani, A.H.: The Application of Solar Energy in Agricultural Systems. Department of Bio-System Mechanical Engineering, Gorgan University of Agricultural Sciences and Natural Resources, Gorgan, Iran (2015)
3. Al-Shamani, A.N., Othman, M.Y.H., Mat, S., Ruslan, M.H., Abed, A.M., Sopian, K.: Design & sizing of stand-alone solar power systems a house Iraq. Technical University, April 2015
4. Legbassi, T.H., Hangnilo, R., Nassara, L., Codjo, G.D.: Etude technico-économique d'une mini-centrale solaire photovoltaïque: cas d'électrification du village de Fètèkou, École Polytechnique d'Abomey – Calavi (2015–2016)
5. Benefits of off grid solar power light systems, 16 March 2015. https://www.sepco-solarlighting.com/blog/benefitsof-off-grid-solar-power-light-systems. Accessed 04 Jul 2018
6. Ghafoor, A., Munir, A.: Design and economics analysis of an off grid PV system for household electrification. Renew. Sustain. Energy Rev. **42**, 496–502 (2015)
7. AMANA, C.R.E.O.H.B.: «CELLULES PHOTOVOLTAIQUES: Etude et comparaison de trois types de cellules,» Université de Cergy-Pontoise. Master 1 Physique. Cellules Photovoltaïques (2009)
8. http://www.leonics.com/support/article2_12j/articles2_12j_en.php

Static-Dynamic Analysis of an LVDC Smart Microgrid for a Saharian-Isolated Areas Using ETAP/MATLAB Software

M. A. Hartani[1,2](✉), M. Hamouda[1], O. Abdelkhalek[2], A. Benabdelkader[2], and A. Meftouhi[2]

[1] Laboratoire de Développement Durable et Informatique LDDI, Faculté des Sciences et de la Technologie, Université Ahmed Draia, Adrar, Algeria
hartani_mohamed@univ-adrar.dz, jhamouda@univadrar.dz
[2] Smart Grid and Renewable Energy Laboratory SGREL, Department of Electrical Engineering, University of TAHRI Mohamed, Bechar, Algeria
othmane.Abdelkhalek@labrier-univ-bechar.dz

Abstract. In this work, a standalone low voltage DC Microgrid LVDC-MG system is studied. This system is destined to supply a residential agricultural zone at the southern of Algeria based on the integrated PV renewable energy resource, the energy storage system ESS, and the backup unit that consists of diesel generator DG. In this regard, the proposed topology is a grid-off DC-coupled system, which regroups four farms linked via power lines and power converters to ensure the bidirectional sharing of energy following a specified load schedule and an energy management policy EMP. The EMP center combines both decentralized controllers such as MPPT, and DC-DC converters closed loops that regulate voltages of the nodal DC buses and states of the energy of the storage devices. In the other hand, a centralized control strategy aims to manage the external shared energy of the DC-MG, considering nodal voltages, states of energy, and the load demand in each farm. Hence, the output signal commands are the PWM to the switching devices of the power converters, references currents, and voltages to the closed control loops, and references power values of the distributed energy resources DERs. Thus, the DC-MG is simulated with ETAP and Matlab software under various tests scenarios and variable conditions, where static and dynamic results are evaluated and discussed to improve the lifetime of the system and the efficiency of the proposed configuration respectively.

Keywords: LVDC microgrid · Optimal load flow · Autonomy unit · Decentralized control · Bidirectional networking · Losses

1 Introduction

Nowadays, Direct current DC-MGs are offering many advantages to traditional and conventional AC grids. Thus, there are high power electrical appliances operate in DC such as the used in industry, buildings and transmission lines, in addition to much digital equipment that is compatible with the DC. In literature, DC-MG is being more

M. Hatti (Ed.): ICAIRES 2019, LNNS 102, pp. 496–505, 2020.
https://doi.org/10.1007/978-3-030-37207-1_53

interested and developed such as the realized projects of Hefei University of Technology Microgrid (Asia, China), Bosch DC Microgrid at California Honda Facility, Burlington DC Microgrid (Canada, North America), and Andaman Island (Indian) Coast Guard Microgrid [1–4]. Furthermore, the available energy in saharian zones is generated by solar and wind sources, which offer a DC power via its DC-DC converters. Hence, the energy storage devices such as battery and supercapacitor have a DC output generation, in addition to its high energy and power densities that can supply transient and permanent load demand avoiding damages of the vital components (sources and loads) when supplying current spikes to the induction motors and the non-linear appliances especially. This device aims to regulate the power balance between the DERs and the load demand, either by injecting or by absorbing the needed or the exceeded energy respectively. Thus, the backup DG is integrated to support renewable sources due to its intermittent nature, which is based only on the available solar potential (sun irradiation and wind speed) [5–7]. In addition, an energy management policy is necessary in such systems to keep the overall network operating in stability, where each DER operates within its specified limits, and with respect to the main variables of the system [8, 9]. Figure 1 shows the proposed framework of the grid-off DC-MG, where the sizing of the DERs, the storage devices, and the backup generator are mentioned.

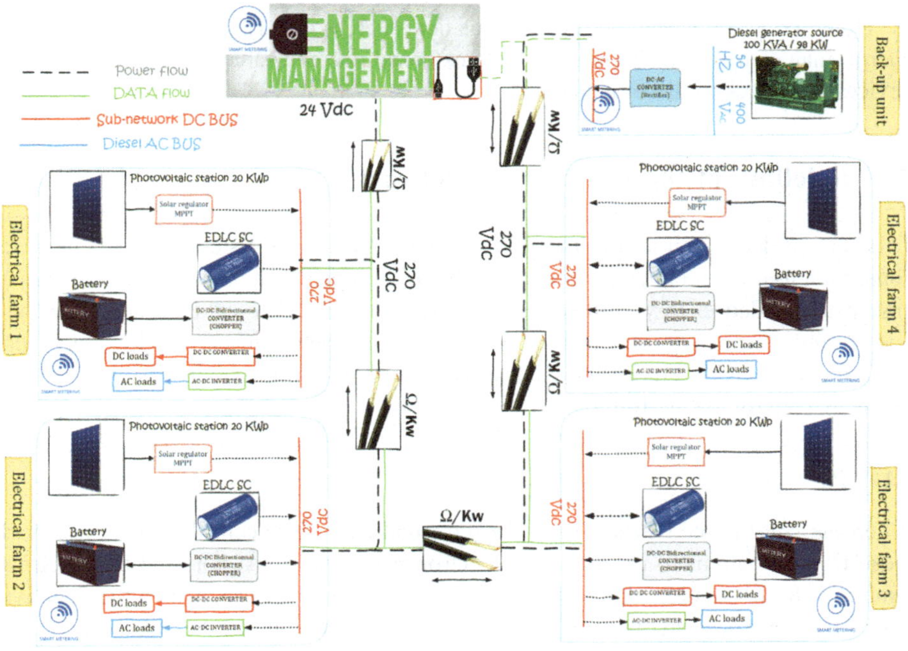

Fig. 1. The electrical circuit of the proposed LVDC-Microgrid system.

As viewed, the treated system regroups four agricultural farms as represented in the Matlab structure, while a composite sub-network in ETAP software represents it. Hence, each sub-network has its own small-scale renewable system to supply its need of energy, while the backup unit copes with the global demand of the whole system following a specified load schedule and management policy. The LVDC-MG is supplied mainly via renewable energies and autonomy storage devices resulting in a high renewable fraction and reducing the use of traditional sources or the grid utility, and minimizing the negative impacts of the environment. In contrast, the big drawback of such systems is the initial coast, which includes the components, the installation, the transport, and the maintenance. For that reason, this work aims to reach the following objectives:

- Model, simulate and control the proposed DC-MG network in Matlab/Etap software.
- Study the static and dynamic regimes of the DC-MG using Etap and Matlab software respectively.
- Evaluate the obtained results under real-variable parameters such as solar irradiation, wind speed, and temperature.
- Test the viability of the proposed energy management policy to manage the shared energy with respect to the boundaries of the DERs, the storage devices, and based on a specified load schedule.

Other works have been presented it this field such as [10–12], while focusing on the energy management of LVDC-MG systems based on renewable energies and storage devices, and backup diesel generator.

2 Modelling of the LVDC-MG Network

This section deals with a brief modelling of the LVDC-MG system. The DC network consists of four interconnected farms, which are equipped with its own sub-networks to ensure its needed energy sufficiently and autonomously. These farms are linked via power cables to ensure the bidirectional networking between them following a centralized energy management policy. The whole structure of the studied network shown in Fig. 1 consists of the following:

2.1 PV

When the solar photon strikes onto the semiconductor solar cell, the electrons knock loose from the atoms to produce hole-electron pairs finally. As forming the electrical circuit of the PV source, the positive electrical conductor is assigned to the negative, where the electrons circulate. Hence, using a pay bass diode to the solar cell is necessary for the darkness where no solar irradiance is available, and the cell works as a diode producing neither voltage nor current. The basic mathematical equation that describes the I-V characteristic of the PV cell is [6–14]:

$$I_{pv} = (I_{pv,N} + K_I.\Delta_T).\frac{G}{G_n} \tag{1}$$

Figure 2 shows the equivalent circuit of the PV source, where the number of series and parallels modules as considered in series and parallel resistances. Real circuit of the PV device includes series resistance R_S for voltage source region operation and parallel resistance R_P for current source region operation.

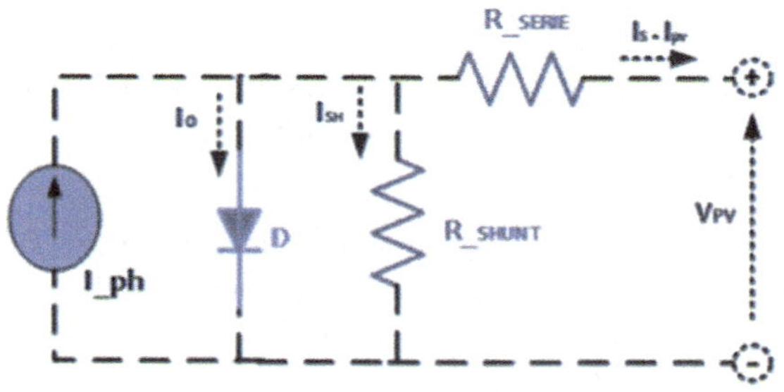

Fig. 2. The representative circuit of the solar PV source.

2.2 Diesel

Usually, diesel generators DGs are used for off-grid systems and isolated networks based on renewable energies. The intermittency and the variation of the environmental variables such as solar radiation, wind speed, and temperature necessitate such backup generator to ensure the minimal service when the renewable energy is not sufficient to fulfill the demand. The general structure of the diesel engine model consists of four main sub-models: the controller, the actuator, the engine, and the flywheel [6–14].

2.3 HESS

In this paper, a hybrid storage unit that consists of chemical batteries and electrical supercapacitor is used. This hybridization aims to cope with both transients and permanents current to the load due to the high power density of the supercapacitor and the high energy density of the battery. Hence, this hybridization can cave the dynamics of the battery against current spikes that can harm it, and which are absorbed and/or injected easily by the EDLC device [15, 16].

2.4 Power Converters

The LVDC-MG, system includes only DC renewable sources, DC storage devices, and AC diesel generator. These sources are connected between them via power converters and power lines. The used converters in this case includes DC-DC converter, AC-DC rectifier, in addition to DC-AC converter for AC loads. The overall efficiency of the system is based on the efficiencies of these converters when transferring and sharing the generated energy [6–14].

3 Power Flow Methods

In this work, DC load flow DC-LF is run in ETAP software to solve power-flow problems such as minimizing losses, improving steady-state voltages, enhanced system reliability, stability, and load ability. In another hand, classical analytical methods are well matched for small systems, while performing unfavorably for expansive, large and complex objective functions. Hence, intelligent techniques for optimal sizing and siting of DGs are characterized by its high speed, good convergence, and well suited for large and complex system scales [17, 18]. The transmission lines ordinarily shape the branches, the nodes are denoted as buses, where the power is either injected or absorbed following the bus types. The gotten results from LF analysis in purely DC systems is the magnitude of the voltage at nodal buses, and the real power flowing in each line. Among the OPF aspects are:

- The total amount of the active power is estimated from the DERs including the renewable sources (PV), the storage devices (battery), or the conventional sources (DG). The generated energy must cover the demand with losses. In addition, the needed energy is shared by the DERs of the DC-MG in a unique ratio to achieve the optimal economic operation.
- Transmission lines should not carry power too close to their stability that affects its thermal limits.
- The operating voltages of the nodal buses should respect the specified boundaries and tolerances.

During the steady-state analysis of the LVDC-MG, the impedances of the lines of the network are assumed as purely resistive. Thus, considering n buses, and using Kirchhoff's current law, the injected current at each bus (i = 1 … n) is equal to the sum of the currents flowing in n − 1 buses, the network equation can be written as follows [19]:

$$I_{DC,i} = \sum_{j=1,j\neq i}^{n} Y_{DC,ij}(V_{DC,i} - V_{DC,j}) \tag{2}$$

With, I_{DC}, i is the injected direct current in the bus I, $Y_{DC,\ ij}$ is the admittance between the i and j buses, and $V_{DC,i}$ is the voltage magnitude in the bus i. Hence, in the unipolar DC-MG, the injected power can be expressed as follow:

$$P_{DC,i} = V_{DC,i} \times I_{DC,i} \tag{3}$$

Hence, Eq. 4 defines the droop buses constraint, while Eq. 9 denotes the active power balance of the DC-MG by:

$$\begin{cases} V_{Gi} = V_{0,i} - (R_{Di} \times I_{Gi}) \\ I_{Gi} = P_{Gi}/V_{Gi} \end{cases} \tag{4}$$

$$P_{Gi} - P_{Di} - P_{DC,i} \tag{5}$$

In this system, we can distinguish three bus types, which are the active power and the voltage magnitude as shown in Table 1.

Table 1. Classification of the bus types used in the LVDC-MG

N°	Type of bus	Active power P	Voltage magnitude |V|
2	Slack bus	Unknown	Known
3	Generator bus PV	Known	Known
4	Load bus PQ	Known	Unknown

4 Simulation Results and Discussion

In this section, the obtained results are discussed separately, where static result are gotten form ETAP software, while dynamic results are gotten from MATLAB software. The next Table 2 summarizes the common initial parameters of the whole DC system.

Table 2. Electrical parameters for simulation tests of the LVDC-MG system

Elements	ID	Electrical parameters	
DERs	Diesel DG	Peak active power (Kwp)/PF/Efficiency (%)/AC Line-Line voltage (V)/Frequency (Hz)	120/0.8/95 400/50
	PV	Peak active power per farm (Kwp)	25
	Main batteries Local batteries	Rated capacity (Ah)/voltage (V)	2030/165 595/165
Loads	*Peak daily consumption*	*Kwh/day - Kwp/day*	*80–800*

4.1 ETAP Software Results

In this part, we present the final values of the nodal voltages, the injected/absorbed currents and powers in each node and breach of the studied LVDC-MG. The static results of ETAP soft are obtained using 6 proposed scenarios based on the availability and the priority of the DERs. Hence, each scenario has its specified state and then we present the possible causes that characterize these scenarios. The Table 3 illustrates the selected scenarios in detail as follow:

Table 3. Energy management scenarios of the LVDC-MG system in ETAP software.

Cases	State	Causes	Cases	State	Causes
PV only	Normal	Sufficient supply to the load + charge the batteries	PV + Local batteries	Normal	Sufficient supply to the load + charge the batteries
Local batteries only	Normal	Sufficient supply to the load	PV + Main storage	Critical	No diesel fuel Critical SOE
Main storage only	Critical	No diesel fuel No PV Critical SOE	PV + Diesel	Critical	Critical SOE
Diesle only	Critical	No PV Critical SOE	Diesel + local batteries	Critical	No PV
Main storage + Diesel	Critical	No PV Critical SOE	Main storage + local batteries.	Critical	No diesel fuel No PV

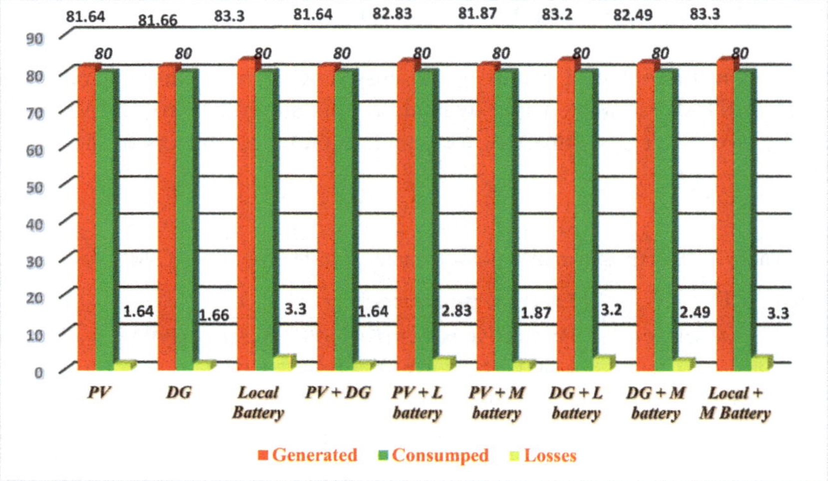

Fig. 3. OPF results of the 5 buses LVDC-MG system

The following curve shows the values in (Kw) of the generated, consumed powers, with the power mismatch of the DC-MG in each scenario and in each branch. The power losses are the sum of the partial losses of the lines and the power converter, where we consider 98% and 97% as the efficiencies of DC-DC and AC-DC converters respectively.

As depicted in the Fig. 3 above, the global generated energy of the DERs of the LVDC-MG varies based on the supplied circuit, where we observe that the power losses varies between [1.312–2.64] p.u. The low power losses values was in 'PV' and 'PV + DG' scenarios, followed by 'DG', than 'PV + Main storage', 'DG + Main storage', 'PV + local batteries', 'DG + local storage', and finally by 'local batteries' and 'Main and local batteries' scenarios. In other hand, the higher the number of the used power converter in each scenario, the low system efficiency of the global system.

4.2 Matlab Software Results

In this part, we aims to simulate a real daily profile of the LVDC-MG following the load schedule, the available energy of the DERs, and the management decisions. The operating parameters of the LVDC-MG system are presented in Fig. 4, which are the operating voltages of each DC BUS inside each farm, the states of charges SOCs of each storage system, the fuel level of the DG, and the states of the storage system inside each farms either ON or OFF following its SOCs. In addition, Fig. 5 illustrates the energy balance of each electrical system, where the instantaneous powers of the PV, battery, and DG are plotted together. The advantage of the MATLAB results is that we can simulate and evaluate more than one management scenario in the same profile, in addition to its transient and permanent regimes.

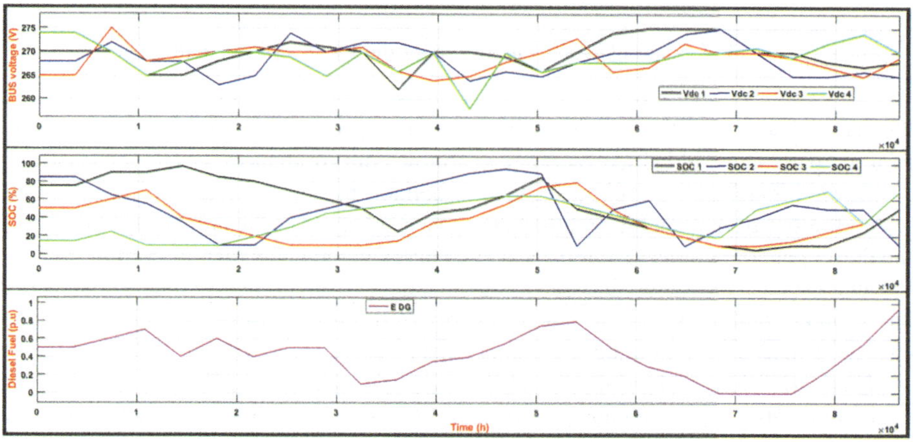

Fig. 4. Simulation variables of the LVDC-MG system: bus voltages, SOC, fuel level

As depicted in the figure above, the three profiles show the variations of the bus voltages and the states of charge inside each electrical farm. The bus voltages was within its exceptibale ranges between: [258–275] V_{DC}. The drop voltage above or below the reference value reflects the availability or the lack of energy at the common bus of the system respectively. Hence, the fuel level of the diesel generator DG is also depicted in the same figure. Moreover, the energy management center receive these variables as inputs in order to get the output management decisions in term of the reference powers of the DERs. Thus, the management center aims to keep the energy balance stable while associate the appropriate DERs to supply the needed energy to the load.

The load demand varies inside each sub-system or farm, while receiving the same solar potential represented by the PV energy. As shown, the energy balance of the whole LVDC-MG system consists of two operating periods. The first is during morning [8 ^{h} – 18 ^{h}], where the energy flowing is represented by $P_{LOAD} = P_{PV} \pm P_{BATTERY} + P_{DG}$. In this balance, the PV has the priority to supply the load, then the

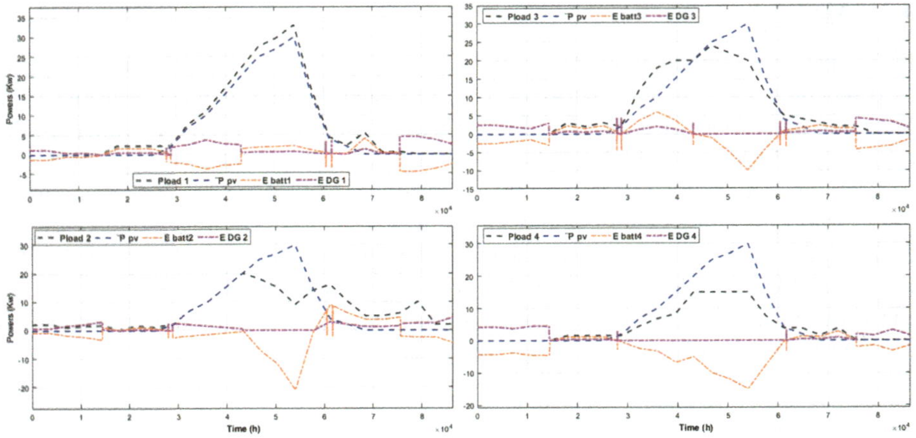

Fig. 5. Matlab software results of the LVDC-MG system

battery and then the DG. The selection of the appropriate DER follows the availability of three parameters classified by order as follow: sun irradiation – SOE – Diesel fuel.

In the other hand, the night balance is defined by: $P_{LOAD} = \pm P_{BATTERY} + P_{DG}$, where the sun is unavailable and the battery is sized to cover the autonomy periods and the diesel generator aims to compensate the energy deference either at morning or night balances. Thus, the diesel has intervened with the battery during the whole period based on the fuel level and the state of energy of the storage devices, while the battery charge or discharge following the management decisions.

From the presented profiles, multiple scenarios can be combined based on the availability of the load demand, the PV, the state of energy, the operating period, the fuel level, and the state of the operating variables such as buses voltages.

5 Conclusion

At the end of this brief study, the proposed LVDC-MG system can be considered as an efficient solution for islanded villages in our Saharan zones especially. The studied system was run on MATLAB and ETAP software to deal with static and dynamic regimes of the system. An energy management policy was implemented to react with the possible scenarios and keep the energy balance stable based on the intermittent DERs and a specified load schedule. From ETAP software, the power losses were evaluated in each management scenarios, while both transient and permanent regimes were discussed via MATLAB software. Thus, the combination of this two software makes the study more comprehensive and integrated with all respects. In general, the evaluated system can be considered for Saharian zones that have high solar potential, where renewable energy is preferred and efficient in addition to the economic investment in such systems.

As a future perspective, we aim to implement smart metering to such renewable island system, in addition, to implement a small-scale experimental test bench.

References

1. Madziga, M., Rahil, A., Mansoor, R.: Comparison between three off-grid hybrid systems (solar photovoltaic, diesel generator and battery storage system) for electrification for Gwakwani village. South Africa. Environ. **5**(5), 57 (2018)
2. Grant, J.A.: Design and Simulation of a DC Microgrid for a Small Island in Belize (Doctoral dissertation) (2018)
3. Silva, F.A.: Clean energy microgrids [book news]. IEEE Ind. Electron. Mag. **12**(2), 79–80 (2018)
4. Samal, S., Hota, P.K.: Design and analysis of solar PV-fuel cell and wind energy based microgrid system for power quality improvement. Cogent Eng. **4**(1), 1402453 (2017). https://doi.org/10.1080/23311916.2017.1402453
5. Ramli, M.A., Bouchekara, H.R.E.H., Alghamdi, A.S.: Optimal sizing of PV/wind/diesel hybrid microgrid system using multi-objective self-adaptive differential evolution algorithm. Renew. Energy **121**, 400–411 (2018)
6. Sofimieari, I., Mustafa, M.W.B., Obite, F.: Modelling and analysis of a PV/wind/diesel hybrid standalone microgrid for rural electrification in Nigeria. Bull. Electr. Eng. Inform. **8**(4), 23–29 (2019)
7. Dragicevic, T., Wheeler, P., Blaabjerg, F.: DC distribution systems and microgrids. Institution of Engineering and Technology (2018)
8. Kumar, J., Agarwal, A., Agarwal, V.: A review on overall control of DC microgrids. J. Energy Storage **21**, 113–138 (2019)
9. Zhang, G., Tian, Z., Tricoli, P., Hillmansen, S., Liu, Z.: A new hybrid simulation integrating transient-state and steady-state models for the analysis of reversible DC traction power systems. Int. J. Electr. Power Energy Syst. **109**, 9–19 (2019)
10. Mohamed Amine, H.: Energy management of a DC hybrid system. In: Second International Conference on Electrical Engineering ICCEEB'2018, Biskra, Algeria, 2–3 December (2018)
11. Mohamed Amine, H.: Proposed DC Microgrid for agricultural farms in south Algeria (Adrar zone). In: Conférence Internationale sur les Matériaux, le Patrimoine et l'Environnement en Zones Arides Université Ahmed DRAIA, Adrar, Algeria, 17–18 February, pp. 34_1–34_8 (2019)
12. Mohamed Amine, H.: Energy management of a hybrid power system for an island agricultural site (Saharian Adrar zone). In: First International Conference on Smart Grids, CIREI'2019, Oran, Algeria, 4–6 March (2019)
13. Brahmi, N., Charfi, S., Chaabene, M.: Optimum Sizing Algorithm for an off grid plant considering renewable potentials and load profile. Int. J. Renew. Energy Dev. **6**(3), 213 (2017)
14. Belila, A., Benbouzid, M., Berkouk, E.M., Amirat, Y.: On energy management control of a PV-Diesel-ESS based microgrid in a stand-alone context. Energies **11**(8), 2164 (2018)
15. Kouchachvili, L., Yaïci, W., Entchev, E.: Hybrid battery/supercapacitor energy storage system for the electric vehicles. J. Power Sources **374**, 237–248 (2018)
16. Jacob, A.S., Banerjee, R., Ghosh, P.C.: Sizing of hybrid energy storage system for a PV based microgrid through design space approach. Appl. Energy **212**, 640–653 (2018)
17. Prakash, P., Khatod, D.K.: Optimal sizing and siting techniques for distributed generation in distribution systems: a review. Renew. Sustain. Energy Rev. **57**, 111–130 (2016)
18. Herbadji, O., Bouktir, T.: Optimal Power flow using firefly algorithm with consideration of FACTS devices "UPFC". Int. J. Electr. Eng. Inform. **7**(1), 12 (2015)
19. Li, C., Chaudhary, S., Dragicevic, T., Vasquez, J.C., Guerrero, J.M.: Power flow analysis for DC voltage droop controlled DC microgrids. In: Proceedings of the 11th International Multi conference on Systems, Signals & Devices, SSD, pp. 1–5. IEEE Press (2014). https://doi.org/10.1109/SSD.2014.6808896

Sizing of a Solar Parking System Connected to the Grid in Adrar

Abdeldjalil Dahbi[1(✉)], Mohammed Boussaid[2], Mohammed Haidas[3], Maamar Dahbi[3], Rachid Maouedj[1], Othmane Abdelkhalek[3], Miloud Benmedjahed[1], Lalla Moulati Elkaiem[2], and Lahcen Abdellah[2]

[1] Unité de recherche en énergies renouvelables en milieu saharien, URERMS, Centre de développement des énergies renouvelables, CDER, 01000 Adrar, Algeria
Dahbi_j@yahoo.fr

[2] Laboratory of Energy, Environment and Information Systems (LEEIS), Department of Material Sciences, University of Ahmed Draia, Street 11, 01000 Adrar, Algeria

[3] Smart Grid and Renewable Energies Laboratory, Tahri Mohammed, University of Bechar, BP 417, Bechar, Algeria

Abstract. This paper presents a sizing study of a solar parking system connected to the grid. This system is installed in Adrar region, south west of Algeria. It is composed from photovoltaic panels, cables, converters. All these elements have been modeled and sized using analytical method. The proposed method is validated by simulation using PVSYST software.

The simulation results were discussed and compared with the analytical results. They proved the validity of the model to make a good system sizing for different applications.

Keywords: Solar parking · Electrical grid · Sizing · Simulation · PV system software

1 Introduction

Nowadays, the electrical energy production presents a very important issue, especially with the environment problems and the cost of the fossil energy. Face to these problems, the renewable energies and especially photovoltaic energy is used to fix them [1]. However, this latter requires a large area in its installation compared to the needed harvested electrical energy. In this context, a solar parking system is proposed and studied. The solar car park is used in order to benefit from the roof area by installing

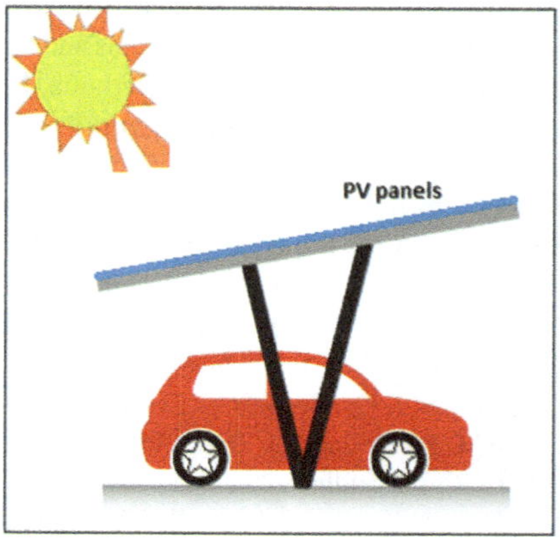

Fig. 1. The solar car park.

M. Hatti (Ed.): ICAIRES 2019, LNNS 102, pp. 506–514, 2020.
https://doi.org/10.1007/978-3-030-37207-1_54

photovoltaic panels, moreover for obtaining shadow and electrical energy. This latter can be used for different applications such as charging electrical cars and grid connection which is the used application in this paper [2]. Algeria has a high solar potential that encourage us to benefit it with an optimization way for a long time according to the literature estimation [3–5]. In this work, the solar car park is studied to be installed in Adrar with a power equals to 31.5 kWc, Fig. 1.

This paper is organized as following: in the second section; the modeling of the whole element of the solar parking system is given. The third section is reserved for the sizing of the car parking system. Then, the simulation of the system installed in Adrar using PVSYST, is presented in the fourth section. At the end, an investigation and evaluation comparison have been done between the analytical and the simulation results.

2 Modeling of the Solar Parking System

The sizing of the PV system passes by many steps, depending basically on the site potential and the required energy.

The estimated energy produced per day is given by:

$$\mathbf{E_c = P * Q * h} \tag{1}$$

Where:

E_c: The consumed energy (Wh/day).
P: Electrical power (W).
Q: Number of the used panels.
h: Time (hour).

2.1 Sizing of the Required Photovoltaic Generator

The peak power of the required panels for installation depends on the site irradiation. It is calculated by applying the following equation [6]:

$$\mathbf{P_C = \frac{E_c}{kI_r} = \frac{E_P}{I_r}} \tag{2}$$

P_c: Peak power in peak watts (Wc).
I_r: Average annual daily irradiation ($kWh/m^2.j$)

The total number of the installed PV modules is calculated by:

$$\mathbf{N} = \frac{\mathbf{P_c}}{\mathbf{P_u}} \tag{3}$$

P_u: unit peak power of the modules watts (W)

The number of the serial PV modules is given by:

$$\mathbf{N_s} = \frac{\mathbf{U_{ond}}}{\mathbf{U_{mod}}} \tag{4}$$

$\mathbf{U_{ond}}$: the input voltage at the terminals of the inverter.
$\mathbf{U_{mod}}$: the unit voltage of the modules

The number of the parallel PV modules is calculated by:

$$\mathbf{N_p} = \frac{\mathbf{P_c}}{\mathbf{N_s P_{mod}}} \tag{5}$$

$\mathbf{P_{mod}}$: the peak power of the PV module

2.2 Cables Sizing

It is very important to use cables that can support the current in both sides DC and AC. This can be achieved by a good sizing of the cable section. Moreover, the isolation must support the environment temperature and the temperature caused by Joule loses. The sections of cables that cause the least potential drop in voltage between the panels and the inverter/charger and between the batteries and the inverter/charger must be determined. The voltage drop must not exceed 3% [7].

The cable section can be obtained by using the following equation:

$$\mathbf{R} = \frac{\mathbf{\Delta U}}{\mathbf{I}} \tag{6}$$

$$\mathbf{S} = \frac{\rho \mathbf{L}}{\mathbf{R}} \tag{7}$$

ΔU: maximum voltage drop in the cable.
I: current circulating in the cable.
R: cable resistance

2.3 Converters Choice

The boost power (P_{-reg}) and the inverter power ($P_{\_ond}$) must be over the peak power (P_c), so it is necessary that:

$$\mathbf{P_{\_reg} \geq P_c} \tag{8}$$

$$\mathbf{P_{\_nd} \geq P_c} \tag{9}$$

2.3.1 Sizing of the Car Parking System

The above equations are used to apply this study on a solar parking. The real characteristics of the installed PV panels in the renewable energy research unit in the Saharian medium are [8]:

Type (Mono crystalline) whose construction parameters are given in the following table (Table 1):

Table 1. Characteristics of the used PV.

Maximum power ($\mathbf{P_{max}}$)	250 W
Open circuit voltage ($\mathbf{V_{oc}}$)	36.99 V
Court-circuit current ($\mathbf{I_{sc}}$)	8.768 A

The peak power which is needed according to the parking area is: $P_c = 31.5\,kW_C$ Hence, the PV module number is:

$$\mathbf{N = \frac{P_c}{P_U} = \frac{31500}{250} = 126\,panels} \tag{10}$$

The number of the serial PV is calculated by:

$$\mathbf{N_s = \frac{U_{ond}}{U_{mod}} = \frac{500}{36.99} = 14\,serial\ connections} \tag{11}$$

The number of the parallel PV is obtained as:

$$\mathbf{N_p = \frac{P_c}{N_s P_{mod}} = \frac{10500}{(14).(250)} = 3\,Parallel\ connections} \tag{12}$$

The cable section S is reached by the following equation:

$$\mathbf{R = \frac{\Delta U}{I} = \frac{0.02.30.75}{8.131} = 0.076\ \Omega} \tag{13}$$

$$\mathbf{S = \frac{\rho L}{R} = \frac{1.610^{-8}.12}{0.076} = 2.5\ \ 10^{-6} m^2 = 2.5\,mm^2} \tag{14}$$

In the following section, the same solar parking system will be presented and simulated using the PV system software, and then the results comparison will be obtained in both cases.

3 Simulation of the Solar Parking

3.1 Simulation of the Solar Parking in Adrar

PVSYST is useful software that allows simulating and verifying projects. It imports weather and geographically data to give results very closer to the real case. Furthermore, it can do a good sizing of the system and its production estimation with optimization.

In this part, the simulation of the solar parking connected to the grid is accomplished using PVSYST. This solar car parking is chosen in Adrar [6].

The sizing of the system using PVSYST has been passed by many steps such as, setting of the localization, PV type, inverter sort, the output power, and the kind of the load…etc., Fig. 2.

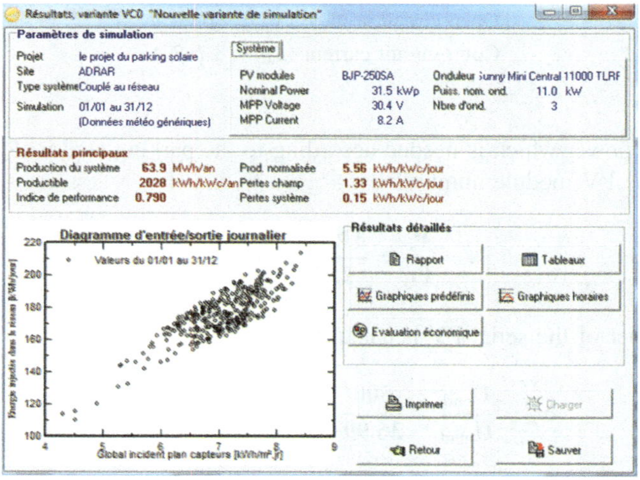

Fig. 2. Parameters setting and simulation in PVSYST.

PVSYST allows presenting different shapes of simulations, graphs, histograms and tables. The other simulation results using PVSYST are shown below (Fig. 3).

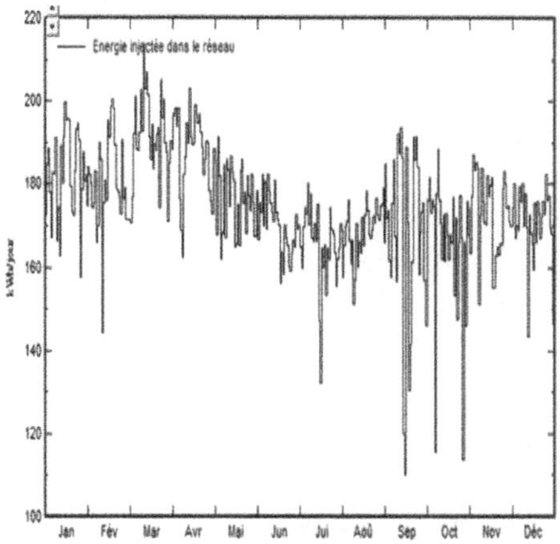

Fig. 3. Daily energy injected to the grid during the year.

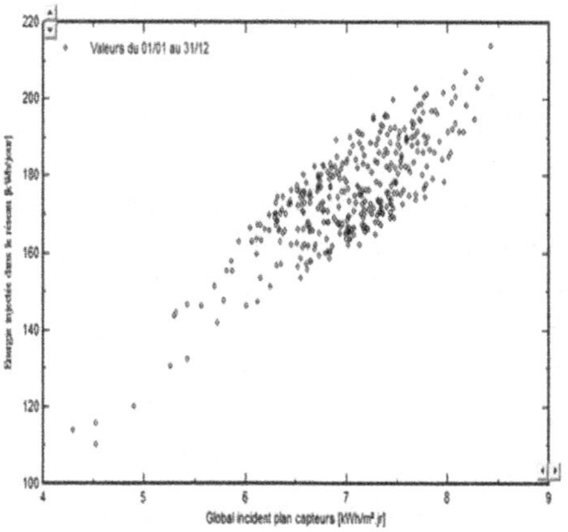

Fig. 4. Diagram of daily input/output.

It is seen that the daily produced energy is variable during the year; it reaches high values from autumn to spring months (September to April). However, it has lower values in summer, because the high temperature in Adrar reduces the PV efficiency (Fig. 4).

The diagram presented in the figure above shows the energy injected to the grid [kWh/day], in function of the global irradiation [kWh/m^2.day] during all the year in Adrar region.

As it is seen in "Fig. 5", the energy injected to the grid increases proportionally with the irradiations.

The table below presents different variables obtained by simulation during the year in Adrar (Table 2).

Table 2. Balance sheet of simulation results during the year in Adrar.

	GlobHor kWh/m^2	DiffHor kWh/m^2	T_Amb °C	GlobInc kWh/m^2	GlobEff kWh/m^2	EArray kWh	E_Grid kWh	PR
Janvier	139.5	20.63	11.55	210.9	206.3	5790	5647	0.850
Février	144.5	27.83	14.82	192.7	188.8	5201	5068	0.835
Mars	200.6	36.59	20.17	234.8	229.6	6097	5939	0.803
Avril	221.3	45.17	23.81	227.1	220.9	5799	5653	0.790
Mai	242.1	57.62	29.14	225.3	218.3	5619	5469	0.771
Juin	240.9	58.44	33.76	215.1	208.3	5248	5099	0.753
Juillet	243.9	59.44	37.68	221.7	214.8	5275	5130	0.735
Août	227.3	59.49	36.76	224.4	218.0	5361	5225	0.739
Septembre	190.8	51.09	31.49	210.2	204.7	5183	5049	0.763
Octobre	164.4	42.37	26.47	207.3	202.7	5282	5146	0.788
Novembre	137.5	23.03	17.34	199.3	195.2	5323	5190	0.827
Décembre	126.4	19.54	12.97	197.7	193.5	5420	5284	0.849
Année	2279.1	501.26	24.72	2566.5	2501.0	65597	63897	0.790

Where:

GlobHor: horizontal globale irradiation
Diffhor: horizontal diffused irradiation
T-Amb: ambiant temperature
GlobInc: Global incident on collectors
GlobEff: Global "effectif", corr. pour IAM et ombrages
EArray: output effective PV energy
E-Grid: Energy injected to the grid
PR: performance Indicator

3.2 Comparison Between Analytic and Simulation Results

The table below presents the comparison between analytic and simulation results

Table 3. Comparison between analytic and simulation results.

Variable	Analytic results	Simulation results
Total number of panels (N)	126	126
Total number of serial panels (N_S)	3	3
Total number of parallel panels (N_P)	14	14
Section (S)	2.5 mm^2	2.5 mm^2

As it has seen according to the Table (3), it is obvious that a very good accuracy is noticed from the comparison between the simulation and the analytical results.

3.3 Experimental Scheme of the Photovoltaic Parking System

According to the analytical and simulation results, it is found that the solar parking is composed of 126 PV modules, each 14 PVs should have a serial connection in order to increase the voltage, the three groups of 14 PVs are connected in parallel for increasing the current [9], these three groups are connected to the grid via three boosts and three inverters, with a total power equals to 31.5 kWc, as it is shown in Fig. 5.

Thanks to the control applied on the inverter, it is noted that this latter insures the grid connection conditions in both sides, (The same voltage magnitude, the same frequency, the same phase delay) [10, 11].

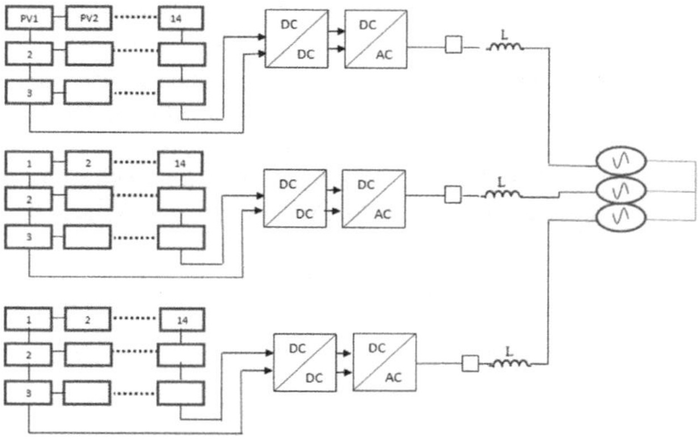

Fig. 5. Diagram of the car park connected to the grid.

4 Conclusion

In this work, a study of solar parking installed in Adrar (Algeria) was achieved. The elements system have been modeled and sized in analytical method. Then, the car parking system connected to the grid is simulated according to Adrar data. After that, the obtained results have been compared; it was found that the analytical and simulation results were similar, which confirm the good sizing of the system. Moreover, the system production and the irradiation are good during the year in Adrar, however, the energy produced decreases in summer months due to the high temperature.

References

1. Rezaei, M., Ghanbari, M.: Optimization of sizing and placement of photovoltaic (PV) system in distribution networks considering power variations of pv and consumers using dynamic particle swarm optimization algorithm (DPSO). Indian J. Fundam. Appl. Life Sci. **5**(S1), 3321–3327 (2015)
2. Coonick, C., BRE National Solar Centre, Gance, D.: A Technical Guide to Multifunctional Solar Car Parks. BRE National Solar Centre (2018)
3. Boussaid, M., Belghachi, A., Agroui, K., et al.: Solar cell degradation under open circuit condition in out-doors-in desert region. Results Phys. **6**, 837–842 (2016)
4. Boussaid, M., Belghachi, A., Agroui, K., Djarfour, N.: Mathematical models of photovoltaic modules degradation in desert environment. AIMS Energy **7**(2), 127–140 (2019)
5. Boussaid, M., Belghachi, A., Agroui, K.: Contribution to the degradation modeling of a polycrystalline photovoltaic cell under the effect of stochastic thermal cycles of a desert environment. Int. J. Control Energy Electr. Eng. (CEEE) **6**, 66–72 (2018)
6. Nacer Eddine, T.: Modélisation et Simulation d'un Système Photovoltaïque, mémoire de master. Université d'El-Oued, Septembre 2015
7. Zeb, K., Islam, S.U., Uddin, W., et al.: An overview of transformerless inverters for grid connected photovoltaic system. IEEE proceedings (2018). 978-1-5386-7939
8. Unité de recherche en énergies renouvelables en milieu saharien, URERMS, Centre de développement des énergies renouvelables, CDER, 01000, Adrar, Algeria (2019)

9. Ghada Bel Hadj Ali: Les installations Photovoltaïques Raccordées au réseau. Tunis, le 27 Octobre 2014
10. Dahbi, A., Nait Said, N., Nait Said, M.S.: A novel combined MPPT-pitch angle control for wide range variable speed wind turbine based on neural network. Int. J. Hydrogen Energy **41**, 9427–9442 (2016)
11. Dahbi, A., Hachemi, M., Nait Said, N., Nait Said, M.S.: Realization and control of a wind turbine connected to the grid by using PMSG. Energy Convers. Manag. **84**, 346–353 (2014)

Power Flow Analyses of a Standalone 5-Buses IEEE DC Microgrid for Arid Saharian Zone (South of Algeria)

M. A. Hartani[1,2,3(✉)], M. Hamouda[1,3], O. Abdelkhalek[2,3], O. Hafsi[2,3], and A. Chakar[2,3]

[1] SGRE & LDDI Laboratoires, Adrar, Algeria
hartani_mohamed@univ-adrar.dz, jhamouda@univadrar.dz
[2] Laboratoire de Développement Durable et Informatique LDDI, Faculté des Sciences et de la Technologie, Université Ahmed Draia, Adrar, Algeria
othmane.Abdelkhalek@labrier-univ-bechar.dz
[3] Smart Grid and Renewable Energy Laboratory SGREL, Department of Electrical Engineering, University of Tahri Mohamed, Bechar, Algeria

Abstract. Fast growing of Algerian population and their non-uniform distribution caused electrical perturbance in grid lines and distribution stations, such as interruptions, voltage/frequency drop, poor power quality, short circuits, and high voltage phenomena. In addition, hard terrains and weather conditions are most impediments in linking Saharian citizens with electricity. For that reason, microgrid networks can be an efficient solution to solve this problem using renewable energies REs due to the high solar potential, and storage system ESS. In this study, we present a prototype of a DC Microgrid destined for residential-agricultural investors in the south of Algeria-Adrar zone. These farms are interconnected through cables to share energy from/to the other farms, which its internal sub-networks to cover its demand independently, while the total demand is covered by a backup unit includes diesel generator and batteries in emergency cases. Thus, an energy management centre EMC controls the functioning of the whole system and managing the shared energy inside the Microgrid. The Microgrid is run in ETAP software, where static and dynamic regimes are studied. From power flow analyses, we can observe that the DC network operates efficiently under stable nodal buses that reflect the availability of the power supply. Hence, the design concept is verified through test scenarios to demonstrate the capability of the proposed microgrid.

Keywords: DC microgrid · Renewable energies REs · Energy management system EMS · Power flow analyses · Standalone system

1 Introduction

This document presents a DC Microgrid structure for feeding a group of agricultural farms in the big south of Algeria [1]. In hand, farms are isolated form the main AC grid and in other hand, the agricultural farms contain residential homes equipped with the needed essential appliance for their works and activities. The principal activities in the

M. Hatti (Ed.): ICAIRES 2019, LNNS 102, pp. 515–523, 2020.
https://doi.org/10.1007/978-3-030-37207-1_55

south Saharian sites focus on water extraction from the ground to irrigate crops, and for daily uses. The DC Microgrid network aims essentially to supply the needed energy continuously and efficiently. The proposed structure regroups four agricultural farms through power links to achieve Load Flow LF and power balance, and data link to exchange sub-networks parameters needed for management control center [2]. The main objectives of this work are:

- Insuring self-sufficiency and continuous electrical supply for the interconnected farms 24 h, and during emergency and critical periods through storage system HESS and the backup diesel genset [3].
- Facilitate agricultural activities of farmers in remote and isolated sites [4].
- Safe renewable electrical supply (DC) as clean, environmentally and friendly [5].
- Stable power balancing between the sub-networks through different LF load flow methods.
- In literature, various methods are used to size such systems including REs and storage devices as optimization methods and genetic algorithm [6].

2 Description of the Proposed DC Microgrid Network

The hybrid-studied network regroups four farms through data and power links with a common backup unit as viewed in Fig. 1. Each separated farm contains PV and batteries for feeding their appropriate load demand [7, 8], where the four farms have a common backup storage system and a diesel source for the emergency cases. The management center insure an optimal load flow for the global network through the measured parameters of each device such as dc bus voltage, frequency, currents, powers, SOC (state of charge) of the storage system, sun irradiations, temperature, losses in buses, branches and cables, and the protective devices state such as fuses and circuit breakers [2, 9, 10].

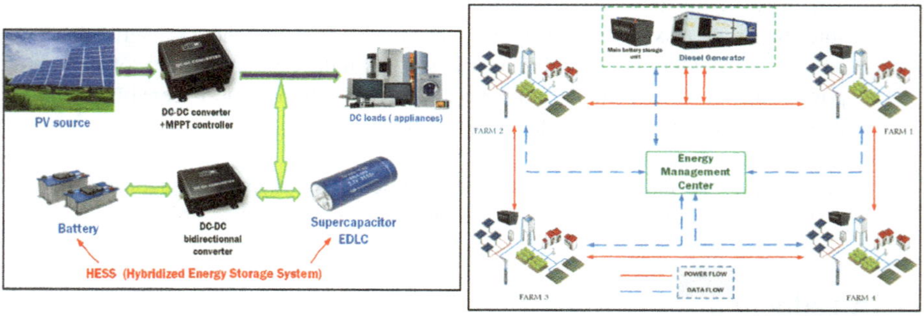

(a) (b)

Fig. 1. Proposed DC Microgrid structure: a-Full circuit, b-sub-network circuit of farms

The treated system links four agricultural farms in the Matlab structure, while a composite sub-network in ETAP software represents it. Hence, each sub-network has its own small-scale renewable system to supply its need of energy, while the backup unit copes with the global demand of the whole system following a specified load schedule and management policy. The DC-MG is supplied mainly via renewable energies and autonomy storage devices resulting in a high renewable fraction and reducing the use of traditional sources or the grid utility, and minimizing the negative impacts of the environment. In contrast, the big drawback of such systems is the initial coast, which includes the components, the installation, the transport, and the maintenance. For that reason, this work aims to reach the following objectives:

- Model, simulate and control the proposed DC-MG network in Matlab/Etap software.
- Study the static and dynamic regimes of the DC-MG using Etap and Matlab software respectively.
- Evaluate the obtained results under real-variable parameters such as solar irradiation, wind speed, and temperature.
- Test the viability of the proposed energy management policy to manage the shared energy with respect to the boundaries of the DERs, the storage devices, and based on a specified load schedule.

2.1 Modelling of the PV Source

The PV source model can be represented by an equivalent circuit, where an ideal current source represents the photovoltaic, series resistance R_S for voltage region operation and parallel resistance R_P for current source region operation respectively [11]. In addition, a pay-bass diode can be added at the output of the PV source to protect the PV against reversed currents. Hence, this equivalent source take into account series and parallel connection of multiple arrays to increase the output generation of I_{PV} and V_{PV}. The mathematical model of the PV can be represented by the following expressions in Eqs. (1–3), where the generation at the maximum point MPP is [12, 13]:

$$\begin{cases} \boldsymbol{I_{pv}} = (\boldsymbol{I_{pv,N}} + \boldsymbol{K_I}.\Delta_T).\frac{G}{G_n} \\ \boldsymbol{P_{pv-MPPT}} = \boldsymbol{V_{MPPT}}.\boldsymbol{I_{MPPT}} \end{cases} \quad (1)$$

The PV array parameters are open-circuit voltage (Voc), short-circuit current (Isc), maximum point voltage (V_{MPP}), maximum point current (I_{MPP}), open-circuit voltage/temperature coefficient (K_V), short circuit current/temperature coefficient (K_I). This information is provided with reference to the nominal condition of temperature and solar irradiation (STC). In addition, The I/V and V/P curves represent the current, voltage and power of the PV at different irradiation (W/m^2). Thus, three main operating points can be distinguished defined by [V_{PV}, I_{PV}], which are the open circuit point [V_{OC}, 0], short-circuit point [I_{SC}, 0], and maximum power point [V_{MPP}, I_{MPP}]. Various MPPT technics can be used to extract the PV power including simple, genetic and developed methods as mentioned in [14].

2.2 Modelling of the Diesel Source

The diesel genset consists of a diesel engine coupled to a synchronous generator. A speed controller maintains the alternating current frequency at the output, and the regulator works by adjusting the fuel flow in order to keep the engine speed and generator speed constant. However, the machine functions as a synchronous compensator and provides reactive energy [15]. Hence, the total fuel cost (Fc) at time and the generation limits of the generator set can be represented by the following expressions in Eqs. (4) and (5) respectively [16]:

$$\begin{cases} F_C = \int_0^{\tau} (aP^2(t) + bP(t) + c)d_t \\ P_{Gen}^{min} \leq P_{Gen} \leq P_{Gen}^{max} \end{cases} \tag{2}$$

2.3 Modelling of the Yrid Storage System Hess

The hybrid storage system includes battery and Electric Double Layer Capacitors EDLC supercapacitor devices, connected in battery semi-active topology. The following Eq. (3) defines the output voltages of the storage devices [13]:

$$\begin{cases} V_{battery} = E = E_0 - K\frac{Q}{Q-i_t}.i_t - R_b.i_t + A_b.\exp(-B.i_t) - K\frac{Q}{Q-i_t}.i^* \\ V_{sc} = \frac{Q_T}{C_T} - R_{sc}.i_{sc} \end{cases} \tag{3}$$

Where E_0 is the battery constant voltage (V), K is the polarization constant (V/Ah), Q is the battery capacity (Ah), i* is the filtered battery current (A), it is the actual battery charge (Ah), A_b is the exponential zone amplitude (V), B is the exponential zone time constant inverse Ah^{-1} and Rb is the battery internal resistance (Ω). Q_T is the total electric charge (C), R_{SC} is the super capacitor module resistance (Ω) and i_{SC} is the supercapacitor module current (A).

2.4 Modelling of Power Onerters

The power converter are used to transfer the generated power of the distributed energy sources to the appropriate loads through a common dc link or bus. In this work, only non-isolated dc-dc converters are used such as boost, and buck-boost converters. The boost converter is connected with the PV to extract its maximum power, where the switching device of the converter is controlled via the MPPT-PWM signals. The HESS consist of a bidirectional converter BDC connected with the battery to control the charge-discharge cycles, the narrow scope of SOC, and the bus voltage stable or nearly stable. Equation (4) is used to linearize the above state-space equations of the Boost converter (5), where X is the steady-state component and D is the steady state or DC component duty-ratio. Hence, the state space averaged model of the bidirectional converter in equilibrium is shown in Eq. (6) [17].

$$\dot{X} = \left[\bar{D}A_{on} + (1-\bar{D})A_{off}\right]X + \left[\bar{D}B_{on} + (1-\bar{D})B_{on}\right]Y \tag{4}$$

$$\left\{ \begin{bmatrix} \dot{i}_L \\ \dot{V}_c \end{bmatrix} = \begin{bmatrix} 0 & -\frac{1-D}{L} \\ \frac{1-D}{C} & -\frac{1}{RC} \end{bmatrix} \begin{bmatrix} i_L \\ V_c \end{bmatrix} + \begin{bmatrix} \frac{V_c}{L} \\ -\frac{i_L}{C} \end{bmatrix} d; \right. \tag{5}$$

$$\left\{ \begin{array}{l} \dot{X} = AX + BY \\ 0 = \begin{bmatrix} -\frac{R_L}{L} & \frac{D}{L} & -\frac{1}{L} \\ -\frac{D}{C_H} & -\frac{1}{R_{in}.C_H} & 0 \\ \frac{1}{C_L} & 0 & -\frac{1}{R_{Load}.C_L} \end{bmatrix} \begin{bmatrix} i_L \\ V_L \\ V_H \end{bmatrix} + \begin{bmatrix} 0 & 0 \\ 0 & \frac{1}{R_{in}.C_H} \\ \frac{1}{R_{Load}.C_L} & 0 \end{bmatrix} \begin{bmatrix} V_L \\ V_H \end{bmatrix} \end{array} \right. \tag{6}$$

R_L is the internal resistance of the inductor, R_{in} is the internal resistance of the input source, R_{load} represents the load resistance, equivalent to V_{load}^2/P_{load}, C_H and C_L are high and low side capacitances respectively, V_L and V_H are low and high side voltages.

The studied DC Microgrid is located in the big southern of Algeria (BECHAR – ADRAR), where: latitude = 27.9716°, longitude = 0.1870°. The full structure of the studied network in Matlab simulink aims to analyze static and dynamic states of the operating variables and to test the viability and efficiency of the management strategies. ETAP software is specifically designed for power system simulation, which facilitate system analysis such as steady state, security assessment, state estimation, optimal power flow, DC/AC load flow. After this description, tests of performance of the selected topology and the efficiency of the network can be achieved through next analysis (Table 1):

Table 1. Matlab and ETAP softwares analysis.

Matlab analysis:	ETAP analysis:
DC/AC load Flow analysis. DC/AC short-Circuit analysis. Energy Management (flowcharts and algorithms). Charge-discharge of storage devices. Transient and permanent analysis	DC/AC Load Flow analysis. DC/AC Short-Circuit analysis. Battery Discharge sizing. Optimal Power Flow analysis. Motor acceleration analysis. Harmonic analysis. Transient stability analysis. Protective device coordination. Switching Sequence Management

3 Results and Discussion

This section represents initial results and discuss it to improve model viability and operating conditions under different scenarios and perturbance. Results of the two softwares are discussed separately.

3.1 ETAP Software Results

Load flow analyses are achieved on ETAP software in order to achieve the following purposes:

1- Off-line method of calculating the voltage and angle at the bus.
2- Solve the set of nonlinear power balance equations.
3- Load flow is root-finding problem, where this problem is converted to optimization problem.

Our five buses DC microgrid shown in Fig. 2 is modelled using ETAP software, where dc load flow LF analysis were made. Thus, 10 load flow scenarios are selected, where the load demand is supplied by the combination between the used DERs of the system, which are PV, local batteries, main storage, and diesel generator. In each case, load flow results include bus voltages, injected and consumed powers; global losses and iteration number are plotted and discussed. Table 2 resumes initial parameters of LF analysis. Hence, the DC loads were supplied via its DC-DC converter to reduce the bus voltage of 270.3 V_{DC} to 24 $V_{DC,}$ which is adaptive with the common DC appliances, and with the used MPPT and PWM regulator in order to control and supply safe DC energy (Table 3).

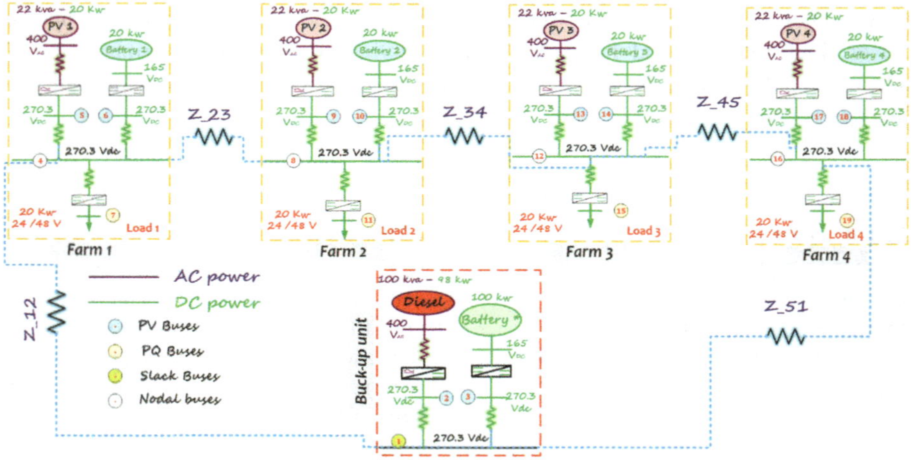

Fig. 2. Proposed DC Microgrid structure: a-Full circuit, b-sub-network circuit of farms

After running load flow analyses of the five buses DC-MG using 10 proposed management scenarios, the energy balance and the losses are summarized in the next table.

Table 2. Load flow line and bus data

	Bus data (pu)		Line data				
Bus No.	Bus type	Initial V_{BUS}	From	To	R_{BUS}	L_{BUS}	Distance (m)
1	SLACK	1	1	4	.05098	.00014	100
2	PV	1	4	8	.05098	.00014	120
3	PV	1	8	12	.05098	.00014	90
4	PV	1	12	16	.05098	.00014	50
5	PV	1	16	1	.05098	.00014	135

Table 3. Load flow results

No.	Scenarios	Parameters					
		Production (pu)	Consumption (pu)	Losses (pu)	Iterations	State	No. of power converter
1	PV only	1.0205	1	0.0205	1	Normal	8/14
2	Local batteries only	1.0413		0.0413		Normal	8/14
3	Diesel only	1.0208		0.0208		Emergency	5/14
4	Main storage only	1.0311		0.0311		Emergency	5/14
5	PV + local batteries	1.0354		0.0354		Critical	12/14
6	PV + Diesel	1.0205		0.0205		Critical	9/14
7	PV + Main storage	1.0234		0.0234		Critical	9/14
8	Diesel + Main storage	1.0311		0.0311		Emergency	6/14
9	Diesel + Local batteries	1.04		0.04		Critical	9/14
10	Main storage + Local batteries	1.0413		0.0413		Critical	9/14

The total load demand of the whole network is about 80 (Kw)–1 (p.u). As seen in Fig. 3, in different scenarios, the operating voltages of the 5 buses were within its limits with a minimum of 0.9997 p.u, which reflects the stability of the power balance, where the totality of the load demand is supplied sufficiently and efficiently by the distributed energy resources DERs under different scenarios. In addition, very few losses was measured during LF analyses, which were varying [1.64–3.3] (Kw) due to the studied sizes of the dc cables with transport distances. The next figure shows the nodal buses of each scenario, where the energy balance were stable under the used management scenarios, except in the 3 emergency cases.

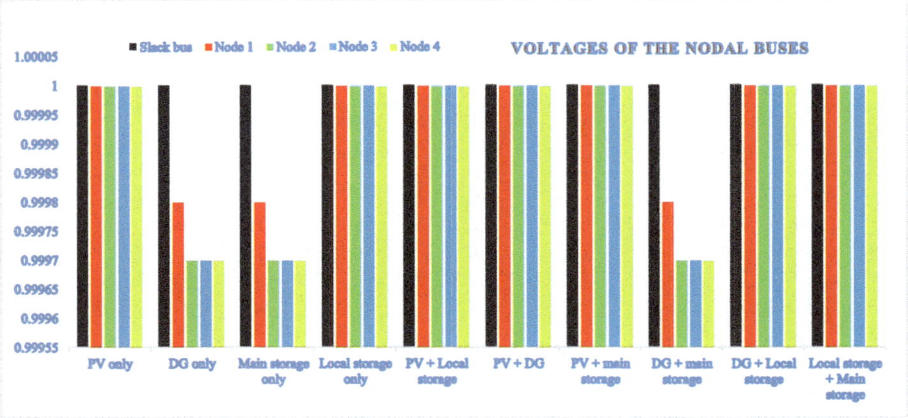

Fig. 3. ETAP results of the nodal buses of the DC-MG system

In these cases, the interconnected farms have been supplied by the backup unit energy in order to cope with their lack of energy. In contrast, and during the rest scenarios, the sub-networks have supplied the loads demand sufficiently and efficiently through the generated energies of the DERs of the studied system.

4 Conclusion

In this brief paper, a dc microgrid network is studied. The microgrid module are presented through mathematical and equivalent circuits of DER and power converters, in addition to management strategies structures. The dc microgrid is tested using Matlab simulink and ETAP softwares in order to study transient and steady state parameters respectively. Thus, load flow tests are carried by ETAP software, where branches, cables, buses, and power flow results are plotted and discussed. In addition, other test are achieved using Matlab simulink, where the PV power conversion chain, storage devices are tested and main parameters are figured and discussed. Therefore, this model is under construction, where backup generator and the energy management of the network are not studied and modelled yet. As a future perspective, next work focus on connecting such sub-systems as a DC Microgrid and modelling the backup unit including diesel generator for the whole microgrid.

References

1. Hatziargyriou, N. (ed.): Microgrids: Architectures and Control. Wiley, Hoboken (2014)
2. Chalise, S.: Power management of remote microgrids considering battery lifetime (2016)
3. Meegahapola, L.G., et al.: Microgrids of commercial buildings: strategies to manage mode transfer from grid connected to islanded mode. IEEE Trans. Sustain. Energy **5**(4), 1337–1347 (2014)

4. Manas, M.: Renewable energy management through microgrid central controller design: An approach to integrate solar, wind and biomass with battery. Energy Reports **1**, 156–163 (2015)
5. Yin, C., et al.: Energy management of DC microgrid based on photovoltaic combined with diesel generator and supercapacitor. Energy Convers. Manag. **132**, 14–27 (2017)
6. Kumar, A., Biswas, A.: Techno-economic optimization of a stand-alone photovoltaic-battery renewable energy system for low load factor situation-a comparison between optimization algorithms. Int. J. Eng.-Trans. A: Basics **30**(10), 1555–1564 (2017)
7. Mahmood, et al.: A power management strategy for PV/battery hybrid systems in islanded microgrids. IEEE J. Emerg. Sel. Topics Power Electron. **2**(4), 870–882 (2014)
8. Diaz, et al.: Intelligent distributed generation and storage units for DC microgrids—a new concept on cooperative control without communications beyond droop control. IEEE Trans. Smart Grid **5**(5), 2476–2485 (2014)
9. Bhattacharjee, A.: Decentralized power management in microgrids (2014)
10. Nunna, H.K., et al.: Multiagent-based distributed-energy-resource management for intelligent microgrids. IEEE Trans. Ind. Electron. **60**(4), 1678–1687 (2013)
11. Pan, L.: Analysis of photovoltaic module resistance characteristics, pp. 1369–1376 (2013)
12. Brahmi, N., et al.: Optimum sizing algorithm for an off grid plant considering renewable potentials and load profile. Int. J. Renew. Energy Dev. **6**(3), 213 (2017)
13. Rivera, O., et al.: Hardware in loop of a generalized predictive controller for a micro grid DC system of renewable energy sources. Int. J. Eng. **31**(8), 1215–1221 (2018)
14. Boutabba, T., et al.: A new implementation of maximum power point tracking based on fuzzy logic algorithm for solar photovoltaic system. Int. J. Eng.-Trans. A: Basics **31**(4), 580–587 (2018)
15. Feddaoui, O.: Contribution à l'Etude des Systèmes Hybrides de Génération: Application aux Energies Renouvelables. Doctoral dissertation, University of Souk Ahras (2014)
16. Rouholamini, M., Mohammadian, M.: Grid-price-dependent energy management of a building supplied by a multisource system integrated with hydrogen. Int. J. Eng.-Trans. A: Basics **29**(1), 40–49 (2016)
17. Ghazanfari, A., et al.: Active power management of multihybrid fuel cell/supercapacitor power conversion system in a medium voltage microgrid. IEEE Trans. Smart Grid **3**(4), 1903–1910 (2012)

A Petri Net Modeling for WSN Sensors with Renewable Energy Harvesting Capability

Oukas Nourredine[1,2] and Boulif Menouar[3(✉)]

[1] Department of Computer Science, Normal Superior School, Kouba, Algiers, Algeria
oukas@ens-kouba.dz

[2] LIMOSE Laboratory, M'Hamed Bougara University of Boumerdes, Independence Avenue, Boumerdes 35000, Algeria

[3] Department of Computer Science, Faculty of Sciences, M'Hamed Bougara University of Boumerdes, Independence Avenue, Boumerdes 35000, Algeria
m.boulif@univ-boumerdes.dz

Abstract. In Wireless Sensor Networks, the Sensor Nodes (SNs) need to operate for a long period of time without any kind of intervention in order to achieve dedicated tasks such as surveillance, automation, monitoring, control and many others. SNs are equipped with non-changeable batteries for power supply. Due to the small dimension of an SN, the energy supply attached to the SN battery has to be very limited in size. The lifetime of the sensor nodes and thus of the overall network greatly depends on these batteries. Because SNs generally operate in harsh conditions, the replacement of batteries is impossible. Hence, we make use of natural sources of renewable energies such as wind, sun, vibrations and alike to provide SNs with permanent harvested power supply. In this paper, we adopt an approach which is relatively poorly investigated in the literature by presenting, to the best of our knowledge, a new Generalized Stochastic Petri Net (GSPN) model to an SN with Energy Harvesting capability. The proposed model allows to determine performance parameters and to extract some experimental results to predict the energy consumption of a sensor node.

Keywords: Wireless sensor network · Renewable energy · Energy harvesting · Energy management · Energy quantization

1 Introduction

A wireless sensor network (WSN) is composed of small, low-cost, and low-power sensor nodes (SNs). An SN can sense events in its vicinity and can receive and/or send event reports to its SN neighbors. WSNs have a wide range of application areas such as military eld, environmental monitoring, and intelligent building systems, healthcare, home automation, and traffic control [1, 2]. In order to be operational, the sensors rely on batteries which are a key issue in the network lifetime.

In most cases, sensor nodes are deployed in harsh and remote areas where a physical access to their location is often almost impossible. This makes the replacement of the battery a tough and risky task [3] and adds an energy constraint that limits the

M. Hatti (Ed.): ICAIRES 2019, LNNS 102, pp. 524–534, 2020.
https://doi.org/10.1007/978-3-030-37207-1_56

functionality of the WSNs. A solution alternative is to equip SNs with an Energy Harvesting system.

Energy harvesting (EH) [4–6] refers to collecting renewable energy from the environment such as sun, wind, body heat, finger strokes, foot strikes or other sources of energy and converting them to electrical energy used to power the nodes and increase their lifetime. This harvested energy can be an alternative energy source for adding-on a principal power source (battery) [6]. EH has significantly improved the lifetime of the whole WSN and has made it possible to add new devices in a much more reliable manner by preventing the perturbation of the energy powering [7].

EH capability triggers issues in all the layers of a network design [8]: in topology control, routing, medium access control, transmission policies, scheduling based congestion control, data cycling and admission control.

Ashraf et al. [8] cites several references for these problems and argues that the stochastic approaches provide a more realistic representation of the underlying processes as energy storage and consumption in wireless nodes are stochastic by nature [8].

Authors in [9] propose a pro le energy prediction model called Pro-Energy for multi-source energy-harvesting WSNs that leverages past energy observations to forecast future energy availability. In [8], the authors adopt an approach that views the problem as a queue control system where the objective is to regulate transmission such that the energy level stays near to a reference value.

Another class of approaches dubbed duty cycling schemes adjusts energy consumption to achieve efficient energy expenditure. Sensor nodes switch to low consumption mode when they are at risk of becoming energy de cient and consume a reasonable amount of energy to achieve higher performance levels when there is a sufficient amount of energy. Among these approaches, channel and queue-aware sleep and wake up scheme maximizes throughput [10, 11]. Another approach [10] of the duty cycles family prevents sensor nodes from working in an energy-negative mode for periods when the real energy harvest drops beneath a given threshold [11].

A Petri Net based modelling of battery charge/discharge process is provided in [12]. The authors consider the retrial phenomenon, breakdowns and the sleeping mechanism to model the behaviour of the network with a sensor-neighbours relationship abstraction.

In this paper, our work is motivated by the lack of Petri Nets modelling approaches for sensor energy consumption with EH capability. Hence, we propose a Generalised Stochastic Petri Nets model that regulates the consumption of energy in order to insure that it is always smaller than the harvested energy and to guarantee by the way the continuity of service of the whole WSN. Our approach contrasts with that of [12]'s as it provides a more realistic description of the system. Indeed, in [12], the authors consider that a sensor enters to the sleeping state if and only if the energy is less than a given threshold. This limitation does not represent the actual behaviour of an SN that can optionally change its sate even if its battery is almost fully charged in order to postpone attaining the maximum cycle count limit.

The rest of this paper is organised as follows: Sect. 2 reviews some principle notions pertaining to renewable energy and solar energy harvesting systems. In Sect. 3, we give a description to the architecture of an SN. In the next section, we give our contribution by presenting a more realistic GSPN model for an SN with EH capability.

In Sect. 5, we derive and discuss our experimental results. Finally, we give our conclusions as well as some directions for future works.

2 Renewable Energies

A renewable energy refers to an energy whose source is not exhausted when used according to. In this section, we present the different sources of renewable energy, then we provide further details for the solar energy.

2.1 Sources of Renewable Energy

There are numerous forms of renewable energy available in our universe such as: thermal, photo-voltaic (solar, light), bio-energy, hydro, tidal, wind, wave, and geothermal [16]. Using renewable energy to feed a sensor with power is called energy harvesting (EH). EH brings a lot of advantages to the operating mode of an SN. We can cite for instance: reducing the dependency on battery power, reducing installation and maintenance costs, increasing the lifetime of the SN as long as the renewable energy is available, eliminating the costs of battery replacements and making the WSN installation very easy [14]. For technical reasons, not all the renewable energies are nowadays suitable for SNs. Figure 1 shows common energy sources for EH and their respective transducers to generate power for SNs.

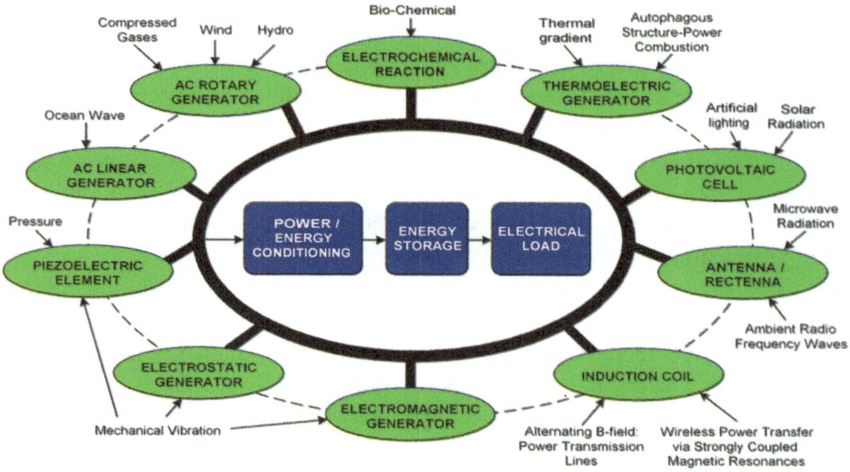

Fig. 1. Sources for energy harvesting. [4, 14].

Renewable energies can be broadly classified in three essential types: thermal energy, radiant energy and mechanical energy (see Fig. 2) [16]. Other classifications exist in the literature for which we refer interested readers to [6].

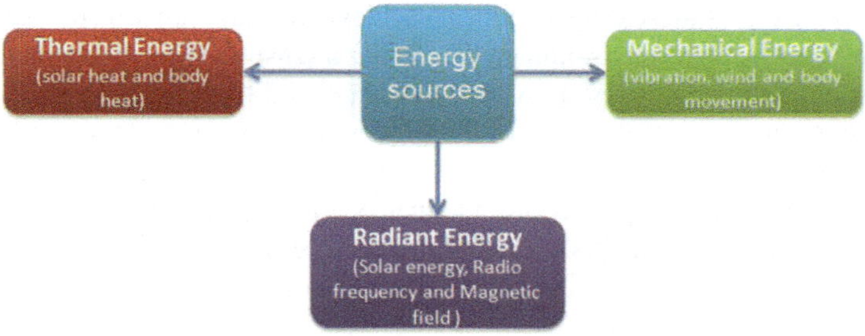

Fig. 2. Appropriate renewable energies for Energy Harvesting.

2.2 Solar Energy Harvesting

Among the renewable energy sources, solar energy (SE) is the most attractive for WSNs, because it is relatively more predictable and it has a higher power density in comparison to other renewable energy sources. SE amount is about 15 mW = cm^3 in direct sun, 150 w = cm^3 outdoor but cloudy day and 6 w = cm^3 indoor [2, 17]. Figure 3 represents a basic solar energy harvesting system which consists of a Solar Panel, a DC-DC converter, a rechargeable battery, a battery charge protection circuit called power management unit and a DC-DC converter control unit [18].

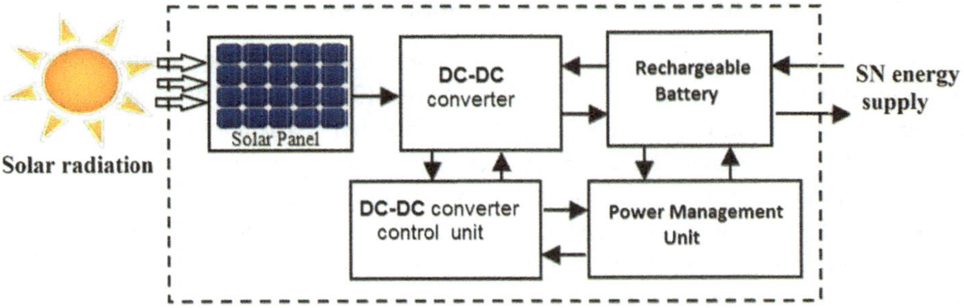

Fig. 3. Basic solar energy harvesting system [18].

There are several of-the-shelf solar EH sensor nodes such as: Solar Biscuit which is a battery-less sensor for environment monitoring, Everlast which is a long-life super-capacitor-operated SN, Heliomote which is an SN extensively used for ecosystem sensing and many others [19].

3 Sensor Node Architecture

3.1 General SN Architecture

An SN comprises a communication or a radio transceiver unit for transmitting and receiving information. The processing unit that contains a processor or a micro-controller and a memory handles these information. To achieve the main function of a sensor that is to capture events from the environment, the SN has a sensing unit which contains sensors and an analogue/digital converter (ADC). The ADC digitizes the analogue signal generated by the sensor and delivers it to the processing unit for further processing. The power unit gets energy from the battery and distributes it to the other system components [4, 7]. Some SNs have a supplement unit such as a mobilizer and/or a GPS in order to procure motion and/or localization capabilities. Figure 4 shows such a typical architecture of a sensor node.

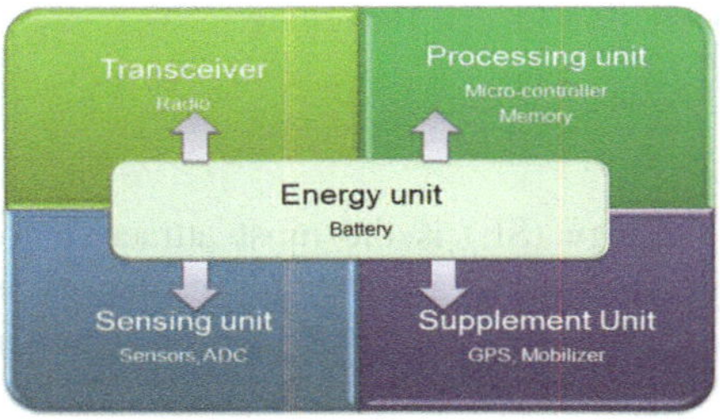

Fig. 4. Typical architecture of an SN

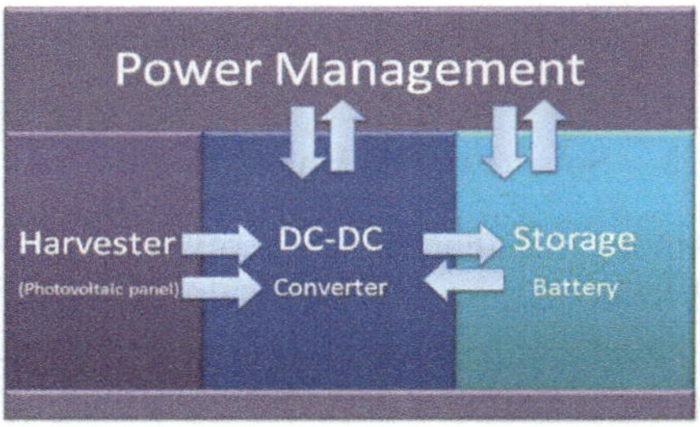

Fig. 5. Architecture of power unit with EH capability

3.2 SN Architecture with Energy-Harvesting Capability

In the last decade, there has been intensive researches that aim at an efficient transforming of renewable energy to electrical energy [3, 17, 20]. The related findings opened new application perspective for WSN. The harvested energy can be directly used by the components of the SN or accumulated and preserved in a rechargeable battery. For example, in the case of solar energy, during daytime, harvested energies are used directly as well as stored for future use. At nighttime, the stored power is utilized to provide energies to the SN. Figure 5 represents the architecture of the power unit with Energy Harvesting capability.

4 A GSPN Model of a Sensor Node with Energy-Harvesting Capability

Petri nets [13] are a graphical modeling tool that is useful for simulating dynamic systems with concurrent, synchronized and conflicting activities. In this section, we use this medium to describe the modus operandi of an SN with EH capability.

This model supposes that an SN transceiver can be in one of following modes:

- Active: the SN does its fundamental functionalities of processing, sensing, receiving and sending.
- listening: the transmitting hardware is turned off.
- Standby: both receiving and transmitting hardware are turned off. The SN enters this sleeping state to save its energy and to increase it by its harvesting capability.

Our model complies with the energy quantization principle. That is, we suppose that the energy is handled with discrete amounts. For this reason, we model the battery energy by quantum levels. We suppose that a quantum is an indivisible value equal to the required amount of energy to send or to receive a message. The level of battery charge can trigger the switching to or from the SN Standby state as follows: Given two charge level thresholds l1 and *l2* with *l1* < *l2*, when the quantum level decreases below the first threshold *l1*, the SN switches to the sleeping mode in order to save energy. When the quantum level enhances to attain *l2* due to energy harvesting, the SN switches to the active mode to resume its duty.

Figure 6 depicts the proposed GSPN where the SN functionalities are all but one represented by timed transitions. The place Battery represents the battery charge level. Each token in this place represents a quantum of energy. The battery has a capacity limit denoted by *C*. When the battery charge becomes less than the threshold l1, the SN enters to the sleeping state immediately (see the immediate transition *Go_sleep*). An SN in the sleeping state is represented by a mark in the place Standby. Even if the SN is sleeping, it consumes energy (see *Fondamental_op* transition) though it cannot listen nor send nor receive (notice the three inhibitor arcs incident to the Standby place). When sleeping, the number of quantums can increase due to energy harvesting. The sensor stays in sleeping state until the number of quantums exceeds the threshold *l2* when the ring of the timed transition Be awake becomes possible. Table 1 gives a short description of each transition.

Notice that even if the model uses a unique quantum value, it can consider different energy consumption rates for the different functionalities by using different values of the transition ring rates.

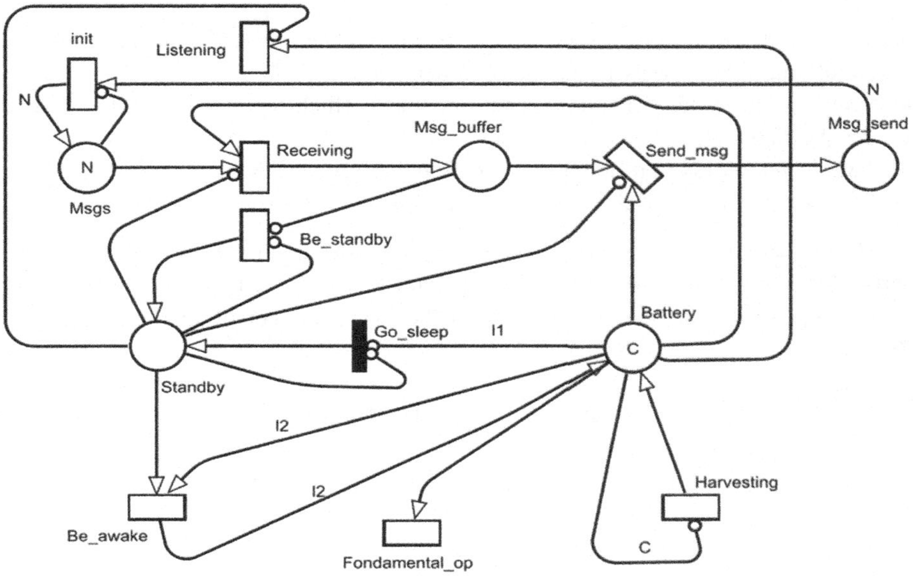

Fig. 6. A GSPN model of an SN with energy harvesting capability.

Table 1. Transition description.

Transition	Type	Description	Firing rate	Value/day
Go_sleep	Immediate	SN enters to the sleeping state if the energy is less than the threshold $l1$	/	/
Init	Timed	For initializing the model	ν	***1***
Receiving	Timed	Represents the receiving of a message	μ	*50*
Send _msg	Timed	Represents the sending of a message	λ	*50*
Be_Standby	Timed	The SN enters to the sleeping sate periodically	ω	*50*
Be_awake	Timed	The SN becomes active	θ	*25*
Listening	Timed	Represents the listen operation. The SN consumes a quantum of energy in each firing	α	*100*
Fondamental_op	Timed	Whatever be the state, the SN consumes energy	β	*10*
Harvesting	Timed	Represents the harvesting operation. The battery energy is increased by a quantum in each firing	δ	*200*

In addition to the Battery place, which is the heart of our model, our Petri net contains four other places. That is:

- *Msgs*: Each token in this place represents a candidate message to be received. The initial number of marks in this place is denoted by *N*.
- *Standby*: This place contains at most one token which represents that the SN is in the sleeping state.
- *Msg_buffer:* A token in this place represents a received message.
- *Msg_send*: A token in this place represents a sent message.

Finally, solving the stationary state system provides us the vector π that allows to derive several formulas of performance parameters. For example, the mean battery charge denoted by $\overline{Battery}$ corresponds to the mean number of marks in the place *Battery*:

$$\overline{Battery} = \sum\nolimits_{i:Mi \in M} Mi(Battery).\pi i \tag{1}$$

We apply Little's law [15] in the equilibrium state. The waiting time *Wait (Msg_buffer)* of a token in the place *Msg_buffer* is equal to the mean number of tokens in that place *Msg_buffer* divided by the mean sending rate $\bar{\lambda}$ (throughput of *Send _msg* transition). So:

$$Wait(Standby) = \frac{\overline{Msg\_buffer}}{\bar{\lambda}}. \tag{2}$$

Where $\overline{Msg\_buffer}$ is the mean number of tokens in the place *Msg_buffer*. That is:

$$\overline{Msg\_buffer} = \sum\nolimits_{i:Mi \in M} Mi(Msg\_buffer).\pi i \tag{3}$$

Mathematical formulas related to the GSPN modeling procure other performance parameters. For other details of these formulas and associated proofs, the reader can refer to [13].

5 Experimental Analysis

In order to trial our model's ability to simulate an actual SN behavior, we conducted our experiments by using the following parameter values. Following common WSN battery features [14], the considered SN battery has a volume of *1* cm^3 that provides *0.8* wh of power. By supposing that a quantum is equal to 26 mwh, the battery capacity *C* is hence equal to *30* quantums. We set the two battery level thresholds to $l1 = 5$ and $l2 = 10$.

We also suppose that the SN consumes in average:

- One quantum per hour when it is doing its fundamental operations of processing and sensing;
- One quantum per message when sending or receiving;
- One quantum per listening operation. The SN conducts a listening operation 5 times per hour.

After having verified the ergodicity of our model, we solve the system in the equilibrium state to derive the performance measures.

Figure 7(a) shows the influence of the harvesting rate on the mean battery charge. It is clear from this figure that the battery charge increases when the harvesting rate increases. That result coincides with the actual circumstances. For example, if the harvesting rate is greater than 40 times per hour, the mean battery charge is greater than 80%. An adequate energy model must make it possible to check if the energy consumption is compensated by the harvesting energy. Figure 7(b) depicts the influence of the harvesting rate on the mean response time. The mean response time decreases when the harvesting rate decreases. Because the mean battery charge is often greater than the threshold *l1* that allows the main sensor to stay activated in most of the time to serve messages quickly.

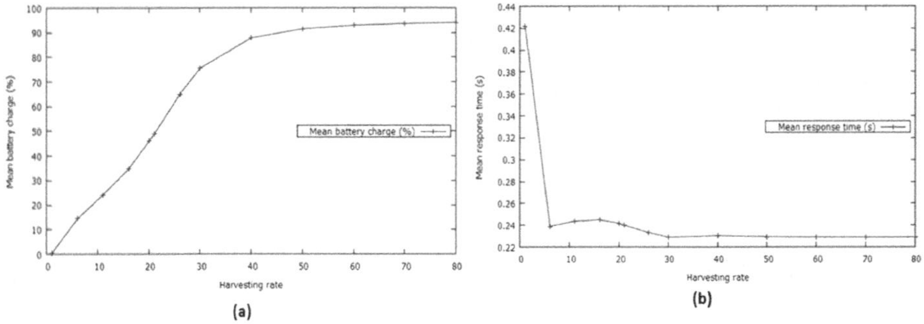

Fig. 7. (a) Battery charge versus harvesting rate. (b) Mean response time versus harvesting rate.

Figure 8(a) depicts the influence of the sleeping rate on the mean battery charge. Increasing the sleeping rate obligate the sensor to be disabled in most of time which conserves energy. Figure 8(b) shows the influence of the sleeping rate on the mean response time. We note that the mean response time decreases when the sleeping rate increases until the value of 25. After this value, the mean response time decreases slowly whereas the sleeping rate increases because this configuration becomes inadequate when we compare the sleeping rate with the awaking rate.

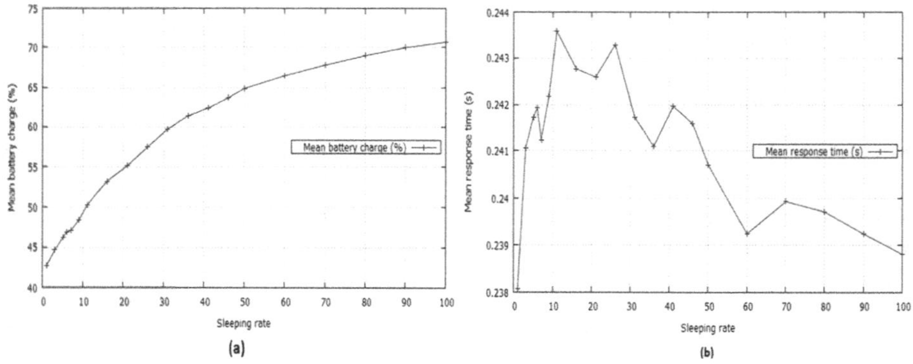

Fig. 8. (a) Battery charge versus sleeping rate. (b) Mean response time versus the sleeping rate

Figure 9(a) depicts the influence of the number of messages present in the system on the mean battery charge. By varying this number, we can see the behavior of the battery charge. Indeed, increasing the number of messages can cause a decrease in the mean battery charge level. If this later drops to a value that is close to the first threshold, the SN will frequently fall in a sleeping mode generating a delay in the response time. In this case, using another type of batteries with better performances can be considered. Figure 9(b) shows that, as expected, the mean response time increases when the number of messages increases.

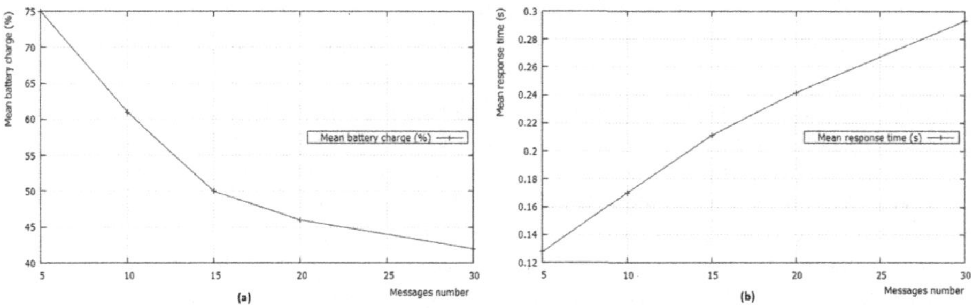

Fig. 9. (a) Mean battery charge versus the number of messages. (b) Mean response time versus the number of messages.

6 Conclusion

In this paper, we give a GSPN modeling for sensor nodes energy management. The energy consumption is modeled by using quantization principles. By using duty cycle approach (sleeping mechanism) to reduce the energy consumption, the proposed model proves to be able to predict the mean level of battery energy and the mean response time. Our model allows the WSN designers to choose the SN components with the most adequate features to achieve the required performances. Some experimental analysis are given to demonstrate the feasibility of the proposed model.

In our future work, we want to investigate the modeling of the sensor nodes by taking into account the difference between message types.

References

1. Othman, M.F., Shazali, K.: Wireless sensor network applications: a study in environment monitoring system. Procedia Eng. **41**, 1204–1210 (2012). International Symposium on Robotics and Intelligent Sensors 2012 (IRIS 2012)
2. Kang, M., Yoon, I., Noh, D.K.: Efficient location service for a mobile sink in solar-powered wireless sensor networks. Sensors **19**, 272 (2019)
3. Kosunalp, S.: An energy prediction algorithm for wind-powered wireless sensor networks with energy harvesting. Energy **139**, 1275–1280 (2017)
4. Basagni, S., Naderi, M.Y., Petrioli, C., Spenza, D.: Wireless sensor networks with energy harvesting. In: Mobile Ad Hoc Networking: Cutting Edge Directions, pp. 701–736 (2013)
5. Akbari, S.: Energy harvesting for wireless sensor networks review. In: Proceedings of the 2014 Federated Conference on Computer Science and Information Systems, pp. 987–992. IEEE (2014)
6. Kausar, A.Z., Reza, A.W., Saleh, M.U., Ramiah, H.: Energizing wireless sensor networks by energy harvesting systems: scopes, challenges and approaches. Renew. Sustain. Energy Rev. **38**, 973–989 (2014)
7. Rashid, A., Khan, F., Gul, T., e Alam, F., Khan, S.A.S., Khalil, F.K.: Improving energy conservation in wireless sensor networks using energy harvesting system. (IJACSA) Int. J. Adv. Comput. Sci. Appl. **9**(1), 354–361 (2018)

8. Ashraf, N., Faizan, M., Asif, W., Qureshi, H.K., Iqbal, A., Lestas, M.: Energy management in harvesting enabled sensing nodes: Prediction and control. J. Netw. Comput. Appl. **132**, 104–117 (2019)
9. Cammarano, A., Petrioli, C., Spenza, D.: Online energy harvesting prediction in environmentally powered wireless sensor networks. IEEE Sens. J. **16**(17), 6793–6804 (2016)
10. Sharma, V., Mukherji, U., Joseph, V., Gupta, S.: Optimal energy management policies for energy harvesting sensor nodes. IEEE Trans. Wirel. Commun. **9**(4), 1326–1336 (2010)
11. Babayo, A.A., Anisi, M.H., Ali, I.: A review on energy management schemes in energy harvesting wireless sensor networks. Renew. Sustain. Energy Rev. **76**, 1176–1184 (2017)
12. Oukas, N., Boulif, M.: Energy-consumption-aware modelling and performance evaluation for eh-wsns. In: Proceedings of the second conference on Informatics and Applied Mathematics (IAM 2019), pp. 57–62. labstic.univ-guelma.dz (2019)
13. Florin, G., Fraize, C., Natkin, S.: Stochastic Petri Nets: properties, applications and tools. Micro-electron. Reliab. **31**(4), 669–697 (1991)
14. Tan, Y.K., Panda, S.K.: Review of energy harvesting technologies for sustainable WSN. In: Seah, W., Kheng Tan, Y. (eds.) Sustainable Wireless Sensor Networks, chap. 2. IntechOpen, Rijeka (2010)
15. Little, J.D.C.: A proof of the queuing formula $l = \lambda.w$ Oper. Res. 9, 383–387 (1961)
16. Thakur, S., Prasad, D., Verma, A.: Energy harvesting methods in wireless sensor network: a review. Int. J. Comput. Appl. **165**(9), 19–22 (2017)
17. Onishi, T., Ogose, S.: Lifetime extension of wireless sensor networks with energy harvesting. J. Signal Process. **22**(2), 77–86 (2018)
18. Sharma, H., Haque, A., Jaffery, Z.A.: Modeling and optimisation of a solar energy harvesting system for wireless sensor network nodes. J. Sens. Actuator Netw. **7**(3), 40 (2018)
19. Kaur, H., Buttar, A.S.: A review on solar energy harvesting wireless sensor network. Int. J. Comput. Sci. Eng. **7**(2), 398–404 (2019)
20. Sudevalayam, S., Kulkarni, P.: Energy harvesting sensor nodes: survey and implications. IEEE Commun. Surv. Tutor. **13**(3), 443–461 (2011)

Robust Residuals Generation for Faults Detection in Electric Powered Wheelchair

S. Tahraoui[1(✉)], M. Z. Baba Ahmed[1], F. Benbekhti[2], and H. Habiba[3]

[1] Department of Electronic Engineering, Universit Hassiba Benbouali de Chlef, Ouled Fares, Chlef, Algeria
s.tahraoui@univ-chlef.dz, zaki.babaahmed@gmail.com
[2] Technology Department, Faculty of Sciences and Technology, Djilali Bounaama Khemis Miliana University, Khemis Miliana, Algeria
f_benbekhti@yahoo.fr
[3] Faculty of Technology, Abou Bekr Belkaid University, B.P 230 Chetouane, Tlemcen, Algeria
houari_habiba@yahoo.fr

Abstract. A diagnostic approach for generation robust residuals to the dynamic model of an electric wheel chair is proposed in this paper. The method formulated here uses the observer presented in this work in order to design a residual generator that allows the detection of faults. The unknown input observer with perfect decoupling allows assessing these faults, which leads to their localization and detection. The present study focuses on faults arising from actuators for linear systems. Finally, the application of the observer algorithm is presented to study the performances of this observer.

Keywords: Electric powered wheelchair · Diagnosis · Robust residual · Detection fault · Observer

1 Introduction

Electric wheelchairs (EWC) belong to the category of under-actuated systems, which have fewer control inputs than the available degrees of freedom. The first electric wheelchair (EWC) was developed in Canada in the early 20th century, but it was not until the 1960s, with the advanced technological discoveries, such as the microprocessors, that it became reliable and was therefore ready to be used. Electric wheelchairs in this category are powered by motors and can be used indoors or outdoors. There are three different categories of chairs, namely the electric wheelchairs with fixed chassis, the electric wheelchairs with a folding chassis and the adjustable electrical wheelchairs [1–3]. In this article, a model-based diagnostic method is proposed for the detection and estimation of faults that can occur in an electric wheelchair (EWC); this system can be considered as a prototype for studying autonomous vehicles. Indeed, the issues raised by the electric wheelchair (EWC) constitute a subject of study in their own right; they provide an excellent basis for the study of more complex mobile systems. Our approach, which is based on the unknown input observer (UIO) with perfect decoupling, aims at designing a residual generator that allows detecting, localizing and

M. Hatti (Ed.): ICAIRES 2019, LNNS 102, pp. 535–545, 2020.
https://doi.org/10.1007/978-3-030-37207-1_57

identifying faults. This operation aims at generating structured residuals in order to locate faults and minimize false alarms. The unknown input observer (UIO) makes it possible to estimate these alarms, and also helps to detect and locate them. This study focuses on the defects of actuators in a linear system. The problem of estimating the state of a system is of considerable practical importance, whether for the implementation of a control law or for the elaboration of a diagnostic strategy. The basic principle of generating residuals, using observers, consists in carrying out an estimation of the outputs of the system from the quantities accessible to measurement, namely the inputs and the outputs. Then, the residual vector may be constructed as the difference between the estimated output and the measured output; this can be achieved by assessing the error on the output. When modeling a system, it is common to use inputs that are not measurable. The term unknown inputs is therefore used to designate such inputs, and the reconstruction of the state of such systems can only be achieved under certain conditions. The observers are then called unknown input observers (UIO) [4]. Therefore, the principle of constructing an observer with unknown inputs is to make the estimation error independent of non-measurable perturbations. Consider a system to be monitored, in which the observer with unknown inputs (UIO) can solve the problem of sensitivity to the various faults and perturbations by introducing their state matrices into the synthesis equations of the observer-based residual generation. In this case, the decision-making requires comparing the fault indicator with the threshold that is obtained empirically or theoretically Chen et al. [5] were first to introduce the usage of unknown input observers (UIOs) in the detection of faults. In this context, the generation of residuals in systems based on linear models has been the subject of several research works using the state observer. Researchers such as Kiyak et al. [6], Sun [7], Khan and Ding [8], Cristofaro and Johansen [9], were particularly interested in this subject. For example, authors like Bagherpour et al. [10], Tahraoui et al. [11] have widely used the unknown input observer (UIO) in the diagnosis of failures in industrial processes and installations. This paper focuses on the problem of diagnosing defects of actuators in an electric wheelchair (EWC) using the unknown input observer (UIO) with a perfect decoupling between unknown inputs.

2 Unknown Input Observer (UIO) with Perfect Decoupling

The principle of the unknown input observer (UIO) consists in generating an estimation error (of the state vector) which tends asymptotically towards zero even in the presence of perturbations. Thus, the generated residual is decoupled from the perturbations because it depends on the estimation error. The unknown input observer (UIO) theory with perfect decoupling consists in making the estimation error independent of the unmeasurable perturbations. This is the theory of the generation of robust residuals. The synthesis of this type of observers is possible for models of systems admitting the form that is properly described by the following state representation:

$$\begin{cases} \dot{\mathrm{x}}(\mathrm{t}) = \mathrm{Ax}(\mathrm{t}) + \mathrm{Bu}(\mathrm{t}) + F_x f(t) + D_x d(t) \\ \mathrm{y}(\mathrm{t}) = \mathrm{Cx}(\mathrm{t}) + F_y f(t) \end{cases} \tag{1}$$

2.1 Observer Structure

The model of the observer with perfect decoupling is given by:

$$\begin{cases} \dot{z}(t) = Mz(t) + Nu(t) + Py(t) \\ \hat{x} = z(t) - L_y y(t) \end{cases} \tag{2}$$

Where $\hat{x} \in R^n$ and $z \in R^n$ are the estimated state vector and the observer state vector, respectively. The matrices M, N, P, Ly are determined so as to obtain a residual r(t). These matrices are determined in such a way that the estimate $\hat{x}(t)$ converges asymptotically towards the real state x(t) of the system, in spite of the influence of the perturbations.

2.2 Observer Synthesis Algorithm

The observer synthesis algorithm can be summarized as follows:

- Rank $CD_x = nd$.
- Calculation of $L_y = -D_x[(CD_x)^T(CD_x)]^{-1}(CD_x)^T$.
- Calculation of $E = I + L_yC$.
- Calculation of $N = EB$.
- Impose M as a Hurwit matrix. For this purpose, one may choose M as a diagonal matrix which shows the eigenvalues that are sought for the observer.

Calculation of $PC = EA - ME$.

Remark: Decoupling is only possible if the rank of the matrix CDx is equal to the number of inputs to be decoupled.

Theoretical calculation of residuals Calculating the transfer matrix linking the defects to the estimation error at the output.

Let: $F = ML_yF_y + PF_y - EF_x$, $\acute{F} = -L_yF_y$

The residual vector is:

$$\begin{cases} e_y(s) = \left[C(sI - M)^{-1}(F + s\acute{F}) - F_y\right]f(s) \\ F = ML_yF_y + PF_y - EF_x \\ \acute{F} = -l_yF_y \end{cases} \tag{3}$$

The error transfer function is:

$$G_f(s) = C(SI - M)^{-1}(F + SF') - F_y \tag{4}$$

The purpose is not only to make the residual generator insensitive to unknown inputs but also to make it as sensitive as possible to faults.

Q(s) Allows structure the residuals in order to facilitate faults location.

Let Q(s) be a proper and stable transfer matrix. Let's generate a residual vector r(s) such that r(s):

$$r(s) = Q(s)e_y(s) = Q(s)G_f(s)f(s) \tag{5}$$

Q(s) makes it possible to structure the residuals in order to be able to localize the defects. This allows generating the signature table which presents the impact of the defects on each residual [12].

3 State Representation of the Electric Wheelchair

The present study was carried out on the dynamic model of the electric wheelchair, which was developed by Boubekeur et al. [13]. The state representation of for the electric wheelchair takes the following form (3):

$$\begin{cases} \dot{\mathbf{x}}(\mathbf{t}) = \mathbf{Ax}(\mathbf{t}) + \mathbf{Bu}(\mathbf{t}) + \boldsymbol{D_x d(t)} \\ \mathbf{y}(\mathbf{t}) = \mathbf{Cx}(\mathbf{t}) \end{cases} \tag{6}$$

This is the nominal form where the disturbances are taken into consideration. The following state representation is therefore obtained:

$$\begin{cases} [\dot{x}] = \begin{bmatrix} 0 & 1 & 0 & 0 \\ 0 & l_1 & 0 & l_2 \\ 0 & 0 & 0 & 1 \\ 0 & l_3 & 0 & l_4 \end{bmatrix} [x] + \begin{bmatrix} 0 & 0 \\ y_1 & y_3 \\ 0 & 0 \\ y_2 & y_4 \end{bmatrix} [u] + \begin{bmatrix} 0 \\ 1 \\ 0 \\ 1 \end{bmatrix} T \\ [y] = \begin{bmatrix} 1 & 0 & 0 & 0 \\ 0 & 1 & 0 & 0 \\ 0 & 0 & 1 & 0 \\ 0 & 0 & 0 & 1 \end{bmatrix} [x] \end{cases} \tag{7}$$

Where: $l_1 = l_4 = -\frac{ac}{a^2 - b^{2\prime}}, l_2 = l_3 = \frac{ac}{a^2 - b^{2\prime}}, y_1 = y_4 = \frac{aR}{a^2 - b^{2\prime}}, y_2 = y_3 = -\frac{aR}{a^2 - b^{2\prime}}$.

A collection of selected papers will be distributed at the opening of the conference in electronic form.

4 Modeling and Generating the Unknown Input Observer (UIO) of the System

4.1 Modeling the System in the Presence of Faults

The model of the system to be monitored is correctly described by the previous state representation (1). The present study assumes that the system is subjected to two actuator defects and one perturbation, in addition to the perturbation that is supposed to be due to the inclined plane. The corresponding state representation has the following form:

$$\begin{cases} \dot{\mathrm{x}}(\mathrm{t}) = \mathrm{Ax}(\mathrm{t}) + \mathrm{Bu}(\mathrm{t}) + F_x f(t) + D_x d(t) \\ \mathrm{y}(\mathrm{t}) = \mathrm{Cx}(\mathrm{t}) \end{cases} \tag{8}$$

Such that:

$$A = \begin{bmatrix} 0 & 1 & 0 & 0 \\ 0 & l_1 & 0 & l_2 \\ 0 & 0 & 0 & 1 \\ 0 & l_3 & 0 & l_4 \end{bmatrix} = \begin{bmatrix} 0 & 1 & 0 & 0 \\ 0 & -1.0537 & 0 & 0.0175 \\ 0 & 0 & 0 & 1 \\ 0 & 0.0175 & 0 & -1.0537 \end{bmatrix}, B = \begin{bmatrix} 0 & 0 \\ y_1 & y_3 \\ 0 & 0 \\ y_2 & y_4 \end{bmatrix} \begin{bmatrix} 0 & 0 \\ 0.9837 & -0.0163 \\ 0 & 0 \\ -0.0163 & 0.9837 \end{bmatrix}$$

$$C = \begin{bmatrix} 1 & 0 & 0 & 0 \\ 0 & 1 & 0 & 0 \\ 0 & 0 & 1 & 0 \\ 0 & 0 & 0 & 1 \end{bmatrix}; D = 0, F_x = \begin{bmatrix} 0 & 0 \\ 0 & 0 \\ 0 & 1 \\ 0 & 0 \end{bmatrix}; D_x = \begin{bmatrix} 0 \\ 1 \\ 0 \\ 1 \end{bmatrix}; F_y = 0; D_y = 0$$

And:

Fx: Action matrix of actuator faults f(t) to be detected
Dx: Action matrix of disturbances d(t), resistant torque T
vi(t): Measurement noise vector.

The new state representation is:

$$\begin{cases} \begin{bmatrix} \dot{x}_1(t) \\ \dot{x}_2(t) \\ \dot{x}_3(t) \\ \dot{x}_4(t) \end{bmatrix} = \begin{bmatrix} \mathbf{0} & \mathbf{1} & \mathbf{0} & \mathbf{0} \\ \mathbf{0} & \mathbf{-1.0537} & \mathbf{0} & \mathbf{0.0175} \\ \mathbf{0} & \mathbf{0} & \mathbf{0} & \mathbf{1} \\ \mathbf{0} & \mathbf{0.0175} & \mathbf{0} & \mathbf{-1.0537} \end{bmatrix} \begin{bmatrix} x_1(t) \\ x_2(t) \\ x_3(t) \\ x_4(t) \end{bmatrix} + \begin{bmatrix} \mathbf{0} & \mathbf{0} \\ \mathbf{0.9837} & \mathbf{-0.0163} \\ \mathbf{0} & \mathbf{0} \\ \mathbf{-0.0163} & \mathbf{0.9837} \end{bmatrix} \begin{bmatrix} c_r \\ c_l \end{bmatrix} \\ \quad + \begin{bmatrix} \mathbf{1} & \mathbf{0} \\ \mathbf{0} & \mathbf{0} \\ \mathbf{0} & \mathbf{1} \\ \mathbf{0} & \mathbf{0} \end{bmatrix} \begin{bmatrix} f_1(t) \\ f_2(t) \end{bmatrix} + \begin{bmatrix} \mathbf{0} \\ \mathbf{1.0220} \\ \mathbf{0} \\ \mathbf{1.0220} \end{bmatrix} d(t) \\ \begin{bmatrix} S_r \\ \dot{S}_r \\ S_l \\ \dot{S}_l \end{bmatrix} = \begin{bmatrix} \mathbf{1} & \mathbf{0} & \mathbf{0} & \mathbf{0} \\ \mathbf{0} & \mathbf{1} & \mathbf{0} & \mathbf{0} \\ \mathbf{0} & \mathbf{0} & \mathbf{1} & \mathbf{0} \\ \mathbf{0} & \mathbf{0} & \mathbf{0} & \mathbf{1} \end{bmatrix} \begin{bmatrix} x_1(t) \\ x_2(t) \\ x_3(t) \\ x_4(t) \end{bmatrix} + \begin{bmatrix} v_1(t) \\ v_2(t) \\ v_3(t) \\ v_4(t) \end{bmatrix} \end{cases} \tag{9}$$

5 Residual Generator Synthesis Based on an Unknown Input Observer (UIO) with Perfect Decoupling

Step 1: Rank $CD_x = nd$
It is first checked whether the system allows obtaining a perfect decoupling; then the rank of CDx is determined. The rank of CDx = 1 is equal to the number of inputs to

decouple d(t). It is possible to construct a fault-sensitive and non-disturbance-sensitive residual generator. The observer with perfect decoupling is given by:

$$\begin{cases} \dot{\mathrm{z}}(\mathrm{t}) = \mathrm{Mz}(\mathrm{t}) + \mathrm{Nu}(\mathrm{t}) + \mathrm{Py}(\mathrm{t}) \\ \hat{\mathrm{x}} = \mathrm{z}(\mathrm{t}) - \mathrm{L_y}\,\mathrm{y}(\mathrm{t}) \end{cases}$$

Remember that the present work aims to determine the matrices M, N, P, Ly such that the estimate $\hat{x}(t)$ converges asymptotically towards the real state x(t) of the system, despite the influence of the perturbations, while checking the following constraints:

$$\begin{gathered} \text{M is a Hurwitz matrix (stable).} \\ ME + PC = EA \\ N = EB \\ ED_x = 0 \\ ML_yF_y + PF_y - EF_x \neq 0 \\ L_yF_y \neq 0 \end{gathered}$$

The observer synthesis algorithm is applied.

Step 2: Calculation of $L_y = -D_x\left[(CD_x)^T(CD_x)\right]^{-1}(CD_x)^T, L_y = \begin{bmatrix} 0 & 0 & 0 & 0 \\ 0 & -0.5 & 0 & -0.5 \\ 0 & 0 & 0 & 0 \\ 0 & -0.5 & 0 & -0.5 \end{bmatrix}$

Step 3: Calculation of E = I + LyC, $E = \begin{bmatrix} 1 & 0 & 0 & 0 \\ 0 & 0.5 & 0 & -0.5 \\ 0 & 0 & 1 & 0 \\ 0 & -0.5 & 0 & 0.5 \end{bmatrix}$

Step 4: Calculation of N = EB, $N = \begin{bmatrix} 0 & 0 \\ \frac{y_1 - y_3}{2} & \frac{y_2 - y_4}{2} \\ 0 & 0 \\ \frac{y_3 - y_1}{2} & \frac{y_4 - y_2}{2} \end{bmatrix} = \begin{bmatrix} 0 & 0 \\ 0.5 & -0.5 \\ 0 & 0 \\ -0.5 & 0.5 \end{bmatrix}$

Step 5: Impose M as a Hurwitz matrix. In this case, it is possible to choose M as a diagonal matrix that shows the eigenvalues that are desired for the observer.

The diagonal matrix M and its eigenvalues are: $\lambda_1 = -1, \lambda_2 = -2, \lambda_3 = -2, \lambda_4 = -1$

$$M = \begin{bmatrix} -1 & 0 & 0 & 0 \\ 0 & -2 & 0 & 0 \\ 0 & 0 & -2 & 0 \\ 0 & 0 & 0 & -1 \end{bmatrix}$$

Step 6: Calculation of PC = EA − ME, C: identité

$$P = \begin{bmatrix} 1 & 1 & 0 & 0 \\ 0 & \frac{l_1-l_3}{2}+1 & 0 & \frac{l_2-l_4}{2}-1 \\ 0 & 0 & 2 & 1 \\ 0 & \frac{l_3-l_1}{2}-\frac{1}{2} & 0 & \frac{l_4-l_2}{2}+\frac{1}{2} \end{bmatrix} = \begin{bmatrix} 1 & 1 & 0 & 0 \\ 0 & 0.4644 & 0 & -0.4644 \\ 0 & 0 & 2 & 1 \\ 0 & 0.0356 & 0 & -0.0356 \end{bmatrix}$$

The unknown-input reconstructor and the output estimation error are obtained as follows:

$$\begin{cases} \begin{bmatrix} \dot{z}_1(t) \\ \dot{z}_2(t) \\ \dot{z}_3(t) \\ \dot{z}_4(t) \end{bmatrix} = \begin{bmatrix} -1 & 0 & 0 & 0 \\ 0 & -2 & 0 & 0 \\ 0 & 0 & -2 & 0 \\ 0 & 0 & 0 & -1 \end{bmatrix} \begin{bmatrix} z_1(t) \\ z_2(t) \\ z_3(t) \\ z_4(t) \end{bmatrix} + \begin{bmatrix} 0 & 0 \\ 0.5 & -0.5 \\ 0 & 0 \\ -0.5 & 0.5 \end{bmatrix} \begin{bmatrix} c_r \\ c_l \end{bmatrix} \\ \qquad + \begin{bmatrix} 1 & 1 & 0 & 0 \\ 0 & 0.4644 & 0 & -0.4644 \\ 0 & 0 & 2 & 1 \\ 0 & 0.0356 & 0 & -0.0356 \end{bmatrix} \begin{bmatrix} S_r \\ \dot{S}_r \\ S_l \\ \dot{S}_l \end{bmatrix} \\ \begin{bmatrix} e_{S_r} \\ e_{\dot{S}_r} \\ e_{S_r} \\ e_{\dot{S}_l} \end{bmatrix} = \begin{bmatrix} S_r \\ \dot{S}_r \\ S_l \\ \dot{S}_l \end{bmatrix} - C \begin{bmatrix} z_1 \\ z_2 + \frac{1}{2}(\dot{S}_r + \dot{S}_l) \\ z_3 \\ z_4 + \frac{1}{2}(\dot{S}_r + \dot{S}_l) \end{bmatrix} \end{cases}$$

5.1 Theoretical Calculation of Residuals

Calculation of the transfer matrix linking the defects to the output estimation error: The error transfer function is:

$$G_f(s) = C(sI - M)^{-1}(F + \acute{F}s) - F_y$$

$$F = ML_yF_y + PF_y - EF_x = \begin{bmatrix} -1 & 0 & 0 & 0 \\ 0 & 0 & 0 & 0 \\ 0 & -1 & 0 & 0 \\ 0 & 0 & 0 & 0 \end{bmatrix}, \acute{F} = -L_yF_y = \begin{bmatrix} 0 & 0 & 0 & 0 \\ 0 & 0 & 0 & 0 \\ 0 & 0 & 0 & 0 \\ 0 & 0 & 0 & 0 \end{bmatrix}$$

The residual vector is written as: r(t)

$$r(s) = \begin{pmatrix} r_1 \\ r_2 \end{pmatrix} = \begin{pmatrix} G_{f11} & G_{f12} \\ G_{f21} & G_{f22} \end{pmatrix} \begin{pmatrix} f_{ac1}(s) \\ f_{ac2}(s) \end{pmatrix}$$

$$= \begin{pmatrix} \frac{-s^3-5s^2-8s-4}{s^4+6s^3+13s^2+12s+4} & 0 \\ 0 & \frac{-s^3-4s^2-5s-2}{s^4+6s^3+13s^2+12s+4} \end{pmatrix} \begin{pmatrix} f_{ac1}(s) \\ f_{ac2}(s) \end{pmatrix}$$

5.2 Table of Signature

The signature table associated with this residual generator is given in Table 1, where "1" indicates the occurrence of a defect f_i affecting the residual r_{ij} and "0" the insensitivity of that residual to the defect.

Table 1. Table of signatures (EWC)

	fac1	fac2
r1	1	0
r2	0	1

According to the signature table, the residuals are insensitive to the disturbances d(t). The structure allows the complete location of the defects. In addition, it is theoretically possible to detect and locate the faulty actuator.

6 Simulation

The residuals generated by the unknown input observer (UIO) with perfect decoupling of the electric wheelchair are evaluated as a function of the values of the nominal parameters of the system as follows [14] (Table 2):

Table 2. The nominal parameters of the system EWC

$J = 16.08$ kg.m^2	$R = 0.17$ m
$J_a = 0.0024$ kg.m^2	$L = 0.57$ m
$J_w = 0.0289$ kg.m^2	$C_a = 0.06$ N.m/rad/s
$\sigma = 0.033$	$C_w = 0.008$ N.m/rad/s
$M = 210.00$ kg	$g = 9.81$ m/s^2
$m_w = 2.00$ kg $b = \sigma R^2\left(\frac{M}{4} + \frac{1}{L^2}J\right)$ $c = \frac{1}{\sigma}C_a + \sigma C_w$	$J_e = \begin{bmatrix} a & b \\ b & a \end{bmatrix}, C_e = \begin{bmatrix} c & 0 \\ 0 & c \end{bmatrix},$ $a = \frac{1}{\sigma}J_a + \sigma\left\{J_w + \left(\frac{M}{4} + m_w\right)R^2 + \left(\frac{R}{L}\right)^2 J\right\}$ $T = \sigma\left(\frac{M}{2} + m_v\right)gR\sin\psi$

The simulation consists in checking the detection of faults. This operation is characterized by the appearance of the signals of the corresponding residuals through the implementation of the unknown input observer that estimates the defects f_{ac1} and f_{ac2}. Indeed, the observer reconstructs the residual r1 or the residual r2 of the system. If the output presents a defect, it will immediately be estimated. Thus, if a residual r(t) deviates from the threshold interval, a defect f will certainly occur. Therefore, with this observer, it is possible to detect and locate the two actuator faults even if they occur simultaneously at the two outputs. In practice, it is considered that the input signal is a driving torque, and a disturbance input is a resistive torque. A measurement noise is added to the measurements $S_r, S_l, \dot{S}_r, \dot{S}_l$ in order to simulate the normal operation. The residuals are evaluated in normal operation and in faulty operation. It is assumed that the two actuator faults are defined as follows:

$$f_{AC1} = \begin{cases} -20 & 10 \le t \le 25 \\ 0 & \text{elsewhere} \end{cases} \quad f_{AC2} = \begin{cases} -35 & 15 \le t \le 20 \\ 0 & \text{elsewhere} \end{cases}$$

Normal operation,

(a) With perturbation
It is modeled using the resistive torque T. Figure 1 illustrates the residuals r1(t) and r2(t). The perturbations are perfectly decoupled because of the observer with perfect decoupling. In practice, the residuals are different from zero due to the measurement noise.

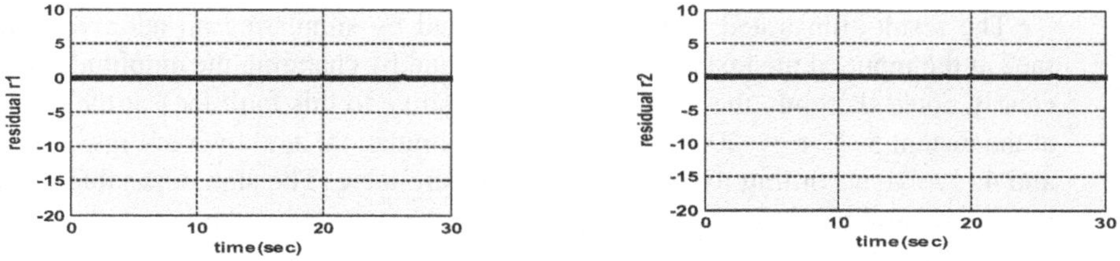

Fig. 1. Residuals r1(t) and r2(t) in the presence of perturbations

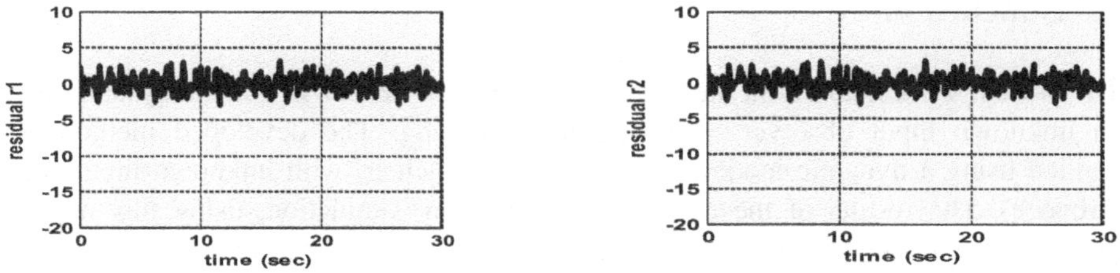

Fig. 2. Residuals r1(t) and r2(t) in the presence of disturbances, with measurement noises

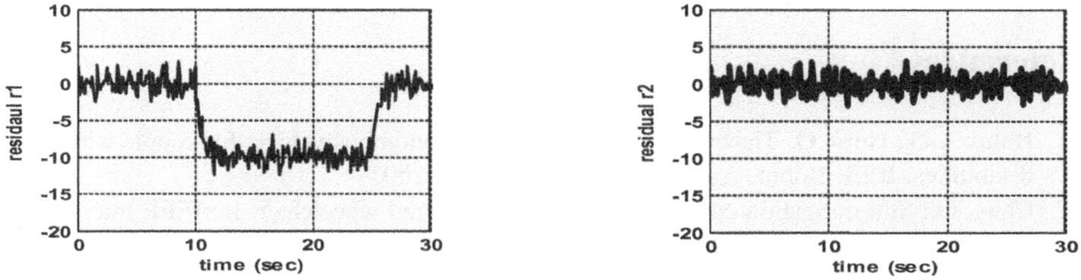

Fig. 3. Residuals r1(t), r2(t) with disturbances, measurement noises and fault fac1

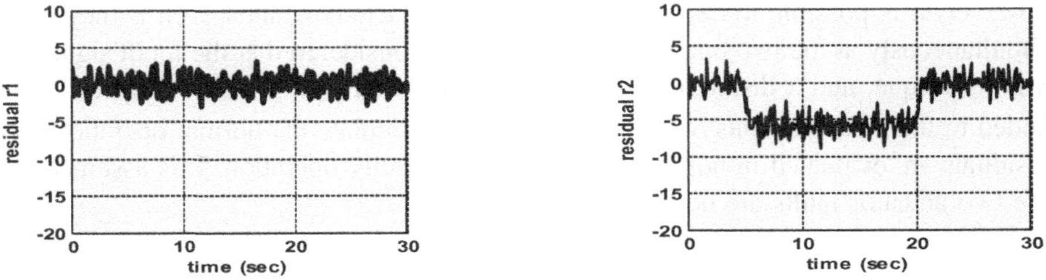

Fig. 4. Residuals r1(t) and r2(t) in with disturbances, measurement noises and fault fac2

(b) Addition of measurement noises
To get closer to reality, random signals are added to the measurements, as shown in Fig. 4.

(c) Faulty operation
The results shown in Fig. 3 are obtained by simulating an additive fault fac1 at the input, at the instant $10 \leq t \leq 25$ For this fault, only the residual r1(t) is sensitive, and therefore detectable, as indicated in the signature table.

The results illustrated in Fig. 4 are obtained by simulating an additive fault fac2 at the input, at the instant $10 \leq t \leq 25$, and by changing the amplitude. It is clearly noted that only the residual r2(t) is sensitive to this fault fac1 at the input, at the instant $5 \leq t \leq 20$. The results of the simulation, as shown in Figs. 1, 2, 3 and 4, evolve according to the previous signature table. The signature table has a localizing structure (two different signatures). The results of the simulation thus confirm the effectiveness of the suggested approach.

7 Conclusion

The present work focused on the detection of actuator defects in a linear system, using an unknown input observer with perfect decoupling. The developed method was applied using a dynamic model of an electric wheelchair, with unknown inputs (disturbance). The results of the residual generation by simulation, using this method, proved to be effective for all the cases considered. It would have been desirable to validate these results through an experimental manipulation on a real electric wheelchair.

References

1. Bourhis, G., Horn, O., Habert, O., Pruski, A.: An autonomous vehicle for people with motor disabilities. IEEE Robot. Autom. Mag. **8**(1), 20–28 (2001)
2. Chuy, O.: Slip mitigation control for an electric powered wheelchair. In: IEEE International Conference on Robotics and Automation (ICRA), Hong Kong, China, 31 May–7 June 2014 (2014)

3. Brahim, K., Chemori, A., desvaux de Marigny, C.: Free move: un fauteuil roulant automatisé de nouvelle génération. Handicap (2014)
4. Ragot, J., Maqui, D.: Diagnostic des systèmes linéaires, Collection Pédagogique d'Automatique. Hermès Science Publications, Paris (2000)
5. Chen, J., Patton, R., Zhang, H.: Design of unknown input observers and robust fault detection filters. Int. J. Control **63**, 85–105 (1996)
6. Kiyak, E., Kahvecioglu, A., Caliskan, F.: Aircraft sensor and actuator fault detection isolation and accommodation. J. Aerosp. Eng. **24**, 46–58 (2011)
7. Sun, X.: Unknown input observer approaches to robust fault diagnosis. Thesis submitted for the Degree of Doctor of Philosophy in the University of Hull by Xiaoyu Sun, M.Sc. Electronics, Shenyang, China and B.Sc. Automation, Shenyang, China (2013)
8. Khan, A.Q., Ding, S.X.: Threshold computation for fault detection in a class of discrete-time nonlinear systems. Int. J. Adapt. Control Sig. Process. **25**, 407–429 (2011)
9. Cristofaro, A., Johansen, T.: A fault tolerant control allocation using unknown input observers. Automatica **50**, 1891–1897 (2014)
10. Bagherpour, E.: Disturbance decoupled residual generation with unknown input observer for linear systems. In: Conference on Control and Fault-Tolerant Systems (SysTol), Nice, France, 9–11 October 2013 (2013)
11. Tahraoui, S., Meghebbar, A., Boubekeur, D.: Unknown input observer based on LMI for robust generation residuals. Turk. J. Electr. Eng. Comput. Sci. **25**, 95–107 (2017)
12. Toscano, R.: Commande et diagnostic des systèmes dynamiques (Modélisation, Analyse, Commande par PID et par retour d'état, Diagnostic). Livre Ellipses Edition. Toscano R. Commande et diagnostic des systèmes dynamiques (Modélisation, Analyse, Commande par PID et par retour d'état, Diagnostic). Ed. by Livre Ellipses (2011)
13. Boubekeur, D., Boumediene, D., Sari, Z., Tahraoui, S.: A control comparative study of an electric powered wheelchair system. Electrotehnică Electronică Automatică **63** (2015)
14. Meliani, S.: Modélisation du Système Pilote-Véhicule dans une tâche de contrôle manuel d'un fauteuil roulant électrique. PhD thesis (2009)

Optimum Synthesis of the PID Controller Parameters for Frequency Control in Microgrid Based Renewable Generations

M. Regad(✉), M. Helaimi, R. Taleb, and A. E. Toubal Maamar

Electrical Engineering Department, Laboratoire Génie Electrique et Energie Renouvelable (LGEER), Hassiba Benbouali University of Chlef, BP. 78C, 02180 Ouled Fares, Chlef, Algeria
m.regad@univ-chlef.dz

Abstract. This work concerns the optimal control of an electric microgrid based on renewable energies. This proposed configuration is consists of two renewable sources (photovoltaic and three wind generators) with conventional sources: diesel generator, fuel cells, and energy storage systems (BESS, BFSS). Due to climate change, renewable energy sources are considered as stochastic and intermittent sources. The fluctuation in frequency and power requires the integration of an adequate control system to reduce the effect of the disturbances on the stability of the electric microgrid. This control is ensured by the integration of PID whose parameters are optimized using an objective function resolved by the application of genetic algorithms. The proposed PID controller gives a favourable performance as illustrated in the results. These obtained results show the high efficiency of the proposed method and robustness of the PID controller.

Keywords: Microgrid · PID controller · Hybrid system · Optimization · Genetic algorithm

1 Introduction

The microgrid based renewable energy sources offer a major possibility of electricity production independently of conventional sources in condition to accept the natural fluctuation [1]. During the last decades, the limit of fossil fuel enforces the use of renewable energy sources which consist an alternative the hybrid system with diesel engine generator, photovoltaic, wind and energy storage system is popularly used for lament rural isolated [2]. The use of hybrid energy system with various renewable energy sources is economical technique optimal solution of electrical production problem. The advantage of hybrid energy system based renewable energy sources such as wind and photovoltaic with a conventional source is the satisfaction of energy demand every time although the intermittent nature of primary sources such as wind and solar on the one hand, the non-correlation of the load that fluctuates over the year and the resources. These same sources of energy can be combined with other sources such as the fuel cell using hydrogen produced by an electrolyzer connected to a renewable energy source. The autonomy of this system will be ensured by an efficient

M. Hatti (Ed.): ICAIRES 2019, LNNS 102, pp. 546–556, 2020.
https://doi.org/10.1007/978-3-030-37207-1_58

storage system for the continuity of the service. The energy storage system plays an important role in the hybrid system through its efficient storage and release of energy in a short time. For stand-alone energy systems the cost of storage represents the greatest constraints of the overall system cost for large power installations. Minimizing the cost of storage and maximizing capacity is the key reason for combining wind and PV systems with a storage system [3]. This hybrid system uses conventional (DG) and another renewable (PV, WTG) sources with a storage system (BESS, BFSS) to compensate for energy when the demand exceeds the capacity of the renewable source. A hybrid system uses static converters (AC-DC-AC) for the exchange of energy between different sources. As a result, the electric power produced has significant fluctuations which make the stability of the microgrid difficult. One of the important control issues associated with a microgrid is the frequency regulation [4]. Several meta-heuristic optimization techniques are often applied for the determination of control system parameters. This article consists of the use of genetic algorithms to optimize the parameters of the PID regulator in order to control the frequency of this hybrid microgrid. AGs as an intelligent optimization method can give a robust response [5]. The principle of genetic algorithms is presented in [6]. Control by the PID controller (proportional, integral and derivative) is often used for several controls scheme, it is efficient and simple to be implemented [7].

2 Configuration Proposed Hybrid System

Figure 1 illustrates the configuration of the studied proposed hybrid system [8]. This system consists of three wind turbine generators (WTG), a photovoltaic (PV), a diesel generator (DEG), two fuel cells (FC), an aqua electrolyzer (AE), an electrochemical battery (BESS), a battery electromechanical (FESS) and static converters (DC and AC). Table 1 present the different parameters of the proposed system.

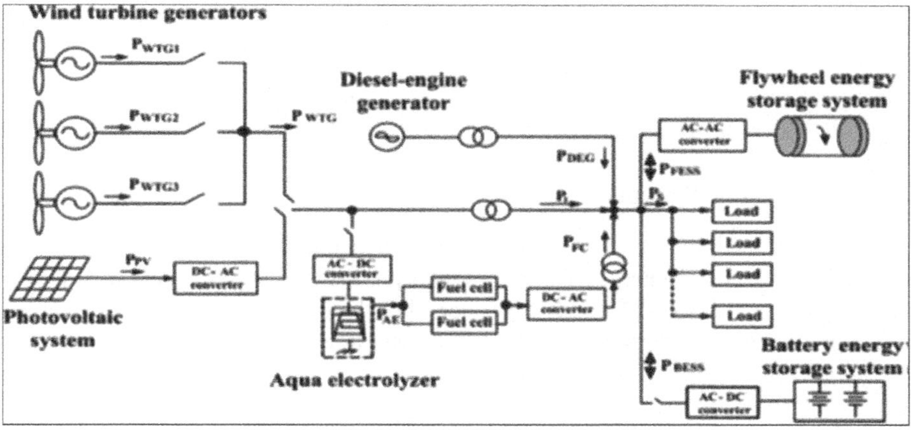

Fig. 1. Proposed configuration of the hybrid energy system.

The use of the storage system is required for the autonomous microgrid. In this system using an electrochemical battery and the electromechanical battery, which is a flywheel storing energy in kinetic form. PV, AE, CF, BESS and FESS require power converters to exchange energy with the AC system studied. Assume that the BESS FESS and have sufficient capacity to store excess energy generated by the generators subsystems. When the power demand increases, the storage system (BESS and/or FESS) can release enough energy to the load connected in a very short time. The diversity of sources and storage system requires the use of electronic converters powers for connecting the various devices of the hybrid system [8].

Table 1. Parameters of hybrid system offers

Component	Gain (K)	The time constant (T)
Wind turbine generator	$K_{WTG} = 1$	$T_{WTG} = 1.5$
Photovoltaic generator	$K_{PV} = 1$	$T_{PV} = 1.8$
Fuel Cell (FC)	$K_{FC} = 0.01$	$T_{FC} = 4$
Diesel engine generator	$K_{DEG} = 0.003$	$T_{DEG} = 2$
Battery energy storage system	$K_{BESS} = 0.003$	$T_{BESS} = 0.1$
Flywheel energy storage system	$K_{FESS} = 0.01$	$T_{FESS} = 0.1$
Aqua Electrolyze (AE)	$K_{AE} = 0.006$	$T_{AE} = 0.5$

2.1 Wind Generators

The wind generation system is based on the conversion of wind energy into electrical energy. This conversion occurs in two steps: The turbine extracts a part of the kinetic energy of wind to convert it into electricity. The generator receives the mechanical energy and energy transforms to be sent to the microgrid. The output of the wind generator depends on wind speed. The wind system can be simplified by a first-order system. The WTG transfer function is represented as [9]:

$$G_{WTG}(s) = \frac{K_{WTG}}{1 + s.T_{WTG}} \tag{1}$$

2.2 Photovoltaic System

The effect PV is a physical process base through which the solar radiation is converted directly into electrical energy. The PV system produces a DC voltage which is converted to alternative using DC-AC converters. For the low-frequency domain analysis, it is represented by a first-order model for a given transfer function as [9].

$$G_{PV} = \frac{K_{PV}}{1 + ST_{PV}} \tag{2}$$

2.3 Fuel Cell

The fuel cell has become a promising solution to the distributed generation of electrical energy. It allows the conversion of chemical energy into electricity through the chemical reaction between hydrogen and oxygen. Fuel cells provide energy with advantages of high efficiency, low pollution and the heat exhaust reusability and water [8]. The fuel cell generator is a higher-order model and a non-linearity. In the field of low-frequency analysis, it is represented by a first-order transfer function given as follows [9]:

$$G_{FC} = \frac{K_{FC}}{1 + ST_{FC}} \quad \text{or} \quad \frac{\Delta P_{FCK}}{\Delta P_{AE}} \tag{3}$$

2.4 Diesel Generator

To ensure continuity of production of electrical energy in an autonomous network it is necessary to add one or more diesel generators. In the system studied classical generator consists of a diesel engine which is coupled to a synchronous generator. The normal operation of the generator is as follows: the torque supplied to the alternator by the motor rotates the rotor of the alternator, which generates phase currents [10] stator. A diesel generator can be simplified by a first-order model as follows:

$$G_{DEG} = \frac{K_{DEG}}{1 + ST_{DEG}} \quad \text{or} \quad \frac{\Delta P_{DEG}}{\Delta f} \tag{4}$$

2.5 Aqua Electrolyzer

Part of the power generated by PV or WTG is absorbed by AE to produce hydrogen for use by the fuel cell. The hydrogen decomposition of water is ensured by the passage of an electric run between two electrodes separated by an electrolyte. The model of the transfer function of an electrolyzer is shown below [2].

$$G_{AE} = \frac{\mathbf{K_{AE}}}{1 + \mathbf{ST_{AE}}} = \frac{\mathbf{\Delta P_{AE}}}{\mathbf{\Delta P_{WTG}}(1 - \mathbf{Kn})} \quad \textbf{or} \quad \frac{\mathbf{\Delta P_{AE}}}{\mathbf{\Delta P_{PV}}(1 - \mathbf{Kn})} \tag{5}$$

2.6 Energy Storage Systems

In order to minimize the operation of the diesel generator and therefore the emissions of gas, it is necessary to use an energy storage system. This enables us to use renewable resources to the maximum possible recharging the batteries by the surplus energy produced by PV and/or WTG. Finally, in the case of a sudden demand for power, storage fills the source function (buffer) until the engine starts and takes the role of the supply of energy [12]. The storage system is constituted by two different types of battery; fuel battery (BESS) and battery flywheel (FESS). Both linear models are presented by the following first-order transfer functions [8]:

$$\mathbf{G_{BESS}} = \frac{\mathbf{K_{BESS}}}{1 + \mathbf{s.T_{BESS}}} \tag{6}$$

$$\mathbf{G_{FESS}} = \frac{\mathbf{K_{FESS}}}{1 + \mathbf{s.T_{FESS}}} \tag{7}$$

2.7 Characteristic of Power Generated and Asked

The power generated by renewable sources depends on solar radiation and wind speed, which is intermittent in nature. The power required by the load also takes the same from the generated power. It can be represented by the following mathematical equation [6]:

$$P = \left(\frac{\varphi\eta\sqrt{\beta}(1 - \mathrm{G(s)}) + \beta}{\beta}\right).\Gamma \tag{8}$$

On the other hand $P = \Gamma.\chi$

P: Represents the solar power, wind or model of the load.
φ: Is the stochastic component of power.
β: Contributes to the average value of the power.
G(s): Is a low passing filter.
η: Is a constant normalizing the generated power or the one requested.
χ: constant for Correspondence per unit [pu].
Γ: Is a switching signal depends on the time with the gain which causes the sudden fluctuation of the average value for the stochastic power.

For wind power settings (8).

$$\begin{array}{l} \varphi \sim \mathrm{U}(-1,1), \quad \eta = 0.8, \quad \beta = 10, \quad \mathrm{G(s)} = \frac{1}{10^4.\mathrm{S}+1} \\ \Gamma = 0.24\mathrm{h(t)} - 0.04\mathrm{h(t-140)} \end{array} \tag{9}$$

Where h(t) is the Heaviside step function. For the generated solar power settings (8) are:

$$\begin{array}{l} \varphi \sim \mathrm{U}(-1,1), \quad \eta = 0.9, \quad \beta = 10, \quad \mathrm{G(s)} = \frac{1}{10^4.\mathrm{S}+1} \\ \Gamma = 0.05\mathrm{h(t)} + 0.02\mathrm{h(t-180)} \end{array} \tag{10}$$

For the power, required settings (8) are:

$$\begin{aligned}
&\varphi \sim \mathrm{U}(-1,1), \eta = 0.8, \beta = 10, \\
&\mathrm{G(s)} = \tfrac{300}{300.\mathrm{S}+1} + \tfrac{1}{1800.\mathrm{S}+1} \\
&\Gamma = \tfrac{1}{\chi}[0.9h(t) + 0.03h(t-110) + 0.03h(t-130) + 0.03h(t-150) \\
&\quad - 0.15h(t-170) + 0.1h(t-190)] + 0.02h(t)
\end{aligned} \tag{11}$$

3 Optimization of PID Parameters by Genetic Algorithm

Genetic algorithms are evolutionary optimization methods inspired by Darwinian natural selection [11]. They are widely used to solve complex optimization problems. Figure 2 shows the flowchart of the implementation of Genetic Algorithms. A Genetic Algorithm evolved a set of known solutions (individual people) that presents a possible solution to the problem. Each individual assigns an adaptation function (fitness function). This function has the function to be optimized. A genetic algorithm starts with an initial population of solutions. Then it creates a sequence of new generations; The best solutions are more likely to be selected, creating a new generation of solutions. The selected individuals are then improved through the application of three basic operators such as selection, crossover and mutation. The algorithm is repeated for a given number of generations and stop when the criteria stop is reached. The optimization of an objective function with a PID based genetic algorithm is as follows:

Step 1: Creation of a population of initial parameters solutions. Each parameter called as a nuisance. A chromosome consists of genes and thus each chromosome has a solution to the problem. K_P, K_I and K_D.
Step 2: Evaluate the objective function (fitness) for each solution of individuals.
Step 3: Remember the best solution for stopping rule verification. If so, finish.
Step 4: Generation "offspring": Offspring is a new chromosome obtained through the selection steps, crossover and mutation.
Step 5: Putting people products in the new population.
Step 6: Replace the old population of individuals by the news.
Step 7: Terminate the program when the termination criterion is reached; otherwise go to step 2.

Figure 2 shows the chart of Algorithms Genetics:

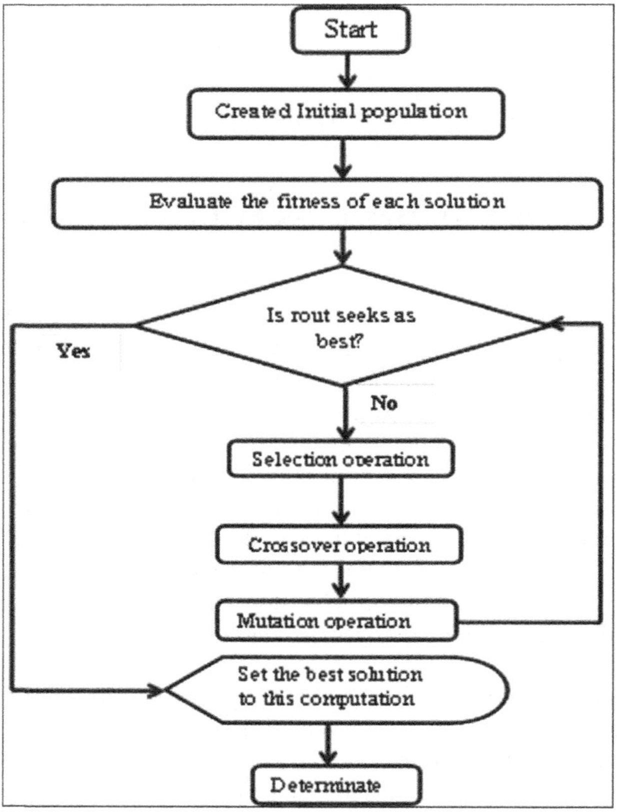

Fig. 2. Flowchart of genetic algorithm

By following the steps of this algorithm we can get Matlab code of Genetic Algorithms to optimize the parameters of a PID controller using an objective function. This code consists of several sub-files (initialization, encoding, decoding, mutation, selection, fitness function, outcome) have the body of the main program. The parameters of Genetic Algorithms are presented in the Table 2.

Table 2. Parameters of AG

Parameter	Value
Population size	40
Number of variables	3
Mutation probability	0.05
Binary number length	30
Number of iteration	100

In order to obtain good performance monitoring frequency of a microgrid, the variation of frequency (.DELTA.f) and the change of the control signal (Δu) to the output of the PID controller should be minimized. Proportional-integral-Derivative (PID) plays an important role in industrial processes have since been introduced. Many control systems use the PID controller for its simplicity and it is proved satisfactory. He still has a wide application in industrial control [12]. There are two popular structures for the implementation of the PID controller; parallel or series. This controller has three parameters (,) which must now be optimized to adjust system control laws while maintaining some control objectives incorporated as performance indices in the time domain. K_P K_I and K_D. In this study, the integral performance index (J) to minimize by appropriate genetic algorithms is the weighted sum of ISTSE (Integral of Squared Time Squared Error Multiplied) And ISDCO (Integral of Squared Deviation of the Output Controller) Given as follows [6].

$$J\_opt = \int_{T_{min}}^{T_{max}} [w(\Delta f)^2 + \left(\frac{1-w}{Kn}\right)(\Delta u)^2]dt \quad (12)$$

4 Results and Discussion

This section presents the simulation results in the time domain of the frequency control by a PID controller for a microgrid. In this study, the three parameters of the PID controller (K_P, K_I and K_D) Are optimized by the application of Genetic Algorithms. Table 3 shows the optimum values of these parameters (Fig. 3).

Table 3. Optimal Parameters PID

PID	K_{P_opt}	K_{I_opt}	K_{D_opt}
	4.6867	4.9998	0.488

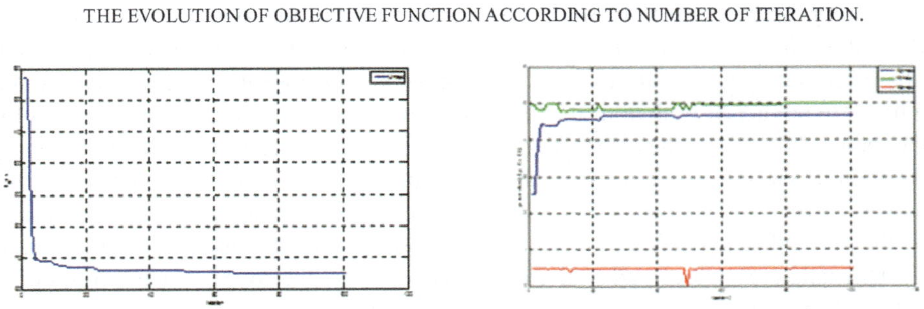

(a) (b)

Fig. 3. (a) Evolution of the objective function based on the number of iteration, (b) Changing PID control parameters based on simulation time.

The results show that the objective function used is stochastic in nature. The same PID parameter values give different values for the same objective function.

GENERATED POWER AND REQUESTED INDEPENDENT REGULATOR PID

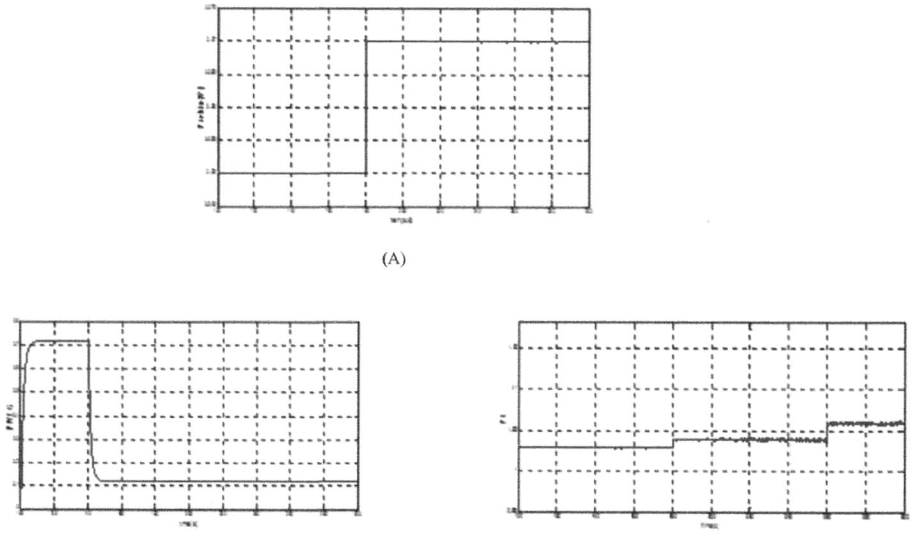

(A)

(b) (c)

Fig. 4. (a), (b) and (c) Stochastic Realization power generated and applied independently of the PID structure

Figure 4 illustrates the profile of the required load and the PV output power and WTG under varying conditions randomly. The power generated by renewable sources of random and intermittent nature. What influences the control of the frequency and power of the microgrid. These results show that the generated power is independent of the controller (Fig. 5).

During the first phase of simulation time the value of the total power produced by PV and reached 0.89 WTG been following the value of solar radiation and wind speed. These resources (wind, sun) are stochastic in nature, have local characteristics and are difficult to predict. It is obvious that this power is not sufficient to cover the demand for the required load. The diesel generator starts producing the rest of the required power. As soon as the renewable source generator can fulfil the demand, the surplus of this power is stored in batteries and/or sent to the electrolyser produces hydrogen absorbed by the fuel cell. The stored power will be used in case of failure of the generated power.

Disruption of power and noticed the frequency is due to the change from solar radiation and wind. Control of the PID controller using frequency parameters optimized by genetic algorithms is robust and responsible to reduce the fluctuations in the frequency and power due to random changes in solar radiation and wind speed.

CHANGE IN FREQUENCY AND POWER BASED ON THE SIMULATION OF TIME

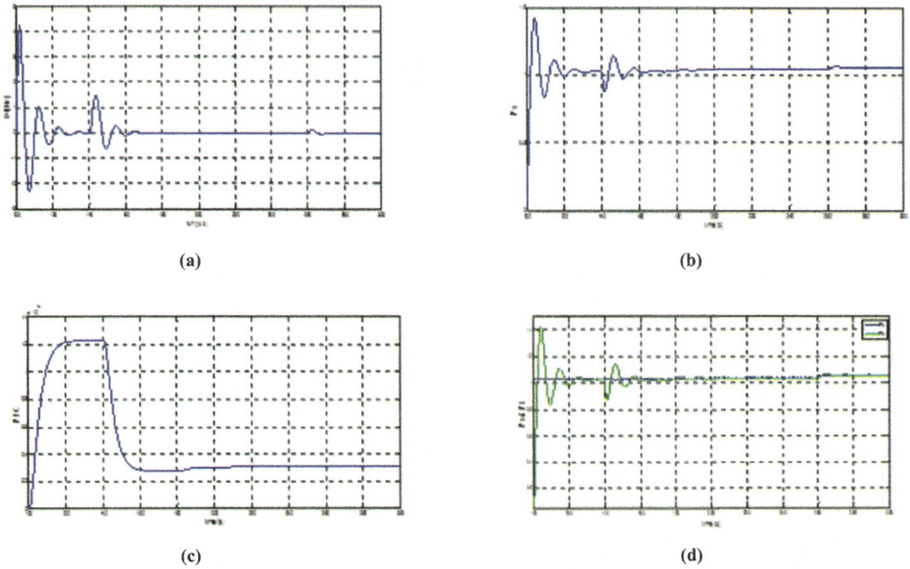

Fig. 5. (a), (b), (c) and (d) Variation of the frequency and power for microgrid with the optimal parameters of the PID controller

5 Conclusion

This item is to develop an optimal control synthesis of the frequency of a hybrid system. The configuration of the proposed system is discussed, the characteristics of major system components such as WTG, PV, FC, DEG, AF, BESS and FESS are presented. The control strategy shows the reliability and efficiency of the proposed system by the simulation of configuration. This strategy is based on the use of the PID controller whose parameters are evaluated by the application of genetic algorithms. The AGs provide good system performance in terms of solution quality and speed of convergence. The proposed PID controller is given a favourable performance as show the illustrated results. The obtained results show the high efficiency of the used method and a robust PID controller. The obtained results show that the proposed method is efficient and robust.

References

1. Gergaud, O.: Modélisation énergétique et optimisation économique d'un système de production éolien et photovoltaïque couplé au réseau et associé à un accumulateur. Diss. École normale supérieure de Cachan-ENS Cachan
2. Lopez, M.: Contribution à l'optimisation d'un système de conversion éolien pour une unité de production isolée (Doctoral dissertation, Université Paris Sud-Paris XI) (2008)

3. Belhamel, M., Moussa, S., Kaabeche, A.: Production d'Electricité au Moyen d'un Système Hybride (Eolien-Photovoltaïque-Diesel). Revue Énergies Renouvelables: Zones Arides, pp. 49–54 (2002)
4. Saini, V., Sathans: Frequency regulation in an AC micro-grid with diverse sources of power using intelligent control technique. J. Autom. Control Eng. **4**(3), 252–256 (2016)
5. Das, D.Ch., Roy, A.K., Sinha, N.: Genetic algorithm based PI controller for frequency control of an autonomous hybrid generation system. In: Proceedings of the International MultiConference of Engineers Scientists, IMECS 2011, Hong Kong, March 2011
6. Pan, I., Das, S.: Kriging based surrogate modelling for fractional-order control of microgrids. IEEE Trans. Smart grid **6**(1), 36–44 (2014)
7. Ross, D., Deguine, E., Camus, M.: Asservissement par PID. rose. eu. org 3 (2010)
8. Lee, D., Wang, L.: Small-signal stability analysis of an autonomous hybrid renewable energy power generation/energy storage system Part I: time-domain simulations. IEEE Trans. Energy Convers. **23**(1), 311–320 (2008)
9. Das, D.Ch., Roy, A.K., Sinha, N.: GA based frequency controller for solar thermal–diesel–wind hybrid energy generation/energy storage system. Int. J. Electr. Power Energy Syst. **43** (1), 262–279 (2012)
10. Kamwa, I., Saulier, B.: Modélisation, simulation et régulation d'un réseau éolien/diesel autonome. Raport NIREQ4340, Varennes, Canada (1989)
11. Forrest, S.: Genetic algorithms: principles of natural selection applied to computation. Science **261**(5123), 872–878 (1993)
12. Alikhani, A., et al.: Optimal PID tuning based on Krill Herd optimization algorithm. In: 2013 3rd International Conference on Control, Instrumentation, and Automation (ICCIA) (2013)

Optimum Dynamic Network Reconfiguration in Smart Grid Considering Photovoltaic Source

Samir Hamid-Oudjana[1(✉)], Mustafa Mosbah[2,3], Rabie Zine[4], and Salem Arif[2]

[1] Unité de Recherche Appliquée en Energies Renouvelables (URAER), Ghardaia, Algeria
samirehamid@yahoo.fr
[2] LACoSERE Laboratory, Department of Electrical Engineering, Amar Telidji University of Laghouat, Laghouat, Algeria
mosbah.mustapha@gmail.com, salem.arif@gmail.com
[3] Algerian Distribution Electricity and Gas Company, Algiers, Algeria
[4] Department of Mathematics, Faculty of Sciences, Moulay Ismaïl University, Meknès, Morocco
rabie.zine@gmail.com

Abstract. Among the advantages of a smart grid is to optimize the power flow by using distribution-management-system (D-M-S) function. Finding of an optimal dynamic configuration is one of the important tasks in D-M-S. The purpose of this paper is to suggest an optimization method based on Genetic Algorithm (GA) to determine the dynamic reconfiguration in one-hour intervals by considering the load variation and generation variation of PV sources in the day. In each hour, the configuration gives minimum active losses under all operating constraints. The GA method is tested on the IEEE 69 bus network and validated on the Algerian network using MATLAB. The proposed method yielded effective results that encourages use it in real time.

Keywords: Distribution network · Dynamic reconfiguration · Optimization · Photovoltaic source

1 Introduction

Electrical power system passes through different steps, production, transmission and distribution. Distribution has now become the most important part. It provides connection among transmission network and electricity consumers [1]. Researchers have given great importance to distribution networks in last years. This is due to introduction of distributed generation (DG) based on renewable sources of energy and non-renewable ones in those networks [2]. Distribution networks operate in lower voltages and present meshed structures, however, they must be operated in a radial structure [3]. A radial structure, i.e. each load bus is connected to the power bus (source station) by a single path [4]. The search of an optimal radial configuration (reconfiguration process) is the procedure on modifies distribution network topology by adjusting the switches

M. Hatti (Ed.): ICAIRES 2019, LNNS 102, pp. 557–565, 2020.
https://doi.org/10.1007/978-3-030-37207-1_59

stat to reduce the chosen objective [5]. The reconfiguration can be invariant on time (static reconfiguration case) or variant during the day according to the load and renewable generation power (dynamic reconfiguration case). Due to the Joule effect, the higher current causes the higher power losses. One of the Smart Grid characteristics is that it can help to minimize these losses [6]. The minimize power losses by dynamic reconfiguration task has a great importance in smart grid and in real time control. A Smart Grid is a modern distribution network, equipped by sensors, communications, control, automation and computer equipment to improve safety and flexibility [7]. Algeria is among the countries that are interested in the development and modernization of its electricity network (Transmission and Distribution network); it has effectively integrated various new technologies, such as programmable logic controllers, optical fiber links, digital protections, smart meters, SCADA system. In this context, a new contribution has been presented in this paper to propose a tool that helps the operator to determine the optimal distribution network configuration in each hour. Various techniques have been published are considered on static distribution network reconfiguration based conventional methods, artificial intelligence methods and meta-heuristic methods [8]. But very few paper are published on Dynamic Distribution Network Reconfiguration (D-DNR) considering the photovoltaic production variation and/or load variation during day, for example: A method for solving the power system reconfiguration problem by introduction of distributed generation based objective of minimizing active losses and improving the voltage profile in the distribution network which has been presented in [9] considering different load levels, by application of Harmony Search algorithm. In paper [10] a method based on the genetic algorithm is presented to evaluate the reconfiguration problem of the distribution network taking into account the effect of load variation and stochastic production of renewable energy sources. According to reference [11], an optimal real-time reconfiguration algorithm is proposed, which uses a classical non-linear optimization technique and guarantees an optimal solution in the shortest possible time, and aims to minimize active losses in each interval time. A method has been proposed in [12] to solve the problem of dynamic reconfiguration of the electrical distribution network considering a variable load, application of the artificial immune network for combinatorial optimization, in order to minimize the losses energy cost in a given period. The study based the reference [13] proposes a method of dynamic reconfiguration that takes into account the initial topological variation in real time. The method combines dynamic topology analysis and network reconfiguration to solve the distribution network optimization problem in real time and in the presence of a fault. Depending on the network state, the optimal configuration is identified to reduce power losses and improve the distribution network voltage profile in real time. The dynamic reconfiguration of the distribution network focused on the Lagrange relaxation method has been proposed in [14]. The objective of this method is to determine the optimal topologies of the distribution network over a specified time interval, in order to minimize the active power losses. A recent research in [15] presents a multi-objective management based on the reconfiguration of the network in parallel by placement of renewable source and the dimensioning, to minimize the active power losses, the annual operating costs (installation costs, maintenance and active power loss) and pollutants gaseous emissions. The variation over time of the wind speed, the solar irradiation and the load is taken into account. The objective of this study is to find the optimal configuration of the network that has had conjunction with the placement and sizing of renewable sources considering multiple criteria. A dynamic reconfiguration approach for the three-phase asymmetric distribution network has been presented in [16]. The network topology is optimized for future periods of time and adapts to load and DG variation while minimizing the daily costs of active losses using mixed integer linear programming method. A recent review of reconfiguration techniques has been presented in [17] for the restoration of service in modern distribution networks based on various practical considerations. This paper present a Genetic Algorithm method based on graphs theory to design an optimal D-DNR taking account the photovoltaic production variation and load variation on real times. D-DNR determine by changing loops switches state to minimize the total real power losses. This study proposed to adapt the GA method to the strategy of balancing the load between medium voltage departures (branches permutation strategy). The proposed method is tested on IEEE distribution network 69 bus and validated on Algerian distribution network, 116 bus.

2 Problem Formulation

2.1 Objective Function

The first concern of the distribution system operator is to minimize active power losses as much as possible [18]. The objective of D-DNR problem is to find the best network configuration with minimal real losses subject to all exigency exploitation constraints. Since many switching combinations in a distribution network exist, the search for an optimal configuration is a NP-hard, non-linear, combinatory and a non-differentiable constraints optimization [19]. The goal is to minimize Eq. 2 in every hour of time. Figure 1 shows an equivalent circuit model of a distribution network in the presence of looping switchers.

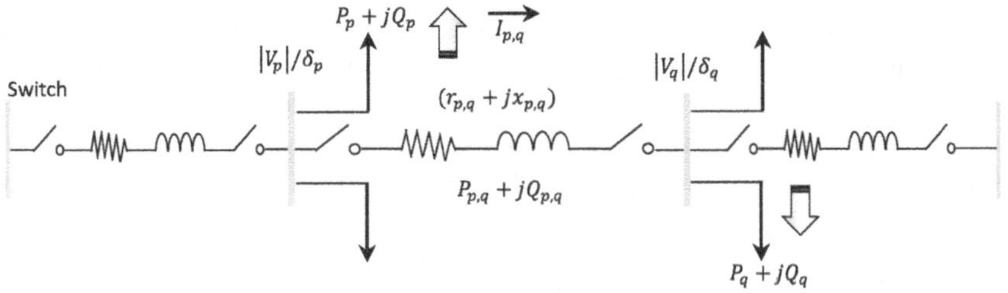

Fig. 1. Two-bus distribution network with one line diagram

$$P^T_{loss}(t) = I^2_{pq}R_{pq} = \frac{S^2_{pq}}{V^2_p}R_{pq} = \frac{P^2_{pq} + Q^2_{pq}}{V^2_p}R_{pq} \tag{1}$$

$$F = \sum_{i=1}^{N_B} P_{loss(i)}(t) \quad (2)$$

where $|V_p|/\delta_p$, is voltage/angle at bus p, (r_{pq}, x_{pq}) are resistance and reactance of branches joining bus p and bus q, (P_{pq}, Q_{pq}) are real and reactive power flow in branch between bus p and bus q. This objective function is subject to equality and inequality constraints (See references [20, 21]).

3 Genetic Algorithm Method

The GA is an optimization technique inspired by natural selection and based on genetics laws (Mendel). The algorithm is based on a set of possible solutions randomly initialized in the search space. Individuals are represented by their design variables or by a coding of these chromosomes. Then evaluate their relative fitness in the base of their performance and create a new population of potential solutions using an evolutionary operations (the selection, crossover and mutation). Repeat this run until locating a solution, this algorithm have been proposed by Holland-John [22].

4 Simulations and Results

In this work, the optimization method chosen is the genetic algorithm; this method is applied to determine the optimal dynamic reconfiguration of distribution networks taking into account the dynamics of load and the dynamics of PV production during the day, under the distribution network constraints. To confirm the efficiency of the developed program, initially is tested on IEEE 69 bus, but with only one consumption (static configuration). After that, a validation on the 116 bus of Algerian network was carried out considering the variation of the load and the PV production during the day. For the case of dynamic reconfiguration, we consider that it is a set of static cases that last one hour of time, for this we necessity the values of the loads and the PV sources for each hour. IEEE 69 bus are generally defined in the literature [23], however the Algerian network consists: 116 bus, 124 branches containing 09 Tai-lines and 09 feeders. The nominal voltage of 116_busses network is 10 kV. The substation is connected to medium voltage network through a 30/10 kV transformer. The voltages limits considered in this paper between 0.95 pu and 1.05 pu. The total limit PV considered in this study is 09 MW, which distributed between the following busses (busses with low voltages): 62, 66, 106, 109, and 75, according to the operator obligation of the Algerian distribution power system $\left(\sum_{i=1}^{NPV} P_{pvi}\right) \leq 10$ MW for distribution networks with a voltage level of 30 kV or 10 kV). The initial configuration of the 116 bus network is assured by opening the switches (116, 117, 118, 119, 121, 122, 123 and 124).

To determine the dynamic configuration of the Algerian network, it is necessary to have the daily values of the load and the PV production. For this reason, the measurements were made on 12/07/2017 to get the load and PV production profiles, the choice of this day is based on the highest consumption in the year. It should be mentioned that the PV source simulated in this study is located at the same location of the studied network (116 bus), called El-Hadjira source, which has been operational since 2017. The dynamics of the load laying the day (12/07/2017) is presented in Fig. 2. The Fig. 3 illustrates the dynamics of El-Hadjira PV production in the same day with a maximum total power output of 9986 kW.

In the process of identifying an optimal configuration, the following steps must be followed: randomly select a network configuration by the GA method; a first test is performed for the feasibility of topological constraints by applying graph theory (see details in Ref [24]), if these constraints are satisfied, a power flow calculation (application of the Newton Raphson method) is used to determine the various electrical parameters, then the technical and security constraints (voltage and thermal limits of the lines) are verified. In the case of an unsatisfied constraint, the penalty of the objective function is necessary (see § 2.4) to exclude unfeasible solutions.

Following the different executions of the program under MATLAB software, the optimal parameters of GA used in this simulation are: population size is 100, maximum iteration is 200, crossover probability is 1, mutation probability is 0.01 and one point crossover.

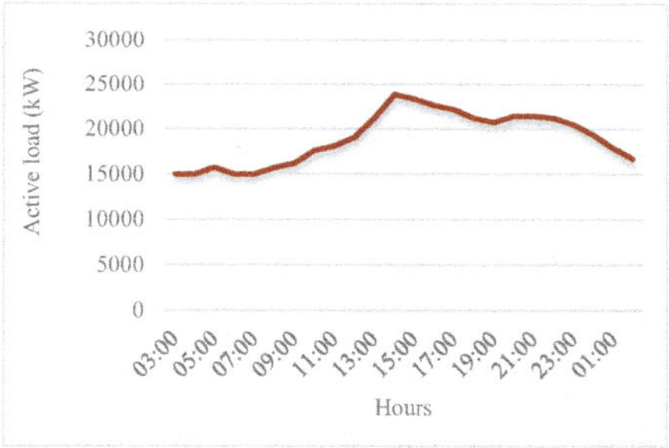

Fig. 2. Dynamic curves of load profile

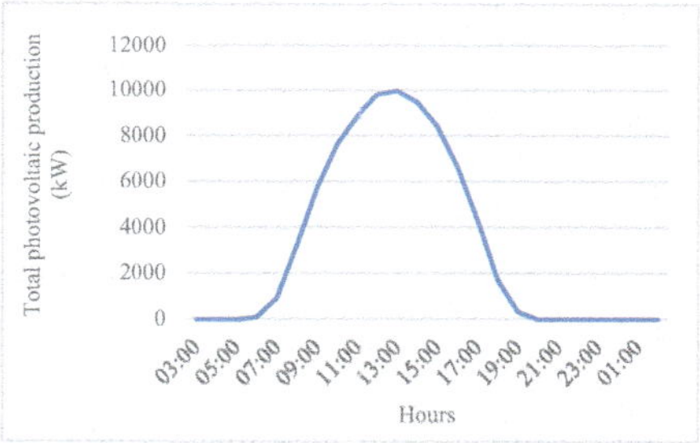

Fig. 3. Dynamic curves of PV sources profile

The simulation results found by the GA method are displayed in the following Fig and tables. Table 1 shows the effectiveness of the proposed method comparable to the results found in previous work. Table 2 illustrates the simulation results of the reconfiguration of the 116 bus distribution network in the presence of PV installed over a 24-h period. Figures 4 and 5 shows respectively the hourly minimum voltages and the hourly active losses before and after the reconfiguration of the 116 bus network in the presence of PV sources in each hour. These curves illustrate the significant improvement in minimum voltages as well as the active losses caused by the reconfiguration of this network in the presence of PV source, particularly during the operation of PV in the day.

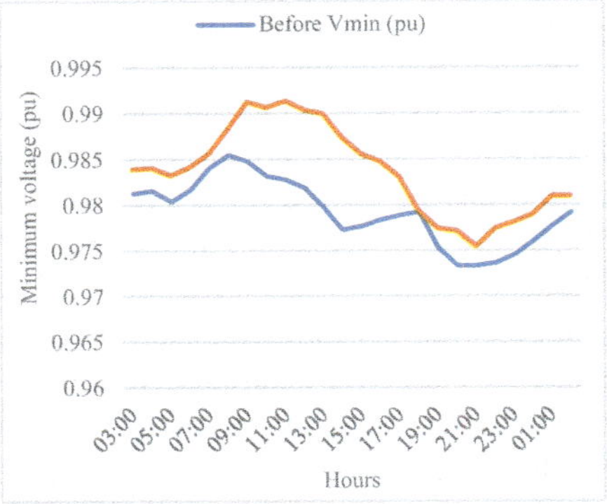

Fig. 4. Hourly minimum voltage for 116_bus

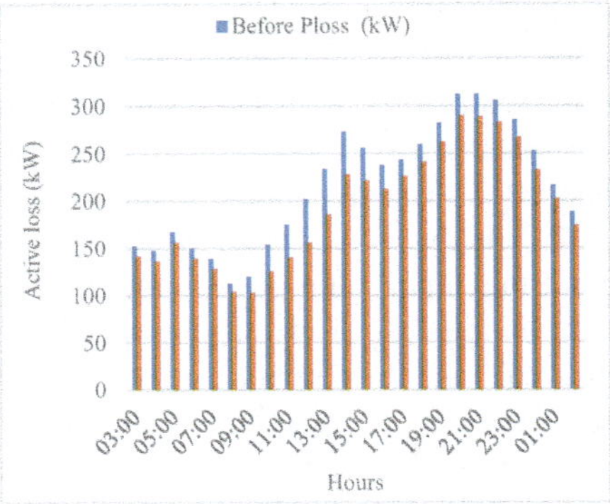

Fig. 5. Hourly total active loss for 116_bus

Table 1. The comparisons results with other works

Test network	Before reconfiguration	Optimization methods	After reconfiguration	Real loss (kW)
69 bus	69 70 71 72 73	GA [25]	9 28 33 34 36	140.6
		PC-GA [26]	No reported	100.95
		F-GA [27]	12 55 61 69 70	99.62
		MHBMO [28]	7 9 14 32 37	139.51
		BBO [29]	14 70 69 58 61	99.58
		ACO [30]	14 56 61 69 70	99.59
		Proposed algorithm	14 70 69 58 61	99.58

The results given in Table 1 show the efficiency and robustness of the proposed method for minimizing active power losses, which encourages its use in practice. Table 2 shows that the configuration changes within each hour, this is because of the variation of the load and the PV production. During the interval of radiation presence between 07:00 and 19:00 (PV source operating period), the active power losses decrease significantly by up to 15% and therefore a significant enhancement in the voltage profile. In the same way, the more power injected by the PV source, the lower the active power losses and the better the voltage profile.

It should be noted that the qualities of smart grid (modern, automated and communicating) make it easy to vary the reconfiguration of its structure during the day. The optimization of the Algerian network configuration during the day is minimized the total losses from 5206.41 kW to 4675.63 kW which represents a reduction of 10.19%, this by no investment, it is only changing the state of the loop switches, which shows the importance of this task in modern distribution network (smart grid).

Table 2. D-DNR determined by GA method in presence PV sources

Hours	Configuration	Open switches									Before reconfiguration		After reconfiguration	
											P_{loss} (kW)	V_{min} (pu)	P_{loss} (kW)	V_{min} (pu)
03:00	1	19	45	68	75	98	99	107	118	121	153.7	0.9814	143	0.9841
04:00	2	19	75	79	83	105	107	116	121	123	148.9	0.9817	137.8	0.9842
05:00	3	19	42	68	79	89	99	107	108	121	168.7	0.9805	157.3	0.9834
06:00	4	29	42	68	75	88	92	105	107	121	151.3	0.9818	140.4	0.9843
07:00	5	19	60	75	79	83	89	103	107	121	139.9	0.9842	129.7	0.9858
08:00	6	19	75	98	103	105	107	118	121	122	113.3	0.9856	104.8	0.9885
09:00	7	29	65	69	75	79	105	121	122	124	120.5	0.9849	104	0.9914
10:00	8	15	36	45	75	88	107	118	121	122	154.8	0.9833	126.5	0.9908
11:00	9	29	31	43	67	79	91	99	105	108	176.4	0.9829	141.5	0.9915
12:00	10	29	36	43	45	67	88	108	118	121	203.4	0.982	157.4	0.9905
13:00	11	15	43	65	67	70	75	79	105	121	235.2	0.9799	187.1	0.9901
14:00	12	29	70	75	79	88	105	121	122	124	274.1	0.9774	229.1	0.9875
15:00	13	29	45	70	79	103	108	121	122	124	256.7	0.9778	223.2	0.9857
16:00	14	29	45	69	75	79	103	107	121	122	238.8	0.9785	213.8	0.9849
17:00	15	29	55	75	98	99	105	107	121	122	244.6	0.979	227.5	0.9832
18:00	16	10	19	68	75	103	105	107	118	121	260.5	0.9794	242.4	0.9795
19:00	17	29	60	75	83	88	89	107	118	121	283.3	0.9754	263.6	0.9776
20:00	18	29	45	68	75	98	99	107	118	121	313.7	0.9734	290.9	0.9773
21:00	19	19	60	75	79	83	99	101	107	121	313.7	0.9734	290.2	0.9756
22:00	20	55	68	75	98	101	103	107	120	121	306.8	0.9737	284.4	0.9776
23:00	21	29	60	68	92	103	105	107	108	121	286.4	0.9746	268.2	0.9783
00:00	22	19	42	75	83	89	103	107	118	121	254.1	0.9761	234.2	0.9791
01:00	23	29	45	68	75	88	92	98	107	121	217.9	0.9778	203.6	0.9811
02:00	24	19	60	75	79	83	89	103	107	121	189.8	0.9793	175	0.9811

5 Conclusion

In this work, a hybrid genetic algorithm-graph theory is proposed in order to optimize D-DNR considering load and photovoltaic production variability during the day with objective to minimize of real power losses under technical, security and topological constraints. The effectiveness of this method is shown in the quality of the results comparable to the works of literature, by testing the algorithm proposed on IEEE distribution network (69 bus) and validating on real distribution network (116 bus). The usefulness of this management method stems from the fact that this reconfiguration does not require any heavy investment, it is only a way of exploiting equipment already existing in distribution network, especially since this task is becoming easier and easier in smart grid. As the future work is to propose to redo programming under industrial software like C^{++}.

References

1. Shefaei, A., Vahid-Pakdel, M., Mohammadi-ivatloo, B.: Application of a hybrid evolutionary algorithm on reactive power compensation problem of distribution network. Comput. Electr. Eng. **72**, 125–136 (2018)
2. Dixit, M., Kundu, P., Jariwala, H.R.: Optimal integration of shunt capacitor banks in distribution networks for assessment of techno-economic asset. Comput. Electr. Eng. **71**, 331–345 (2018)
3. Merlin, P.: Search for a minimal-loss operating spanning tree configuration for an urban power distribution system. In: Proceedings of 5th PSCC, pp. 1–18 (1975)
4. Shirmohammadi, D., Hong, H.W.: Reconfiguration of electric distribution networks for resistive line losses reduction. IEEE Trans. Power Delivery **4**(2), 1492–1498 (1989)
5. de Assis, L.S., Vizcaı, J.F., Usberti, F.L., Lyra, C., Cavellucci, C., Von Zuben, F.J.: Switch allocation problems in power distribution systems. IEEE Trans. Power Syst. **30**(1), 246–253 (2015)
6. Jumar, R., Maaß, H., Hagenmeyer, V.: Comparison of lossless compression schemes for high rate electrical grid time series for smart grid monitoring and analysis. Comput. Electr. Eng. **71**, 465–476 (2018)
7. Ding, F., Loparo, K.A.: Hierarchical decentralized network reconfiguration for smart distribution systems—Part I: problem formulation and algorithm development. IEEE Trans. Power Syst. **30**(2), 734–743 (2015)
8. Mosbah, M., Arif, S., Mohammedi, R.D., Oudjana, S.H.: A genetic algorithm method for optimal distribution reconfiguration considering photovoltaic based DG source in smart grid. In: International Conference in Artificial Intelligence in Renewable Energetic Systems, pp. 162–170. Springer (2018)
9. Rao, R., Ravindra, K., Satish, K., Narasimham, S.V.L.: Power loss minimization in distribution system using network reconfiguration in the presence of distributed generation. IEEE Trans. Power Syst. **28**(1), 317–325 (2013)
10. Zidan, A., El-Saadany, E.F.: Distribution system reconfiguration for energy loss reduction considering the variability of load and local renewable generation. Energy **59**, 698–707 (2013)
11. Masteri, K., Venkatesh, B.: Real-time smart distribution system reconfiguration using complementarity. Electr. Power Syst. Res. **134**, 97–104 (2016)
12. Souza, S.S., Romero, R., Pereira, J., Saraiva, J.T.: Artificial immune algorithm applied to distribution system reconfiguration with variable demand. Int. J. Electr. Power Energy Syst. **82**, 561–568 (2016)
13. Wen, J., Tan, Y., Jiang, L., Lei, K.: Dynamic reconfiguration of distribution networks considering the real-time topology variation. IET Gener. Transm. Distrib. **12**(7), 1509–1517 (2018)
14. Kovački, N.V., Vidović, P.M., Sarić, A.T.: Scalable algorithm for the dynamic reconfiguration of the distribution network using the Lagrange relaxation approach. Int. J. Electr. Power Energy Syst. **94**, 188–202 (2018)
15. Hamida, I.B., Salah, S.B., Msahli, F., Mimouni, M.F.: Optimal network reconfiguration and renewable DG integration considering time sequence variation in load and DGs. Renew. Energy **121**, 66–80 (2018)
16. Zhai, H., Yang, M., Chen, B., Kang, N.: Dynamic reconfiguration of three-phase unbalanced distribution networks. Int. J. Electr. Power Energy Syst. **99**, 1–10 (2018)

17. Abu-Elanien, A.E., Salama, M., Shaban, K.B.: Modern network reconfiguration techniques for service restoration in distribution systems: A step to a smarter grid. Alexandria Eng. J. **57**(4), 3959–3967 (2018)
18. Mohammedi, R.D., Zine, R., Mosbah, M., Arif, S.: Optimum network reconfiguration using Grey Wolf Optimizer. TELKOMNIKA (Telecommun. Comput. Electron. Control) **16**(5), 2428–2435 (2018)
19. Zine, R., et al.: Optimum distribution network reconfiguration in presence DG unit using BBO algorithm, pp. 180–189 (2018)
20. Mosbah, M., et al.: A genetic algorithm method for optimal distribution reconfiguration considering photovoltaic based DG source in smart grid. In: 2th International Conference in Artificial Intelligence in Renewable Energetic Systems, pp. 162–170. Springer International Publishing AG 2019 (2018)
21. Mosbah, M., Arif, S., Mohammedi, R.D., Zine, R.: Optimal reconfiguration of an Algerian distribution network in presence of a wind turbine using genetic algorithm. In: 1st International Conference in Artificial Intelligence in Renewable Energetic Systems, pp. 392–400. Springer International Publishing AG 2018 (2018)
22. Holland, J.H.: Adaptation in Natural and Artificial Systems: An Introductory Analysis with Applications to Biology, Control, and Artificial Intelligence. MIT press, Cambridge (1992)
23. Baran, M.E., et al.: Optimal capacitor placement on radial distribution systems. IEEE Trans. Power Delivery **4**(1), 725–734 (1989)
24. Mosbah, M., Zine, R., Arif, S., Mohammedi, R.D.: Optimum distribution network reconfiguration in presence DG unit using BBO algorithm, pp. 180–189 (2018)
25. Hong, Y.-Y., Ho, S.-Y.: Determination of network configuration considering multiobjective in distribution systems using genetic algorithms. IEEE Trans. Power Syst. **20**(2), 1062–1069 (2005)
26. Qin, Y., Wang, J., Gui, W.: Particle clonal genetic algorithm using sequence coding for solving distribution network reconfiguration. In: The 9th International Conference for Young Computer Scientists, ICYCS 2008, pp. 1807–1812 (2008)
27. Liu, L., Chen, X.: Distribution network reconfiguration based on fuzzy genetic algorithm, pp. 66–69 (2000)
28. Niknam, T.: An efficient multi-objective HBMO algorithm for distribution feeder reconfiguration. Expert Syst. Appl. **38**(3), 2878–2887 (2011)
29. Kouzou, A., Mohammedi, R.D., Hellal, A.: An efficient biogeography-based optimization algorithm for smart radial distribution power system reconfiguration. In: 2015 First Workshop on Smart Grid and Renewable Energy (SGRE), pp. 1–7 (2015)
30. Swarnkar, A., Gupta, N., Niazi, K.R.: Adapted ant colony optimization for efficient reconfiguration of balanced and unbalanced distribution systems for loss minimization. Swarm Evol. Comput. **1**(3), 129–137 (2011)

Optimal Location and Size of Wind Source in Large Power System for Losses Minimization

Mustafa Mosbah[1,2(✉)], Rabie Zine[3], Samir Hamid-Oudjana[4], and Salem Arif[1]

[1] LACSERE Laboratory, Department of Electrical Engineering, Amar Telidji University of Laghouat, Laghouat, Algeria
mosbah.mustapha@gmail.com
[2] Algerian Distribution Electricity and Gas Company, Algiers, Algeria
[3] Department of Mathematics, Faculty of Sciences, Moulay Ismaïl University, Meknès, Morocco
rabie.zine@gmail.com
[4] Unité de Recherche Appliquée en Energies Renouvelables (URAER), Ghardaia, Algeria
samirehamid@yahoo.fr

Abstract. This paper presents a Genetic Algorithm approach for optimal integration of Wind based on Distributed Generation (WDG) in smart grid taking in to count technical and security constraints. The optimal integration that is to say, look for the optimal location and size of the WDG to be integrated into the network. The objective function considered in this study is minimize active power losses. The proposed method is applied on IEEE 14 bus using MATLAB software.

Keywords: Wind · Distributed Generation · Location and size · Loss reduction · Power system · Optimal Power Flow · Genetic Algorithm

1 Introduction

The world electricity demand has permanently and quickly developed. This evolution has been contributed by population increasing, economic activity development, household equipment evolution, life modernization and comfort [1]. The fact to introduce other sources (e.g. Distributed Generation, DG) in bus load can cause an impact on the transmission system. The location and size of DG can play an important role in the power system exploitation and operation. The DG placement at non suitable location can provoke in transmission system negative impacts [2]. The efficient solution to avoid this impact passes throw optimal integration of DG. In literature, many different types of techniques varying from meta-heuristic approaches [3] have been used to determine the optimal location and size of DG, for example, the research [3] presented a technique based an Optimal Power Flow (OPF) calculation with DG unit, bat the location is predetermined on bus 7, 10 and 30, the bi-objective function is fuel cost and power losses minimization. The paper [4] used continuation power flow method

M. Hatti (Ed.): ICAIRES 2019, LNNS 102, pp. 566–574, 2020.
https://doi.org/10.1007/978-3-030-37207-1_60

and voltage stability index, in order to determine the location vulnerable to voltage, to decide which bus has to be targeted (candidate bus), and then DG location. Another study proposed in [5] to show the influence of DG on gas emission and fuel cost. In this study the used optimization technique is interior-point method, the DG location is chosen before. Zhu et al. [6] had formulated of Distribution Optimal Power Flow (DOPF) into consideration DG units, with an objective function to minimize total power losses. The quadratic programming method is used to resolve this problem. Harrison and Wallace [7] developed a method based on OPF in a distribution network and considered DG as a negative load. The DPOPF algorithm (Distributed and Parallel OPF) proposed to solve the OPF with DG units of renewable energy in the transmission system [8]. Authors of [9] also presented Particle Swarm Optimization technique for solving multi type DG unit sizing and placement problem in distribution networks. A multi-objective optimization (minimization power losses and number of DG) has been proposed, in order to deduce the adequate location and optimal size of DG, using Non-Linear Programming technique [10]. The work in [11] presented Mixed Integer Non-Linear Programming (MINLP) approach for determine optimal location and number of DG in hybrid electricity market. Reference [12] developed application of Multi-Objective Particle Swarm Optimization (MOPSO) with the aim of determine optimal location and size of DG and shunt capacitor simultaneously with considering load uncertainty in distribution network. The multi-objective optimization includes three objective functions: improving voltage, decreasing active power losses and voltage stability. Another study proposed a multi-objective index-based approach to optimally determine size and location of multi DG units in distribution system with non-unity power factor considering different load models [13]. In paper [14] presented a Genetic Algorithm method for optimal location and sizing of Photovoltaic based-DG unit. The objective function is minimize total power losses in transmission power system. In this work, a metaheuristic technique based on the genetic algorithm was applied to determine the optimal location and size of wind based on distributed generation, while minimizing active losses. The developed program is applied on the IEEE 14 node network under MATLAB language.

2 Problem Formulation

2.1 Objective Function

Optimal location and size of WDG are defined by active power loss minimization in power network with system operating constraints.

In order to avoid the negative effect of power losses, the location of delivered power should be optimal by even installed WDG-unit, to balance load with production at every moment. So our problems consist to optimize the size of WDG and their location in power system. The fitness function of active losses can be expressed as:

$$P_{loss}^{T} = I_{pq}^{2} R_{pq} \tag{1}$$

$$F = Min\sum_{i=1}^{NB} P_{loss(i)} \tag{2}$$

where V_p is voltage at bus p, R_{pq} is resistance of line connecting bus p and bus q, I_{pq} is current through the branch between bus p and bus q, NB is number of lines.

The objective function, subject to set of equality and inequality constraints that should be satisfied while achieving the minimization of active power loss.

2.2 Equality Constraints

This constraints represent active, reactive power balance equations. The power balance equation in transmission system in presence of distributed generation units with renewable and non-renewable energy units can be expressed as follows:

$$\begin{cases} P_G + P_{WDG} = P_D + P_L \\ Q_G + Q_{WDG} = Q_D + Q_L \end{cases}$$

where (P_G, Q_G), are the total active and reactive power of conventional generator, respectively, (P_D, Q_D) the total active and reactive power of load, respectively, (P_L, Q_L) is the total active and reactive power losses, respectively, P_{DG} is active power of DG (DG source modeled as photovoltaic power).

2.3 Inequality Constraints

Voltage limit:

$$V_{imin} \leq V_i \leq V_{imax} \quad for\ i = 1.\ldots..N \tag{5}$$

Line thermal limit:

$$S_k \leq S_{kmax} \quad for\ k = 1.\ldots\ldots NB \tag{6}$$

Real power generation limit:

$$P_{Gimin} \leq P_{Gi} \leq P_{Gimax} \quad for\ i = 1.\ldots NG \tag{7}$$

The DG source limit:

$$0 \leq \sum_{i=1}^{NDG} P_{DGi} \leq 0.3 * \sum_{i=1}^{Nbus} P_{Di} \quad for\ i = 1.\ldots..NDG \tag{8}$$

2.4 Preserving Solution Feasibility

Note that the control variables are generated in their permissible limits using strategist preservation feasibility (perform a random value between the minimum and maximum value), while for the state variables, including the voltages of load bus, the power flowing in distribution lines, it appealed to penalties functions that penalize solutions that violate these constraints. The introduction of penalty in the objective function, transforms the optimization problem with constraints in an optimization problem without constraints [14], so it is easier to deal, in this case the Eq. 2 shall be replaced by:

$$F_p = Min \sum_{i=1}^{NB} P_{loss(i)} + P_f \tag{9}$$

$$P_f = k_v.\Delta V + k_s.\Delta S \tag{10}$$

$$\Delta V = \sum_{i=1}^{NL} \left(V_{Li} - V_{Li}^{lim}\right)^2 \tag{11}$$

$$\Delta S = \sum_{i=1}^{NB} \left(S_{li} - S_{li}^{lim}\right)^2 \tag{12}$$

where k_v and k_s, are penalty factors, in this study, the values of penalty factors have been considered 10.000.

3 Applied Approach

3.1 Genetic Algorithm (GA)

GA are stochastic optimization algorithms found on the natural selection mechanism of a generation. Their operation is extremely simple. We start by an initial population of potential solutions (chromosomes) chosen randomly. We evaluate their relative performance (fitness). In the base of their performance we create a new population of potential solutions by using simple evolutionary operations: The selection, crossover and mutation. We repeat this cycle until we find a satisfied solution. GA have been initially developed by John Holland.

3.2 OPF with WDG

Optimal Power Flow Considering Wind Distribution Generation (OPF-WDG) has already been raised by formulas of Eq. (1). The state variables vector χ consisting of, slack bus real power P_{G1}, load bus voltages V_{L1}, reactive power outputs the all conventional generator Q_{Gi} transmission line power flow S_{l1}. Hence, χ can be expressed as:

$$\chi^T = \left[P_{G1}, V_{L1} \ldots V_{LNL}, \delta_2 \ldots \delta_N, Q_{G1} \ldots Q_{GNG}, S_{l1} \ldots S_{lNl}\right] \tag{13}$$

The vector v of control variables consisting, renewable generator active power outputs (size) P_{WDG} and distribution generation location (location)L_{WDG}. The other control variables (P_G,V_G, T) are considering in OPF function. Hence, v can be expressed as:

$$v^T = \left[L_{WDG1} \ldots \ldots L_{WDGn}, P_{f_{WDG1}} \ldots P_{f_{WDGn}}, P_{WDG1} \ldots P_{WDGn}\right] \tag{14}$$

The objective of optimal power flow in presence DG unit is minimize a selected fitness function via optimal settings of control variables vector. The Fig. 1 illustrate chromosome structure applied in this study.

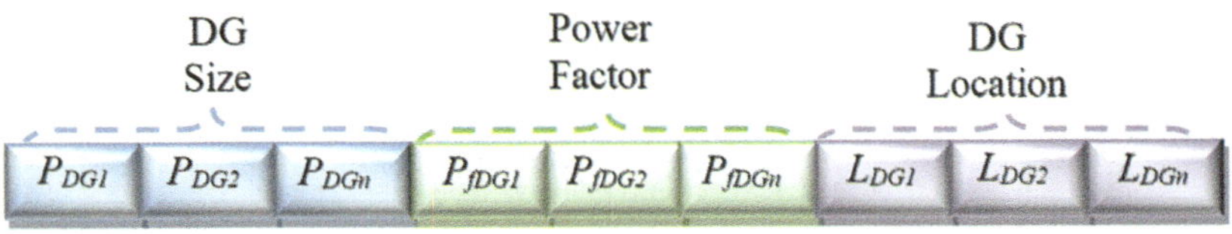

Fig. 1. Chromosome structure

OPF role is to provide different control variables values, namely, conventional generators active power, generators voltage, transformers tap settings, transforms angle control and FACTS devices, to minimize an objective function, considering technical, security, economic and environmental constraints. OPF challenge is able also to determine optimal size and location of WDG. For that, an algorithm based on OPF function and coupled with genetic algorithm method is proposed. The aim of GA method is to define optimal size and location of WDG. We should mention that, we have used OPF function implanted in MATPOWER software, and we added a new control variable (WDG size, WDG location). Figure 2 presents combination strategy of classic OPF with WDG-unit.

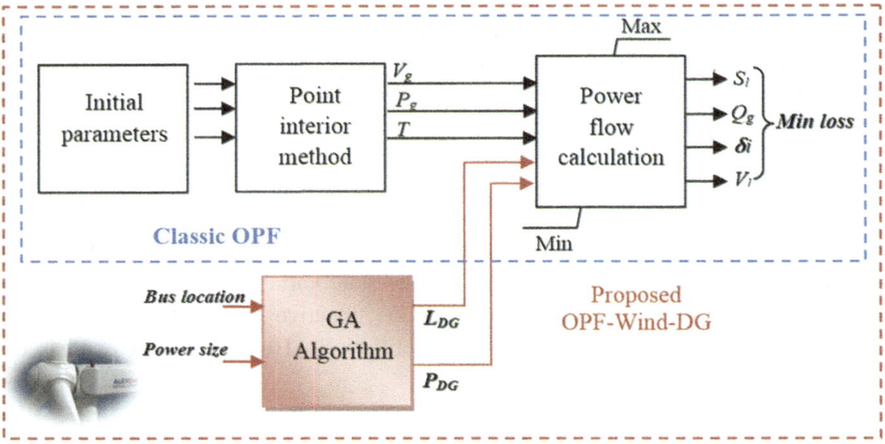

Fig. 2. Proposed model based WDG-unit integrated in classic OPF

4 Simulations and Results

The test network IEEE 14_bus, consists of 6 generators, 20 transmission lines and 4 on load tap changing transformers. The total active and reactive power absorbed by load is 259 MW and 73.5 MVAr. Figure 3 illustrate the IEEE 14_bus power system topology.

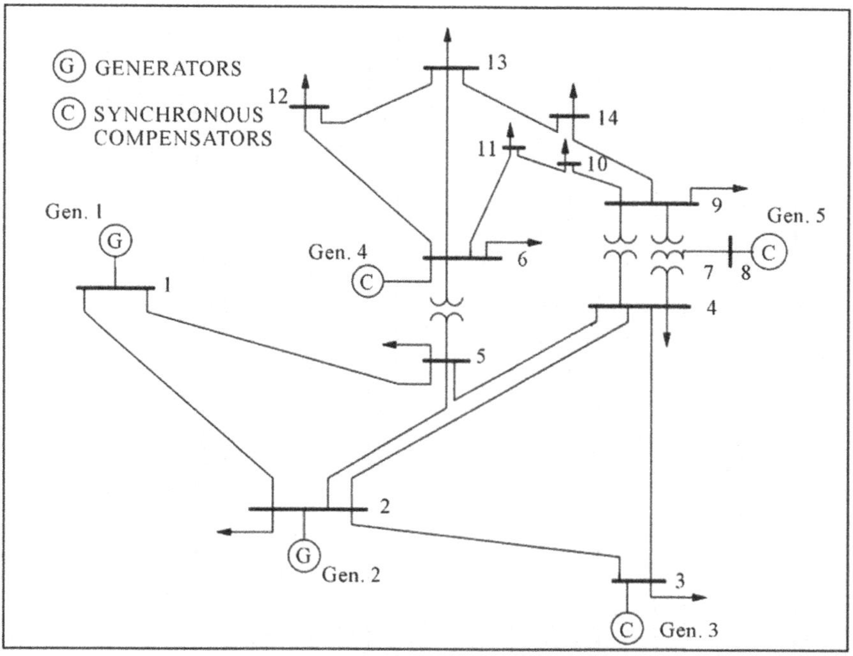

Fig. 3. IEEE 14_bus power system topology

In this study, we made an application of optimal power flow considering wind turbine based distributed generation using genetic algorithm technique. The inferior and superior voltage magnitude limits for all generator buses (PV-bus) are 0.9 pu–1.1 pu, and voltage limits for load buses (PQ-bus) are 0.95 pu–1.05 pu. It is known that all load buses have been considered as candidate for WDG location. You have to know that the values of control variables are generated in their acceptable limits using randomly strategy, in an interval of 0 to 50 MW. This work, we had considered that the generation power by WDG are negative loads, such as the WDG active power for each load bus is limited by a minimum value 0 (no generation power by WDG) and a maximum value. The distributed generation are modeled as wind source with power factor comprised between 0.8 and unity (capable injecting P and Q). The application of proposed technique to transmission power system has been examined on seven case studies: case 1 (OPF without WDG), case 2 (OPF with 1 WDGs), case 3 (OPF with 2 WDGs), case 4 (OPF with 3 WDGs), case 5 (OPF with 4 WDGs), case 6 (OPF with 5 WDGs) and case 7 (OPF with 6 WDGs).

Number of iteration chosen for this case is 50, with 100 populations of GA algorithm. The probability of mutation is 0.01, and crossover probability is 0.9. After convergence algorithm, the results obtained by genetic algorithms method are represented as following.

Table 1 shown the different parameters system for different cases. The comparison of voltage profile in various cases simulations is presented in Fig. 4. Figure 5 illustrate active load bus, total load for different cases and negative loads in bus 4, 9, 12 and 13.

Table 1. Active loss with and without WDG, location and size

Variables	Limits		Case 1	Case 2	Case 3		Case 4			Case 5				Case 6					Case 7					
	Inferior	Superior	OPF Without WDG	OPF With 1 WDG	OPF With 2 WDG		OPF With 3 WDG			OPF With 4 WDG				OPF With 5 WDG					OPF With 6 WDG					
P_{g1} (MW)	0	332.4	194.33	81.01	1.22		44.43			0.00				0.19					0.00					
P_{g2} (MW)	0	140	36.72	17.49	11.86		16.06			25.10				31.72					26.66					
P_{g3} (MW)	*	*	28.74	19.93	96.18		54.15			91.89				93.14					95.27					
P_{g6} (MW)	*	*	00.00	00.00	00.00		00.00			0.00				0.00					0.00					
P_{g8} (MW)	*	*	8.50	97.65	100.0		00.00			0.00				15.40					0.00					
Q_{g1} (MVAr)	-990	990	0.00	0.00	10.00		0.00			0.00				0.00					0.00					
Q_{g2} (MVAr)	-40	50	23.69	2.08	-0.83		-40.00			-34.02				-30.27					-24.65					
Q_{g3} (MVAr)	-6	24	24.13	40.00	22.25		13.27			0.54				3.61					5.33					
Q_{g6} (MVAr)	-6	40	11.55	-6.00	-6.00		-6.00			-6.00				-6.00					-6.00					
Q_{g8} (MVAr)	-6	24	8.27	8.24	8.32		-6.00			-6.00				-6.00					-6.00					
V_1 (pu)	0.9	1.1	1.060	0.959	0.950		0.954			0.947				0.946					0.946					
V_2 (pu)	0.9	1.1	1.041	0.946	0.943		0.940			0.940				0.940					0.940					
V_3 (pu)	0.9	1.1	1.016	0.940	0.948		0.940			0.940				0.940					0.940					
V_6 (pu)	0.9	1.1	1.060	0.990	0.989		1.033			1.026				1.034					1.021					
V_8 (pu)	0.9	1.1	1.060	0.940	0.940		1.007			1.036				1.032					1.005					
WDG location bus			**	12	9	12	4	9	13	4	7	10	13	7	9	11	13	14	4	5	9	10	12	14
WDG size (MW)			**	48.77	3.59	49,01	48.93	48.75	48.69	34.44	48.89	33.17	26.59	45.76	13.39	21.47	36.66	2.20	28.29	23.84	36.59	7.69	15.80	25.57
WDG size (MVAr)			**	34.08	1.61	34.6	21.19	30.14	26.54	13.02	32.10	22.45	16.42	34.13	10.04	16.10	17.16	1.65	1.51	4.58	26.55	5.571	11.12	17.90
Total real power loss (MW)			9.287	5.860	2.866		2.026			1.112				0.960					0.737					
Total load (MW)			259	210.22	206.39		112.62			115.88				139.48					121.19					

According to simulation results presented in Fig. 4, the voltage profile is affected by integration of WDG. In the first case before integration of DG-unit, the active power losses is 9.287 MW. After integration of WDG, in case 2 the total losses (TL) have become 5.86 MW, case 3 is 2.866 MW, case 4 is 2.026 MW, case 5 is 1.112 MW, case 6 is 0.96 MW and in case 7 have become 0.737 MW.

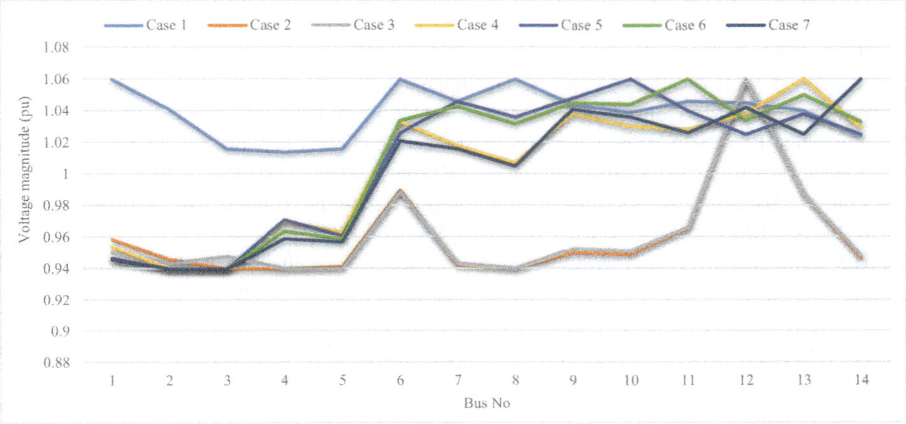

Fig. 4. The voltages profile with and without WDG

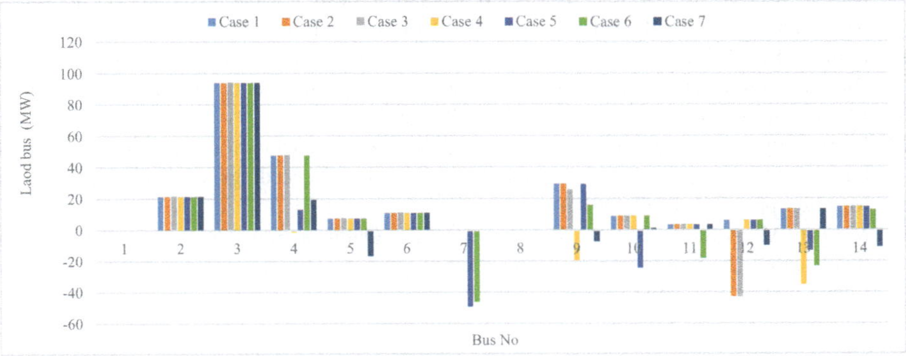

Fig. 5. Load actives buses with and without WDG

5 Conclusion

The optimization method based on genetic algorithm was used for integration of one and two wind based DG source in term of optimal location and size in power system, this is bay calculating the optimal power flow, including technical and security constraints. From this work, we have found that the integration of WDGs has proven its effectiveness better than without DG by minimizing of total power losses and acceptable voltage profile, and this integration provides relief overload transmission lines through the local production of wind source.

References

1. Pham, T.: Energiewende and competition in Germany: diagnosing market power in wholesale electricity market. Econ. Policy Energy Environ. (2015)
2. Mosbah, M., Khattara, A., Becherif, M., Arif, S.: Optimal PV location choice considering static and dynamic constraints. Int. J. Emerg. Electr. Power Syst. **18**, 1–13 (2017)
3. Canard, J.F.: Impact de la génération d'energie dispersee dans les réseaux de distribution, Doctorat de l'Institut National Polytechnique de Grenoble. Laboratoire d'Electrotechnique de Grenoble (2000)
4. Momoh, J.A., Boswell, G.: Value-based Implementation of distributed generation in optimal power flow. IEEE Trans. Power Syst. 7803–9255 (2005)
5. Yingchen, L., et al.: Optimal power flow of receiving power network considering distributed generation and environment pollution. IEEE Trans. Power Syst. 4244–4813 (2010)
6. Zhu, Y., Tomsovic, K.: Optimal distribution power flow for systems with distributed energy resources. Electric Power Energy Syst. **29**, 260–267 (2007)
7. Harrison, G.P., Wallace, A.R.: Optimal power flow evaluation of distribution network capacity for the connection of distributed generation. IEEE Proc. Gener. Transm. Distrib. **152**, 115–122 (2005)
8. Lin, S.Y., Chen, J.F.: Distributed optimal power flow for smart grid transmission system with renewable energy sources. Energy **56**, 184–192 (2013)
9. Satish, K., Vishal, K., Barjeev, T.: Optimal placement of different type of DG sources in distribution networks. Electr. Power Energy Syst. **53**, 752–760 (2013)
10. Masoud, E.: Placement of minimum distributed generation units observing power losses and voltage stability with network constraints. IET Gener. Transm. Distrib. **7**, 813–821 (2013)
11. Kumar, A., Gao, W.: Optimal distributed generation location using mixed integer non-linear programming in hybrid electricity markets. IEEE Trans. Power Syst. 1751–8687 (2009)
12. Zeinalzadeh, A., et al.: Optimal multi objective placement and sizing of multiple DGs and shunt capacitor banks simultaneously considering load uncertainty via MOPSO approach. Electr. Power Energy Syst. **67**, 336–349 (2015)
13. El-Zonkoly, M.: Optimal placement of multi-distributed generation units including different load models using particle swarm optimisation. In: Proceedings of IEEE PES Winter Meeting, pp. 1751–8687 (2010)
14. Mosbah, M., Arif, S., Zine, R., Mohammedi, R.D., Oudjana, S.H.: Optimal size and location of PV based DG-unit in transmission system using GA method for loss reduction. J. Electr. Eng. **17**(37), 2577–2594 (2017)

Compounds and Materials in Renewable Power Systems

Comparison of the Impacts of SVC and STATCOM on the Stability of an Electrical Network Containing Renewable Energy Sources

Kadri Abdellah(✉) and Makhloufi Salim

Laboratoire Energie, Environnement et Systèmes d'Information, Faculté des Sciences et Technologies, Université Ahmed Draia Adrar, Alger, Algeria
kadridet@gmail.com, makhlofi_s@yahoo.fr

Abstract. Currently and in the coming decades the subject of integrating FACTS into a power network will have more importance within the scientific community. This is mainly due to the liberalization of the electricity sector and advances in power electronics [1].

This work deals with the impact of the integration of renewable energy into the electricity network, and comparisons between two types of FACTS (SVC, STATCOM), with particular reference to transient stability. This work shows that the introduction of SVC (Static Var Compensator), and STATCOM (Static Synchronous Compensator) allowed the improvement of the stability of the network in the presence of voltage fault.

Keywords: Renewable energy · Transient stability · SVC · STATCOM · Electricity network

1 Introduction

Modern power system is a complex network comprising of numerous generators, transmission lines, variety of loads and transformers. As a consequence of increasing power demand, some transmission lines are more loaded than was planned when they were built. With the increased loading of long transmission lines, the problem of transient stability after a major fault can become a transmission limiting factor [2].

This paper deals with the problem of static voltage stability in electrical networks. Basic notions of integrating FACTS to solve instability and voltage collapse have been presented.

The purpose of this work is to analyze the temporary stability of an electrical network containing photovoltaic generator and wind turbine in case of voltage fault in the presence of a FACTS device (SVC or STATCOM).

M. Hatti (Ed.): ICAIRES 2019, LNNS 102, pp. 577–584, 2020.
https://doi.org/10.1007/978-3-030-37207-1_61

2 Studied System

The studied network is modeled thanks to the software PSAT, it is represented in Fig. 1. It consists of three conventional generators, slack bus, PV generator, a wind turbine, solar source, 14 transmission lines, 11 static loads and three transformers. The base power is 100 MVA and the base voltage is 13.8 kV.

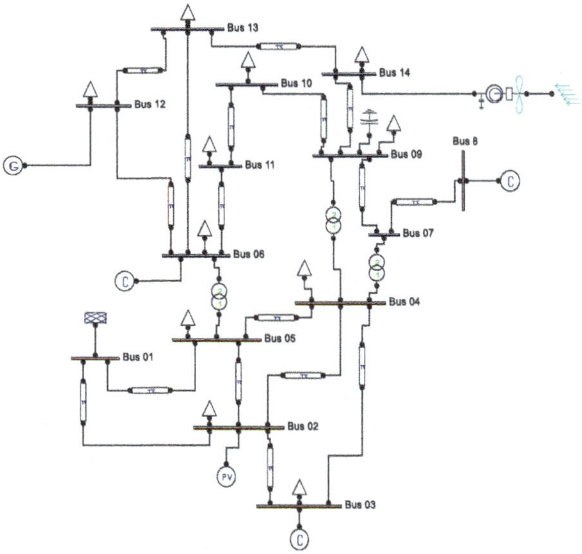

Fig. 1. Studied network

3 Static Var Compensator (SVC)

Static VAR systems are applied by utilities in transmission applications for several purposes. The primary purpose is usually for rapid control of voltage at weak points in a network. Installations may be at the midpoint of transmission interconnections or at the line ends. Static VAR Compensators are shunting connected static generators/absorbers whose outputs are varied so as to control voltage of the electric power systems. In its simple form, SVC is connected as Fixed Capacitor Thyristor Controlled Reactor (FC-TCR) configuration as shown in Fig. 2 [3].

The SVC is connected to a coupling transformer that is connected directly to the AC bus whose voltage is to be regulated. The effective reactance of the FC-TCR is varied by firing angle control of the anti-parallel thyristors. The firing angle can be controlled through a PI (Proportional + Integral) controller in such a way that the voltage of the bus, where the SVC is connected, is maintained at the reference value.

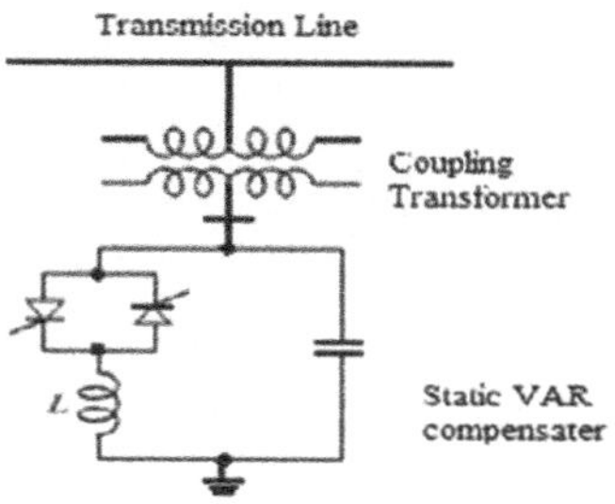

Fig. 2. Static VAR compensator of SVC [3].

4 Static Synchronous Compensator (STATCOM)

The STATCOM is based on a solid state synchronous voltage source which generates a balanced set of three sinusoidal voltages at the fundamental frequency with rapidly controllable amplitude and phase angle. The configuration of a STATCOM is shown in Fig. 3.

Basically it consists of a voltage source converter (VSC), a coupling transformer and a DC capacitor. Control of the reactive current and hence the susceptance presented to power system is possible by variation of the magnitude of output voltage (V_{VSC}) with respect to bus voltage (V_B) and thus operating the STATCOM in inductive region or capacitive region [3].

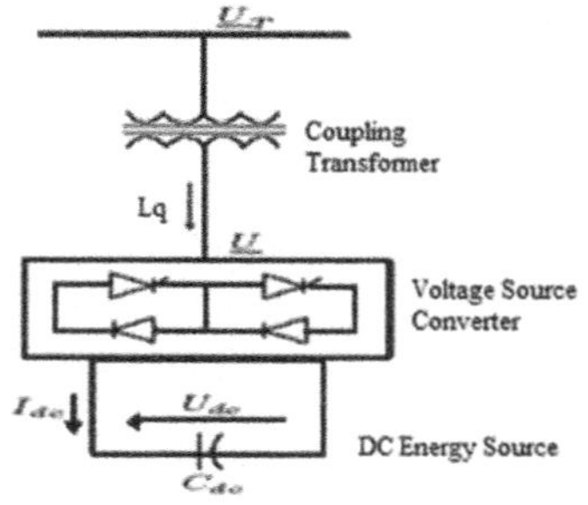

Fig. 3. Static synchronous compensator (STATCOM) [3].

5 Network Test Settings

See Tables 1, 2, 3 and 4.

Table 1. Parameters of the wind turbine and Photovoltaic generator

Wind turbine

parameters	Meaning
Apparent power S (MVA)	4
Nominal voltage U (KV)	2.3
Nominal frequency F (Hz)	50
Radius of the Wind turbine m	R=35
Speed multiplier gain	G=90
Density of air (kg/m^3)	ρ=1.225
Stator resistance (*pu*)	R_s =1.6
Rotor resistance (*pu*)	R_r =0.4
Magnetization inductance (pu)	X_m =0.055
Stator inductance (pu)	X_s =0.15
Rotor inductance (pu)	X_r= 0.023
Inertia of the tree (*pu*)	J=0.01
Number of pole pairs	P=3

Photovoltaic generator

Activer power P (MW)	40
Nominal voltage U (KV)	13.8
Nominal frequency F (Hz)	50
Reference voltage	1.045
max current (p.u)	1.2
min current (p.u)	0.8

Table 2. Generator parameters

Generator parameters	Generator 1	Generator 2	Generator 3	Generator PV	Slack bus
U (KV)	139.88	147.66	150.42	144.21	146.28
F (Hz)	50	50	50	50	50
δ [deg]	−12.7250	−14.2209	−13.3596	−4.9826	0.0000
P [pu]	0.0000	0.0000	0.0000	0.4000	2.3239
Q [pu]	0.2508	0.1273	0.17.62	0.4356	−0.1655

Table 3. Parameters (a) SVC and (b) STATCOM

(a)SVC

Apparent power S (MVA)	100
Nominal voltage U (KV)	13.8
Nominal frequency F (Hz)	50
Reference voltage	1.00
Bmin (p.u)	-1.00
Bmax (p.u)	1.00

(b) STATCOM

Apparent power S (MVA)	100
Nominal voltage U (KV)	13.8
Nominal frequency F (Hz)	50
Reference voltage	1.00
max current (p.u)	1.2
min current (p.u)	0.8

Table 4. Transformer parameters

Transformer	S (MVA)	First voltage	Secondary voltage	R (p.u)	X (pu)
T1	100	69	13.8	0	0.17615
T2	100	69	13.8	0	0.17615
T3	100	69	13.8	0	0.17615

6 Simulation Résultats

The simulation is run for four cases as follows:

6.1 Results Without Fault

In this case, we integrate a renewable source with four conventional generators. A very slight disturbance of the voltage at the bus levels (V_{BUS08}–V_{BUS14}) is shown in the figure below (Fig. 4).

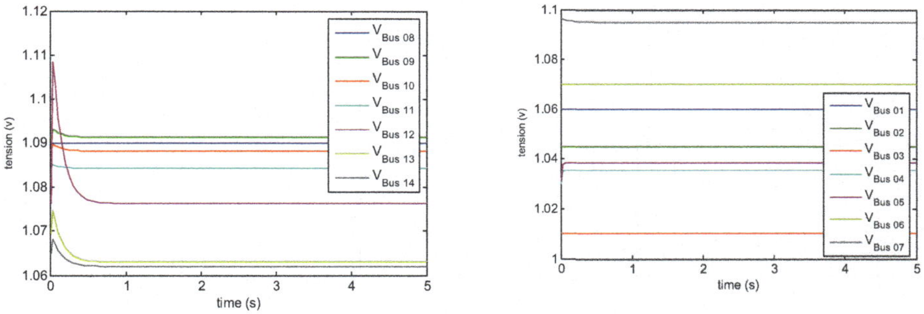

Fig. 4. Tensions to the bus (Without default and FACTS)

6.2 Results with Voltage Fault

A three-phase fault has been applied on the bus 11 at time t = 3 s and deleted at time t = 3.25 s. This causes breakdowns in the lines. The simulation results are shown in Fig. 5. There is a voltage drop (short circuit) in the buses (4, 5, and 7 to 14) as shown in the figure below.

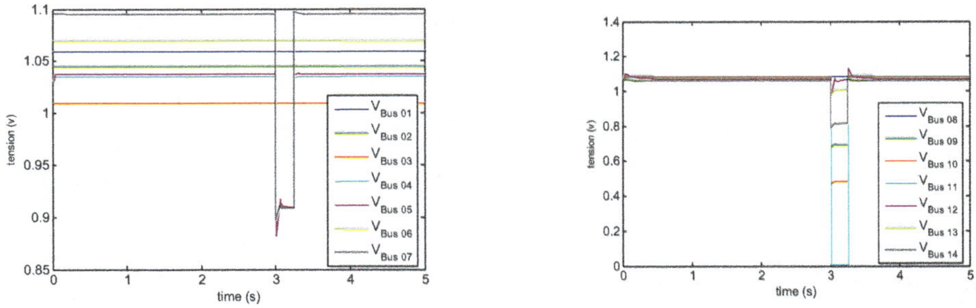

Fig. 5. Voltage to the bus a fault and without FACTS.

6.3 Simulation Results with Integration of the SVC

The same fault as in the previous paragraph is applied to the bus 11 in the presence of the SVC reactive energy compensator at the same bus. The simulation results are shown in Fig. 6. There is a marked improvement in the stability of voltage of the buses 7 to 14 (Fig. 7).

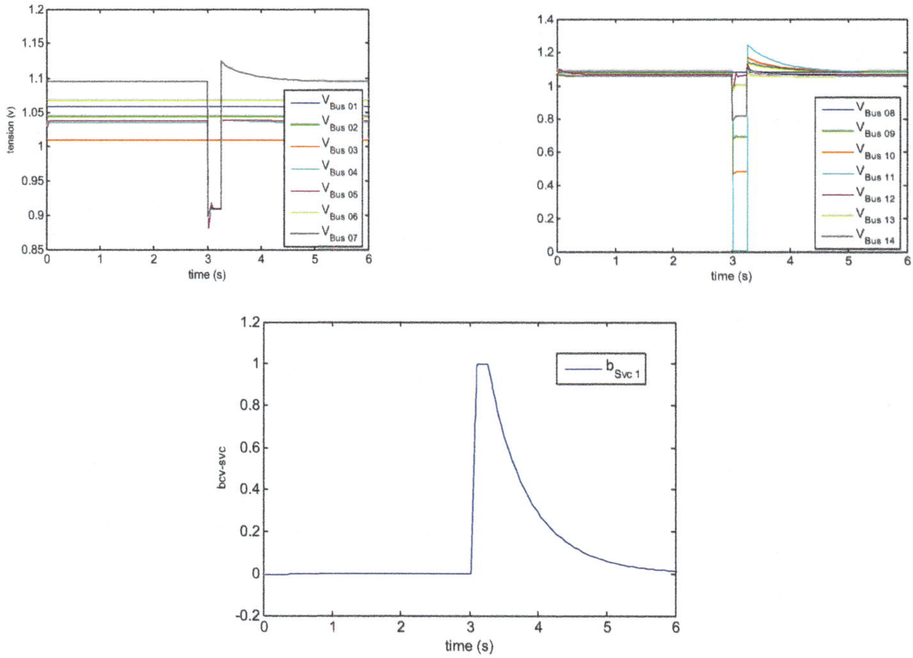

Fig. 6. Voltage to the bus a fault and SVC

6.4 Simulation Results with Integration of the STATCOM

The same fault as in the previous paragraph on Bus 11 is applied in the presence of the STATCOM. The simulation results are shown in Fig. 6. There is a marked improvement and rapid return to stability of voltage of the buses 7 to 14.

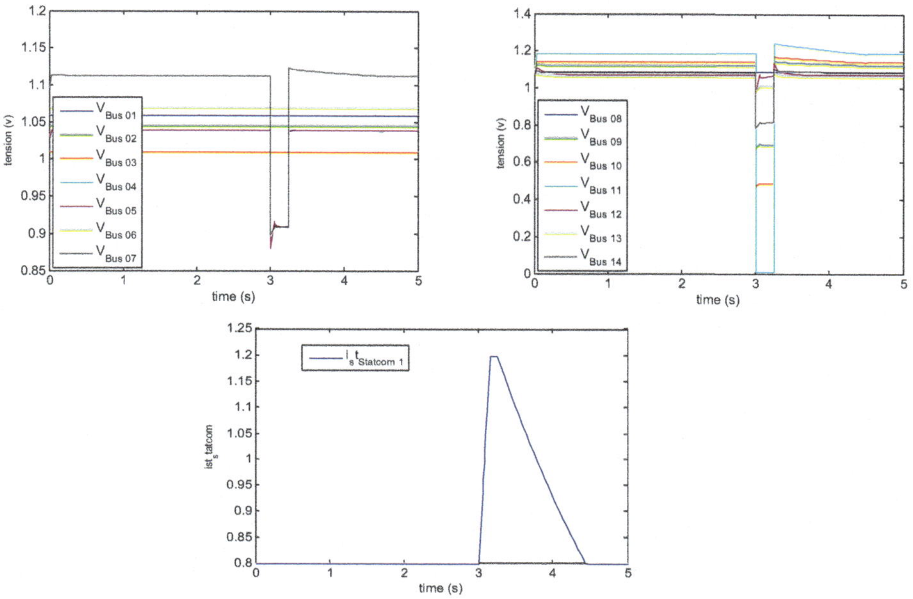

Fig. 7. Voltage to the bus a fault and STATCOM

7 Conclusion

In this work, we studied the stability of a network system containing a renewable energy sources in the presence of a fault. We used PSAT simulation tool running under MATLAB environment. After applying a three-phase fault on one bus, we found that the network became completely unstable. For this reason, we used compensation mechanism connected to the network. We proposed to install a flexible system (SVC and STATCOM). The results obtained show that this FACTS system gives good results for the improvement of the transient stability of the studied network with a slight superiority to STATCOM over SVC.

References

1. Feguir, S., Badi, F.: Placement Optimal de FACTS dans un Réseau électrique par les Méthodes Méta heuristiques, thèse de Master, L'Université Echahid Hamma Lakhdar d'El Oued, Faculté de Technologie 2017/2018

2. Mihalic, R., Zunko, P., Povh, D.: Improvement of transient stability using unified power flow controller. IEEE Trans. Power Deliv. **11**(1), 485–491 (1996)
3. Bisen, P., Shrivastava, A.: Comparison between SVC and STATCOM FACTS devices for power system stability enhancement. Int. J. Emerg. Technol. **4**(2), 101–109 (2013). Department of Electrical & Electronics Engineering, Oriental College of Technology, Bhopal, (MP)

Simulation of Electromagnetic Systems by COMSOL Multiphysics

S. Khelfi[1,2(✉)], B. Helifa[1,2], I. K. Lefkaier[1,2], and L. Hachani[1,2]

[1] ENS of Laghouat, LPM Laboratory, University Amar Telidji, Laghouat, Algeria
s.khelfi@ens-lagh.dz, helifa@yahoo.fr, l.hachani@lagh-univ.dz

[2] Génie Mécanique Laboratory, University of Laghouat, University Amar Telidji, Laghouat, Algeria

Abstract. Non-destructive techniques are used widely in the metal industry in order to control the quality of materials. Eddy current testing is one of the most extensively used non-destructive techniques for inspecting electrically conductive materials at very high speeds that does not require any contact between the test piece and the sensor. The present work aims at identifying the Non-destructive testing (NDT) by eddy currents (EC) in its various modeling as well as experimental aspects, the benchmark problems TEAM Workshop 15-1 have been considered to validate a COMSOL-Multiphysics 3D-resolution using a 3D electromagnetic formulation with Whitney edge elements. Our calculations' findings using COMSOL Multiphysics software are highly reliable and in harmony with the experimental data, this definitely permits us to examine numerous other cases in a perfect way and create a database ready for a study pertinent to inverse problems. The realized experimental setup can be used for Automatic sorting of defective materials in some industrial applications, which can reduce losses and costs and increase quality and efficiency.

Keywords: Non-destructive control · Eddy current · Edge element method 3D · Defective materials · COMSOL Multiphysics

1 Introduction

The modeling of an actual configuration of CND-CF can not generally be obtained analytically and uses numerical methods. Among them, the finite element method (FEM) which allows to take into account complex geometries of probes and parts [1].

Today a wide range of digital tools is available. It is based on the implementation of theoretical models using different mathematical tools including finite element technique. Numerical simulation makes it possible to study the functioning and the properties of a modeled system as well as to predict its evolution. It is very interesting to have a simulation environment that includes the possibility of adding different physical phenomena to the studied model. It is in this philosophy that Comsol Multiphysics has been developed. It is a modular finite element numerical computation software

M. Hatti (Ed.): ICAIRES 2019, LNNS 102, pp. 585–589, 2020.
https://doi.org/10.1007/978-3-030-37207-1_62

allowing to model a large variety of physical phenomena characterizing a real problem. It will also be a design tool thanks to its ability to manage complex 3D geometries [2].

2 Use of COMSOL Multiphysics in ECT

The release of the software used for the study is the 5.3 a release.

2.1 Formulation

The system of equations in magnetic vector and electric scalar potential to solve is:

$$\iiint_D [\nabla \times (\nu\nabla \times \vec{A}) - \nabla(\nu\nabla.\vec{A})]dxdydZ + j\omega \iiint_D [\sigma(\vec{A} + \nabla V)dxdydz =$$
$$\iiint_D \vec{J_s}dxdydz \qquad (1)$$
$$\iiint_D \nabla.[-j\omega\sigma(\vec{A} + \nabla V)dxdydz = 0$$

$\vec{A}$ and V are respectively the magnetic vector potential and electric scalar potential, ν is the magnetic reluctivity and σ the electrical conductivity of the conductive plate [3].

2.2 Geometry

Geometric and electromagnetic characteristics are shown in Table 1.

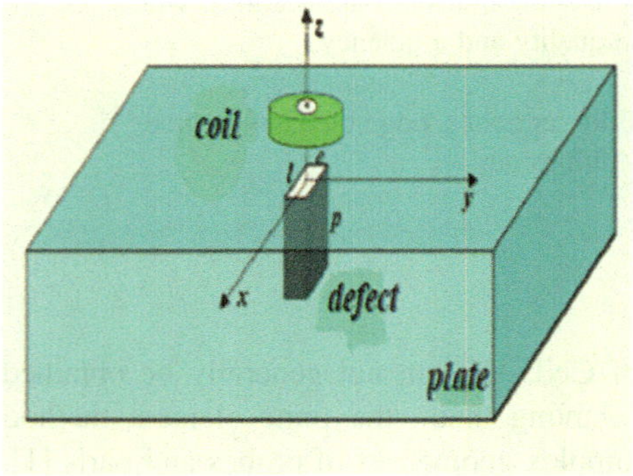

Fig. 1. Sensor-crack system.

Table 1. Geometric and electromagnetic characteristics of the system

Conductive plate	Length: 120 mm
	Width: 103 mm
	Thickness: 20 mm
	Electrical Conductivity: 3.85×10^7
	Magnetic Permeability: $4\pi \times 10^{-7}$
Defect	Length: 14 mm
	Width: 1 mm
	Depth: 5 mm
Coil	Coil inner radius
	Coil outer radius
	Coil length
	Lift off
	Resonance frequency: fc = 717k

2.3 Boundary Condition

In all the boundaries of the study domain, the magnetic induction B is supposed tangent and then the magnetic insulation condition (nxA = 0) is imposed (default condition in COMSOL Multiphysics).

2.4 Mesh

The mesh is generated with tetrahedral elements. The entire mesh consisting of 25214 domain elements, 4091 boundary elements, and 419 edge elements (Fig. 2).

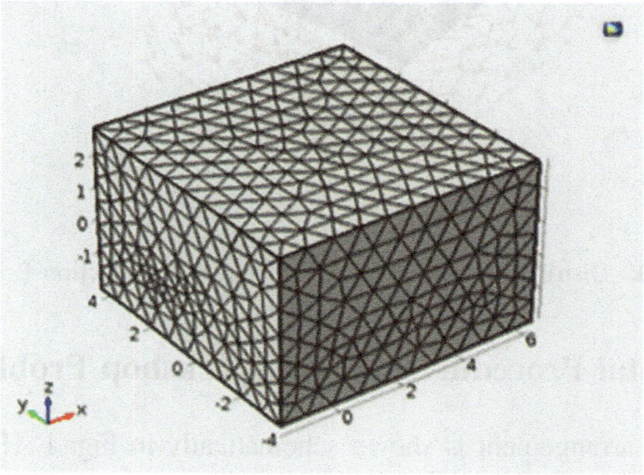

Fig. 2. Mesh

2.5 Resolution

The impedance variation ΔZ is a complex number. The real part is computed with the Joule Losses (J_L) in the conductive media and the imaginary part is computed with the magnetic energy (W_M) in the whole meshed domain (Fig. 3).

$$Z = R + Xi$$

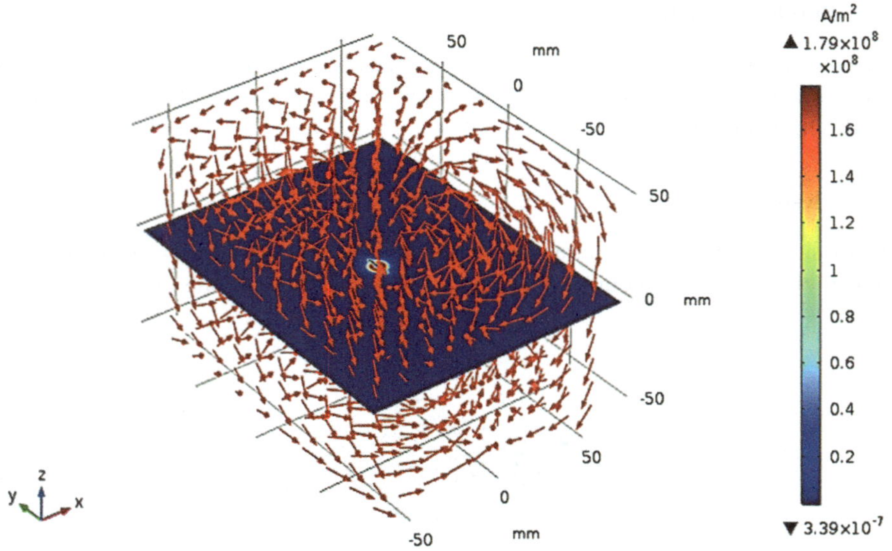

Fig. 3. Distribution of the vectors of magnetic induction B in 3D.

3 Experimental Procedures (Team Workshop Problem 15)

The experimental arrangement is shown schematically in Fig. 1. Here, a circular air-cored coil is scanned, parallel to the x-axis, along the length of a rectangular slot in an aluminum alloy plate. Both the frequency and the coil lift-off are fixed and ΔZ is

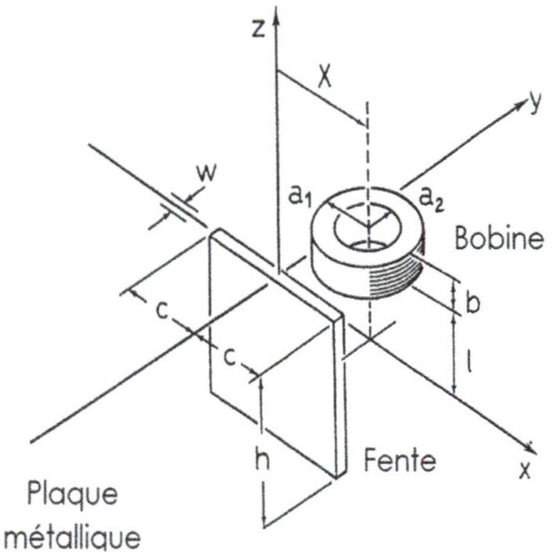

Fig. 4. Detail of TEAM Workshop Pb N ° 15-1

Table 2. Parameters of the Problem Benchmark TEAM Workshop Pb N ° 15-1

Bobine	
Rayon externe (a_1)	$12,40 \pm 0,05$ mm
Rayon interne (a_2)	$6,15 \pm 0,05$ mm
Hauteur (b)	$6,15 \pm 0,1$ mm
Nombre de spires (N)	3790
Lift-off (l)	0,88 mm
Spécimen	
Conductivité (σ)	$(30,6 \pm 0,20).10^6$ S/m
Épaisseur	$12,22 \pm 0,02$ mm
Défaut	
Longueur ($2c$)	$12,6 \pm 0,02$ mm
Profondeur (h)	$5,00 \pm 0,05$ mm
Largeur (w)	$0,28 \pm 0,01$ mm
Autres	
Fréquence	900 Hz
Épaisseur de peau(δ) à 900 Hz	3,04 mm

measured as a function of coilcenter position. The parameters for this test experiment are listed in Table 2 (Fig. 4).

4 Validation

In this section we compared the experimental results with the numerical results of 3D simulations (Figs. 5 and 6).

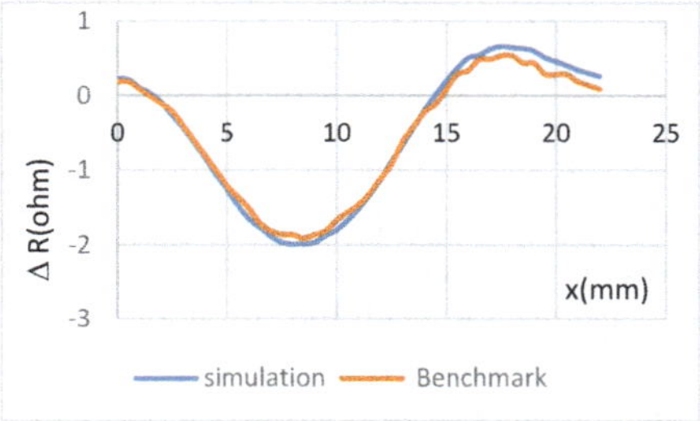

Fig. 5. Variation of the resistance according to the displacement of the sensor

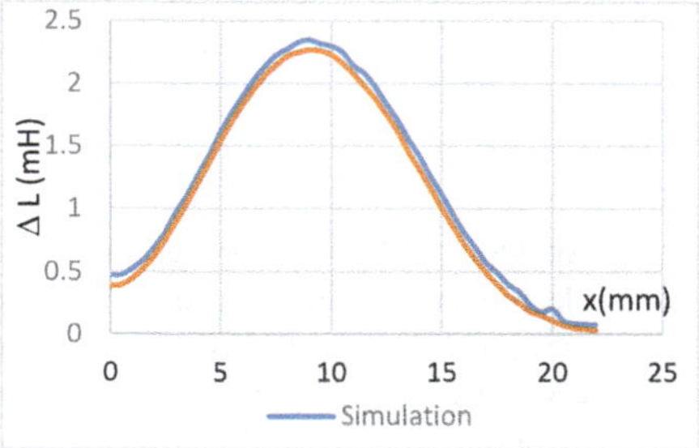

Fig. 6. Variation of inductance as a function of sensor displacement

5 Conclusion

This work presents a simulation of a 3D ECT problem with the COMSOL Multiphysics 5.3a software. The use of the software to solve this type of problem is detailed and explained. The results of our COMSOL Multiphysics calculations are in very good agreement with the experimental data, this definitely permits us to examine numerous other cases in a perfect way.

References

1. Lakhdari, A.-E.: Etude de modélisation de capteurs en CND par Courant de Foucault: Application à la détection des fissures, Magister's thesis, University Mohamed Khider – Biskra, 22 Mai 2011
2. Fekiri, N.: Transferts de chaleur dans des éléments micro et nano structurés: Simulation et expérimentation par thermographie infrarouge et microscopie SThM, Internship report, 05 Septembre 2013
3. Helifa, B., Zaoui, A., Feliachi, M., Lefkaier1, I.K., Boubenider, F., Cheriet, A.: Simulation du CND par courants de Foucault en vue de la caractérisation des fissures débouchantes dans les aciers austénitiques

The Use of Nanofluids in Electrocaloric Refrigeration Systems

B. Kehileche[1(✉)], Y. Chiba[2], N. Henini[1], and A. Tlemçani[1]

[1] Faculty of Technology, Department of Electrical Engineering, University of Medea, Médéa, Algeria
kehileche@yahoo.fr
[2] Faculty of Technology, Department of Mechanical Engineering, University of Medea, Médéa, Algeria
Chiba.younes@univ-medea.dz

Abstract. Electrocaloric cooling is based on electrocaloric materials (ferroelectrics) exhibiting electrocaloric effect, detected in a temperature change of the electrocaloric materials due to the variation of the electric field. Electrocaloric cooling could represent potentially an alternative to classic refrigeration. Electrocaloric cooling is the best refrigeration solution, applied to a electrocaloric materials made by the ferroelectrics materials, placed between CHEX and HHEX and exposed to alternative polarization/depolarization cycle. To vehiculate the refrigerant between CHEX and HHEX, a liquid refrigerant is used: water or a water/ethylene glycol mixture, but innovative liquid refrigerant could be adopted, such as water/nanoparticles in order to enhance the COP and the temperature span ΔT of the nanofluids. In this work we report the simulation results on a active electrocaloric regenerators refrigerator made in parallel plate, utilizing water and nanoparticles (CuO and Al_2O_3) as liquid refrigerant. The results are reported in terms of COP; power and temperature span ΔT. In this paper, for electrocaloric cooling system, a decoupled multiphysics numerical approach (Electric, Fluid Flow, and Heat Transfer) is developed using COMSOL multiphysics.

Keywords: Electrocaloric refrigeration · Nanofluid · COMSOL multiphysics

1 Introduction

As a result of population growth, climate change and in order to meet the growing needs in industrial refrigeration and air conditioning, research is now focusing on alternative refrigeration technologies. Indeed, environmental requirements limit conventional technologies, including thermodynamic techniques [1–3]; one of the problems regarding the classic refrigeration is that refrigeration is usually based on a dangerous liquid as refrigerant fluid. Research on future technologies of refrigeration turned to other principles: refrigeration electrocaloric based on nanofluids (water/nanoparticles (CuO and Al_2O_3); the operating principle is based on the electrocaloric effect (ECE), which is related to a change of temperature in the electrocaloric material (ECM) due to a variation of the electric field [4, 5]. The thermodynamic cycle

M. Hatti (Ed.): ICAIRES 2019, LNNS 102, pp. 590–597, 2020.
https://doi.org/10.1007/978-3-030-37207-1_63

(four processes: (1) adiabatic polarization, (2) isochamps heat removal, (3) adiabatic depolarization, and (4) isochamps heat absorption) is applied to a electrocaloric materials made by the ferroelectrics materials, which is both nanofluids and AER. The AER is placed between CHEX (depolarization of electrocaloric materials) and HHEX (polarization of electrocaloric materials). To vehiculate the nanofluids between HHEX and CHEX water and nanoparticles (CuO and Al_2O_3) is used.

The results were carried out utilizing a 2-D model solved with COMSOL multiphysics; fort two different nanofluids (Al_2O_3 and CuO).

2 Description of Electrocaloric Refrigeration Cycle

Figure 1 shows a schematic diagram of an AER cycle.

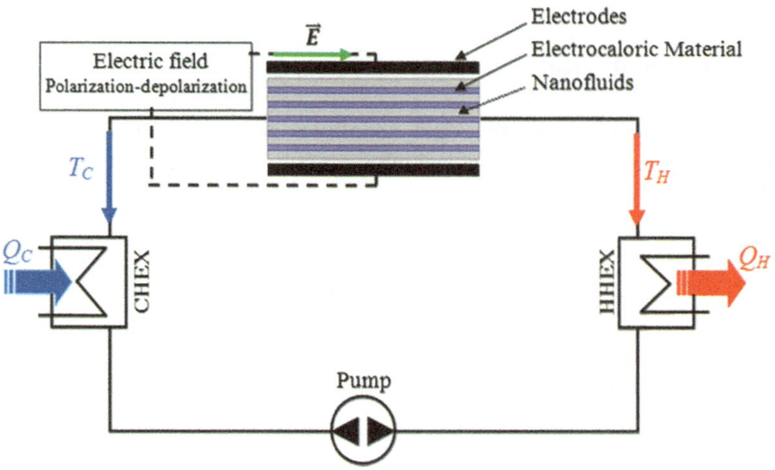

Fig. 1. Schematic diagram of the AER cycle setup.

Four phases of AER cycle [1–5]:

(1) The electrocaloric materials is polarized;
(2) The heat is exchanged using water and nanoparticles that crosses the electrocaloric regenerator from the CHEX to HHEX;
(3) The electrocaloric materials is then depolarized;
(4) The cooled down electrocaloric materials absorbs heat from the water and nanoparticles that crosses it from the HHEX to CHEX (Table 1);

Table 1. Geometric parameters of the regenerator

Parameters	Not.	Valeurs	Unit
Length	L	50	mm
Width	l	20	mm
Height of the ECM plate	e_S	1	mm
Gap between plate	e_f	0.3	mm
Number of layers	N_s	14	
Electrocaloric materials	$BaTiO_3$		
The Curie temperature	T_C	297	K
Electric field	E	10;40	kV/cm
Flow velocity	u	0,06	m/s
Frequency	f	1	Hz
Time Polarization/depolarization	$t_{Pol/Dépol}$	2	s

3 Mathematical Model

The governing equations of AER and nanofluids mathematical model consists of coupled partial differential equations, which were solved with the commercial software COMSOL Multiphysics [9–13].

Maxwell's relations:

$$\left(\frac{\partial S}{\partial E}\right)_T = \left(\frac{\partial P}{\partial T}\right)_E \tag{1}$$

The velocity distribution in the fluid is determined by navier stokes equations:

$$\rho_f\left(\frac{dU}{dt} + (U.\nabla)U\right) - \mu_f\nabla^2 U + \nabla p = 0 \tag{2}$$

$$\nabla.U = 0 \tag{3}$$

the standard heat transfer equation:

For the solid domains:

$$\rho_{p,s}\frac{\partial T_s}{\partial t} - k_s\nabla^2 T_s = 0 \tag{4}$$

For the fluid domains:

$$\rho_f c_{p,f}\left(\frac{\partial T_f}{\partial t} + (U.\nabla)T_f\right) - k_f\nabla^2 T_f = 0 \tag{5}$$

Nanofluids density [5]:

$$\rho_{nf} = (1 - \varnothing)\rho_f + \varnothing\rho_s \tag{6}$$

Specific heat capacity:

$$(\rho C_p)_{nf} = (1 - \varnothing)(\rho C_p)_f + \varnothing(\rho C_p)_s \tag{7}$$

The viscosity of the nanofluid is determined by Brinkman relations:

$$\mu_{nf} = \frac{\mu_f}{(1 - \varnothing)^{2.5}} \tag{8}$$

Thermal Conductivity of Nanofluids by Maxwell-Garnetts:

$$\frac{k_{nf}}{k_f} = \frac{k_s + 2k_f - 2\varnothing(k_f - k_s)}{k_s + 2k_f + 2\varnothing(k_f - k_s)k_f} \tag{9}$$

Table 2 shows the thermophysical properties of water and nanoparticles.

Table 2. Thermophysical properties of base fluid and nanoparticles

Thermophysical properties	Water	Al_2O_3	CuO
Density ρ[kg m^{-3}]	999.2	3970	6350
Specific heat capacity C_p [$Jkg^{-1}K^{-1}$]	4183	765	535
Thermal conductivity k [$WK^{-1}m^{-1}$]	0.6	36	69

4 Resultats

AER was composed in three combination of electrocaloric materials and liquid refrigerant:

(1) electrocaloric materials and water;
(2) electrocaloric materials and water/nanoparticles Al_2O_3;
(3) electrocaloric materials and water/nanoparticles CuO.

Figure 2 shows the evolution of the powers for the three composed combinations of electrocaloric materials and water/nanoparticles, as a function T[K] and flow velocity; respectively. Al_2O_3 is the liquid refrigerant the highest power. The employment of water/nanoparticles (Al_2O_3-CuO) as liquid refrigerant takes to increasing the coefficients of performance and power. If the AER works with Al_2O_3 as liquid refrigerant, the effect of concentration Al_2O_3 particles in water translates into cooling powers higher than of CuO. Moreover, one can observe that the cooling powers has a slight maximum in correspondence of T > Tcurie. Since from Fig. 2 we observed that water/nanoparticles (Al_2O_3-CuO) is one of the best nanofluids.

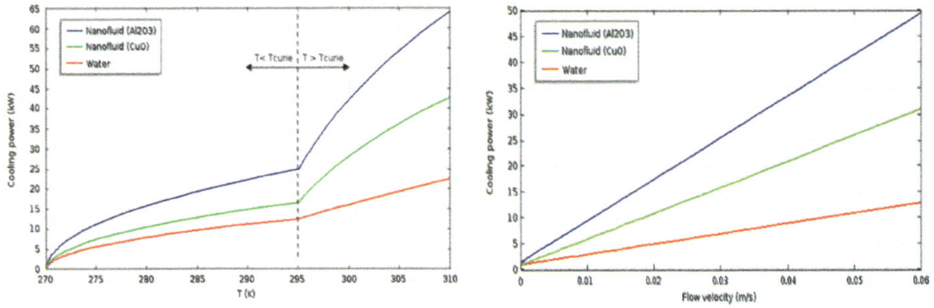

Fig. 2. Variation of cooling powers as function of the temperature and flow velocity

Figure 3 shows the evolution of the COP for the three composed combination of electrocaloric materials and water/nanoparticles, as a function the time and the temperature; respectively. The liquid refrigerant with Al_2O_3 are the ones to which the greatest coefficients of performance are attributed, with maximum COP of 20 for the AER working with water/nanoparticles (Al_2O_3).

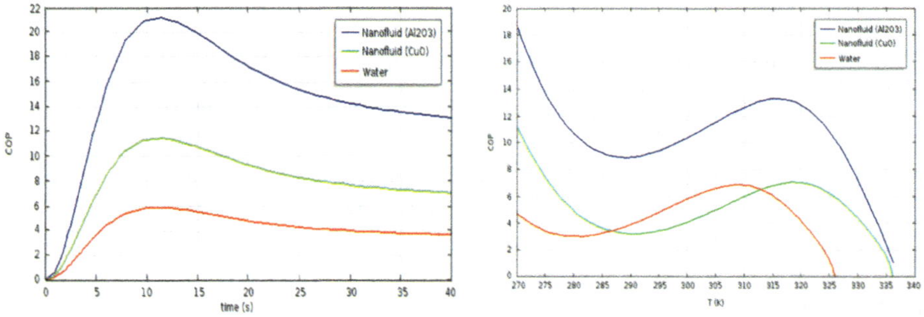

Fig. 3. Variation of the coefficients of performances as function of time and the temperature

Figure 4 shows impact of electric field on the evolution of the temperature span ΔT under electric field; 10 to 40 kV/cm, all the curves exhibit sharp increasing trends with electric field; we observed that water/nanoparticles (Al_2O_3-CuO) is one of the best combinations, the electric field is imposed so as to avoid the breakdown of the electrocaloric material.

Figure 5 shows variation of the temperature span ΔT as function of the solid and fluid thickness for regenerator AER. Figure 5 illustrates clearly the positive impact of water/nanoparticles (Al_2O_3-CuO) of the temperature span and negative impact on the solid and fluid thickness.

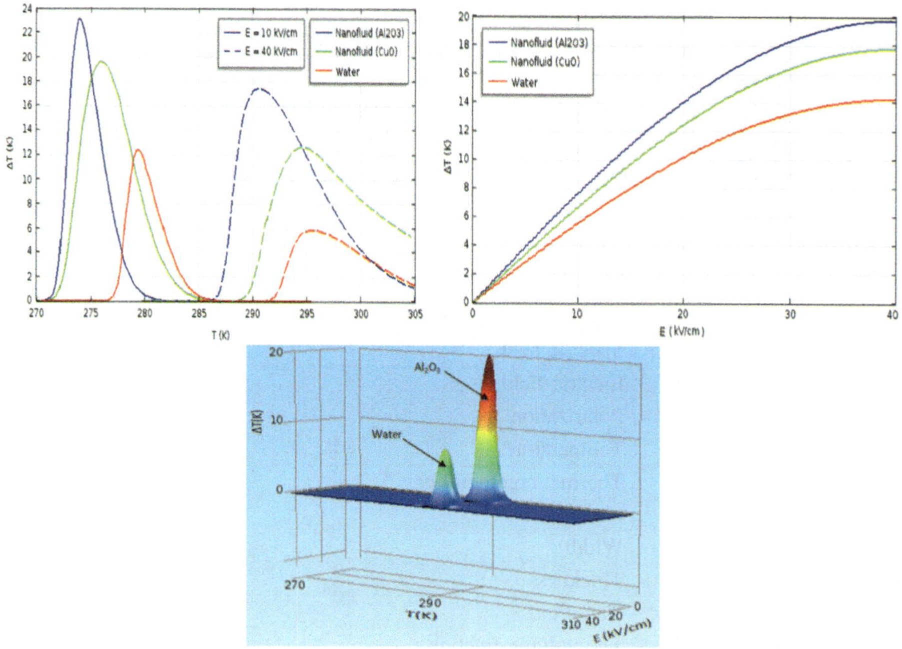

Fig. 4. Variation of temperature span ΔT as function of the temperature and electric field

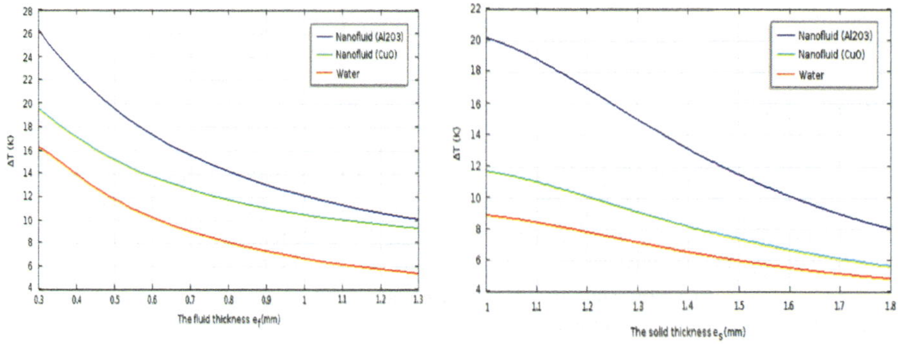

Fig. 5. Shows variation of the temperature span ΔT as function of the solid and fluid thickness

5 Conclusions

In this work we studied, the cooling power; coefficient of performance and the temperature span ΔT of AER employing water/nanoparticles (Al_2O_3-CuO) as liquid refrigerant. We observed that the best results both in terms of cooling power, coefficient of performance and the temperature span ΔT are given by $BaTiO_3$ and water/nanoparticles (Al_2O_3-CuO) as liquid refrigerant. The effect of using water/nanoparticles (Al_2O_3-CuO) is positive in terms of energy efficiency. The concentration of Al_2O_3 nanoparticles results in the cooling powers and COP higher than working with only water as liquid refrigerant and slightly greater than working with water/nanoparticles (CuO).

In conclusion, the employment of water/nanoparticles (Al_2O_3-CuO) as liquid refrigerant and elecrocaloric materials $BaTiO_3$ of AER enhances the energy efficiency both in terms of power, COP and the temperature span ΔT.

Nomenclature

Symbols	Meaning	Unit
COP	Coefficient of performance	
C_p	Specific heat	J kg^{-1}°C^{-1}
E	Electric field	V m^{-1}
P	Polarization	C m^{-2}
T	Temperature	K
k	Thermal conductivity	$Wm^{-2}K^{-1}$
L	Length	Mm
l	Width	Mm
e	Thickness	Mm
nf	Nanofluid	
∅	The volume fraction	%
μ_f	Viscosity	$kgm^{-1}s^{-1}$
ρ	Density	kgm^{-3}
CHEX, HHEX	Cold and Hot heat exchanger	

References

1. Ozbolt, M., Kitanovski, A., Tusek, J., Poredo, A.: Electrocaloric refrigeration: thermodynamics, state of the art and future perspectives. Int. J. Refrig **40**, 174–188 (2014)
2. Correia, T., Zhang, Q.: Electrocaloric Materials: New Generation of Coolers, pp. 1–14. Springer, Heidelberg (2014)
3. Scot, J.F.: Electrocaloric Materials. Cavendish Laboratory, Cambridge University (2011)
4. Aprea, C., Greco, A., Maiorino, A., Masselli, C.: Electrocaloric refrigeration: an innovative, emerging, eco friendly refrigeration technique. J. Phys. (2017)
5. Chiba, Y., Smaili, A., Sari, O.: Enhancements of thermal performances of an active magnetic refrigeration device based on nanofluids. Mechanika **23**, 31–38 (2017)
6. Mugica, I., et al.: Exergy analysis of a parallel-plate active magnetic regenerator with nanofluids. Entropy **19**, 464 (2017)
7. Hernández, D., et al.: Analysis of working nanofluids for a refrigeration system. Dyna **83**, 176–183 (2016)

8. Senthilkumar, A., et al.: Effectiveness study on $Al2O_3$-TiO_2 nanofluid heat exchanger. Int. J. Eng. Robot Technol. 73–81 (2016)
9. Aprea, C., Greco, A.: A comparison between electrocaloric and magnetocaloric materials for solid state refrigeration. Int. J. Heat Technol. **35**, 225–234 (2017)
10. Lionte, S., Vasile, C., Siroux, M.: Approche multiphysique et multi-échelle d'un régénérateur magnétothermique actif. Institut National des Sciences Appliques INSA (2015)

Robust Speed Sensorless Fuzzy DTC Using Simplified Extended Kalman Filter for Dual-Star Asynchronous Motor (DSIM) with Stator Resistance Estimation

A. Cheknane(✉), K. Kouzi, H. Sayaf, and I. Benhamida

Laboratory of Semiconductors and Functional Materials,
Amar Telidji University, BP 37G Ghardia Road, 03000 Laghouat, Algeria
{abd.cheknane,k.kouzi}@lagh-univ.dz

Abstract. Direct torque control seems DTC good solution for problems as robustness and dynamics, encountered in vector control with rotor flux oriented. Current researches are directed towards improving the feature of this technique, by resolving the main problems being the evolution of the switching frequency, the ripples on the torque, the flux and the current and stator resistance deviation. In this work, we develop novel design of the DTC scheme based on simplified EKF witch estimate both the mechanical speed and resistance stator. Simulation results were performed in matlab show best performance of the suggested scheme.

Keywords: Dual-stator induction machine · Direct torque fuzzy controlled · Direct torque control · Extended Kalman filter

1 Introduction

The Multi-phase machine could be an interesting alternative for the speed variable control because it gives various advantages over classical three-phase Machine. In a multi-phase machine drive system, more than three-phase windings are housed in the same stator of the electric machine and the current per phase in the machine is thereby reduced. Most commonly, two sets of three-phase windings are spatially phase shifted by 30°. However, model and control strategy are widely found in recent literature, but very few papers deal with DTFC technology for six-phase engines. Therefore, in this work, a study is reserved for the (DTFC) strategy of the double stator asynchronous motor (DSIM) and its speed setting. These machines require double three-phase power and have been used in many applications for their benefits [1, 2].

The advantages of this method include a very fast and accurate response to the torque control signal and an available comparative simplicity of the control algorithm with a few input variables such as motor torque, flux magnitude and flux Sector DTC for induction machines (IM). It simplifies the overall drive technology, because the controller is directly connected to the drive circuitry of switches, which eliminates the

M. Hatti (Ed.): ICAIRES 2019, LNNS 102, pp. 598–609, 2020.
https://doi.org/10.1007/978-3-030-37207-1_64

modulator. Provided the torque and flux controllers produce good results, DTC is rather robust to the motor parameters and external disturbances. The other advantage of DTC-based controllers is the ability to avoid coordinate transformation in calculations and eliminate current regulation and voltage modulation blocks [3].

The established disadvantages of DTC are highlighted at startup and in low velocity or high waviness regions and the slow response to beneficial torque changes. To improve performance, fuzzy logic is successfully implemented in various control systems and electrical drive control. It is therefore necessary to test it in DTC [4].

The purpose of using kalman filter in this work is the minimization of the sensors (speed sensor) for the cut and the compensation of the DTC estimator (estimate the variation of resistance), and also test the robustness of the control and the variation of parameters of machine (DSIM). The Kalman Filter is used as an estimator and filters the noise of modeling and measurements. The results of the simulation show good performance and robustness. The error between real and estimated speed is within a few rad/s.

2 Modeling of Double Stator Induction Machine

The DSIM offers several advantages for industrial drive systems over conventional three-phase drive systems, such as improving reliability, minimizing torque pulsations, reducing magnetic flux harmonics and reducing power for the cascaded H-bridge multi-level inverters [3, 5].

The DSIM model is derived from some assumptions, as it is known under the name "simplifying hypotheses".

DSIM modeling and control in the original reference is very difficult. For this reason, it is necessary to obtain a simplified model of this machine. The DSIM model is decomposed into two main sub-models noted (ds1-qs1) and (ds2-qs2) for the stator side and a sub-model noted (dr-qr) for the rotor side.

All sub models are expressed in the synchronous coordinate system. Stator and rotor flux expressions are

The state space representation of the machine is as follow [5]:

$$\frac{\mathrm{dX}}{\mathrm{dt}} = \mathrm{AX} + \mathrm{B} \quad \text{where} \quad \mathrm{X} = \left[\phi_{\mathrm{ds1}}\phi_{\mathrm{ds2}}\phi_{\mathrm{qs1}}\phi_{\mathrm{qs2}}\phi_{\mathrm{dr}}\phi_{\mathrm{qr}}\right]^T; \quad U = \left[v_{ds1}v_{ds2}v_{qs1}v_{qs2}\right]^T; \tag{1}$$

The matrices A and B are given

$$A=\begin{bmatrix} -\left(\frac{R_{s1}}{L_{s1}}-\frac{R_{s1}L_a}{L_{s1}^2}\right) & \frac{R_{s1}L_a}{L_{s1}L_{s2}} & w_s & 0 & \frac{R_{s1}L_a}{L_{s1}L_r} & 0 \\ \frac{R_{s2}L_a}{L_{s1}L_{s2}} & -\left(\frac{R_{s2}}{L_{s2}}-\frac{R_{s2}L_a}{L_{s2}^2}\right) & 0 & w_s & \frac{R_{s2}L_a}{L_{s2}L_r} & 0 \\ -w_s & 0 & -\left(\frac{R_{s1}}{L_{s1}}-\frac{R_{s1}L_a}{L_{s1}^2}\right) & \frac{R_{s1}L_a}{L_{s1}L_{s2}} & 0 & \frac{R_{s1}L_a}{L_{s1}L_r} \\ 0 & -w_s & \frac{R_{s2}L_a}{L_{s1}L_{s2}} & -\left(\frac{R_{s2}}{L_{s2}}-\frac{R_{s2}L_a}{L_{s2}^2}\right) & 0 & \frac{R_{s2}L_a}{L_{s2}L_r} \\ \frac{R_rL_a}{L_{s1}L_r} & \frac{R_rL_a}{L_rL_{s2}} & 0 & 0 & -\left(\frac{R_r}{L_r}-\frac{R_rL_a}{L_r^2}\right) & (w_s-w) \\ 0 & 0 & \frac{R_rL_a}{L_{s1}L_r} & \frac{R_rL_a}{L_rL_{s2}} & -(w_s-w) & -\left(\frac{R_r}{L_r}-\frac{R_rL_a}{L_r^2}\right) \end{bmatrix},$$

$$B=\begin{bmatrix} 1 & 0 & 0 & 0 \\ 0 & 1 & 0 & 0 \\ 0 & 0 & 1 & 0 \\ 0 & 0 & 0 & 1 \\ 0 & 0 & 0 & 0 \\ 0 & 0 & 0 & 0 \end{bmatrix} \tag{2}$$

The fundamental mechanical equation of the DSIM is expressed as

$$C_{em}-C_r=J\frac{d\Omega_r}{dt}+K_f\Omega_r \tag{3}$$

Electromagnetic torque established by the DSIM can be articulate affording flux and currents as

$$C_{em}=P\frac{L_m}{L_m+L_r}\left[(i_{qs1}+i_{qs2})\phi_{dr}-(i_{ds1}+i_{ds2})\phi_{qr}\right] \tag{4}$$

where P denotes the poles pairs number.

3 Direct Torque Control

The idea of DTC is to directly select stator voltage vectors according to the torque and flux errors. The governing equation for torque in this scheme is due to the interaction of stator and rotor fields. Torque and stator flux linkage are computed from measured motor terminal quantities i.e. stator voltages and current. An optimal voltage vector for the switching of VSI is selected among the six nonzero voltage vectors and two zero voltage vectors by the hysteresis control of stator flux and torque. A three-phase VSI has eight possible combinations of six switching devices. which have a well defined state: ON or OFF in each configuration. So all the possible configurations can be identified with three bits (Sa, Sb, Sc), one for each inverter leg. The bit is set to 1 if the top switch is closed and to 0 when the bottom switch is closed. In order to prevent short circuit of the supply, the state of the upper switch is always opposite to that of the lower on [6].

3.1 Operation and Sequences of a Three Phase Voltage Inverter

The inverter voltage switches must be controlled to maintain the flow and torque of machine. The vector of the stator voltage Vs can be written in the form

$$\overrightarrow{\mathrm{V_s}} = \sqrt{\frac{2}{3}\mathrm{U_{c0}}\left[\mathrm{S_a} + \mathrm{S_b}e^{j\frac{2\pi}{3}} + \mathrm{S_c}e^{j\frac{4\pi}{3}}\right]} \tag{5}$$

$S_i = 1$ means that the high switch is closed and the low switch is open $(V_i = +U_0/2)$, and $S_i = 0$ means that the switch 'high switch is open and the down switch is closed $(V_i = -U_0/2)$.

We will try to control the flow and torque via the choice of voltage vector that will be done by a configuration of the switches. Since

In each modulation period, one arm does not switch twice. Figure 2 shows the representation in the complex plane of the six non-zero voltage vectors that can be generated by a two-level three-phase voltage inverter [7].

3.2 Torque and Flux Estimator

The feedback flux and torque are calculated from the machine terminal voltages and currents. The components of stator flux is given as

$$C_{kemi} = p\left(\phi_{is\alpha} i_{s\beta} - \phi_{is\beta} i_{s\alpha}\right) \tag{6}$$

3.3 Estimation of the Stator Flux

The estimation of the flow can be carried out on the basis of measurements of the stator quantities current and voltage of the machine.

$$\phi_{s1,s2}(t) = \int_0^t (V_{s1,s2} - \mathrm{R_{s1,s2} i_{s1,s2}})\mathrm{dt} \tag{7}$$

We obtain from this equation the components α and β of the vector ϕ_s

$$\phi_{s1,s2\alpha} = \int_0^t (V_{s1,s2\alpha} - R_{s1,s2} i_{s1,s2\beta})dt \quad \phi_{s1,s2\beta} = \int_0^t (V_{s1,s2\beta} - R_{s1,s2} i_{s1,s2\beta})dt \tag{8}$$

3.4 Electromagnetic Torque Estimation

The electromechanical torque can be estimated from the estimated fluxes, and from measured stator currents

$$C_{em} = p\left(\phi_{s\alpha} i_{s\beta} - \phi_{s\beta} i_{s\alpha}\right) \tag{9}$$

It can be noted in this equation that the accuracy of the equation depends on the flow estimation quality and the accuracy of the stator current measurement.

3.5 Torque and Flux Controller

The instantaneous values of flux and torque are calculated from stator variables by using flux and torque estimator.

The command stator flux and torque magnitude are compared with their respective estimated values and the errors are Processed by the hysteresis band controllers. The flux loop controller has two levels of digital output according to the Following equations [8]

$$\Delta\phi_s = \text{abs}\left(\phi_{ksiref} - \phi_{ksi}\right) \tag{10}$$

if $\phi_{ksi} \leq \phi_{ksiref} - \Delta\phi_s \quad F_{ki} = 1;$ *if* $\phi_{ksi} \geq \phi_{ksiref} + \Delta\phi_s \quad F_{ki} = 0;$ *if* $\phi_{ksi} = \phi_{ksiref}$ $F_{ki} = F_{ki-1};$

The torque controller loop has three levels of digital output according to the following equations

$$\Delta C_{em} = \text{abs}\left(C_{emref} - C_{kemi}\right) \tag{11}$$

if $C_{kemi} < C_{emref} - \Delta C_{em} \quad F_{ki} = 1;$ *if* $C_{kemi} > C_{emref} - \Delta C_{em} \quad F_{ki} = -1;$
if $C_{kemi} = C_{emref} \quad F_{ki} = F_{ki-1} = 0;$

4 Basic Structure of a Fuzzy Speed Controller of the Double Stator Induction Machine

In this work, we will illustrate the principles of fuzzy controllers, their design and their use in direct torque control of the DSIM. To the principle of fuzzy logic controllers is based on the techniques of artificial intelligence, The block diagram of fuzzy logic control is mainly depicted in Fig. 1.

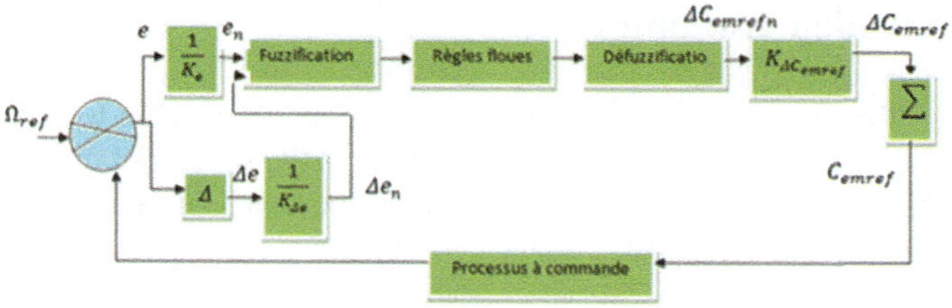

Fig. 1. Synoptic diagram of a fuzzy PI.

With e and Δe are the error and the variation of the error, respectively

$$e(k) = \Omega_{ref} - \Omega_r(k), \Delta e(k) = e(k) - e(k-1) \tag{12}$$

These variables can be normalized as follows

$$e_n = \frac{e}{K_e}, \Delta e_n = \frac{\Delta e}{K_{\Delta e}}, \Delta C_{emrefn} = \frac{\Delta C_{emref}}{K_{\Delta C_{emref}}} \tag{13}$$

Or:

K_e, $K_{\Delta e}$, $K_{\Delta C_{emref}}$: Are normalization gains that can be constant or variable.

The fuzzy rules make it possible to determine the variable of output of the regulator according to input variable using the matrix Mac-Vicar, which groups together in this case 49 rules [9].

This inference matrix is established from a perfect knowledge of the behavior of the system, as well as the knowledge of the objective of the control to be achieved. We take as defuzzification criterion the method of the center of gravity presented above, so the generated control action is given by the following expression

$$\Delta C_{emrefn} = x_{Gr} = \frac{\sum_{i=1}^{49} \mu_{ci} x_{Gi} S_i}{\sum_{i=1}^{49} \mu_{ci} S_i} \tag{14}$$

The action of the command that gives us the reference electromagnetic torque to impose on the machine can be expressed by

$$C_{emref} = C_{emref} + K_{\Delta C_{emref}} \Delta C_{emrefn} \tag{15}$$

5 The Extended Kalman Filter

The Kalman filter was developed by R. E. Kalman in 1960. Due to advances in the development of digital computing, the Kalman filter is the subject of much research and application. Kalman filtering has been applied in the fields of aerospace, navigation, manufacturing and many others.

5.1 Representation of the System

The kalman filter presented in the previous section assumes evolution and observation equations are linear, which is not the case of nonlinear systems (our DSIM case). To extend the estimation to nonlinear systems, we have used the extended Kalman filter, this filter is linearized simply the equations thanks to a development of Taylor.

Starting from the following nonlinear stochastic equations [10]

$$X_k = f(X_{k-1}, u_{k-1}, w_{k-1}) \tag{16}$$

With the following exit equation

$$Z_k = h(X_k, v_k) \tag{17}$$

The state function f and the observation function h are two differentiable functions. w_k et v_k still represent process and measurement noise.

5.2 Extended Kalman Filter Algorithm

In EKF, the system is linearized at each moment k. The Jacobian matrices A, W, H, V are defined as:

- A is the Jacobian matrix of partial derivatives of f with respect to X
- W is the Jacobian matrix of the partial derivatives of f with respect to w
- H is the Jacobian matrix of partial derivatives of h with respect to X
- V is the Jacobian matrix of partial derivatives of h with respect to V

The algorithm of the EKF is summarized by the equations given in the Fig. 2 [11].

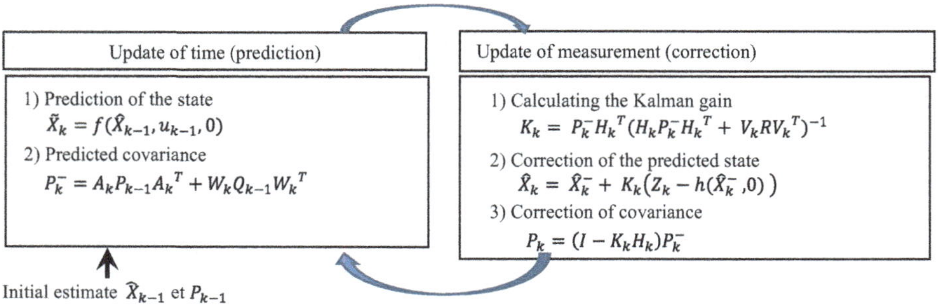

Fig. 2. Algorithm of the EKF

6 Simulation Results of Direct Torque Control

To show the effectiveness of the proposed DTC algorithm of the DSIM with EKF, various tests were performed at different dynamic operating conditions. The parameters of the test motor are given in Table 1. In the first part, we will present the simulation results of the speed control by a Fuzzy PI of the DTC controlled DSIM.

A-First test: Variation of speed (Fig. 3)

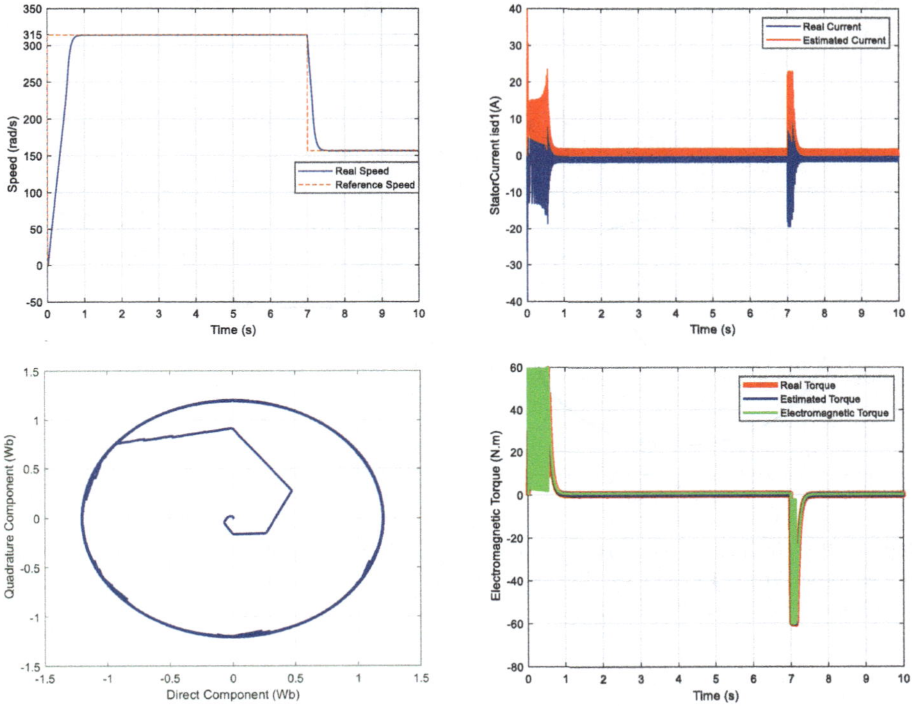

Fig. 3. Performance of the DTFC control of the DISM followed by a speed variation at t = 7 s from 315 rad/s to 157 rad/s.

B-Second Test: Variation of Charge (Fig. 4)

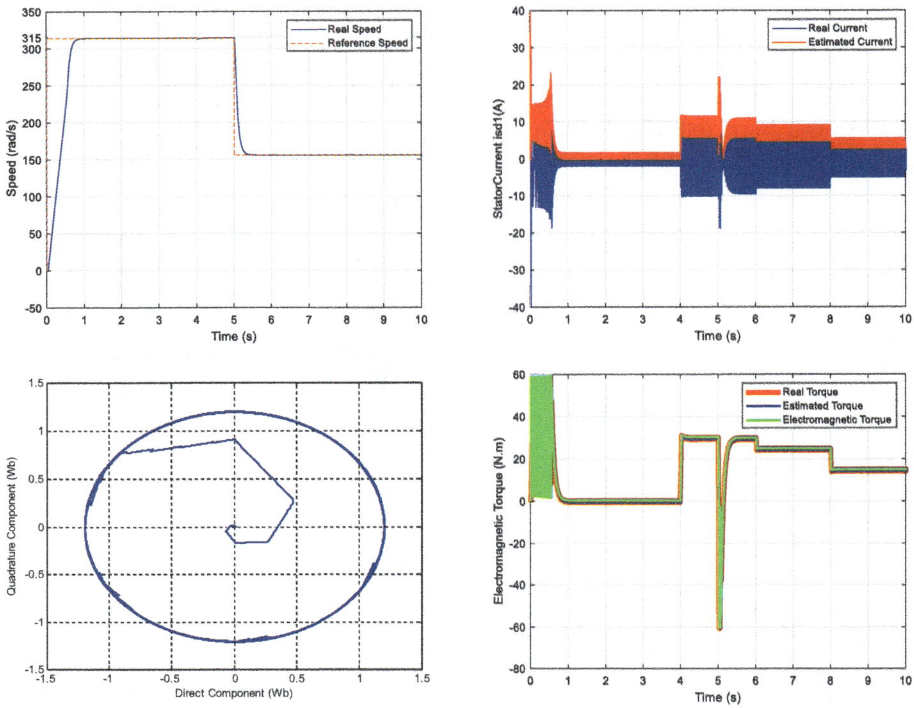

Fig. 4. Performance of the DTFC control of the DSIM followed by a variation of the load (Cr = 30 Nm at t = 4 s) at (Cr = 25 Nm at t = 6 s), (Cr = 15 Nm at t = 8 s) and variation of the speed at t = 5 s 315 rad/s to 157 rad/s.

Table 1. DSIM parameters

P_n	1.5 MW
V_n	220/380 V
P	2
R_{s1}, R_{s2}	3.72 Ω
R_r	2.12 Ω
L_{s1}, L_{s2}	0.022 H
L_r	0.006 H
L_m	0.3672 H
J	0.0662 kg m^2
K_f	0.001 N.m.s/rad

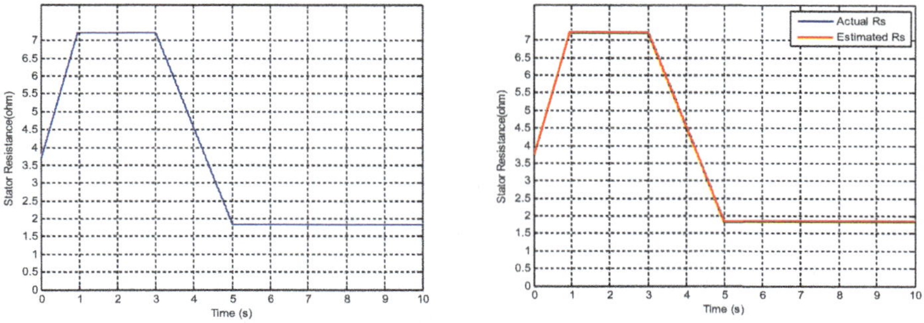

Type-2- exponential variation type, the deviation of the resistance as a function of the exponential shape temperature

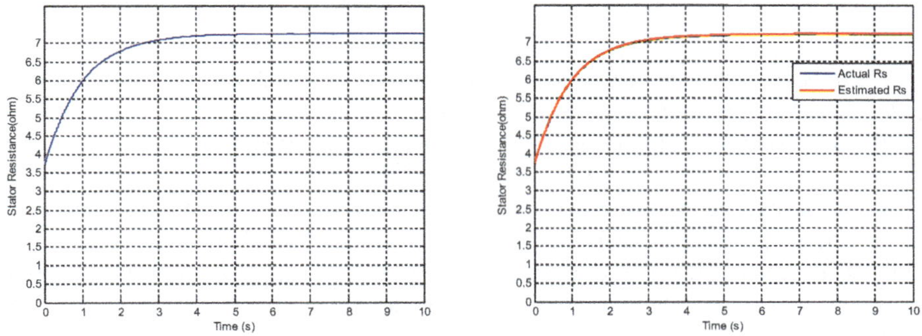

Estimation of the speed by EKF

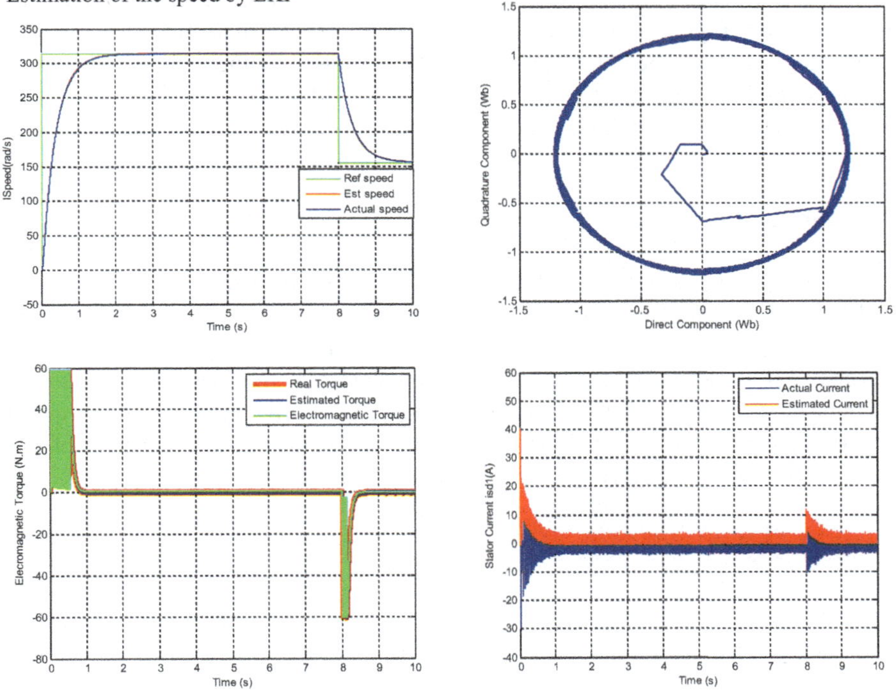

Fig. 5. DTFC sensorless speed performance using extended kalmen filter

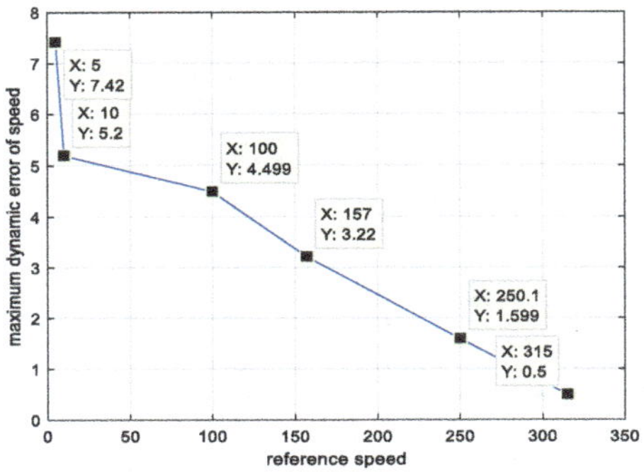

Fig. 6. Speed maximum dynamic error at rated load operation Versus reference speed.

It can be seen that the magnitudes of the torque, the speed, the flux and the current have a better accuracy and less insensitivity to the deviation of the speed and the load.

In the second part, simulation results, estimation of stator resistance and speed by EKF.

In this work, we proposed two types of variation of stator resistance, Type-1- the stator resistance deviation, which is supposed to increase linearly by 100% between 0 s and 1 s, remains twice its nominal value between 1 s and 3 s, then the resistance decreases linearly by 50% between 3 s and 5 s to regain a value of 50% of its value from 5 s.

The performance of the presented scheme in terms of the speed estimation accuracy are also tested. Figure 6 show the maximum of the speed dynamic error versus the speed reference. It can be noted that the speed estimation is sufficiently accurate, in fact the dynamic error is 7.42% at a very low speed of (5 rad/s) and doesn't exceed 0. 5% at the real speed (315 rad/s) (Fig. 5).

7 Conclusion

In this study novel structure of DTC algorithm based on simplified extended filter KALMAN for dual star asynchronous motor drive was well explained. The performance of the suggested scheme has been simulated under various variations of stator resistance. It is determined from the simulation results that the EKF estimator of mechanical speed and stator resistance has restored the drive system stability and has enhanced the robustness of the DTC drive of the DSIM against large changes of speed consign and stator resistance during operation of the DSIM.

References

1. Zaidi, E., Marouani, K., Bouadi, H., Kassel, A.E., Bentouhami, L., Merabet, E.: Fuzzy sliding mode method for speed regulation of a dual star induction machine drive fed by multi-level inverters. In: 2018 International Conference on Applied Smart Systems (ICASS), pp. 1–7 (2018)
2. Hadiouche, D., Razik, H., Rezzoug, A.: On the modeling and design of dual-stator windings to minimize circulating harmonic currents for VSI fed AC machines. IEEE Trans. Ind. Appl. **40**, 506–515 (2004)
3. Yahia, M., Katia, K., Khalilsp, M., Saddam, B.: Direct torque control to improve the performances of the DSIM powered by indirect matrix converter. In: 2018 International Conference on Communications and Electrical Engineering (ICCEE), pp. 1–5 (2018)
4. Boukhalfa, G., Belkacem, S., Chikhi, A., Benaggoune, S.: Direct torque control of dual star induction motor using a fuzzy-PSO hybrid approach. Appl. Comput. Inform. (2018)
5. Zaimeddine, R., Berkouk, E.: Direct torque control of double-star induction motors. In: Proceedings of the 5th WSEAS International Conference on Software Engineering, Parallel and Distributed Systems, World Scientific and Engineering Academy and Society (WSEAS) (2006)
6. Merabet, E., Amimeur, H., Hamoudi, F., Abdessemed, R.: Self–tuning fuzzy logic controller for a dual star induction machine. J. Electr. Eng. Technol. **6**, 133–138 (2011)
7. Casadei, D., Serra, G., Tani, A.: Implementation of a direct torque control algorithm for induction motors based on discrete space vector modulation. IEEE Trans. Power Electron. **15**(4) (2000)
8. Bose, B.K.: Modern Power Electronics and AC Drives. Pearson Education (2001)
9. Zadeh, L.A.: The role of fuzzy logic in the management of uncertainty in expert systems. Fuzzy Sets Syst. **11**(1–3), 197–198 (1983)
10. Aydogmus, O., Talu, M.F.: Comparison of extended-kalman- and particle-filter-based sensorless speed control. IEEE Trans. Instrum. Measur. **61**(02), 402–410 (2012)
11. Lin, G., Wan, Q.: Estimation of rotor resistance of induction motor based on extended Kalman filter. In: Jin, D., Lin, S. (eds.) Advances in CSIE, AISC, vol. 169, pp. 193–198. Springer, Heidelberg (2012)

Impact of SiC/Si Technology on the Conducted EMI Generated by a Buck Converter

K. Saci(✉), S. Khelladi, A. Hadjadj, and A. Bensaci

Laboratory of Analysis, Control of Energy Systems and Electrical Networks (LACoSERE), University Amar Telidji, Laghouat, Algeria
k.saci@lagh-univ.dz

Abstract. This paper presents a comparative study of conducted common mode (CM) and differential mode (DM) EMIs generated by a DC-DC BUCK converter based on silicon (Si) or Silicon Carbide (SiC) Diodes and/or MOSFETs. Using Si technology as a reference, the prediction of EMI waveforms is carried out with simulation in both time and frequency domains using LTspice software. The paper shows the great influence of the use of SiC Diodes on the conducted EMIs in the frequency range [10 MHz–100 MHz]. The obtained results show that the switching cell equipped with the SiC MOSFET has a faster switching and a faster current response (High di/dt and dv/dt) generating an increase in CM and DM disturbance. Using SiC technology for both the MOSFET and Diode shows a reduction of conducted EMI for frequencies greater than 20 MHz. The association of Si MOSFET and Sic Diode reduces conducted EMIs generated by the converter.

Keywords: Buck converter · Differential mode (DM) · Common mode (CM) · Conducted EMI · Silicon (Si) · Silicon carbide (SiC)

1 Introduction

The technological progress of semiconductors that are used in components such as Silicon Carbide (SiC) or Gallium Nitride GaN. SiC and GaN power devices are introduced to increase both the switching frequency and the efficiency of power converters. Increase power converters is crucial in the sense of efficiency, EMI performance and the size of the converter. All these progress involve the development and use of large-gap semiconductor in the switching frequency and efficiency results in reducing the size of the converter. These benefits are presented in many papers [1–3]. Beside the above cited benefits of using SiC devices in power converters, there is also the increase of the conducted and radiated electromagnetic interference (EMI). This increase is due to the high switching speed (high di/dt and dv/dt) compared to other conventional components in power electronic converters [4]. In [5], the conducted and radiated EMI of Si or SiC components for a boost converter are presented. These results show that the use of SiC diode has a great influence on the radiated EMI. The comparison results of the conducted EMI performance of Si and SiC MOSFETs in a CCM PFC boost converter in [6], show that there is not a significant difference. In [4, 7], The conducted EMI using SiC JFET and Si MOSFET are compared. The results

M. Hatti (Ed.): ICAIRES 2019, LNNS 102, pp. 610–616, 2020.
https://doi.org/10.1007/978-3-030-37207-1_65

study shows that the conducted perturbations measured at the LISN are larger with the SiC JFET than with the Si MOSFET.

The aim of this paper is to discuss through simulation the EMI performances of a buck converter when using silicon (Si) or silicon carbide (SiC) Diodes and/or MOSFET. The comparison results of the differential mode (DM) and common mode (CM) emission levels for the Si and SiC components are presented in the simulation results section.

2 Buck Converter Circuit and Description of Procedures

The simulation model of the DC-DC buck converter used for comparison purposes is shown in Fig. 1. The buck converter model was proposed in [8]. The parameters of this converter are listed in Table 1. The active components are a Silicon Carbide Schottky Diode (CPW4-1200S020), Silicon Diode (RFUH30TS6D), a Silicon Carbide Power MOSFET (C3M0065100 K), and a Si MOSFET (IRFP450). Advances in computer technology enabled Spice software such as LTspice to achieve faithful and realistic simulations of waveforms within switching power supplies (SMPS). Accordingly, the simulation of the DC/DC converter circuit fed through a line-impedance stabilization network (LISN) is carried out in LTspice software, which contains a library proposing relatively accurate switches models based on semiconductor physical equations. these models are dedicated to the simulation of genuine MOSFETs and diodes. Thus, the prediction of conducted disturbance levels generated by the DC-DC converter is performed with precision to obtain the CM and DM noise levels, in both time and frequency domain. The simulation was performed with four different MOSFET-Diode combinations. First we used Si for both MOSFET and Diode, the second case uses a Si MOSFET and a SiC Diode, a third case using a SiC MOSFET and Si Diode and finally, we used the SiC MOSFET-Diode combination.

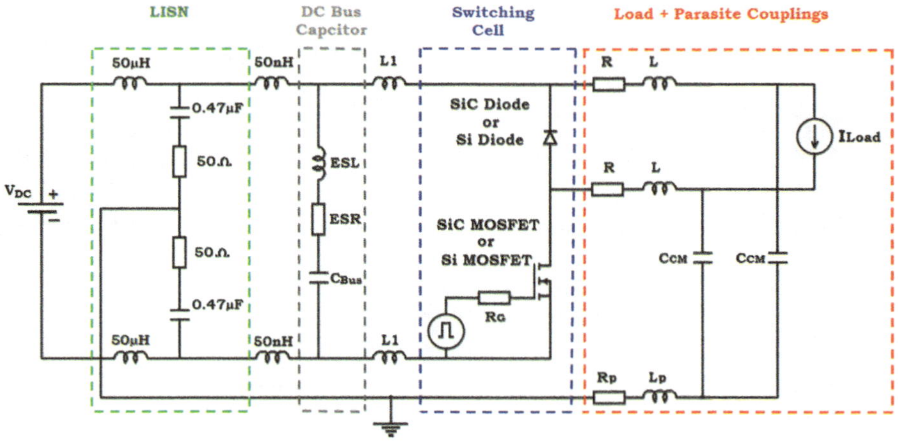

Fig. 1. Considered model of the converter connected to the LISN.

Table 1. Simulation parameters

V_{DC}	300V	ESL	50nH
I_{Load}	10A	L_1	50nH
f switching	20 KHz	L	20nH
Duty Cycle	50%	R	20mΩ
Rg	50Ω	Ccm	1nF
C_{Bus}	100µF	Lp	10nH
ESR	200mΩ	Rp	20mΩ

3 Simulation Results

3.1 Switching Characteristics

In order to compare the EMI of the converter, Drain-source voltage (Vds) and drain current (Id) were simulated to compare the switching characteristics of the devices. The same conditions were applied for all cases simulated. Figures 2, 3, 4 and 5 shows the turn-on and turn-off characteristics of the MOSFETs.

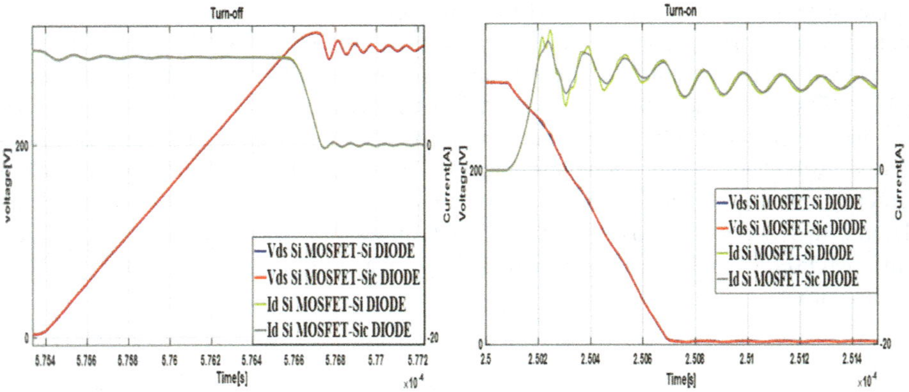

Fig. 2. Turn-on and turn-off MOSFET switching.

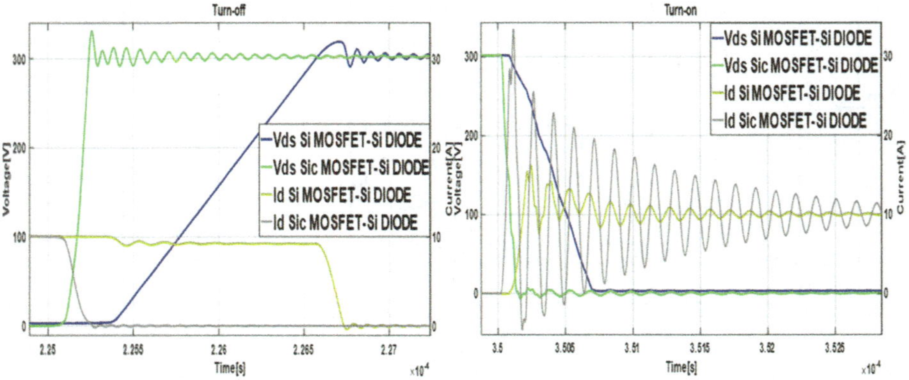

Fig. 3. Turn-on and turn-off MOSFET switching.

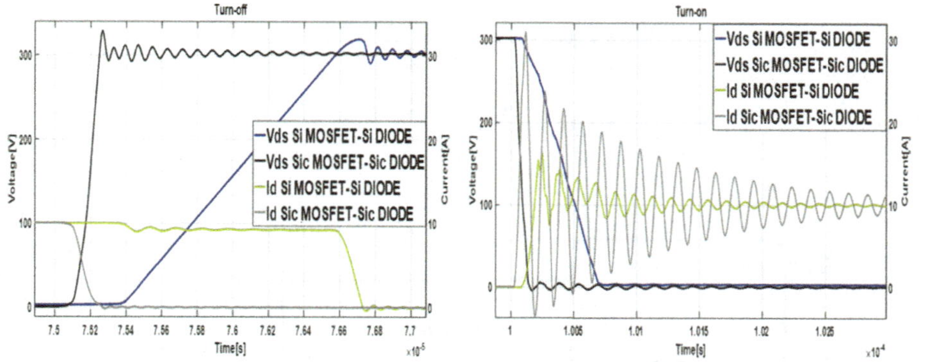

Fig. 4. Turn-on and turn-off MOSFET switching.

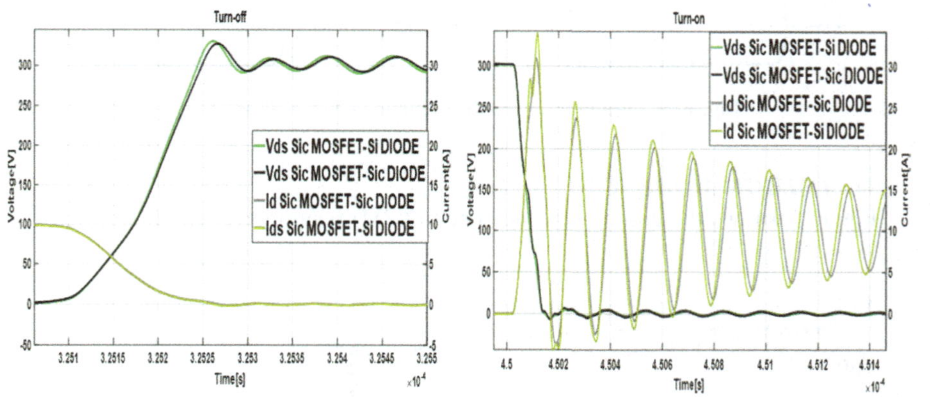

Fig. 5. Turn-on and turn-off MOSFET switching.

The figures shows that there is not a significant difference either for the reverse recovery times or of the turn-off times when using Si or SiC Diodes. The switching time is lower for the SiC MOSFET than for the Si MOSFET. Also, we notice that the current Id response for the SiC MOSFET is faster than that of Si MOSFET. Nevertheless, the use of the SiC Diode causes a reduction in oscillations amplitudes during the turn-on switching. A lowering in oscillations during the turn-off time is remarkable when SiC technology is used for both, MOSFET and Diode.

3.2 Conducted EMI

A comparison of the CM and DM voltages (VCM, VDM) spectra for the four cases described above has been performed. VCM and VDM are calculated with Eqs. (1) and (2) using the total noise voltages seen at the LISN output ports, which comprise the voltages of the line to ground and the neutral to the ground as shown in Fig. 6. Then the VCM and VDM spectrums are obtained using LTspice Fast Fourier Transform (FFT).

$$VCM = \frac{(V_{LISN_{L-G}} + V_{LISN_{N-G}})}{2} \quad (1)$$

$$VDM = \frac{(V_{LISN_{L-G}} - V_{LISN_{N-G}})}{2} \quad (2)$$

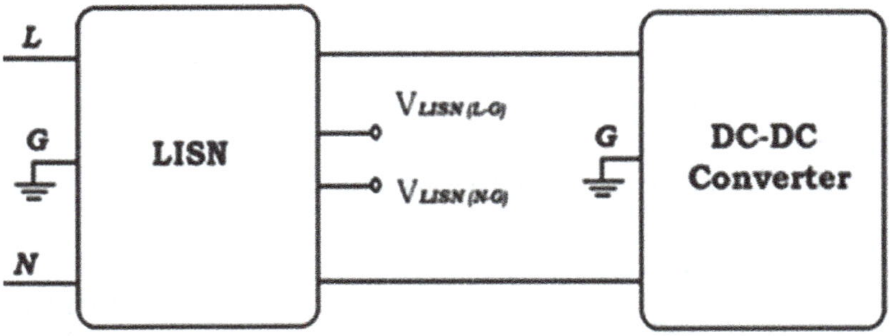

Fig. 6. Configuration for the conducted EMIs Simulation.

1) ***COMMON MODE (CM)***

Figure 7 shows the common mode spectra (CM) resulting for the four cases.

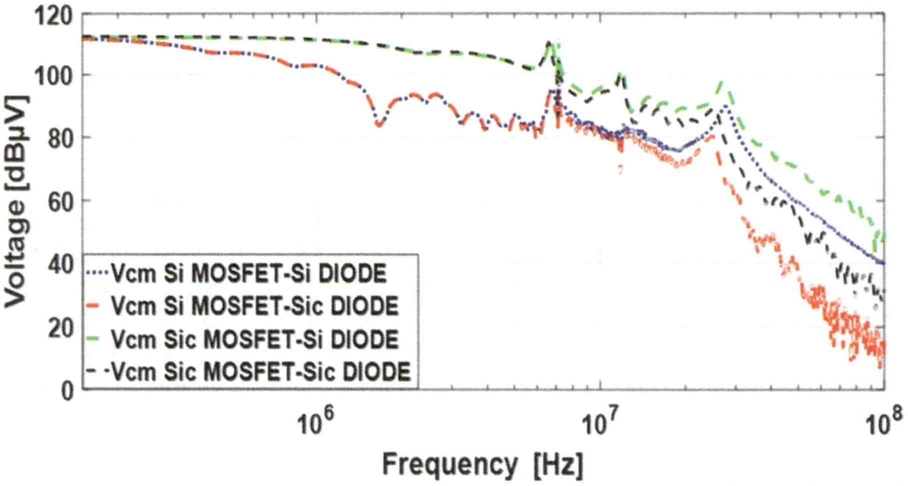

Fig. 7. Comparison of the CM voltages in frequency domain

2) ***DIFFERENTIAL MODE (DM)***

Figure 8 shows the differential mode spectra (DM) resulting for the four cases.

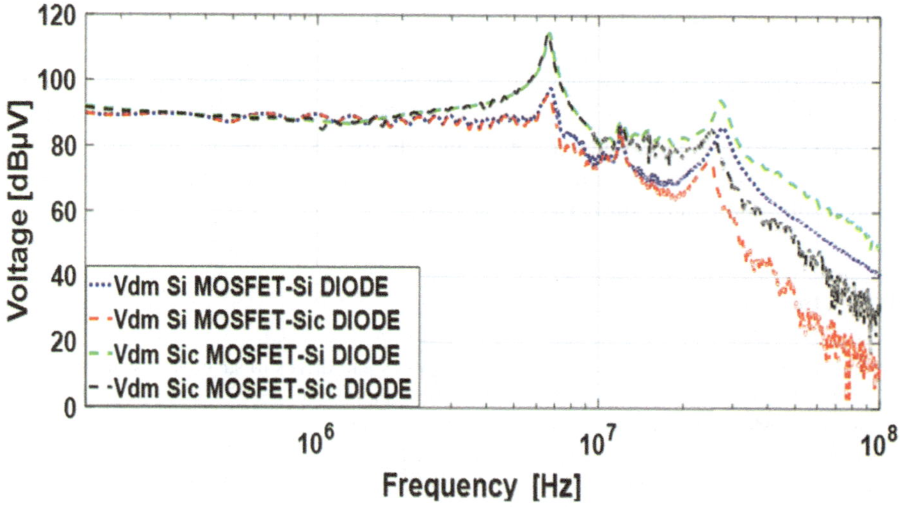

Fig. 8. Comparison of the DM voltages in frequency domain

- From the figures, it can be seen that when using a Si or SiC Diode, there is not a significant difference of noises levels (CM/DM) in the range of 150 kHz to 10 MHz.
- It is seen that the amplitude of the spectrum of the SiC MOSFET is higher than that for the Si MOSFET in the frequency range 400 kHz to 100 MHz for common mode (CM) disturbances and from the frequency of 3 MHz for differential mode (DM).
- In cases where SiC Diodes are used, there is a remarkable difference in the amplitude of the spectrum for frequencies greater than 10 MHz, the SiC Diodes cause a reduction of conducted EMI in the given range of frequency comparing to the same converter using Si Diodes.
- In the range greater than 20 MHz, the CM & DM conducted EMI levels when using SiC technology for both MOSFET and Diode are lower than the reference values when using Si technology.

4 Conclusions

The results show that some oscillations appeared when using both Si and SiC Diodes but they are smaller with the SiC Diode due to its much lower reverse recovery charge and time. This is also reflected on the CM and DM spectrums which are lower than the reference value when using Si diode in the range from 10 MHz to 100 MHz. On the other hand, the comparisons between the performances of the SiC and Si components in the converter show that the switching times of the SiC MOSFETs are shorter and with a faster current response than those of the Si MOSFET. These factors generate a raise of common mode disturbances in the frequency range 400 kHz to 100 MHz and an increase in differential mode disturbances from the frequency of 3 MHz.

The reverse recovery charge and time near zero in SiC Diodes with the switching speed (dv/dt) and current variations (di/dt) of the MOSFET are key factors that decide on the conducted EMI performance in a switching cell. This been said, more detailed and expanded investigations on this subject are to be accomplished to understand the EMI behavior of dc-dc converters regarding the used switches technology. In future works, the objective is to realize the proper experimental test setup to validate simulation results and extend furthermore the study.

References

1. Fujihira, T., et al.: Impact of SiC on power supplies and drives to save energy and materials. In: PCIM Asia 2017; International Exhibition and Conference for Power Electronics, Intelligent Motion, Renewable Energy and Energy Management, pp. 1–8 (2017)
2. Kostov, K., Rabkowski, J., Nee, H.: Conducted EMI from SiC BJT boost converter and its dependence on the output voltage, current, and heatsink connection. In: 2013 IEEE ECCE Asia Downunder, pp. 1125–1130 (2013)
3. Stevanovic, L.D., Matocha, K.S., Losee, P.A., Glaser, J.S., Nasadoski, J.J., Arthur, S.D.: Recent advances in silicon carbide MOSFET power devices. In: 2010 Twenty-Fifth Annual IEEE Applied Power Electronics Conference and Exposition (APEC), pp. 401–407 (2010)
4. Rondon, E., Morel, F., Vollaire, C., Schanen, J.: Impact of SiC components on the EMC behaviour of a power electronics converter. In: 2012 IEEE Energy Conversion Congress and Exposition (ECCE), pp. 4411–4417 (2012)
5. Bogónez-Franco, P., Sendra, J.B.: EMI comparison between Si and SiC technology in a boost converter. In: International Symposium on Electromagnetic Compatibility - EMC EUROPE, pp. 1–4 (2012)
6. Kavak, H., Iskender, I., Jahi, A.: Conducted EMI performance comparison of Si and SiC MOSFETS in a CCM boost PFC converter for MIL-STD-461F CE102. Poznan Univ. Technol. Acad. J. Electr. Eng. **95**, 87–94 (2018)
7. Rondon-Pinilla, E., Morel, F., Vollaire, C., Schanen, J.: Modeling of a buck converter with a SiC JFET to predict EMC conducted emissions. IEEE Trans. Power Electron. **29**(5), 2246–2260 (2014)
8. Marlier, C.: Modélisation des perturbations électromagnétiques dans les convertisseurs statiques pour des applications aéronautiques, thesis, Lille 1 (2013)

Detection of Defects Using GMR and Inductive Probes

Touil Dalal Radia[1(✉)], Daas Ahmed[2], Helifa Bachir[1], and Lefkaier Ibn Khaldoun[1]

[1] Materials Physics, LPM Laboratory, University de Amar Telidji, Laghouat, Algeria
d.touil@lagh-univ.dz, helifa@yahoo.fr, lefkaierk@gmail.com

[2] Laboratory of Process Engineering, University Laghouat, Bp37G03000, Laghouat, Algeria
daaskossay@gmail.com

Abstract. Eddy current testing (ECT) is one of the most extensively used non-destructive techniques for inspecting electrically conductive materials at very high speeds that does not require any contact between the test piece and the sensor. However, the characterization of a crack is not easy to obtain, and for this purpose, other eddy current evaluation methods are still under investigation.

This work introduce a new eddy current testing technique for surface on near surface defect detection in nonmagnetic metals using giant magnetoresistive (GMR) sensors. It is shown that GMR-based eddy-current probes are more sensitive than the Inductive probes to determinate the cracks dimensions that were machined on aluminum plates.

Keywords: Giantmagneto-resistorsensor · Eddy current testing (ECT) · Non-destructive testing · Crack detection

1 Introduction

ECT is widely used to inspect the presence of cracks in metallic structures [1]. Those materials are important to monitor how some previously known cracks and characterize theme, also localize the very small defects less than 100 μm and inside maximum in those materials. The delicate measurement of crack's location is quite difficult and has been the field of study of several researchers [2]. To develop the performance of the inspection of metallic structures using ECT [4], the probe with best properties [3, 4] excitation signals methods [5] and signal processing techniques [6] are still under investigation. ECT is generally used with high-frequency magnetic fields, above kilohertz order, or search coils [7]. High-frequency magnetic fields are suitable for identifying surface defects. In order to apply ECT to examining defects in depths, it is necessary to use a low-frequency magnetic field. However, it is difficult to sense a weak signal due to a defect by using a low-frequency magnetic field because the sensitivity

M. Hatti (Ed.): ICAIRES 2019, LNNS 102, pp. 617–622, 2020.
https://doi.org/10.1007/978-3-030-37207-1_66

of the search coil is not very high [4]. For this reason, other sensitive sensors for weak magnetic fields are desired. Recently, in our system we testing defects in samples by using a low-frequency magnetic field, an MR and inductive sensor, and a lock-in amplifier [8, 10]. This system has sensitivity of sub-nanotesla order in a non-shielded environment. In this study, we measured the length of cracks machined in plate of aluminum; we use two types of sensors inductive and GMR probes in the same condition of experience (frequency 20 kHz) and we use the (same coil) for the excitation with the GMR.

2 The Experimental Setups

The experimental setup is depicted in Fig. 1, including a pancake coil with Ferrite pot (external radius: 9 mm, internal radius: 4.7 mm, number of turns: 175 and height: 2.2 mm) manufactured by SCIENSORIA, an MR sensor (Giant magneto-resistor).type AAH004 00 manufactured by NVE is inserted along the coil axis, with the sensing direction perpendicular to the plate. With an active area about 100 by 200 mm in the middle of the layout [10]. Their characteristics of coil and MR are presented in Tables 1 and 2.

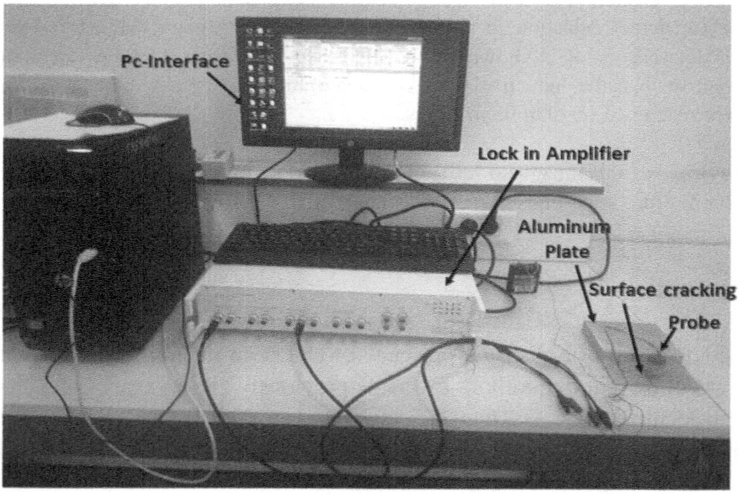

Fig. 1. Measuring system.

Table 1. Geometric parameters of the coils and their pot

	Dimension coil
Inside radius	4.7 mm
Outside radius	9 mm
Length of coil	2.2 mm
Number of turns	175
Number of layers	14
Diameter of wire	0.14 mm
	Ferrite pot size
Material	T6
Core diameter	4.6 mm
Internal ring diameter	9.1 mm
Internal height	2.5 mm
Permeability	$\mu_r = 4000 \pm 25\%$

Table 2. Geometric parameters of the GMR sensor

	Dimension GMR NVE AAH004 00
Saturation	1.5 mT = 15 Oe
Sensitivity	32–48 V/T/V
Linear range	0.15 and 0.75 mT
Resistance	2 K ± 20% (ohms)
Power supply	9 V
Package	MSOP8

A lock-in amplifier is also source of alternating current (AC source) [9], arranged as shown in Fig. 1. The applying coil are for applying an alternating magnetic field to the sample. The coil is connected to the lock-in amplifier. The lock-in amplifier is connected with the MR sensor. The signal detected from the sample, which includes amplitude, is referred to that of the alternating current source. In this experiment, the general procedure consists in scanning the area including cracks to determinate the Length. Firstly, we scanning with GMR probe depicted in Fig. 2. The same measurement was done with a Inductive probe, the scanning with the same excitation coil, which depicted in Fig. 3. In this way. Depicting the variations of the tension recorded as measured on the scanned area. This variations useful information about the crack's characterization.

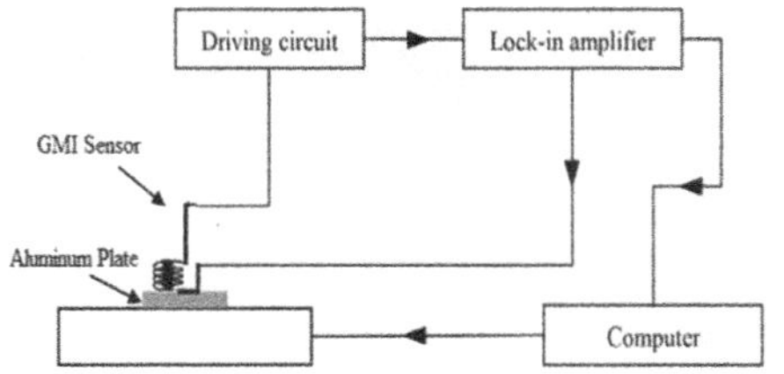

Fig. 2. Schematic of the experimental setup for the ECT system with Inductive Probe.

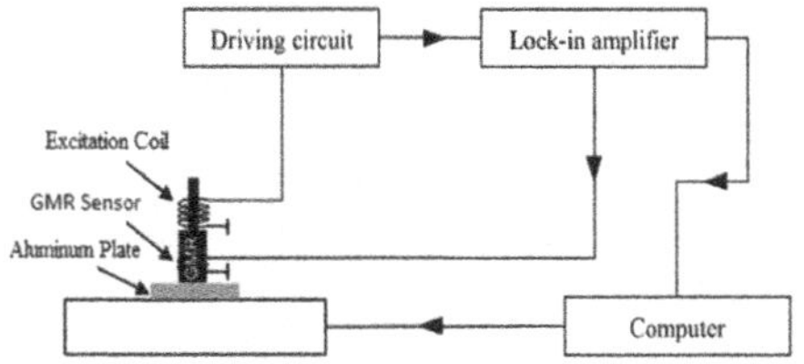

Fig. 3. Schematic of the experimental setup for the ECT system With GMR Probe.

Figure 4 depicts the internal configuration of the giant Magneto-resistor sensor AAH004-00 produced by non-volatile electronics. Four giant magneto-resistors are connected in a bridge configuration, with two of them magnetically shielded. Figure 5 depicts Schematization of the measurement system explained the Block diagram of the experimental setup.

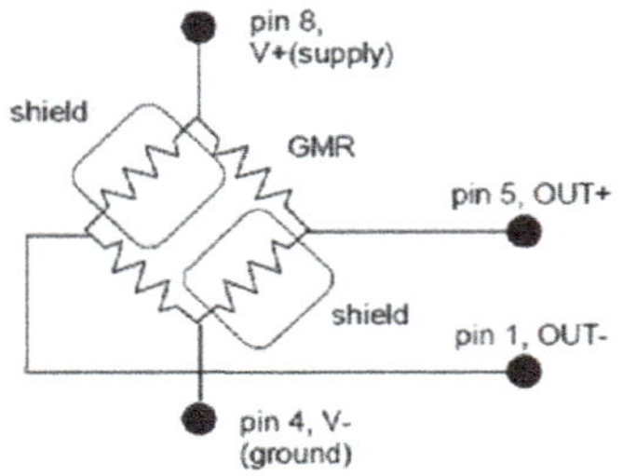

Fig. 4. Giant magneto-resistor bridge sensor.

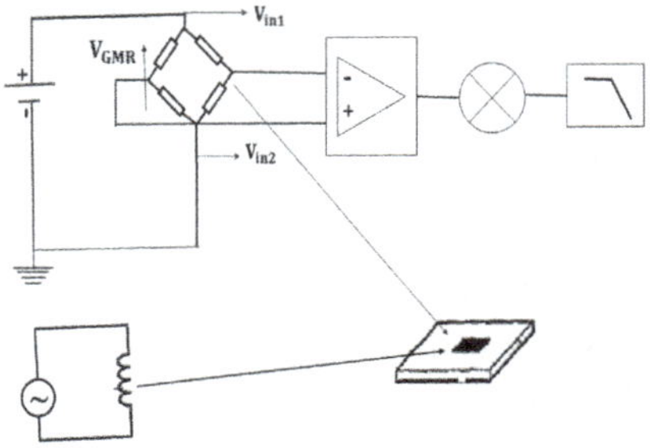

Fig. 5. Schematization of the measurement system.

3 Results and Discussion

ECT exploits the phenomenon of electromagnetic induction. A time varying current passing through an excitation coil will produce electromagnetic field. If an electrically conductive material is in the proximity of this electromagnetic field, an eddy current will be induced in the material. If a flaw exists in the testing sample, the amplitude and the distribution of the eddy current will be changed. Figure 6 depicts the scan tests that were performed to evaluate the conditions of the material. Three surface cracks were machined in the aluminum plate crack 1, 2 in plate 1 (thickness = 4 mm) crack 3 in plate 2 (thikness = 20 mm), and their dimensions are presented in Table 3. The scanning tests were performed along of the cracks in a direction of the sensitivity of GMR sensor with depicts in Fig. 7 because it is the region where the maximum perturbation of the x component of the magnetic field (Bx) occurs.

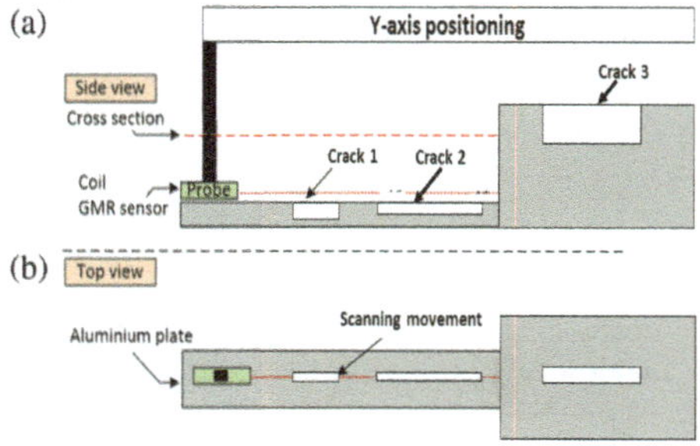

Fig. 6. Representation of the performed scan: (a) Side view; (b) Top view.

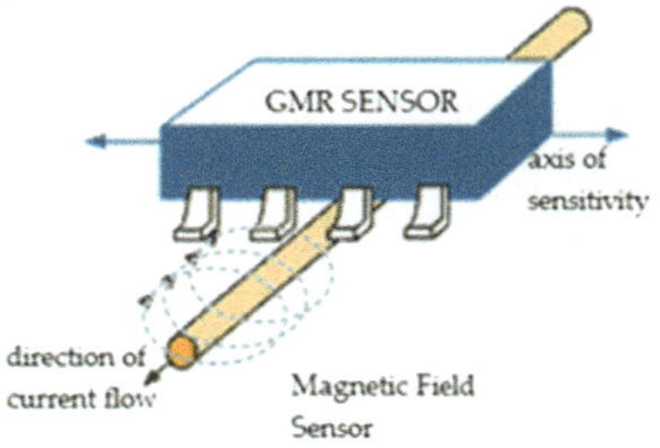

Fig. 7. Representation of the sensitivity direction of GMR Sensor

Table 3. Dimension of the cracks in plate

	Width	Length	Depth
Crack 1	0.5 mm	10 mm	2.75 mm
Crack 2	0.5 mm	15 mm	1.5 mm
Crack 3	1 mm	14 mm	5 mm

A large increase in the output voltage of the GMR and inductive sensors circuit was observed when the sensor was moved on the top of the crack, after the further movement of the sensor the output voltage came back to the nearly previous value. Figures 8, 9, and 10 show variation of the sensors output as a function of the position of the sensor. The largest change in the sensor output was observed with GMR sensor then the inductive sensor. If the materials do not have any crack, no significant variation in the sensor output was measured. Because all magnetic flux lines pass through the material due to its higher permeability.

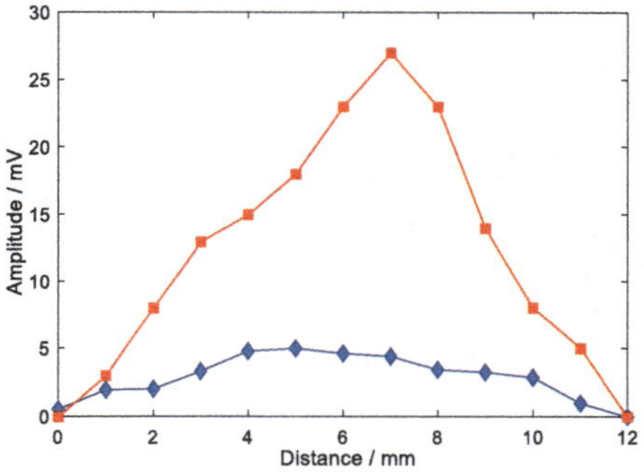

Fig. 8. GMR Inductive output voltage (crack 1).

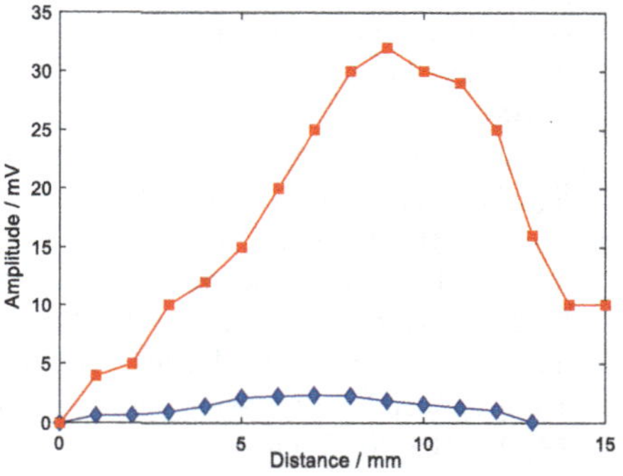

Fig. 9. GMR Inductive output voltage (crack 2).

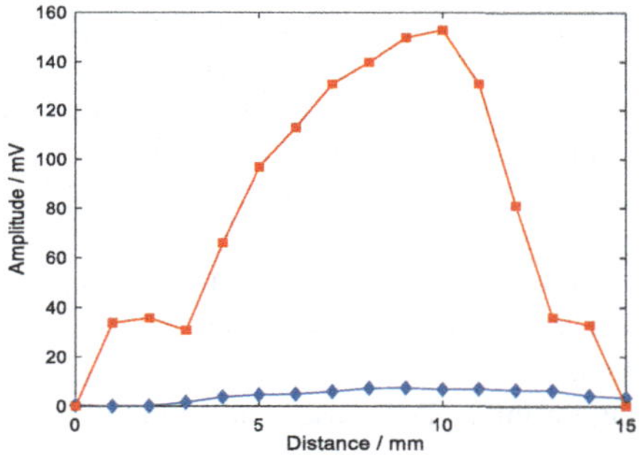

Fig. 10. GMR Inductive output voltage (crack 3).

4 Conclusion

Experimental tests were performed to evaluate surface cracks in aluminum plates, using mono-frequency excitation signal to compare both Inductive and GMR sensors were used for the detection. The analyses of the results were presented in the tension. For a 4 mm of aluminum plate thickness allowed the detection of surface and cracks with different depth (1.5 mm, 2.7 mm) and the second aluminum plate with 20 mm including crack of (5 mm). The results obtained confirmed that the GMR sensor-based eddy-current probes are more sensitive than the Inductive probes to determinate the cracks dimensions. Therefore, it is easy to obtain information about characterization of cracks (Depth, Length, Width, position…). So the GMR sensor give good results with the ferrite coil and we observed them in other works before. As future work, we are going to extend this technique to characterize very small defects less than 100 μm inside in non-ferromagnetic metallic structures.

References

1. Pasadas, D.J., Ribeiro, A.L., Ramos, H.G., Feng, B., Baskaran, P.: Eddy current testing of cracks using multi-frequency and noise excitation. In: 2018 IEEE International Instrumentation and Measurement Technology Conference (I2MTC), Houston, TX, USA, pp. 1–6 (2018)
2. Pattabiraman, M., Nagendran, R., Janawadkar, M.P.: Rapid flaw depth estimation from SQUID-based eddy current nondestructive evaluation. NDT E Int. **40**(4), 289–293 (2007)
3. Yamada, H., Hasegawa, T., Ishihara, Y., Kiwa, T., Tsukada, K.: Difference in the detection limits of flaws in the depths of multi-layered and continuous aluminum plates using low-frequency eddy current testing. NDT E Int. **41**(2), 108–111 (2008)
4. Yusa, N., Sakai, Y., Hashizume, H.: An eddy current probe suitable to gain information about the depth of near-side flaws much deeper than the depth of penetration. NDT E Int. **44**(1), 121–130 (2011)
5. Simm, A., Theodoulidis, T., Poulakis, N., Tian, G.Y.: Investigation of the magnetic field response from eddy current inspection of defects. Int. J. Adv. Manuf. Technol. **54**(1–4), 223–230 (2011)
6. Tsukada, K., Kiwa, T.: Magnetic property mapping system for analyzing three-dimensional magnetic components. Rev. Sci. Instrum. **77**(6), 063703 (2006)
7. Pasadas, D.J. Ribeiro, A.L., Rocha, T.J., Ramos, H.G.: Open crack depth evaluation using eddy current methods and GMR detection. In: 2014 IEEE Metrology for Aerospace (MetroAeroSpace), Benevento, Italy, pp. 117–121 (2014)
8. Zeng, Z., et al.: EC-GMR data analysis for inspection of multilayer airframe structures. IEEE Trans. Magn. **47**(12), 4745–4752 (2011)
9. Hea, D., Wanga, Z., Kusanoa, M.: Evaluation of 3D-Printed titanium alloy using eddy current testing with high-sensitivity magnetic sensor. NDT and E Int. **102**, 90–95 (2019)
10. Espina-Herna, J.H., Pacheco, E.R., Caleyo, F.: Rapid estimation of artificial near-side crack dimensions in aluminium using a GMR-based eddy current sensor. NDT E Int. **51**, 94–100 (2012)

Fault Ride-Through Improvement of an Offshore DFIG Wind Turbine

Kouider Khaled(✉) and Bekri Abdelkader

Department of Electrical Engineering, University of Tahri Mohammed, Bechar, Algeria
khouiledkhaled@gmail.com, bekriabdkader@yahoo.fr

Abstract. Fault ride through (FRT) is recognized as one of the most exciting topics in inspecting wind turbine performance. In this paper, the behavior of a DFIG offshore wind farm under fault circumstances is reviewed. Three case studies basing on the 3-phase fault location are taken into account. In this situation, the crowbar protection is primordial to prevent any over-currents in the rotor windings and to short the Rotor side converter (RSC) from the system. However, the use of this protection individually is insufficient and causes a lack of rotor speed and reactive power control. Here, the need for the internal model control (IMC) for the grid side converter (GSC) is essential to intensify the fault ride through capabilities of the DFIG. The simulation is performed using MATLAB/SIMULINK. As the results show, the use of the IMC for the grid side converter with the crowbar protection boosts both the Fault ride through and the performance of the DFIG.

Keywords: Grid side converter (GSC) · Rotor side converter (RSC) · Doubly fed induction generator (DFIG) · Internal model control (IMC) · Wind turbine · Fault ride through (FRT)

1 Introduction

Day after day, renewable energies do not stop growing and breaking new records in the global energy markets. Several factors push up the world to find new clean and harmless energy sources. Due to its significant potential in the energy markets, wind energy is considered one of the most relevant renewable sources [1]. The last statistics from [2] proves that wind energy leads all the other renewable energies with a more than 590 MW as the total capacity and a new installed capacity of 51.3 MW only in 2018 (46.3 MW onshore, 4.5 offshore). The DFIG was introduced to the wind power generation system due to many advantages, such as economic benefits, the partial scale conversion, lightweight mechanism structure, and many other factors. The grid is suddenly affected by many tips of disturbances, such as faults, voltage dips, and so on, and since the DFIG is directly connected to the grid via its stator. Thus, this machine will be susceptible to these abrupt circumstances. For the grid code requirements, the DFIG should support these fault conditions and remain connected to the grid, which is known as the fault ride through (FRT). Because of a large wind farm of 315 MW disconnection and leads to a non-fault ride through, Australia was suffered from a

M. Hatti (Ed.): ICAIRES 2019, LNNS 102, pp. 623–631, 2020.
https://doi.org/10.1007/978-3-030-37207-1_67

Blackout on 28 September 2016 [3]. The crowbar protection is a crucial part to preserve the Rotor side converter (RSC) and to ensure the fault ride through obligation as well [3].

Many kinds of research are carried out about the fault ride through (FRT), a crowbar protection system application to A DFIG wind system is well discussed in [4], Enhanced Fault Ride–Through Ability of DFIG using g Superconducting Fault Current Limiter [5], a nonlinear dynamic modeling for fault ride-through capability of DFIG-based wind farm is performed by [6]. The detailed configuration and control laws are well examined and established in various works such as [7–9]. In this paper, the crowbar protection will be employed for the RSC converter to ensure high protection against large rotor currents overshoots. The whole system used in this paper is well displayed in Fig. 1. To ensure high performance and robust response against these faults, the Internal Model Control (IMC) is used for the Grid side converter (GSC). Eventually, this paper will be divided into three main parts: 1-mathematical modeling of the wind system, 2-The overall control of the Back-to-Back converter, 3-Simulation results and discussion.

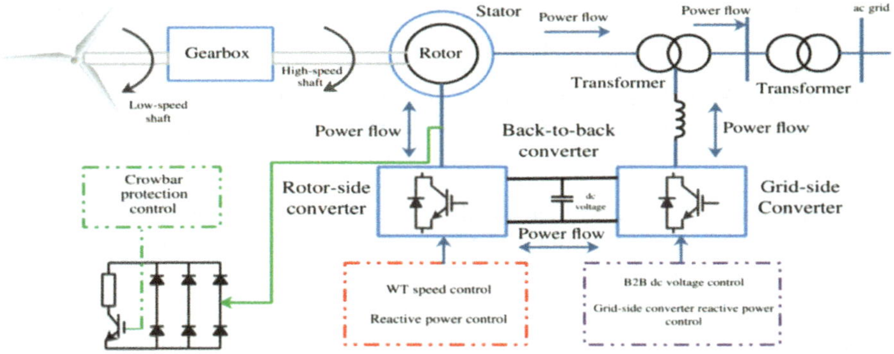

Fig. 1. DFIG wind turbine system with crowbar protection.

2 DFIG Wind System Modeling

In this section, the mathematical modeling of the overall DFIG wind system will be examined.

2.1 DFIG Mathematical Modeling

Basing on the *dq* synchronous reference frame, the analytical model of the machine can be re-written employing the stator and rotor windings equations as follows:

$$
\begin{aligned}
u_{sd} &= R_s i_{sd} + \frac{d\psi_{sd}}{dt} - \omega_s \psi_{sq} \\
u_{sq} &= R_s i_{sq} + \frac{d\psi_{sq}}{dt} + \omega_s \psi_{sd} \\
u_{rd} &= R_r i_{rd} + \frac{d\psi_{rd}}{dt} - \omega_r \psi_{rq} \\
u_{rq} &= R_r i_{rq} + \frac{d\psi_{rq}}{dt} + \omega_r \psi_{rd}
\end{aligned} \tag{1}
$$

After that, the Eq. (2) represents the stator and rotor flux with L_m, L_s, L_r are the mutual, stator and rotor inductances:

$$
\begin{aligned}
\psi_{sd} &= L_s i_{sd} + L_m i_{rd} \\
\psi_{sq} &= L_s i_{sq} + L_m i_{rq} \\
\psi_{rd} &= L_r i_{rd} + L_m i_{sd} \\
\psi_{rq} &= L_r i_{rq} + L_m i_{sq}
\end{aligned} \tag{2}
$$

Moreover, the electromagnetic torque:

$$
T_{em} \frac{3}{2} p(i_{rd} i_{sq} + i_{rq} i_{sd}) \tag{3}
$$

3 Back-to-Back (B2B) Modeling and Control

One of the most dominant interfaces in wind power application is the Back-to-Back (B2B) converter. The topology of this Voltage Source Converter (VSC) is well discussed and detailed in several references like [7, 8], [10, 11]. As shown in the Fig. 1, this converter consists of two Voltage Sources converters, one for the machine (Rotor Side converter) (RSC), and the other is the grid side converter (GSC).To ensure high performance and accurate control of the DFIG wind system, this converter should be commanded appropriately. In the next subsections, both vector control strategies for the two converters will be concisely reviewed.

3.1 RSC Control

The primary role of the rotor side converter is to adjust both rotor speed and stator reactive power. In a DFIG wind system, the stator active and reactive power can be commanded independently using suitable rotor currents injections. Basing on the *dq* synchronous reference frame, and corresponding to the comprehensive development of the classic vector control method for the RSC converter in [7], we can see that I_{dr} control the torque, while the reactive power is directly affected by I_{qr}.

$$T_e = -\frac{3}{2}p\left(\frac{-L_m\psi_{qs}}{L_m + L_{ls}}\right)i'_{dr} \tag{4}$$

$$Q_s = \frac{3}{2}\left[v_{ds}\left(\frac{\psi_{qs} - L_m i'_{dr}}{L_{ls} + L_m}\right)\right] \tag{5}$$

3.2 GSC Control

The grid side converter GSC is the second part of the back-to-back converter which links the DFIG with the grid. Many types of research are examining the vector control strategies for the GSC of the DFIG like in [9], [12–14]. In this paper, we will use the Internal Model Control as an additional controller for the GSC controller to enhance the fault ride through and to ensure high grid voltage and currents control. In this paper, we will use the Internal Model Control as an additional controller for the GSC controller to enhance the fault ride through and to establish high grid voltage and currents control.

The 2-degree of freedom IMC controller applied in this paper is well described in [7], the final development of the system equation of this controller yields:

$$M_{gsc}(s) = \frac{1}{G_{gsc} + L_s + r} \tag{6}$$

$$v_{gsc}(s) = v'_{gsc}(s) - i_{gsc}(s)G_{gsc} \tag{7}$$

$$F_{gsc}(s) = \alpha_{gsc}L + \frac{(r + G_{gsc})\alpha_{gsc}}{s} \tag{8}$$

$$Kp_{gsc} = \alpha_{gsc}L \text{ and } Ki_{gsc} = (r + G_{gsc})\alpha_{gsc}$$

α_{gsc} is the bandwidth of the closed loop system of i_{gsc}.

Then, the two-degree-of-freedom GSC currents IMC control loop is well presented in Fig. 2.

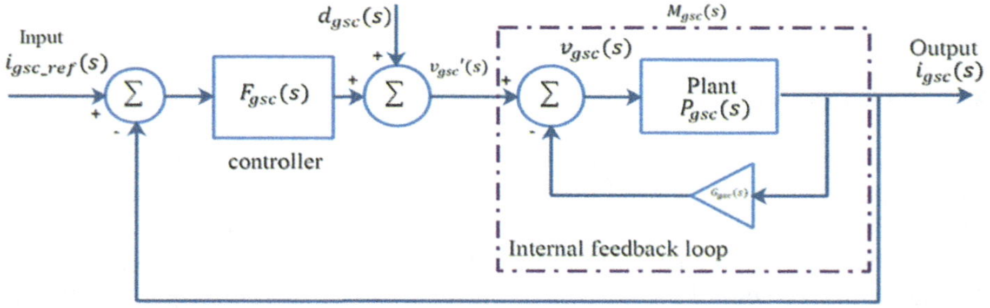

Fig. 2. Two degree-of-freedom for GSC currents IMC control loop.

Finally, the complete GSC converter control including the IMC controller is exposed in Fig. 3. The MATLAB/SIMULINK model of the GSC with the IMC is well described in the appendix in the associated Supplementary Material.

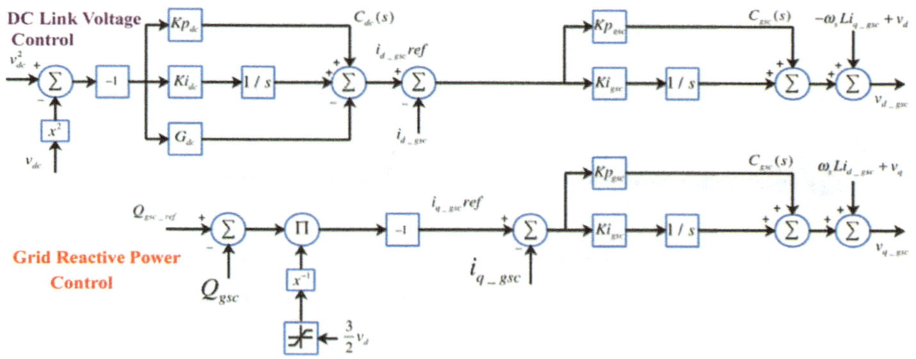

Fig. 3. Configuration diagram of GSC controllers.

4 Results and Discussion

The simulation was implemented using the MATLAB/SIMULINK Software Package. To investigate the behavior of an offshore DFIG wind farm under fault conditions, we take three case studies into account, as indicated in Fig. 4. All the data handled in the simulation are mentioned in the Appendix included in the (Supplementary Material).To study the performance of the applied Internal Model Control (IMC) control for the grid side converter (GSC), a three phases-fault is tested at 8 s and cleared after 85 ms in each case. Before that, the DFIG can generate 0.71 pu of active power with a wind speed of 11 m/s.

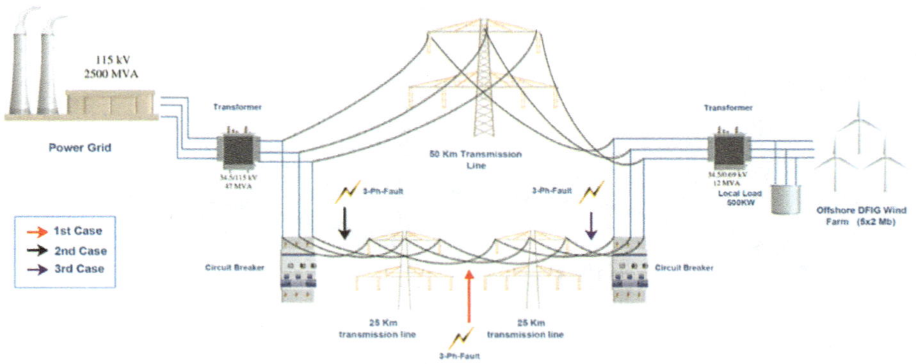

Fig. 4. The overall diagram of the simulation case studies.

The Figs. 5, 6, 7, 8 and 9 show all the most important electrical parameters of the DFIG such rotor voltages and currents, DC link voltage, stator active and reactive power, and finally the on/off crowbar trigger function.

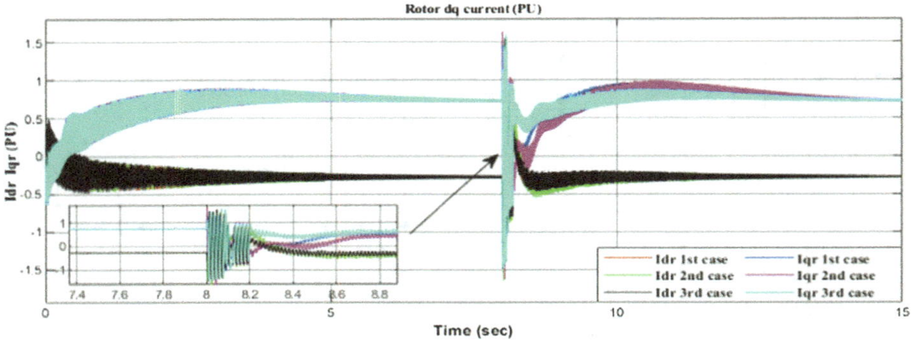

Fig. 5. The *dq* components of the rotor current.

The Fig. 5 shows the *dq* components of rotor current. In case 1 and 2 the magnitude of I_r still less than the crowbar triggering current and the turbine remains connected, while in case 3 the magnitude rise above the I_{crow} causing that the crowbar protection intervention and disconnect the turbine from the grid.

As demonstrated in Fig. 6, the rotor voltage has a massive augmentation in amplitude just after the fault application. In addition to that, in the three cases, we can see that the system is suffering from two interruptions (8.054 s–8.075 s) and from (8.11 s–8.17 s), inducing a lack of RSC control. After fault clearance, the rotor voltage regains gradually.

The Fig. 7 shows the active and reactive power driven from the stator. Obviously, the reactive power is well controlled near to 0 pu, when the active power approaches

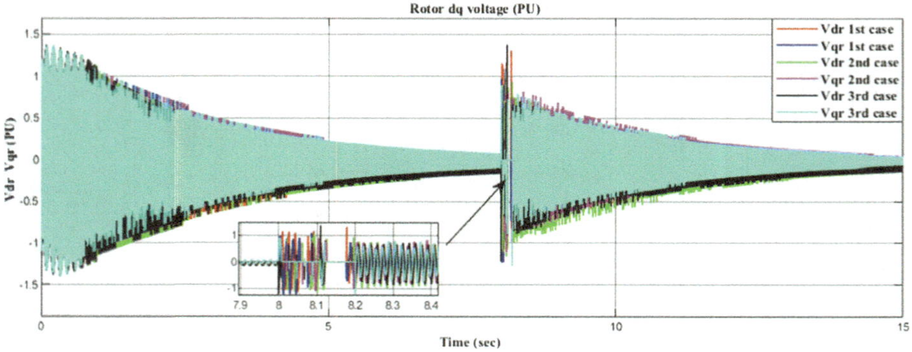

Fig. 6. The *dq* components of the rotor voltage.

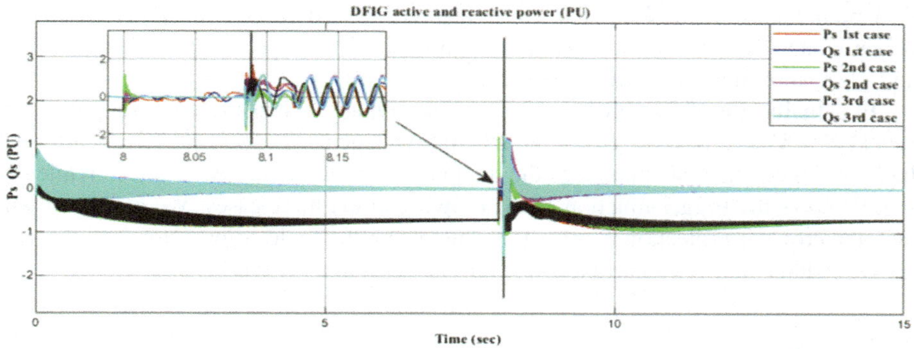

Fig. 7. Active and reactive power delivered from the stator.

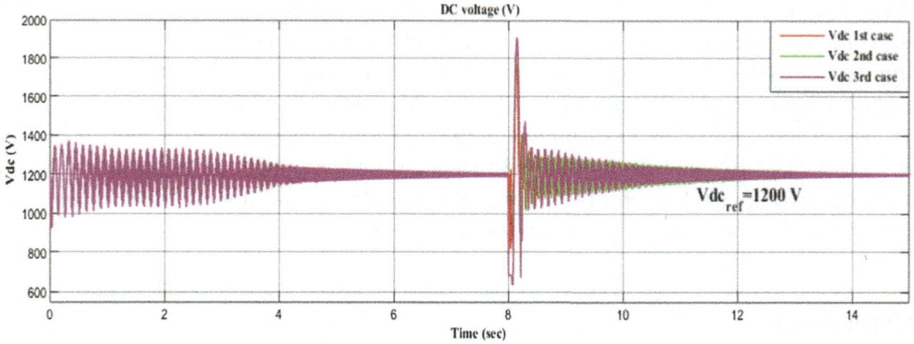

Fig. 8. The DC link voltage evolution.

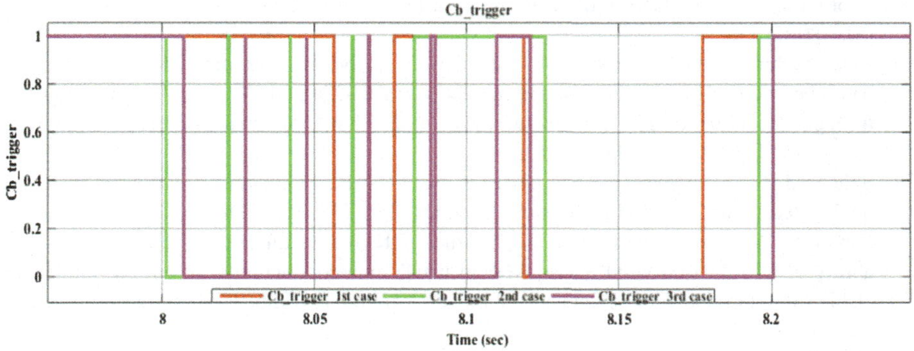

Fig. 9. The crowbar trigger function.

0.71 pu in the pre-fault. When the fault is applied, we can observe that there is a loss in power control because of the interruptions in the rotor voltage. After the fault, both P_s and Q_s restore their rated values rapidly.

Eventually, Fig. 8 establishes the DC link voltage. When the crowbar protection acts ($V_{dc} > V_{crowb}$) or ($I_r > I_{crow}$), the DC link voltage starts discharging, and after the fault clearance, the GSC maintains the dc link voltage to its reference value. While the Fig. 9 shows the trigger function of the crowbar in the three cases. We can figure out that, the crowbar protection is activated when using the logic comparator with the pre-set Dc voltage and rotor current to prevent any damage for the RSC.

5 Conclusion

In this work, a dynamic model of a grid-connected DFIG together with crowbar protection is achieved. After that, we briefly review the complete vector control of both RSC and GSC including the IMC control technique in Sect. 2. In the last part, the simulation of the three cases based on the fault location is performed. Basing on the simulation results, we can conclude that the fault ride through of the tested wind farm is highly affected by the 3-phase fault location. Also, since the FRT is a high necessity, the associated controllers must be well exploited, and their parameters must precisely be chosen. Finally, the results showed that the crowbar protection is not enough and must be accompanied by the IMC controller to help the machine to ride through the fault or at least, to reduce any severe currents fastly.

References

1. Global Wind Energy Council (GWEC): Global Wind Report—Annual Market Update; Global Wind Energy, Council (GWEC), Brussels, Belgium, p. 61, April 2019
2. Duong, M.Q., Leva, S., Mussetta, M., Le, K.H.: A comparative study on controllers for improving transient stability of DFIG wind turbines during large disturbances. Energies **11** (3), 480 (2018)
3. Romphochai, S., Pichetjamroen, A., Teerakawanich, N., Hongesombut, K. (eds.): Coordinate operation of fuzzy logic voltage regulator and Bi-2212 SFCL for enhancing fault ride through capability of DFIG wind turbines. In: 2017 International Electrical Engineering Congress (iEECON), pp. 1–4 (2017)
4. Salles, M.B.C., Hameyer, K., Cardoso, J.R., Grilo, A.P., Rahmann, C.: Crowbar system in doubly fed induction wind generators. Energies **3**(4), 738–753 (2010)
5. Sahoo, S., Mishra, A., Chatterjee, K., Sharma, C.K. (eds.): Enhanced fault ride—Through ability of DFIG-based wind energy system using superconducting fault current limiter. In: 2017 4th International Conference on Power, Control & Embedded Systems (ICPCES), pp. 1–5 (2017)
6. Döşoğlu, M.K.: Nonlinear dynamic modeling for fault ride-through capability of DFIG-based wind farm. Nonlinear Dyn. **89**(4), 2683–2694 (2017)
7. Anaya-Lara, O., Campos-Gaona, D., Moreno-Goytia, E., Adam, G.: Offshore Wind Energy Generation Control, Protection, and Integration to Electrical Systems, p. 307. Wiley, Hoboken (2014)

8. Abu-Rub, H., Malinowski, M., Al-Haddad, K.: Power Electronics for Renewable Energy Systems, Transportation and Industrial Applications, p. 827 (2014)
9. Abad Jsp, G., Rodriguez, M.A., Marroyo, L., Iwanski, G.: Power Electronics and Electric Drives for Traction Applications, p. 684. Wiley, Hoboken (2016)
10. Teodorescu, R., Liserre, M., Rodriguez, P.: Grid Converters for Photovoltaic and Wind Power Systems, p. 407. Wiley, Hoboken (2011)
11. Wu, B., Lang, Y., Zargari, N., Kouro, S.: Power Conversion and Control of Wind Energy System, p. 480. Wiley, Hoboken (2011)
12. Kerrouche, K., Mezouar, A., Belgacem, K.: Decoupled control of doubly fed induction generator by vector control for wind energy conversion system. Energy Procedia **42**, 239–248 (2013)
13. Chowdhury, M.A., Hijazin, I., Hosseinzadeh, N., Pota, H.R.: Modeling-and-Analysis-With-Induction-Generators. CRC Press, p. 486 (2015)
14. Ali, M.H.: Wind Energy Systems; Solutions for Power Quality and Stabilization, p. 287. CRC Press, Boca Raton (2012)

Experimental EMC Qualification Test of an EMI Filter for a DC-DC Converter Intended to Smart Grid Applications

S. Khelladi[1](✉), K. Saci[1], A. Hadjadj[1], A. Ales[2], Z. Chebbat[2], and A. Layoune[2]

[1] Laboratory (LACoSERE), University Amar Telidji, Laghouat, Algeria
s.khelladi@lagh-univ.dz

[2] Laboratory Electromagnetic Compatibility, Ecole Militaire Polytechnique, Algiers, Algeria
Achour.ales@gmail.com

Abstract. This paper presents a practical approach of design & implementation of an EMI-filter for high frequency and high power DC-DC converter, qualifying to the EMC standard CISPR 11. The proposed design method is based on direct experimental identification of the filter elements taking into account the topology of the filter and the technology of its components. The proposed filter has been designed and implemented for the high frequency-power DC-DC converter, which operates at switching frequency of 10 kHz. Measurements made on this converter, without and with filter, show the effectiveness of the proposed design method, they also indicate that the lower saturation induction of the chosen magnetic materiel for the filter core inductors degrade the attenuation of the filter at high frequencies.

Keywords: Electromagnetic interference (EMI) · Experimental identification · DC-DC converter · EMI filter · Common mode (CM) · Differential mode (CM) · Core induction

1 Introduction

Electromagnetic interference (EMI), it has been the most occurring and major problem in power electronics converters, rapid changes in voltage and current within the switched mode power converters, makes these equipments the main source of radiated and conducted EMI to other nearby equipments. EMI is generally conductive in nature. The conducted emissions are mainly reduced by the EMI filters together with proper design of the circuit. It can also be combined with other solutions like the slowing down the dv/dt during the transitions of the power semi-conductor components and/or by acting on the converter controls. The EMI filters are made from coupled inductors combined with capacitors; the choice of the filter topology depends on network and load impedances. Generally, the CM and DM filters are used for power converters. The passive components have a strong impact on filter efficiency [1, 2]. The measurement method, according with EMC standard (EN 55011 or CISPR 11), is based on the utilization of the line impedance stabilizing network (LISN) in frequency band from

M. Hatti (Ed.): ICAIRES 2019, LNNS 102, pp. 632–640, 2020.
https://doi.org/10.1007/978-3-030-37207-1_68

0.15 MHz to 30 MHz [3]. A procedure for designing EMI filters will be presented in this study. It is based on the experimental analysis of conducted EMI induce by the DC-DC converter, and the identification of the adequate filter elements taking in account filter topology and technology of components namely the characteristics of the appropriate magnetic materials.

2 Implementation of the Experimental Measuring Bench

The design of EMI filters is based on the spectrum of emissions to be eliminated. For this purpose, an experimental test bench for conducted disturbances has been realized in accordance with CISPR 11, based on the utilization of the line impedance stabilizing network (LISN) in frequency band from 0.15 MHz to 30 MHz [4]. The test bench realized is illustrated in Fig. 1. The DC-DC converter studied is a series chopper assembled in an aluminum box, to guard against external radiated noise. It comprises a switching cell composed of a controlled power switch MOSFET (IRFP350) and a freewheeling diode (MUR650) placed in parallel with the load. The converter powered through a LISN by a DC bus of 100 V and operating at a switching frequency fsw = 10 kHz with a duty cycle of 0.3, is loaded by R = 10 Ω and L = 1 mH.

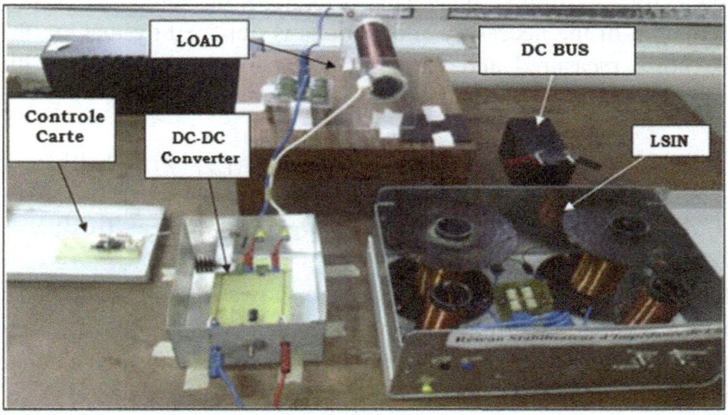

Fig. 1. Test bench made according to the EMC standard.

The measurement results show that the two spectra of the CM and DM conducted emissions shown in Fig. 2 far exceed the standard in the frequency range 0.15 MHz to 3.5 MHz.

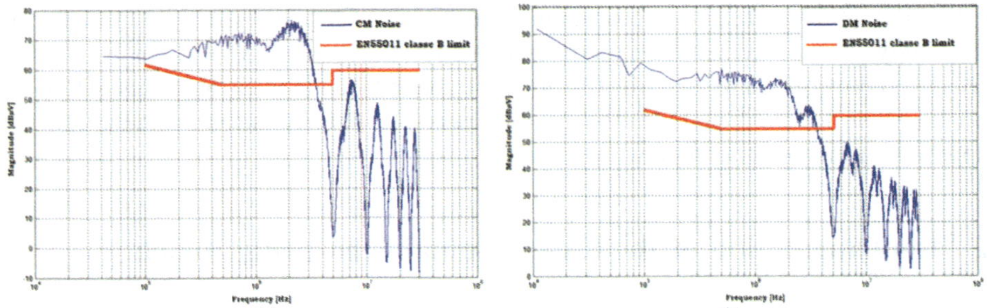

(a) (b)

Fig. 2. EMI noise spectra. (a) Common mode, (b) Differential mode.

3 EMI Filter Design

In order to reduce the conducted emissions induced by the power converter, a T-structure filter (C-L-C) is required, thus the CM and DM filters are used in cascade as indicated in Fig. 3. The CM filter uses a coupled inductors L_{CM} and two capacitors C_Y connected to the ground and the DM filter uses an inductor L_{DM} and two capacitors C_X. The EMI filter is supposed to guarantee two main roles: maximum power adaptation of victim-source sense, and power mismatch in the opposite direction (source-victim).

The calculation of the necessary attenuations is obtained by making the difference between the noise measured and the limit given by the standard as illustrated by Eqs. (1) and (2).

$$Att_{req,DM}(f) = A_{DM_{dB\mu V}}(f) - Limit_{\mathrm{EN55011}_{dB\mu V}}(f) + 6\,\mathrm{dB\mu V} \tag{1}$$

$$Att_{req,CM}(f) = A_{CM_{dB\mu V}}(f) - Limit_{\mathrm{EN55011}_{dB\mu V}}(f) + 6\,\mathrm{dB\mu V} \tag{2}$$

The attenuations required for both modes are shown in Figs. 4 and 5. The slopes of ±40 dB/decade allow to determine the cutoff frequencies f_{CM}, and f_{DM} and f_{CMH} and f_{DMH} for an LC filter structure. The CM and DM filter elements are calculated from the cutoff frequencies f_{CM} and f_{DM}. Frequencies f_{CMH} and f_{DMH} are used to determine the limit values of the parasitic elements of the passive components to be used for the realization of the filter [5]. The EMI filter is a passive filter that requires no power

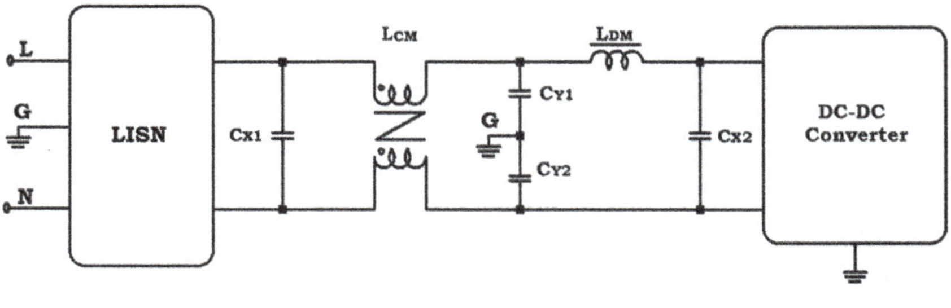

Fig. 3. EMI filter structure.

supply, so it is necessarily stable, its role is to mitigate the disturbances generated in the power chain. The identification of the parameters is carried out by experimental measurements.

4 Identification of EMI Filter Elements

4.1 Common Mode Filter

The attenuation of the CM filter is given by Eq. (1). According to the trace of the attenuation required for the CM filter shown in Fig. 4, the cutoff frequencies are defined as follows: The low cutoff frequency of the filter F_{CM} = 61 kHz, The high cutoff frequency of the filter F_{CMH} = 3.05 MHz and the frequency of intersection of the two slopes F_r = 424 kHz.

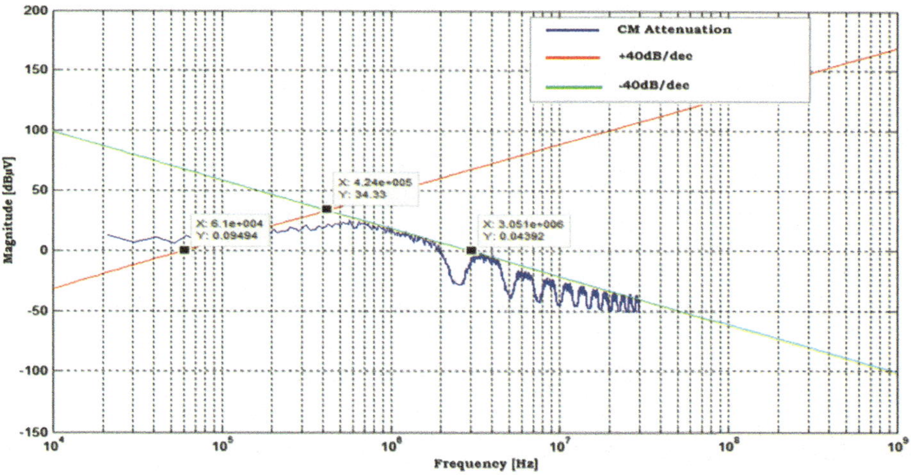

Fig. 4. CM attenuation requirement.

There is an infinity of couples (L, C) that satisfy the relation (3). However, because of technological constraints, and other related to the practical aspect, such as the imperfections of the coil (inter-turn capacitors) and the winding direction, it is generally preferred to set the value of the capacity to C_Y = 1.5. nF to deduce L_{CM} inductance from Eq. (3).

$$L_{CM/DM} = \left(\frac{1}{2\pi F_{CM/DM}}\right)^2 \cdot \left(\frac{1}{2C_{Y/X}}\right) \quad (3)$$

Thus we calculate the inductance of CM filter where L_{CM} = 2.3 mH, the series element of the capacitor C_Y is L_Y = 8 nH obtained by the LCR-meter module at the frequency of 200 kHz, the cables inductance: L_{CABLES} = 45 μH, the capacitive parallel element of the L_{CM} inductance is Cp = 60 pF.

4.2 Differential Mode Filter

According to Eq. (2) the attenuation of the DM filter is shown in Fig. 5. We obtain the low cutoff frequency F_{DM} = 60 kHz. By choosing a value C_X = 1,5nF for the capacitors of the DM filter and taking into account the capacitors of the CM filter, the equivalent differential mode capacitance Cx is of value 2.25 nF and the inductance of the DM filter is of value L_{DM} = 6.2 mH according to Eq. (3).

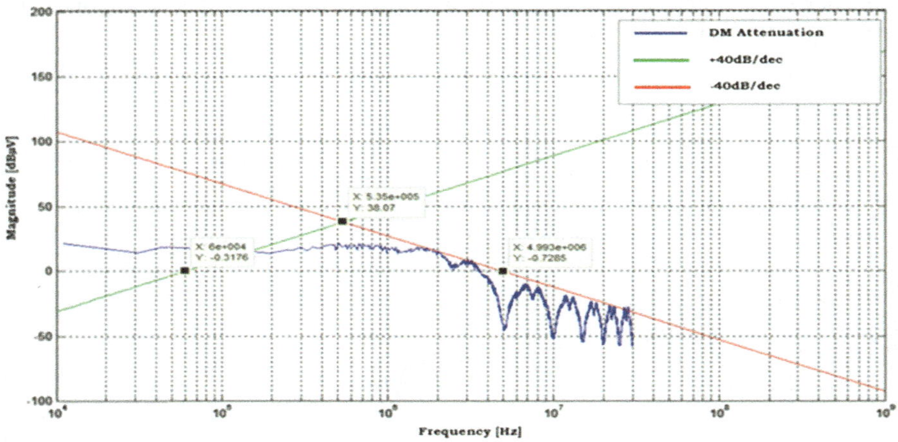

Fig. 5. DM attenuation requirement.

4.3 Choice of Magnetic Material

For the intended application and the chosen filter structure, a material having high permeability over a high frequency range is required to provide a proper common mode filtering. The material available at the laboratory meeting the desired criteria is the N30 ferrite manufactured by EPCOS shown in Fig. 6. The dimensions of the magnetic core are illustrated in Table 1. In order to ensure the validity of the manufacturer's data, an experimental test was carried out to verify the limits of the permeability by the injection method illustrated in Fig. 7. The objective of this test is to inspect the working frequency range of the material, through the phase shift between the current and the voltage absorbed by the toroid. To do this, the wound ferrite core is connected in series with a resistance of 50 Ω which keeps its resistive appearance over a wide frequency range. The circuit is excited by a sinusoidal voltage using a low frequencies Generator (GBF) of variable frequency up to 10 MHz.

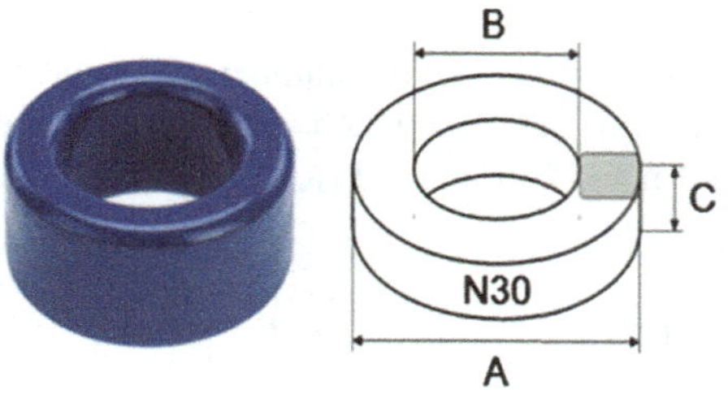

Fig. 6. Geometrical dimension of the magnetic core.

Table 1. Magnetic core size definition.

Core Type	A (O.D) mm	B (I.D) mm	C (Height) mm	AL (±25%) nH/T^2
N 30	25.3 mm	14.8 mm	10 mm	4260

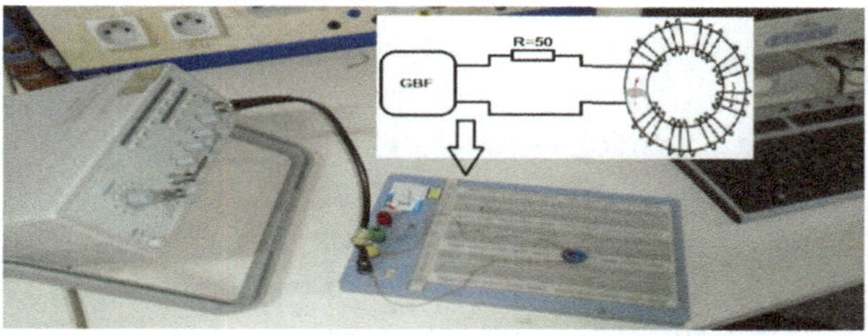

Fig. 7. Experimental test bench for the injection method.

Figure 8a Shows the impedance argument for a frequency range from 10 kHz to 3 MHz. It can be seen that the relative permeability of the material decreases from about 0.8 MHz, which is consistent with the manufacturer's data shown in Fig. 8b.

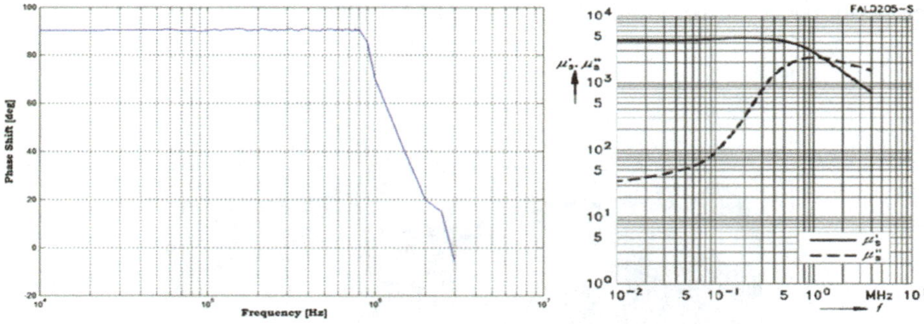

(a) (b)

Fig. 8. Variation of the complex permeability of ferrite N30 with frequency. (a) Test result by injection method. (b) According to the manufacturer [6].

In order to achieve the common mode and differential mode inductance with the chosen magnetic material (ferrite N30), the number (n) of turns of the two inductances can be approximated with Eq. (4) from which we get the number of turns n = 24 turns for L_{CM} inductance, and n = 37 turns for L_{DM}.

$$L_{CM/DM} = n^2 A_L \quad \Rightarrow n = \sqrt[2]{\frac{L_{CM/DM}}{A_L}} \tag{4}$$

5 Hardware Implementation, Experimental Results and Discution

The EMI filter is placed on the PCB and which comprises X - capacitors, Y - capacitors, the L_{DM} and L_{CM} inductors are realized on the same magnetic core (N30) as shown in Fig. 9.

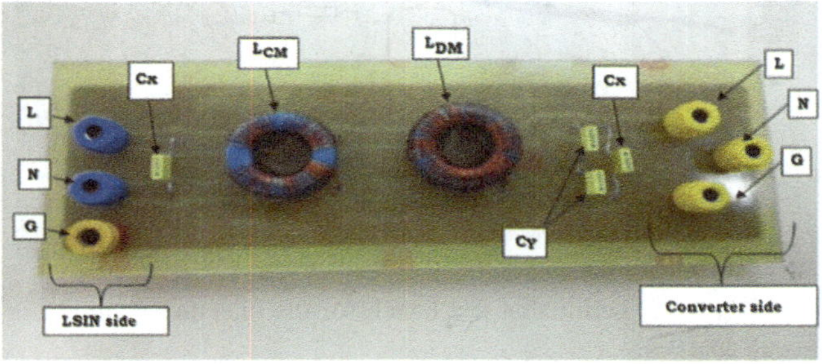

Fig. 9. The realized EMI filter

The realized EMI filter is inserted into the energy conversion chain. Figure 10 shows the final test bench in accordance with the EMC standard CISPR 11 to evaluate the performance of the designed and realized EMI filter.

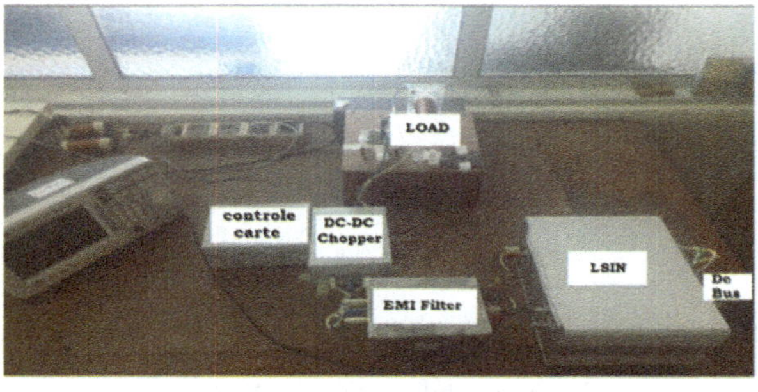

Fig. 10. Final test bench of conducted EMI.

5.1 Discusion of Test Results

In this section the DM and CM noise waveforms of the input line voltage of the conversion chain with and without the EMI filter are presented in Fig. 11. Figure 11a shows the level of the DM emission spectrum after inserting the filter. Note that the filtering is quite good (the level is below the level imposed by the standard), until the frequency of 0.8 MHz where a fluctuation occurs (non-compliant emission level with the limit of the standard). Similarly, the attenuation of CM disturbances illustrated in Fig. 11b is quite good up to about 1 MHz where we notice the appearance of two peaks of disturbances at 0.5 MHz and 0.8 MHz. On these points, we note that the realized EMI filter is a little less efficient due to the limited saturation induction of the ferrite core N30 and this has been confirmed by the complex permeability measurements both tested and those provided by the manufacturer which show that the cutoff frequency of the material is about 0.8 MHz.

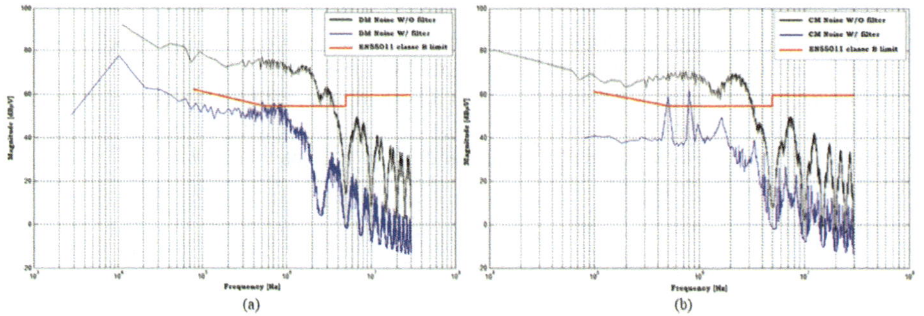

Fig. 11. DM & CM emission levels with and without filter. (a) Differential mode, (b) Common mode.

6 Conclusions

During this research paper, an experimental prediction of the electromagnetic CM & DM disturbances allowing the design and the implementation of an EMI filter was carried out by the realization of an experimental measuring bench in accordance with the CISPR 11 standard. The proposed EMI filter is successfully implemented satisfying the requirements of the specifications. However, the experimental prediction also shows that the performances of the implemented EMI filter cannot be maintained on more than 40 dB to 60 dB per decade on frequency because of the complexity of the elaborated structure and the constraints of weight and volume. On the other hand, the limited performance of the chosen magnetic material, which results in a degradation of the attenuation of the filter at certain frequencies (0.5 MHz and 0.8 MHz), results from the low saturation induction of the material (N30) which imposes a short frequency range (up to 0.8 MHz) leading the magnetic components of the EMI filter to lose their induction at critical filtering frequencies. As perspectives, one can optimize the choice and technology of the components, namely the characteristics of the appropriate magnetic materials that guarantee a smooth operation in the frequency range without degrading the performance of the magnetic cores. Hybridization of the EMI filter is also an effective solution to the previous difficulties.

References

1. Ashish, T., Jayapal, R., Venkatesh, S.K., Singh, A.: Design & implementation of a practical EMI filter for high frequencyhigh power DC-DC converter according to MIL-STD-461E, India (March 2017)
2. Kotny, J.L., Duquesne, T., Idir, N.: Design of EMI filters for DC-DC converter, pp. 1–6 (2010)
3. IEC: CISPR 11:2003 | IEC Webstore | electromagnetic compatibility, EMC, smart city. https://webstore.iec.ch/publication/12025. Accessed 30 June 2019
4. Kotny, J.L., Duquesne, T., Idir, N.: EMI filter design using high frequency models of the passive components. In: 2011 IEEE 15th Workshop on Signal Propagation on Interconnects (SPI), pp. 143–146 (2011)
5. Kotny, J.-L., Duquesne, T., Nadir, I.: Determination DES parametres parasites hf d'un filtre cem pour convertisseur a base de composants SiC. Presented at the Conference: 17ème Colloque International et Exposition sur la Compatibilité ÉlectroMagnétique, Clermont-Ferrand, France (2014)
6. Ferrites and Accessories, SIFERRIT, Material N30, p. 6 (September 2006)

Comparison of Different Extraction Methods for the Simulation of Thin-Film PV Module

Bouchra Benabdelkrim[1,2(✉)] and Ali Benatillah[2]

[1] Department of Material Sciences, Institute of Science and Technology, University of Ahmed Draia Adrar, Adrar, Algeria
benaekbouchra@gmail.com

[2] Laboratoire Energie, Environnement et Systèmes d'Information (LEESI), Université Ahmed Draia Adrar, Adrar, Algeria
benatiallah.ali@gmail.com

Abstract. This paper investigates and compares three different methods commonly employed in solving current–voltage equation of single diode and two diode solar PV models using manufacturer's data sheet based on three parameter estimation methods are employed: an iterative method, method of the slope at point and an analytical method based on the Lambert W function. These three existing mathematical methods are implemented to estimate the parameters of thin-film PV module under changing environmental conditions.

The results reveal that no single method performs best in all the metrics and there will always be a trade-off in the choice of the method based on the user's focus.

The present work can be a potential tool for researchers and designers working in the area of photovoltaic systems, to make decisions related to the selection of the best possible method for the extraction of the characteristic parameters of thin-film PV modules.

Keywords: Thin film PV modules · Numerical methods · Analytical methods · Parameter extraction · Performance I–V curves

1 Introduction

During the last years the international market of thin-film photovoltaic (PV) modules has been increasing considerably mainly due to their simple and low-cost manufacturing process. The various thin-film technologies reduce the amount of light absorbing material that is necessary to produce a solar cell.

Therefore, to ensure the maximum use of the available solar energy by a PV power system, it is important to study its behaviour through modeling, before implementing it in reality. The mathematical model of the PV device is very useful in studying various PV technologies and in designing several PV systems along with their components for application in practical systems.

Some researchers have compared the algorithm used in extracting the parameter of solar PV models. For example, the extraction of the parameters of the single-diode solar cell model using experimental I–V characteristics of Si and Multi-junction solar

M. Hatti (Ed.): ICAIRES 2019, LNNS 102, pp. 641–649, 2020.
https://doi.org/10.1007/978-3-030-37207-1_69

cells has been performed by Appelbaum and Peled [1]. In another research, Ciulla et al. [2] compared the I–V and P–V curves at various temperatures and irradiance for a generic PV panel for five different algorithm models which are: Hadj Arab et al. [3] model, De Blas et al. [4], Lo Brano et al. [5], Villalva et al. [6] and De Soto et al. [7] models.

Extensive studies have been conducted to determine the series resistance (Rs) and parallel resistance (Rsh). Some authors neglect Rp to simplify the model as the value of this resistance is generally high [8], and sometimes, the Rs is neglected, as its value is very low [9]. The neglect of Rs and Rp has significant impact on the model accuracy. Several algorithms have been proposed to determine both Rs and Rsh through iterative techniques [10]. In [11], Rs and Rsh are evaluated by using additional parameters which can be extracted from the current versus voltage curve of a PV module.

This present study tends to contribute also along this direction by including other metrics in addition to the accuracy of the models for an encompassing comparison of parameter extraction models. This study therefore evaluates the performances of three extraction methods (iterative method, method of the slope at point and the Lambert W function). These three methods applies the one diode and two diode models and based only on the manufacturer datasheet of ST40 thin film PV panel, whose the objective is to predict the behavior of Shell ST40 panel under real environmental conditions. The Shell ST40 belongs as a copper indium diselenide (CIS) thin film technology, their module size (W × L) is 0.328 × 1.293 m^2.

2 Materials and Methods

2.1 Lambert W-Function Method

A PV cell of current equation mathematically solved by the Newton's Raphson method is difficult to employ the Large PV structure. This equation gives the relation involving the output current (I) and terminal voltage (V) under the environment condition as:

$$I = I_{pv} - I_0.\left(\exp\left(\frac{V + R_s.I}{V_T}\right) - 1\right) - \left(\frac{V + R_s.I}{R_{sh}}\right) \tag{1}$$

Where

I_{pv} Photocurrent
I_0 Cell saturation current
R_{sh} Shunt resistance
R_s Series resistance
V_T the thermal voltage (V_T = a.Ns.k.T/q)
N_s Number of cells in series
a Ideal factor of the PV diode
q Electron charge (1.60281 × 10^{-19} C)
k Boltzmann's constant = 1.38066 × 10^{-23} J/K
T Cell operating temperature

When reach the level of entire PV structure it is difficult to solve Newton's method because in large PV structure all cell is individually described by a only one equation, so task becomes complicated, rising the convergence issues. To overcome this limitation Lambert W Function is used to the explicit results of current and voltage equations (Fig. 1). The Lambert W Function W(x) is defined as Eq. (3) [13].

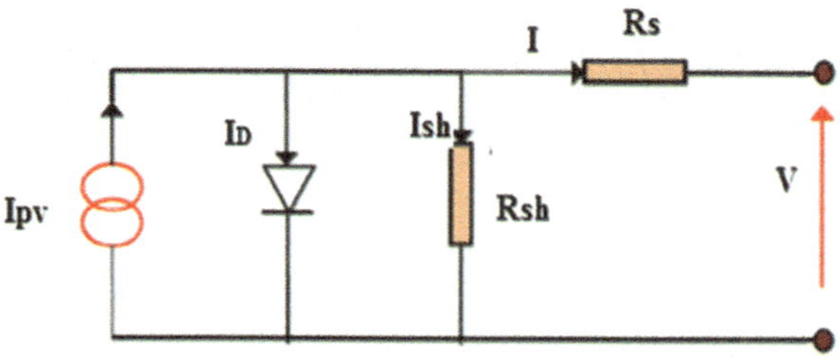

Fig. 1. PV-cell equivalent-circuit models: single-diode model [12].

If we make the following change of variable:

$$x = \frac{V_{mp} + I_{mp}.R_S}{a.N_s.V_T} \tag{2}$$

The analytical solution based on the use of the Lambert W function, which is the solution of the equation:

$$f(x) = x.e^x \tag{3}$$

Then the series and parallel resistances can be written as follows:

$$R_s = \frac{x.a.N_s.V_T - V_{mp}}{I_{mp}} \tag{4}$$

$$R_{sh} = \frac{x.a.N_s.V_T}{I_{PV} - I_{mp} - I_0[exp(x) - 1]} \tag{5}$$

Where x's expression is given in Eq. (6). The value obtained by (6) is substituted in (4) and (5) to deduce the values of R_{sh} and R_S.

$$x = \text{Lambert W}\left[\frac{V_{mp}.(2I_{mp} - I_{PV} - I_0)\left[exp\left(\frac{V_{mp}.(V_{mp}-2a.V_T)}{(a.N_s.V_T)^2}\right) - 1\right]}{a.N_s.V_T.I_0}\right] + 2\frac{V_{mp}}{a.N_s.V_T} - \left(\frac{V_{mp}}{a.N_s.V_T}\right)^2 \tag{6}$$

The mathematically relation the Eq. (3) is applied to Eq. (1) in order to obtained the equation for the cell gives the explicit results of current and voltage equation. By solving Eq. (1) with the Lambert W method the equation of the output current as the function of output voltage as given in Eq. (7) [14].

$$I = -\frac{V}{R_s + R_{sh}} - \frac{LambertW\left(\frac{R_{sh}\left(R_s.I_{pv} + R_s.I_0 + V\right)}{a.V_T(R_s + R_{sh})}\right)}{R_s} + \frac{R_{sh}\left(I_0 + I_{pv}\right)}{R_s + R_{sh}} \tag{7}$$

Where:

$$I_0 = I_{pvn}.\exp\left(-\frac{V_{ocn}}{a.N_s.V_{Tn}}\right) - 1 \tag{8}$$

The ideality factor (a) is calculated by:

$$a = \left(K_V - \frac{V_{oc}}{T_n}\right) / a.N_s.V_{Tn}\left(\frac{K_i}{I_{pvn}} - \frac{3}{T_n} - \frac{E_g}{k.T_n^2}\right) \tag{9}$$

2.2 Method of the Slope at Point

The difference given by this method in comparison of the previous method is in the manner of calculating the series resistance. It is based on the fact that the series resistance influences remarkably the slope of the characteristic curve I-V in the vicinity of the point (*Voc,* 0).

$$\frac{dI}{dV} = -I_0\left[\exp\left(q\frac{V + IR_s}{aN_skT} - 1\right)\right]\frac{q}{aN_skT}\left(1 + R_s\frac{dI}{dV}\right) \tag{10}$$

$$R_s = -\frac{dV}{dI}\bigg|_{V_{oc}} - \frac{1}{\frac{I_0q}{aN_skT}\exp\left(q\frac{V_{oc}}{aN_skT}\right)} \tag{11}$$

The slope $M = \frac{dV}{dI}$ ($I = 0$) at the point (*Vco,* 0) is deduced geometrically from experimental data (Fig. 2).

The photocurrent I_{pv} of the PV cell is directly proportional to the solar insolation. The output current I of the cell is equal to photo generated current $I_{pv,}$ minus diode current I_D, minus shunt current I_{Sh}.

$$I = Ipv - ID - Ish \tag{12}$$

As $R_{sh} \approx \infty, I_{sh} \approx 0$.

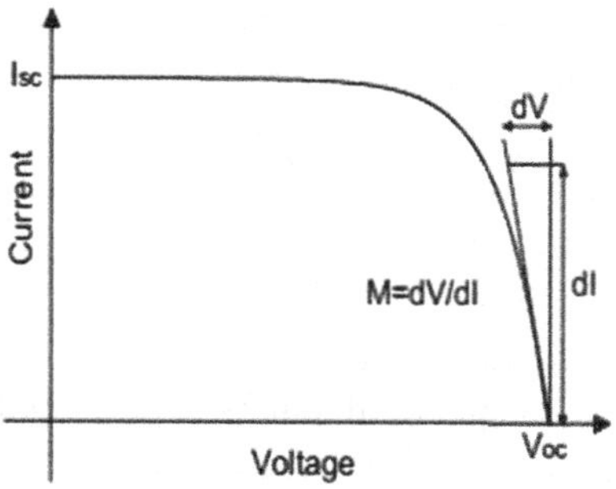

Fig. 2. The slope calculation at the open circuit voltage point

$$I = Ipv - LD \tag{13}$$

Thus the equations for the I-V characteristics of the PV cell are:

$$I = I_{pv} - I_0\left[exp\left(q\frac{V + IR_s}{aN_skT} - 1\right)\right] \tag{14}$$

Hence, to solve Eq. (14) the value of voltage (Voc) and current (Isc) at different temperature T1 & T2 is the Eq. (15) [15]:

$$I_0(T) = I_{0(T1)}\left(\frac{T}{T_1}\right)^3 \exp\left[\left(\frac{qE_g}{a.K}\right)\left(\frac{1}{T_1} - \frac{1}{T}\right)\right] \tag{15}$$

Here, the energy gap of the material is defined as Eg

Where

$$I_{0(T1)} = \frac{I_{pvn}}{\exp\left(\frac{V_{ocn}}{aV_{Tn}}\right) - 1} \tag{16}$$

$$I_{PV} = [I_{PVn} + K_i.(T - T_n)].G/G_n \tag{17}$$

Where:

$$I_{sc} = I_{PV} \tag{18}$$

$$K_i = \left[I_{sc(T2)} - I_{sc(T1)}\right]/(T2 - T1) \tag{19}$$

2.3 Iterative Method

This method appears an improved modeling approach for the two-diode model of photovoltaic (PV) module. The main contribution of this method is the simplification of the current equation. Furthermore the values of the series and parallel resistances are computed using a simple and fast iterative method. The both reverse saturation currents I_{01}, I_{02} are set to be equal in magnitude

$$I_{01} = I_{02} = \frac{I_{sc} + K_i.\Delta T}{\exp\left(\frac{q.(V_{oc} + K_v.\Delta T)}{kT.(a_1 + a_2)/p}\right) - 1} \tag{20}$$

Diode ideality factors a_1 and a_2 represent the diffusion and recombination current components, respectively. Ishaque [10] put $\frac{a_1 + a_2}{p} = 1$ and $a_1 = 1$, it follows that variable p can be chosen to be $p \geq 2.2$. This generalization can eliminate the ambiguity in selecting the values of a_1 and a_2. Equation (20) can be simplified in terms of p as:

$$I = I_{pv} - I_{01}.\left(\exp\left(\frac{V + R_s.I}{V_{T1}}\right) - 1\right) - I_{02}.\left(\exp\left(\frac{V + R_s.I}{V_{T2}}\right) - 1\right) - \left(\frac{V + R_s.I}{R_{sh}}\right) \tag{21}$$

The R_s and R_{sh} are calculated simultaneously, similar to the procedure proposed in [10]. From Eq. (21) at maximum power point condition, the expression for R_{sh} can be rearranged and rewritten as:

$$R_{sh} = \frac{V_{mp} + R_s.I_{mp}}{I_{pv} - I_{01}.\left(\exp\left(q\frac{V_{mp} + R_s.I_{mp}}{k.T}\right) + \exp\left(q\frac{V_{mp} + R_s.I_{mp}}{(p-1)k.T}\right) + 2\right) - \frac{P_{max,e}}{V_{mp}}} \tag{22}$$

The initial conditions for both resistances are given below

$$R_{s0} = 0, R_{p0} = \frac{V_{mp}}{I_{sc,STC} - I_{mp}} - \frac{V_{0c,STC} - V_{mp}}{I_{mp}} \tag{23}$$

3 Results and Discussion

The equations of the previous section were implemented in MATLAB environment to simulate, evaluate and test the three methods mentioned above for ST40 PV modules. The datasheet parameters specified under STC are already given in Table 1.

Table 1. Specification of the PV modules

Modules	Isc(A)	Voc(V)	Imp(A)	Vmp (V)	$Ki_{(Isc)}$ (mA/°C)	$Kv_{(Voc)}$ (mV/°C)	Ns
Thin-Film Shell ST40	2.68	23.3	2.41	16.6	0.35	−100	36

Figure 3 show the I-V characteristics compared with measured data extracted from ST40 PV module's datasheet, for different irradiance levels at 25 °C and for different temperature levels and at 1 kW/m^2. It is noted that the I-V characteristics of the method of the slope at point show good agreement with the measured data, with the exception of iterative method and the W-function method around V_{OC} for low irradiance.

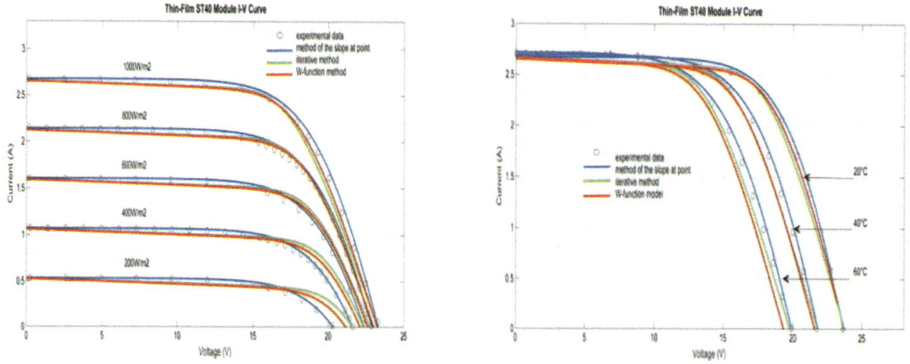

Fig. 3. The I-V characteristics of ST40 module at varying irradiance and temperature.

In the other hand, It is observed that at temperatures around STC, the models have similar behavior for all method. However, as the temperature increases, the Lambert W function method's characteristics tends to a slight deviate from the other methods.

Table 2 show the parameters estimated for ST40 PV module. The values these parameters (R_S, R_{sh}, a, I_0 and I_{PV}) are estimated using three methods. Certainly, the similarity of the results between these methods is noteworthy and the differences have no appreciable influence on the simulated I-V characteristics at STC.

Table 2. The estimated parameters of ST40 using three models at STC.

Modules	*Thin-Film (ST40)*		
Models	Method of the slope at point	Iterative method	*W-function method*
Ipv	2.68	2.657	2.68
a_1	2	1	1.23
a_2	–	1.3	–
Rs	0.839	1.71	1.435
Rsh	–	198.941	174.531
$Io_1 = Io_2$	9.122e − 6	3.075e − 11	3.415e − 09

To provide thorough evaluation, data corresponding to the above mentioned panels are taken from manufacturer's datasheet and I-V curves are matched with the simulation results obtained using three models. Further, to know the quality of the curve fit between these models values to the experimental data, statistical analysis is carried out by measuring Individual Absolute Error (IAE) values. The IAE values are calculated by using the mentioned formula.

$$\text{Individual Absolute Error: } (IAE) = |\mathrm{I}_{\text{measured}} - \mathrm{I}_{\text{estimated}}| \tag{24}$$

Figure 4 show the absolute errors of I–V characteristic for ST40 PV module. The absolute errors of method of the slope at point are less than that of the other methods at all the conditions.

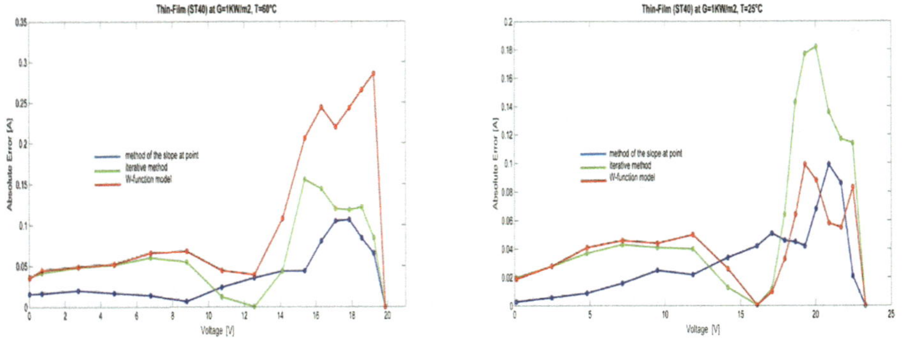

Fig. 4. Absolute error for Thin-Film ST40.

4 Conclusion

In this work, three parameter estimation models existing in the literature are described and have been verified by simulation and measured data, which were extracted from datasheet I-V characteristics. The methods that the most accurate method of parameter estimation given in the PV modules' datasheets of the ST40 module is method of the slope at point. The differences between the three estimation methods have no appreciable influence on the simulated I-V characteristics under varying environmental conditions. In particular, excellent accuracy exhibited at high irradiance and low temperature conditions for all models.

References

1. Appelbaum, J., Peled, A.: Parameters extraction of solar cells – a comparative examination of three methods. Sol. Energy Mater. Sol. Cells **122**, 164–173 (2014)
2. Ciulla, G., Lo Brano, V., Di-Dio, V., Cipriani, G.A.: Comparison of different one-diode models for the representation of I-V characteristic of a PV cell. Renew. Sustain. Energy Rev. **32**, 684–696 (2014)
3. Hadj Arab, A., Chenlo, F., Benghanem, M.: Loss-of-load probability of photovoltaic water pumping systems. Sol. Energy **76**, 713–723 (2004)
4. De-Blas, M.A., Torres, J.L., Prieto, E., Garcia, A.: Selecting a suitable model for characterizing photovoltaic devices. Renew. Energy **25**, 371–380 (2002)

5. Lo-Brano, V., Orioli, A., Ciulla, G., Di-Gangi, A.: An improved five-parameter model for photovoltaic modules. Sol. Energy Mater. Sol. Cells **94**, 1358–1370 (2010)
6. Villalva, M.G., Gazoli, J.R., Filho, E.R.: Comprehensive approach to modeling and simulation of photovoltaic arrays. IEEE Trans. Power Electron. **20**, 1198–1208 (2009)
7. De Soto, W., Klein, S.A., Beckman, W.A.: Improvement and validation of a model for photovoltaic array performance. Sol. Energy **80**, 78–88 (2006)
8. Ulapane, N.N.B., Dhanapala, C.H., Wickramasinghe, S.M., Abeyratne, S.G., Rathnayake, N., Binduhewa, P.J.: Extraction of parameters for simulating photovoltaic panels. In: 2011 6th International Conference on Industrial and Information Systems, Kandy, Sri Lanka, August 2011
9. Tan, Y.T., Kirschen, D.S., Jenkins, N.: A model of PV generation suitable for stability analysis. IEEE Trans. Energy Convers. **19**(4), 748–755 (2004)
10. Ishaque, K., Salam, Z., Taheri, H.: Simple, fast and accurate two-diode model for photovoltaic modules. Sol. Energy Mater. Sol. Cells **95**(2), 586–594 (2011)
11. Chan, D.S.H., Phang, J.C.H.: Analytical methods for the extraction of solar-cell single- and double-diode model parameters from I-V characteristics. IEEE Trans. Electron Devices **34** (2), 286–293 (1987)
12. Chatterjee, A., Keyhani, A., Kapoor, D.: Identification of photovoltaic source models. IEEE Trans. Energy Convers. **PP**, 1–7 (2011)
13. Chatzigeorgiou, I.: Bounds on the lambert function and their application to the outage analysis of user cooperation. IEEE Commun. Lett. **17**(8), 1505–1508 (2013)
14. Piazza, M.C.D., Vitale, G.: Photovoltaic Sources: Modeling and Emulation. Springer, London (2013)
15. Walker, G.: Evaluating MPPT converter topologies using a MATLAB PV model. J. Electr. Electron Eng. Aust. **21**, 8 (2001)

Identification of the Common Mode Impedance of a DC-DC Buck Converter According to the System Earthing Arrangement

Djelloul Bensaad[1,2](✉), A. Hadjadj[1], A. Ales[2], S. Khalidi[1], and K. Saci[1]

[1] LACoSERE Laboratory, Ammar Telidji University, Laghouat, Algeria
bensaadd@gmail.com

[2] Laboratoire Systemes Électrotechniques, École Militaire Polytechnique, Algiers, Algeria
Achour.ales@gmail.com

Abstract. In this paper, we make an identification comparison of the common mode impedance of a DC grid consisting of a single converter, this identification between two types of grounding systems widely is used in the industrial sector. The study shows the impact of the earthing arrangement on the identification models, in normal operation, and under an electrical fault, knowing that the sizing of the filters is based on the knowledge of these impedances. We validated the results with Pspice in a frequency range up to 30 MHz; this work provides analytical models for predicting the effect of common mode interference in a DC network at the operating phase.

Keywords: Modeling · Common mode · Grid DC · EM disturbances · System earthing arrangements · Filter

1 Introduction

The evolution of the renewable energy reveals electrical microgrids with direct current used in the industrial and domestic distribution that is being connected to the main energy network or which are autonomous networks [1]. The efficiency of these DC networks has presented them as an optimal choice in the distribution of electrical energy, and research has a primary goal of making DC networks similar to those of the AC network, regarding techniques and applications [2–5]. Any electrical installation must meet requirements such as the safety of people and equipment, the quality of energy to provide and continuity of service with efficiency and reliability and as main axes the two types DC and AC share as examples the systems of earthing and protection [6].

The study of pole-to-pole (PP) or pole-to-pole (PG) faults is the major concern of engineering to prevent, predict and protect installations and guarantee continuity of service [7, 8]. And between the quality and efficiency that is achieved by the integration of load power electronics on the distribution networks are connected via power converters [9], these loads are nonlinear and with the activity of power electronics

M. Hatti (Ed.): ICAIRES 2019, LNNS 102, pp. 650–659, 2020.
https://doi.org/10.1007/978-3-030-37207-1_70

switching generating noise sources in the form of harmonics and other electrical and electromagnetic noises which interfere with the electrical installation and reduce its quality and even interrupt their operation and cause material damage [10, 11].

Acquired efficiency by the progress of power electronics puts us in confrontation with another challenge requires studies that identify sources of disturbance in DC networks, many studies focus on electromagnetic interference but the majority ignores the influence of the diversity of earthing systems on electronic systems and, in general, these studies are limited to the design phase when the neutral is connected to ground.

This work combines two axes, the System Earthing Arrangements (SEA), and the identification of disturbance sources in a DC network. In [12] Fault Detection, DC, Power Systems Based on Impedance Characteristics of Modules. The identification of impedances is of great importance in filter optimization, and also provides a better understanding of the disturbance and current propagation paths. In differential mode, the pole (+) (−) conductors are the propagation path, and the protection wire (PE), chassis, and metals are common mode propagation paths through stray capacitances. The above is just in a normal operating mode, but the occurrence of an insulation fault changes the system behavior and creates new paths, which requires in-depth identification to have an early prediction for the design phase and also engineering after design in the implementation of devices in networks. Electrical installation to react under a fault of electrical insulation according to the (SEA), this work will concentrate on two types of earthing IT (Unearthed (or impedance-earthed)) and TN (Exposed-conductive parts connected to neutral) [1].

This study organized in:

Firstly, section focuses on the identification of the common-mode impedance of normal function [13, 14]. Secondly gives an idea of the influence of malfunction of a converter on according to (SEA). This malfunction is simulated by a short-circuit (insulation defect) between one of the conductors (L-N) and the ground (chassis). Behavioral models are developed in and are based on generic electrical systems given by passive components associated with voltage or current sources [15]. This study offers the Engineer a prediction of electromagnetic disturbance by mathematical models which allows better optimization of the filters of the DC network [18] or in the protection system. More than provides an idea of electromagnetic interference (EMI) in the event of insulation defects for two types of (SEA).

2 General Consideration

2.1 Common Mode Impedance

The DM and CM signals (currents and voltages) are expressed in terms of input signals (currents and voltages) of the model shown in Fig. 2. The structure of the converter can be represented by a generic quadruple (see Fig. 1(a)) [14].

$$\begin{pmatrix} i_{dm} \\ i_{cm} \end{pmatrix} = [M_i].\begin{pmatrix} i_1 \\ i_2 \end{pmatrix}, [M_i] = \begin{pmatrix} \frac{1}{2} & -\frac{1}{2} \\ 1 & 1 \end{pmatrix} \quad (1)$$

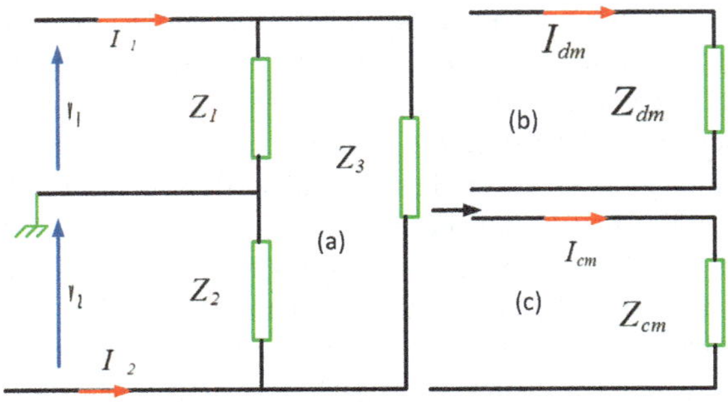

Fig. 1. Electrical diagram of a buck converter (a) in (DM/CM) basis, (b) in DM basis with decoupling, and (c) in CM basis with decoupling mode [14]

$$\begin{pmatrix} v_{dm} \\ v_{cm} \end{pmatrix} = [M_v].\begin{pmatrix} v_1 \\ v_2 \end{pmatrix}, [M_v] = \begin{pmatrix} 1 & -1 \\ \frac{1}{2} & \frac{1}{2} \end{pmatrix} \tag{2}$$

According to Fig. 1(a), the DM and CM signals can be expressed as follows:

$$\begin{pmatrix} v_{dm} \\ v_{cm} \end{pmatrix} = [Z].\begin{pmatrix} i_{dm} \\ i_{cm} \end{pmatrix} \tag{3}$$

$$[Z] = \begin{pmatrix} \left(\frac{Z_1Z_3}{Z_1+Z_3} + \frac{Z_2Z_3}{Z_2+Z_3}\right) & 0.5.\left(\frac{Z_1Z_3}{Z_1+Z_3} + \frac{Z_2Z_3}{Z_2+Z_3}\right) \\ 0.5.\left(\frac{Z_1Z_3}{Z_1+Z_3} + \frac{Z_2Z_3}{Z_2+Z_3}\right) & 0.25.\left(\frac{Z_1Z_3}{Z_1+Z_3} + \frac{Z_2Z_3}{Z_2+Z_3}\right) \end{pmatrix} \tag{4}$$

Thus, the DM and CM impedances are, respectively, given by (4), and (5) according to the impedance of the circuit in Fig. 1(a).

$$Z_{dm} = \frac{Z_1Z_3}{Z_1+Z_3} + \frac{Z_2Z_3}{Z_2+Z_3} \tag{5}$$

$$Z_{cm} = 0.25.\left(\frac{Z_1Z_3}{Z_1+Z_3} + \frac{Z_2Z_3}{Z_2+Z_3}\right) \tag{6}$$

The matrix [Z] has some non-diagonal terms (Z12 and Z21) corresponding to the coupling DM and CM. It should be noted that they are equal. For the case of the symmetrical system Z1 = Z2, the coupling terms become zero: Z12 = Z21 = 0. This results in a perfect decoupling between DM and CM. In this case, the electrical representation of the system becomes as shown in Fig. 2(b) and (c), and this approach will follow in this work according to:

$$[Z] = \begin{pmatrix} Z_{dm} & 0 \\ 0 & Z_{cm} \end{pmatrix} \quad (7)$$

The selected topology is a model structure of Buck converters sees Fig. 2.

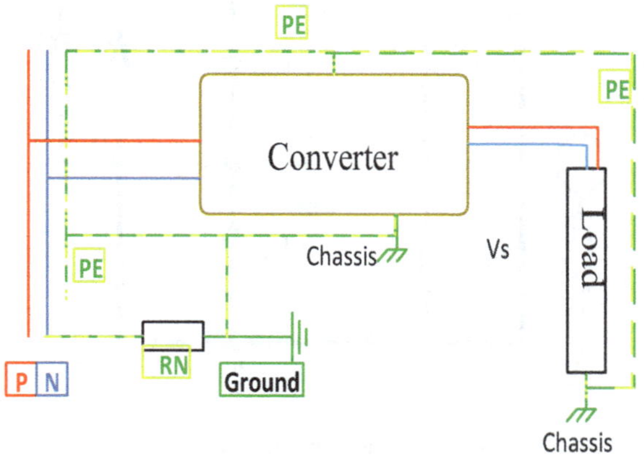

Fig. 2. Electrical circuit converter

In Fig. 2:

PE: Wire protection.
P: The plus.
N: The minus.
RN: (Resistance connected minuses to the ground).

2.2 TN Earthing System

The system is characterized by the direct neutral to ground point connection all the masses bonded to the PE protection wire. In a DC system called (−V) neutral and (+V) phase. Protection is reacted by overload. In our case eliminates RN (liaison resistance to the earth), see Fig. 2.

2.3 IT Earthing System

In the IT system ((Unearthed (or impedance-earthed)) the ground point at two possibilities, it is separately neutral or is connected by an important resistance value (RN). The protection is reacted only to the second insulation fault (PG). The first defect modifies the IT regime in TN, but the device remains in function. The installation is endowed with a permanent isolation controller.

3 Computing of a Common Mode (MC)

The identification of common mode impedances by the injection of small signals, we simulate sources of pole/ground voltages approach of the study and as follows (Fig. 3):

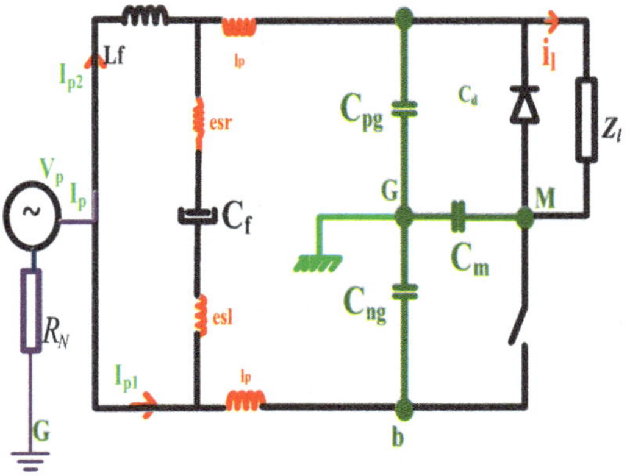

Fig. 3. Converter under the test of MC impedance

- Identification of the converter in normal operation in two systems TN and IT and another study under insulation fault similar to the previous one see Figs. 5 and 6.

3.1 Identification of MC in Normal Function

Figure 5 shows a buck converter under test for impedance indemnification in common mode. The Cpg, Cng, Cm represent the parasitic capacitances between neutral/chassis (G), phase/chassis (G) and the floating capacity according to the state of the switch in our case the (MOSFET).

$$Z_{mc} = \frac{v_{cm}}{i_{cm}} \tag{8}$$

The formula (8) represents the impedance of the converter in two modes: the online mode and the offline mode. The parallelism of the parasitic capacitors is in series with half the impedance of the filter. The addition of the resistance Rn depends on the neutral regime if the regime is TI Rn is a significant value otherwise it is zero.

$$ZIT_{mc} = \frac{1}{(C_{ng} + C_{pg}).p + Z_m} + \frac{1}{2}(Z_{lf} + Z_{lp}) + R_N \tag{9}$$

$$ZTN_{mc} = \frac{1}{(C_{ng} + C_{pg}).p + Z_m} + \frac{1}{2}(Z_{lf} + Z_{lp}) \tag{10}$$

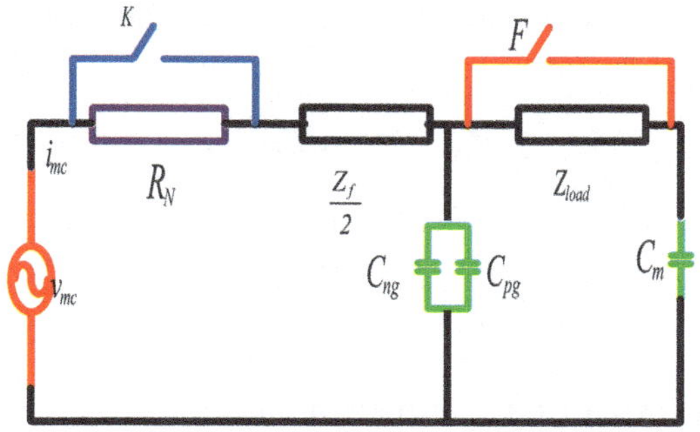

Fig. 4. Electrical model of converter's MC impedance

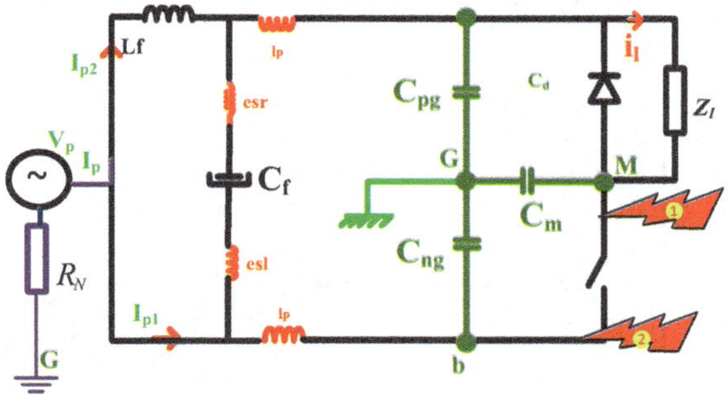

Fig. 5. Converter under insulation fault

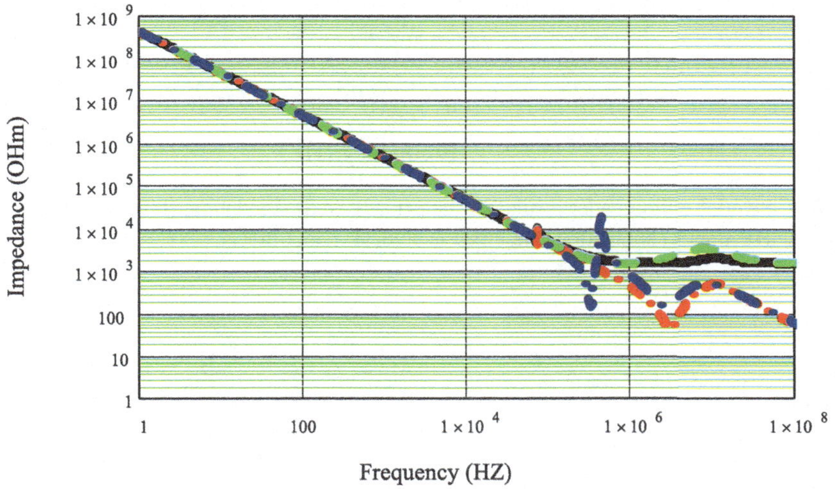

Fig. 6. Converter in normal function, (red) simulation, (blue) analytical model in the IT (online). The (Black) simulation, the (green) analytical model in TN (offline)

The impedance Zm is dependent on the state of the switch see function (10) (Fig. 4).

$$Z_m = Z_{C_M} + F.Z_{load} \tag{11}$$

$$\begin{cases} F = 0; \text{ (online mode)} \\ F = 1; \text{ (offline mode)} \end{cases} \tag{12}$$

3.2 Identification (MC) Under Insulation Fault

In this phase of the study, the behavior of the converter will be observed under a lack of isolation compared to the regime of the earth. Knowing that in the IT regime the protection of this reacts except that the second fault. Figure 7 shows the location of the two presumed defects.

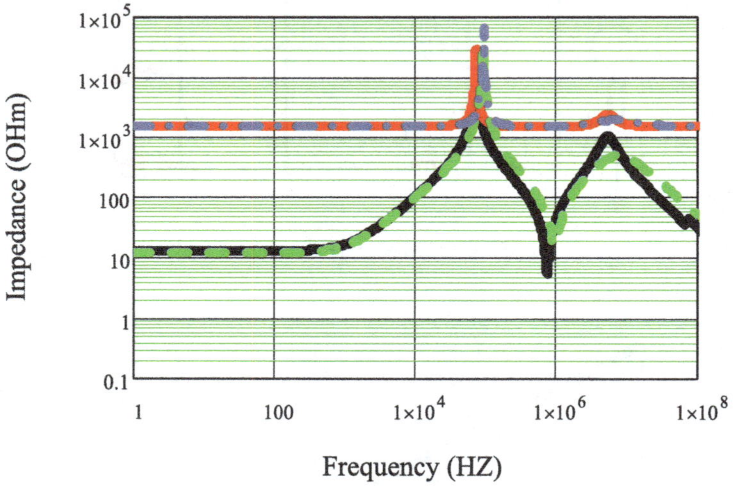

Fig. 7. Converter offline under insulation fault (load), (red) simulation, (blue) analytical model in the IT. The (black) simulation, the green analytical model in TN

$$ZIT_{-fault-minus-sour} = \frac{1}{(C_{ng} + C_{pg}).p + Z_m} + \frac{1}{2}(Z_{lf} + Z_{lp}) \tag{13}$$

$$Z_{mc-general} = \frac{1}{(C_{ng} + K.C_{DS} + C_{pg}).p + Z_m} + \frac{1}{2}(Z_{lf} + Z_{lp}) + R_N \tag{14}$$

$$\begin{cases} K = 0; \text{ (}\textit{normal}\text{ mode)} \\ K = 1; \text{ (}\textit{defect}\text{ mode)} \end{cases}$$

4 PSpice Validation

The circuit is given in Fig. 1 is simulated by PSpice. Converter components such as inductances capacitors and even the load are also characterized by Impedance Analyzer (Agilent 4294A) to introduce parasitical elements on the simulated circuit. An AC-source $v_p(t)$ is introduced to inject frequencies point by point from few Hertz up to 100 MHz (Fig. 1). The DM impedance Z_{DM} is performed for each injected frequency as given by (1). Simulation parameters are given below (Fig. 8):

- Switching frequency f_{sw} = 10 kHz,
- Duty cycle α = 0.5,
- Filter inductor L_f = 30 µH, pre (parallel resistive element) = 2 kΩ, pce (parallel capacitive, element) = 2 pF,
- Filter capacitor C_f = 100 µF, sre (series resistive element) = 230 mΩ, sle (series inductive element) = 30 nH,
- Smoothing inductor L_o = 1.61 mH, pre = 119.5 kΩ, pce = 79.6 pF.
- Resistive load R_l = 12.5 Ω,
- Parasitical inductors layout L_p = 10 nH.

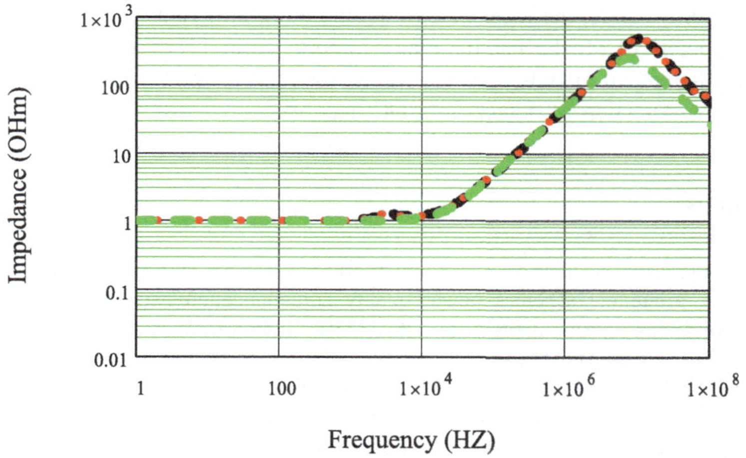

Fig. 8. Converter offline/online under insulation (filter) fault, (red) and simulation, (black), the (green) analytical model in TN.

5 Results

The results obtained sense a similarity of the capacitive aspect of the two types of neutral but:

In the IT regime, the resistance limits the minimum value in both cases, healthy and faulty.

And in the earthed neutral mode, the minimum value is the resistance of connecting cables.

The impedance models in different cases can be used in protection algorithms and in fault detection software.

This work also adds an optimal prediction of filter sizing in unfavorable cases.

6 Conclusion

This paper provides an analytical approach for calculating the common mode impedance of a practical case introducing the neutral regime influence as in the industrial field. The results obtained are validated by PSpice Simulation in both operating modes, in a normal operating mode without any abnormality, and in an unfavorable case where the converter has an electrical insulation fault.

For MC impedance, note that:

In the healthy mode the impedance depends on the parallelism of the parasitic capacitances between NG, PG and the load impedance which is in series with the floating parasitic capacitance and the PG resistor which has an important impact, in the IT neutral system, or the resistance of the earth connection of the cable - pole in TN.

Two resonances are found, the first with the inductance of charge, the second with the inductances of the filter and the inductance of the parasitic line.

In degrading mode, the resistance limits the minimum value in IT system and the parasitic capacity of Mosfet has the major importance in TN system.

The mathematical models obtained can be used in the optimization of the filters, as well as in the detection of defects in a DC network.

References

1. Kumar, D., Zare, F., Ghosh, A.: DC microgrid technology: system architectures AC grid interfaces grounding schemes power quality communication networks applications and standardizations aspects. IEEE Access. **5**, 12230–12256 (2017)
2. Baran, M.E., Mahajan, N.R.: DC distribution for industrial systems: opportunities and challenges. IEEE Trans. Ind. Appl. **39**(6), 1596–1601 (2003)
3. Justo, J.J., Mwasilu, F., Lee, J., Jung, J.W.: AC-microgrids versus DC-microgrids with distributed energy resources: a review. Renew. Sustain. Energy Rev. **24**, 387–405 (2013)
4. Lotfi, H., Khodaei, A.: AC versus DC microgrid planning. IEEE Trans. Smart Grid **8**(1), 296–304 (2017)
5. Bahramirad, S., Reder, W., Khodaei, A.: Reliability-constrained optimal sizing of energy storage system in a microgrid. IEEE Trans. Smart Grid **3**(4), 2056–2062 (2012)
6. Li, H., Li, W., Luo, M., Monti, A., Ponci, F.: Design of smart MVDC power grid protection. IEEE Trans. Instrum. Meas. **60**, 3035–3046 (2011)
7. Saleh, K.A., Hooshyar, A., El-Saadany, E.F.: Hybrid passive-overcurrent relay for detection of faults in low-voltage DC grids. IEEE Trans. Smart Grid **8**, 1129–1138 (2017). ISSN 1949-3053
8. Emhemed, S., Burt, G.M.: An advanced protection scheme for enabling an LVDC last mile distribution network. IEEE Trans. Smart Grid **5**(5), 2602–2609 (2014)

9. Thomas, D.: Conducted emissions in distribution systems (1 kHz–1 MHz). IEEE Electromagn. Compat. Mag. **2**(2), 101–104 (2013)
10. Frantz, G., Frey, D., Schanen, J.-L., Bishnoi, H., Mattavelli, P.: EMC models for power electronics: from converter design to system level. In: Energy Conversion Congress and Exposition (ECCE), 15–19 September 2013. IEEE (2013)
11. Ales, A., Gouichiche, Z., Karouche, B., Moussaoui, D., Schanen, J.-L., Roudet, J.: The accurate input impedances of a DC-DC converters connected to the network. In: 2015 IEEE 15th International Conference on Environment and Electrical Engineering (EEEIC), Rome, 10–13 June 2015 (2015)
12. Feng, X., Ye, Z., Liu, C., Zhang, R., Lee, F.C., Boroyevich, D.: Fault detection in DC distributed power systems based on impedance characteristics of modules. In: Proceedings of Conference Record IEEE Industry Applications Society Annual Meeting, vol. 4, pp. 2455–3462, 8–12 October 2000 (2000)
13. Ales, A., Schanen, J.L., Moussaoui, D., Roudet, J.: Impedances identification of DC/DC converters for network EMC analysis. IEEE Trans. Power Electron. **29**(12), 6445–6457 (2014)
14. Foissac, M., Schanen, J.L., Vollaire, C.: Black box EMC model for power electronics converter. In: Proceedings IEEE Energy Conversion Congress and Exposition, pp. 3609–3615, 20–24 September 2009 (2009)
15. Tarateeraseth, V., Hu, B., See, K.Y., Canavero, F.G.: Accurate extraction of noise source impedance of an SMPS under operating conditions. IEEE Trans. Power Electron. **25**(1), 111–117 (2009)

Direct Torque Controlled Doubly Fed Induction Motor Supplied by WG and Based on ANN

Fethia Hamidia[1(✉)], Amel Abbadi[1], Oumsaad Benbouabdllah[2], and Younes Chiba[3]

[1] Electrical Engineering Department, LREA Laboratory, Yahia Feres University, Medea, Algeria
fe_hamidia@yahoo.fr
[2] Ingénieur études lignes électrique, Société de travaux d'électrification, Médéa, Algeria
[3] Department of Mechanical Engineering, Yahia Feres University, Medea, Algeria

Abstract. Faced with a growing demand for energy and the exhaustion of fossil fuels in the more or less long term, various alternatives have been envisaged as renewable energies.

Renewable energy is a type source of energy that renews itself quickly enough to be considered inexhaustible on a human scale. Renewable energies come from regular or constant natural phenomena (sun, wind…).

Renewable sources of energy, allowing decentralized production of electricity, can help to solve the problem of electrification of isolated site where a large number of individuals are devoid of any energy supply, thus being unable to satisfy any need even minimal and improve its living conditions. Utilizing universally applicable, low maintenance energy sources, the wind solution most often represents the ideal economic and technological choice for the regions or isolated installations.

In this work, we will discuss the application of wind for pumping system based on the use artificial neural network to track the maximum power point and to control DC voltage, the simulation show the effectiveness the proposed method.

Keywords: DTC · DFIM · WIND generator · MPPT · DC/DC converter · ANN

1 Introduction

Wind energy is, for many reason, one of the most promising renewable energy sources [1]. Wind energy systems are the fastest growing technology. Among these wind technologies, many systems of different types have been designed and developed while extending an experiment in this area dating back several centuries. Nowadays, the most known and used form of wind technology is the wind turbine, a machine that gets energy from the wind to generate an electric current. The chain of conversion of wind

M. Hatti (Ed.): ICAIRES 2019, LNNS 102, pp. 660–668, 2020.
https://doi.org/10.1007/978-3-030-37207-1_71

energy into electrical energy integrates various electrical components. In order to maximize the efficiency of this energy conversion, many solutions have been discussed both at the level of the generator to be used and power electronics. Most devices use synchronous or asynchronous machines but studies are also carried out in order to develop special machines. The electronic power interface between the wind turbine and the load is generally intended to control the generator in order to extract a maximum of wind power. Agricultural watering needs are usually greatest [2]. Wind energy is used in various isolated site applications such as pumping. Its use in pumping has responded quickly especially in rural and Saharan regions. For wind power generation several structures are used with vertical or horizontal axes and several methods are available at fixed or variable speeds. While for the generator system, the researchers use the different electrical machines for energy conversion as permanent synchronous machine. In addition, induction motor is used more and more for pumping systems. The low cost of the engine, the low maintenance requirements [3]. However, lately the researchers are interested in the doubly fed asynchronous machine for pumping system.The introductions of new use of DTC control based on artificial techniques in the renewable domain are very promising. First i dea of direct torque control was developed in 1986. This method of controlling that has been progressed during past decade, provides a fast torque response and also it is robust against machine parameter variations [4–6]. Currently, artificial intelligence techniques are also widely used in many fields such as industrial process control, image processing, diagnostics, medicine, space technology, computer data management systems and pumping system. Among all intelligent techniques, Artificial Neural Networks (ANN) seems to have the greatest impact in the field of power electronics and in the control of electrical machines, which is evident from the large number of publications produced in Literature.

The main objective of this work is so to continue and improve the research in the framework of production by wind and wind energy. Develop control methods to improve the efficiency and production of electrical energy. We present and discuss the application of direct torque control on the doubly fed induction motor supplied with wind system, then to track the maximum power point and to regulate the DC voltage needed to feed our system, an artificial neural network controllers are used.

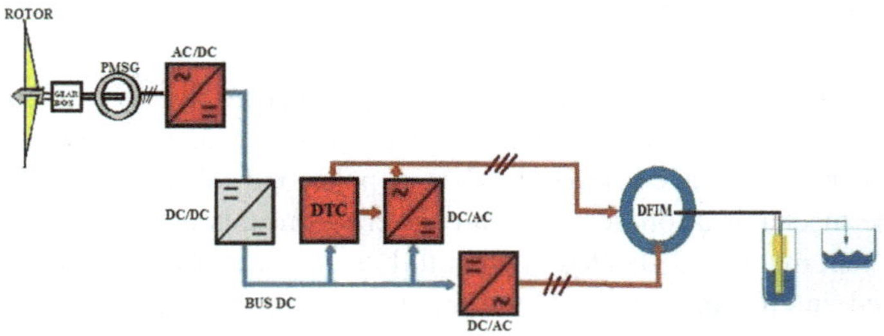

Fig. 1. Proposed intelligent pumping system

2 Wind Generator

A wind turbine is a device that transforms part of the kinetic energy of the wind (fluid in motion) into mechanical energy available on a transmission shaft and then into electrical energy via a generator. In this section, we are mainly interested in modeling and controlling the wind turbine. Where, an aerodynamic study of the turbine is presented, in order to know these main operating parameters, namely the power and torque coefficients, and the BETZ limit.

Artificial control strategies are described in the third and fourth section in order to control the turbine power of the turbine or to track the maximum power point. The last part of this section will be devoted to the control of the duty cycle of transistor to get the DC voltage (= 500 V).

The maximum available power of a site for a given wind speed is given by the following relation.

$$P_{\omega} = \frac{1}{2}\rho S V_{\omega}^{3} \tag{1}$$

With $S = \pi R^2$

The mechanical power recovered by the wind turbine can be written:

$$P_T = \frac{1}{2}\rho S V_{\omega}^{3} C_p(\lambda) \tag{2}$$

$$\lambda = \frac{R\Omega}{V_{\omega}} \tag{3}$$

$$T_T = \frac{1}{2}\rho R^3 V_{\omega}^{2} C_r(\lambda) \tag{4}$$

The wind torque depends to three variables: the wind speed, the rotation speed of the turbine shaft and wedge angle ß. [7]

$$T_g(V_{\omega}, \omega, \beta) = \frac{1}{2}\rho\pi R^3 V_{\omega}^{2} C_r(\lambda, \beta) \tag{5}$$

The development of magnetic materials has enabled the construction of synchronous machines with permanent magnets at costs that become competitive. Machines of this type have a large number of poles and make it possible to develop considerable mechanical torque. There are several concepts of synchronous machines with permanent magnets dedicated to wind turbine applications. As is shown in Fig. 1, the proposed intelligent pumping system, we use permanent magnet synchronous generator (PMSG).

3 Artificial Neural Network Control

3.1 MPPT Control

We propose in this part, a system of control of the output voltage of the wind system. Indeed, in most cases, such as wind pumping, for example our inverter (Fig. 4) must be powered by a constant voltage U_{dc}. The system comprises a wind generator, PMSG, a rectifier and DC-DC converter with its MPPT control, In this work, we propose an intelligent neuronal technique to control duty cycle of the switching transistor D1, as shown in Fig. 2.

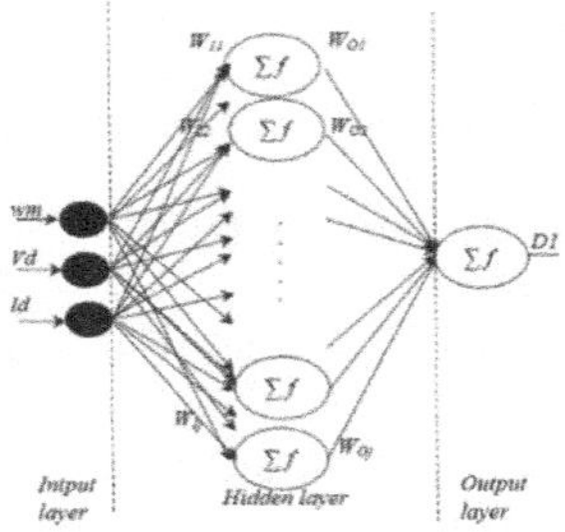

Fig. 2. Proposed intelligent MPPT appraoch

The caracteristic of the MPPT controller is presented as follow:

- Learning algorithm:**back-propagation algorithm**
- Input/hidden/output layer: **3/15/1**
- Input: three inputs presented by the current, voltage output of the rectifier connected to wind generator as presented in Fig. 2 and wind speed
- output: duty cycle of the switching transistor of DC-DC converter.
- activation functions: tansig for all neuron layers
- goal: the sum squared error falls under 2e−4 after 100 iterations.

3.2 DC Voltage Control

A static DC/DC converter equipped with an output voltage control system whose block diagram is shown in Fig. 4, this boost implements the regulation of the DC voltage on the DC bus. Thus, the DC bus is regulated to get a constant voltage value needed to supply our doubly fed induction motor (U_{dc}). We propose an artificial neural network ANN methods to control the output voltage

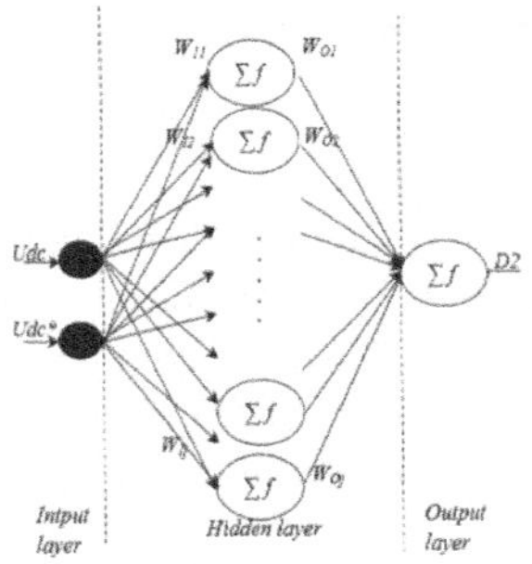

Fig. 3. Proposed neuro DC voltage regulator

The caracteristic of the DC voltage controller is presented as follow:

- Learning algorithm:**back-propagation algorithm**
- Input/hidden/output layer: **2/25/1**
- Input: two inputs presented by DC voltage (U_{dc} and U^*_{dc}) as presented in Fig. 3.
- output: duty cycle of the switching transistor of DC-DC converter
- activation functions: tansig for all neuron layers
- goal: the sum squared error falls under 5e−2 after 126 iterations.

4 Dual Star Induction Motor

Currently, the asynchronous motor is increasingly used in pumping systems. This type of engine is appreciated for its robustness, its low cost of purchase and maintenance. His order is more difficult than for other machines especially those with direct current. In addition, the efficiency of wind pumping systems makes the latter more interesting, even with the additional cost of the inverter. In recent years, the appearance of the inverter to control the speed of these motors has allowed their use in wind pumping applications. ($\sigma = 1 - \frac{M_{sr}^2}{L_s L_r}$)

$$\begin{aligned} \frac{dI_s}{dt} &= \frac{R_s}{\sigma L_s} I_s + \frac{M_{sr} R_r}{\sigma L_s L_r} I_r + \frac{1}{\sigma L_s} V_s - \frac{M_{sr}}{\sigma L_s L_r} V_r \\ \frac{dI_r}{dt} &= \frac{R_r}{\sigma L_r} I_r + \frac{M_{sr} R_s}{\sigma L_s L_r} I_r + \frac{1}{\sigma L_r} V_r - \frac{M_{sr}}{\sigma L_s L_r} V_s \end{aligned} \tag{6}$$

5 Direct Torque Control

In order to have an adjustable flow rate of the pump, it is necessary to have an asynchronous motor control system. The vector control by orientation of the rotor flow has the disadvantage of being relatively sensitive to variations in the parameters of the machine, which is why in the mid-1980s, a direct control strategy of the torque of the asynchronous machine appeared under the name of DTC (Direct Torque Control), this method was proposed by Takahashi and Depenbrock. The principle of the DTC control consists of the voltage vectors selection among the eight vectors delivered by the voltage inverter to control the torque and the stator flux after the determination of the components of the stator flux vector,

$$\phi_s = \int_0^t (V_s - Ri_s)dt \tag{7}$$

$$V_s(S_a, S_b, S_c) = \frac{2}{3} U_{dc}\left(S_a + S_b e^{\frac{j2\pi}{3}} + S_c e^{\frac{j4\pi}{3}}\right) \tag{8}$$

Subsequently, the instantaneous torque error is calculated and applied to a dual band hysteresis regulator which will generate at its output a three-level variable (−1.0, 1), to define the desired direction of evolution for the electromagnetic torque (as shown in Fig. 4). Regarding the stator flux, its module is calculated from the values of its coordinates, and then its error will be applied to a single band hysteresis regulator that will provide a binary variable, defining the evolution of the desired flux.

S_a, S_b, S_c, represent the state of the three phase legs 0 meaning that the phase is connected to the negative and 1 meaning that the phase is connected to the positive leg [7–10].

The stator current space vector is calculated from measured currents i_a, i_b, i_c:

$$i_s = \frac{2}{3}\left(i_a + i_b e^{\frac{j2\pi}{3}} + i_c e^{\frac{j4\pi}{3}}\right) \tag{9}$$

The component α and β of vector φ_s can be obtained

$$\begin{aligned} \phi_{s\alpha} &= \int_0^t (V_{s\alpha} - Ri_{s\alpha})dt \\ \phi_{s\beta} &= \int_0^t (V_s - Ri_{s\beta})dt \end{aligned} \tag{10}$$

Stator Flux amplitude and phase angle are calculated in expression:

$$\begin{aligned} \phi_s &= \sqrt{\phi_{s\alpha}^2 + \phi_{s\beta}^2} \\ \angle\phi_s &= arctg\frac{\phi_{s\beta}}{\phi_{s\alpha}} \end{aligned} \tag{11}$$

The DTC command has the advantage of having a simple implementation without resorting to mechanical sensors and fast dynamic response, however it presents problems for low speeds and the need to have stator flux estimators and electromagnetic torque

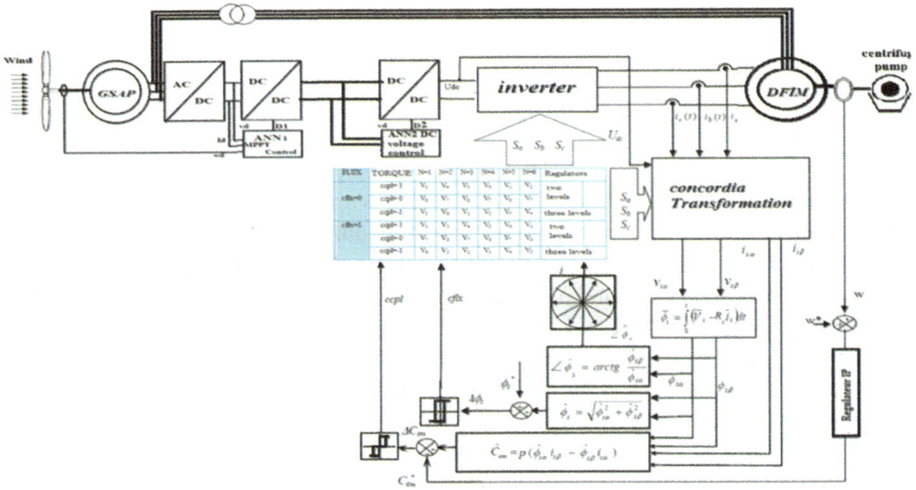

Fig. 4. Schematic diagram of DTC-DFIM supplied by wind generator

6 Simulation Results

The wind pumping system studied consists of a wind generator, PMSG, rectifier, a DC-DC boost converter with its MPPT control, a DC-DC adapter, an inverter and a centrifugal pump driven by DFIM as shown below and to confirm the effectiveness of this proposed system, simulations are performed in this section by using MATLAB/SIMULINK.

Figure 5 presents the speed, electromagnetic torque, current and flux obtained while starting up the induction motor initially under no load then connecting the nominal load. The speed reaches quickly its reference value without overshoot, however when the nominal load is applied at 0.2–0.4 s a little overtaking is noticed and the command reject rapidly the disturbance.

The results obtained from the closed-loop drive using the ANN controller to provide the maximum power point and also to track the U_{dc} voltage signal are shown in Fig. 5. Random step changes in the speed reference are applied. The response for a drive run-up is shown in Fig. 6. These results show that satisfactory closed-loop performance is obtained.

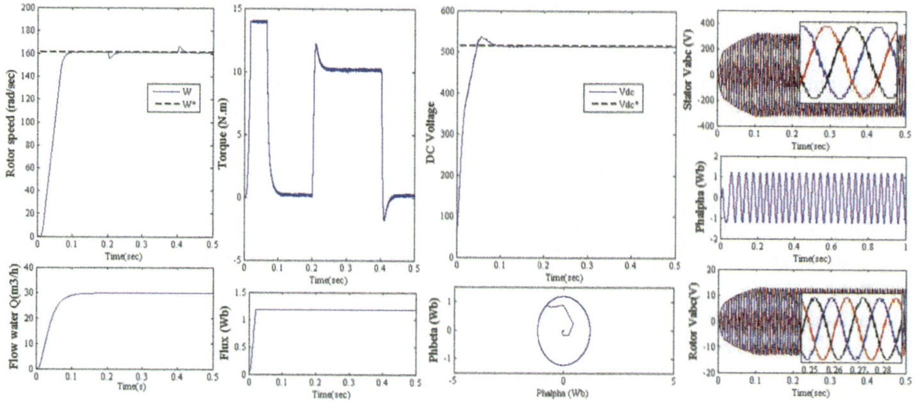

Fig. 5. Torque, flux, DC volatge, current and rotor speed responses using DTC-DFIM supplied by wind gnerator

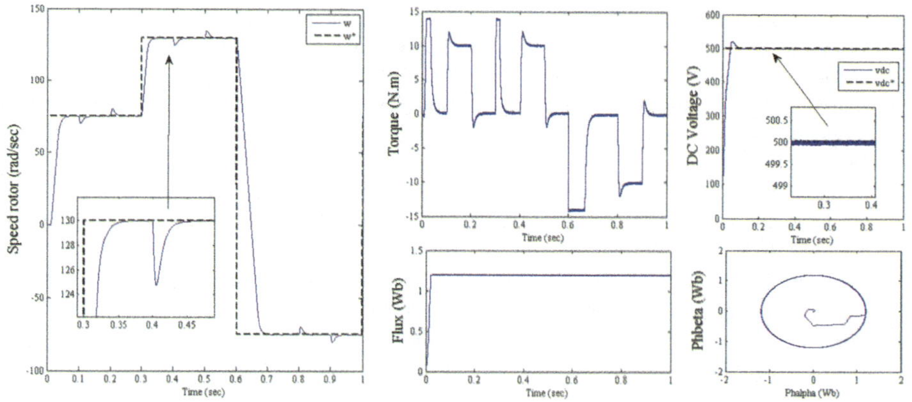

Fig. 6. Torque and rotor speed responses using DTC-DFIM supplied by wind gnerator (with speed and electromagnetic variation)

7 Conclusion

For an isolated and windy site, the use of wind energy for pumping water can be indispensable and highly competitive with other energy sources. We presented in this work a method of electric wind pumping from a wind turbine. So from current waveforms responses obtained from wind generator, it can be possible to power and to feed the AC motor and maintained the output voltage at a predetermined value. To get better performance, this paper proposes and discusses the application of one the most known artificial neural network named neural network, so the obtained results were very successful and confirm the validity of proposed technique

References

1. Thakur, R.K., Argawal, V.: Application of internal computation technique to fixed speed wind energy conversion system. IEEE Trans. Sustain. Energy Tehnol. 1093–1098 (2008)
2. Bouzid, A.M., et al.: Performance analysis of an isolated self-excited induction generator driven by a wind turbine under balanced/unbalanced loads. In: International Conference on Automatic and Mechatronics, pp. 322–327 (2011)
3. Bouzeriaa, H., Fethaa, C., Bahib, T., Abadliab, I., Layateb, Z., Lekhchinec, S.: Fuzzy logic space vector direct torque control of PMSM for photovoltaic water pumping system. Energy Proc
4. Abouda, S., Nollet, F., Chaari, A., Essounbouli, N., Koubaa, Y.: Direct torque control of induction motor pumping system fed by a photovoltaic generator. In: International Conference on Control, Decision and Information Technologies (CoDit), pp. 404–408 (2013)
5. Kianinezhad, R., Alcharea, R., Nahid, B., Betin, F., Capolino, G.: A novel direct torque control (DTC) for six-phase induction motors with common neutrals. In: International Symposiim on Power Electronics, Electrical Drives, Automation and Motion, SPEEDAM, pp. 107–112 (2008)
6. Seifossadat, Gh.: A direct torque control (DTC) method for dual three phase induction motors using a fuzzy inference system. In: International Symposium on Power Electronics Electrical Drives Automation and Motion, June 2008
7. Hamidia, F., Abbadi, A., Tlemçani, A., Boucherit, M.S.: Dual star induction machine supplied by photovolatic panel based on fuzzy type 2. Nonlinear Dyn. Syst. Theory (NDST) **18**(4), 359–371 (2018)
8. Hamidia, F., Abbadi, A., Boucherit, M.S.: Wind water pumping system based on ANN. In: 6th International Conference on Control Engeneering & Information Technology, Istambul, Turkey (2018)
9. Merzoug, M.S., Naceri, F.: Comparison of field-oriented control and direct torque control for permanent magnet synchronous motor. World Academy
10. Hamidia, F., Abbadi, A., Tlemçani, A.: Improved pumping system supplied by double photovoltaic panel. Rev. Roum. Sci. Tech.-Electrotechn **64**, 87–93 (2019)

Transformerless PV Three Level NPC Central Inverter

Mohammed Yassine Dennai(✉), Hamza Tedjini, and Abdelfettah Nasri

SGRE Laboratory, Tahri Mohamed University, Bechar, Algeria
dennaimoh@yahoo.fr, tedjin_h@yahoo.fr

Abstract. Generally inverter is developed with centre on high reliability, low cost and mass-production for transformed electrical energy from the PV module to the grid. Basically the DC/AC inverters are used in grid connected photovoltaic energy production systems as the power processing interface between the electric grid and the photovoltaic energy source. The energy injected into the electric grid by the photovoltaic system depends on the amount of power extracted from the photovoltaic power source and the efficient processing of the DC/AC inverter power. Our future energy sources are renewable energy sources, one of the renewable energy is solar energy, so there is need to convert these energy forms into electrical energy efficiently. Generally solar to electrical energy conversion efficiency is very less. The conversion process has different stages at each stage there will be some energy loss. In solar system dc-ac conversion is one of the stages in these paper three level NPC converter is used. In this paper explain the basic idea about conversion DC voltage of PV in AC voltage by use The three level NPC inverter.

Keywords: Photovoltaic system (PV) · Three level NPC converter · SPWM · THD

1 Introduction

Recent year renewable energy sources, like wind, tides, solar, geothermal heat, biomass etc., are advantageous for electrical power generate due to the friendly environmental nature and infinite existence. Recent year energy policies and global warming have become a main topic on the international agenda. Today developed countries are trying to mitigate the greenhouse effect [1]. In this circumstance, generation of photovoltaic power has an important role to play due to the fact that it is a green source. Solar photovoltaic array generates DC power at its maximum by using boost converter with while power inverter converts this DC power to AC power and feeds to utility grid. Requirement of renewable energy sources is rapidly increasing because of the high cost and limited conventional energy sources as gas, oil, coal and multilevel inverters are more reliable to interface with renewable energy sources than a conventional two-level inverter for a high power application in aspects of better quality of power [2, 10], further more multilevel Inverters have become more popular over the years in application of

M. Hatti (Ed.): ICAIRES 2019, LNNS 102, pp. 669–678, 2020.
https://doi.org/10.1007/978-3-030-37207-1_72

electric power with the secure of less disturbances outputs and the possibility to function at higher switching frequencies other than ordinary two-level inverter [2].

2 The Different Photovoltaic Sectors

2.1 Crystalline Silicon Solar Cell

Crystalline silicon photovoltaics is the most widely used photovoltaic technology. Crystalline silicon photovoltaics are modules built using crystalline silicon solar cells (c-Si), developed from the microelectronics technology industry. Crystalline silicon solar cells have high efficiency, making crystalline silicon photovoltaics an interesting technology where space is at a premium (Fig. 1).

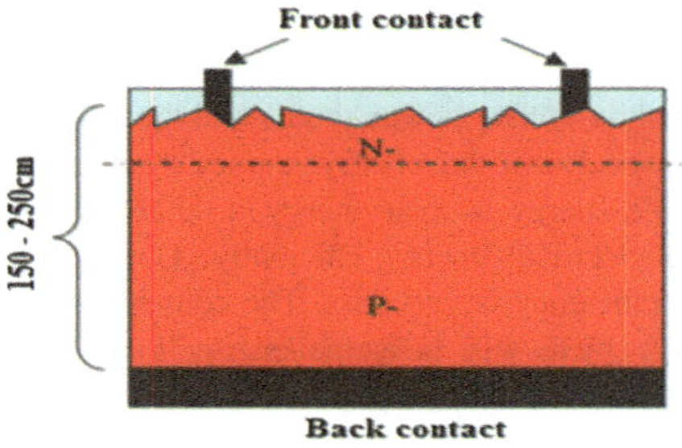

Fig. 1. Schematic diagram of a cell based on crystalline silicon.

2.2 High-Efficiency Multijunction Solar Cells

Today, most inorganic photovoltaic cells consist of a single PN junction. In this junction, only photons whose energy is equal to or greater than the forbidden band of the material (denoted E g in eV) are capable of creating electron-hole pairs. In other words, the photovoltaic response of a single junction cell is limited. Only the proportion of the solar spectrum whose photon energy is greater than the absorption gap of the material is useful, the energy of the lower photons is therefore not usable.

On the other hand, even if the energy of the photons is sufficient, the probability of meeting an electron is low. Thus, most photons pass through the material without having to transfer their energy. A first response to limit losses has long been known from the technological point of view, it is sufficient to use multi-level systems, stacking junctions with decreasing gaps, (Fig. 2). Thus it is possible to exploit almost all the solar spectrum with very high conversion efficiencies [11].

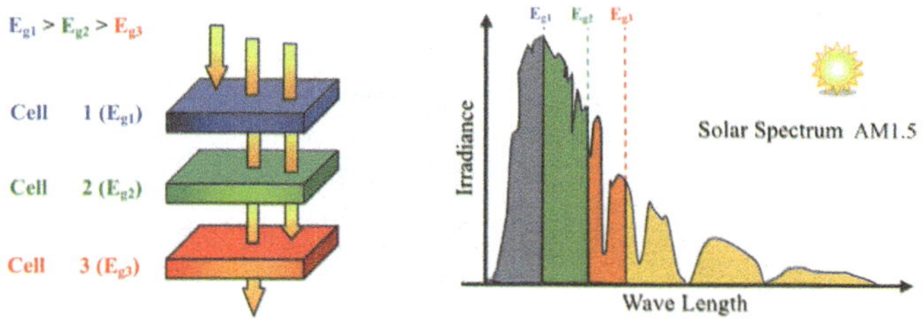

Fig. 2. Schematic diagram of a cell based on crystalline silicon. [11]

2.3 New Photovoltaic Technologies

Other research has focused on an organic and inorganic photovoltaic cell model (hybrid cells) that offers many advantages over traditional cells. Thus, a nanocrystal line cell has been developed that mimics photosynthesis of plants. Used for a long time in dye lasers or optical materials, the efficient use of organic dye molecules in photovoltaics was discovered in 1991 by Michael Graetzel. Pigmented organic molecules (dye) absorb light and release electrons. The electrons are transported to the anode by a porous layer of titanium dioxide (TiO2), an inorganic semiconductor material. At the anode, the electrons are directed to an external circuit where their passage produces electrical energy [12, 13] (Fig. 3).

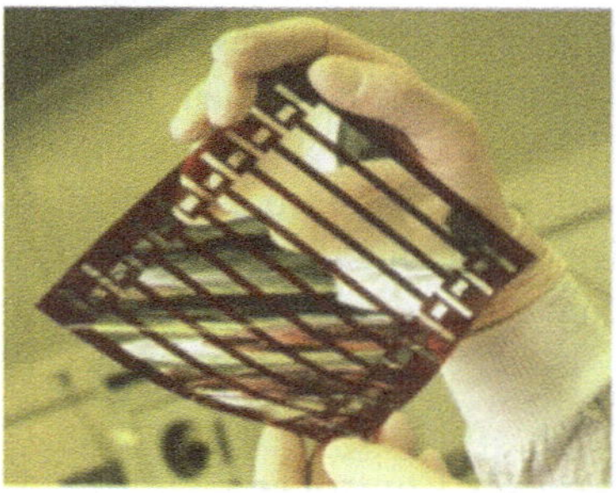

Fig. 3. All organic solar cell on flexible substrate [11]

3 The PV-Load Generator Connection via a DC-AC Adapter Stage

There are different management topologies for these installations. Nevertheless, all of these approaches rely on grid-connected GPVs through inverters that transfer and shape electrical solar energy. The progress made in recent years in the development of inverters dedicated to photovoltaics has made it possible to significantly change these management systems. there are mainly three UPS architectures giving good technical solutions: the central inverter, strings inverters and inverters integrated into the panels, (Fig. 4).

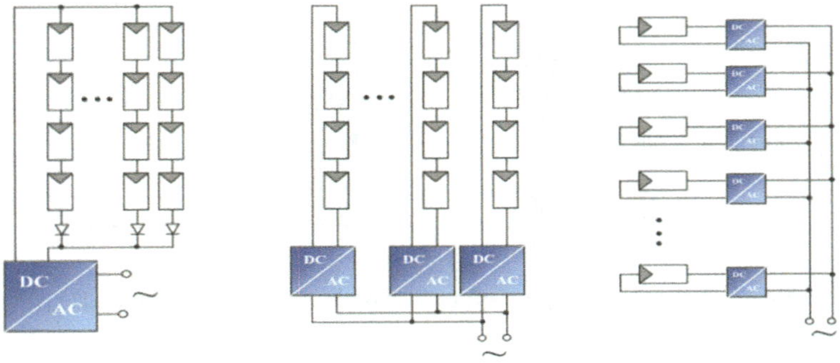

a) Central Inverter b) String Inverter c) PV integrated inverter

Fig. 4. Schematic diagrams of different topologies of PV installations connected to an electrical network [11].

In this case we used the central inverter, the various solar panels for a large installation are mounted in rows to form a string, which is itself coupled in parallel with non-return diodes to several others, Fig. 4a. The high power PV generator thus structured is connected on the DC side to a single inverter. This central inverter is highly energy efficient at reduced costs. The main function of this device is to create an AC voltage from a DC voltage most compatible with the network and to constantly examine the presence or absence of the network to allow the injection of current. The conventional structure of the inverter is often a bridge circuit for connecting each of the two input poles to each of the two output poles by means of power switches.

Solar array consists of solar module connected in series and parallel. A single solar module is constructed with a number of cells depending upon the capacity of module. The cells operate at a low voltage, ranging from less than 1 V (PV cell) to 3 or 4 V (Li-Ion cell). These low voltages do not interface well to existing high power systems, so the solar cells are series connected to create a battery or a PV module with a higher voltage terminal [8, 9].

3.1 Maximum Power Tracker Controller

The algorithm implemented in the first MPPT commands was relatively simple. Indeed, the capabilities of microcontrollers available at the time were low and applications, especially for the space had much less constraints in temperature and illumination variation than terrestrial applications. Applied initially to photovoltaics, its principle was first described by A. F. Boehringer. This command is based on an adaptive control algorithm, keeping the system at its maximum power point (PPM). The latter is described in Fig. 5 and can be implemented entirely in digital.

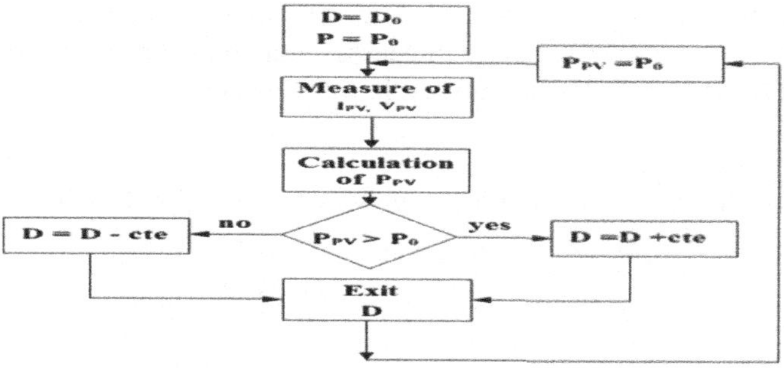

Fig. 5. Principle of MPPT control

4 Multilevel Inverter Topology

The basic three types of multilevel topologies are [3, 4]:

- Diode clamped multilevel inverters
- Flying capacitors multilevel inverter or capacitor clamped multilevel inverter
- Cascaded inverter with separate DC sources.

4.1 Three-Level Diode Clamped Inverter

Usually simple two-level inverters are generally used to generate an AC voltage from a DC voltage. Working of the two-level inverter can create only two different output voltages for the load is +Vdc/2 and −Vdc/2 (This is consider when the inverter is fed with Vdc). The basic circuit diagram of 3-level diode clamped inverters as illustrate in Fig. 6 does not depend on just two levels of voltage to generate an AC signal. When we get the smoother waveform, the more voltage levels generated in the inverter. However if many levels of voltage will make raise the complexity of the circuit, with more apparatus and a more complications the controllers for the inverter are required and being used [6].

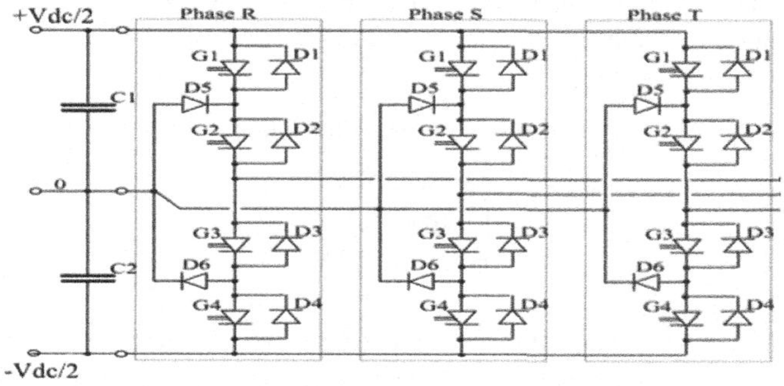

Fig. 6. Circuit arrangement for three-level diode clamped inverter for single phases.

In neutral point clamp topology or diode clamp topology, multilevel inverter diodes have been apply as a clamping device to generate required output voltage levels [5]. In 3-level diode clamp inverter, the duty cycle for switches ON/OFF is done shown in Table 1 [5].

Table 1. Switching modes of 3-level inverter for phase R.

Magnitude of voltage (VDC)	No. of switches to be on/off			
	S1	S2	S3	S4
+Vdc/2	On	On	Off	Off
0	Off	Off	On	Off
−Vdc/2	Off	Off	On	On

4.2 Control Technique: Sinusoidal Pulse Width Modulation (SPWM)

For the switching techniques inverters that use pwm techniques have a dc input voltage that is always constant in magnitude. The inverter will acquire this as an input voltage and output ac where the frequency and magnitude can be controlled as per the requirement. There are various ways by which pwm can be implemented to make the output to be ac. A widespread technique known as spwm (sinusoidal pulse width modulation) has been used and discussed here. In order to get output as a sinusoidal waveform for 7-level inverter at a definite frequency a sinusoidal control signal (vsine) at the definite frequency is compared with a (m-1) triangular waveform (vtri). Here, m is number of inverter level (2, 3, 4, …, etc.). Usually the inverter will use than the frequency of the triangle wave as the switching frequency which is essentially kept constant [5–7] (Figs. 7 and 8).

5 Simulation and Results

The studied system is presented in Fig. 9; The output power of solar array depends upon solar irradiance (in KW/sq.), cell temperature and area of array. Here total 300 modules have been used (connected in series) each having 54 solar. The rating of solar panel is tabulated in Table 2.

Table 2. Solar panel rating.

Voltage	32.9 V
Current	8.21 A
Power	270 W

The PV arrays output is gives as input to the inverter and Fig. 9 shows the I-V Characteristics of the solar cell here Alternate phase opposition displacement Sinusoidal Pulse Width Modulation Scheme is used to generate the control signals for the inverter Fig. 11 shows the modulation scheme diagram (Fig. 10).

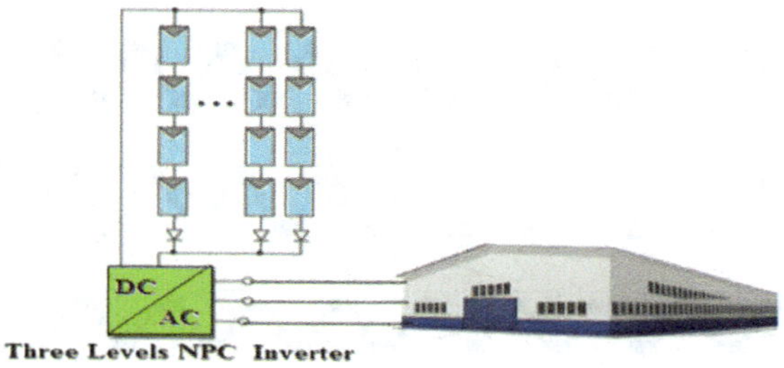

Fig. 7. Studied system.

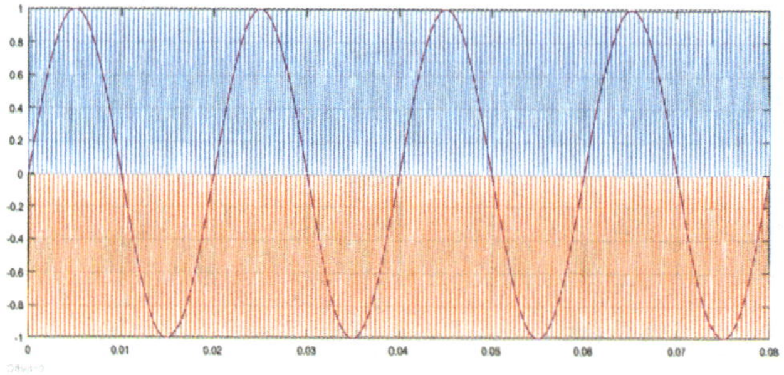

Fig. 8. Comparison of carrier signals and reference for SPWM generation.

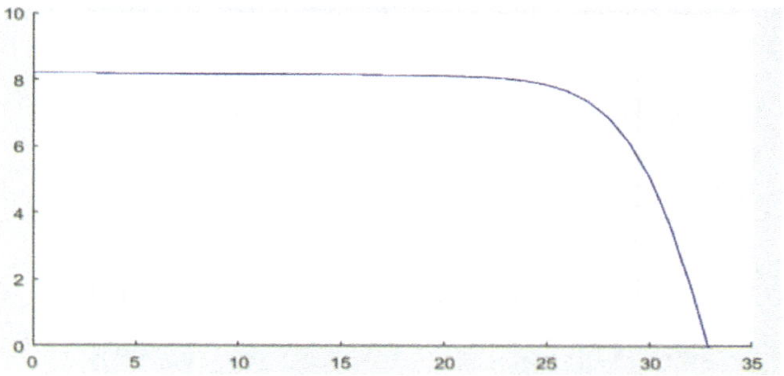

Fig. 9. Current and voltage (I-V) characteristic of the solar cell.

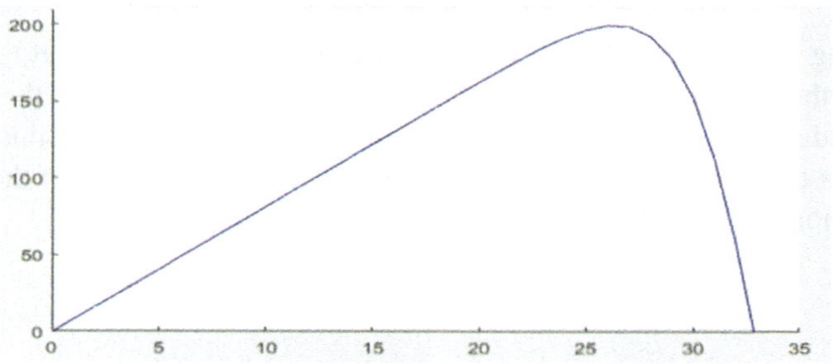

Fig. 10. Power and voltage (P-V) characteristic of the solar cell.

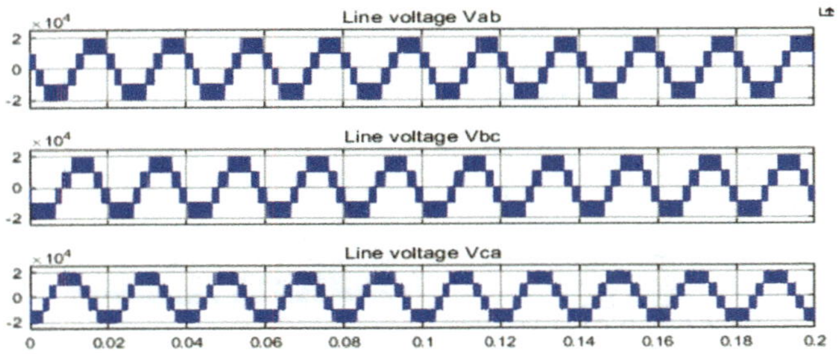

Fig. 11. Output line voltage waveform of three level NPC converter.

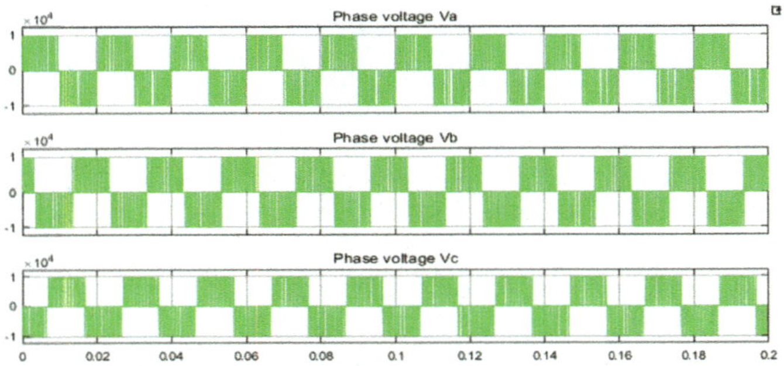

Fig. 12. Waveform of phase voltage across the switches.

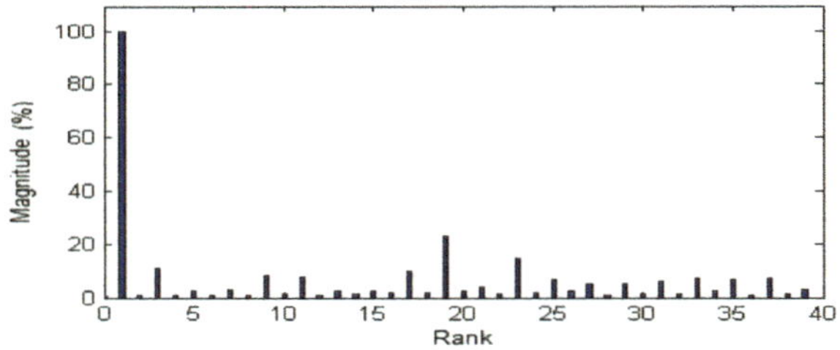

Fig. 13. Total harmonic distortion of the three level NPC converters for phase displacement sinusoidal pulse width modulation, for switching frequency f s = 2 kHz, reference wave frequency f r = 50 Hz.

According to the spectral analysis of each one of these results, that the increase of the levels of the converter makes it possible to have a better quality of the signal; and a more sinusoidal look; and following several scientific researches that this quality does not change in case of more than nine levels, the major disadvantage of these structures remains economic (Figs. 12, 13 and 14).

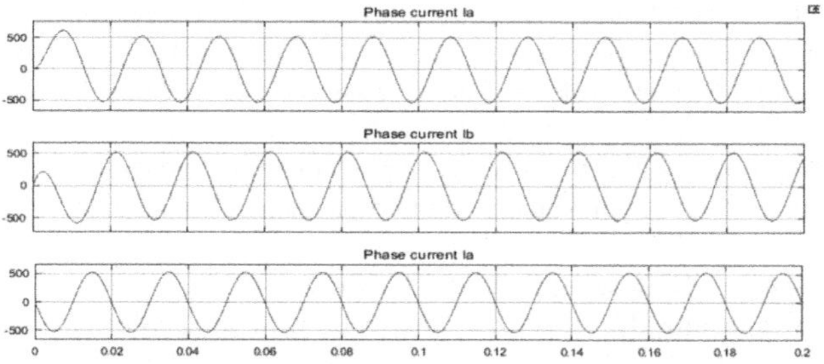

Fig. 14. Load current waveform of 3 phase for R = 10 ohms, L = 50e-3 H.

6 Conclusion

The paper has existing the model of the 3-level diode clamped inverter for solar application along by using SPWM modulation techniques and simulation is done by using MATLAB. By doing different analysis on basis of MATLAB results. Here a complete design of solar power system in open loop has been done to feed required power in grid system with appropriate number of solar module connected in series . Also different level of inverters have been used to reduce the harmonics and filter, to achieve sinusoidal waveform by NPC three level inverter.

References

1. Jiang, Y., Qahouq, J.A.A., Batarseh, I.: Improved solar PV cell Matlab simulation model and comparison. In: Proceedings of 2010 IEEE International Symposium on Circuits and Systems (ISCAS), 30 May 2010–2 June 2010, pp. 2770–2773 (2010)
2. Rodriguez, J., Lai, J.S., Peng, F.Z.: Multilevel inverters: a survey of topologies, controls, and applications. IEEE Trans. Ind. Electron. **49**(4), 724–738 (2002)
3. Renge, M.M., Suryawanshi, H.M.: Five-level diode clamped inverter to eliminate common mode voltage and reduce dv/dt in medium voltage rating induction motor drives. IEEE Trans. Power Electron. **23**(4), 1598–1607 (2008)
4. Jain, K., Chaturvedi, P.: MATLAB-based simulation & analysis of three-level SPWM inverter. Int. J. Soft Comput. Eng. (IJSCE) **2**(1), 56–59 (2012). ISSN 2231-2307
5. Ibrahim, Z.B., Hossain, M.L., Bugis, I.B., Lazi, J.M., Yaakop, N.M.: Comparative analysis of PWM techniques for three level diode clamped voltage source inverter. Int. J. Power Electron. Drive Syst. (IJPEDS) **5**(1), 15–23 (2014)
6. Jain, K., Chaturvedi, P.: Matlab - based simulation & analysis of three - level SPWM inverter. Int. J. Soft Comput. Eng. (IJSCE) **2**(1), 56–59 (2012)
7. Kedareswari, M.: Reduction of THD in diode clamped multilevel inverter employing SPWM technique. Int. J. Sci. Res. Publ. **3**(6), 1–4 (2013)
8. Walker, G.R., Sernia, P.C.: Cascaded DC-DC converter connection of photovoltaic modules. IEEE trans. Power electron. **19**(4), 1130–1139 (2004)
9. Wenham, S.R., Green, M.A., Watt, M.: Applied Photovoltaics (1994)
10. Jiang, Y., Qahouq, J.A.A., Batarseh, I.: Improved Solar PV Cell Matlab Simulation Model and Comparison (2010). 978-1-4244-5309-2/10/$26.00 ©2010 IEEE

11. Petibon Stéphane, M.: Nouvelles architectures distribuées de gestion et de conversion de l'énergie pour les applications photovoltaïques. Délivré par l'Université Toulouse III - Paul Sabatier, Doctoral School: GEET; Systems Analysis and Architecture Laboratory, Janvier 2009
12. Koide, N., Islam, A., Chiba, Y., Han, L.: Improvement of efficiency of dye-sensitized solar cells based on analysis of equivalent circuit. Photochem. Photobiol. A: Chem. **182**(3), 296–305 (2006)
13. Ito, S., Murakami, T.N., Comte, P., Liska, P., Grätzel, C., Nazeeruddin, M.K., Grätzel, M.: Fabrication of thin film dye sensitized solar cells with solar to electric power conversion efficiency over 10%. Thin Solid Films **516**(14), 4613–4619 (2008)

Spin-Orbit Coupling's Effect on the Electronic Properties of Heavy Elements-Based Compounds

M. Abane(✉), M. Elchikh, and S. Bahlouli

University of Sciences and Technology USTO-MB, Oran, Algeria
mahdi.bvb38@gmail.com, m.elchikh@gmail.com, bsoumia56@yahoo.fr

Abstract. The topological insulating is the topic studied in the last days, these compounds have a lot of properties as electronics, optics and till superconducting, among compounds that will be a topological insulating or topological semimetals are: the semi-Heuslers that contain heavy elements of general formula XYZ, this study was about one of them is: ScPdBi. We studied its structural properties to investigate its crystallographic phase: ferromagnetic (FM) or anti-ferromagnetic (AFM), we found that it is an AFM compound and we classified its structural properties parameters. After that, we studied its electronic properties (band structure and DOS) without spin-orbit coupling (SOC) where we found that this compound has a metallic character. Finally, we applied the SOC where we found different results according the electronic properties where the inversion band appeared so the topological non-trivial nature appeared; this compound is from the most important candidate for the superconducting applications found by an external effect about it as hydrostatic pressure. The calculations were by the first principle based in DFT implemented in ELK code; the exchange potential calculated by general gradient approximation (GGA).

Keywords: Heavy fermions · Topological insulating · Semi-Heuslers · Spin-orbit coupling

1 Introduction

The spin-orbit coupling (SOC) [1] effect is more important because give us a different characters specially in electronic properties as we will show in this work, our study was about one of the important compounds in industry and intelligent materials actually [2, 3], that is the heavy elements compounds that can be appeared as a semi-Heusler compound of general formula XYZ, in this work we studied the structural properties of one of these compounds: ScPdBi and we will show its crystallographic phase if it is a ferromagnetic or anti-ferromagnetic, then we did an electronic calculations: band structure and DOS without SOC for compare with those are with and the appearance of topological insulating (Ti) character [4, 5]. Finally; we applied the SOC where we showed a different character according the band structure, the DOS confirmed the result. We used the first principle based by the density functional theory (DFT) [6] with general gradient approximation (GGA) [7, 8] implemented in ELK code. Our results are compatible with literature.

M. Hatti (Ed.): ICAIRES 2019, LNNS 102, pp. 679–683, 2020.
https://doi.org/10.1007/978-3-030-37207-1_73

Finally, we can confirm after interpreting results found that the effect of spin-orbit coupling is more important to specify a topological insulating character of heavy fermion-based compounds that it is a good candidate for the superconducting applications. We did the spin-orbit coupling because our compound has heavy atoms that have a grand electronic correlation.

2 Compound Studied

2.1 Details

The compounds studied is ScPdBi, it is a semi-Heuslers compound, with space group 216 (P $\bar{4}$3m) (Fig. 1).

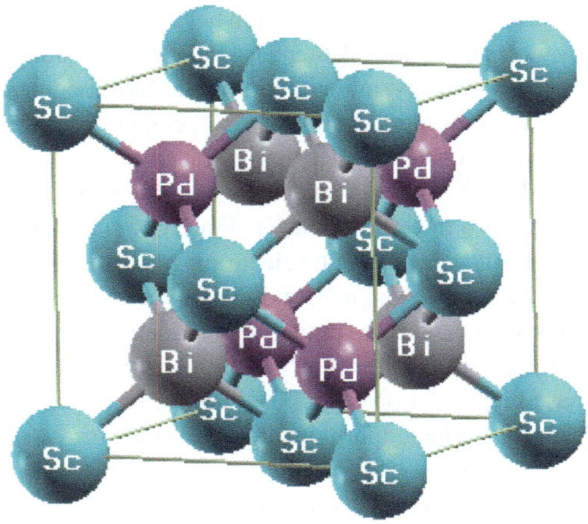

Fig. 1. ScPdBi mesh.

2.2 Calculations Methods

The study was carried out by a first principle calculation based on the density functional theory (DFT) implemented in the ELK code, and using the linear potential augmented linear waveform method (FP- LAPW). The exchange and correlation potential is processed by the generalized gradient approximation (GGA).

3 Results and Discussion

3.1 Structural Properties

We made many calculations about different mesh parameter in the both phases: ferromagnetic (FM) and anti-ferromagnetic (AFM), then we plotted the volume-energy curve (Fig. 2), where we find that the most stable phase is AFM; so it is AFM compounds. The structural properties parameters are is Table 1.

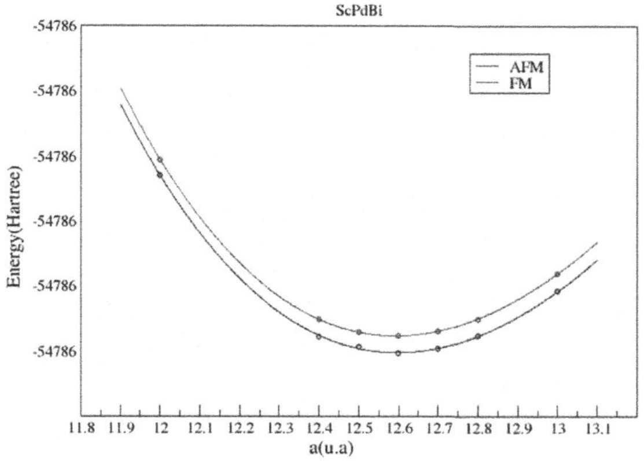

Fig. 2. Volume-Energy curve of ScPdBi

Table 1. Structural properties parameters of ScPdBi

Compounds	A(A°)	B_0(GPa)	B'_0
ScPdBi AFM	6,664	66,914	4,452

3.2 Electronic Properties

We calculated the band structure and DOS for the AFM compounds without Spin-Orbit Coupling, we found that ScBiPd AFM has a metallic character; where the spin up (a) and spin down (b) of band structure (Fig. 3) are the same.

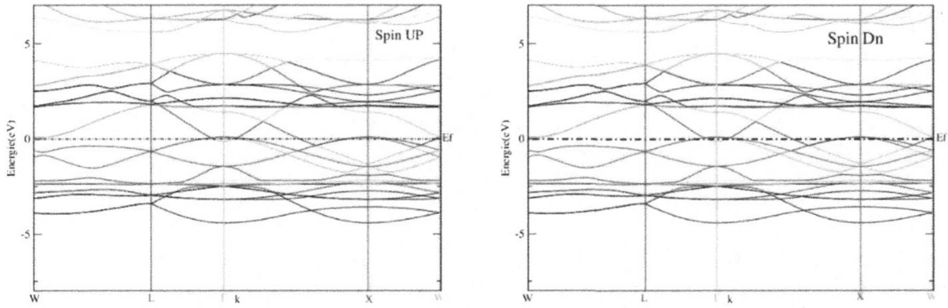

(a) (b)

Fig. 3. Band structure (a): spin Up, (b): spin Dn of ScPdBi

The results found by band structure are confirmed by TDOS (Fig. 4), where we show that is the same character in spin Up and spin Dn because the compound is antiferromagnetic.

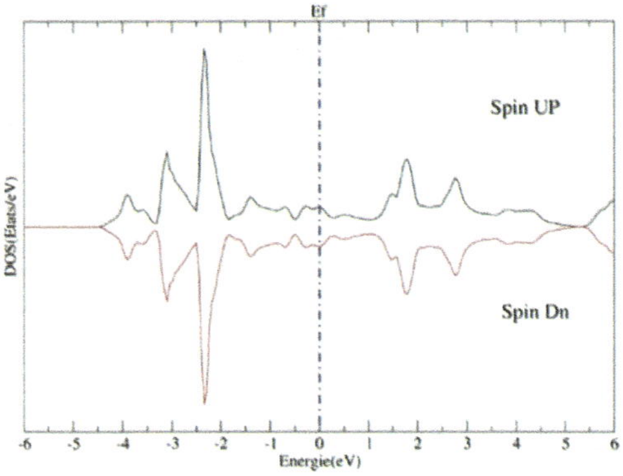

Fig. 4. Total DOS (spin (Up and Dn)) of ScPdBi

3.3 Spin-Orbit Coupling Effect

We did the same study about ScPdBi previous but we aplied the SOC, so we found that the topological nature appeared in band structure (Fig. 5) as a topological non-trivial with $\Delta E = E(\Gamma 6) - E(\Gamma 8) = -0.21$ eV (Fig. 5b). As we show that is a band inversion and states surface because the spin Up and spin Dn of band here gave different results; so this compound can be as topological insulating or topological semimetal.

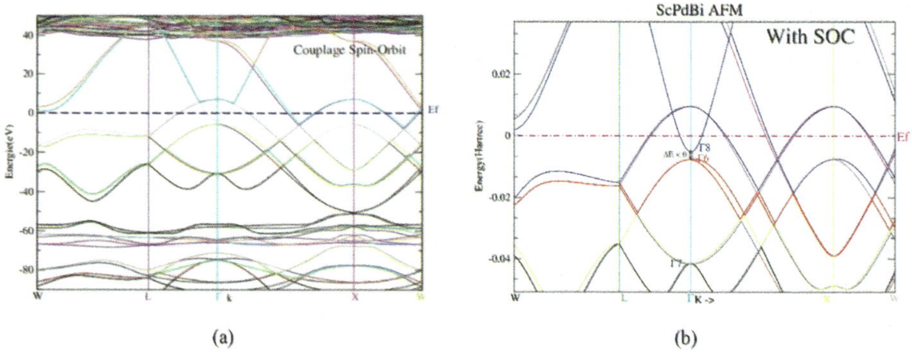

Fig. 5. Band structure (spin (Up and Dn)) with SOC of ScPdBi, (a): general band structure, (b) specify band structure of Γ states: Γ8 in blue, Γ6 in red and Γ7 in black.

The previous results found by band structure are confirmed by the TDOS with a spin-orbit coupling (SOC) where we show that in spin Up and spin Dn of TDOS aren't the same but there is a band inversion appeared (Fig. 6).

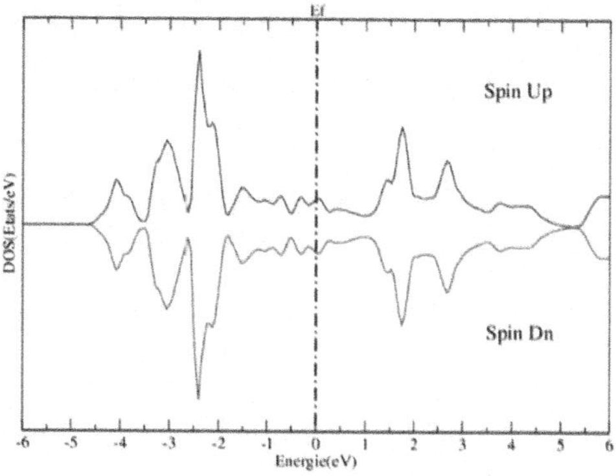

Fig. 6. Total DOS (spin (Up and Dn)) with SOC of ScPdBi

4 Conclusion

The study of structural stability has shown us that this compound is antiferromagnetic. We then adjusted the mesh parameter variation curve as a function of energy by a Murnaghan state equation [9], which allowed us to calculate the mesh parameter of this compound. Our results found are in good agreement with the literature.

The study of the electronic properties shows us that this material is as topological insulating or topological semimetal compound with a topological nature non-trivial appeared by SOC effect that give us a supply fuel to investigate the band structure. Under the effect of hydrostatic pressure and therefore a good candidate for superconducting applications [10].

References

1. Gao, Y., Guo, X., Lu, J.: Effect of the spin-orbit coupling on the topological phase of half Heusler compounds. Int. J. Sci. **2**(8) (2015)
2. Fu, L., Kane, V.L., Mele, E.J.: Phys. Rev. Lett. **98**, 106803 (2007)
3. Culver, Dimitrie: Phys. E **44**, 860 (2012)
4. Moore, J.E.: Nature **464**, 194 (2010)
5. Qi, X.-L., Zhang, S.-C.: Phys. Today **63**(1), 33 (2010)
6. Hohenberg, P., Kohn, W.: Phys. Rev. **134**, 3583 (2011)
7. Perdew, J.P., Burke, K., Ernzerhof, M.: Phys. Rev. Lett. **77**, 3865 (1996). Comput. Mater. Sci. 592–599
8. Wu, Z., Cohen, R.E.: Phys. Rev. B **73**, 235116 (2006)
9. Birch, F.: Phys. Rev. **71**, 809 (1947)
10. Ishikawa, M., Jorda, J.L., Junod, A.: In superconductivity in d- and f-Band Metals (2015)

Selective Control Approach for DFIG Powered by Parallel Inverters

Dris Younes[1(✉)], Benhabib Mohamed Choukri[2], and Meliani Sidi Mohammed[2]

[1] Laboratoire d'automatique de Tlemcen (LAT), Université de Tlemcen, Tlemcen, Algeria
younes.dris@univ-tlemcen.dz

[2] Manufacturing Engineering Laboratory of Tlemcen, Université de Tlemcen, Tlemcen, Algeria

Abstract. The use of power electronics in the wind energy application is an essential part. In this work, parallel inverters are used to supply the DFIG with the goal to meet the high current demand. The disadvantage of this topology is the large number of power electronic devices (IGBTS) that lead to high power switching losses. The purpose of this paper is to use a control algorithm, named "selective control", to address this weakness. This technique makes it possible to activate or deactivate the inverters used according to the amplitude of the current demand. In other hand a maximum power tracking MPPT technique is used to extract the maximum power from the wind. The proposed system is simulated and tested by using the SimPowerSystem toolboxes in MATLAB. The results show the high effectiveness of the proposed control in terms of pursuing maximum power and reducing THD.

Keywords: Wind energy · Parallel inverter · DFIG · Selective control · Maximum power tracking MPPT · SimPowerSystem

1 Introduction

The future in the energy sector is to develop a clean and alternative energy source, which could regenerate naturally and meet the world's energy demand. Wind energy is one of the most useful renewable energies in the world. It has provided an economical solution, because it reduces the cost of electricity and creates jobs (Herbert et al. 2007). Wind turbines can operate in many applications (agriculture; Industry, commerce and residence). The global cumulative installed wind capacity has increased exponentially from approximately 23.9 GW in 2001 to 539 GW in 2017 (REN21 2018).

Variable-speed wind energy conversion systems (WECS) using doubly fed induction generator (DFIG) have become in the Foreground of the wind generator used in the world, because of their high-energy efficiency, controllability and improved power quality. The main advantage of DFIG is the size of the converter, which is considered as a small fraction, it is estimated by 20% to 30% of the overall power system. The power converter of the (DFIG) should have the ability to control the reactive power Q and the active power P injected to the grid regardless of the wind speed. For that, the self-commutated converter

M. Hatti (Ed.): ICAIRES 2019, LNNS 102, pp. 684–692, 2020.
https://doi.org/10.1007/978-3-030-37207-1_74

normally adopts pulse width-modulated (PWM) control methods, which is widely used in (WECS). This converter is based on semiconductor devices, which have the ability to turn ON/OFF and can operate at high switching frequencies, such as insulated gate bipolar transistor (IGBT), MOSFET, integrated gate-commutated thyristor (IGCT) and silicon carbide MOFET (Ma et al. 2015).

There are many proposed topologies, which use the IGBTs device. In (Chemidi et al. 2015), a matrix converter is used to control the DFIG in both modes of operations (sub-synchronous and super-synchronous). However, the major disadvantage of matrix converter that is more complicated in terms of control (Kolar et al. 2011). The simplicity of conventional back-to-back converter proposed in (Pena et al. 1996) make it very useful in (WECS). It is composed of two inverters sharing a common DC link. In this type of structure, the choice of inverter topology, depends on the power produced by the wind turbine, the grid voltage and IGBT characteristics. For low voltage a two-level inverter is enough as mentioned in (Teichmann and Bernet 2005). But for high voltage applications, it is necessary to use a multilevel inverter (Narimani et al. 2014). A lot of research is done to develop this type of inverter which is divided into three categories; neutral-point diode clamped structure, flying capacitor clamped structure, and cascaded converter cells (Ke 2015). However, all the previous structures are adopted for high voltage but they cannot operate in high current applications, because the current supported by one IGBT is limit.

As a solution of this weakness, this paper proposes a structure, which is composed of two parallel inverters with a common DC-link. The aim of this structure is to withstand high current intensity. To control this converter, a selective algorithm is proposed to minimize the switching losses of the IGBT.

2 System Description

The proposed WECS is composed of both mechanical, electrical and control parts. The mechanical part consists of three blades and a gearbox, because designing a low speed multipolar DFIG is not technically feasible. The mechanical part transforms wind kinetic energy to mechanical energy. The electrical part is made of a DFIG that transforms the energy captured by the mechanical part into electrical energy. This generator can produce power at constant voltage and constant frequency though the rotor speed varies (Pena et al. 1996). The stator winding of the DFIG is connected directly to the grid while the rotor winding is connected to the grid through the proposed parallel inverter, this converter is made from two parts: rotor side converter (RSC) and grid side converter (GSC). RSC is connected to the DFIG rotor and the GSC is connected to the grid. Concerning the control part, a decoupled d-q vector control technique is employed for the control of the rotor circuit. For controlling the RSC to achieve a maximum power, a maximum power point tracking (MPPT) technique is used to estimate the speed reference Ω_{ref}. Then a speed controller adjusts the speed of the DFIG to achieve the desired speed. Next, a current controller is used to control the active power P and the reactive power Q injected to the grid via the stator of the DFIG. The GSC is used to keep the voltage at the DC bus voltage constant regardless of the magnitude and direction of the rotor power. Finally, a proposed selective control is employed to turn ON/OFF converters of the rotor circuit. The control is done according to the power transferred to or from the grid. Figure 1 shows the configuration of the DFIG system connected to the grid utility.

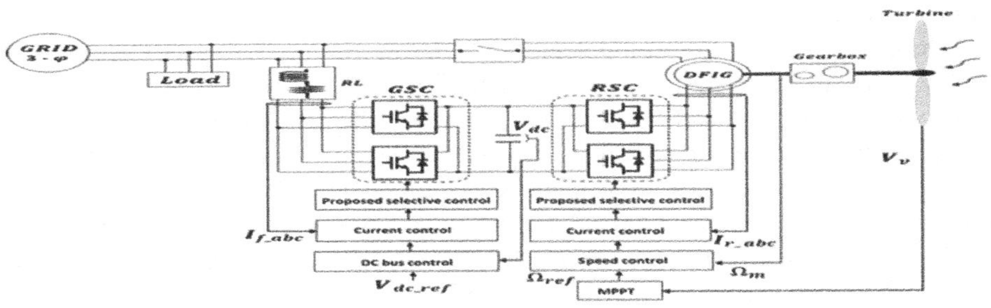

Fig. 1. Wind turbine system connected to the grid utility.

3 Control of the DFIG

In this article, an independent control of the active and the reactive power is done by the decoupled vector control approach. The back-to-back converters are controlled independently which allows us to split into two parts the control of WTCS: the RSC control and the GSC control.

3.1 RSC Control

By considering the network voltage constant, neglecting resistance of the stator winding and aligning the stator flux with the d axis. The model of the DFIG in synchronous frame will be simplified as below:

$$\begin{cases} V_{sd} = 0 \\ V_{sq} = \omega_s \varphi_{sd} \\ V_{rd} = R_r I_{rd} + \sigma L_r \frac{di_{rd}}{dt} - \sigma L_r \omega_r i_{rq} \\ V_{rq} = R_r I_{rq} + \sigma L_r \frac{di_{rq}}{dt} + \sigma L_r \omega_r i_{rd} + g \frac{MV_{sq}}{L_s} \end{cases} \tag{1}$$

The active power, reactive power and the electromagnetic torque of DFIG stator will be expressed by following equations:

$$\begin{cases} P_s = \frac{3}{2} V_s I_{sq} \\ Q_s = \frac{3}{2} V_s I_{sd} \\ C_{em} = -\frac{3}{2} \frac{Mp}{L_s} I_{rq} \varphi_{sd} \end{cases} \tag{2}$$

From the previous equations, it is obvious that the active and reactive power are decoupled while, they are controlled by the quadrature and direct current respectively. For that a proportional-integral (PI) controller is used. The direct axis reference current is chosen to keep the reactive power zero, while the quadrature axis reference is selected to extract the maximum power. This can be achieved for a particular wind speed, so an MPPT technique with a speed controller are used. There is a lot of research to develop the MPPT algorithm for WECS (Abdullah et al. 2012). In this paper, a simple MPPT algorithm is used, which is called Tip speed ratio (TSR) control (Loucif et al. 2014). A proportional-integral (PI) is used as a speed controller. Figure 2 shows the structure of the RSC control

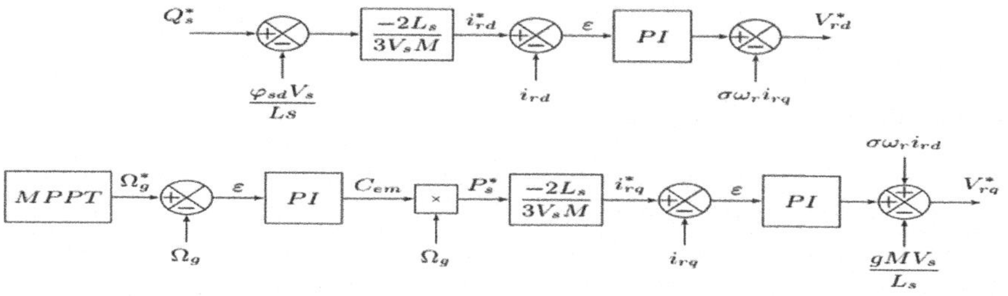

Fig. 2. RSC control schema blocks.

3.2 GSC Control

The main purpose of the GSC is to ensure a constant voltage at the DC link and guarantee a unity power factor at the grid side. The control of the GSC is based on the following real power from the grid side to the DC link. RSC. By applying the voltage vector control (Gaillard 2010), the voltage source will be oriented on the quadrature axis. The model of the filter will be simplified as below:

$$\begin{cases} V_{fd} = -R_f I_{fd} - L_f \frac{dI_{fd}}{dt} + \omega_s L_f I_{fq} \\ V_{fq} = -R_f I_{fq} - L_f \frac{dI_{fq}}{dt} - \omega_s L_f I_{fd} + V_{sq} \\ \frac{dV_{dc}}{dt} = \frac{1}{C}(I_{RSC} + I_{GSC}) \end{cases} \quad (3)$$

$$\begin{cases} P_f = \frac{3}{2} V_{sq} I_{fq} \\ Q_f = \frac{3}{2} V_{sq} I_{fd} \end{cases} \quad (4)$$

The direct and the quadrature current acts on reactive and active power respectively. According to (Gaillard 2010), if the set of power losses in the capacitor, inverter and RL filter are neglected, the active filter and the capacitor power is expressed by the following equations:

$$\begin{cases} P_f = P_{cap} + P_{RSC} \\ P_{cap} = V_{dc} I_{cap} \end{cases} \quad (5)$$

Figure 3 shows the structure of the GSC control. The filter current and the DC bus voltage are controlled by the outer loop and the inner loop respectively. The outer loop use an IP controller, it reduces the overshoot at the DC link voltage. This type of controller is discussed in (Loucif et al. 2015).

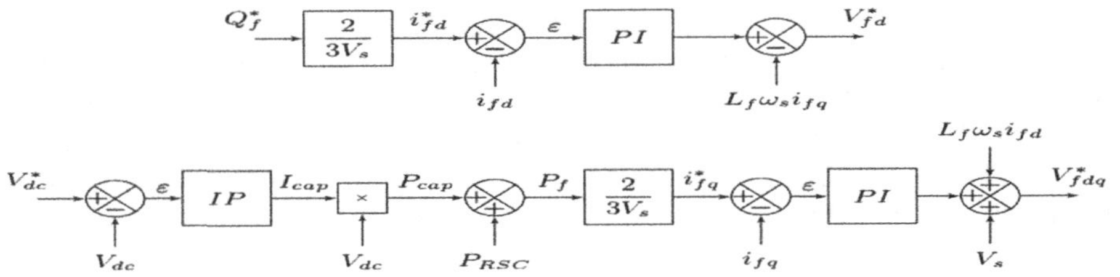

Fig. 3. GSC control schema blocks.

3.3 Proposed Selective Control

With the parallel back-to-back inverter, it is possible to use a low power IGBTs for high power. In WECS, the produced power depends on the wind speed variations. So, the two inverters are not needed all the time. Thus, the contribution of this paper lies in using a selective technique. The aim of this algorithm is to turn on/off the inverters of the parallel structure without affecting the operation of the WECS. The same working principle of the selective technique is applied to control the RSC and the GSC. If the exchanged current is less than the limit of the used IGBTs, the selective control turns off one of the inverters of the concerned side converter. The algorithm of selective control is shown in Fig. 4, the main advantage of the proposed control is to reduce the power loss.

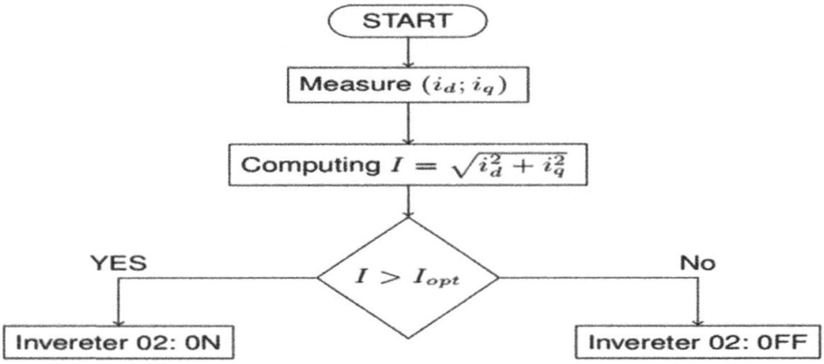

Fig. 4. Selective control diagram.

4 Simulation Results

In this section, the WTCS is developed and simulated in Simulink environment and SimPowerSystems MATLABs block set. This simulation is done by using a discrete solver with a fixed step. The DFIG studied in this paper is rated at 2 MW and its parameters are given in the appendix. The DC bus voltage is set at 1200 V. The grid voltage is considered balanced where the RMS value between phases is 690 (V) and the frequency is 50 (Hz). The simulation is performed with a variable wind speed, varying between 7.8 m/s and 10.2 m/s as shown in Fig. 5.

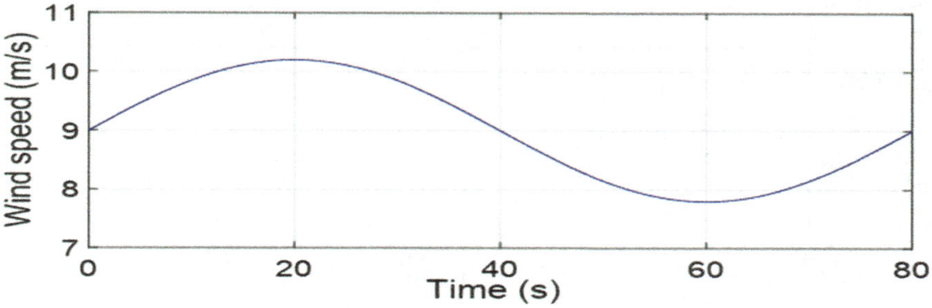

Fig. 5. Wind speed profile.

Figures 6 and 7 shows the performance of the MPPT used where the Cp is varied around the max and the measured speed track the reference speed under different wind speed.

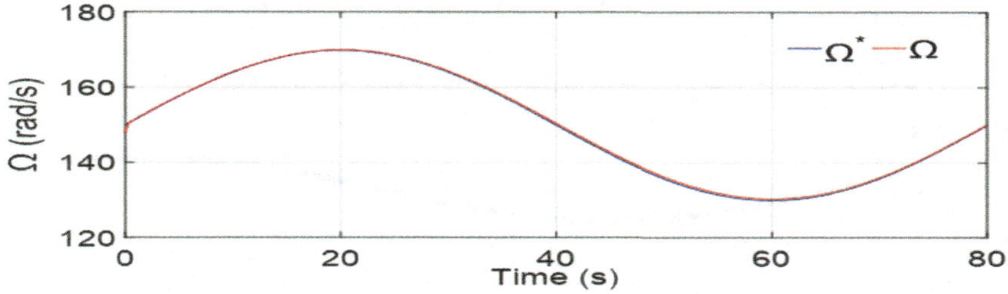

Fig. 6. Generator speed variation according to the wind profile.

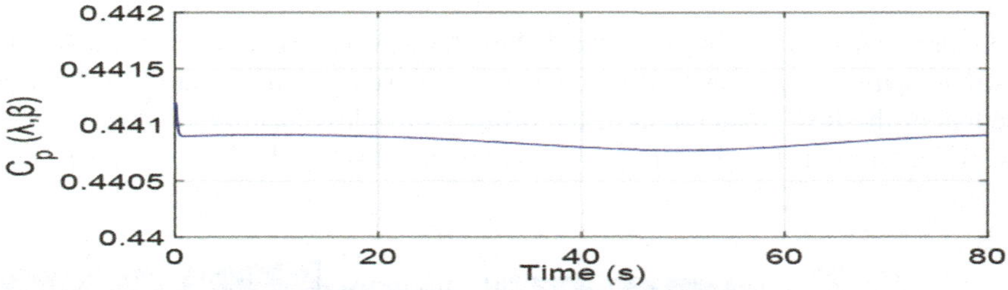

Fig. 7. Power coefficient Cp using the MPPT technique.

Figure 8 shows the efficiency of the IP controller in elimination of the over shoot. Where it is less than 10% of the DC bus voltage and the DC link is quickly stabilized at the reference value (1200 V).

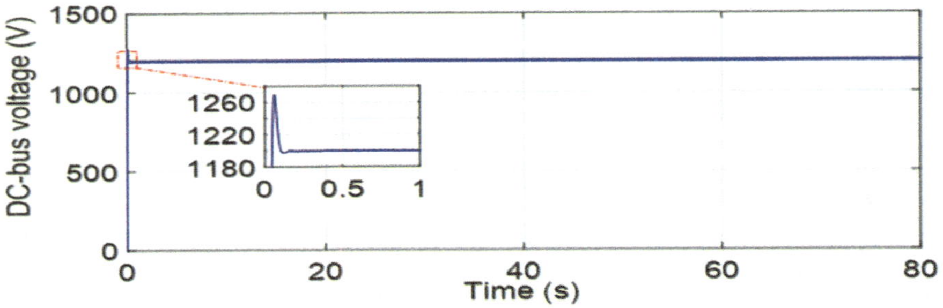

Fig. 8. DC bus voltage.

Figure 9 presents the power exchanged to the grid. When the mechanical speed is less than the synchronous speed (sub synchronous mode), the rotor absorb power from the grid but if the generator speed is greater than the synchronous speed (super synchronous mode) the rotor produces power.

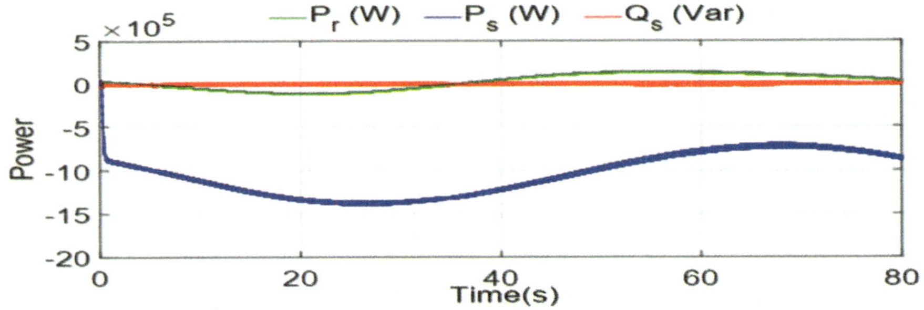

Fig. 9. Exchanged power between the DFIG and the grid.

Figure 10 shows the current controlled by the selective technique at the RSC. If the maximum current is greater than 1200 (A) the proposed selective control activates both inverters of the RSC, the current of the inverter stays less than the limit of the IGBT. Else, if the current is less than the previous limit, the RSC works with one inverter only.

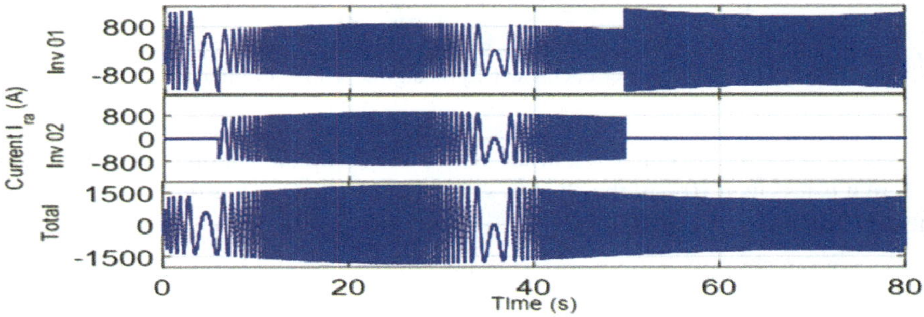

Fig. 10. Rotor currents inverter using proposed selective control.

Figure 11 presents the current controlled by the selective technique at the GSC. The current limit is set to 100 (A).

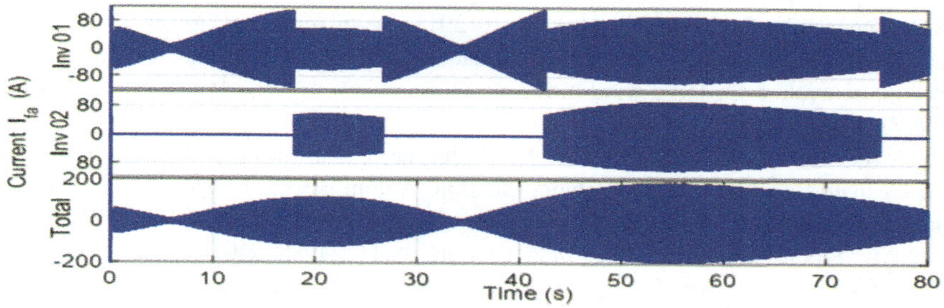

Fig. 11. Filter currents inverter using proposed selective control.

The efficiency performance of the proposed control is shown in Fig. 12. The THD is lower when both inverters are OFF. In all others states, the THD is less than 5% which corresponds to the international laws (Jana et al. 2017).

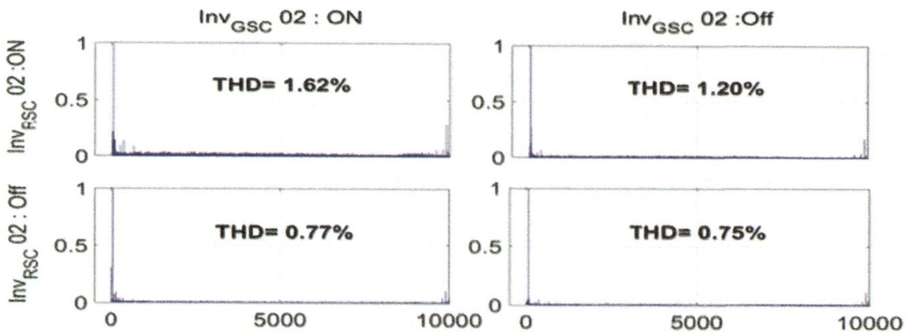

Fig. 12. Frequency analyses of the stator current.

5 Conclusion

This paper presented a study of a WTCS system, based on DFIG. A parallel Back-to-Back is used in order to increase the current limit of the conventional converter. The TSR technique is found capable to drive the DFIG to track the maximum power from the available wind regime. A direct current control is applied with a proposed selective technique to decrease the THD. The simulation results show the stability of the system and the good performance of the proposed control. The parallel converter is one of the solutions to high wind power applications. The proposed selective control has proven its ability to reduce power losses in the system by reducing the THD.

References

Herbert, G.M.J., Iniyan, S., Sreevalsan, E., Rajapandian, S.: A review of wind energy technologies. Renew. Sustain. Energy Rev. **11**, 1117–1145 (2007)

REN21: Renewables 2018 global status report (2018). www.ren21.net. Accessed 22 Jan 2019

Ma, K., Tutelea, L., Boldea, I., Ionel, D.M., Blaabjerg, F.: Power electronic drives, controls, and electric generators for large wind turbines - an overview. Electric Power Components Syst. **43**(12), 14061421 (2015)

Chemidi, A., Meliani, S.M., Benhabib, M.C.: Performance analysis of DFIG wind power system fed by matrix converter. Electrotehnic, Electron. Autom. **63**(1), 78–87 (2015)

Kolar, J.W., Friedli, T., Rodriguez, J., Wheeler, P.W.: Review of three-Phase PWM ACAC converter topologies. IEEE Trans. Ind. Electron. **58**(11), 4988–5006 (2011)

Pena, R., Clare, J.C., Asher, G.M.: Doubly fed induction generator using back-to-back PWM converters and its application to variable-speed wind-energy generation. IEE Proc. Electric Power Appl. **143**(3), 231–241 (1996)

Teichmann, R., Bernet, S.: A comparison of three-level converters versus two-level converters for low-voltage drives, traction, and utility applications. IEEE Trans. Ind. Appl. **41**(3), 855–865 (2005)

Narimani, M., Wu, B., Cheng, Z., Zargari, N.R.: A new nested neutral point-clamped (NNPC) converter for medium-voltage (MV) power conversion. IEEE Trans. Power Electron. **29**(12), 6375–6382 (2014)

Ke, M.: Power Electronics for the Next Generation Wind Turbine System. Research Topics in Wind Energy. Springer, Heidelberg (2015). https://doi.org/10.1007/978-3-319-21248-7

Abdullah, M.A., Yatim, A.H.M., Tan, C.W.: A review of maximum power point tracking algorithms for wind energy systems. Renew. Sustain. Energy Rev. **16**, 3220–3227 (2012)

Loucif, M., Boumediene, A., Mechernene, A.: Maximum power point tracking based on backstepping control of wind turbine. Electrotehnic Electron. Autom. **62**(3), 103–109 (2014)

Gaillard, A.: Système éolien basé sur une MADA: contribution à l'étude de la qualité de l'énergie électrique et de la continuité de service. Ph.D. thesis, University of Nancy-I, France (2010)

Loucif, M., Boumediene, A., Mechernene, A.: Nonlinear sliding mode power control of DFIG under wind speed variation and grid connexion. Electrotehnic, Electron. Autom. **63**(3), 103–109 (2015)

Jana, J., Saha, H., Bhattacharya, K.D.: A review of inverter topologies for single-phase grid connected photovoltaic systems. Renew. Sustain. Energy Rev. **72**, 1256–1270 (2017)

Efficiency of Polyaniline/(ZnO, Cds) Junctions Doped by Ionic Liquid in Photovoltaic Properties

A. Benabdellah[1(✉)], M. Debdab[1], Y. Chaker[1], B. Fetouhi[1], and M. Hatti[2]

[1] Synthesis and Catalysis Laboratory, Ibn Khaldoun University of Tiaret, Tiaret, Algeria
abdelkader.benabdellah@univ-tiaret.dz, benalk2000@gmail.com

[2] UDES, Unité de Développement des Equipement Solaires, EPST-CDER, Bou Ismail, Tipasa, Algeria

Abstract. Polyaniline doped by $[EtOHMIM^{+}][HSO_4^{-}]$ ionic liquid as solvent PANI (ES $[SO_4^{-}]$) was synthesized by chemical polymerization method using ammonium persulphate as oxidizing agent. An entirely approach has been to made heterojunctions by evaporation vacuum method of PANI (EB & ES $[SO_4^{-}]$) onto n-type (ZnO, Cds) thin films. I-V characteristics of these junctions were studies at room temperature 303 K by using Keithley electrometer 6517 A. High rectifying ratio and low ideality factor of these junctions shows in results are comparable with those reported for inorganic semiconductors junctions. The efficiency of these heterojunctions in the photovoltaic properties were studied in the dark and under illumination indicates that the Ag/Cds - ES $[SO_4^{-}]$/Al may be preferred over the Ag/ZnO - ES $[SO_4^{-}]$/Al heterojunctions solar cell which was found to deliver short circuit current density Jsc = 200 μA/cm^2, open circuit voltage Voc = 400 mV, and solar cell efficiency η = 3.8% under AM 1.5 simulated solar light with the intensity of 100 mW/cm^2.

Keywords: Polyaniline · Ionic liquid · Heterojunctions · Photovoltaic properties · Evaporation vacuum method

1 Introduction

Polymer photovoltaic cells have shown great potential as a means to harvest solar energy in a highly processable and cost-effective manner [1]. Among of these conducting polymers, polyaniline occupies a prominent place in the study and applications [2, 3]. It combines the lightness, low cost and modularity of polymers with the electrical conduction properties of metals. Polyaniline can be doped by protonation of emeraldine base (PANI-EB) with conventional or organic acids [4]. This encouraged us to undertake its synthesis and study in organic media as ionic liquid. In recent years, we have seen the development of the great use of ionic liquids in the preparation of conductive polymers [5, 6]. Ionic liquids have unique properties as a solvent and

M. Hatti (Ed.): ICAIRES 2019, LNNS 102, pp. 693–699, 2020.
https://doi.org/10.1007/978-3-030-37207-1_75

electrolyte [7]. These have a high ionic conductivity (>10^{-4} S/cm), an electrochemical range > 1 V during which the ionic liquid is neither oxidized nor reduced.

Organic semiconductors, including conjugated polymers and small molecular organic materials [8, 9], have recently attracted much attention and show great promise in organic light emitting diodes (OLEDs), field-effect transistors (OFETs) and the ongoing effort to lower the cost of solar cells [10–12]. In recent years, the development of thin film plastic solar cells, using polymer-fullerene or polymer-polymer bulk heterojunctions as an absorber (and transport layer at the same time), has made significant progress [13, 14]. Polyaniline (PANI) is one of the most important feature promising conducting polymers that makes PANI an interesting as sensitive p-n junction layer for the solar cell is the variation of its electrical and optical properties at room temperature [15, 16]. The typical structure of these devices consists of a composite of two materials with donor and acceptor properties, respectively, sandwiched between two electrodes.

In this research, polyaniline salt (ES $[SO_4^-]$) doped by $[EtOHMIM^+][HSO_4^-]$ ionic liquid as solvant was synthesized by chemical polymerization method. An effort has been to made heterojunctions between PANI (EB & ES $[SO_4^-]$) onto n- type (ZnO, Cds) thin films by evaporation vacuum method. I-V characteristics of PANI (EB & ES $[SO_4^-]$)/n-type (ZnO, Cds) junctions have been recorded at room temperature by a Keithley electrometer 6517 A was used to measure current density (J) as a function of applied voltage in range ± 1 V and in steps of 0.05 V. The current density-voltage (J-V) characteristics have been measured for solar cell devices prepared from the organic polymer polyaniline PANI (EB & ES $[SO_4^-]$) onto n- type (ZnS, TiO_2) thin films. The studied J-V characteristics were used to determine solar cell parameters such as (V_{oc}), (J_{sc}), (I_{max}), (FF), (η).

2 Experimental Procedures

2.1 Materials and Instruments

Aniline (98.6%), Ammonium persulfate ($[(NH_4)_2S_2O_8]$, APS), Ammonium hydroxide (NH_4OH) was obtained from Sigma Aldrich (99.3%). All the chemical and reagents were used as received without further purification. n-type wafers of (ZnO (99.99%) and Cds (99.995%) as obtained from Alfa Aesar) with <100 nm> and resistively of (5–30 Ω. cm) have been used as thin films for the deposition of PANI (EB & ES $[SO_4^-]$). The thickness is measured by a Quartz Crystal Thickness monitor (CTM-200).

2.2 Preparation of ES $[SO_4^-]$ by Chemical Method

Preparation and characterization of ($[EtOHMIM^+]$ $[HSO_4^-]$ viscous ionic liquid as-synthesized) followed the literature [17]. Distilled before use at 70 °C. Doping of polyaniline by chemical means was carried out using $[EtOHMIM^+]$ $[HSO_4^-]$ ionic liquid as solvent. The reactions as follows: an aqueous solution of ammonium persulfate (0.1 M) was adding drop by drop into a (0.05 M) of aniline solution dissolved in (5 ml) of $[EtOHMIM^+]$ $[HSO_4^-]$ at a temperature about 3 °C to prevent heating of the solution. After stirring for 1 h, a dark green precipitate is obtained, filtered, washed with distilled water, transferred to a beaker and mixed with methanol to remove all oligomeric impurities. After standing overnight, the filtered precipitate, dried under vacuum at 60 °C for 8 h, gives a green powder of emeraldine salt ES $[SO_4^-]$.

2.3 Preparation of PANI (EB & ES $[SO_4^-]$/n-Type (ZnO, Cds) Junctions for Solar Cell Devices

Thin films of n-type (ZnO, Cds) of were deposited using a vacuum evaporation technique at 2×10^{-5} Torr. The evaporation is performed at a temperature 150 °C for 3 min to obtain a thin films of n-type (ZnO, Cds) with a thickness of 300 nm. Thin films of n-type (ZnO, Cds) were deposited on bulk samples of PANI (EB & ES $[SO_4^-]$) in the form of pellets (diameter 12 mm and thickness of approx 1 mm) prepared under a load of 6 tonnes. The thickness is measured by using a Quartz Crystal Thickness monitor (CTM-200). Electrical contacts have been provided by a conducting paste of silver. For ensuring a good adhesion of PANI (EB & ES $[SO_4^-]$) bulk samples, the n-type (ZnO, Cds) thin films were ultrasonically cleaned in acetone and then de-ionized water, after they were placed on a hotplate for 4 h with temperature of 363 K. The photovoltaic properties of PANI (EB & ES $[SO_4^-]$/n-type (ZnO, Cds) junctions for solar cell devices were measured under illumination using a Bentham 605 solar simulator fitted with a xenon lamp.

3 Results

3.1 I-V Characteristics Between PANI (ES $[SO_4^-]$)/n-Type (ZnO, Cds)

Whatever the n-type (ZnO, Cds) thin films, we found that PANI (ES $[SO_4^-]$) has the best specific conductivity compared PANI (EB), for this we have choosing only PANI (ES $[SO_4^-]$) in fabrication and study of our junctions. I-V characteristics of PANI (ES $[SO_4^-]$)/n-type (ZnO, Cds) junctions at room temperature are shown in Fig. 1. Junctions characteristics indicate asymmetric and non-linear curves, the forward current increases exponentially in the lower voltage region and linearly in the higher voltage region, which indicates the formation of the depletion layer at PANI (ES $[SO_4^-]$)/n-type (ZnO, Cds) interface and show the rectifications effect.

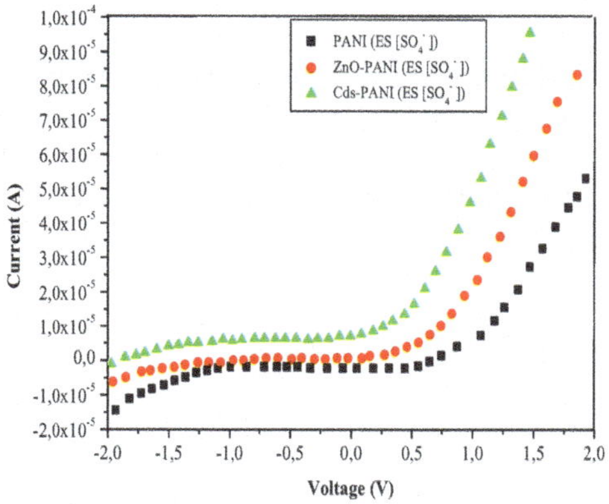

Fig. 1. I-V characteristics of PANI (ES $[SO_4^-]$)/n-type (ZnO, Cds) junctions.

3.2 J-V Characteristics Between PANI (ES $[SO_4^-]$)/n-Type (ZnO, Cds)

Solar cell parameters that, open circuit voltage (V_{oc}), short circuit current density (J_{sc}), maximum current (I_{max}), fill factor (FF) and conversion efficiency (η) are summarized in Table 1.

Table 1. Solar cell parameters of PANI (ES $[SO_4^{-2}]$)/n-type (ZnO, Cds) measured both in dark and under illumination intensity.

Solar cells devices	V_{oc} (V)	J_{sc} (μA/cm^2)	I_{max} (μA)	FF	η (%)
Ag/ZnO - ES $[SO_4^{-2}]$/Al	0.2	400	120	0.18	0.8
Ag/Cds - ES $[SO_4^{-2}]$/Al	0.4	2000	640	0.48	3.8

J-V characteristics of PANI (ES $[SO_4^-]$)/n-type (ZnO, Cds) solar cell have measured both in dark and under illumination intensity in 100 mW/cm^2 are shown in Figs. 2 and 3.

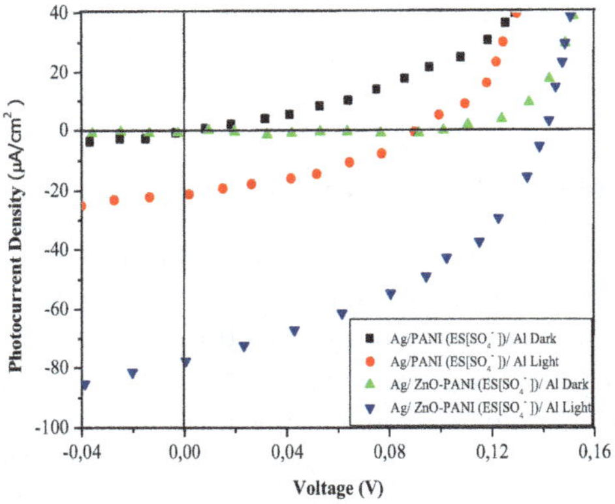

Fig. 2. J-V Curves for Ag/ZnO - ES $[SO_4^-]$/Al tested in the dark and under AM 1.5 solar simulated illumination intensity is 100 mW/cm^2.

This can be further justified by the fact that the n-type (ZnO, Cds) thin films used in this work is of n-type while the PANI (ES $[SO_4^-]$) bulk simples are considered as the hole transporting layer [18]. The forward-bias mode, positive voltage on the PANI (ES $[SO_4^-]$) region clearly demonstrates that we have formed a rectifying contact with a current rectification-ratio at 0.2 V.

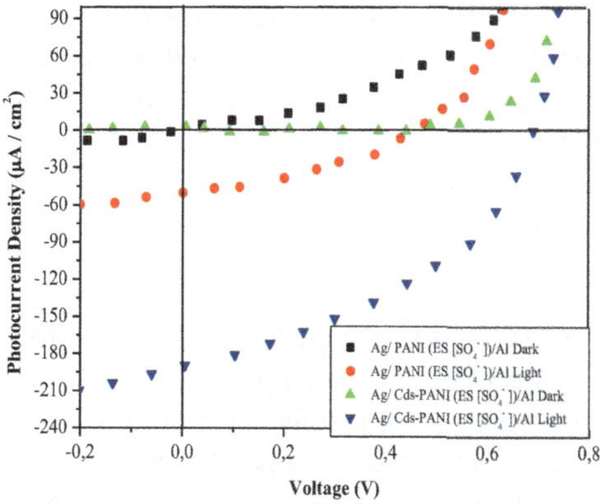

Fig. 3. J-V Curves for Ag/Cds - ES $\left[SO_4^-\right]$/Al tested in the dark and under AM 1.5 solar simulated illumination intensity is 100 mW/cm^2.

4 Discussion

In this work all tested junctions of PANI (ES $[SO_4^-]$)/n-type (ZnO, Cds) show rectifying behavior, with rectifying ratio (forward to reverse current ratio) of about 100. Assuming the thermionic emission theory, the ion of applied voltage (V) is given by [19]:

$$J = J_s\left(e^{qV/nKT} - 1\right) \tag{1}$$

Where *Js* is the reverse saturation current density, *q* electron charge, *n* ideality factor, *k* Boltzmann's constant and *T* is the temperature measured in Kelvin. *Js* is given by the following relation:

$$J_s = A^* T^2 e^{-\phi/kT} \tag{2}$$

Where $A*$ is Richardson constant, and ϕ is the barrier height. The solar conversion efficiency η is given by the formula:

$$\eta = FF \times V_{OC} J_{SC} / P_{in} \tag{3}$$

Where P_{in} is the power of the incident light, and *FF* was calculated by the formula:

$$FF = (J_{\max} V_{\max} / J_{SC} V_{OC}) \tag{4}$$

Atypical solar conversion efficiency of 3.8% has been obtained, which is of small value as compared with aluminum/polyaniline/GaAs metal-insulator semiconductor solar cell which was found to give efficiencies in the region of 1% [20]. The value of 0.4 V for the open circuit voltage obtained in this work compares well with the value 0.65 V obtained for Pani/n-Si solar cell devices [21]. The low value of *FF* is associated with a high series resistance and a high shunt resistance. High values for R_S may originate from electrode contact resistance and high R_{sh} is related to morphology of PANI (ES $[SO_4^-]$) layer deposited on n-type (ZnO, Cds); a poor absorber morphology limiting the electron hopping transport [22, 23].

5 Conclusion

Junctions between PANI (ES $[SO_4^-]$) doped with $[EtOHMIM^+]$ $[HSO_4^-]$ ionic liquid and n- type (ZnO, Cds) substrates have been prepared by evaporation vacuum method. I-V characteristics results of these junctions shows redressment effect with high rectifying ratio and low ideality factor are comparable with those reported for inorganic semiconductors junctions. The Solar cell efficiency under AM 1.5 simulated solar light with the intensity of 100 mW/cm^2 found about 3.8% is obtained for Ag/Cds-ES $[SO_4^-]$/Al with ES $[SO_4^-]$. The low efficiency obtained in this work was caused by the morphology of the polymer film.

Acknowledgements. This work is partially of a research project supported by CNEPRU (code: B00L01UN140120150002) of research direction (Algerian Ministry of Higher Education and Scientific Research). The authors also gratefully acknowledge the helpful comments and suggestions of the reviewers, which have improved the presentation.

References

1. Nechtschein, M., Devreux, F., Genoud, F., Vieil, E., Pernaut, J.M., Genies, E.: Polarons, bipolarons and charge interactions in polypyrrole: physical and electrochemical approaches. Rev. Synth. Metals **15**(1), 59–78 (1986)
2. Payerne, R.: Structure et propriétés électroniques a différentes échelles de systèmes modèles de polymères conducteurs et semi-conducteurs. Thèse de l'Université Joseph Fourier, Grenoble I, Français, pp. 74–76 (2004)
3. Sultan, A.S., Adewole, J.K., Al-Ahmed, A., Nazal, M., Javaid Zaidi, S.M.: Preparation and performance evaluation of SPEEK/polyaniline composite membrane for direct methanol fuel cell. Int. Polym. Process. **32**(1), 41–49 (2017)
4. Thorat, S.M., Kulkarni, M.V.: Morphology tuning of conducting polyaniline via static, liquid-liquid interfacial polymerization process and its application for optical PH sensing. Int. Polym. Process. **30**(2), 243–247 (2015)
5. Benabdellah, A., Belarbi, E.-H., Ilikti, H., Hatti, M., Fettouh, B.: Electrical properties of PANI/chalcogenide junctions doped with ionic liquids anions. J. Tenside Surfactants Deterg. **49**(3), 241–246 (2012)
6. Fetouhi, B., Benabdellah, A., Belarbi, E.H., Ilikti, H., Benabdallah, T.: Junction characteristics system based on composite organic semiconductors: polystyrene/polyaniline doped by [BMIM] $[BF_4]$ ionic liquid. J. Tenside Surfactants Deterg. **51**(6), 541–546 (2014)
7. Chauvin, Y., Lecocq, V., Graille, A., Santini, C.C., Baudouin, A., Basset, J.M., Arzel, L., Bouchu, D., Fenet, B.: Synthesis and characterization of ionic liquids based upon 1-butyl-2, 3-dimethylimidazolium chloride/ZnCl2. Revue Of New Journal of Cemistry **29**, 700–706 (2005)

8. Lee, S.-Y., Choi, G.-R., Lim, H., Lee, K.-M., Lee, S.-K.: Electronic transport characteristics of electrolyte-gated conducting polyaniline nanowire field-effect transistors. Rev. Appl. Phys. Lett. **95**, 013113 (2009)
9. Cao, Y., Smith, P., Heeger, A.J.: Counter-ion induced processibility of conducting polyaniline and of conducting polyblends of polyaniline in bulk polymers. Rev. Synth. Metals **88**, 91–97 (1992)
10. Burn, P.L., Lo, S.-C., Samuel, I.D.W.: The development of light-emitting dendrimers for displays. Adv. Mater. **19**, 1675–1688 (2007)
11. Nam, S., Jang, J., Kim, K., Yun, W.M., Chung, D.S., Hwang, J., Kwon, O.K., Chang, T., Park, C.E.: Solvent-free solution processed passivation layer for improved long-term stability of organic field-effect transistors. J. Mater. Chem. **21**, 775–780 (2011)
12. Williams, E.L., Jabbour, G.E., Wang, Q., Shaheen, S.E., Ginley, D.S., Schiff, E.A.: Conducting polymer and hydrogenated amorphous silicon hybrid solar cell. Appl. Phys. Lett. **87**, 1–3 (2005)
13. McEvoy, A.J., Gratzel, M., Wittkopf, H., Jestel, D., Benemann, J.: Nanocrystalline electrochemical solar cells. In: IEEE First World Conference on Photovoltiac Energy Conversion, USA, 5–9 December, pp. 1779–1782 (1994)
14. Cao, Y., Smith, P., Heeger, A.J.: Counter-ion induced processibility of conducting polyaniline and of conducting polyblends of polyaniline in bulk polymers. Synth. Met. **48**, 91–97 (1992)
15. Nicolas-Debarnot, D., Poncin-Epaillard, F.: Polyaniline as a new sensitive layer for gas sensors. Anal. Chim. Acta **475**, 1–15 (2003)
16. Hoppe, H., Glatzel, T., Niggemann, M., Schwinger, W., Schaeffler, F., Hinsch, A., Lux-Steiner, MCh., Sariciftci, N.S.: Efficiency limiting morphological factors of MDMO-PV: PCBM plastic solar cells. Thin Solid Films **511**, 587–592 (2006)
17. Chaker, Y., et al.: Synthèse et caractérisation de nouveaux liquides ioniques et de poly (liquides ioniques) à base d'imidazolium. Thesis USTO (Algeria), pp. 69–75 (2017)
18. Yu, G., Gao, J., Hummelen, J.C., Wudl, F., Heeger, A.J.: Polymer photovoltaic cells: enhanced efficiencies via a network of internal donor-acceptor heterojunctions. J. Sci. **270**, 1789–1791 (1995)
19. Talib, R.A.: Preparation of conducing polymr (PAni) by chemical method, Study some of physical proprtis and its application as ammonia gas sensor. MSc. thesis, Basrah University, Basrah, Iraq, pp. 125–134 (2009)
20. Shaheen, S.E., Brabec, C.J., Padinger, F., Fromherz, T., Hummelen, J.C., Sariciftci, N.S.: 2.5% efficient organic plastic solar cells. Appl. Phys. Lett. **78**, 841–843 (2001)
21. Wang, W., Schiff, E.A.: Polyaniline on crystalline silicon heterojunction solar cells. Appl. Phys. Lett. **91**, 133504 (2007)
22. Levitsky, I.A., Euler, W.B., Tokanova, N., Xu, B., Castracane, J.: Hybrid solar cells based on porous Si and copper phthalocyanine derivatives. Appl. Phys. Lett. **85**, 6245–6247 (2004)
23. Riedel, I., Martin, N., Giacalone, F., Segura, J.L., Chirvase, D., Parisi, J., Dyakonov, V.: Polymer solar cells with novel fullerene-based acceptor. Thin Solid Films **451–452**, 43–47 (2004)

Correction to: The Fire Risk in Green Building Caused by Photovoltaic Installations

Hadj Miloua

Correction to:
Chapter "The Fire Risk in Green Building Caused by Photovoltaic Installations" in: M. Hatti (Ed.):
***Smart Energy Empowerment in Smart and Resilient Cities*, LNNS 102, https://doi.org/10.1007/978-3-030-37207-1_49**

In the original version of the book, the following belated correction has been incorporated: The name of the author has been corrected from "Miloua Hadj" to "Hadj Miloua" in the frontmatter, backmatter and Chapter 49.
The book and the chapter have been updated with the change.

The updated version of this chapter can be found at
https://doi.org/10.1007/978-3-030-37207-1_49

M. Hatti (Ed.): ICAIRES 2019, LNNS 102, p. C1, 2020.
https://doi.org/10.1007/978-3-030-37207-1_76

Author Index

M. Hatti (Ed.): ICAIRES 2019, LNNS 102, pp. 701–703, 2020.
https://doi.org/10.1007/978-3-030-37207-1

The manufacturer's authorised representative in the EU is Springer Nature Customer Service Centre GmbH, Europaplatz 3, 69115 Heidelberg, Germany. If you have any concerns regarding our products, please contact ProductSafety@springernature.com

Printed and bound by CPI Group (UK) Ltd, Croydon, CR0 4YY
30/09/2026
02225643-0016